Mineral Resource Reviews

Mineral Resource Reviews is the official book series of the Society for Geology Applied to Mineral Deposits (SGA). It publishes selected conference and symposium proceedings, monographs, Short Course Volumes, In-depth Reports and Field Guides by researchers and professionals in the fields of mineral resources; their study, development and environmental impacts. The series will focus on providing in-depth descriptions, analyses and reviews of deposit types, metallogenic regions, ore forming processes as well as the analytical/investigative techniques needed to study them. High quality scientific contributions from experts in mineral deposit geology are invited for publication in this series.

Rob J. Bowell • Charles R. M. Butt
Editors

Geology, Geochemistry and Formation of Supergene Mineral Deposits in Deeply Weathered Terrain

Springer

Editors
Rob J. Bowell
SRK Consulting (UK) Limited
Cardiff, South Glamorgan, UK

Charles R. M. Butt
ARRC
CSIRO Mineral Resources
Bentley, WA, Australia

ISSN 2365-0559 ISSN 2365-0567 (electronic)
Mineral Resource Reviews
ISBN 978-3-031-75732-7 ISBN 978-3-031-75733-4 (eBook)
https://doi.org/10.1007/978-3-031-75733-4

This Springer imprint is published by the registered company Springer Nature Switzerland AG
The registered company address is: Gewerbestrasse 11, 6330 Cham, Switzerland

Contents

About the Editor

Rob J. Bowell is a Corporate Consultant with SRK Consulting, an adjunct professor in the Department of Geological Sciences and Geological Engineering at Queen's University, Kingston, Ontario, Canada, and a research fellow of the Queen's Facility for Isotope Research. He has 35 years of experience in the global mining industry. His background is in exploration and mining in tropical and deeply weathered terrain, including extensive work in Equatorial and Southern Africa, South America, and Southeast Asia. He holds degrees in chemistry, geology, and chemical engineering and a PhD in geochemistry on the occurrence of oxide gold deposits in Ashanti, Ghana.

He has worked for TCL, AGC, and MEL prior to joining SRK. For the last 30 years, he has worked in mining consulting in the fields of due diligence, financial and technical audits, process chemistry, environmental geochemistry, environmental engineering, and mineralogy. Specializes in the application of chemistry and mineralogy to solve engineering problems. Has been involved in the discovery of copper projects that have been developed into mines in DRC, Namibia, and Zambia and discovered Gold and Diamond deposits in the Dodoma region of Tanzania and in northern Mozambique. He is currently involved in the development of the Kuene Copper district in northern Namibia.

He has authored/co-authored over 250 papers, edited monographs, and reports, including one on Arsenic as Reviews in Mineralogy & Geochemistry, Volume 79, and edited and contributed to volumes on applied chemistry/geochemistry for Nova Publishing series on chemical elements. He is deputy chief editor for Geochemistry, Exploration, Environment, and Analysis and past president of the Association of Applied Geochemists.

Charles R. M. Butt is an exploration geochemist and regolith geologist with over 50 years research experience in deeply weathered terrain. He has mainly concentrated on the development of exploration procedures for gold, nickel, platinum-group elements, uranium and base metals, and in the geology of secondary mineral deposits, particularly surficial uranium, nickel-cobalt laterite and supergene gold. Both themes have required an understanding of the processes of chemical weathering during the formation of intensely weathered and lateritic regolith, and the subsequent influences of climatic, tectonic and other environmental changes on their evolution, a focus of this volume.

Charles was formerly Deputy Director, CRC for Landscape Evolution and Mineral Exploration, and Chief Scientist, CSIRO Division of Exploration and Mining, based in Perth, Western Australia. He is now an Honorary Research Fellow with CSIRO Mineral Resources. He has led many industry-funded research projects in Australia, and delivered courses and field workshops to industry, university extension and government in Australia, Canada and several countries in South America and Africa. He was awarded the Gold Medal of the Association of Exploration (now Applied) Geochemists, Gibb Maitland Medal (Geological Society of Australia) and A.B Edwards Medals (Australian Journal of Earth Sciences). He was elected Fellow of the Australian Academy of Technological Sciences in 2004.

Charles Butt has authored/co-authored over 300 papers, reports and other articles, including reviews of oreforming processes related to lateritic weathering. Edited books include Regolith exploration geochemistry in tropical and subtropical terrains (CRM Butt and H Zeegers, Handbook of Exploration Geochemistry 3.). He was Associate Editor of the Journal of Geochemical Exploration (1976-1999) and Geochemistry: Exploration, Environment, Analysis (2000-2010), and Guest Editor for special issues of these journals, the Australian Journal of Earth Sciences and Elements.

Introduction

Rob J. Bowell and C. R. M. Butt

Abstract

Supergene mineral deposits form a significant proportion of certain global mineral resources. For some metals, such as aluminium, iron, niobium, nickel and cobalt, supergene deposits predominate, providing 50–100% of global supply. For others, they are regionally important, e.g., oxidized zones on sediment-hosted copper-cobalt deposits of the African Copperbelt; or relatively minor surface enrichments that provide cash flow in the early stages of mine development, exemplified by lateritic and saprolitic gold overlying orogenic quartz-sulfide-carbonate systems and epigenetic systems in more juvenile terranes. Some are niche deposit types of local importance, such as palaeochannel uranium, or by-products of multi-commodity deposits that supplement the yield of the principal ore.

1 Significance as Global Mineral Resources

Supergene mineral deposits form a significant proportion of certain global mineral resources. For some metals, such as aluminium, iron, niobium, nickel and cobalt, supergene deposits predominate, providing 50–100% of global supply. For others, they are regionally important, e.g., oxidized zones on sediment-hosted copper-cobalt deposits of the African Copperbelt; or relatively minor surface enrichments that provide cash flow in the early stages of mine development, exemplified by lateritic and saprolitic gold overlying orogenic quartz-sulfide-carbonate systems and epigenetic systems in more juvenile terranes. Some are niche deposit types of local importance, such as palaeochannel uranium, or by-products of multi-commodity deposits that supplement the yield of the principal ore (Table 1).

These deposits are the products of chemical weathering, either the residues left after intense leaching, or secondary concentrations of mobilized elements. For many, the initial weathering was under lateritic conditions in humid savanna climates. Some have subsequently been modified under later conditions ranging from equatorial rainforest to hyper-arid and from tropical to glacial. Further modifications can be due to tectonic changes (whether uplift or down-warping) altering drainage conditions or initiating erosion or burial. Some of the modifications are major, others minor or insignificant. Other deposits, of course, such as evaporites and porphyry copper, have not undergone lateritic weathering, but formed under different climatic

R. J. Bowell (✉) · C. R. M. Butt
SRK Consulting (UK) Limited, Cardiff, South Glamorgan, UK
e-mail: rbowell@srk.co.uk

R. J. Bowell, C. R. M. Butt (eds.), *Geology, Geochemistry and Formation of Supergene Mineral Deposits in Deeply Weathered Terrain*, Mineral Resource Reviews, https://doi.org/10.1007/978-3-031-75733-4_1

Table 1 Supergene mineralization; commodity type, estimated global supply and bedrock source. (Sourced from SPG Global and inference from published literature)

Commodity	Type	% Supply	Regolith host	Bedrock protore
Aluminium (bauxite)	Lateritic and karst	100	Lateritic residuum	Felsic to mafic igneous and sedimentary
Niobium	Lateritic	90	Lateritic residuum Saprolite	Carbonatite; alkaline intrusives
Nickel	Ni-Co laterite	75	Plasmic horizon Saprolite	Olivine-rich ultramafic (mostly serpentinized)
Cobalt	Ni-Co laterite	20	Fe-Mn -rich plasmic horizon and saprolite	Olivine-rich ultramafic (mostly serpentinized) carbonate-hosted Cu-Co sulfides
	'Copperbelt'	70	Mn-Co duricrust 'caps' Saprolite; oxidized zone	
Copper	Gossan	<10	Gossan, supergene sulfide	Massive and vein sulfides in volcanic rocks
	Exotic Cu	<10	Saprock/saprolite	Supergene sulfides and oxides in palaeo-gravels, alluvium and karst infill
Zinc	Gossan Non sulfide	25	Non sulfide deposits	Siliciclastic and carbonate host rocks
Lead	Gossan Non sulfide	5	Non sulfide deposits	Carbonate
Iron	Martite-Goethite Channel(CID); Detrital DID	60	Residual Detrital and allochthonous	Banded iron formation
Gold	Lateritic regolith Gossan	<10 <5	Residual to transported	Most lithologies; orogenic and epithermal Au deposits
Kaolinite	Lateritic and sedimentary	80	Residual residuum to ferricrete Saprolite sedimentary	Weathered felsic silicate rocks, particularly granitoid intrusives
REE	Lateritic regolith Ion adsorption	10	Residual Residual	Carbonatite Granite
Uranium	Hydromorphic	<5	Allochthonous. calcrete; channel sediments	Granitic terrain

and tectonic conditions. All are subject to continuing modification under present surficial environments. Detrital or placer deposits are generally transported materials that have undergone little or no in-situ alteration in deeply weathered environments and are hence considered outside the scope of this work.

The mineral assemblages of many of the rocks, including primary mineralization, that constitute the protore of supergene deposits were formed at elevated temperature and pressure. When exposed to surface conditions, they are unstable and alter to re-establish equilibrium with the surrounding environment, typically adjusting to lower temperatures and pressures, and higher oxygen and moisture environments. The rate at which this occurs is very much dependent on the prevailing climate and geomorphological setting, each of which may change with time. Deciphering and characterizing these effects are essential in determining the genesis of supergene mineral deposits, and understanding the processes that can cause depletion or enrichment of elements, resulting in the formation of potentially economic mineral deposits.

The objective of this volume, therefore, is to discuss the characteristics and genesis of supergene deposits in the context of their weathering history, evaluating the chemical and mineralogical transformations that have occurred under changing climatic, tectonic and geomorphological settings.

The first two chapters outline the context under which supergene mineral deposits form and become enriched to form economic ore. Chapter "The Geochemistry of Supergene Pro cesses" encompasses much of the basic science of the mineralogy and chemistry of weathering, addressing mineral stability, chemical reactions involving major rock-forming minerals; formation and weathering of secondary minerals, enrichment and depletion; physical weathering and groundwater processes. Chapter "The Role of Regolith Formation and Evolution in the Development of Supergene Mineral Deposits" describes how the lateritic regolith that hosts supergene deposits has formed and evolved over long time periods under changing climatic and tectonic conditions. The deposits may occur in present-day environments quite different from those under which they formed, as illustrated by weathering and geomorphological histories of selected regions.

The remaining chapters include descriptive information of deposit types and styles, describing each deposit type, in turn, indicating global distribution, principal characteristics and formation, followed by individual or grouped case studies set in different weathering and/or geomorphological environments or locations. The deposit types are arbitrarily grouped into four categories:

1.1 Deposits on Massive and Matrix Sulfides

These deposits are dominantly residual enrichments of base metals and gold hosted by ferruginous gossans overlying Fe-rich base metal sulfide assemblages (chapter "In Situ Supergene Base Metal Deposits"), or as so-called 'non-sulfide' residual and translocated deposits enriched in Cu, Zn and Pb carbonates and hydro-silicates (calamines) derived from Fe-poor sulfides (chapter "Supergene Non-Sulfide Base Metal Deposits"). Both styles of mineralization were of considerable historical significance, with numerous mines in countries surrounding the Mediterranean Sea, and extending eastwards to Iran (Persia), India and beyond. Similar deposits were found, in the Americas and, latterly, in southern Africa and Australia. The bright colours of secondary minerals such as malachite and azurite, as well as native Au, Cu and Ag, in gossan and other outcrops attracted the attention of early miners in the Neolithic to early Copper-Bronze Ages (3500–1500 BC) and led to the development of the metallurgy of Sn, Cu, Pb, Fe and other metals (Boyle 2024). Gossans are now more important as exploration guides than ore. 'Non-sulfide' deposits have less spectacular minerals but were important sources of Zn, Pb and, in places, Ag and Cu. They are found dominantly in semi-arid to arid terrain, occurring as direct pseudomorphic replacements of primary sulfides, and as distal Zn-rich mineralization, derived by leaching and transport of Zn by percolating groundwater, and precipitated as replacements in wall rocks, or in cavities caused by tectonic uplift or karst development.

1.2 Deposits on Disseminated Primary Mineralization

These are lateritic deposits with residual and absolute enrichments in saprolite or ferruginous lateritic residuum or both. They include largely residual accumulations of primary and secondary Nb, rare earth elements (REEs), Ti and phosphate minerals on carbonatites (chapters "Carbonatite Hosted Residual REE Deposits" and "Regolith Hosted Niobium Deposits in Carbonatite Complexes in Brazil"), formed by the near total dissolution of the primary igneous calcite and dolomite that comprise more than 30% of the fresh rock. In many, the surface lateritic horizons are not preserved and the resources are dominantly hosted by saprolite; the Nb deposit at Seis Lagos, Brazil, however, is reported as being wholly within ferruginous lateritic duricrust. Deposits described herein focus on worldwide resources of the 'critical' REE, and niobium, the latter describing the Brazilian deposits which supply over 90% of the global market. Associated phosphate and titanium accumulations are discussed briefly. Chapter "Supergene Ores of the Central African Copperbelt" describes the globally important Cu-Co and associated Mn-Co deposits in the Central African Copperbelt. Although the potential resources of primary mineralization are large, the bulk of production to date has been from the high-grade supergene enrichments in the oxidized and transition zone, similar to base metals in gossans, and Co-rich Mn duricrusts ('Cobalt caps').

Lateritic and supergene gold deposits are widely distributed in areas of low relief in predominantly Archaean-Proterozoic shields in South America, Africa, India and Australia (chapter "Supergene Gold Deposits in Intensely Weathered Regolith"). Related deposits occur in tropically weathered dissected terrains of higher relief, in India, SE Asia and the circum-Pacific arcs. Low-grade 'lateritic' Au deposits occur in the ferruginous upper horizons of the regolith, referred to as ferruginous gravels and duricrusts (FGD) due to their initially indeterminate origin—they may vary laterally from residual to transported within a seemingly continuous unit. Despite Au leaching, weathered mineralization in saprolite may be relatively enriched compared to the bedrock source due to density variations and contribute to the mineable resource even if the primary mineralization is uneconomic. Supergene chemical enrichment in saprolite has only been recognized in arid zones in southern Australia subject to groundwater salinization. Although supergene deposits only contribute a small proportion of global Au production, the research leading to their discovery and understanding of their genesis has led to significant advances in regolith geology and geochemistry of intensely weathered lateritic terrain.

1.3 Deposits on 'Unmineralized' Bedrock

These are deposits developed on bedrock that that would not generally be considered to be mineralized, although having characteristics that lead to enrichment of one or more elements during weathering. They include bauxite, Ni-Co(-Sc) laterites, iron ore, kaolinite and other clays and ion adsorption REE deposits. Bauxites (chapter "Lateritic Bauxite Deposits") contribute essentially all of the world's supply of aluminium. 'Lateritic' bauxites have formed over a wide range of felsic and mafic igneous and sedimentary rocks in regions subject to intense leaching, generally under humid, tropical climates, although some possibly developed under more temperate conditions with different seasonality, similar to present Mediterranean climates. 'Karst' bauxites formed over limestones and dolomitic rocks under similar conditions by solution of the carbonates followed by intense weathering of aluminous silicate minerals in interbedded sedimentary and volcanic units. Bauxites are globally widespread, either exposed and still actively forming or being modified, or as 'fossil' deposits, buried by later sediments. The latter demonstrates ages of the deposits and the previous extent of intense weathering in higher latitudes, especially in Europe and Asia. The geology of bauxites has been well documented. However, the vast resources of the Fouta Djallon sub-province in Guinea are less well-known. Chapter "Lateritic Bauxite Deposits" describes these as the principal examples.

Over the past 20 years, Ni-Co(-Sc) laterites (chapter "Nickel-Cobalt Laterite Deposits") have become the major source of Ni, surpassing sulfide deposits. They have formed only on olivine-rich ultramafic rocks, including both ophiolites in island arc settings (e.g., Antilles and western Oceania), and komatiitic volcanic rocks and ultramafic units in layered intrusions in stable shields (e.g., Amazonian Craton, Yilgarn Craton). There are three broad and mineralogically-defined ore types (oxide, hydrous Mg silicate and clay silicate), the formation of which is due to the climatic and/or tectonic histories of the deposits.

Supergene iron ore deposits (chapter "Iron Ore") have formed from banded iron formations (BIF), distinctive finely-layered sedimentary rocks composed of alternating bands of Fe-rich magnetite and Fe-poor chert, with associated silicates and carbonates as gangue minerals. Martite-Goethite (M-G) ores form by residual lateritic enrichment of BIF and mimetic enrichment of BIF by Fe beneath the water table, followed by further enrichment by lateritic weathering following exposure. These ores form major deposits in Brazil, Australia, India and South Africa. Detrital ores resulting from erosion, redeposition and further weathering of M-G ore form smaller deposits. In addition, and almost exclusively in Australia, Channel Iron Deposits (CID) occur as ferruginous granules in units of considerable thickness and length in meandering paleochannels carved through the Precambrian basement. CID either rests on a lateritic surface or, less commonly, overlie fresh bedrock. Their genesis is uncertain but may include detrital components and post-depositional hydromorphic enrichment. Iron is ultimately derived from BIF, but impurities such as Al indicate contributions from a variety of bedrock lithologies.

Residual and sedimentary kaolinite deposits (chapter "Supergene Origin of Kaolin Deposits") occur on felsic igneous and sedimentary rocks in similar regions to bauxites, extending into less intensely weathered regions characterized by ferruginous lateritic residuum. Erosion of these regoliths and redeposition of kaolinite and other clays in lakes and organic continental swamps gives rise to sedimentary kaolin. Some of the most important deposits have formed by a multi-stage process involving both steps, followed by further kaolinization of associated feldspathic sediments by microbial nitrification and acidification, with some of the largest deposits associated with coal. Kaolin deposits are widely distributed, most abundantly in China, the Americas, Europe, Australia and Japan. In recent years, China has become the largest producer, surpassing the USA. The origin of the highly prized China clays of SW England has remained contentious, although recent studies suggest hydrothermal kaolinization of igneous feldspar, followed by weathering of partially altered meta-stable feldspar under humid tropical conditions.

Ian-adsorption rare earth element (REE) deposits have lowest cut-off grade of any currently producing REE deposit type (ca. 0.1 wt. % total oxides) and are the main global producers of heavy REE (chapter "Ion Adsorption Mineralisation in Regolith-Hosted REE Deposits"). They are formed via breakdown of igneous, metamorphic and sedimentary rocks during weathering, which releases the REE from rock-forming and accessory REE-bearing mineral phases. The REE are precipitated at horizons in the profile marked by sharp pH or redox contrast, with the REE adsorbed onto clay minerals, principally kaolinite and halloysite. To date, most deposits are known from SE Asia, especially China, with others in Madagascar and Brazil. They are commonly associated with granitic and felsic bedrock in which resistant magmatic REE-bearing minerals have been subjected to high-temperature alteration to phases subject solution by acidic weathering. The most favourable environments are similar to those causing bauxitization and kaolinization during intense lateritic weathering.

1.4 Hydromorphic Sedimentary Deposits

Although not a product of humid tropical weathering, the largest (Yeelirrie) and the most numerous other deposits and occurrences of U in palaeochannels sediments are in the regolith-dominated terrain of the northern Yilgarn Craton in Western Australia. These deposits are hosted by some of the youngest sedimentary units of the regolith, including near-surface calcretes and associated clastic units, and mostly evaporitic sediments in playas, and formed under quite different climatic conditions to the thick regolith that characterizes the landscape. In contrast, similar deposits in Namibia occur in eroded terrain with widespread surficial aeolian sediments with little residual regolith. Under the present semi-arid to arid conditions in both regions, U, V and K, integral components of carnotite, are released from primary minerals (e.g., feldspar, biotite) in granitic rock and/or, in the Yilgarn, secondary phases in residual and transported regolith. The released metals are transported to valleys by alkaline groundwaters where evaporative concentration of dissolved components leads to the precipitation of carbonates, Mg-rich smectites, opaline silica and carnotite.

2 Regolith Nomenclature

The regolith has long been largely ignored by geologists, but the increasing importance of exploration for minerals that the regolith conceals and, the topic of this volume, the supergene ores that it hosts, has brought the subject increasing attention. One barrier to understanding the geology of the regolith has been the absence of consistent terminology. Regolith nomenclature has largely been derived from disciplines such as geology, soil science, geomorphology and geography, and also from casual observations by travellers. In many instances, the original terms were in any case not clearly defined or did not encompass the wide range of possible variations, so when applied to the regolith, may have been used inaccurately or not fully understood. In addition, local observations and traditions, or industry-wide conventions, can lead to misunderstanding or ambiguity. These issues were addressed by The Regolith Glossary (Eggleton 2001), which, for the most part, has been followed in this book. However, the terminology in the Glossary was developed for regolith developed over predominantly silicate rocks. It can be difficult to apply where the bedrock is composed dominantly of non-silicate minerals, such as carbonates or sulfides, and in regolith later modified by weathering under changed climatic and/or tectonic conditions. The latter may have accumulations of secondary carbonates, sulfates, silicates and oxides, which in some instances form supergene mineralization. In Fig. 1, the generalized lateritic regolith profile over silicate rocks adopted by Eggleton (2001) and widely used in recent literature is compared with a commonly depicted profile of a 'mature' gossan formed by the oxidation of massive and matrix sulfides (after Taylor and Thornber 1992). In the silicate profile, each horizon may be considered to have developed from a precursor similar to that immediately underlying it. The gossan profile is simpler, as Fe-sulfides generally alter to Fe oxides, either directly or via intermediate supergene sulfides in a transition zone above the weathering front. The Fe oxides persist through the profile, outcropping at surface in erosional terrain; they may retain the sulfide crystal fabric and hence can be considered as saprolite. Where the transition zone contains abundant secondary carbonates, sulfates and other minerals, however, these are likely to be the product of later weathering events as their precipitation destroys the original sulfide fabric and imposes a new one—but this is not seen in gossans. Similar newly-imposed fabrics can be present in the regolith over silicate rocks, as discussed in chapter "The Role of Regolith Formation and Evolution in the Development of Supergene Mineral Deposits".

3 Environmental Impacts of Mining Supergene Deposits

All human activities produce waste and have the potential to affect our environment, including the land, water and atmospheric resources in contact with mining and metallurgical operations (Bowell et al. 2023). The characteristics and magnitude of any impact depend on several factors such as deposit geology, nature of the regolith, type of mining and processing implemented, sensitivity of the local environment to the operation, and proximity to human settlements or sensitive environmental settings.

Environmental challenges associated with mineral extraction and processing include impacts on land, air, water, soil, flora, fauna and socio-economic fabric. Mines can generate large land disturbances and can also result in sterilization of land through deposition of barren, nonreactive rock piles during production and after operation. Other mine waste piles can be chemically reactive and interact with precipitation and runoff, leading to metal and sulfate release into surface and groundwater. Impacts on soils may

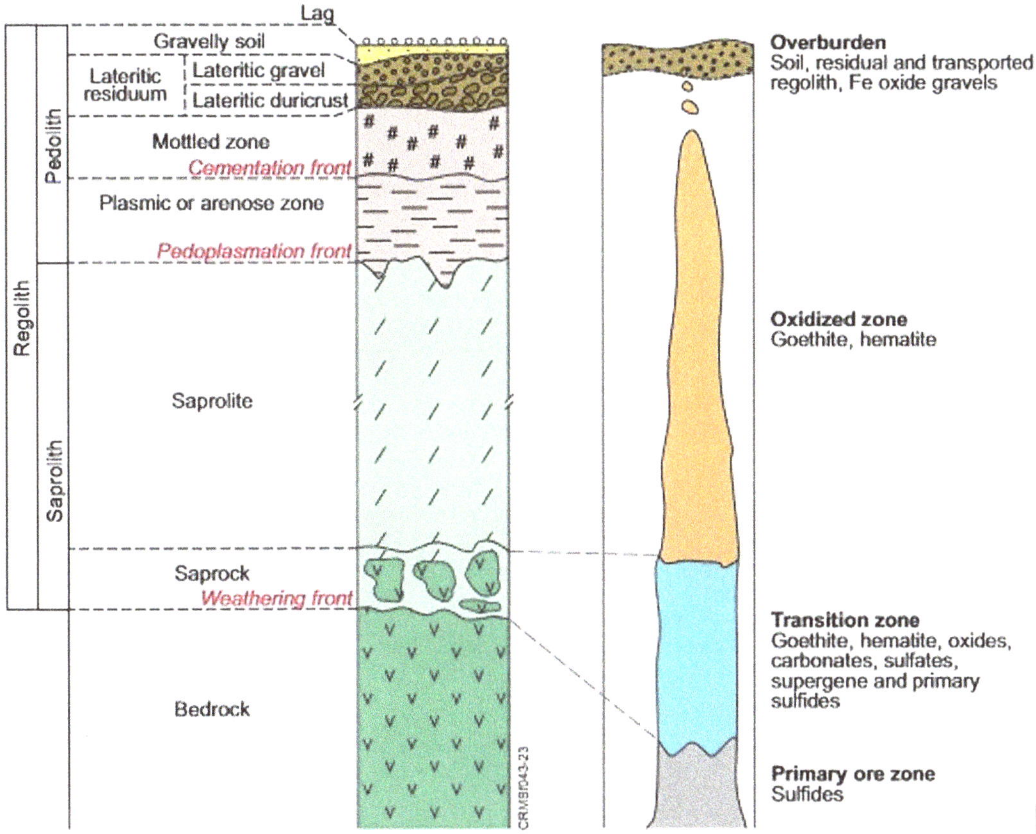

Fig. 1 *Left*: The generalized terminology of a lateritic regolith profile developed over silicate rocks (After Eggleton 2001), compared with *Right*: a 'mature' gossan profile formed by the oxidation of massive and matrix sulfides in the same environment. (After Taylor and Thornber 1992)

result directly from weathering of mine waste or through interaction with affected waters. Interaction with the atmosphere may occur with release of aerosol or airborne metals or minerals affecting air quality that may even lead to acidification of rain. Associated with the large 'footprint' of most open-cut mines (e.g., Fe ore), high stripping ratios, high amounts of waste (e.g., very low Au grades), leached soils (bauxite, kaolinite, ion adsorption REE) and waste from processing (gold, bauxite), abandoned mine pits, rarely backfilled, scar the landscape. Red mud derived from the Bayer process for refining aluminium represents an environmental and economic problem: the annual production is over 120 million tonnes and the global inventory is estimated at 4 billion tonnes. Intense weathering may lead to the formation of infertile, leached soils with a low exchange capacity. This causes difficulty with revegetating mine sites, particularly those associated with Ni-Co laterites due to nutrient-poor nature of serpentine soils (high Mg, Ni, Cr and very low N, P, K, Ca) and, in places, erosion in high relief terrain. Similar issues are discussed in relation to mining, processing, waste-disposal and contamination due to uncontrolled mining of some ion adsorption REE deposits in southern China (chapter "Ion Adsorption Mineralisation in Regolith-Hosted REE Deposits") but could well apply also to artisanal mining in many other countries.

Discussion of these environmental impacts is outside the scope of this volume but further information can be found in Brundtland 1987; Fleury and Davies 2012; Mudd 2010; Gorman and Dzombak 2018; Alda 2019; Silveira et al. 2021; Bowell et al. 2023, and references contained therein.

References

Alda M (2019) Corporate sustainability and institutional shareholders: the pressure of social responsible pension funds on environmental firm practices. Bus Strateg Environ 28:1060–1071

Bowell RJ, Williams CR, Merry EJ, Carpenter A, Bertrando K, Parshley JV (2023) Mitigation of mining effects on the environment. SEG Discovery #135, October:27–43

Boyle RW (2024) In: Wilson GC, Butt CRM, Garrett RG, Robinson HA (eds) A history of geochemistry and cosmochemistry. Prehistory to the end of the classical period. Cambridge Scholars Publishing, Newcastle upon Tyne. (circa 600 pp, in press)

Brundtland GH (1987) Our common future: report of the World Commission on Environment and Development. Geneva, UN-Dokument A/42/427

Eggleton RA (ed) (2001) The regolith glossary. Surficial geology, soils and landscapes. CRC LEME, Perth. 144 pp. See also: https://publications.csiro.au/publications/#publication/PIlegacy:3643. http://crcleme.org.au/Pubs/Monographs/RegolithGlossary.pdf

Fleury AM, Davies B (2012) Sustainable supply chains—minerals and sustainable development, going beyond the mine. Resour Policy 37:175–178

Gorman MR, Dzombak DA (2018) A review of sustainable mining and resource management: transitioning from the life cycle of the mine to the life cycle of the mineral. Resour Conserv Recycl 137:281–291

Mudd GM (2010) Global trends and environmental issues in nickel mining: sulfides versus laterites. Ore Geol Rev 38:9–26. https://doi.org/10.1016/j.oregeorev.2010.05.003

Silveira NCG, Martins MLF, Bezerra ACS, Araújo FGS (2021) Red mud from the aluminium industry: production, characteristics, and alternative applications in construction materials—a review. Sustainability 13. https://doi.org/10.3390/su132212741. (Open access)

Taylor GF, Thornber MR (1992) Gossan formation and gossan surveys. In: Butt CRM, Zeegers H (eds) Regolith exploration geochemistry in tropical and subtropical terrains, Handbook of exploration geochemistry, 4. Elsevier, Amsterdam, pp 115–202

The Geochemistry of Supergene Processes

Rob J. Bowell and M. I. Leybourne

Abstract

Supergene (or weathering) processes, occurring near the Earth's surface, are the geochemical phenomena that describe the alteration of primary minerals in contact with water and atmospheric gases to form more thermodynamically stable secondary minerals. Supergene alteration involves a suite of chemical reactions, including oxidation, hydration, dissolution and precipitation, predominantly influenced by climatic conditions, lithology, topography and hydrology. These reactions result in the decomposition of primary minerals, the release of metals into solution and their transport and reprecipitation as secondary minerals, governed by factors such as redox potential, pH, complexation and adsorption-desorption, driven by factors such as temperature, precipitation, organic matter input and microbial activity. Silicate weathering plays a crucial role in regulating atmospheric CO_2 concentrations by its consumption during mineral dissolution. Selective leaching and preferential enrichment mechanisms lead to the concentration of economically valuable elements and minerals, including copper, nickel, cobalt, aluminium, iron, gold and kaolinite, to form supergene ore deposits. An understating of these processes has implications for ore genesis, exploration targeting strategies, mining and metallurgy and environmental management. Due to the importance of oxygen during weathering, the maximum age of such products is about 2.2 billion years, i.e. after the 'great oxidation event'. Climate has and continues to play an important role in the development of supergene mineral deposits as it exerts control over precipitation, evaporation and temperature as well as concentration levels of reactive gases in the atmosphere.

1 Introduction

The secondary or supergene dispersion of chemical elements is a function of the active chemical, biological and physical processes during weathering, lateritization, soil formation and landscape modification (Tardy 1981; Trescases 1992; McQueen 2009; Gupta 2011). Supergene processes refer to the weathering and alteration of rocks and minerals that occur at or near the surface of the Earth by chemical, physical and biological mechanisms that result from exposure

R. J. Bowell (✉)
SRK Consulting, Cardiff, Wales, UK
e-mail: rbowell@srk.co.uk

M. I. Leybourne
Queen's Facility for Isotope Research (QFIR), Queen's University, Kingston, ON, Canada
e-mail: m.leybourne@queensu.ca

R. J. Bowell, C. R. M. Butt (eds.), *Geology, Geochemistry and Formation of Supergene Mineral Deposits in Deeply Weathered Terrain*, Mineral Resource Reviews, https://doi.org/10.1007/978-3-031-75733-4_2

to the atmosphere, water and other environmental factors (McQueen 2009).

The high temperatures, abundant rainfall and intense biological activity characteristic of humid tropical environments contribute to the rapid transformation of the Earth's surface. Consequently, supergene processes are best expressed in tropical rainforests and humid savannas, which are known for their biodiversity and exceptionally high weathering rates.

Weathering can lead to the formation of economically important mineral deposits from otherwise uneconomic protores of metals such as the Al, Nb, Cu, Ni, Au and rare earth elements (REEs). For example, Cu is commonly found as chalcocite derived from Cu leached from primary sulfide minerals and subsequently deposited in a more concentrated form as supergene sulfides below the water table (Chavez 2021); e.g. chalcopyrite ($CuFeS_2$) can be upgraded to chalcocite (Cu_2S), Eq. (1):

$$2CuFeS_2 + 6O_2 = Cu_2S + 3SO_4^{2-} + 2Fe^{3+} \quad (1)$$

Gold can also be concentrated in the supergene zone either by release from oxidizing sulfide minerals and formation of secondary Au or by the residual concentration of native gold (Bowell et al. 1993; Boyle 2003). Weathering can also lead to the formation of supergene deposits on essentially unmineralized bedrock. Bauxite (the main ore of aluminium), e.g. is formed by the weathering of a range of rocks including granite, basalt and sediments containing aluminosilicate minerals. In this chapter, we review the main weathering processes operating in supergene environments.

2 Age of Oxidation and Initiation of Chemical Weathering

The development of extensive secondary oxide minerals is considered to have been initiated from approximately 2.45–1.85 Ga, with a peak at around 1.9 Ga (Poulton et al. 2021; Large et al. 2022). The initiation of an oxygen-rich atmosphere is dated as starting around 2.45 Ga. Anomalous spikes in Mo and Re concentrations in marine shales of that age indicate oxidation (Anbar et al. 2007) at or around 2.30 Ga, based on reduction in the extent of mass-independent fractionation of S isotopes (Farquhar et al. 2000). During this period, the oxygen content in near-surface continental and marine environments, as well as the atmosphere, increased significantly (Holland 2006). This event, termed the Great Oxidation Event, led to an exponential increase in mineral diversity (Hazen et al. 2008; Sverjensky and Lee 2010; Hazen and Morrison 2022). Prior to this, mineral diversification in Archean rocks was relatively limited (Hazen and Ferry 2010; Hazen and Morrison 2022; Hazen et al. 2008). Rock units, such as the conglomerates of the Witwatersrand basin in South Africa, showed evidence of low oxygen fugacity with the presence of relatively pristine but rounded grains of pyrite and uraninite, indicating water transport but no surface oxidation (Frimmel 2005; Hazen et al. 2008). Geological, isotopic and chemical evidence suggests that biologically produced oxygen molecules (O_2) accumulated in the Earth's atmosphere as it evolved from reducing with little to no free oxygen to oxidizing with abundant free oxygen (Hazen et al. 2008; Sosa Torres et al. 2015).

Over time, prolonged reaction of rainwater enriched in CO_2, SO_2, H_2, H_2S and CO occurred within dominantly volcanic materials (Berner and Berner 1996; Sverjensky and Lee 2010). Numerical modelling of mineral reactions for enstatite, ferrosilite, diopside, anorthite, albite and magnetite indicates that Na–Mg–Ca–HCO_3 water was produced with the formation of minerals such as pyrite, kaolinite, chalcedony, siderite, calcite and lizardite (Sverjensky and Lee 2010).

Dissolved iron in these reactions is predicted to reach a molar value around $10^{-4.3}$ (approximately 3 mg/L), and siderite saturation is therefore indicated. The full reaction is shown in Fig. 1c as a function of oxygen fugacity (log f_{O2}) and pH. Changes in oxygen fugacity are relatively small, and the oxidation state of the system does not appear to change. The pH, by comparison, evolves from acidic to alkaline. Minerals other than kaolinite, pyrite and silica, and minor amounts of siderite (depending on the extent of weathering), are not expected. This situation could presumably have prevailed for a prolonged period,

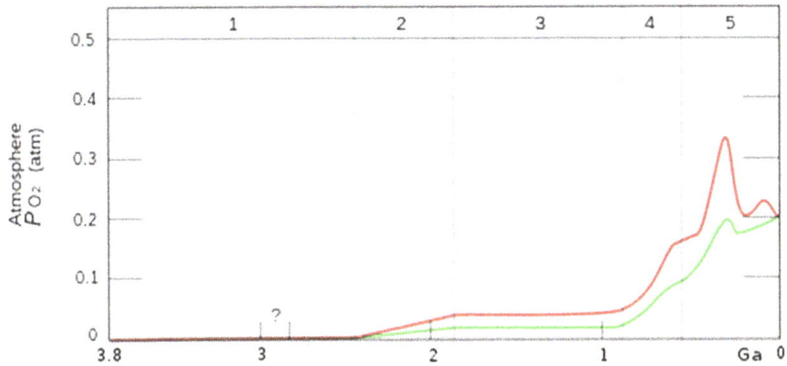

Fig. 1 Oxygen build-up in the Earth's atmosphere. Red and green lines represent the range of the estimates where time is measured in billions of years (Ga). (Modified after Holland 2006). The stages indicated above the plot are: stage 1 (3.85–2.45 Ga)—practically, there is no O_2 in the atmosphere, and the oceans were also largely anoxic (with the possible exception of O_2 in the shallow oceans); stage 2 (2.45–1.85 Ga)—O_2 produced, rising to values of 0.02 and 0.04 atmospheres, but absorbed in oceans and seafloor sedimentary rocks; stage 3 (1.85–0.85 Ga)—O_2 starts to out gas from the oceans but is absorbed by land surfaces, and there is no significant change in oxygen level; and stages 4 and 5 (0.85 Ga–present)—other O_2 reservoirs filled, and O_2 gas accumulates in the atmosphere

until the rise of oxygenic photosynthesis from the development of cyanobacteria-evolved porphyrin-based photosynthesis (Cardona et al. 2015). Such reactions would initially have been small scale, with any available oxygen consumed in the near-surface environment (e.g., in the formation of banded iron formations), and only gradually would oxygen fugacity have increased in Earth's atmosphere (Fig. 1).

The increasing oxygen level eventually depleted the reducing capacity of ferrous Fe and hydrogen sulfide, and atmospheric methane impacted the development of widespread microbial mats (Schirrmeister et al. 2013). Subsequently, surviving archaea adapted to changes in atmospheric composition, which, combined with a global glaciation, led to a rise in eukaryotic organisms and the evolution of multicellular life forms (Kerr 1999; Hoffman 2011).

Following the Great Oxidation Event, large-scale hematite-rich banded iron formations were deposited globally (Hoffman 2011; Hofmann et al. 2009; Olejarz et al. 2021). The occurrence of red bed sedimentary rocks indicates that there was sufficient oxygen to oxidize Fe to its ferric state, representing a marked contrast to sandstones deposited under anoxic conditions, which are typically green (Eriksson and Cheney 1992; Johnson et al. 2014). The release of reactive free oxygen, potentially toxic to the anaerobic biosphere, may have initiated the extinction of early molecules and organisms. Changes in sulfate sulfur isotope ratios over this time have been interpreted to indicate a decrease in the size of the biosphere of more than 80% associated with changes in nutrient supplies (Hodgskiss et al. 2019). It has been estimated that once oxygen production started, it took only 1% oxygen in the atmosphere to sustain weathering rates similar to the present day, suggesting that perhaps as much as 90% of all iron formations preserved in the rock record were formed between 2.5 and 1.9 Ga (Trendall and Blockley 2004; Holland 2006). With the development of microbial life, biologically mediated supergene minerals could form (Dove 2010; Walker et al. 1981; Warke et al. 2020).

3 Mineralogy and Chemistry of the Weathering Environment

3.1 Chemical Weathering

Chemical weathering is one of the fundamental processes that have shaped the Earth's surface over long periods. It involves the alteration and decomposition of rocks, minerals and other geological materials through chemical reactions through the interaction of the composite mineralogy with chemical agents, principally water, oxygen and carbon dioxide (Trescase 1992). A considerable number of experimental and field studies have been undertaken to assess these processes (e.g., Steidtmann 1908; Goldich 1938; Thornber 1975a, b, 1983, 1985, 1992; Stoch and Sikora 1976; Berner 1981, 1984, 1990; Thornber and Wildman 1984; Thornber et al. 1981; Siever and Woodford 1979; Bowell 1993; Velbel 1993; White 1995; Kowalewski and Rimstidt 2003; Boyle 2003; Holland 2005; Meunier 2005; Leybourne and Goodfellow 2007, 2010; Dixon and von Blanckenburg 2012; West 2012).

Chemical weathering occurs when minerals become unstable on exposure to surface conditions (Berner 1981). The changes that take place are highly specific to the mineral and the environmental conditions. Some minerals, like quartz, are virtually unaffected by chemical weathering, whereas others, such as feldspar (especially Ca-rich), are more readily altered. In general, the degree of chemical weathering is greatest in warm and wet climates and least in cold and dry climates (Nesbitt and Young 1982). The important characteristics of surface conditions that lead to chemical weathering are the presence of water (in the air, at the ground surface and as groundwater), the abundance of oxygen and the presence of carbon dioxide, which produces weak carbonic acid when combined with water (Berner and Caldeira 1997). That process, which is fundamental to most chemical weathering, can be shown as follows:

$$H_2O + CO_2 \longrightarrow H_2CO_3 \quad \text{Water + carbon dioxide} \longrightarrow \text{carbonic acid} \tag{2}$$

$$H_2CO_3 \longrightarrow H^+ + HCO_3^- \quad \text{Carbon dioxide} \longrightarrow \text{hydrogen ion + carbonate ion} \tag{3}$$

Here, water (e.g., rain) and CO_2 in the atmosphere combine to form carbonic acid. Subsequently, carbonic acid dissociates to form hydrogen (acidity) and bicarbonate ions. The amount of CO_2 in air is only sufficient to make very weak carbonic acid, but there is typically much more CO_2 in the soil, so water that percolates through the soil zone can become more acidic, i.e. lower pH.

3.2 Goldich's Reaction Series

Minerals differ widely in their resistance to chemical weathering. The mineral series based on the observations of Goldich (1938), depicted in Fig. 2, summarizes the relative resistance to weathering of the common rock-forming silicate minerals and indicates that minerals crystallized at the highest temperatures, under the most

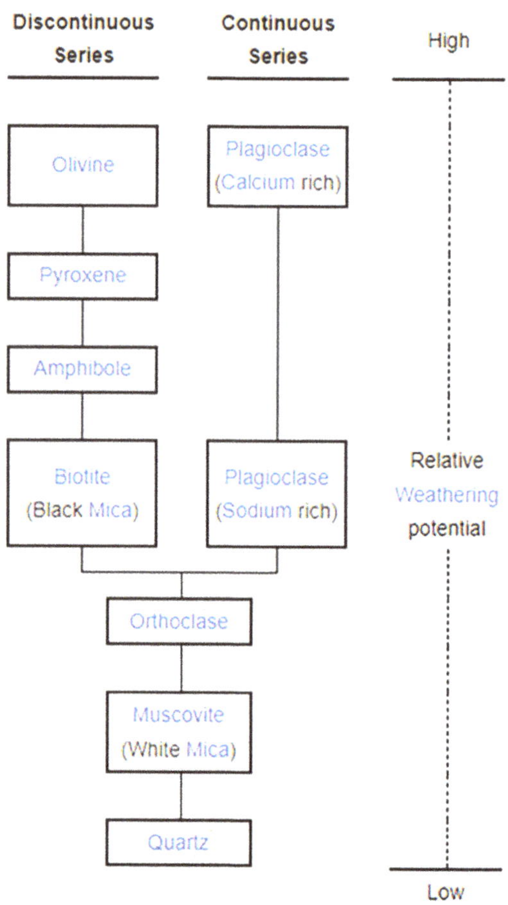

Fig. 2 Goldich's reaction series. (Modified from similar schemes by Goldich 1938; Parker 1970)

anhydrous conditions, are more readily weathered than those that crystallized last from lower temperature, more water-rich magmas (Bhatti et al. 1994; Brantley 2003).

The Goldich dissolution series follows the same trend as Bowen's reaction series, with the minerals that are first to crystallize also the first the undergo chemical weathering (Bowen 1922, 1956). The Bowen's reaction series dictates that during fractional crystallization, olivine and Ca-rich plagioclase are the first to crystalize out of a melt, after which follow pyroxene, amphibole, biotite, Na-rich plagioclase, orthoclase, muscovite and, finally, quartz. This order is controlled by the temperature of the melt and its composition. Because the surface of the Earth is a low-temperature environment compared to the zones of igneous rock formation, Fig. 3 also reflects the relative stability of minerals, with the ones at the bottom being the most stable and the ones at the top being the quickest to weather (Parker 1970). This is not the only control on the weathering rate, which is dependent on both intrinsic (qualities specific to the minerals) and extrinsic (qualities specific to the environment) variables (Berner 1981).

Climate is a key extrinsic variable, controlling the water-to-rock ratio, pH and alkalinity, all of which impact the rate of weathering (Goldich 1938). The Goldich dissolution series is in part based on earlier work by Steidtmann (1908) who determined that the order of mineral ion release in response to weathering for the major elements was:

$$CO_3{}^{2-} > Mg^{2+} > Ca^{2+} > Fe^{2+} > Na^{+} > K^{+} > SiO_2 > Fe^{3+} > Al^{3+}$$

Climate or rather climatic agents such as water availability and mass, as well as temperature changes, control the water-to-rock ratio that in turn generates pH, redox and alkalinity, all of

Fig. 3 Karst features: (**a**) failed cave system, Otjikoto Lake, Namibia (field of view, approximately 50 m); (**b**) aragonite 'pig's ear', Lucky Hill Mine, Sarawak, Malaysia (field of view 0.5 m); (**c**) 'Organ pipes', Otavi Mountains, Namibia (field of view 30 m); and (**d**) Fairy Cave and shrine, Sarawak, Malaysia (field of view 40 m)

which control the rate of mineral weathering (Berner 1981). The mineral evolution observed in the weathering regime is a function of changes in temperature, availability of oxidants and acidity, as well as changes in water chemistry (Parker 1970). Although Goldich's original order of mineral weathering potential was qualitative, later work has placed more quantitative estimates on chemical weathering (Berner 1981; Thornber 1983, 1985; White 1995; Kowalewski and Rimstidt 2003).

Chemical weathering of minerals in bedrock leads to the formation of secondary minerals, which constitute the weathering products of the parent minerals. Secondary minerals, the minerals formed in the weathering environment, are important representatives of the prevailing geochemical conditions of that environment (Pacheco and Van der Weijden 2014). Secondary weathering of minerals in igneous rocks can be classified mainly as iron oxide minerals, salts and phyllosilicate minerals. The chemistry of secondary minerals is controlled in part by the chemical and mineralogical composition of the parent rock and its minerals.

Mafic rocks tend to contain higher proportions of Mg and ferric and ferrous Fe, which can lead to secondary minerals with high abundance of these cations, including serpentine, Al-, Mg- and Ca-rich clay minerals, carbonate minerals and Fe oxide minerals, such as hematite (Stoch and Sikora 1976; Nesbitt and Young 1982; Li et al. 2016). Felsic rocks have relatively higher proportions of K and Na, which lead to secondary minerals rich in these ions, including Al-, Na- and K-rich clay minerals such as kaolinite, montmorillonite and illite (Sequeira Braga et al. 2002; Meunier 2005). Experimental work has shown some limitations of the Goldich dissolution series, most notably variations in mineral weathering rates of different minerals that overlap and are not as pronounced as presented in Fig. 3 (Kowalewski and Rimstidt 2003; White and Brantley 2003).

4 Mechanisms of Tropical Weathering

Intense tropical heat, followed by heavy rainfall, leads to thermal expansion and contraction of rocks, causing them to crack and disintegrate (Tardy 1981). This process, exfoliation, is particularly prevalent in areas with high diurnal temperature variations. The frequent cycles of wetting and drying also contribute to the physical breakdown of rocks through the action of erosion and abrasion. In addition, the high density of flora in many tropical environments provides a wide range of flora and fauna that actively participate in weathering processes.

Rainwater and surface waters penetrate along structural zones of weakness in rocks such as fractures, joints, vein margins and cleavage planes. The scale can vary from centimetres to several metres, and the rate of water ingress can vary considerably. The penetration of water into massive crystalline rock is comparatively much slower, and the scale is restricted to a few millimetres. Water penetration into the minerals is orders of magnitude lower, a few microns or less, and is largely actioned by diffusion except with highly soluble minerals such as gypsum or calcite. Thus, the initial pace of weathering is slow and controlled by micro features. Penetrating solutions spread outward into mineral grains in a tube-like pattern and, as they do, expand mineral-water interaction and increase the rate of dissolution. Dissolved ions decrease in concentration away from the weathering front (Robertson and Butt 1993). The warm and wet climate of the humid tropics accelerates chemical reactions, leading to the dissolution, decomposition and alteration of minerals. Rainwater, rich in CO_2, forms carbonic acid that reacts with minerals, such as feldspar and limestone, leading to their dissolution.

Plant roots can accelerate weathering by penetrating into rocks, exerting pressure and causing mechanical disintegration. Burrowing animals, such as termites, ants and earthworms, facilitate the breakdown of organic matter, promoting the release of CO_2 (resulting in carbonic and bicarbonate acids; Eqs. 1, 2), nutrients and the development of fertile soils (Drever 1994; Guo et al. 2018; Deng et al. 2022). In addition to water-rock reactions, organisms have a role in the acceleration of mineral weathering, in particular by the production of soil organic acids (Shotyk 1992). These not only generate protons that can replace cations in the mineral lattice but also contain functional groups that can detach chelate metals. The principal acid formed is carbonic acid.

5 Types of Chemical Weathering

5.1 Dissolution in Water

Dissolution occurs when minerals or rocks dissolve in water or other solvents as chemical bonds within the minerals are broken down. Weathering reactions can involve the conversion of a primary mineral to a secondary mineral (e.g. feldspar to clay), partial weathering of a mineral releasing ions into solution or the complete dissolution of a mineral. Calcite, for example, will dissolve in weak acid, to produce calcium and bicarbonate ions:

$$\begin{aligned} &CaCO_3 + H^+ + HCO_3^- \longrightarrow Ca^{2+} + 2HCO_3^- \\ &\quad \text{Calcite} + \text{hydrogen ions} + \text{bicarbonate} \longrightarrow \text{calcium ions} + \text{bicarbonate ions} \end{aligned} \tag{4}$$

Calcite is the major component of limestone, and under surface conditions, limestone will dissolve to varying degrees (depending on which minerals it contains, other than calcite), to produce karst features (Maher and Chamberlain 2014; Waltham et al. 2005). These features occur on a highly variable scale up to tens of metres (Fig. 3).

Mineral dissolution reactions are classified as either congruent or incongruent. Congruent dissolution refers to minerals that dissolve completely into their constituent ions, e.g. calcite. By contrast, incongruent dissolution applies to minerals that only partially dissolve and leave a residual solid weathering product. For example, feldspar is altered by hydrolysis to clay minerals (Stoch and Sikora 1976). The incongruent dissolution of plagioclase feldspar can be written:

$$\begin{aligned} &CaAl_2Si_2O_8 + 2H_2CO_3 + \tfrac{1}{2}O_2 \\ &\quad \longrightarrow Al_2Si_2O_5(OH)_4 + Ca^{2+} + 2CO_3^{2-} \\ &\text{Plagioclase} + \text{carbonic acid} \longrightarrow \text{kaolinite} + \text{dissolved calcium} + \text{carbonate ions} \end{aligned} \tag{5}$$

Reaction 5 is representative of endmember calcium plagioclase (anorthite). Reactions can be similarly written for the solid solutions between the Ca and Na plagioclase endmembers and between the Na- and K-feldspar endmember minerals. In reaction 5, the products are the clay mineral kaolinite and dissolved calcium and carbonate ions.

A variation of dissolution is hydrolysis, which involves the reaction of minerals with water, leading to the formation of new phases. The hydrolysis of silicate minerals and the oxidation of Fe in ferromagnesian silicate minerals all serve to create softer and weaker weathered rocks that are more susceptible to mechanical weathering (White 1995; Meunier 2005). Silicate minerals undergo hydrolysis reactions, with the formation of different product minerals, depending on the composition of the primary mineral, fluid chemistry, pH and redox state. For example, pyroxene can be converted to clay minerals chlorite or smectite, and olivine can be converted to serpentine, brucite, tremolite or talc (Nesbitt and Young 1982).

5.2 Carbonation

Carbonation involves the reaction of minerals with carbon dioxide (CO_2) dissolved in water and leads to the binding of free protons (H^+) (Sverdrup 1990). To counter excess free protons, some minerals buffer or consume (neutralize) the generated acid. Acid-neutralization reactions result from mineral buffering of H^+ in aqueous systems. This buffering is commonly accompanied by the precipitation of secondary minerals (Nordstrom and Alpers 1999). These buffering reactions can reduce the impact of protons through weathering or sulfide oxidation by forming an inhibitory surface coating on the reactive sulfide minerals. Under acidic conditions, carbonate minerals (e.g., calcite, dolomite and magnesite) readily dissolve and provide bicarbonate alkalinity, which results in neutralization of acid and precipitation of metal hydroxide phases.

Carbonate minerals vary in their neutralizing reaction rate, i.e. calcite > dolomite = magnesite> ankerite > siderite. Both siderite and ankerite are Fe-bearing, which limits their neutralizing capacity because the Fe^{2+} in these minerals produces additional acidity owing to redox changes during hydrolysis reaction, i.e. Fe^{3+} is unstable in water and precipitates as $Fe(OH)_3$, liberating three moles of protons for every mole of Fe^{3+} lost from solution. The neutralizing order of the carbonate minerals is controlled, in part, by equilibrium mass-action constraints and by kinetic limitations (Morse 1983). Carbonate minerals (especially calcite) are not, as is commonly thought, the only source of acid naturalization in nature; silicate and hydroxide minerals also have neutralizing capacity, albeit at rates typically less than carbonate minerals.

The weathering of silicate minerals to neutralize acidity has been examined in many studies (Sverdrup 1990; Bhatti et al. 1994; Moss and Edmunds 1992; Kwong and Lawrence 1997; Bufe et al. 2021). Thus, silicate and hydroxide minerals must also be assessed in considering the buffering capacity of a lithology or regolith. From soil acidification studies, Sverdrup (1990) divided the most common minerals into six groups according to the pH dependency of their dissolution rate (Table 1). From these relative weathering rates, minerals in the slow to very slow neutralizing categories are kinetically inhibited from providing much neutralizing capacity. Even for minerals in the intermediate and fast mineral weathering groups, these minerals will not be practical neutralizing materials unless they occur in excess of ~10% (Sverdrup 1990).

5.3 Oxidation

Oxidation is an important weathering process and occurs when minerals react with oxygen, leading to a change in their chemical composition. Oxidation reactions occur through the breaking of bonds within a mineral and can include the loss or gain of electrons by components resulting in altered mineral structures. Iron oxidation from the weathering of ferromagnesian silicate minerals starts with the dissolution of the Fe (Nesbitt and Young 1982). Fayalitic olivine in the presence of carbonic acid is converted to aqueous ferrous Fe, carbonate and silicic acid:

$$\begin{aligned} &Fe_2SiO_4 + 4H_2CO_3 \rightarrow 2Fe^{2+} + 4HCO_3{}^- \\ &\quad + H_4SiO_{4(aq)} \\ &\text{Olivine} + (\text{carbonic acid}) \\ &\quad \rightarrow \text{dissolved Fe} + \text{dissolved carbonate} \\ &\quad + \text{silicic acid} \end{aligned} \tag{6}$$

In the presence of water, the dissolved Fe is then quickly converted to ferric hydroxide:

$$\begin{aligned} &2Fe^{2+} + 3H_2O + 4e - \rightarrow 2Fe(OH)_{3(s)} + 3H^+ \\ &\quad \text{Dissolved Fe} + \text{water} + \text{electrons} \\ &\quad \rightarrow \text{ferric hydroxide mineral} \\ &\quad + \text{protons(acid)} \end{aligned} \tag{7}$$

Equation 6 is for fayalite (the Fe endmember of the olivine solid solution with the Mg endmember forsterite (Mg_2SiO_4)), but it could apply to other ferromagnesian silicate minerals, including pyroxene, amphibole and biotite. Iron sulfide minerals (in particular, pyrite and pyrrhotite) are oxidized similarly (Eq. 7). The oxidation of ferrous Fe (Fe^{2+}) to ferric Fe (Fe^{3+}) results in the precipitation of ferric (oxy)hydroxide phases, as ferric iron is unstable in solution (Eq. 7; Figs. 6, 7). Ferric hydroxide phases are significant because they form an important attenuation sponge for many trace elements from solution (Bowell 1994; Leybourne and Johannesson

Table 1 Mineral groups according to neutralization potential (Sverdrup 1990)

Group name	Typical minerals	Buffering pH range[a] (s.u.)	Approx. NP[b] range	Relative reactivity[c]
1. Dissolving	Calcite, aragonite, dolomite, magnesite, aragonite, portlandite and brucite	6–11.2	7.8–14.8	1.0
2. Fast weathering	Anorthite, nepheline, olivine, garnet, jadeite, leucite, spodumene, kutnohorite diopside, siderite and wollastonite	5.5–11	2.8–0.6.2	0.6
3. Intermediate weathering	Epidote, zoisite, enstatite, hypersthene, augite, hedenbergite, hornblende, glaucophane, tremolite, actinolite, anthophyllite, serpentine, chrysotile, talc, chlorite and biotite	4.8–7.3	1.7–5.8	0.4
4. Slow weathering	Albite, oligoclase, labradorite, vermiculite, montmorillonite, manganite, goethite, gibbsite and kaolinite	2.4–5.1	0.5–2.9	0.02
5. Very slow weathering	K-feldspar, ferrihydrite and muscovite	2.2–4.1	0.2–0.6	0.01
6. Extremely slow weathering	Quartz, hematite, rutile and zircon	3.3–3.5	<0.01	0.004

[a]Buffering pH range evaluated by crushing 5 g of pure mineral and mixing with 5 mL of distilled water and left to react for 30 minutes. The pH of the distilled water was 3.4 s.u.
[b]NP range assessed as equivalent buffering potential of 10 g of pure mineral to calcite and titrated with hydrochloric acid. So for example, 10 g of portlandite ($Ca(OH)_2$) was found to have the equivalent capacity to neutralize HCl acid as 14.8 g of calcite, whereas 10 g of hornblende was required to buffer HCl acid to a similar pH to only 3.1 g of calcite
[c]Calculated from Sverdrup's eq. (1990), see below and based on 100% mono-mineral sample

2008). The results of this process are illustrated in Fig. 4, which shows gossan in an open pit, originally a massive sulfide deposit (Fig. 4a), weathered to ferric Fe oxide minerals. Also shown are ferric hydroxide, Cu carbonate and sulfate minerals formed by groundwater dissolution and mobilization of metals (Fig. 4; e–f).

A particular type of oxidation takes place in areas where the rocks have elevated levels of sulfide minerals, especially pyrite (FeS_2) and pyrrhotite ($Fe_{1-x}S$, where $x = 0 - 0.1235$). Pyrite reacts with water and oxygen, as follows:

$$4FeS_2(s) + 15O_2(g) + 14H_2O(l) = 4Fe(OH)_3(s) + 8SO_4{}^{2-}(aq) + 16H^+(aq)$$
$$\text{Pyrite} + \text{oxygen} + \text{water} \rightarrow \text{Fe ions} + \text{sulfate} + \text{hydrogen ions} \quad (8)$$

Understanding the controls on weathering is important to ensure adequate characterization of all aspects of the deposit and potential lithological and alteration types. However, it is not just the primary element dispersion that is important, but for many deposits, the zonation produced by sulfide oxidation is also critical, e.g., for mineral resources (Fig. 5).

Sulfide oxidation can increase the loading of metals, sulfate and acidity in the receiving environment. Oxidation of ferrous iron and hydrolysis of ferric iron at pH > 2 provide the additional source of acidity through the reactions:

$$4FeS_2 + 16O_2 = 4Fe^{2+} + 8SO_4{}^{2-} \text{ and}$$
$$4Fe^{2+} + 10H_2O + O_2 = 4Fe(OH)_{3(s)} + 8H^+ \quad (9)$$

For each mole of pyrite oxidized, most of the available hydrogen ions are precipitated as ferric Fe hydroxide so that only a portion of the hydrogen is released as acidity. Note that oxidation of Fe^{2+} is the main acid-producing reaction. Some hydrogen ions can also be sequestered as oxidized metal-sulfate minerals, some of which are highly soluble and, as such, can represent an instantaneous source of acidic, metal-sulfate-rich water

Fig. 4 Field evidence of sulfide oxidation: (a) gossan profile in open pit, Mina San Miguel, Huelva, Spain; (b) gossan (in red) overlying massive sulfide, Xeros, Cyprus; (c) lateritized gossan (orange), Justice Pit, Ashanti, Ghana; (d) gossan zone (orange-black-stained rock) in ore pillar, Plaka Mine#80, Lavreotiki, Greece; (e) copper-bearing aragonite (blue) precipitated from groundwater interaction with limestone level 29, Tsumeb Mine, Namibia; and (f) chalcanthite, woodwardite and other Cu sulfates precipitated from sulfide-water interaction, 150 fathom workings, Botallack, Cornwall

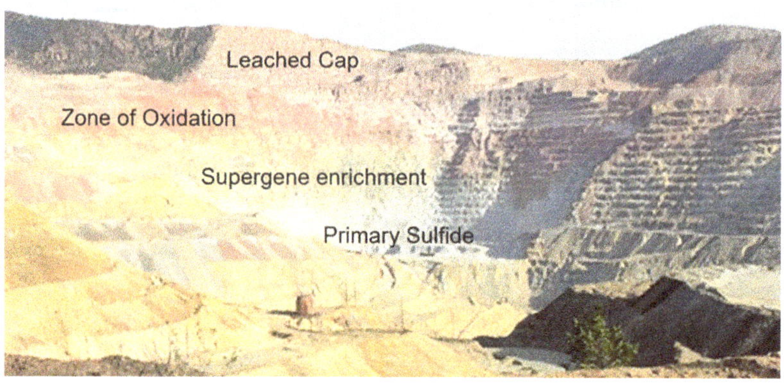

Fig. 5 Oxidation zoning in a porphyry copper deposit (Santa Rita pit, New Mexico). Labels indicate the standard supergene weathering profile of such deposits in semi-arid to arid environments. Supergene enrichment has upgraded the Cu content of Cu sulfide minerals and increased the economic value of the mineralization

upon dissolution and hydrolysis. For example, melanterite ($Fe^{2+}SO_4.7H_2O$) is readily soluble as:

$$4Fe^{2+}SO_4.7H_2O + O_2 = 4FeO.OH + 4SO_4{}^{2-} + 8H^+ + 22H_2O \quad (10)$$

Other metal-sulfate minerals formed through oxidation of sulfide minerals are less soluble including alunite ($KAl_3(SO_4)_2(OH)_6$) and jarosite ($KFe_3(SO_4)_2(OH)_6$). Sulfide oxidation and formation of ferric hydroxide phases can be visualized through the use of an Eh-pH diagram as shown in Fig. 6 (Sato 1992).

The textural features of sulfide oxidation are characteristic and have been used effectively in the mapping and exploration of sulfide mineral deposits through surface expressions (Locke 1926; Blanchard 1968; Thornber and Taylor 1992; Taylor 2011). The development, genesis

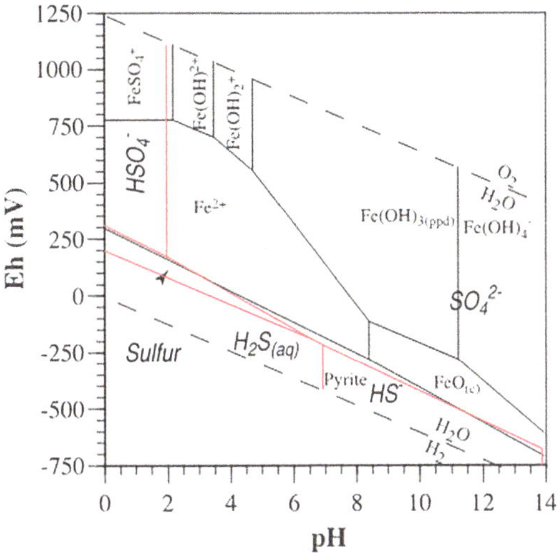

Fig. 6 Eh-pH diagram for the Fe-S-O-H system at 298 K. Sulfur speciation, with $[SO_4] = 10^{-2}$, $[HCO_3] = 10^{-3}$, $[Cl] = 10^{-1}$ and $[Fe] = 10^{-3}$. Square brackets = activity in moles/L. Black lines are for Fe speciation, and red lines are for S speciation

and properties of such features, termed gossans, are discussed in more detail in chapter "Gossans: In Situ Supergene Base Metal Deposits".

5.4 Alkaline Reactions

In highly weathered or alkaline environments, even silica can become mobile and result in the formation of bauxite-rich regolith. This is discussed further in chapter "Lateritic Bauxite Deposits".

The weathering of feldspar results in the formation of calcite (Blum 1994):

$$\begin{aligned} &CaAl_2Si_2O_8 + 4OH^- + 5H_2O + CO_2 \\ &= \begin{matrix} 2\left(Al(OH)_4{}^-\right) + 2\left(H_3SiO_4{}^-\right) \\ + CaCO_3 \text{Anorthite calcite} \end{matrix} \end{aligned} \tag{11}$$

5.5 Acidic Reactions

In contrast, in highly acid environments, it is cations that become more mobile, and silica is retained; therefore, many gossans have a silica framework (Blanchard 1968; Boyle 2003). These reactions typically occur at pH < 5 in equatorial regolith.

$$\begin{aligned} &CaAl_2Si_2O_8 + 8H + +2H_2O \\ &= \begin{matrix} Ca^{2+} + 2\left[Al^{3+} --- RX\right] + 4H_2O + 2H_4SiO_4 \\ \text{Anorthite soluble organic chelate + quartz} \end{matrix} \end{aligned} \tag{12}$$

5.6 Transformation

Where minerals have adaptable structures, for example, phyllosilicate minerals, chemical weathering can occur through ion exchange as 'solid-state' reactions. In these, cations can be exchanged with those in solution, but the Al-Si structure of the mineral is preserved (Oliva et al. 2003; Meunier 2005), e.g. in the alteration of muscovite to vermiculite (White 1995). Such a reaction occurs through a mechanism as follows:

$$\begin{aligned} &KAl_2(AlSi_3O_{10})(OH)_2 + Me^{2+} + Al^{3+} \\ &\quad + Si(OH)_4{}^{4-} + 6\,H_2O \\ &\quad = (Me)Al_4Si_4O_{10}(OH)_2 \cdot 8H_2O + O_2 \\ &\quad + K^+ \end{aligned} \tag{13}$$

where Me = Mg, Fe^{2+} etc.

This type of weathering is common in humid tropical climates but can also occur in any climate during the weathering of mafic volcanic rocks (Li et al. 2016). An important economic example of solid-state weathering is the alteration of serpentine minerals in ultramafic rocks to a 'garnierite' phase in the Ni laterites (Trescases 1979; chapter "Nickel-Cobalt Laterite Deposits") by a mechanism such as:

$$\begin{aligned} &Mg_3Si_2O_5(OH)_4 + Ni^{2+} \\ &= \begin{matrix} (Mg_2, Ni)Si_2O_5(OH)_4 + Mg^{2+} \\ \text{Serpentine + nickel} \end{matrix} \\ &= \text{'garnierite' + magnesium} \end{aligned} \tag{14}$$

6 Rock-Water Interaction

6.1 Introduction

Groundwater moves through an aquifer contacting more rock than would surface water in a stream, and consequently, greater mineral-water reaction may potentially occur although the minerals comprising most aquifers are relatively inert or react very slowly (Hem 1986; Leybourne and Goodfellow 2007).

In some areas, aquifer rocks or sediments include minerals that could potentially contaminate the water with elements that might make the water less than ideal for human consumption or agricultural use. Examples include Cu, As, Hg, Pb, Cd, F, Na and B. Such contamination may occur because the aquifer rocks contain high abundances of the element in question (e.g., Leybourne and Cameron 2008) or some hydrological or hydrogeochemical features cause concentrations to build up to significant levels (Hem 1986). The features that contribute to the dispersion or retention of an element during weathering are functions of a variety of factors including aqueous dissolution, speciation, complexation, redox state and pH. Dissolution during weathering is related to the instability of a mineral compound with respect to water chemistry and can lead to:

- Aqueous interactions of dissolved constituents and interactions along the flow path that may lead to ion exchange, precipitation or change in aqueous speciation.
- Mineral-water interactions along the flow path that may lead to saturation with respect to a mineral compound or interaction with minerals that leads to saturation or attenuation, such as adsorption.

6.2 Chemical Speciation

A chemical element can exist in solution in different species as a function of the redox potential, pH and overall chemistry of the solution (Fig. 7). For example, Mn, depending on water chemistry, can occur as various ionic species—cations, anions or neutral ions (Fig. 8). Elemental complexes have significant control over mobility and the propensity for adsorption to mineral surfaces. For example, rare earth element complexation results in fractionation across the REE

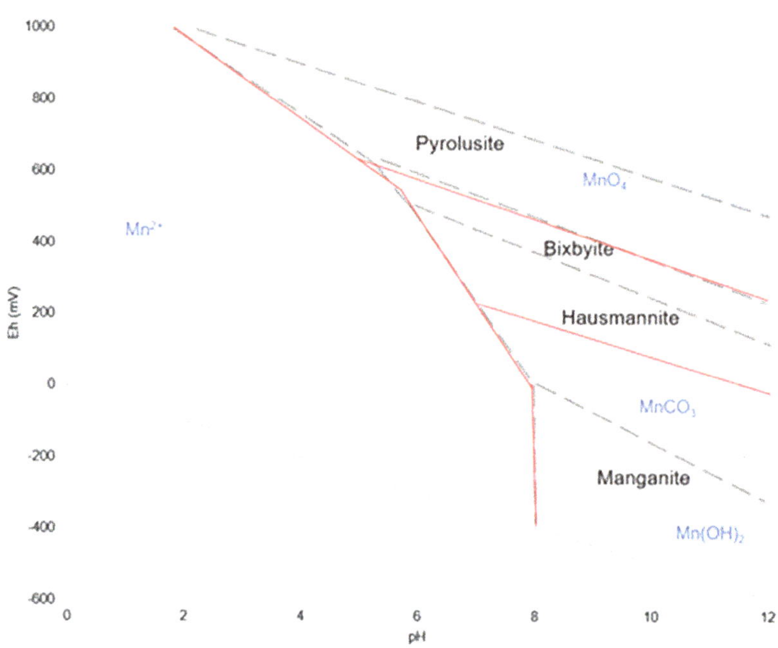

Fig. 7 Eh-pH diagram Mn-C-O-H system at 298 K. Mn = $[10^{-4}]$ M and $CO_3^{2-} = [10^{-2}]$ M. The speciation of some elements such as Mn and Fe is highly sensitive to the redox potential of the water with which they are in contact, whereas others like Cu are controlled more by the pH

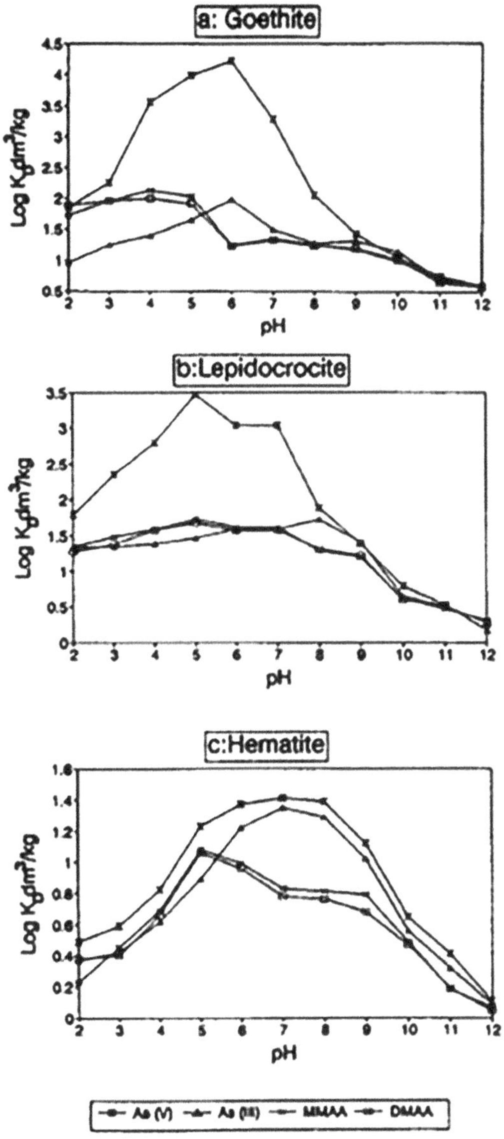

Fig. 8 Adsorption of arsenic species as a function of pH. Plots of log distribution coefficient (Ko, dm^3kg^{-1}). $[As] = 10^{-6}$ M. (**a**) Goethite, (**b**) lepidocrocite and (**c**) hematite. MMAA is monomethyl arsenic acid, and DMAA is di-methyl arsenic acid (From Bowell 1994)

as a function of pH and available ligands. Thus, at the pH of seawater, complexation with carbonate is important, especially for the heavy REE, which preferentially form $HREE(CO_3)_2^-$ complexes, compared to the light REE which are preferentially adsorbed to Fe- and Mn- oxide mineral surfaces and are lost from seawater. Conversely, at low pH (<4), humic acid-REE complexes enhance REE adsorption to mineral surfaces (Tang and Johannesson 2010). Similarly, Au has low solubility in most waters because it forms $AuOH(H_2O)^0$ complexes, which limits Au solubility to a few parts per trillion (ppt) (Leybourne et al. 2000). However, Au solubility is enhanced in the presence of thiosulfate ($S_2O_3^{2-}$), halides and cyanide (CN^-) (Vlassopoulos and Wood 1990; Gray et al. 1992; Bowell et al. 1993; Leybourne et al. 2000).

6.3 Biological Controls

Microbial communities within soils play crucial roles in mediating chemical processes that govern nutrient cycling, organic matter decomposition, and the transformation of contaminants (Johnson and Hallberg 2003). Microbial-mediated chemical processes in soils encompass a diverse array of transformations, including organic matter decomposition, nitrogen fixation, nitrification, denitrification, phosphorus solubilization and metal mobilization. These processes are driven by microbial enzymatic activities, metabolic pathways and interactions with soil minerals and organic compounds. The cycling of carbon, nitrogen and phosphorus is intricately linked to microbial metabolism, with microorganisms acting as catalysts for the conversion of organic substrates into simpler compounds and vice versa. Microbial decomposition of organic matter releases nutrients such as nitrogen and phosphorus, which are subsequently utilized by plants and other soil organisms.

Moreover, microbial communities contribute to the transformation and detoxification of pollutants through processes such as biodegradation, sorption and immobilization. Microorganisms can degrade a wide range of contaminants, including hydrocarbons, pesticides and heavy metals, thereby mitigating the environmental impacts of pollution.

The activity and diversity of soil microbial communities are influenced by various factors, including soil physicochemical properties, land management practices, climate and vegetation cover. Changes in these factors can alter microbial community composition and activity, with cascading effects on soil chemical processes and ecosystem functioning.

Biological oxidation is an important mechanism for the passive attenuation of sulfate and other nutrient-rich species, such as ammonia (Johnson and Hallberg 2003). Much of the reaction occurs in the root zone or rhizosphere with the generation of organic acids and interaction with minerals. Some microbial-mediated reactions make use of energy generated in mineral breakdown or formation as part of their metabolism. For example, some divalent metals (e.g., Fe, Zn, Pb) can be removed by precipitation as sulfide minerals following microbial sulfate reduction. Using Fe as an example, a simple, complete reaction can be written as follows:

$$Fe^{2+} + H_2S_{(g)} + 2HCO_3^- \rightarrow FeS_{(s)} + 2H_2O + 2CO_{2(g)} \quad (15)$$

Mackinawite FeS or, alternatively, greigite (Fe_3S_4) may form (Skousen et al. 2017). Both are generally precursors to pyrite. Thus, precipitating metals as a sulfide phase typically repeats the cycle that placed the metals in the deposit originally, reversing the oxidation reaction that liberated them (Rose 2010). Some microbes also enhance sulfide mineral oxidation, especially at low pH conditions. For example, *Acidithiobacillus ferrooxidans* enhances the oxidation of pyrrhotite, pyrite and chalcopyrite (e.g., Vera et al. 2013; Kocaman et al. 2016).

6.4 Chemical Adsorption and Precipitation

Chemical adsorption is a fundamental process occurring at the mineral-water interface. The chemical adsorption of major ions and trace elements to mineral surfaces has significant environmental implications (Bowell et al. 2020). Adsorption influences the mobility, bioavailability and fate of trace elements in natural systems. For example, the adsorption of transition metals to mineral surfaces can decrease their mobility, preventing them from leaching into groundwater or being transported in surface waters. Adsorption can help mitigate the environmental impact of contaminated sites and reduce the potential for human exposure to toxic metals. Adsorption onto mineral surfaces can also affect nutrient cycling in soils and sediments. Trace elements such as phosphorus or micronutrients can be adsorbed onto mineral surfaces, making them less available for uptake by plants or microorganisms. This can impact the productivity and nutrient balance of ecosystems.

Understanding the adsorption behaviour of trace elements can aid in the development of remediation strategies for contaminated sites. Commonly, this requires site-specific data to be calculated to determine the efficiency and pathway of attenuation (Bowell 1994; Bowell et al. 2015, 2020). By characterizing the adsorption capacity and mechanisms of minerals, it is possible to design efficient methods to remove or immobilize trace elements in soils, sediments or water. In tropical soils, adsorption of As species, for example by different Fe oxyhydroxide minerals, can be very different, and the soil mineralogy as well as As speciation will determine the extent to which As can be attenuated (Fig. 8).

Chemical adsorption and absorption of trace elements to mineral surfaces can occur through several mechanisms, including surface complexation, ion exchange and precipitation (Chapman et al. 1983; Dzombak and Morel 1990; Stumm and Morgan 1992). Surface complexation involves the formation of specific chemical bonds between trace elements and functional groups on mineral surfaces (Stumm and Morgan 1992). These functional groups may include hydroxyl (-OH), carboxyl (-COOH) or amino (-NH2) groups, which are present on the surface of many minerals. The strength and stability of these bonds depend on factors such as the mineral composition, pH and the nature of the trace element.

Trace elements may also be sorbed onto mineral surfaces by ion exchange. Here, trace elements replace other ions that are already present in the mineral lattice. This exchange occurs due to differences in the ionic charges and sizes of the ions involved. For example, in clay minerals, cations such as Na, K or Ca can be exchanged with transition metals such as Ag, Pb, Cu, Ni or Zn. Several factors influence the adsorption of trace elements to mineral surfaces. These include the mineralogy and surface properties of the minerals, pH, temperature, ionic strength of the solution and the activities of the trace elements in the solution. The mineral composition determines the availability and reactivity of surface functional groups for adsorption. For example, minerals rich in Fe and Al oxides tend to have a higher affinity for trace elements due to the presence of hydroxyl groups on their surfaces. The critical parameters that influence the speciation of trace elements and the surface charge of minerals are pH and Eh or reduction-oxidation potential (Baas-Becking et al. 1960). As shown for As schematically in Fig. 9.

Changes in pH can alter the surface charge of minerals, thereby influencing the electrostatic interactions between the mineral surface and the trace elements (Bowell 1994). Additionally, variations in pH can affect the solubility and speciation of trace elements, which in turn influence their adsorption behaviour. The activity of trace elements in solution can also influence their

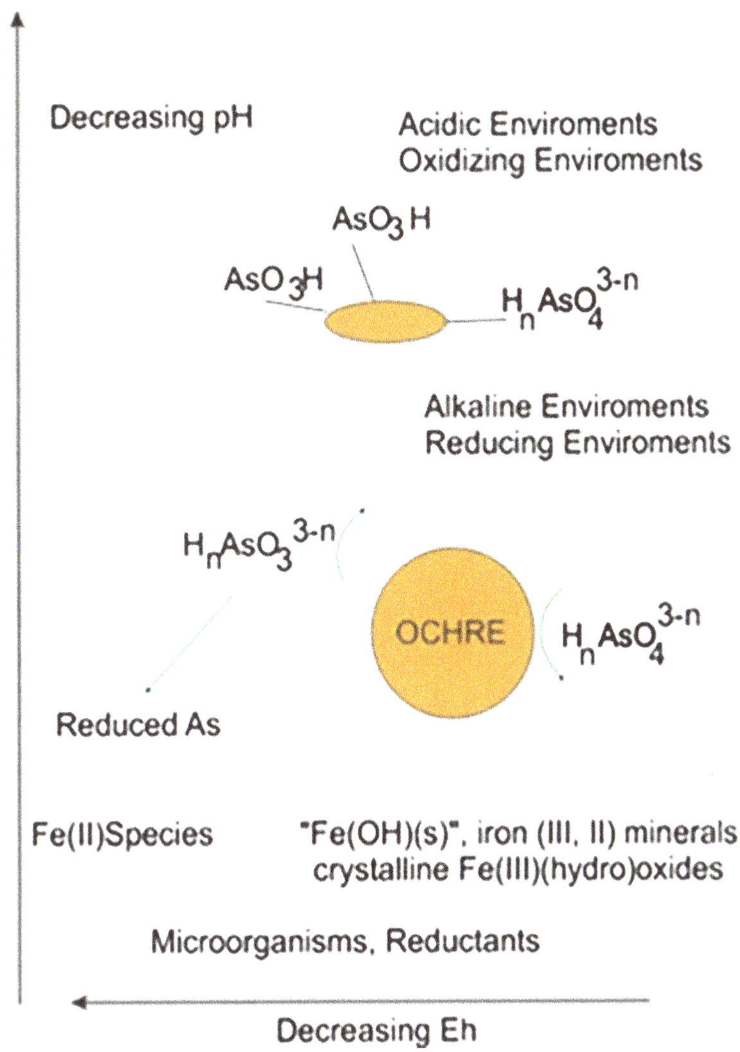

Fig. 9 Schematic representation of arsenic attenuation as a function of Eh and pH

adsorption onto mineral surfaces. At low activities, trace elements may compete for available adsorption sites on mineral surfaces, leading to greater adsorption efficiency. However, at higher activities, the adsorption sites may become saturated, resulting in reduced adsorption capacity.

Precipitation or co-precipitation occurs when trace elements are incorporated into the crystalline structure of minerals during their formation. These processes can happen through the precipitation of secondary minerals in the presence of trace elements or the co-precipitation of trace elements. The solubility and speciation of the trace elements in the fluid phase play a crucial role in determining their propensity for precipitation onto mineral surfaces. Mineral precipitation in the vadose zone is influenced by several factors, including the chemical composition of the infiltrating water, temperature, pressure, pH and the presence of dissolved ions. Where there is potential for competing ions, then a multitude of minerals can form dependent on the pH and Eh conditions at the point of precipitation, e.g. at the Tsumeb mine, Namibia (Bowell 2014).

However, in some deposits, it is not just as oxide zone minerals that attenuation can occur but also as sulfide minerals, replacing the oxide zone assemblage. For example, in the case of galena replacing pyromorphite, Breithaupt (1863) identified in mineral deposits in Germany and the UK (Fig. 10a). Another example is that of tennantite replacing azurite at the Tsumeb mine, Namibia (Fig. 10b).

When water percolates through the oxide zone, it interacts with the surrounding rock and soil, leading to the dissolution of primary minerals. Subsequently, as the water moves vertically or laterally, changes in chemical conditions can trigger mineral precipitation.

Mineral precipitation occurs when the concentration of dissolved ions in the water exceeds their solubility limit. Saturation occurs when the dissolved ions are in equilibrium with the mineral phase, whereas supersaturation arises when the ion concentration surpasses the equilibrium level. Supersaturation can be caused by evaporation, cooling, changes in pH or mixing of different fluids.

Precipitation is preceded by nucleation, a mechanism by which small insoluble particles aggregate with each other or else form an interface with a surface, such as a mineral or within the

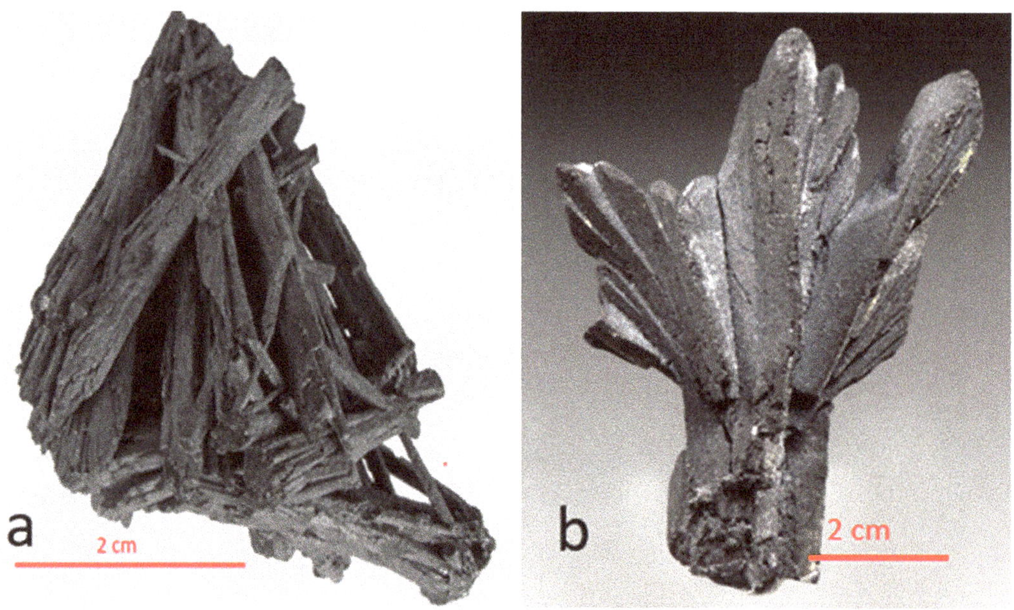

Fig. 10 (**a**) Galena pseudomorph of pyromorphite from Wheal Hope, Perranzabuloe, Cornwall, and (**b**) tennantite pseudomorph of azurite from the Tsumeb mine, Namibia

pore of an aggregate. Nucleation is the initial step in mineral precipitation, where dissolved ions cluster together to form solid nuclei which act as the foundation for further crystal growth. The rate of nucleation is influenced by factors such as temperature and ion concentration as well as suitable sites for precipitation. The nuclei grow through the addition of more ions from the surrounding solution. The growth rate depends on factors such as temperature, ion concentration, pH and the availability of space for crystal expansion. The resulting crystals can range in size from microscopic to macroscopic, with their morphology determined by environmental conditions.

7 Climate Controls on Chemical Weathering

Chemical weathering is controlled by water availability and temperature (Williams 1990; Pacheco et al. 2014; Li and Yang 2010; Perri 2020; Deng et al. 2022), i.e., where there is water, there can be chemical weathering and vice versa (Trescases 1992). Soil type and thickness are related to the intensity of chemical weathering, which is controlled by contact with water and the ambient temperature.

Chemical weathering of silicate minerals has been cited as being an important control of Earth's climate over geological timescales (Deng et al. 2022). As discussed, climate is a critical control on regolith development, with thicker more evolved terrains in environments with higher temperatures and greater precipitation. The temperature dependency of silicate weathering has been demonstrated in several laboratory and field-based studies (White 1995; Kowalewski and Rimstidt 2003). More recently, studies have attempted to assess the processes on a global scale to quantify the importance of climate on silicate weathering. The role of silicate weathering in mountain terrains has even been postulated to present a larger potential CO_2 sink than the Amazon rainforest per square kilometre (Berner 1991). The rate of weathering has been linked to plant productivity (Oeser and von Blanckenburg 2020; Olejarz et al. 2021).

As the present is a key to the past, one approach to test the link between environmental forcing factors and silicate weathering is to focus on an abundant modern dataset that covers a large environmental gradient (Willenbring and von Blanckenburg, 2010). A recent study by Deng et al. (2022) assessed siliciclastic sediment weathering by determining the chemical index of alteration (CIA) and weathering index of Parker (WIP), as these are generally considered the most appropriate proxies to assess chemical weathering and element mobility (Parker 1970; Nesbitt and Young 1982; McLennan 1993). The CIA is the proportion of immobile Al_2O_3 versus labile oxides (CaO, Na_2O and K_2O) in the silicate fraction, whereas the WIP is the sum of molar proportions of labile elements (Ca, Mg, Na and K) weighted by their susceptibility to weathering (Deng et al. 2022). Both indicate the intensity of silicate weathering and track feldspar hydrolysis reactions. Deng et al. (2022) amassed a compiled database of major element geochemistry for siliciclastic sediment from modern rivers across six continents to generate CIA-WIP calculation ($n = 3828$). Analysis was focused on silt- and clay-sized sediments to minimize mineralogical differentiation (Fig. 11).

The results demonstrate stronger weathering in the humid tropics for, as expected, the low-latitude (<10°) regions are characterized by high CIA (mostly 80–100), whereas weathering index values at latitudes of >60° are closer to those of fresh rocks (CIA as low as ~50) (Fig. 12). These trends are consistent with latitude regardless of continent, indicating a strong global control. Although this observation is not new (Brantley and Lebedeva 2011; Misra and Froelich 2012); Deng et al. (2022) is the first study to demonstrate this on a global scale. Nesbitt and Young (1982) used a similar approach, although using a smaller dataset, to propose from the CIA that changes in the latitude of deposition of siliciclastic sediments (and thus climate) were a major control on the geochemistry of Proterozoic sedimentary rocks. The study by Deng et al. (2022) went further and evaluated the major controls of silicate weathering intensity. They defined four major controls on weathering:

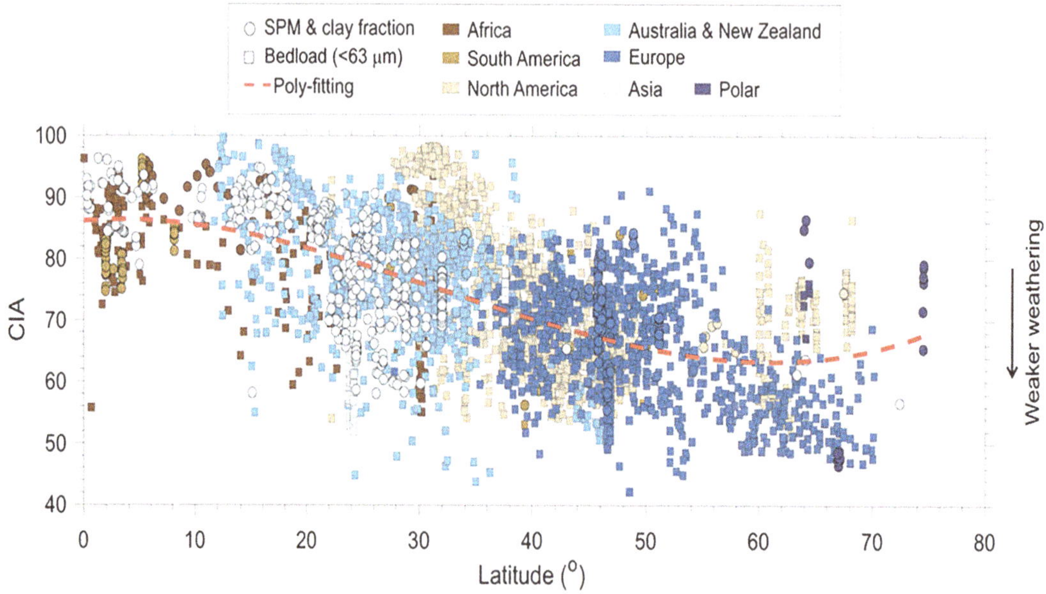

Fig. 11 Latitudinal distribution of CIA in fine-grained sediments. (After Deng et al. 2022)

climate, geomorphology, lithology and land cover. To identify the key variables governing the weathering indices, they compared correlation coefficients to the CIA (Fig. 12).

The weathering indices show a relatively poor correlation to land cover factors ($|R| \leq 0.12$) and lithological factors ($|R| \leq 0.20$). Plant-associated enhancement of weathering rates related to soil matrix or organic matter is separated from plant productivity. Ice as an agent is limited to high latitudes or high altitudes only. The nature of source rocks has a strong control over sand-sized coarse-grained sediments where unaltered primary silicates may still be present, whereas in finer-grained sediments that contain abundant clay minerals, chemical weathering is more dominant.

Temperature control of weathering intensity is reflected in the high correlation coefficient between mean annual temperature (MAT) and CIA ($R = 0.60$); however, the correlation to precipitation (MAP) is poor (0.21). This reflects the complexity of water sources in weathering and the cycling of solutes within the water and also their dispersion. Consequently, temperature can be considered a global control on chemical weathering, and by extension, any change in climate will influence rates of chemical weathering (chapter "The Role of Regolith Formation and Evolution in the Development of Supergene Mineral Deposits").

Nickel laterites are the products of the pervasive weathering of Ni-bearing silicates in ultramafic rocks exposed to humid tropical to subtropical environments. These deposits contain ~60% of the world's Ni resources and account for over 60% of annual global Ni production (chapter "Nickel-Cobalt Laterite Deposits"). Formation of such deposits involves a number of factors including geology, tectonics, topography, geomorphology and climate. From an evaluation of paleoclimate and modern climate data, the most favourable conditions are precipitation 1000–3100 mm pa. and alternating seasons with mean cold month temperatures between 15 and 27 °C and mean warm month temperatures between 22 and 31 °C (Thorne et al. 2012). The Çaldağ nickel laterite deposit of western Turkey is developed on a Cretaceous ultramafic complex that was exposed in the early Eocene, and the majority of the deposit formed during the Eocene (Thorne et al. 2012). By contrast, the development of the Bitincke deposit in Albania occurred only in the early Eocene and then stopped when the mean cold month temperature dropped below the range noted above (Fig. 13).

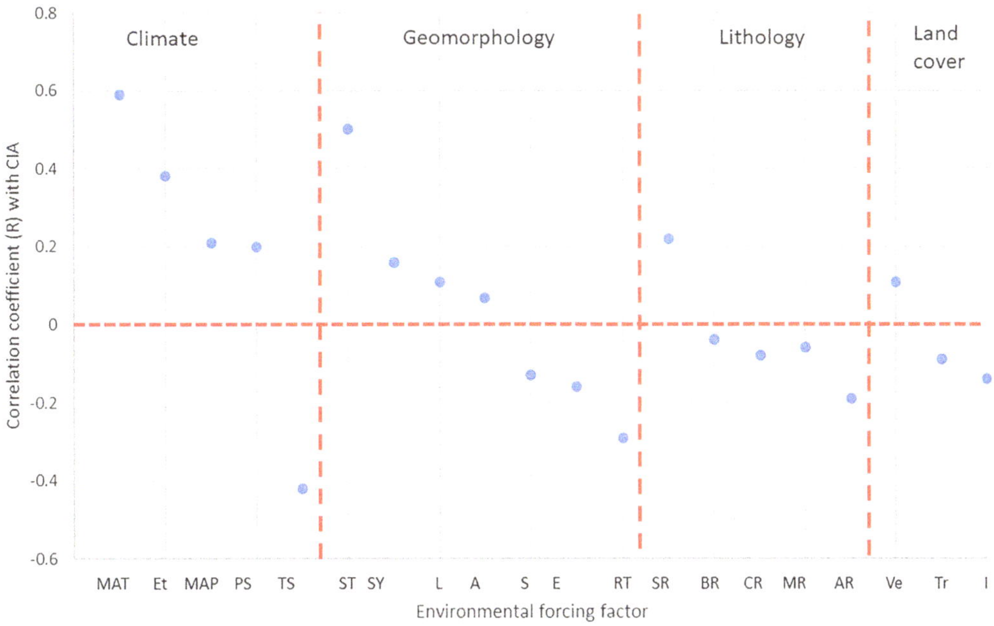

Fig. 12 Correlation coefficients (R) between each environmental forcing factor ($n = 20$) and CIA. (Modified after Deng et al. 2022). Correlations with a p value of >0.001 are shown as open symbols. The order of forcing factors in each category is sorted by correlation coefficient with CIA. Negative values mean a negative correlation. Most |R| are below 0.5, indicating a subordinate control of these factors on weathering intensity on a global scale. Mean annual temperature (MAT) is the only factor with |R | > 0.5. Climatic metric: MAT—mean annual temperature (°C), Et—actual evapotranspiration (m/yr), MAP—mean annual precipitation (m/yr), PS—precipitation seasonality (coefficient of variation, %) and TS—temperature seasonality (standard deviation, °C); geomorphic metric: ST—upland hillslope soil thickness (m), SY—modelled sediment yield (t/km^2/yr) using BQART model 29, L-flow length (km), A—drainage area (km^2), S—mean local slope (o), E—mean elevation (m) and RT—upland hillslope regolith thickness (m); lithological metric: SR—clastic sedimentary rock area (%), BR—basic rock area (%), CR—carbonate rock area (%), MR—metamorphic rock area (%) and AR acidic—intermediate rock area (%); and land cover metric: Ve—vegetation cover area (%), Tr—tree cover area (%) and I—ice and snow cover area (%)

Applying these data to other deposits and considering regional palaeoclimates could be used to determine periods of intense weathering in the geological record and identify regions with suitable geology that might also host Ni laterite deposits.

8 Conclusion

This chapter presents an overview of the geochemistry of supergene processes. It demonstrates supergene zones at the Earth's surface are produced under primarily predominantly oxidizing conditions. The processes require oxygen so that, although in places weathering can occur to depths of hundreds of metres, they are typically most active above the water table. They are functions of the primary and secondary mineralogy, oxygen availability and pH.

Chemical weathering is best expressed in the regoliths of the humid tropics that represent the residual component of bedrock. Such regoliths are enriched in relatively immobile elements which, in addition to Al, Fe and Si include many of the transition metals and REE hosted by resistant minerals.

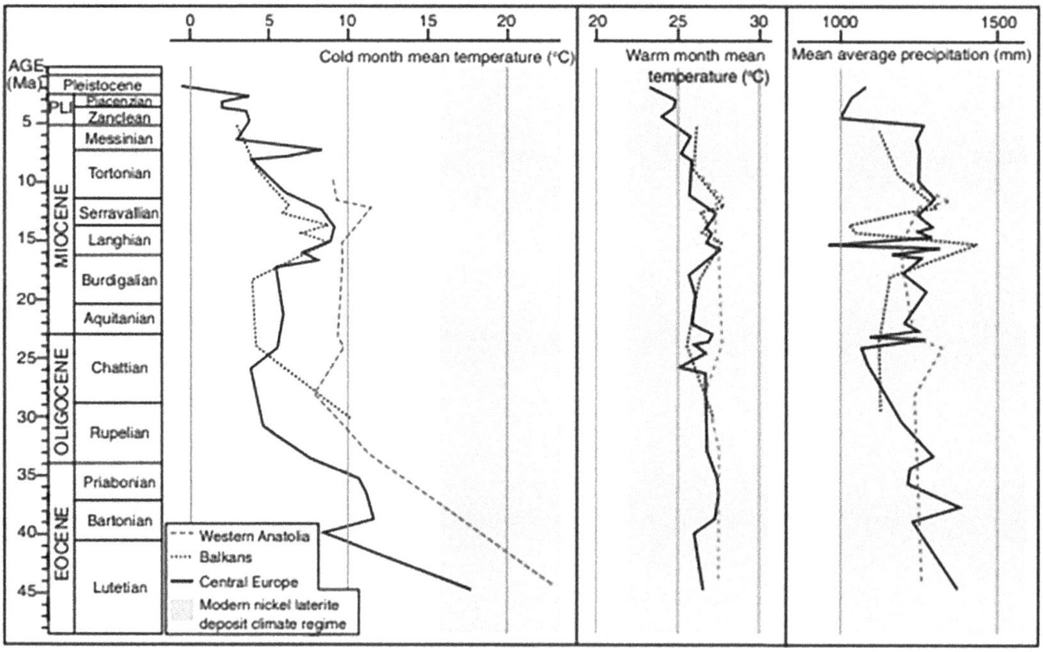

Fig. 13 Paleoclimatic data for Çaldağ (western Anatolia, Turkey), Bitincke (Albania) and Central Europe. Climatic conditions of modern-day nickel laterite deposit formation are shown as shaded. PLI = pliocene. (After Thorne et al. 2012)

Weathering can lead to the enrichment of some elements to the point that potentially economic ores are formed. Sulfide minerals are highly susceptible to oxidation supergene processes, as they form under reducing conditions and so are unstable above the water table. Due to the importance of oxygen in the breakdown of sulfide minerals, the maximum age of such products is generally around 2.2 billion years, i.e., after the 'Great Oxidation Event'. Climate has and continues to play an important role in the development of chemical weathering as it exerts control over precipitation, evaporation and temperature as well as concentrations of reactive gases in the atmosphere.

Acknowledgements The author would like to thank Mr. Stephen Phillips for expert drafting in this chapter. The input of authors who kindly approved the use of their published material in this chapter is greatly appreciated. Thank you to the reviewers: Charles Butt Chris Benn, Robert Thorne and Mehrooz Aspandiar who graciously gave their time and knowledge to greatly improve this chapter.

The Royal Society is thanked for approving the publication of Fig. 1, Elsevier is thanked for approving the reproduction of Fig. 8, the Geological Society of America is thanked for approving the publication of Fig. 13 and Springer Nature is thanked for approving the reproduction of Figs. 11 and 12.

References

Anbar AD, Duan Y, Lyons TW, Arnold GL, Kendall B, Creaser RA, Kaufman AJ, Gordon GW, Scott C, Garvin J, Buick R (2007) A whiff of oxygen before the great oxidation event? Science 317:1903–1906

Baas-Becking LGM, Kaplan IR, Moore D (1960) Limits of the natural environment in terms of pH and oxidation-reduction potentials. J Geol 68:243–284

Berner RA (1981) A new geochemical classification of sedimentary environments. J Sediment Petrol 51: 359–336

Berner RA (1984) Sedimentary pyrite formation: an update. Geochim Cosmochim Acta 48:605–615

Berner RA (1990) Global biogeochemical cycles of carbon and sulfur and atmospheric O2 over phanerozoic time. Chem Geol 84:159

Berner RA (1991) A model for atmospheric CO2 over phanerozoic time. Am J Sci 291:339–376

Berner EK, Berner RA (1996) Global environment: water, air and geochemical cycles. Prentice Hall, Upper Sattle River. 376 pp
Berner RA, Caldeira K (1997) The need for mass balance and feedback in the geochemical carbon cycle. Geology 25:955–956
Bhatti TM, Bigham JM, Vuorinen A, Tuovinen OH (1994) Alteration of mica and feldspars associated with the microbiological oxidation of pyrrhotite and pyrite. In: Alpers CN, Blowes DW (eds) Environmental geochemistry of sulfide oxidation, ACS symposium, American Chemical Society Series, volume 550, pp 90–107
Blanchard R (1968) Interpretation of leached outcrops. Carson City, NV, Nevada State Bureau of Mines Bulletin 66, 180 p
Blum AE (1994) Feldspars in Weathering, In: Parsons, I. (eds) Feldspars and their Reactions. NATO ASI Series, vol 421. Springer, Dordrecht. https://doi.org/10.1007/978-94-011-1106-5_15
Bowell RJ (1993) Mineralogical and geochemical characteristics of tropical rain forest soils, Ashanti region, Ghana. Chem Geol 106:345–358
Bowell RJ (1994) Arsenic sorption by Iron oxyhydroxides and oxides. Appl Geochem 9:279–286
Bowell RJ (2014) Hydrogeochemistry of the Tsumeb deposit: implications for arsenate mineral stability. Rev Mineral Geochem 79:589–630
Bowell RJ, Gize AP, Foster RP (1993) The role of fulvic acid in the supergene mobility of gold in tropical rain forest soils. Geochim Cosmochim Acta 57:4179–4190
Bowell RJ, Luetcke R, Luetcke MR (2015) Mineralogy of the Gallagher vanadium property and Manila mine, Cochise County, Arizona. Rocks and Minerals 90:338–352
Bowell RJ, Donkervoort L, Griffiths R, Bailey AS, Clarkson B, Amy Prestia A (2020) Adsorption of arsenic, and other elements in alluvium: implications for exploration and environmental geochemistry in the Great Basin. In: Vision for discovery. Proceedings GSN symposium, Reno, May 2020
Bowen NL (1922) The reaction principle in petrogenesis. J Geol 30(3):177–198
Bowen NL (1956) The evolution of the igneous rocks. Dover, New York, pp 60–62
Boyle DR (2003) Preglacial weathering of massive sulfide deposits in the Bathurst Mining Camp: economic geology, geochemistry, and exploration applications. In: Goodfellow WD, McCutcheon SR, Peter JM (eds) Massive sulfide deposits of the Bathurst Mining Camp, New Brunswick, and Northern Maine, Economic geology monograph, 11. Society of Economic Geologists, Littleton
Brantley SL (2003) Reaction kinetics of primary rock-forming minerals under ambient conditions. In: Holland HD, Turekian KK (eds) Treatise on geochemistry. Ch. 5.03, Elsevier, Amsterdam
Brantley SL, Lebedeva M (2011) Learning to read the chemistry of regolith to understand the critical zone. Annu Rev Earth Planet Sci 39:387–416
Breithaupt A (1863) Die Paragenesis der Mineralien. Informationszentrum der Bergakad, Freiberg, 292 p
Bufe A, Hoviu N, Emberson R, Rugenstein JKC, Galy A, Hassenruck-Gudipati HJ et al (2021) Co-variation of silicate, carbonate and sulfide weathering drives CO2 release with erosion. Nat Geosci 14(4):211–216
Cardona T, Murray JW, Rutherford AW (2015) Origin and evolution of water oxidation before the last common ancestor of the cyanobacteria. Mol Biol Evol 32:1310–1328
Chapman BM, Jones DR, Jung RF (1983) Processes controlling metal ion attenuation in acid mine drainage streams. Geochimica et Cosmochimica Acta 47 (11):1957–1973
Chavez WX (2021) Weathering of copper deposits and copper mobility: mineralogy, geochemical stratigraphy, and exploration implications. SEG Discovery, 2021:1–5
Deng K, Yang S, Guo Y (2022) A global temperature control of silicate weathering intensity. Nat Commun 13:1781
Dixon JL, von Blanckenburg F (2012) Soils as pacemakers and limiters of global silicate weathering. Compt Rendus Geosci 344:597–609
Dove PM (2010) The rise of skeletal biominerals. Elements 6:37–41
Drever JI (1994) The effect of land plants on weathering rates of silicate minerals. Geochim Cosmochim Acta 58:2325–2332
Dzombak D, Morel F (1990) Surface complexation Modeling: hydrous ferric oxide. J Wiley, New York
Eriksson PG, Cheney ES (1992) Evidence for the transition to an oxygen-rich atmosphere during the evolution of red beds in the lower Proterozoic sequences of southern Africa. Precambrian Res 54:257–269
Farquhar J, Bao T, Thiemens M (2000) Atmospheric influence of earth's earliest sulfur cycle. Science 289: 756–759
Frimmel HE (2005) Archaean atmospheric evolution: evidence from the Witwatersrand goldfields, South Africa. Earth-Sci Rev 70:1–46
Goldich S (1938) A study in rock-weathering. J Geol 46: 17–58
Gray DJ, Butt CRM, Lawrance LM (1992) The geochemistry of gold in lateritic terrains. In: Butt CRM, Zeegers H (eds) Regolith exploration geochemistry in tropical and subtropical terrains. Handbook of exploration geochemistry: 4. Elsevier, Amsterdam, pp 461–482
Guo Y et al (2018) Revisiting the effects of hydrodynamic sorting and sedimentary recycling on chemical weathering indices. Geochim Cosmochim Acta 227: 48–63
Gupta A (2011) Weathering in the tropics. In: Tropical geomorphology. Cambridge University Press, Cambridge, pp 61–81
Hazen RM, Ferry JM (2010) Mineral evolution: Mineralogy in the fourth dimension. Elements 6:9–12
Hazen RM, Morrison SM (2022) On the paragenetic modes of minerals: a mineral evolution perspective. Am Mineral 107:1262–1287

Hazen RM, Papineau D, Bleeker W, Downs RT, Ferry JM, McCoy TJ, Sverjensky DA, Yang H (2008) Mineral evolution. Am Mineral 93:1693–1720
Hem JD (1986) Study and interpretation of the chemical characteristics of natural water, USGS water supply paper 2254, 3rd edn, American Chemical Society Series, volume 550, 272 p
Hodgskiss MSW, Crockford PW, Peng Y, Wing BA, Horner TJ (2019) A productivity collapse to end Earth's Great Oxidation. Proc Natl Acad Sci USA 116(35):17207–17212
Hoffman PF (2011) A history of Neoproterozoic glacial geology, 1871–1997. In: Arnaud E, Halverson GP, Shields-Zhou G (eds) The geological record of Neoproterozoic glaciations. Geological Society, London, Memoirs, pp 17–37
Hofmann A, Bekker A, Rouxel O, Rumble D, Master S (2009) Multiple sulphur and iron isotope composition of detrital pyrite in Archaean sedimentary rocks: a new tool for provenance analysis. Earth Planet Sci Lett 286: 436–445
Holland HD (2005) Sedimentary mineral deposits and the evolution of Earth's near-surface environments. Econ Geol 100:1489–1509
Holland HD (2006) The oxygenation of the atmosphere and oceans. Philos Trans R Soc Biol Sci 361:903–915
Johnson DB, Hallberg KB (2003) The microbiology of acidic mine waters. Res Microbiol 154:466–473
Johnson JE, Gerpheide A, Lamb M, Fischer P, Woodward W (2014) O_2-constraints from Paleoproterozoic detrital pyrite and uraninite. Geol Soc Am Bull 126:813–830
Kerr RA (1999) Early life thrived despite earthly travails. Science 284(5423):2111–2113
Kocaman AT, Cemek M, Edwards KJ (2016) Kinetics of pyrite, pyrrhotite, and chalcopyrite dissolution by Acidithiobacillus ferrooxidans. Can J Microbiol 62: 629–642
Kowalewski M, Rimstidt JD (2003) Average lifetime and age spectra of detrital grains: toward a unifying theory of sedimentary particles. J Geol 111:427–439
Kwong YTJ, Lawrence R (1997) Mineralogical changes during NP determinations and their implications. In: Proceedings ICARD'97, Vancouver, pp 435–447
Large RR, Hazen RM, Morrison SM, Gregory DD, Steadman JA, Mukherjee I (2022) Evidence that the GOE was a prolonged event with a peak around 1900 Ma. Geosyst Geoenviron 1:100036
Leybourne MI, Cameron EM (2008) Source, transport, and fate of rhenium, selenium, molybdenum, arsenic, and copper in groundwater associated with porphyry-Cu deposits, Atacama Desert, Chile. Chem Geol 247: 208–228
Leybourne MI, Goodfellow WD (2007) Br/Cl ratios and O, H, C, and B isotopic constraints on the origin of saline waters from eastern Canada. Geochim Cosmochim Acta 71:2209–2223
Leybourne MI, Goodfellow WD (2010) Geochemistry of surface waters associated with an undisturbed Zn-Pb massive sulfide deposit: water-rock reactions, solute sources, and the role of trace carbonate. Chem Geol 279:40–54
Leybourne MI, Johannesson KH (2008) Rare earth elements (REE) and yttrium in stream waters, stream sediments, and Fe-Mn oxyhydroxides: fractionation, speciation, and controls over REE + Y patterns in the surface environment. Geochim Cosmochim Acta 72:5962–5983
Leybourne MI, Goodfellow WD, Boyle DR, Hall GEM (2000) Form and distribution of anthropogenic gold mobilized in surface waters and sediments from a gossan tailings pile, Murray Brook massive sulfide deposit, New Brunswick, Canada. Appl Geochem 15:629–646
Li C, Yang S (2010) Is chemical index of alteration (cia) a reliable proxy for chemical weathering in global drainage basins? Am J Sci 310:111–127
Li G et al (2016) Temperature dependence of basalt weathering. Earth Planet Sci Lett 443:59–69
Locke A (1926) Leached outcrops as guides to copper ores. Baillere, Tindall and Cox, London, 175 p
Maher K, Chamberlain CP (2014) Hydrologic regulation of chemical weathering and the geologic carbon cycle. Science 343:1502–1504
McLennan SM (1993) Weathering and global denudation. J Geol 101:295–303
McQueen K (2009) Regolith geochemistry. In: Scott K, Pain C (eds) Regolith science. Springer/CSIRO, Dordrecht/Melbourne, pp 73–104
Meunier A (2005) Clays. Springer, Berlin, p 265
Misra S, Froelich PN (2012) Lithium isotope history of Cenozoic seawater: changes in silicate weathering and reverse weathering. Science 335:818–823
Morse JW (1983) The kinetics of calcium carbonate dissolution and precipitation. In: Reeder RJ (ed) Reviews in mineralogy, vol 11. Mineralogical Society of America, pp 227–264
Moss PD, Edmunds WM (1992) Processes controlling acid attenuation in an unsaturated zone of a Triassic sandstone aquifer UK, in the absence of carbonate minerals. Appl Geochem 7:573–583
Nesbitt HW, Young GM (1982) Prediction of some weathering trends of plutonic and volcanic rocks based on thermodynamic and kinetic considerations. Geochim Cosmochim Acta 48:1523–1534
Nordstrom DK, Alpers CN (1999) Geochemistry of acid mine waters. In: Plumlee GS, Logsdon MJ (eds) The environmental geochemistry of mineral deposits. Reviews in economic geology, 6A. Society of Economic Geologists, Littleton, pp 133–160
Oeser RA, von Blanckenburg F (2020) Do degree and rate of silicate weathering depend on plant productivity? Biogeosciences 17:4883–4917
Olejarz J, Iwasa Y, Knoll AH et al (2021) The Great Oxygenation Event as a consequence of ecological dynamics modulated by planetary change. Nat Commun 12:3985
Oliva P, Viers J, Dupré B (2003) Chemical weathering in granitic environments. Chem Geol 202:225–256
Pacheco FAL, Van der Weijden CH (2014) Modeling rock weathering in small watersheds. J Hydrol 513:13–27

Pacheco J, Vasconcelos V, Santos FC (2014) Climate change governance, cooperation and self-organization. Phys Life Rev 11:73–86

Parker A (1970) An index of weathering for silicate rocks. Geol Mag 107:501–504

Perri F (2020) Chemical weathering of crystalline rocks in contrasting climatic conditions using geochemical proxies: an overview. Palaeogeogr Palaeoclimatol Palaeoecol 556:109873

Poulton SW, Bekker A, Cumming VM, Zerkle AL, Canfield DE, Johnston DT (2021) A 200-million-year delay in permanent atmospheric oxygenation. Nature 592(7853):232–236

Robertson IDM, Butt CRM (1993) Atlas of weathered rocks. CSIRO Australia, Division of Exploration Geoscience report no. 390R (unpaginated). Revised and reprinted as open file report 1, CRC for Landscape Evolution and Mineral Exploration, Perth https://crcleme.org.au/Pubs/OFRSindex.html

Rose AW (2010) Advances in passive treatment of coal mine drainage. In: Proceedings America Society of Mining and Reclamation, 2010 pp 847–887. https://doi.org/10.21000/JASMR10010847

Sato M (1992) Persistency-field eh-pH diagrams for sulfides and their application to supergene oxidation and enrichment of sulfide ore bodies. Geochim Cosmochim Acta 56:3133–3156

Schirrmeister BE, de Vos JM, Antonelli A, Bagheri HC (2013) Evolution of multicellularity coincided with increased diversification of cyanobacteria and the Great Oxidation Event. Proc Natl Acad Sci USA 110(5):1791–1796

Sequeira Braga MA, Paquet H, Begonha A (2002) Weathering of granites in a temperate climate (NW Portugal): granitic saprolites and arenization. Catena 49:41–56

Shotyk W (1992) Chapter 9: Organic soils. In: Developments in Earth surface processes, vol 2. Elsevier, Amsterdam, pp 203–224

Siever R, Woodford N (1979) Dissolution kinetics and the weathering of mafic minerals. Geochim Cosmochim Acta 43(5):717–724

Skousen J, Zipper CE, Rose AW, Ziemkiewicz PF, Nairn R, McDonald LM, Kleinmann RL (2017) Review of passive systems for acid mine drainage treatment. Mine Water Environ 36:133–115

Sosa Torres ME, Saucedo-Vázquez JP, Kroneck PMH (2015) Chapter 1, section 2: the rise of dioxygen in the atmosphere. In: Kroneck PMH, Sosa Torres ME (eds) Sustaining life on planet earth: metalloenzymes mastering dioxygen and other chewy gases, Metal ions in life sciences, vol 15. Springer, pp 1–12

Steidtmann E (1908) A graphic comparison of the alteration of rocks by weathering with their alteration by hot solutions. Econ Geol 3:381–409

Stoch L, Sikora W (1976) Transformations of micas in the process of kaolinitization of granites and gneisses. Clay Clay Miner 24:156–162

Stumm W, Morgan L (1992) Chemistry of the solid-water interface. Wiley-Interface, New York

Sverdrup HU (1990) The kinetics of base cation release due to chemical weathering. Lund University Press, Lund. 246 p

Sverjensky DA, Lee N (2010) The Great Oxidation Event and mineral diversification. Elements 6:31–36

Tang J, Johannesson KH (2010) Rare earth elements adsorption onto Carrizo sand: influence of strong solution complexation. Chem Geol 279:120–133

Tardy Y (1981) Le Cycel l'Eau, Climats, Palèoclimats et Gèochimie Globale, Masson, Paris. 338 p [in French]

Taylor R (2011) Gossans and Leached Caps: Field Assessment. Springer, 146 p

Thornber MR (1975a) Supergene alteration of sulfides, I: a chemical model based on massive nickel sulfide deposits at Kambalda, Western Australia. Chem Geol 15:1–14

Thornber MR (1975b) Supergene alteration of sulfides, II: a chemical study of the Kambalda nickel deposits. Chem Geol 15:117–144

Thornber MR (1983) Mineralogical and electrochemical instability of nickel-iron sulfides—pentlandite and violarite. J Appl Electrochem 13:253–267

Thornber MR (1985) Supergene alteration of sulfides, VII: distribution of elements during gossan-forming process. Chem Geol 53:279–301

Thornber MR (1992) The chemical mobility and transport of elements in the weathering environment. In: Butt CRM, Zeegers H (eds) Handbook of exploration geochemistry, v. 4: regolith exploration geochemistry in tropical and subtropical terrains. Elsevier, Amsterdam/New York, pp 79–96

Thornber MR, Taylor GF (1992) The mechanism of sulfide oxidation and gossan formation. In: Butt CRM, Zeegers H (eds) Handbook of exploration geochemistry, v. 4: Regolith exploration geochemistry in tropical and subtropical terrains. Elsevier, Amsterdam/New York, pp 119–138

Thornber MR, Wildman JE (1984) Supergene alteration of sulfides. VI. The binding of cu, Ni, Zn, co and Pb with gossan (iron-bearing) minerals. Chem Geol 44:399–434

Thornber MR, Allchurch PD, Nickel EH (1981) Variations in gossan geochemistry at the perseverance Nickel Sulfide deposit, Western Australia: a descriptive and experimental study. Econ Geol 76:1764–1774

Thorne RL, Roberts D, Herrington R (2012) Climate change and the formation of nickel laterite deposits. Geology 40:331–334

Trendall AF, Blockley JG (2004) Precambrian iron-formation. In: Eriksson PG, Altermann W, Nelson DR, Mueller WU, Catuneanu O (eds) Evolution of the hydrosphere and atmosphere, Developments in Precambrian geology, vol 12, pp 359–511

Trescases J-J (1979) Remplacement progressif des silicates par les hydroxydes de fer et de nikel dans les profils d'alteration tropicals des roches ultrabasiques. Sci Geol Bull 32:181–188. [in French]

Trescases J-J (1992) Chemical weathering. In: Butt CRM, Zeegers H (eds) Handbook of exploration geochemistry, Regolith exploration geochemistry in tropical and subtropical terrains, vol 4. Elsevier, pp 25–41

Velbel MA (1993) Temperature dependence of silicate weathering in nature: How strong a negative feedback on long-term accumulation of atmospheric CO2 and global greenhouse warming? Geology 21:1059–1062

Vera M, Schippers A, Sand W (2013) Progress in bioleaching: fundamentals and mechanisms of bacterial metal sulfide oxidation--part A. Appl Microbiol Biotechnol 97:7529–7541

Vlassopoulos D, Wood SA (1990) Gold speciation in natural waters: I. Solubility and hydrolysis reactions of gold in aqueous solution. Geochim Cosmochim Acta 54:3–12

Walker JCG, Hays PB, Kasting JF (1981) A negative feedback mechanism for the long-term stabilization of Earth's surface temperature. J Geophys Res 86:9776–9782

Waltham T, Bell FG, Culshaw MG (2005) Sinkholes and subsidence: karst and cavernous rocks in engineering and construction. Springer, Berlin

Warke MR, Di Rocco T, Zerkle AL, Lepland A, Prave AR, Martin AP, Ueno Y, Condon DJ, Claire MW (2020) The Great Oxidation Event preceded a Paleoproterozoic snowball Earth. Proc Natl Acad Sci USA 117(24): 13314–13320

West AJ (2012) Thickness of the chemical weathering zone and implications for erosional and climatic drivers of weathering and for carbon-cycle feedbacks. Geology 40:811–814

White AF (1995) Chapter 9. Chemical weathering rates of silicate minerals in soils. In: Chemical weathering rates of silicate minerals. De Gruyter, Berlin, pp 407–462

White AF, Brantley SL (2003) The effect of time on the weathering of silicate minerals: why do weathering rates differ in the laboratory and field? Chem Geol 202:479–506

Willenbring JK, von Blanckenburg F (2010) Long-term stability of global erosion rates and weathering during late-Cenozoic cooling. Nature 465:211–214

Williams PA (1990) Oxide zone geochemistry. Ellis Horwood, London. 286 p

The Role of Regolith Formation and Evolution in the Development of Supergene Mineral Deposits

C. R. M. Butt and R. R. Anand

Abstract

The majority of supergene mineral deposits discussed in this volume are the product of intense weathering in humid tropical environments, mostly having formed in a regolith 30- > 200 thick over long time periods on stable continental landmasses. Deposits such as bauxite, Ni laterite and kaolinite in near-equatorial regions are currently forming under humid savanna to rainforest climatic conditions that have changed little throughout the Cenozoic or even earlier. They have mostly been modified by tectonic events during this period, which have affected drainage characteristics and, in turn, the nature and rate of weathering, in places upgrading the supergene resource. Conversely, in regions such as the Caribbean and SE Asia-western Oceania, high rates of tectonic activity may also result in partial or even total destruction of the deposit. Similar supergene deposits in higher latitudes have had a more complex formation history, evolving as climates have changed to dry savanna, Mediterranean, temperate or near-polar. These changes, in places also coupled with tectonic activity, have induced chemical reactions that may upgrade the deposit and/or modify its mineralogy. This chapter reviews the processes of deep weathering and the formation of intensely weathered and lateritic regoliths and their evolution due to climatic and tectonic change. These processes are discussed in terms of the formation and distribution of supergene deposits, and the landscapes in which they occur, in different regions of the world.

C. R. M. Butt (✉) · R. R. Anand
CSIRO Mineral Resources, ARRC, Kensington, WA, Australia
e-mail: Charles.Butt@csiro.au; Ravi.Anand@csiro.au

1 Introduction

Supergene mineral deposits occur in a wide range of tectonic, climatic and geological environments across the globe. Most are autochthonous, products of residual weathering of a protore, with only a few being wholly allochthonous, with the ore elements precipitated distant from their source. Most deposits have formed and evolved over long periods, quite commonly under different climatic and/or tectonic conditions to those in which they currently occur. For many supergene deposits, successive changes in weathering environment may have resulted in mineralogical and chemical upgrading to enhance their economic value; conversely, such changes may cause their destruction. In contrast, deposits such as evaporites in playas and sabkhas, and uranium in palaeochannels, are the product of their present or geologically recent environments.

R. J. Bowell, C. R. M. Butt (eds.), *Geology, Geochemistry and Formation of Supergene Mineral Deposits in Deeply Weathered Terrain*, Mineral Resource Reviews, https://doi.org/10.1007/978-3-031-75733-4_3

Sulfide oxidation occurs under many weathering environments, resulting in broadly similar supergene products, consisting of supergene sulfides and/or a range of sulfate, carbonate and oxide minerals. For many massive and disseminated sulfide deposits, the primary and supergene zones may both be economic ores. The supergene component is commonly minor, for a long period of intense weathering, with little erosion necessary for the supergene ores to develop into a significant proportion of the total resource. This is the case for major bodies of massive sulfides in a few locations, such as the Iberian Pyrite Belt, Bathurst Camp in Canada and Broken Hill in Australia. Conversely, in low grade disseminated sulfides protores, such as some Cu-Mo porphyries, and many non-sulfide base metal deposits, only the supergene zone has sufficient grades and tonnages to be economic. These zones of supergene enrichment (and depletion) are, of course, the surface expression of concealed sulfide mineralization and include gossans and other 'caprocks' sought during exploration (chapter "Gossans: In Situ Supergene Base Metal Deposits"; Taylor and Thornber 1992). Unweathered, disseminated, non-sulfide primary mineralization (e.g., some Nb-REE-phosphate carbonatites, Zn-, Pb- or Mn-carbonate-rich sediments) may similarly be economic, but many only form a mineable resource where supergene enrichment has occurred during periods of intense weathering. Additionally, some important supergene/secondary deposits (e.g., Ni-Co laterite, bauxite and some Fe ores) are formed from a protolith that would not at present be considered economic ore.

The intense weathering responsible for the formation of these supergene deposits has occurred under lateritic conditions in humid savannas, especially on shields and other stable landmasses. Further weathering under changed tectonic and climatic conditions caused modification to and, in some cases, upgrading of, the supergene mineralization. The tectonic changes include uplift or down-warping affecting drainage conditions within the regolith and the landscape as a whole and initiating erosion or burial. Modifications due to changes in climate have occurred under rainfall conditions ranging from rainforest to hyper-aridity, and temperatures from tropical to glacial. Some have been eroded and redeposited during formation, with post-depositional weathering causing further enrichment. This chapter describes lateritic regolith formation and supergene enrichment under successive climatic regimes and associated erosional and depositional processes, as a general introduction to the more specific genetic mechanisms responsible for the formation of supergene deposits of different commodities.

2 Occurrence and Distribution of Intensely Weathered Lateritic Regoliths

Lateritic regoliths, including ferruginous gravels and duricrust, lateritic bauxites and/or thick saprolite are widespread in intertropical regions, mainly on the continental landmasses between 35^0N and 35^0S (Fig. 1). They are exposed at surface or buried by a mostly thin, later cover of sediments. The total thickness of the residual regolith and sediments overlying basement rock varies from a few metres to over 150 m and is controlled by lithology, structural features and landscape position. In places, the overlying sediments may also be weathered. The overall distribution, shown by the occurrence of mainly buried supergene deposits and deep regolith remnants, is much wider and extends into temperate regions at higher latitudes and, in the northern hemisphere, to recently glaciated regions north of the Arctic Circle (Fig. 2).

A Scottish surgeon, Francis Buchanan, travelling in the Malabar Coast region of SW India (present-day Karnataka and Kerala), observed the widespread use of a soft, ferruginous mottled clay that was cut as blocks from shallow quarries and within a few days hardened into bricks suitable for building. He termed this material 'laterite' after *later*, Latin for brick (Buchanan 1807). Laterite bricks have been used for over 1000 years and remain an important construction material, not only in southern India but across similar humid tropical regions in SE Asia,

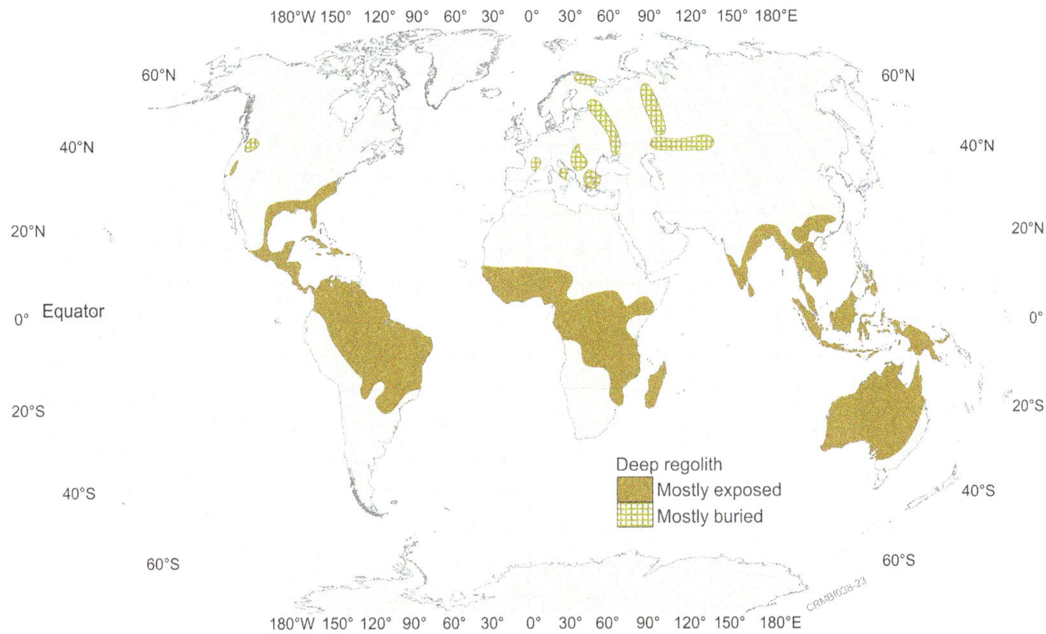

Fig. 1 Global distribution of intensely weathered regolith. (Revised after Bardossy and Aleva 1990). Winkel Tripel projection

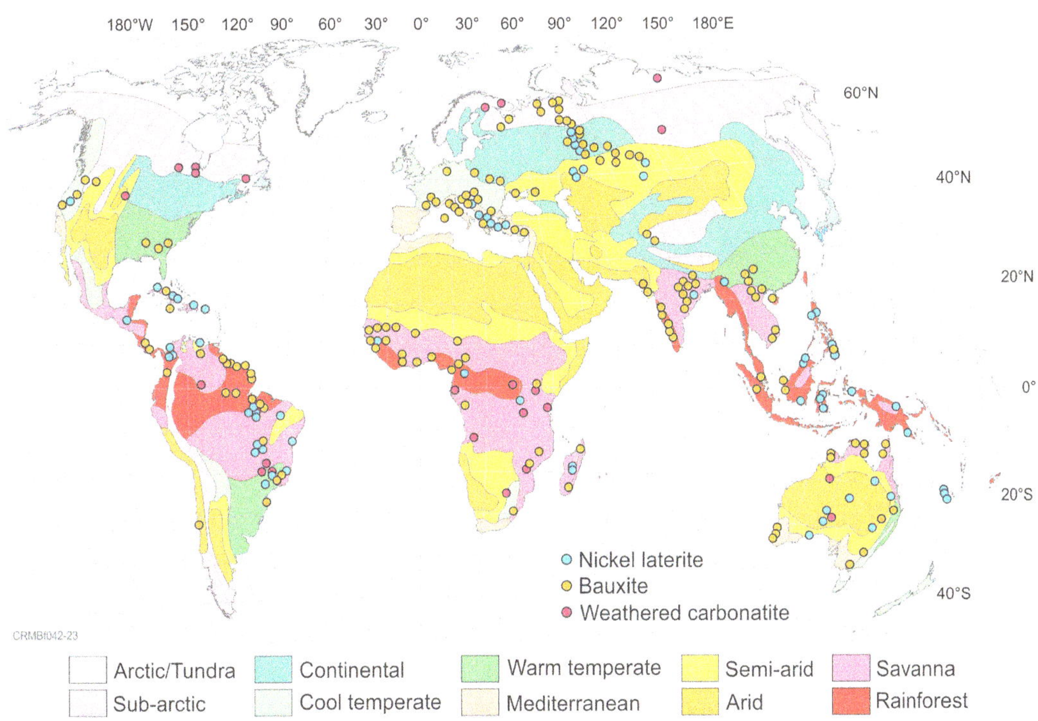

Fig. 2 Global climate map and distribution of lateritic bauxite, Ni laterite and supergene Nb-P-REE deposits on carbonatite. (Winkel Tripel projection)

including Thailand, Laos, Cambodia and Vietnam. Buchanan described laterite from the type locality as "*... diffused in great masses, without any appearance of stratification, and placed over the granite that forms the basis of Malayala. It is full of cavities and pores, and contains a very large quantity of iron in the form of red and yellow ochres*". From the present authors' observations, material from the type area is a mottled saprolite merging to vermiform mottled clay, with the mottles related to tree roots. It is not a duricrust, nor a friable pisolitic or nodular lateritic residuum.

Buchanan's original description was based on some relatively casual observations with no intent to be a precise definition. Nonetheless, the term 'laterite' has been extensively applied to a variety of mostly hard, ferruginous regolith components having a wide range of compositions and genetic origins, resulting in considerable confusion and misunderstanding when used without qualification. Following Eggleton (2001), 'laterite' is herein generally used as an adjective, and thus a lateritic regolith refers to "a vertical sequence of regolith facies" showing (or reasonably interpreted to have if hidden) some or all of the following, from the bottom up bedrock, saprock, saprolite, plasmic zone, mottled zone or ferruginous saprolite, lateritic residuum (lateritic duricrust, lateritic gravel) (Fig. 3). Saprolite commonly comprises over 80% of the total thickness of the regolith. In some locations, the 'bedrock' for all, or the upper part, of the profile may be lacustrine, alluvial or colluvial sediment overlying older basement rocks. There are many variations in the components of profiles, controlled by factors including lithology, topographic position and climatic history. During erosional episodes, the upper units may have been removed, commonly to saprolite. The Fe in lateritic residuum is mainly a relative enrichment and it is commonly observed that Fe contents are lower over granites and other felsic rocks than over mafic or ultramafic rocks. Terms such as

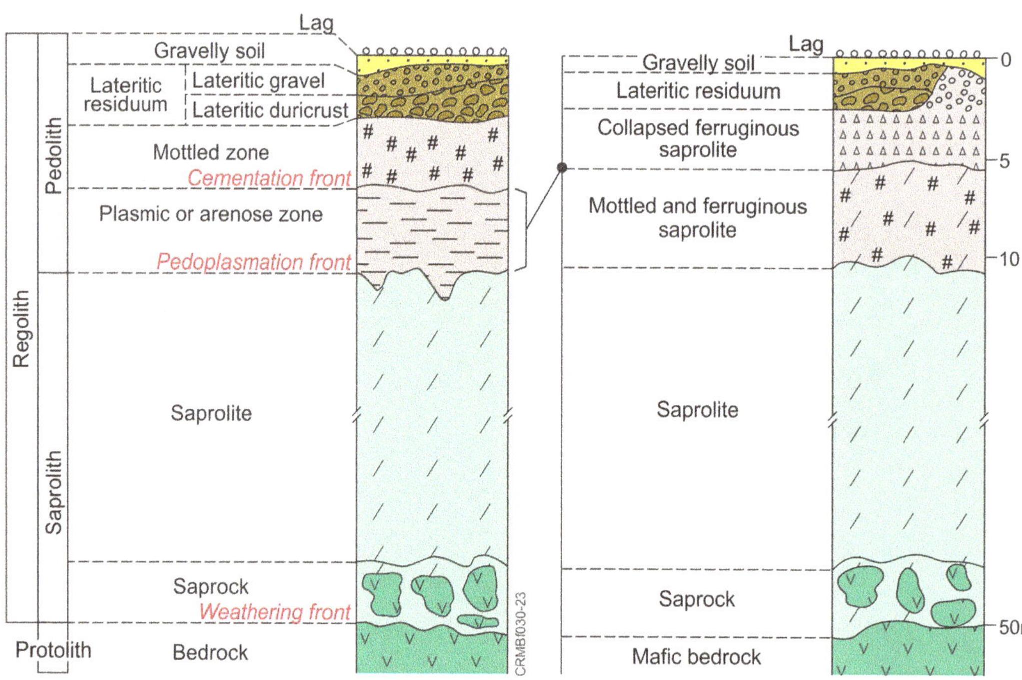

Fig. 3 Structure of a lateritic regolith. (From Butt et al. 2000; Anand and Paine 2002; Anand and Butt 2010) using the terminology recommended by Eggleton 2001 and references cited therein. NB The thickness of the pedolith is exaggerated to portray the different constituent horizons; the saprolith commonly comprises 70–90% of the profile. The right column represents a variant most commonly on mafic rocks

'bauxitic laterite' and 'lateritic bauxite' refer to highly aluminous facies of lateritic residuum. There is no definitive Al content because bauxite is an economic term, referring to an ore of aluminium. It is important to distinguish between lateritic duricrusts and ferricrete. Ferricrete is an indurated, Fe oxide-cemented colluvial or alluvial sediment that may include quartz clasts and ferruginous nodules, pisoliths and mostly saprolitic lithic fragments derived from the erosion of residual lateritic regolith. The iron in the cement is an absolute enrichment, derived laterally in solution. Ferricrete can easily be confused with residual lateritic duricrust, in both appearance and occurrence in the landscape, but they have quite different origins. This confusion extends to 'ferricrete' being used synonymously for residual lateritic duricrust by some authors. These ferruginous materials are commonly the most conspicuous components of intensely weathered lateritic regoliths and the landscapes in which they occur and can be the subject of much controversy regarding their mode of formation. Anand et al. (2019) suggest using the informal term *ferruginous gravel and duricrust* (FGD) to refer to the uppermost horizons of ferruginous nodular and pisolitic gravel and/or duricrust where it is unclear whether components are residual, semi-residual or transported.

3 Formation and Modification of Lateritic Regoliths

3.1 Overview

Lateritic regoliths are considered to form under the following environmental conditions (Table 1):

1. *Seasonally humid tropical climates*: Savanna climates are generally considered the most favourable for lateritic weathering, although this is not universally accepted, especially in the higher latitudes (Taylor et al. 1992). Since the impact of temperature is largely kinetic, higher temperatures merely enable reactions to occur more rapidly, so deep weathering may occur in cooler climates, but over longer periods. However, due to the different seasonality and the lower evaporation rates, winter rainfall in cooler Mediterranean climates is possibly more effective than a similar amount of summer rainfall in savanna climates.
2. *Moderate relief*: The relief must be sufficient to allow adequate drainage and leaching of soluble weathering products—but not so great as to promote rapid erosion.
3. *Long periods of tectonic stability*. Estimates of the net rate of regolith development (the difference between surface erosion and descent of the weathering front) in humid tropical regions vary from about 20–50 m/Ma (Nahon and Tardy 1992) to as little as 1.3–3.7 m/Ma (Boeglin and Probst 1998), so that thick, lateritic regoliths may require several million years to form. On geomorphological evidence, Budel (1982) suggested the characteristic low-relief landscapes of most deeply weathered regions may require 10^7 years to develop. Tectonic uplift during this time will promote instability and, hence, erosion of the regolith. A thick regolith and significant supergene enrichment can only develop if the net rate of weathering, as expressed by the downward progress of the weathering front, exceeds the rate of surface lowering by erosion and profile collapse.

Table 1 Conditions for intense lateritic weathering

Requirement	Rate of chemical weathering (*descent of weathering front*) > Rate of physical erosion (*surface lowering)*
Weathering rates:	5–50 mm/1000 years (5–50 m/million years)
Conditions	Tectonic stability (*minimizes erosion rate*) Moderate relief (*balances slow erosion rate with removal of dissolved products*) Humid tropical climate (*maximizes water-rock interaction*)

Under these conditions, regolith profiles develop progressively, with each horizon formed from a progenitor that resembles that currently underlying it (McFarlane 1976). The full sequence of horizons (Fig. 3) is thus not present until considerable time has passed. Some young lateritic regoliths, such as Ni laterites in Sulawesi, have probably formed in essentially a single weathering episode since exposure in the late Miocene (chapter "Nickel-Cobalt Laterite Deposits"), but the majority occur in terrains that have been subjected to several different episodes, and the products vary or have been modified accordingly. In mid- or lower-slope positions, the upper horizons may incorporate or, in some cases, be composed entirely of, detrital colluvial material eroded from higher in the landscape. Deeper horizons may similarly contain components derived in solution or in suspension from higher in the profile and/or upslope.

The result of prolonged chemical weathering under stable, humid conditions, with little erosion of the insoluble products, is to produce soil underlain by increasing thicknesses of saprolite. Leaching is promoted by organic acids derived from decaying vegetation. Over dominantly silicate rocks, the original fabric of the parent rock is preserved by resistant primary minerals, with stable neo-formed clays and Fe oxides commonly pseudomorphing aluminosilicate and ferromagnesian minerals, respectively, and weathering is essentially isovolumetric. Over carbonate-dominated rocks such as limestones and carbonatites, however, nearly complete dissolution of the major minerals results in a loss of volume, collapse of the saprolite and the only primary features retained are structures such as veins, faults and lithological contacts. An effect of the loss of mobile constituents is relative accumulation of less mobile ones, such that the uppermost saprolite consists of the most resistant primary minerals, e.g. quartz, zircon, chromite, rutile, pyrochlore, gold and stable secondary minerals such as kaolinite, Al and Fe oxides and anatase. The now thick, relatively porous saprolite retains groundwater, with the water-table fluctuating seasonally. This provides a site of considerable chemical activity, especially the redox reactions of Fe and Mn, which may dissolve and reprecipitate seasonally as the groundwater rises and falls. In this zone, Fe oxides accumulate as secondary segregations such as coatings and nodules, and the high acidity these oxidation reactions promote, in turn, dissolution of otherwise stable minerals, including clays such as kaolinite, releasing Si and Al. Some Al co-precipitates with Fe oxides as a precursor to further enrichment and the formation of bauxites. With time, collapse of the highly weathered upper horizons and compaction of Fe oxide segregations lead to the development of the characteristic, near-surface ferruginous horizon, beneath a highly leached soil. The overall process is one of chemical wasting that, together with continuing slow erosion, results in an overall lowering of the landscape with time (Fig. 4). A concomitant effect is the slow reduction of relief, as some of the eroded material is deposited on lower slopes and valleys.

3.2 Regolith Profile Structure and Terminology

The mechanisms of profile formation summarized above refer to the process applying particularly to weathering of silicate rocks in a freely-draining, upland position. Intense weathering results in a regolith comprising a number of horizons formed by structural and mineralogical modifications of the bedrock. Different physical and chemical conditions characterize the various parts of a residual profile. Thus, in the slightly weathered rock at the base of the profile, mineral weathering will take place in micro-fissures and narrow solution channels, and the capillary water in such spatial restricted volumes may be expected to be close to equilibrium with the primary mineral. In these circumstances, the initial weathering products may be closely related to the primary minerals in both composition and structure. Higher in the profile, the close compositional relationship between primary mineral and weathering product may be lost. In the upper horizons and soil, biological weathering assumes a much greater significance as organic ligands

Fig. 4 Profile formation and landscape reduction

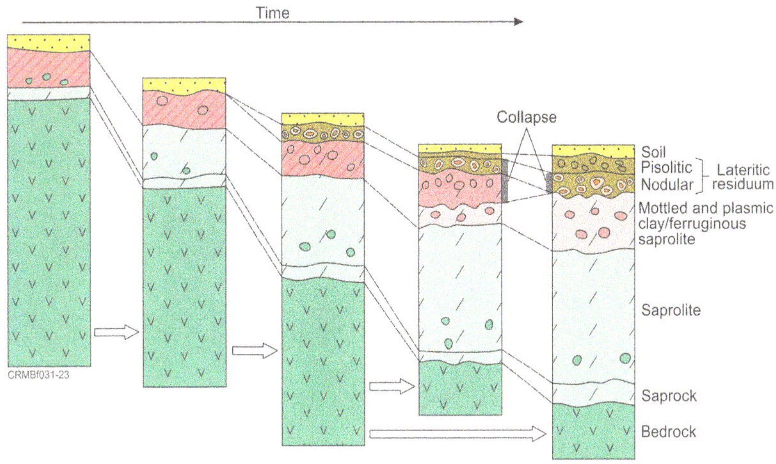

derived from decomposing organic matter and the direct activities of soil microbes or plant roots supplement the mainly inorganic processes occurring at depth (Anand et al. 2019). The resultant 'typical' regolith is illustrated in Fig. 3. Above the unweathered bedrock (*protolith*), a lateritic profile has two major components, the *saprolith* and the *pedolith*, distinguished by their fabrics. The base of the saprolith is the boundary with fresh rock *(weathering front),* and the boundary between saprolith and pedolith is the *pedoplasmation front*. The saprolith, consisting of *saprock* overlain by *saprolite*, is weathered bedrock in which the structure and fabric (texture) of the parent rock have been retained by pseudomorphic replacement of the primary minerals. The weathering front and the saprock-saprolite boundary may be sharp or gradational; they may vary markedly in depth and separation (a few centimetres to several metres), over short distances laterally, with only minor lithological changes. Saprock is compact, slightly weathered rock of low porosity with less than 20% of the weatherable minerals altered (Fig. 5), whereas saprolite (Fig. 6a) is more porous and has more than 20% of the weatherable minerals altered. Saprolite progressively loses its primary fabric upwards as porosity increases due to continued dissolution of remaining primary and some secondary minerals, leading to collapse and settling. Where collapse is extreme, settling can distort the fabric and structure. For example, collapse has resulted in a pseudo-sedimentary fabric in ferruginous saprolite developed on serpentinized dunite at Mt. Keith, W. Australia, and changed the apparent dip of more resistant chert wall-rock (Fig. 6b). Mass balance calculations show that the collapsed saprolite has lost about 75% of its original rock column (Brand and Butt 2001). Similar extreme collapse also occurs over carbonatites, due to dissolution of the dominant dolomite and other carbonate minerals, resulting in significant relative enrichment of resistant minerals and immobile elements (chapters "Carbonatite Hosted Residual REE Deposits" and "Regolith Hosted Niobium Deposits in Carbonatite complexes in Brazil").

The *pedolith* is the upper part of the profile, above the pedoplasmation front, that has been subjected to surficial and soil-forming processes. These result in loss of the parent rock fabric and the development of new fabrics through non-isovolumetric weathering, collapse, the precipitation of introduced materials and mechanical processes such as shrinking and swelling of clay. The principal horizons of the pedolith are the *plasmic* and *arenose* zones, *mottled zone* and *lateritic residuum*. (The plasmic and arenose zones are not present in all profiles. On some mafic rocks, for example, the saprolite more commonly becomes increasingly ferruginous and brecciated, before merging into a lateritic gravel or duricrust). The plasmic zone is a mesoscopically homogeneous horizon of profiles developed on quartz-poor rocks, composed dominantly of clay or silty clay. It has neither the

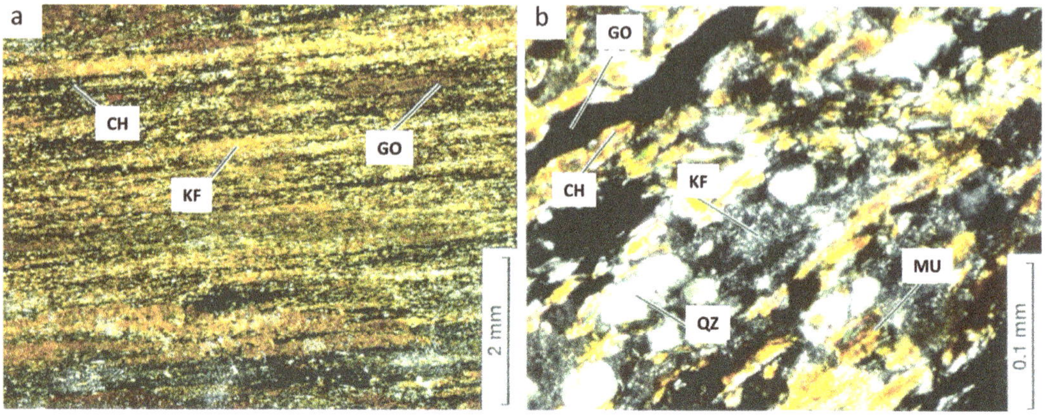

Fig. 5 Saprock-saprolite transition, Reedy mine, Western Australia. (**a**) quartz-chlorite schist. Kaolinized feldspar (KF), stained yellow by goethite, with brown stain (GO) extending from goethite after magnetite or pyrite. (**b**) quartz-chlorite muscovite schist. Completely kaolinized feldspar (KF), chlorite (CH) partially altered to smectite and unaltered muscovite (MU) and quartz (QZ). Goethite lenses (GO) along cleavage. From: Robertson and Butt 1991. (Photos: I.D.M. Robertson)

Fig. 6 Saprolite. (**a**) Saprolite exposed in the Tanami mine pit, Northern Territory, Australia, showing preservation of bedrock structure (depth approx. 30 m.). (**b**) change in the apparent dip of weathered chert wall-rock due to collapse and settling of adjacent ferruginous saprolite on serpentinized olivine dunite, Mt. Keith, Western Australia. (Photos: C.R.M. Butt)

fabric of saprolite nor any significant development of secondary segregations such as mottles, nodules and pisoliths. The arenose zone is the equivalent of the plasmic zone on quartz-rich rock and is composed of angular quartz, with a grain-supported fabric. The mottled zone is that part of a weathering profile having macroscopic segregations distinguished by colour, usually due to precipitation of authigenic Fe oxyhydroxides. Although typically part of the pedolith, mottling may also occur in the saprolite. Mottles may be sharply defined or diffuse; some represent ferruginization of the matrix, with little change in fabric, and others may form secondary structures, with a new fabric. Some mottling may also be due to bleaching of ferruginous clays or saprolite. Lateritic residuum is the upper ferruginous zone of a lateritic profile, composed dominantly of secondary oxides and oxyhydroxides of Fe (goethite, hematite, maghemite), hydroxides of aluminium (e.g. gibbsite, boehmite) and kaolinite, with or without quartz. It consists of a cemented duricrust and/or loose pisolitic or nodular gravel. Lateritic residuum has a broad genetic and compositional relationship with the substrate. It is residual rather than in situ due to minor lateral movement (generally 5–>50 m) and vertical collapse during its formation. Lateritic duricrust may be massive but, more commonly, it consists of various secondary segregations such as nodules and pisoliths, open and infilled vermiform voids, or angular fragments of ferruginous saprolite, older duricrust and vein quartz, accumulated by vertical collapse.

3.3 Mineralogy, Geochemistry and Profile Formation

Mineral Stability The composition and formation of the regolith depend largely on the relative stabilities of the constituent minerals that comprise the bedrock and, in turn, the weathering horizons successively derived from it (Table 2; Fig. 7). Especially over silicate rocks, each horizon may be considered to have developed from a precursor similar to that immediately underlying it, as depicted in Fig. 4.

In the humid, oxidizing conditions favouring intense lateritic weathering, the least stable minerals are usually salts, carbonates and sulfides; thereafter, in some respects, the list is the reverse of Bowen's reaction series, with major phases that crystallize at high temperatures and/high pressures being less-stable under surface conditions than those formed under lower temperatures and pressures. The chemical weathering mechanisms involved are discussed in chapter "The Geochemistry of Supergene Processes". The subsequent distributions of elements in the regolith reflect their chemical mobilities following the mineral transformations that occur in the principal horizons. Elements commonly leached or retained at each step are shown in Table 3. This is, of course, an oversimplification: K, for example, is considered a relatively mobile alkali metal, but is essentially 'immobilized' if hosted by resistant minerals such as muscovite. Some primary minerals are very resistant to weathering: muscovite and talc may persist to the upper saprolite or higher, then weather to kaolinite, whereas others, including quartz, zircon, rutile and chromite, become residually concentrated in lateritic residuum, although generally showing signs of some corrosion. In places, minerals generally regarded as 'weatherable' may persist high in the regolith due to unusual compositions or local effects, such as metamorphic welding reducing porosity and access by weathering fluids. An example is essentially fresh ultramafic rocks that contain secondary olivine observed by the author (CRMB) in intensely weathered terrain, Mt. Keith region, Western Australia.

Table 2 Relative stability of common rock-forming minerals in the weathering environment (developed from Goldich 1938)

Principal minerals	Similar minerals	
Halite	other salts	Increasing stability →
Gypsum		
Calcite	dolomite, aragonite	
Sulfides		
Olivine		
Pyroxene		
Amphibole		
Plagioclase		
Biotite	glauconite	
K Feldspar		
Chlorite	vermiculite	
Smectite		
Muscovite	illite, sericite, talc	
Kaolinite	halloysite	
Quartz	hyalite	
Al oxides &. oxy-hydroxides		
Fe oxides & oxy-hydroxides		
Zircon	Monazite, spinel, anatase	

Secondary minerals formed low in the regolith may become unstable in upper horizons—for example, smectites may weather to kaolinite in the upper saprolite. Kaolinite in turn weathers to release silica and alumina which may be leached, or precipitate as amorphous phases, or as

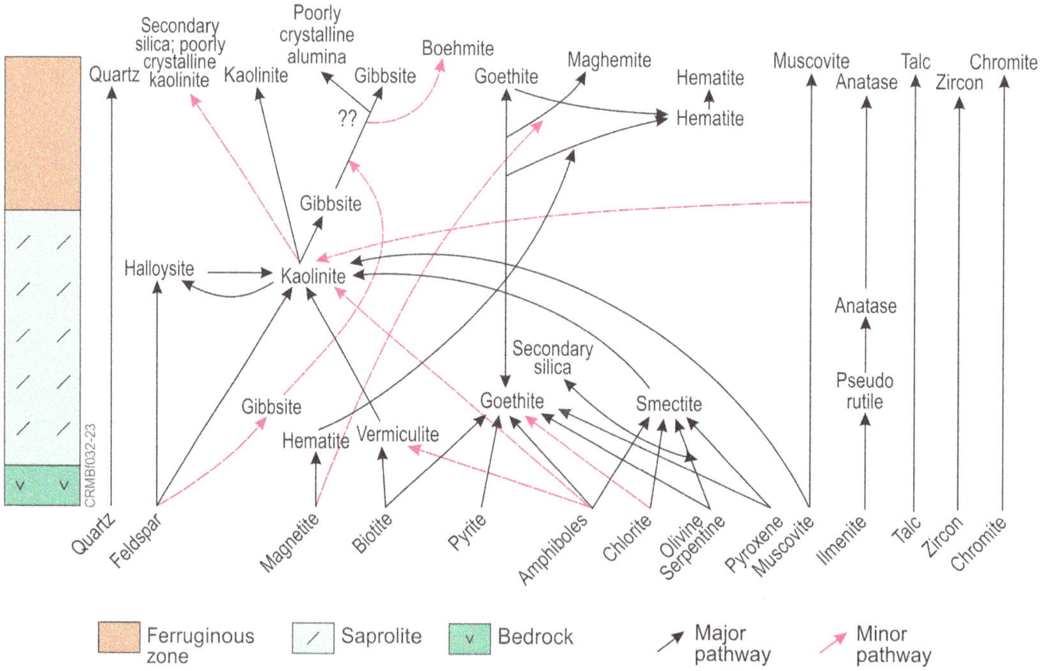

Fig. 7 Pathways of formation of secondary minerals in lateritic regoliths. (After Anand and Paine 2002; Anand and Butt 2010). During weathering, most primary minerals are altered to a mineral assemblage more stable in the surficial environment than bedrock. Kaolinite, halloysite and smectite are generally the most abundant minerals in saprolite, with Fe oxide abundances increasing towards the top of the profile. The Fe oxide mineralogy (goethite, hematite, maghemite) is controlled by weathering conditions rather than the original primary minerals from which Fe was derived

secondary silica and gibbsite respectively. Indeed, no mineral is entirely unaffected by weathering, no element is entirely leached from any regolith horizon and no element is entirely immobile.

Weathering Front and Saprock Weathering commences along mineral boundaries, cleavages, shears, joints and fractures. Locally, it may extend along shears for tens of metres below the depth at which most rock appears fresh. Sulfides are some of the most unstable minerals in humid, oxidizing environments and are only found high in the profile if preserved within vein quartz or resistant lithorelicts. This is consistent with observations that S has been strongly leached from the deepest levels of the regolith and appears to be the element most susceptible to weathering. Many elements hosted by sulfides (e.g., Cd, Co, Cu, Mo, Ni, Zn) are also commonly strongly leached from deep in the profile, although a proportion is retained in Fe oxides or other secondary minerals such as jarosite and other sulfates derived from the sulfides. Carbonates are similarly highly susceptible to weathering, especially in acidic environments, hence elements dominantly or significantly hosted by them, such as Ca, Mg, Mn and Sr, are strongly leached. Calcium and Sr, for example, are, thus, commonly reduced to very low concentrations throughout most of the regolith. Iron oxides commonly pseudomorph the sulfides, whereas dissolved carbonates leave a void. Limestones and carbonatites may be almost entirely dissolved, resulting in collapse of the remnant primary and secondary minerals, or forming karst and cave systems.

Lower Saprolite Weathering in saprock and the lower saprolite most strongly affects ferromagnesian minerals and feldspars. Hydrolysis and acid

Table 3 Element mobility during intense lateritic weathering. (After Butt et al. 1991, 2000; Butt and Zeegers 1992, Anand and Butt 2010, and references therein)

Host minerals	Leached	Partly retained in secondary minerals
Released at weathering front:		
Sulfides	As, Au, Cd, Co, Cu, Mo Mo, Ni, Zn, S	Ag, As, Cu, Ni, Pb, Sb, Zn (Fe oxides, sulfates arsenates, carbonates, alunite-jarosite. Some Ag, Au, Cu as native metals)
Carbonates	Ca, Mg, Mn, Sr	
Released in the lower saprolite:		
Aluminosilicates	Ca, Cs, K, Na, Rb	Si, Al, (kaolinite), Ba (barite)
Ferromagnesian minerals (olivine, pyroxene, amphiboles, chlorite, biotite)	Ca, Mg	Fe, Ni, Co, Cr, Ga, Mn, Ni, Ti, V (Fe & Mn oxides)
Released in the upper saprolite:		
Aluminosilicates (muscovite)	K, Rb, Cs	Si, Al (Kaolinite)
Ferromagnesian minerals (chlorite, talc, amphibole)	Mg, Li	Fe, Ni, Co, Cr, Ga, Mn, Ti, V, (Fe & Al oxides; anatase)
Smectites	Ca, Mg, Na	Si, Al (kaolinite)
Accessory phases (bastnäsite, eudialyte, apatite, allanite, titanite)	REE	REE adsorbed on clays[a]
Released in mottled and ferruginous zones		
Aluminosilicates (muscovite, kaolinite)	K, Rb, Cs	Si, Al (kaolinite)
Iron oxides, gold	Trace elements; Ag	

Retained in stable minerals:
Au, B, Cr, Fe, Hf, K, Nb, Rb, REE, Th, Ti, V, W, Zr
[a]REE in ion adsorption deposits are largely derived from accessory phases (e.g., as listed above) formed by hydrothermal alteration (see chapter "Ion Adsorption Mineralisation in Regolith-Hosted REE Deposits")

weathering of olivine, pyroxene, amphibole and biotite results in replacement by goethite, and the progressive loss of Mg and Si. Minor and trace elements, such as Co, Cu, Mn and Ni, may be partially retained, adsorbed to goethite. In poorly-drained environments, some Mg and Si may be retained in smectite. Serpentinization of olivine- and pyroxene-rich rocks generally increases their resistance to weathering, but the products are similar. As feldspars weather, Na, Ca and Sr are leached, with Si and Al retained as kaolinite and halloysite forming pseudomorphs. Smectites may be intermediate products but are usually only retained in less freely drained or arid environments. In addition, K, Rb and Cs will be lost if hosted by orthoclase or biotite but, if present mainly in muscovite, concentrations are maintained or residually increased through much of the regolith. Barium, commonly hosted by feldspars, is released early during weathering but is reprecipitated as barite, which remains stable through the regolith and may only be partly leached even close to the surface.

Mid to Upper Saprolite The final alteration of all but the most resistant primary minerals occurs in these zones; in addition, less-stable secondary minerals, such as smectites, are also destroyed. Serpentine and chlorite are progressively weathered through the zone; but talc and, in particular, muscovite may persist through to the mottled clay zone and, in places, to the lateritic residuum. Ferromagnesian minerals are the principal hosts for transition metals, such as Co, Cu, Ni and Zn in sulfide-poor mafic and ultramafic rocks, and are retained higher in the profile than if hosted by sulfides. Nevertheless, they become leached from the upper horizons and reprecipitate with secondary Fe–Mn oxides in the mid to lower saprolite.

Mottled and Ferruginous Zones Most remnant major primary minerals, except quartz, are commonly finally destroyed in these horizons, as is the primary fabric of the rocks. These horizons, in particular, demonstrate one of the principal characteristics of lateritic regoliths, namely domination by Si, Al and Fe, resident in kaolinite, quartz, Fe oxides (hematite and goethite), and, in places, gibbsite. The abundances and distributions of these elements broadly reflect their source lithologies. Thus, kaolinization of saprolites is strongest over the felsic rocks but only tends to be well developed in the upper saprolite over mafic and ultramafic rocks. Although some dissolution can occur in ferruginous zones, quartz is commonly only slightly weathered and most is left as a residual, resistant mineral, most abundantly over felsic igneous rocks and quartzose sedimentary rocks. Secondary quartz characteristically accumulates as massive silcrete over dunite (including serpentinized dunite), probably reflecting an environment with impeded drainage and a lack of reactive alumina to form clays such as kaolinite. The ferruginous horizon, *lateritic residuum* is commonly most strongly developed over ultramafic and mafic rocks, tending to be thinner, less ferruginous and less strongly cemented over other lithologies and almost absent over granitoids. Accumulation, partial dissolution and recrystallization of Fe oxides to form lateritic residuum results in the development of a variety of secondary structures such as pisoliths and nodules composed dominantly of hematite, goethite, maghemite, kaolinite, and/or gibbsite. They may be cemented (ferruginous duricrust) or loose (ferruginous gravel). Massive duricrusts may form by Fe-Al oxides replacing or cementing kaolinitic saprolite or plasmic clays. The presence of vermiform voids, ferruginized paleo-rhizospheres, roots, organic matter, burrows and microbes suggest the structures are formed in a biologically active environment (Anand et al. 2019).

Released Si and Al either reprecipitate as silica, gibbsite or kaolinite lower in the profile, e.g. in the mottled clay horizon, or are leached. The distributions of several minor and trace elements are controlled wholly or in part by those of these major elements, due to substitution or coprecipitation. Thus, Cr, As, Ga, Sc, V and Au tend to accumulate with Fe oxides; Cr, mainly derived from ferromagnesian minerals, is also associated with neo-formed kaolinite. Many immobile elements also tend to concentrate with Fe oxides in the upper horizons although, for most, no chemical interactions are involved. Thus, the distributions of Cr, K, Zr, Hf, Th, Nb, Ta, W, REE, Ti and V relate wholly or in part to the stability of their primary and/or secondary host minerals (e.g., Zr and Hf in zircon; Ti in rutile and anatase; Cr in chromite; K in muscovite; REE in zircon and monazite) or to their relative chemical immobility during weathering (V, Ti). The abundances of these elements tend to increase upwards through the profile due to the gradual loss of other components, with marked accumulation in lateritic residuum, within which lateral dispersion by colluvial action can occur during the course of profile evolution. During its development, lateritic residuum is overlain by a humus-rich ferruginous soil. In many areas, however, the lateritic regolith is no longer forming, the original soil, and possibly some of the ferruginous residuum, have been eroded, and residual regolith is overlain by a variable thickness of transported cover.

3.4 The Influence of Lithology and Topography on Regolith Development

Although the formation of intensely weathered lateritic profiles may be portrayed as a single, continuing essentially vertical process, there are also lateral variations. On similar bedrock, the characteristics of the regolith profile developed under given climatic and tectonic conditions also depend topographic position, which affects surface and subsurface drainage. Consequently, in a section across a valley, a catenary association

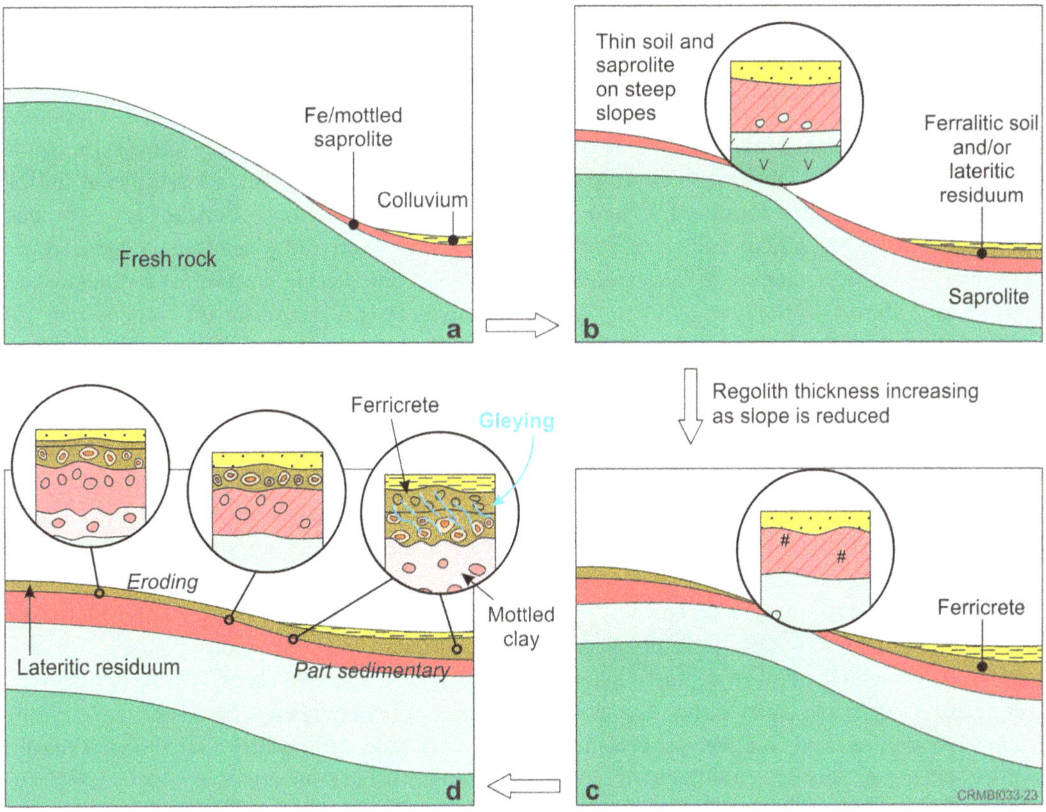

Fig. 8 Landscape reduction and toposequence development. As weathering proceeds, erosion from uplands, colluvial and alluvial sedimentation and chemical wasting gradually reduce the relief. Iron leached from uplands precipitates as oxides and oxy-hydroxides in colluvium and alluvium on lower slopes and valleys, cementing the sediments, and dehydrating and hardening to form ferricrete when exposed. Reducing conditions may develop in water-logged valleys

or toposequence of profiles may develop (Fig. 8). Saprolite formation and the broadly residual accumulation in the pedolith described above continue throughout but particularly apply to upland positions. On upper and mid-slopes, however, the regolith tends to be thinner, with less residual accumulation due to leaching by a greater through-put of groundwater and greater surface erosion. Conversely, absolute enrichment may also occur, due to accumulation of components derived from upslope in solution, in suspension or, at surface, as detritus. As the relief is low, with slow surface runoff and stream and river flow rates, colluvial and alluvial sedimentation takes place on lower slopes and valley floors. Here, the water-table is shallower, and groundwater is less oxidizing and less efficient at removing soluble weathering products. This may favour the formation of smectites in the lower saprolite, especially over mafic rocks, but the total depth of weathering may be greater than elsewhere in the toposequence (Mabbutt 1980). The total profile, including in situ, residual and transported units, tends to be deeper. Absolute enrichment, especially of laterally derived Fe oxyhydroxides and, in places, Mn oxides, may accumulate in both transported and residual upper horizons of the regolith (Maignien 1966; McFarlane 1976; Anand and Paine 2002) that may later become indurated as ferricrete.

The uppermost horizons may form a continuum from essentially residual (as lateritic residuum) in upland sites to entirely transported (as ferricrete) in valleys (Fig. 8). On slopes, clearly residual lateritic residuum may transition upwards to similar overlying ferruginous units having evidence of lateral transport. Low in the landscape, Fe eroded or leached from adjacent hill slopes or updrainage may be deposited or precipitate as oxides in local to distal sediments or in underlying saprolite, forming ferricrete. As the origin of these ferruginous units is difficult to determine, these are referred to informally as FGD (Sect. 2 above).

Weathering may also be episodic, with numerous interruptions due to climatic changes and, even in the most stable cratons, tectonic activity. Even if such changes are quite minor, they may accelerate or slow erosion and alter rates of weathering. The uppermost horizons of the profile are again most likely to be affected, eroded high in the landscape and buried in lowlands. Thereafter, weathering may recommence, with the upper horizons reformed, developing from saprolite or within colluvial and alluvial sediments on slopes and valley floors. Apparently 'complete' unified and mature profiles in a toposequence may thus have multiphase origins.

Regolith profiles developed on different lithologies commonly appear quite similar, mainly because the assemblage of dominant minerals (i.e., Fe (-Al) oxyhydroxides, kaolinite and quartz) is restricted compared to those of fresh rocks. Nonetheless, within a region, profiles over different lithologies may vary markedly according to their topographic setting, primary mineralogy and porosity affecting the manner in which weathering reactions occur. In general, the depth of weathering is greater on the more mafic and most porous rocks, but there are many exceptions, especially where low-grade metamorphic alteration has formed more stable phases such as talc, amphibole and serpentine. In the Yilgarn Craton of Western Australia, for example, mafic and porous felsic rocks are generally the most deeply weathered (60–70 m), whereas profiles over more massive rocks including granite and dolerite are shallower (35–45 m). Weathering is typically much deeper over sulfide-rich, strongly sheared and/or faulted rocks, extending 50 m or more than in adjacent rocks (Anand and Butt 2010).

Bedrock composition can impose characteristic features on the regolith formed from it. Thus, lateritic residuum is commonly most strongly developed over Fe-rich mafic and ultramafic rocks but may never have formed on some Fe-poor felsic rocks (Anand and Paine 2002). Instead, the uppermost horizon is a mottled sandy clay overlying plasmic or arenose zones and saprolite. Ferruginization and silicification are typical of saprolite on olivine-rich or serpentinized dunites, whereas saprolite on peridotites is more or less clay-rich—giving rise to different types of Ni laterite, commonly following later modification and enrichment (chapter "Nickel-Cobalt Laterite Deposits"). Gossans and other resistant materials such as silicified weathered dunite ('birbirite') may be the only remnants of a paleoregolith cover in the North African and Arabian deserts (Butt 2014; Taylor and Thornber 1992; Budel 1982; Augustithis 1965).

The nature and thickness of the regolith on limestones, dolomites, carbonate-rich clastic sediments and carbonatites varies depending on the proportion of non-carbonate minerals. The regolith is better developed and thicker where the carbonate content is low. Pure limestones and dolomite rocks may outcrop, whereas terra rossa soils and karst bauxites are considered to have formed by weathering of intercalated aluminosilicate minerals, concentrated residually following carbonate dissolution. Non-sulfide Zn-Pb deposits in such terranes are the product of oxidation of sphalerite and galena, forming secondary carbonate minerals instead of Fe-oxide-rich gossans.

3.5 Post-Lateritic Modification of Intensely Weathered Regoliths

Global Climate Changes Intensely weathered lateritic regoliths are generally considered to have formed under seasonally humid climates of the tropical savannas. However, they are globally widespread and present on most continental landmasses throughout much of their present range of latitudes and in some tectonically active belts (Fig. 1). This is exemplified by the global distribution of major supergene deposits of lateritic bauxite, Ni laterite and REE-Nb-P carbonatite (Fig. 2). The regolith in these regions has been successively formed, partially eroded, modified and reformed over long periods during the Phanerozoic due to tectonic activity and climate changes. Weathering commenced as early as the Carboniferous in ancient shields such as the Archaean Yilgarn Craton in Western Australia, or as late as the upper Miocene in recently emergent terrains, most importantly in the Caribbean and in SE Asia and Oceania (e.g., Indonesia, New Caledonia, Philippines).

Recent isotopic data, biostratigraphic studies and advances in plate tectonics have confirmed that temperate to warm humid climates prevailed at high latitudes in the mid-late Mesozoic and early Cenozoic, providing conditions conducive to intense weathering in Australia and, exemplified by bauxites and weathered carbonatites, across northern Europe and Asia. Noting, of course, that there are vastly different climatic conditions across each of the present continents so uniform timing of a particular weathering phenomenon is not expected. Nonetheless, the most significant global climatic events associated with lateritic weathering were probably:

1. *Late Jurassic to mid Cretaceous*, with some of the most compelling examples being in south east Europe where bauxite, Ni-Co laterites and associated lateritic regoliths are preserved within and/or beneath sediments of known stratigraphic ages, and gossan formation in Spain and Portugal (chapter "Gossans: In Situ Supergene Base Metal Deposits").
2. *Palaeocene-early Eocene (66–48 Ma)*, including the Eocene Climatic Optimum.
3. *Miocene Climatic Optimum (20–14 Ma).* Numerous other short-lived or more local weathering events, indicated by $^{40}Ar/^{39}Ar$ dating and other techniques, continue to the present.

Despite the different ages that can be ascribed to lateritic regoliths in different locations and bedrock lithologies, there is a remarkable similarity in their structure and mineralogical and chemical compositions. Accordingly, the conditions and mechanisms of formation were probably similar initially, and the same processes continued when comparable weathering re-occurred at a later date—even if partial erosion had taken place in the interim.

Post-lateritic Modifications Modifications to Intensely weathered regoliths and landscapes are due to tectonic and climatic changes. Many modifications have been minor because the profound alterations that occurred during weathering have resulted in the regolith being composed of only the most stable secondary minerals (e.g., kaolinite, gibbsite, Fe oxides and the most resistant primary minerals such as quartz (Fig. 7). These minerals are largely unaffected by changes in conditions, and hence many of the principal chemical, mineralogical and petrographic properties of the regolith are retained. However, they have significance where re-weathering has led to the transformation or further concentration of some components to produce economic supergene deposits.

Both tectonic activity and climatic change commonly result in partial erosion of the regolith, partly offset where lateritic residuum in upland plains and slopes has been indurated to resistant duricrusts. The resultant plateau landforms are broadly similar across different climatic zones, so that deeply weathered landscapes in semi-arid regions have many similarities to those in the humid tropics. Nevertheless, landforms alone can be misleading, because indurated ferricrete is similarly resistant to erosion, to produce plateau landforms by topographic inversion (Butt and Bristow 2013; Sawadogo et al. 2020).

The effects of these changes are summarized in Table 4. Uplift, whether epeirogenic or orogenic, leads to lowering of the water table and dehydration of near-surface horizons, including irreversible hardening of lateritic residuum and ferricrete. Conversely, down-warping may result in waterlogging, imposition of reducing conditions and less leaching in valleys, and burial of residual regolith by colluvial and alluvial sediments. Tectonic plate movements will also slowly modify weathering regimes by drifting into different climatic zones (Fig. 9).

Table 4 Effects of changes from the conditions of formation on the lateritic regolith

Tectonic activity	
Uplift	Lowering of the water-table
	Irreversible dehydration and hardening of lateritic residuum
	Increased leaching of upper horizons under more oxidizing conditions, continued weathering at depth
	Rejuvenation of rivers and increased headward fluvial erosion
	Increased surface erosion, especially upper slopes, slowly removing surface duricrust and, more rapidly, underlying friable units
	Ongoing weathering leading to redevelopment of upper horizons
	Deposition of erosion products as ferruginous sediments and overlying silty clay in lower slopes and valleys. Cementation of sediments by Fe oxyhydroxides to form ferricrete
Down-warping	Decrease in fluvial erosion, increased sedimentation in valleys
	Waterlogging of lower parts of the landscape and imposition of reducing conditions
Plate movements	Drift of landmasses into different climatic zones
	Uplift or down-warping, especially close to plate boundaries
Climatic change	
More humid climate	Increased leaching, deeper soil development, continued intense weathering at depth
	Chemical solution and destruction of ferruginous duricrusts
	Decreased erosion due to thicker vegetation.
Less humid climate	Decreased leaching; weathering at depth less intense, with persistence of smectites in lower saprolite
	Irreversible dehydration and hardening of lateritic residuum
	Increased erosion on slopes due to reduced vegetation and greater run-off
	Deposition of erosion products as ferruginous sediments and overlying silty clay in lower slopes and valleys
Semi-arid or arid climate	Decreased leaching
	Irreversible dehydration and hardening of lateritic residuum
	Increased erosion due to reduced vegetation and greater run-off
	Retention and precipitation of silica, alkaline earths and alkalis in silcretes, clays, calcretes, salts
	Salinization of groundwaters, especially in valleys
	Increased erosion
Mediterranean and temperate climates	Continued leaching—during cooler winter months. Rainfall more effective due to less run-off, reduced and evaporation and uptake by vegetation; weathering reaction rates are slower
	Leaching at depth less intense, with persistence of smectites in lower saprolite erosion reduced due to persistence of vegetation throughout year

In humid climates, regolith modifications are largely those related to intensified leaching due to better drainage following uplift or increased rainfall due to a change to wetter climates (e.g., rainforest). Desilication and gibbsite formation cause upgrading of bauxites and, over ultramafic rocks, mobilization of Ni to form or upgrade some Ni laterites (chapter "Nickel-Cobalt Laterite Deposits"). Uplift promotes leaching of Ni from oxide zones and its precipitation as hydrous Ni-silicates in veins and fissures in underlying saprolite and saprock. Pseudomorphs of these structures are not reported from saprolite in oxide zones, indicating their late formation—i.e., they were not present as the precursor to the oxide zone (Sect. 3.3). Upgrading of bauxite may also occur when lateritic regoliths formed under savanna climates are later exposed to a Mediterranean climate. The annual rainfall is similar, but the seasonality of precipitation differs, with the winter rainfall of Mediterranean regions being more effective due to lower evaporation and greater infiltration and biological activity over longer periods. In the present equatorial

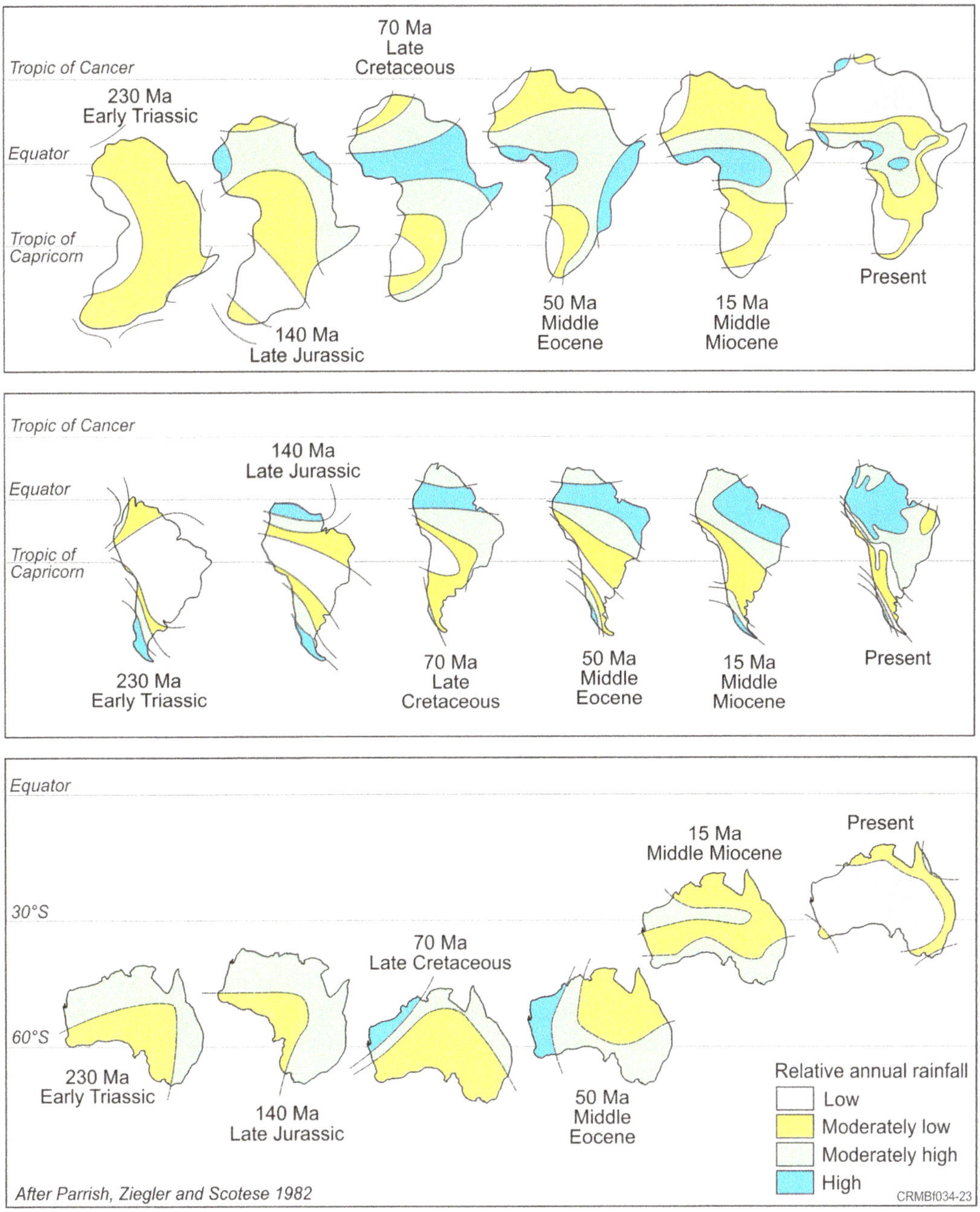

Fig. 9 Changes in relative annual rainfall over time in Africa, S. America and Australia, after Parrish et al. 1982. Whereas the S. American continent has been in much the same position since the Late Cretaceous, Africa and Australia have rotated and drifted north into different climatic zones

rainforests of South America and Africa, however, lateritic plateau landforms formed under earlier seasonal and drier conditions have, in places, been degraded and partially eroded to a 'half-orange' landscape of circular rounded hills with a relief of 30–50 . The hills have lateritic stone line regolith profiles, whereas slopes and valleys have quartz stone line profiles (Lecomte 1988).

Overall, modification is probably greater in drier climates, due to more pronounced variations in temperature and moisture contents both diurnally and seasonally. Continued weathering at the base of the profile results in rather different mineralogy, especially the development of 2:1 lattice clays (e.g. smectites) even in dry savannas. In semi-arid and arid environments, smectite formation, retention and accumulation of alkali metals, alkaline earth elements and silica significantly alter the bulk chemical and mineralogical composition of some regolith horizons, and groundwaters can have greatly increased total dissolved solids contents. These have economic significance where there is significant chemical (hydromorphic) mobility, for example of U and V to form calcrete U deposits (chapter "Channel and Playa Uranium Deposits"), and a range of non-marine brines and evaporites rich in halides, sulfates, borates and nitrates. The regolith phases of non-sulfide base metal deposits, such as the Zn and Pb carbonates and sulfates smithsonite, hydrozincite, cerussite and anglesite, have a strong potential for trace element accumulation under semi-arid conditions due to the mobility of Zn and Pb under these conditions (chapter "Supergene Non-Sulfide Base Metal Deposits"). For these deposits, tectonic uplift may enable groundwater to oxidize sulfide mineralization and transport mobilized metals to precipitation sites.

Late-stage silicification, generally as quartz rather than opaline silica, of weathered felsic and dunitic ultramafic rocks, and some dolomites may preserve or, when intense, destroy the bedrock fabric in saprolite and dilute the concentrations of other components. The precipitated quartz is effectively inert to any further weathering, which may proceed either above or below the silicified horizon. At the Onca and Puma Ni laterite deposits in Amazonia, Brazil, some oxide and hydrous silica ores occur beneath a protective silica cap rock. Golightly (2010) suggested that silica was deposited in an earlier arid phase, and that the ores formed under later humid conditions. However, the silicification may instead be a product of poor drainage during an earlier phase of lateritization, so these deposits may be the result weathering re-continuing after relief inversion, without significant climate change.

4 Geomorphology, Landscape Development and Weathering History

4.1 Plateau Landforms and Geochronology

Most lateritically weathered continental landscapes have been tectonically stable since the Palaeozoic or earlier, undergoing mainly gentle vertical epeirogenic movements. These have given rise to landscapes characterized by one or more elevated, remnant plateaux, commonly capped by ferruginous duricrusts or silcrete. The seeming continental-scale continuity of these plateaux, especially in Africa and South America, was recognized by King (1962), but their origin and geomorphological significance have been debated since. The initial hypothesis was that the different plateau levels depict the long-term evolution of the landscape, the result of repeated periods of scarp retreat, weathering and uplift. According to this model, denudation starts at the continental margins, the ultimate base level, and proceeds towards the interior. Episodic uplift, erosion and scarp retreat, followed by quiescent periods of intense weathering, result in stepped diachronous landscapes in which the surface with the highest elevation is the oldest and those with lower elevations are progressively younger. These ideas were based on identifying surfaces by correlating plateaux having similar elevations and geomorphological characteristics.

As discussed by Vasconcelos and Carmo (2018), this concept has been challenged by proponents of models that considered that uplift has been essentially continuous and that the erosional and weathering processes that mould the landscape are in dynamic equilibrium with their tectonic and climatic environments. Vasconcelos and Carmo (2018) argued that if the weathering surfaces represented by the plateaux can be dated objectively by geochronological techniques, and that the ages obtained are consistent with the geomorphologically correlated surfaces, then the hypothesis that they are direct evidence for episodic landform evolution is supported. Numerous studies of methods for dating regolith materials have been applied in recent years, with palaeomagnetic, numerical radiogenic isotope and fission track techniques having the greatest application over the relevant time scales (Pillans 2008). The techniques are based on determining the date of crystallization of specific minerals as proxies for the surface, such as K-bearing Mn oxides (K-Ar, $^{40}Ar/^{39}Ar$) and Fe oxides ((U-Th)/He), palaeomagnetism). Broader age estimates are given by stable isotope determinations of O in Fe oxides and kaolinite, and fission track analysis of apatite can reveal the thermochronology and denudation history of landscape evolution. These studies have generally supported the geomorphological correlations, although many discrepancies in interpretation remain. For example, regional variations in the timing and amount of uplift, in erosion rate and in duration and intensity of weathering can result in erosion surfaces of different ages, at different elevations, eroding at different rates.

Dating of surfaces expressed as plateaux can apply not only to the stable cratons of the tropics and sub-tropics but also to younger, active terrains on plate margins, such as the Caribbean and parts of SE Asia and Oceania, where complete lateritic regoliths are preserved. Remnants of intensely weathered regoliths are also present at higher latitudes in the northern hemisphere, mostly buried by younger rocks. Age dating of these occurrences has mainly been through stratigraphic correlation rather than more objective laboratory-based analysis of key minerals. The distribution of lateritic regoliths on each continent are discussed below in terms of the influence of their geological and geomorphological expression and tectonic and climatic histories on the nature and distribution of supergene mineral deposits.

4.2 Regional Geomorphological and Weathering Histories

4.2.1 Africa

Following the break-up of the Gondwana continent and opening of the Equatorial Atlantic Ocean in the late Cretaceous, the African continent has drifted northward about 15^0 and rotated anti-clockwise (Fig. 9). This movement influenced oceanic currents and hence the climate, including temperature and rainfall patterns. During this time, weathering has been essentially continuous on emergent terrain, whereas complex regolith and sedimentological sequences have developed on low-lying regions due to periodic flooding during successive cycles of uplift and down-warping. Lateritic regoliths now occur principally from the semi-arid Sahel in the north, through the humid savannas and rainforest of West, Central and East Africa, and Madagascar, to the dry savannas of southern Mozambique and NE South Africa. A few weathered remnants in the Sahara (Budel 1982) and the deserts of SW Africa show that lateritic weathering was more extensive in the past. Epeirogenic "basin and swell" tectonics since the Early Jurassic (Burke and Gunnell 2008) have given rise to a topography characterized by several elevated plateaux, commonly capped by lateritic duricrusts, that King (1962) correlated across the continent. King considered the highest and thus oldest to be Cretaceous in age and termed it the Gondwana surface. King also equated similar high-level plateaux in South America to the Gondwana surface.

West Africa Plateaux capped by Al-Fe-rich duricrusts have been described, named and mapped from Senegal, across southern Mali, Burkina Faso and northern Côte d'Ivoire to eastern Niger (e.g., Michel 1973; Tardy 1997) at elevations from 1400–850 m near Fouta Djallon, Guinea, to the low pediments of the alluvial plains in the valleys of the present major rivers (Fig. 10a). Until recently, most such duricrusts were interpreted as lateritic residuum by many authors (e.g., Tardy 1997; Zeegers and Lecomte

1992). However, others (e.g., Maignien 1966; Thomas 1994) and more recent observations have recognized that, even on some of the higher plateaux, many duricrusts are sedimentary, consisting of ferricrete composed of detrital cobbles, including pisolitic lateritic duricrust, cemented by Fe oxides, and represent instances of inverted topography (Sawadogo et al. 2020; Chardon et al. 2018; Butt and Bristow 2013; Bamba et al. 2002). The surfaces are undulating, not flat, which, coupled with vertical tectonic movements, means that correlation by elevation alone is unreliable. Burke and Gunnell (2008) concluded that the bauxitic African, Dongol-Sigon and Labé surfaces were of the same age and equate to the African surface of King (1962). The Gondwana surface is not recognized. After 34 Ma, domal uplift of the West African swell (Fig. 10b) raised the African surface to different elevations, resulting in the stepped plateau topography. They also considered the lower surfaces to be essentially co-eval, formed on an undulating terrain disrupted by further uplift of the swell and consequent erosion to form the discontinuous 'Glacis' surface. Over time, parts of the original African surface have been elevated by over 1 km. Contemporaneous down warping in northern Mali, Burkina Faso and Niger formed a basinal depression. The tectonic activity resulted in major changes to erosional base levels at regional and local scales.

Numerical radiogenic $^{40}Ar/^{39}Ar$ dating of cryptomelane in Mn-rich lateritic regolith at Tamboa in northern Burkina Faso has been used to estimate timing of different phases of weathering and correlate these with the erosion surfaces (Beauvais and Chardon 2013; Beauvais et al. 2008; Colin et al. 2005; Gunnell 2003). The results are summarized in Table 5, compared with similar data from East Africa. The data suggest that the combined African surface formed at 59–45 Ma. Subsequent clusters of ages have been related to the other surfaces at 29–24, 18–11, 7–5 and 3 Ma, separated by erosional episodes triggered by climate change and/or uplift. This assumed relationship between the age clusters determined by $^{40}Ar/^{39}Ar$ dating and the formation of the plateau topography is at variance with the conclusions of Burke and Gunnell (2008) that the Glacis surfaces are of the same age, and has yet to be proven. These age estimates also do not equate to dating of the surfaces themselves as discussed by Vasconcelos and Carmo (2018).

Climates in West Africa are considered to have been humid savanna to savanna through most of the Cenozoic, possibly becoming drier as the continent drifted north. The Al-rich, bauxitic composition of the duricrust on the African surface is postulated to be due to weathering at quite low, pre-swell elevations under very humid rainforest climates, although it is probable that continued weathering since that time has led to further leaching and upgrading. Later surfaces are more ferruginous due to less humid climate, and lower ongoing leaching.

The principal supergene deposits in West Africa are the rich and extensive bauxites developed from Silurian shale and Ordovician sandstone in Guinea and southwestern Mali, and Fe ores on Archaean BIF across Guinea, Liberia, Côte d'Ivoire and Sierra Leone, with the largest resource at Simandou in SE Guinea. Bauxite at Fouta Djallon has formed on the African surface at elevations ranging from 200–250 m to 850 m (chapter "Lateritic Bauxite Deposits") the Fe ores are probably at similar geomorphological levels. Supergene gold occurs in lateritic residuum and saprolite developed on Proterozoic Birrimian rocks in Côte d'Ivoire, Mali, Burkina Faso and northern Ghana, but deposits are mostly small relative to associated bedrock resources, with development (and discovery) limited by the extensive ferricrete cover on the lower post-African surfaces. Exceptions are the Ity and Yatela deposits, where supergene enrichment over carbonate-rich primary mineralization forms the principal ore (chapter "Supergene Gold Deposits in Intensely Weathered Regolith").

Equatorial, East and Southern Africa Plateaux landforms are generally less evident in these regions, although numerous plateau levels or erosion surfaces at different elevations have been mapped in Uganda. Geomorphological

Table 5 Jurassic to Holocene climates, weathering processes and plateau surfaces in West, Eastern and Equatorial and Africa. (***Uganda***: after Taylor and Howard 1998; Burke and Gunnell 2008; Roberts et al. 2012; McFarlane 1976. ***Zambia-DRC***: after De Putter et al. 2015; Decree et al. 2010; Burke and Gunnell 2008. ***Burkina Faso/West Africa***: after Beauvais et al. 2008; Burke and Gunnell 2008; Butt and Bristow 2013; Chardon et al. 2018)

		West Africs			Eastern and Equatorial Africa				
Period/ Epoch	Date Ma	Climate Burkina Faso	Process West Africa	Surface West Africa	Climate Uganda	Process Uganda	Climate Zambia,-DRC	Process Zambia, DRC	Surface East Africa
Holocene Pleistocene	2.6	Dry savanna Savanna	↑ erosion Ferruginization/ lateritization		Savanna Arid	⍰ lateritization Erosion	Savanna Arid *Kalahari sand*	↑ lateritization	
Pliocene	5.3	Dry savanna Savanna	↑ erosion Ferruginization/ lateritization	Bas glacis	Warm arid Savanna	↑ erosion Lateritization	Arid—Desert Savanna	↑ erosion Lateritization	Undefined
Miocene	10 23	Dry savanna-savanna Humid savanna	↑ erosion Ferruginization/ lateritization ↓ erosion	Moyen glacis Haut glacis	Savanna-drying Humid savanna Rainforest	↔ lateritization ↓ ↑ local erosion	Savanna Dry savanna *Congo desert*	Lateritization ↑ erosion	Miocene
Oligocene	33	Savanna Humid savanna	↑ erosion Ferruginization/ lateritization	Relief intermediaire	Dry savanna; cooling	↑ lateritization ↓ local erosion	Savanna–humid savanna	? Lateritization Lateritization	
Eocene Paleocene	56 66	Humid savanna Rainforest	↑ erosion Lateritization	African (includes Labe & Dongol-Sigon surfaces) (pre 30 Ma)	Humid savanna	↔ lateritization erosion	Semi-arid; dry savanna *Congo lake*	↑ erosion ↓↓	African (includes Buganda surface) (pre 30 Ma)
Cretaceous	100 145			Gondwana (cretaceous-Jurassic)	Savanna Humid savanna	↔ erosion Lateritization	Semi-arid Dry savanna "thermal optimum"	Erosion	Gondwana (cretaceous-Jurassic)
Jurassic	201				Semi-arid; dry savanna	Erosion	Semi-arid	Erosion	
Permo-Triassic	299				Temperate Glacial	Erosion Sedimentation	Temperate Glacial	Erosion Sedimentation	

↔: rifting; ⍰ warping; ↓ uplift; ↓ down warp

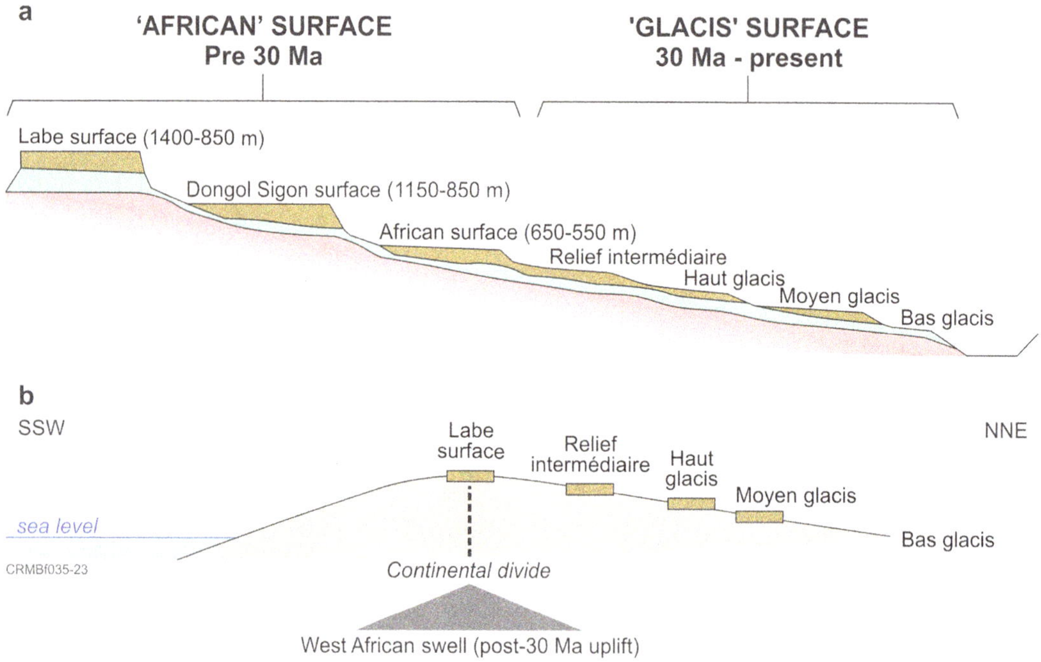

Fig. 10 Depiction of plateau landscapes in West Africa. Each plateau level was originally considered to be of a distinct relative elevation and age (Michel 1973; Tardy 1997), but (**a**) reinterpreted as two surfaces by Burke & Gunnell (2008). The grouped 'African' and 'Glacis' surfaces (**b**) are regarded as co-eval, with their different elevations due to the domed structure of the swell. Duricrusts on the higher African surfaces are bauxitic, whereas those on the Glacis surfaces are ferruginous

models, summarized by McFarlane (1976), recognize one to four different surfaces, with the variation in elevations attributed to undulations due to differential weathering of contrasting lithologies or to tectonic activity associated with the East African Rift (Fig. 11). The surface materials are generally referred to as 'laterite', with no distinctions between lateritic residuum, ferricrete and ferruginous sediments. McFarlane concluded that there were two surfaces, namely the Buganda Surface, a component of the African Surface, and the Miocene Surface (Table 5), attributing the different elevations to rifting, warping, volcanism and swell tectonics. Her studies also recognized the presence of detrital materials (ferricrete) on some surfaces, probably contemporaneous with lateritic residuum. West and south of the Rift, in Zambia, the eastern Democratic Republic of the Congo (DRC) and the Congo Basin, the relief is generally more subdued and plateau landforms less evident. The weathering and tectonic history of the Congo Basin and surrounding uplands suggests that lateritic deep weathering was widespread in the west (Cameroon and Angola Domes) and Eastern DRC in the late Cretaceous-Palaeocene (75–47 Ma), terminated by uplift of the Central African Atlantic Swell and East African Dome that elevated the planation surface and initiated erosion, with a later event at 34 Ma (Guillocheau et al. 2018). The Congo Basin, however, was at or below sea level from the Palaeocene, and subject to sedimentation by shallow marine and terrestrial (lacustrine and marsh) sediments. The basin was not fully emergent until the late Miocene (11.5 Ma), a result of further uplift. As in West Africa, the surface ferruginous duricrusts have generally been identified as lateritic residuum. However, ferricrete very similar to that in West Africa has been observed on upland plateaux in Zambia by the present authors and could be widespread (Fig. 12) (Anand and Butt 1998).

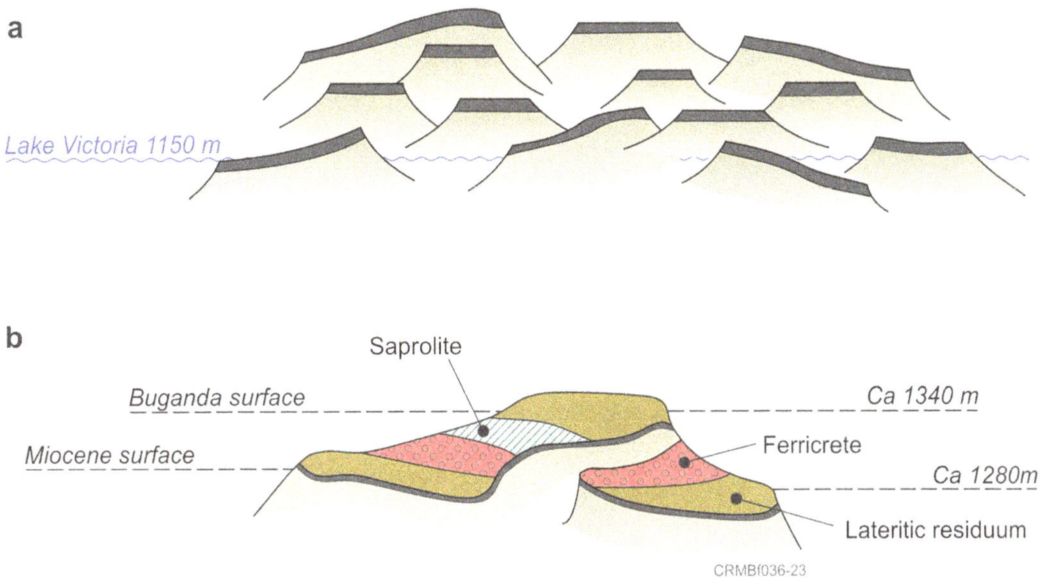

Fig. 11 Plateau surfaces in Uganda. (**a**) Lateritic surfaces developed at many elevations, to below the surface of Lake Victoria, possibly due to rifting, warping, volcanism & swell tectonics. (**b**) Interpreted as two surfaces, the upper Buganda surface, part of the African surface, and a lower Miocene surface, both modified by erosion and deposition. (After McFarlane 1976)

Fig. 12 Ferricrete boulders on upland plateaux consisting of cobbles and pebbles cemented by a ferruginous matrix. (**a**) ferruginous nodules and lithic fragments, and vein quartz, Fouli, Burkina Faso. (**b**) quartz pebbles and lithic fragments, Solwezi, Zambia. (Photos: C.R.M. Butt)

$^{40}Ar/^{39}Ar$ dating of supergene Mn deposits in the DRC (De Putter and Ruffet 2020; De Putter et al. 2015) confirms the pre-45 Ma weathering event and indicates several others, principally at 35 Ma, 24 Ma, 13–11 Ma and 7–3 Ma, similar to those noted by Guillocheau et al. (2018). These suggest weathering under savanna climates, separated by drier periods, from the mid-Eocene to late Miocene. During the Miocene, a desert extended south from the Congo Basin to the Kalahari, but thereafter the climate became increasingly humid due to the northward drift of the continent (Fig. 9) with rainforests in the Congo Basin and in the near-coastal uplands.

Supergene mineralization includes important Cu-Co-Mn deposits in the Copperbelt of DRC and Zambia (chapter "Supergene Ores of the Central African Copperbelt"), Ni-Co laterites in Burundi (Musangati), and Madagascar (Ambatovy), and important residual P-REE-Nb enrichments overlying carbonatites in Angola (Longonjo), DRC (Lueshe), Gabon (Mabounié), Kenya (Mrima) and South Africa (Zankopsdrift, Glenover) (chapter "Carbonatite Hosted Residual REE Deposits"). Bauxites and Mn deposits occur at high elevations in South Africa. Surficial U deposits are widespread in Namibia, including Langer Heinrich, the only calcrete U deposit to have been mined (chapter "Channel and Playa Uranium Deposits"). Whereas the formation of the high-level bauxites possibly commenced on the African surface in the Eocene, the supergene Cu-Co-Mn deposits are younger, dating from humid periods following uplift in the mid-Miocene to Pliocene. Aridity in the late-Pliocene to Quaternary (~2 Ma) has led to induration of Mn-Co and Fe-rich duricrusts and silicification typical of dolomitic bedrock, giving partial protection from later erosion. Enrichment of ore elements in the Copperbelt, northern Namibia and over carbonatites has been enhanced by the dissolution of the carbonate-rich host rocks, generation of voids in the karstic terrain and long residence times that accommodates mineral precipitation. Non-sulfide base metal deposits in Namibia are also a product of deep weathering under former humid conditions, but now under climates ranging from semi-arid in the north (Tsumeb) to desert in the south (Skorpion)—see chapter "Supergene Non-Sulfide Base Metal Deposits".

4.2.2 India, SE Asia and Oceania

Distribution of Lateritic Regoliths Lateritic regoliths occur extensively across India, from the lowland plains of the SW coast, as plateau remnants in the mountain range that form the Western and Eastern Ghats and extensive plateaux over the late Cretaceous Deccan Trap basalts of the centre and northeast. This regolith is discontinuous and partially or wholly eroded in areas of high relief, especially in the far north, and largely absent from the alluvial terrain of the Indo-Gangetic Plain. It is also largely absent in the semi-arid to desert regions in the NW; in Rajasthan, where the gossans of the Khetri copper belt may be the only remnants of past lateritic weathering. Tropical soils (acrisols) derived from intensely weathered and lateritic regoliths are present across the monsoonal (humid savanna and rainforest) regions of NE India, extending to Bangladesh, Myanmar (Burma), SW China and throughout SE Asia, northern Australia and the islands of Indonesia, Philippines and western Oceania.

$^{40}Ar/^{39}Ar$ dating of supergene Mn deposits in the Western Ghats and adjacent coastal lowlands in SW India has identified major periods of intense weathering at approximately 53–44, 39–22 and 14–10 Ma (Jean et al. 2020; Bonnet et al. 2016), with the earlier dates applying both to the highlands and lowlands. These authors infer from the data that there were three successive palaeo-landscapes, with slow denudation commencing from ca. 44 Ma, and accelerating after ca 22 Ma., presumably following uplift. The topography of the Ghats is now characterized by plateau remnants capped by lateritic bauxite, whereas in central India the plateaux are more continuous, especially over the Deccan Traps, but with pronounced escarpments at their margins. The dates of intense weathering correspond to the periods when the Indian subcontinent was drifting northwards through the equatorial zone during and after the Eocene climatic optimum. During this time, the climate has changed from rainforest to monsoonal (humid savanna) over much of the sub-continent, but with semi-arid to desert conditions in the northwest.

Supergene Deposits There are numerous large bauxite deposits in India, principally developed in high plateaux (900–1300 m) and on lower (50–350 m) coastal regions (Bhukte et al. 2020). The largest deposits are in the plateaux underlain by the Deccan trap basalts. Smaller deposits on Archaean to Proterozoic rocks are found on the western coastal lowlands and dissected terrain in the western and eastern Ghats. Supergene Mn deposits and minor supergene gold, as well as the more mundane ferruginous saprolite material quarried for brick-making, occur in the lowlands and coastal tracts in the SW. Few supergene deposits have been discovered to date in much of SE Asia, apart from small Ni laterites in Myanmar (Tagaung Taung) and Vietnam, as well as quarrying for lateritic saprolite for construction. However, ion-adsorption REE mineralization developed over granites has been discovered in several SE Asian countries, most especially in southern China and Myanmar (chapter "Ion Adsorption Mineralisation in Regolith-Hosted REE Deposits"). REE enrichment occurs in kaolinite-halloysite in the mid-upper saprolite of mostly hydrothermally-altered muscovite granite. Deposits in China and Myanmar are, to date, the only ones in production, with China accounting for about 90% of global HREE supply.

The SE Asian island nations in the equatorial to sub-tropical zone are rich in important supergene deposits, especially Ni-Co laterites, with Indonesia, Philippines and New Caledonia being major global producers. They mostly exploit high-grade hydrous silicates ores, formed by lateritic weathering of ophiolitic ultramafic rocks exposed by uplift commencing in the late Miocene. Porphyry Cu-Au-Mo deposits occur in Irian Jaya-Papua New Guinea, the Philippines, Solomon Islands and Fiji, with many having significant supergene mineralization developed under humid tropical conditions, coupled with uplift, since emplacement in the late Miocene and continuing to the present.

4.2.3 South and Central America

Distribution of Lateritic Regoliths East of the Andes, the Amazonian Craton was relatively stable tectonically during the Phanerozoic, except for epeirogenic activity. This led to the formation of extensive basins on the Craton during the Palaeozoic, surrounding the mostly emergent Guiana and Brazilian Shields. After the break-up of Gondwana in the Cretaceous, the continent of South America has largely retained its latitudinal position. The Amazonian Craton continued to straddle the equator, and the Amazon river developed in a newly-formed eastward-flowing drainage basin between the Shields. The separation from Africa, the opening of the Atlantic and the eventual uplift of the Andes along the western margin fundamentally altered the geology, geomorphology and climate of the entire continent. The Andes did not fully develop until the mid-Miocene (10–15 Ma), reaching as high as 4 km in southern Peru (Sundell et al. 2019) and ~ 7 km on the Argentina-Chile border. In the late Miocene, relative uplift of the Shields and down-warping of the basins caused further erosion and sedimentation cycles that continue to the present; intermittent marine incursions into the Amazon and other basins occurred until the Miocene. Climates throughout much of the continent north of about 30^{0}S have remained warm and mostly humid since the Paleocene. However, the elevation of the Andes had a significant orographic effect, with extreme aridity at high elevations. Periods of aridity also occurred in Amazonia during the Pliocene. Lateritic regoliths are present across the humid savannas and rainforests of this region, developed on the Precambrian rocks of the Shields and on the Palaeozoic to Cenozoic sediments of the basins and coastal plains. As in Africa, the present topography is characterized by several plateau levels. King (1956) described four distinct surfaces in SE Brazil, since dated by Vasconcelos & Carmo (2018): Gondwana/post-Gondwana (Mesozoic, 70–~50 Ma), Sul-Americana, (Early Cenozoic), Velhas (Late Cenozoic, 13–10 Ma), and Paraguaçu (Pliocene, < 4 Ma). These have also been recognized throughout the region by other authors. Théveniaut & Freyssinet (2002) equated paleomagnetic dates of lateritic bauxites at different elevations in French Guiana and Suriname to these surfaces, with broadly similar results (Table 6). Lateritic regoliths also extend to the younger terrains of northern Colombia, Central America to southern Mexico, and to the Caribbean islands.

Table 6 Palaeomagnetic and U-Th/He dating of lateritic bauxite, French Guiana and Suriname. (After Théveniaut and Freyssinet 2002; Heller et al. 2022; Ansart et al. 2022)

Epoch	Duricrust/soil	Climate	Planation surface	Elevation (m)
Pleistocene 2.5–0 Ma	Latosols, podzols	Rainforest	na	0–1000
Pliocene–Pleistocene 5–2 Ma	Bauxitic	Savanna—rainforest	na	< 250
Late Miocene 15–5 Ma	Ferruginous	Savanna—rainforest	Cayenne 3	0–20
Miocene 13, 10 Ma	Ferruginous ± aluminous/ bauxitic	Savanna	Late Velhas Cayenne 2,	50–150 (incl: Moengo)
Oligocene—Early Miocene Erosional hiatus; 35–15 Ma	Ferruginous	Savanna	Early Velhas	200–400
Palaeocene-Eocene 60, 50, 40–35 Ma	Bauxitic	Savanna	Sul-Americano Cayenne 1	400–700 Bakhuis
Late Cretaceous 90–60 Ma	Ferruginous	Savanna	Gondwana	1000

na not applicable—weathering effects at several elevations

Supergene Deposits Amazonia and the adjacent Shields in the NE of the continent host a wide range of important supergene resources. The Archaean Carajás Domain, Brazil, is one of the world's most richly endowed mineral provinces, including rich Fe, Mn and Cu-Au deposits, each with a substantial supergene component. Also in Brazil, major supergene Fe ores occur in the Quadrilátero Ferrífero in Minas Gerais, where weathering reaches depths of 150–400 m. Lateritic bauxites are widely developed on rocks from Precambrian to Pliocene in age in the Guianas and Venezuela in the north to south of the Amazon in Brazil. These are developed on plateau surfaces at different elevations and ages, as noted above. In Amazonia, many bauxites are buried by the yellowish, homogeneous, silty Belterra Clay, an enigmatic surface regolith unit up to 25 m thick. Postulated origins of this material include lacustrine sedimentation of erosional detritus from the Andes and aeolian dust from the Sahara, but it is now generally considered they correspond to yellowish latosols derived by degradation of former lateritic bauxite (Negrão et al. 2018). This origin is similar to that proposed for silty clays developed over lateritic regoliths in rainforest terrains in Gabon (Lecomte (1988). Bauxite has also developed on alkaline rocks of the Poços de Caldas complex in southern Brazil. This resembles a karst bauxite due to the solubility of the bedrock, resulting in pinnacles of unweathered rock, with adjacent bauxite-filled solution cavities. There is little or no saprolite developed between bedrock and bauxite ore.

Nickel-Co laterite deposits developed over ultramafic units of Archaean and Proterozoic layered intrusive rocks are widespread across savanna to humid savanna regions in Central Brazil (Barro Alto and Niquelandia, Goias), and in more humid regions on the Atlantic Coast and Amazonia. Important deposits also occur in similar climates on Colombia (Cerro Matoso) and Guatamala (Eximbal), formed on Cretaceous ophiolites on the northern extension of the Andean Cordillera into Central America. Huge reserves of Nb-REE-P occur in the thick lateritic regolith developed on carbonatite complexes (e.g., Araxá, Catalão) in the Alto Paranaíba Alkaline Province (APIP) in Central Brazil and in Brazilian Amazonas (Seis Lagos)—see chapters "Carbonatite Hosted Resid ual REE Deposits" and "Regolith Hosted Niobium Deposits in Carbonatite Complexes in Brazil". The carbonatites of the APIP were intruded in Precambrian rocks at about 85 Ma, exhumed by rapid erosion to be exposed between

66 Ma (Araxa) and 44 Ma (Catalão 1) under arid to semi-arid conditions (Conceição et al. 2024), and later, under more stable tectonic conditions, subjected to episodic deep weathering in warm and humid climates similar to the present.

Somewhat unusually, several small, high-grade supergene non-sulfide Zn deposits occur in a similar humid tropical climate in the Amazonas region of northern Peru. These are hosted by carbonate rocks with well-developed karst features. The deposits have formed during successive episodes of uplift, erosion and weathering, with an evolving regolith in which suggests that smithsonite has pseudomorphically replaced primary sulfides at the base, but in places has altered is hydrozincite in surface horizons. (chapter "Supergene Non-Sulfide Base Metal Deposits"). The ion-adsorption REE deposit at Serra Verde Goiás, Bazil, occurs in the upper saprolite developed over biotite granite (chapter "Ion Adsorption Mineralisation in Regolith-Hosted REE Deposits"). It is likely to be the first of several discoveries in the region.

Significant supergene Cu-Au-Mo mineralization is present over porphyry intrusions along the length of the Andes into Central America. In many instances, the economic viability of a mine depends on the size and quality of the enrichment. Weathering has been dated as occurring intermittently since 44 Ma at different locations on the Cordillera, probably under a wide variety of climatic and tectonic conditions, although semi-arid conditions have been considered the most important for enrichment (Hartley and Rice 2005).

North of continental South America, major deposits of Ni laterite and bauxite have developed on the Caribbean island arc, mainly in Cuba, the Dominican Republic and Jamaica. The Ni laterites in the Dominican Republic are mainly hydrous silicate deposits developed on ophiolitic peridotites. These were emplaced from the Cretaceous to the early Miocene, and exposed to at least four erosion and weathering cycles, triggered by punctuated uplift commencing in the Miocene. Uplift in Cuba, however, was by gradual epeirogenic movement, resulting in oxide and clay Ni laterites. Jamaica hosts some of the world's largest karst bauxite deposits, formed by weathering of Miocene volcanic ash and associated sediments (25–8.5 Ma), deposited on the free-draining mid-Eocene-Lower Miocene White Limestone.

Supergene deposits unrelated to deep weathering include low grade calcrete U deposits in the Andes and Patagonia of Argentina, and minerals such as nitrates, borates, halides and carbonates of, K, Li, Mg, Na in salinas. The latter occur dominantly in the hyper-arid Andean plateau of the Altiplano, Atacama and Puna, at altitudes of 2500–4500 m. Most recently, the so-called 'lithium triangle', extending from the Salar de Uyuni in Bolivia to the Salar de Atacama in Chile and Salar del Hombre Muerto, NW Argentina, has come into prominence. These evaporite deposits have accumulated in closed basins and their formation is closely related to the uplift of the Andes and the associated orographic change to extreme aridity. Recharge of the salinas with Li-rich waters derived from volcanic fluids is ongoing (Godfrey and Alvarez-Amado 2020).

4.2.4 North America

Distribution of Lateritic Regoliths Isolated lateritic plateau remnants are present in southern Oregon and northern California, developed particularly over ophiolitic ultramafic rocks in the Coast Range of the American Cordillera, where they have been exploited as Ni laterites. More extensive residual regoliths remain in unglaciated regions east of the Cordillera, where Cenozoic erosion has been low. They are mainly represented by saprolite developed on crystalline and sedimentary rocks in the present are mostly warm temperate regions of south eastern USA in a belt extending from eastern Pennsylvania in the north through Virginia, South Carolina and west to Arkansas. Ferruginized (i.e., lateritic) surface horizons are uncommon, but equivalent alumina-rich horizons occur as both residual and associated sedimentary bauxites, dated stratigraphically as Palaeocene-early Eocene. Intensely weathered saprolite, probably formed during the late Eocene to Miocene, occurs in several locations in north western USA (Wyoming, Idaho), and deeply weathered carbonatites are known in glaciated terrain in Ontario, Canada, at latitude ca. 50^{0}N. Ferruginous pisoliths have been recorded beneath glacial overburden in Wisconsin, USA (Smith 1987). These occurrences indicates that an intensely weathered regolith, presumably formed under humid tropical conditions, was previously widespread across North America prior to erosion in the late Cenozoic.

Supergene Deposits Bauxite has been mined, mainly in Arkansas, since 1888 (Bramlette 1936). It occurs as residual deposits derived from mid-Cretaceous nepheline syenite and from shallow marine and terrestrial Palaeocene sediments of the Mississippi River Embayment of the Atlantic Coastal Plain. The sediments contain erosional detritus from the syenites and other alkaline rocks. Some mostly smaller deposits are semi-residual and transported. All are overlain by Palaeocene to lower Eocene lignitic sediments. Sedimentary kaolinite deposits in Georgia are of similar age and form the largest kaolin resource in the USA They occur in a NE-SW trending belt on the Atlantic Coastal Plain and were probably derived from the deeply weathered metamorphic rocks and granites of the southern Appalachians and deposited in near-shore fluvial, deltaic to marine environments (chapter "Supergene Origin of Kaolin Deposits").

Residual kaolinite was developed on the Cretaceous Thatuna granodiorite NW Idaho by intense weathering in a humid, warm to hot climate during the Miocene. Contemporaneous eruption of the Columbia River basalts led to damming of drainages to form lakes. Erosion of the granite saprolite resulted in the deposition of clay-rich lacustrine sediments of the Latah Formation that constitute the 'clay belt' of northern Idaho. This was subject to small-scale mining for fire clay from about 1910, but more recent interest is in the saprolite as a source of high purity kaolinite (Kerr and Horn 2020).

Small deposits of Ni laterite occur as isolated lateritic plateau remnants in southern Oregon and northern California, developed particularly over Cretaceous ophiolitic ultramafic rocks in the Coast Range (Cumberlidge and Chace, 1968). The date of weathering is assumed to be Eocene-Miocene.

REE enrichments on the Eocene Bear Lodge carbonatite in NE Wyoming are probably due to the Miocene weathering causing kaolinization of granite in NW Idaho (Andersen et al. 2017). Similarly weathered carbonatites rich in P, Nb and REE have been located in glaciated terrain at Cargill Lake, Martinson Lake, and Martison Lake, Ontario, Canada, at latitude ca. 50^{0}N. The saprolite is overlain by pre-glacial, well-sorted, silica-kaolinite sand, organic clay and peat deposits of late Cretaceous to Early Tertiary age. (Porter 2012a, b).

The southwestern states, especially Arizona, host significant porphyry Cu deposits, including the Morenci, Bingham Canyon and Resolution Mines. These exploit supergene enrichment zones of malachite, azurite chalcocite, covellite and native copper developed over mostly low grade sulfide protores. When first described, the original grades were over 20% Cu, whereas in modern mines, grades are 1% Cu or less. A description of the original mineralization of the Copper Queen Mine, Bisbee, Arizona stated that malachite "*. . . was abundant, occurring in large and beautiful masses with azurite and calcite in limestone caverns. The walls of these caves were covered with velvety moss-green malachite, and sparkled with the blue crystals of azurite, while from the roofs hung translucent stalactitic draperies of calcite, delicately banded and tinted with the salts of copper*" (Ransome 1904).

The Sierra Mojada Mine in northern Mexico has been an important source of Ag for nearly 150 years, derived from supergene Ag halides, native Ag and acanthite, and associated Zn and Pb minerals.

4.2.5 Australia

Distribution of Lateritic Regoliths Parts of the north and west of the Australian continent have been emergent since the Neoproterozoic and subjected to weathering and erosion since that time. The adjacent, now inland, basins were infilled with sediments during the Phanerozoic and have since remained largely undeformed. The only significant tectonic activity has been accretion and uplift along the eastern seaboard commencing at the end of the Proterozoic, and intraplate uplift in South Australia, largely ended by 250 Ma. In consequence, the continent is very flat, with a mean elevation of about 325 m and local relief nowhere more than 1500 m.

Present day southern Australia was situated at and below latitude 60^0S at the time of the initial break-up of Gondwana, with separation not completed until about 34 Ma. No part of the continent was above about 30^0 S until the early Eocene (50 Ma; Fig. 9). The Australian plate has since drifted 20^0 NE and continues to do so at about 7 m per century. Despite this mid-latitude position, lateritic regoliths are widespread across the continent, but except for the humid tropical belt in the far north, and Mediterranean and temperate climates in the southwest and southeast, they occur predominantly in warm semi-arid regions across the centre of the continent. Here, the landscape is generally of low relief and the residual regolith is commonly overlain by locally derived sediments from 1 to >30 m thick, which are commonly also weathered. Whether exposed or buried, the residual regolith may be partially or wholly eroded in places but, overall, residual and transported regolith covers in excess of 70% of the land area. Due in part to the overall flatness of the topography, plateau landforms are less prominent than in other similarly weathered continents and attempts to cross-correlate plateau levels and develop denudation chronologies have seen little success. The present land surface of the West Australian Craton, for example, is similar to that exhumed after the early Permian glaciation. Thick residual regolith is less abundant in the temperate southeastern margin of the continent due to uplift and subsequent erosion. However, plateau remnants are preserved in uplands, and saprolite of partially eroded profiles is locally abundant.

Palaeoclimatic data suggest that warm to temperate climates prevailed for several periods during the Mesozoic and early Cenozoic even in the most southerly parts of Australia (Martin 2006 and references therein). Recent oxygen isotope data has shown that warm conditions were present at near-polar latitudes from the Late Jurassic to the Palaeocene, based on estimates of ocean and sea surface temperatures, with the latter as high as 25–45 °C in the tropics and 10–23 °C in polar latitudes (van Dyck et al. 2020; Vickers et al. 2019). Biostratigraphic analyses of cores recovered from West Antarctica show the presence of terrestrial rainforest-type vegetation at latitude 82^0S during the mid-Cretaceous (Klages et al. 2020). The data are interpreted to imply a strongly seasonal, temperate climate, with high atmospheric CO_2 concentrations, that supported dense vegetation able to survive up to four months of complete polar night darkness. Palynological and palaeontological evidence suggests warm to tropical humid conditions supported rainforest in South Australia in the early and mid Eocene at latitude 60–65^0S, with similar conditions returning in the late Oligocene and mid Miocene (McGowran and Hill 2015; Mao and Retallack 2019). The rest of the continent had broadly similar climates, becoming warmer to the north. As the northward drift continued during the Cenozoic, intervening periods were cooler and drier, with semi-arid conditions generally prevailing after the mid-Miocene. This paleoclimatic history has a strong influence on regolith development. The long-established view of the paleoclimatic history of the West Australian Craton since the Cretaceous is summarized by Anand and Paine (2002) and Anand and Butt (2010). This perspective considers that this period comprises two distinct phases (1) Cretaceous to mid-Miocene generally humid, sub-tropical to tropical conditions, with lateritic weathering peaking in the late Cretaceous and early-mid Miocene and (2) mid-Miocene to Present: drier, semi-arid to arid climates.

This long history of repeated warm to tropical, humid climates is reflected by geochronological evidence for intense weathering as early as the

Carboniferous (Pillans 2005), pre-dating Permian glaciation on the Yilgarn Craton, and currently active in the humid savanna regions of the north and northeast. One way to constrain the regolith history in more detail is direct dating of regolith minerals such as Mn oxides, hematite, goethite, alunite and kaolinite. However, the ages obtained for these minerals are more likely to represent the termination of a prior weathering event rather than the peak of such an event. Based on paleomagnetic dating from across the Yilgarn Craton, Pillans (2005) found that two main regolith hematite formation events are preserved, a pervasive one between 20 and 10 My and earlier event restricted to the eastern Yilgarn at about 60 Ma. Vasconcelos et al. (2013) investigated the geochronology of both goethite and Mn oxides from Channel Iron (CID) and Detrital iron Deposits (DID) in the Pilbara region of Western Australia. They found that detrital fragments (source material) from both CID and DID had ages in the range 65–44 My, and authigenic material at 35–16 My. Importantly, detrital and authigenic Mn oxides and goethite have similar ranges, despite two different, independent techniques being used, $^{40}Ar/^{39}Ar$ dating for Mn oxides and U (U-Th)/He dating for goethite (Vasconcelos et al. 2013). In contrast, the lateritic bauxites further south in the Darling Range are dated from 10 to 1.3 My (Pidgeon et al. 2004; Wells et al. 2019). However, (U–Th)/He dating of ferruginous nodules and pisoliths in the arid, eastern Yilgarn show that they are of Miocene age (18.6–11.5 Ma; Anand et al. 2021). The younger ages in the Darling Range and SW coastal areas, and older ages in the interior Yilgarn, indicate that Mn oxide formation and Al enrichment may still be ongoing in the Darling Range, probably because rainfall remains relatively high (600- > 1000 mm p/a) under the present Mediterranean climate. The results from the eastern Yilgarn are comparable to authigenic goethite cortices in the Miocene (18–5 My) goethite-rich CID of the Hamersley region further north, with younger ages tending to be deeper in the profile (Heim et al. 2006) and reported elsewhere in Africa and Brazil (Monteiro et al. 2014; Dequincey et al. 2002). The goethite ages in ferruginous nodules and pisoliths are thought to be younger than the underlying saprolite (Paleocene-Eocene) and were formed in different environmental conditions than the underlying saprolite (Anand et al. 2021). In Queensland, weathering ages of 20, 16–12, 6–4 and 2,2–0.8 Ma have been recorded (Vasconcelos and Conroy 2003; Li and Vasconcelos 2002).

Supergene Deposits The extensive regolith cover over much of Australia hosts a large endowment of supergene mineral deposits. These include major resources of bauxite in the SW and NE, with smaller bodies in the SE, Fe ore in the Pilbara region in the NW, supergene and alluvial gold in most goldfields, as well as Ni-Co laterites, carbonatite-derived REE, kaolinite, and supergene Cu on porphyries and some massive sulfides.

Major bauxite provinces are located along the northern coast of the Arnhem peninsula in the Northern Territory (NT) and western Cape York Peninsula, Queensland (Qld), and in the Darling Range, of SW Western Australia (WA). Each province has numerous deposits, the largest of which support significant mining operations. The Gove, NT and Weipa, Amrun, Qld are amongst the world's highest grade deposits, at 49–53% Al_2O_3, and are developed on feldspathic sediments of Cretaceous to early Cenozoic age. In contrast, most Darling Range bauxite is developed on Archaean granite (Huntly-Willowdale), whereas Worsley-Mt Saddleback is formed on mafic volcanic rocks. Those in the Darling Range such low reactive silica (generally 1–2%) that they remain profitable despite having the lowest grade bauxite (27–30% Al_2O_3) of any commercial operation globally. Much smaller bauxite resources occur in New South Wales and Tasmania. (Britt 2018).

Australia is a major producer of Fe ore, dominantly supergene martite-goethite Fe ores derived from BIF protore from the Pilbara, WA and the Middleback Range, S Australia. Smaller deposits are developed over BIFs in the Archaean Yilgarn Craton, WA. In addition, very large Channel Iron Deposits (CID), also in the Pilbara, comprised of

Fe-rich fluviatile and lacustrine Miocene sediments support significant mining operations (chapter "Iron Ore"). Australia has large resources of lateritic Ni and Co. Oxide (Cawse, Ravensthorpe) and clay silicate (Murrin Murrin) ores formed over komatiitic ultramafic rocks in the Yilgarn Craton; the large undeveloped Wingellina oxide deposit in the Proterozoic Musgrave Province of central Australia formed on a mafic-ultramafic intrusion. These deposits are in present-day semi-arid environments in terrains of low relief. Smaller oxide and hydrous silicate deposits have formed in higher relief terrain over Precambrian to Devon ophiolitic ultramafic rock in NE Qld and other localities in Qld and NSW, but mine production has been small. These and some oxide-rich deposits over Alaskan-type, mafic to ultramafic complexes also have potential for PGE and Sc (chapter "Nickel-Cobalt Laterite Deposits").

Supergene and surficial deposits have been an important component of the success of the Australian gold industry from its beginning in the nineteenth century. Large alluvial/placer deposits in palaeochannels ('deep leads') in Victoria and, to a lesser extent Tasmania and WA, historically produced substantial quantities of gold. Most of the ore consisted of detrital hypogene gold ranging from large nuggets to fine particles of a few tens of micrometres or less, although there is clear evidence for chemical weathering, such as pitting and Ag-depletion rims on grain surfaces and crystal boundaries (Butt 2022; Butt et al. 2020). Similarly, Au enrichments in the regolith overlying hypogene deposits commonly comprise partially weathered residual primary grains and some secondary grains; secondary Au grains only appear to be dominant in saprolitic deposits in areas having saline groundwater (chapter "Supergene Gold Deposits in Intensely Weathered Regolith"). Although supergene deposits are mainly small, they can provide an early cash-flow during the development of bedrock resources.

Gossans developed over massive sulfide bodies such as Broken Hill, NSW, and Mt. Morgans, Queensland, formed the principal ore in the early stages of mostly small-scale mining of such deposits (chapter "Gossans: In Situ Supergene Base Metal Deposits"). Subsequently, however, the primary sulfide mineralization provided the basis for the development of major long-term mining operations.

Potentially economic supergene deposits unrelated to intense lateritic weathering are present in semi-arid to arid regions across Australia, essentially as evaporite deposits in paleochannels and playas. The principal commodities being sought include U, K, Li and B. Apart from some reduced U deposits in S. Australia, none has yet been exploited. Numerous surficial U deposits have formed during the past 1 My in granitic terrain in the northern Yilgarn Craton, WA and Arunta Block, NT. Uranium, V and K have precipitated as carnotite under oxidizing conditions in valley calcrete and associated alluvial and lacustrine sediments in palaeodrainages and playas (chapter "Channel and Playa Uranium Deposits"). In S. Australia, U has precipitated as coffinite and uraninite under reducing conditions in palaeodrainages at depths of 150 m or more. Some of these have been mined by in situ leaching. Coffinite and other reduced species are the main minerals in the Mulga Rock deposits in Western Australia, hosted in part by lignitic and organic-rich silt and sand sediments (Douglas et al. 2011); Hou et al. 2017). Some playas in palaeochannels in WA also contain significant resources of sulfate of potash (K_2SO_4) in hypersaline brine. The most promising are in regions of low rainfall ($<\sim$350 mm pa) and high evaporation rates (3000- $>$ 4000 mm pa) that have led to concentration of salts in groundwater and allow the efficient functioning of evaporation ponds. None is yet in production.

4.2.6 Northern Europe and Asia

Distribution of Lateritic Regoliths Although lateritic regoliths and associated supergene deposits occur dominantly in the intertropical zone, there are many occurrences in higher latitudes in the northern hemisphere, suggesting a much wider distribution of such regoliths in the past, but now largely eroded (Fig. 2). Intensely weathered saprolite and lateritic regolith are known in now mostly temperate regions from Central Asia to Western Europe, north to beyond the Arctic Circle (Fig. 1). Many of these occurrences have been studied because of their economic potential, especially with bauxite and/or Ni-Co laterites occurring in France, Hungary, Turkey, Greece, the Balkans, Russia and Kazakhstan. Many of these are blind, developed on early Cenozoic and older basement and concealed by younger sedimentary rocks, which constrain their dates of formation. In Greece, lateritic weathering occurred during the Upper Jurassic to Middle Cretaceous; intermittent erosion resulted in bauxites and Ni-Co laterites occurring as residual deposits on basement rocks, as detrital sedimentary deposits and as residual deposits on emergent sedimentary rocks (Kalatha and Economou-Eliopoulos 2015; Eliopoulos and Economou-Eliopoulos 2000; Valeton et al. 1987). Similar deposits were formed during later phases of weathering in the early to mid-Eocene and the mid-Miocene (Thorne et al. 2012).

Further north, lateritic Al-Fe-rich saprolite is preserved as a weathered inter-flow unit 10–15 m thick in the Antrim plateau basalts in Northern Ireland, and kaolinitic saprolite occurs as scattered remnants, in places 30–100 m thick, in Scotland and across Scandinavia and northern Russia, preserved in locations where Pleistocene glacial ice was stagnant and not erosive. Examples include districts in Lapland in Finland (Hirvas 1991; Sarala 2011) and the adjacent Kola Peninsula in Russia.

Oxygen isotope data from pedogenic siderite in Scotland, northern Siberia and Alaska, and from kaolinite in saprolite at Buchan, NE Scotland, indicate mean sea surface and groundwater temperatures of 10–23 °C and 23 ± 5 °C, respectively during the Palaeocene-early Eocene (van Dyck et al. 2020; Hall et al. 2015). These dates correspond closely to the formation of the Al-Fe-rich saprolite in the Antrim basalts at 60 Ma (Cooper 2004) when Antrim and NE Scotland were at 45–50^0 N. The temperature estimates are similar to those proposed by Thorne et al. (2012) which provide optimal conditions for the formation of Ni-Co laterites on ultramafic rocks and hence similar lateritic regoliths on other lithologies. A later phase of kaolinization in Scotland occurred under cooler climates in the mid-Miocene. A controversial conclusion based on K-Ar dating of authigenic illite in saprolite suggests that large regions of southwestern Scandinavia were subject to intense deep weathering in the Late Triassic, ca. 220–200 Ma (Fredin et al. 2017). However, Palaeocene to Miocene dates, similar to those in Northern Ireland, Scotland and SE Europe are more probable for remnant saprolite in northern Norway, Sweden, Finland and Russia, which were at or below 60^{0}N at this time.

Supergene Deposits Karst and lateritic bauxites are widespread from France, including the type locality at Les Baux in Provence, and in Hungary, Albania, Turkey and Greece. Nickel laterites also formed in part of the same area of SE Europe, mainly in Serbia, Albania, Greece and Turkey, developed over ophiolitic ultramafic rocks of Jurassic-Cretaceous age. These bauxite and No-Co laterite deposits are mostly small by world standards but can be mined economically, mainly for domestic use. Larger Ni laterite deposits occur in the Urals (Russia) and Kazakhstan (Freyssinet et al. 2005). The 'linear' deposits in the Urals contain hydrous Ni-Mg silicate minerals in deep structural settings in areas of strong relief, although whether truly 'lateritic' is perhaps debatable. The 'areal' deposits in Kazakhstan have clay silicate ores and developed in areas of low relief. Numerous, mainly karst, bauxite occurrences have been recorded in NE Russia (Fig. 2), many concealed by later sedimentary rocks, although most are uneconomic to mine. Even the weathered inter-basaltic beds in Antrim were considered as a potential Al resource during the second World War. Residual and transported kaolin deposits formed by these weathering events are present in several countries in Europe, including Germany, Ukraine, Turkey and Spain (chapter "Supergene Origin of Kaolin Deposits"). Non-sulfide base metal deposits are widespread through the Mediterranean region of north Africa and southern Europe from Spain and Morocco to Iran. These have a seasonally climate that favours their formation (Hitzman et al. 2003).

Residual P-REE-Nb deposits have formed on the Sokli and Lovazero carbonatites in Finland and the Kola Peninsular, Russia, respectively, both ca 67^0N. Carbonatite-derived residual Nb-REE ores indicate similarly preserved regoliths further east in Russia at Tomtor (71^0N) and Chuktukon (59^0N). The huge and complex supergene deposit at Tomtor is considered to have been formed in several stages (Lapin et al. 2016; Tolstov et al. 2023):

1. Hydrothermal alteration, followed during the Devonian-Carboniferous (390–340 My) by lateritic weathering, This resulted in residual accumulation of REE and Nb in a combined hypogene-supergene alteration profile 300–400 m thick.
2. Erosion and redeposition of the REE-Nb-phosphate-enriched horizon of the residual regolith as a distinct lacustrine sedimentary unit.
3. Burial by carbonaceous Permian sediments. Subsequent exposure, erosion and re-burial of the sediments led to further epigenetic and diagenetic alterations and enrichment of the ore minerals under variety of chemical conditions. These caused Fe and Mn to be leached and the formation of a bleached horizon (mean 12–15 m thick) rich in REE minerals, including introduced HREE. The sequence was finally buried by Jurassic marine sediments.

5 Conclusion

With the exception of evaporites in present-day semi-arid to hyper-arid regions, supergene mineral deposits are the products of the interplay of intense, lateritic weathering of mostly specific bedrock lithologies, modified by subsequent climatic and/or tectonic changes. Some deposits are essentially 'fossil' whereas others are still evolving under their ambient environment. While it has been argued that some supergene deposits are Proterozoic in age (e.g., certain Fe ores, Morris 1983), the majority formed in the Phanerozoic, the oldest possibly being Carboniferous (Murrin Murrin Ni-Co laterite, Australia; Tomtor Nb-REE, Russia). The most 'productive' period globally was probably the late Cretaceous to early Eocene, with evidence for intense chemical weathering having occurred in latitudes then between about 60^0S to 60^0N. This is shown by the presence of lateritic regoliths, including supergene deposits such as bauxite, kaolinite and Nb-P-REE-enriched saprolite on carbonatite, together with palaeontological and isotopic evidence from sedimentary rocks, from equatorial to sub-polar latitudes. Accordingly, supergene deposits are present in all continents, possibly except Antarctica, although their distribution is obviously biased towards regions that have had appropriate climatic regimes (tropical rainforest, savanna, and, in places, Mediterranean) and stable tectonic settings for the longest periods. Erosion resulting from changes to arid and glacial climates and from tectonism, during the Cenozoic in particular, has led to losses, especially in higher latitudes. Lithological control is paramount for the formation of supergene deposits of some specific commodities, such as carbonatites for Nb and some REE and P deposits, ultramafic rocks for Ni-Co laterites and BIF for the majority of Fe ores. Supergene Cu and Au deposits may develop over a variety of lithologies, but only where there is specific primary bedrock enrichment, either of the native metal (Au) or as sulfides (e.g., porphyry Cu, Mo) or other minerals (e.g., non-sulfide Zn, Pb). Bauxites have formed from a range of felsic and mafic lithologies, whereas kaolinites are most commonly derived from felsic rocks. There are numerous locations where economic bauxite and kaolinite deposits formed over granitic and other felsic rocks occur together, in the same profile, or in close proximity, such as in a toposequence or proximal sedimentary basins. In southeastern China, bauxite and kaolinites are also present in the same districts as ion-adsorption REE deposits. This implies each type of deposit is formed under very similar weathering conditions of free drainage and low pH groundwaters in humid climates that induce strong leaching of all but the most resistant minerals. In the case of REE deposits, the mineralogy of the minor and accessory minerals is a critical factor, with most deposits forming over calc-alkaline granitic rocks that have undergone hydrothermal alteration that has replaced resistant REE host minerals with more soluble phases. The key factors controlling the specific locations of these deposits within their regolith-landform context have yet to be elucidated.

Acknowledgements The authors thank their colleagues and numerous other geoscientists for sharing their work and insights into the nature and formation of the regolith in Australia and many other countries, including valuable visits to field locations, from which the concepts described in this chapter have been derived. Comments by Charles Beard and Paul Morris assisted in improving the manuscript; Angelo Vartesi drafted the figures to his usual high standard.

References

Anand RR, Butt CRM (1998) Regolith-landform features of the Kabwe and Mumbwa areas. Report prepared for Billiton Development (Zambia) Limited (unpublished)

Anand RR, Butt CRM (2010) A guide for mineral exploration through the regolith in the Yilgarn Craton, Western Australia. Aust J Earth Sci 57:1015–1114

Anand RR, Paine M (2002) Regolith geology of the Yilgarn Craton, Western Australia: implications for exploration. Aust J Earth Sci 49:3–162

Anand RR, Hough R, Salama W, Aspandiar MF, Butt CRM, González-Álvarez I, Metelka V (2019) Gold and pathfinder elements in ferricrete gold deposits of the Yilgarn craton of Western Australia: a review with new concepts. Ore Geol Rev 104:294–355

Anand RR, Wells M, Lintern M, Schoneveld L, Danišík M, Salama W, Noble RRP, Metelka V, Reid N (2021) The (U-Th)/He chronology and geochemistry of ferruginous nodules and pisoliths formed in paleochannel environments at the Garden Well gold deposit, Yilgarn craton of Western Australia: implications for landscape evolution and geochemical exploration. Minerals 2021(11):679. https://doi.org/10.3390/min11070679

Andersen AK, Clark JG, Larson PB, Donovan JJ (2017) REE fractionation, mineral speciation, and supergene enrichment of the Bear Lodge carbonatites, Wyoming, USA. Ore Geol Rev 89:780–807. https://doi.org/10.1016/j.oregeorev.2017.06.025

Ansart C, Calmels D, Allard T, Roig JY, Coueffe R, Heller B, Pinna-Jamme R, Nouet J, Reguer S, Vantelon, D, Gautheron C (2022) (U-Th)/He geochronology constraints on lateritic duricrust formation on the Guiana shield. Front Earth Sc 10. https://doi.org/10.3389/feart.2022.888993

Augustithis SS (1965) Mineralogical and geochemical studies of the platiniferous dunite-birbirite-pyroxenite complex od Yubdo/Birbir, Ethiopia. Chemi der Erde Bd24:159–196

Bamba O, Parisot J-C, Grandin G, Beauvais A (2002) Ferricrete genesis and supergene gold behaviour in Burkina Faso, West Africa. Geochem Explorat Envir Analy 2:3–13

Bárdossy G, Aleva GJJ (1990) Lateritic bauxites. Developments in economic geology, 27. Elsevier, Amsterdam. 624 pp

Beauvais A, Chardon D (2013) Modes, tempo, and spatial variability of Cenozoic cratonic denudation: the West African example. Geochem Geophys Geosyst 14:1590–1608. https://doi.org/10.1002/ggge.20093

Beauvais A, Ruffet G, Hénocque O, Colin F (2008) Chemical and physical erosion rhythms of the West African Cenozoic morphogenesis: the ^{39}Ar-^{40}Ar dating of supergene K-Mn oxides. J Geophys Res 113:F04007. https://doi.org/10.1029/2008JF000996

Bhukte PG, Daware GT, Masurkar SP, Mahendiran P, Janbandhu K, Rao KR, Singh U, Puttewar SP, Agnihotri A (2020) Geochemical, mineralogical and petrological characteristics of lateritic bauxite deposits formed on Deccan Trap basalt with reference to high-level and coastal (low level) deposits of Maharashtra. J Geol Soc India 95:587–598. https://doi.org/10.1007/s12594-020-1485-1

Boeglin J-P, Probst J-L (1998) Physical and chemical weathering rates and CO_2 consumption in a tropical lateritic environment: the upper Niger basin. Chem Geol 148:136–156

Bonnet NJ, Beauvais A, Arnaud N, Chardon D, Jayananda M (2016) Cenozoic lateritic weathering and erosion history of Peninsular India from ^{40}Ar/^{39}Ar dating of supergene K–Mn oxides. Chem Geol 446:33–53. https://doi.org/10.1016/j.chemgeo.2016.04.018

Bramlette MN (1936) Geology of the Arkansas bauxite region. Information Circular 8, Arkansas Geological Survey, Little Rock, 66 pp

Brand NW, Butt CRM (2001) Weathering, element distribution and geochemical dispersion at Mt. Keith, Western Australia: implication for nickel sulfide exploration. Geochem Explorat Envir Analy 1:391–407

Britt AF (2018) Australian Resource Reviews: Bauxite 2017. Geoscience Australia, Canberra. https://doi.org/10.11636/9781925297720

Buchanan F (1807) In: A journey from Madras through the countries of Mysore, Kanara and Malabar. East India Co., London. Volume 2: pp. 436–461, 559; volume 3: 66, 89, 251, 258, 378

Budel J (1982) Climatic geomorphology. (trans: L. Fischer and D. Busch). Princeton University Press, Princeton, 443 pp

Burke K, Gunnell Y (2008) The African erosion surface: continental-scale synthesis of geomorphology, tectonics, and environmental change over the past 180 million years. Memoir 201. Geological Society of America, Boulder, USA. 66 pp

Butt CRM, Lintern MJ, Anand RR (2000) Evolution of regoliths and landscapes in deeply weathered terrain - implications for 40 39 40 39 4 3 geochemical exploration. Ore Geol Rev 16:167–183. https://doi.org/10.1016/S0169-1368(99)00029-3

Butt CRM, Gray DJ, Lintern MJ, Robertson IDM, Taylor GF, Scott KM (1991) Gold and associated elements in the regolithdispersion processes and implications for exploration. Final report. AMIRA project 241A. Re-issued as CRC LEME Open file report 57, 1998, 64 pp https://crcleme.org.au/Pubs/OFRSindex.html

Butt CRM, Zeegers H (eds) (1992) Regolith exploration geochemistry in tropical and subtropical terrains. Handbook of exploration geochemistry, 4. Elsevier, Amsterdam. 607 pp

Butt CRM (2014) Discussion of 'Silicified serpentinite—a residuum of a Tertiary palaeo-weathering surface in The United Arab Emirates'. Geol Mag 151:1144–1146. https://doi.org/10.1017/S0016756813001143

Butt CRM (2022) Alluvial gold at Avoca, Victoria — evidence for supergene alteration of hypogene gold grains. Aust J Mineral 23:5–15

Butt CRM, Bristow APJ (2013) Relief inversion in the geomorphological evolution of sub-Saharan West Africa. Geomorphology 185:16–26. https://doi.org/10.1016/j.geomorph.2012.11.024

Butt CRM, Hough RM, Verrall M (2020) Gold nuggets: the inside story. Ore Energy Res Geol 4–5. https://doi.org/10.1016/j.oreoa.2020.100009

Chardon D, Grimaud J-l, Beauvais A, Bamba O (2018) West African lateritic pediments: landform-regolith evolution processes and mineral exploration pitfalls. Earth Sci Rev 179:124–146. https://doi.org/10.1016/j.earscirev.2018.02.009

Colin F, Beauvais A, Ruffet G, Hénocque O (2005) First $^{40}Ar/^{39}Ar$ geochronology of lateritic manganiferous pisolites: implications for the Palaeogene history of a West African landscape. Earth Planet Sci Lett 238(2005):172–188. https://doi.org/10.1016/j.epsl.2005.06.052

Conceição FT, Vasconcelos PM, Navarro GRB, Farley KA (2024) $^{40}Ar/^{39}Ar$ and (U-Th)/He constraints on emplacement, exhumation, and weathering of alkaline-carbonatite complexes in the Alto Paranaíba Igneous Province (APIP), Brazil. Gondwana Res 130: 116–130. https://doi.org/10.1016/j.gr.2024.01.010

Cooper MR (2004) Antrim lava group, Palaeogene extrusive igneous rocks, Northern Ireland. In: Mitchell, WI (ed). The geology of Northern Ireland-our natural foundation. Geological Survey of Northern Ireland, Belfast. https://earthwise.bgs.ac.uk/index.php/Geology_of_Northern_Ireland:_our_natural_foundation Webpage last edited 29 September 2017.

Cumberlidge JT, Chace FM (1968) Geology of the Nickel Mountain, Riddle, Oregon. In JD Ridge (ed) Ore deposits of the United States, 1933–1967 (Graton-Sales volume). American Institute of Mining, Metallurgical and Petroleum Engineers, 2, 1650–1671

De Putter T, Ruffet G (2020) Supergene manganese ore records 75 Myr-long Campanian to Pleistocene geodynamic evolution and weathering history of the Central African Great Lakes region—tectonics drives, climate assists. Gondwana Res 83:96–117. https://doi.org/10.1016/j.gr.2020.01.021

De Putter T, Ruffet G, Yans J, Mees F (2015) The age of supergene manganese deposits in Katanga and its implications for the Neogene evolution of the African Great Lakes region. Ore Geol Rev 71:350–362. https://doi.org/10.1016/j.oregeorev.2015.06.015

Decrée S, Deloule E, Ruffet J, Dewaele S, Mees F. Marignac, C, Yans J, De Putter T (2010) Geodynamic and climate controls in the formation of Mio-Pliocene worldclass cobalt and manganese oxidized ores in the Katanga (DRCongo). Mineral Dep 45:621–629. https://doi.org/10.1007/s00126-010-0305-8

Dequincey O, Chabaux F, Clauer N, Sigmarsson O, Liewig N, Leprun JC (2002) Chemical mobilizations in laterites; evidence from trace elements and 238U-234U-230Th disequilibria. Geochimica Cosmochimca Acta 66:1197–1210

Douglas GB, Butt CRM, Gray DJ (2011) Geology, geochemistry and mineralogy of the lignite-hosted Ambassador palaeochannel uranium and multi-element deposit, Gunbarrel Basin, Western Australia. Mineral Deposita 46:761–787. https://doi.org/10.1007/s00126-011-0349-4

Eggleton RA (ed) (2001) The regolith glossary. Surficial geology, soils and landscapes. CRC LEME, Perth (144 pp. See also: https://publications.csiro.au/publications/#publication/PIlegacy:3643. http://crcleme.org.au/Pubs/Monographs/RegolithGlossary.pdf

Eliopoulos DG, Economou-Eliopoulos M (2000) Geochemical and mineralogical characteristics of Fe–Ni- and bauxitic-laterite deposits of Greece. Ore Geol Rev 16:41–58

Fredin OA et al (2017) Reply to 'The inheritance of a Mesozoic landscape in western Scandinavia'. Nat Comm 8:1503. https://doi.org/10.1038/s41467-017-01468-6

Freyssinet P, Butt CRM, Morris, RC, Piantone P (2005) Ore-forming processes related to lateritic weathering. In: Hedenquist JW, Thomson JFH, Goldfarb RJ, Richards JP (eds) Economic geology 100th

anniversary volume. Economic Geology Publishing Company, New Haven, Connecticut, p 681–722
Godfrey L, Alvarez-Amado F (2020) Volcanic and saline lithium inputs to the Salar de Atacama. Minerals 10: 201. https://doi.org/10.3390/min10020201
Goldich SS (1938) A study in rock-weathering. J Geol 46: 17–58. https://doi.org/10.1086/624619
Golightly JP (2010) Progress in understanding the evolution of nickel laterites. In RJ Goldfarb, EE Marsh and T Monecke (eds) The challenge of finding new mineral resources: global metallogeny, innovative exploration, and new discoveries volume II: Zinc-Lead, Nickel-Copper-PGE, and Uranium Special Publication 15, Society of Economic Geologists, Inc. pp. 451–475
Guillocheau F, Simon B, Baby G, Bessin P, Robin C, Dauteuil O (2018) Planation surfaces as a record of mantle dynamics: the case example of Africa. Gondwana Res 53:82–98
Gunnell Y (2003) Radiometric ages of laterites and constraints on long-term denudation rates in West Africa. Geology 31:131–134
Hall AM, Gilg A, Fallick AE, Merritt JW (2015) Kaolins in gravels and saprolites in north-east Scotland: Evidence from stable H and O isotopes for Palaeocene–Miocene deep weathering. Palaeogeography, Palaeoclimatol Palaeoecol 424:6–16. https://doi.org/10.1016/j.palaeo.2015.02.019
Hartley AJ, Rice CM (2005) Controls on supergene enrichment of porphyry copper deposits in the Central Andes: a review and discussion. Mineral Deposita 40: 515–525. https://doi.org/10.1007/s00126-005-0017-7
Heim JA, Vasconcelos PM, Shuster DL, Farley KA, Broadbent G (2006) Dating paleochannel iron ore by (U-Th)/He analysis of supergene goethite, Hamersley province, Australia. Geology 34:173–178. https://doi.org/10.1130/G22003.1
Heller BM, Riffel SB, Allard T, Morin G, Roig J-Y, Couëffé R, Aertgeerts G, Derycke A, Ansart C, Pinna-Jamme R, Gautheron C (2022) Reading the climate signals hidden in bauxite. Geochim Cosmochim Acta 323: 40–73. https://doi.org/10.1016/j.gca.2022.02.017
Hirvas H (1991) Pleistocene stratigraphy of Finnish Lapland. Geological survey of Finland, Bulletin 354. Geologian Tutkimuskeskus, Espoo. 125 pp
Hitzman MH, Reynolds NA, Sangster DF, Allen CR, Carman CE (2003) Classification, genesis, and exploration guides for non-sulfide zinc deposits. Econ Geol 98:685–714
Hou B, Keeling J, Li Z (2017) Paleovalley-related uranium deposits in Australia and China: a review of geological and exploration models and methods. Ore Geol Rev 88:201–234. https://doi.org/10.1016/j.oregeorev.2017.05.005
Jean A, Beauvais A, Chardon D, Arnaud N, Jayananda M, Mathe PE (2020) Weathering history and landscape evolution of Western Ghats (India) from 40Ar/39Ar dating of supergene K–Mn oxides. J Geol Soc 177: 523–536. https://doi.org/10.1144/jgs2019-048
Kalatha S, Economou-Eliopoulos (2015) Framboidal pyrite and bacterio-morphic goethite at transitional zones between Fe–Ni-laterites and limestones: evidence from Lokris, Greece. Ore Geol Rev 65:413–425. https://doi.org/10.1016/j.oregeorev.2014.10.012
Kerr SB, Horn AD (2020) Technical report Bovill project: pre-feasibility study prepared for I-Minerals Inc. Millcreek Mining Group, Salt Lake City. https://www.imineralsinc.com/assets/docs/I-Minerals%20Bovill%20PFS%20Report_04062020.pdf. Accessed 2023
King LC (1956) A geomorfologia do Brasil oriental. Rev Bras Geogr 18:147–265. (in Portuguese)
King LC (1962) Morphology of the earth: a study and synthesis of world scenery. Oliver and Boyd, London. 699 pp
Klages JP et al (2020) Temperate rainforests near the south pole during peak cretaceous warmth. Nature 580:81–86. https://doi.org/10.1038/s41586-020-2148-5
Lapin AV, Tolstov AV, Kulikova IM (2016) Distribution of REE, Y, Sc, and Th in the unique complex rare-metal ores of the Tomtor deposit. Geochem Int 54: 1061–1078
Lecomte P (1988) Stone line profiles: importance to geochemical exploration. J Geochem Explor 30:35–61
Li J-W, Vasconcelos P (2002) Cenozoic continental weathering and its implications for the palaeoclimate: evidence from $^{40}Ar/^{39}Ar$ geochronology of supergene K–Mn oxides in Mt Tabor, central Queensland. Australia Earth Planet Sci Lett 200:223–239
Mabbutt JA (1980) Weathering history and landform development. J Geochem Explor 12:96–116
Maignien R (1966) Review of research on laterites. UNESCO, Paris. 148 pp
Mao X, Retallack G (2019) Late Miocene drying of central Australia. Palaeogeogr Palaeoclimatol Palaeoecol 514: 292–304. https://doi.org/10.1016/j.palaeo.2018.10.008
Martin HA (2006) Cenozoic climatic change and the development of the arid vegetation in Australia. J Arid Environ 66:533–563. https://doi.org/10.1016/j.jaridenv.2006.01.009
McFarlane MJ (1976) Laterite and landscape. Academic Press, London/New York. 151 pp
McGowran B, Hill RS (2015) Cenozoic climatic shifts in southern Australia. Trans Roy Soc S Aust 139:19–37. https://doi.org/10.1080/03721426.2015.1035215
Michel P (1973) Les bassins des fleuves Sénégal et Gambie. Etude géomorphologique: Memoire ORSTOM 63. 752 pp
Monteiro HS, Vasconcelos PM, Farley KA, Spier C, Mello CL (2014) (U–Th)/he geochronology of goethite and the origin and evolution of cangas. Geochimica Cosmochimca Acta 131:267–228. https://doi.org/10.1016/j.gca.2014.01.036
Nahon D, Tardy Y (1992) The ferruginous laterites. In: Butt CRM, Zeegers H (eds) Regolith exploration geochemistry in tropical and subtropical terrains. Handbook of exploration geochemistry 4. Elsevier, Amsterdam, p 41–55

Negrão LBA, Costa ML, Poellmann H (2018) The Belterra Clay on the bauxite deposits of Rondon do Pará, Eastern Amazon. Brazilian Journal of Geology 48:473–484. https://doi.org/10.1590/2317-4889201820180128

Parrish JT, Ziegler AM, Scotese CR (1982) Rainfall patterns and the distribution of coals and evaporites in the Mesozoic and Cenozoic. Palaeogeogr Palaeoclimatol Palaeoecol 40:67–101

Pidgeon RT, Brander TH, Lippolt HJ (2004) Dating of laterite from the Darling Ranges. Aust J Earth Sci 51: 901–916

Pillans B (2005) Geochronology of the Australian regolith. In: Anand RR, de Broekert P (eds) Regolith landscape evolution across Australia. CRC LEME, Perth, pp 41–61. http://crcleme.org.au/RegLandEvol/Geochron_of_%20Aust_Regolith.pdf

Pillans B (2008) Regolith dating methods: a guide to numerical dating techniques in Australia CRC LEME, Perth. 46 pp. http://crcleme.org.au/Pubs/guides/RegDatMeth.html

Porter TM (2012a) Cargill, Ontario Canada. Porter GeoConsultancy Pty Ltd., Adelaide. http://www.portergeo.com.au/database/mineinfo.asp?mineid=mn1390

Porter TM (2012b) Martison, Ontario, Canada. Porter GeoConsultancy Pty Ltd., Adelaide. http://www.portergeo.com.au/database/mineinfo.asp?mineid=mn1392

Ransome FL (1904) The geology and ore deposits of the Bisbee quadrangle, Arizona. United States Geological Survey, Washington. Professional Paper 21, 168 pp

Roberts EM, Stevens NJ, O'Connor PM, Dirks PHGM, Gottfried MD, Clyde WC, Armstrong RA, Kemp AIS, Hemming S (2012) Initiation of the western branch of the East African Rift coeval with the eastern branch. Nat Geosci 5:289–294. https://doi.org/10.1038/ngeo1432

Sarala P (2011) Pre-glacial weathering of the Vuotso area. In: P. Sarala and V.J. Ohala (Editors), Geochemical and indicator mineral exploration methods and ongoing projects in the glaciated terrains in northern Finland. Excursion Guide, 25th International Applied Geochemistry Symposium. Viorimiesyhdistys—Finnish Association of Mining and Metallurgical Engineers, Espoo. Serie B92–11. 72 pp

Sawadogo B, Bamba O, Chardon D (2020) Landform-regolith mapping in the West African context. Ore Geol Rev 126. https://doi.org/10.1016/j.oregeorev.2020.103782

Smith RE (1987) Some conceptual models for geochemistry in areas of pre-glacial deep weathering. In: R.G. Garrett (editor) Geochemical Exploration 1985. J Geochem Explor 28:337–352

Sundell KE, Saylor JE, Lapen TJ, Horton BK (2019) Implications of variable late Cenozoic surface uplift across the Peruvian Central Andes. Nat Sci Rep 9:1–12

Tardy Y (1997) Petrology of laterites and tropical soils. A.A. Balkema, Rotterdam. 408 pp

Taylor GF, Thornber MR (1992) Gossan formation and gossan surveys. In: Butt CRM, Zeegers H (eds) Regolith exploration geochemistry in tropical and subtropical terrains. Handbook of exploration geochemistry, vol 4. Elsevier, Amsterdam, pp 115–202

Taylor G, Eggleton RA, Holzhauer C, Maconachie L, Gordon M, Brown MC, McQueen KG (1992) Cool climate lateritic and bauxitic weathering. J Geol 100:669–677. https://doi.org/10.1086/629620

Taylor RG, Howard KWF (1998) Post-palaeozoic evolution of weathered landsurfaces in Uganda by tectonically controlled deep weathering and stripping. Geomorphology 25:173–192. https://doi.org/10.1016/S0169-555X(98)00040-3

Théveniaut H, Freyssinet P (2002) Timing of lateritization on the Guiana shield: synthesis of paleomagnetic results from French Guiana and Suriname. Palaeogeogr Palaeoclimatol Palaeoecol 178:91–117

Thomas MF (1994) Geomorphology in the tropics. Wiley, Chichester. (460 pp)

Thorne RL, Roberts S, Herrington R (2012) Climate change and the formation of nickel laterite deposits. Geology 40:331–334. https://doi.org/10.1130/G32549.1

Tolstov AV, Cherenkov VG, Baranov LN (2023) Genesis and age of the ore thickness of the Tomtor deposit of niobium and rare-earth elements (northeast Siberian Platform). Geol Ore Deposits 65(Suppl 2):S276–S286. https://doi.org/10.1134/S1075701523090076

Valeton M, Biermann M, Reche R, Rosenberg F (1987) Genesis of nickel laterites and bauxites in Greece during the Jurassic and cretaceous, and their relation to ultrabasic parent rocks. Ore Geol Rev 2:359–404

van Dyck J, Fernandez A, Bernasconi SF, Caves Rugenstein JK, Passey SR, White T (2020) Spatial pattern of super-greenhouse warmth controlled by elevated specific humidity. Nat Geosci 13:739–744. https://doi.org/10.1038/s41561-020-00648-2

Vasconcelos PM, Carmo IO (2018) Calibrating denudation chronology through $^{40}Ar/^{39}Ar$ weathering geochronology. Earth Sci Rev 179:411–435. https://doi.org/10.1016/j.earscirev.2018.01.003

Vasconcelos PM, Conroy M (2003) Geochronology of weathering and landscape evolution, Dugald River valley, NW Queensland, Australia. Geochim Cosmochim Acta 67:2913–2930. https://doi.org/10.1016/S0016-7037(02)01372-8

Vasconcelos PM, Heim JA, Farley KA, Monteiro H (2013) $^{40}Ar/^{39}Ar$ and (U–Th)/He-$^{4}He/^{3}He$ geochronology of landscape evolution and channel iron deposit genesis at Lynn Peak, Western Australia. Geochimica et Cosmochimica Acta 117(2):83–31. https://doi.org/10.1016/j.gca.2013.03.037

Vickers ML, Bajnai D, Price GD, Linckens J, Fiebig J (2019) Southern high-latitude warmth during the Jurassic–Cretaceous: new evidence from clumped isotope thermometry. Geology 47:724–728. https://doi.org/10.1130/G46263.1

Wells MA, Danišík M, McInnes BIA, Morris PA (2019) (U-Th)/he-dating of ferruginous duricrust: insight into laterite formation at Boddington, WA. Chem Geol 522: 148–161. https://doi.org/10.1016/j.chemgeo.2019.05.030

Zeegers H, Lecomte P (1992) Seasonally humid tropical terrains (savannas). In: Butt CRM, Zeegers H (eds) Regolith exploration geochemistry in tropical and subtropical terrains. Handbook of exploration geochemistry, vol 4. Elsevier, Amsterdam, pp 203–240

Gossans: In Situ Supergene Base Metal Deposits

Rob J. Bowell and M. I. Leybourne

Abstract

The oxidation of sulfide-rich base metal mineral deposits results in residual, generally in situ, materials rich in iron oxyhydroxide and oxide minerals. Historically, these decomposed rocks have commonly been termed gossan. Gossans form through the oxidation of Fe sulfide minerals, typically pyrite and pyrrhotite. When pyrite is exposed to air and water, it undergoes a chemical reaction that produces Fe oxyhydroxide, oxide and sulfate minerals and sulfuric acid. The distinctive reddish-brown to yellow colour of gossans is due to the formation of goethite, jarosite, and Fe-hydroxy-sulfate phases. Acidic oxidizing solutions then oxidize the remaining sulfide minerals as well as primary silicate minerals. Neutralization and metal attenuation lead to formation of numerous secondary minerals. A portion of the metal content is generally precipitated lower down in the mineralized structure, commonly termed an enrichment zone, with 'gossan' generally referring to the upper leached zone. Miners historically used gossan as an indicator of potentially economic base and/or precious metal ore bodies in the underlying weathered rock and primary sulfide minerals. Although no longer targeted as ores themselves, gossans are still sought in modern exploration as a guide to concealed sulfide deposits.

R. J. Bowell (✉)
SRK Consulting, Cardiff, Wales, UK

Department of Geological Sciences and Geological Engineering, Queens University, Kingston, ON, Canada
e-mail: rbowell@srk.co.uk

M. I. Leybourne
Department of Geological Sciences and Geological Engineering, Queens University, Kingston, ON, Canada

1 Introduction

When sulfide mineralization is exposed to oxygen and water during chemical weathering, uplift and erosion, the sulfide minerals become unstable and react. The reaction products of this typically involve a residual concentration of the least soluble components such as Fe oxyhydroxide and sulfate minerals, and the more soluble components are mobilized. In acidic waters, this may include many trace metals (Thornber 1992). The residual material left at the surface composed of Fe oxyhydroxide minerals is termed a gossan, a term derived from old Cornish slang for weathered residual remains of a Cu-tin deposit (Pryce 1778; Buckley 2005).

The process that initiates gossan development is the oxidation of sulfide minerals, chiefly pyrite and the liberation of protons. This requires oxygen and water along with a catalyst such as

R. J. Bowell, C. R. M. Butt (eds.), *Geology, Geochemistry and Formation of Supergene Mineral Deposits in Deeply Weathered Terrain*, Mineral Resource Reviews, https://doi.org/10.1007/978-3-031-75733-4_4

Acidithiobacillus ferrooxidans for ferrous Fe to be present (Nordstrom and Alpers 1999). The general equation for pyrite oxidation is summarized below:

$$4FeS_2(s) + 15O_2(g) + 14\ H_2O \rightarrow 4Fe(OH)_3(s) + 8SO_4^{2-}(aq) + 16H^+(aq) \quad (1)$$

The net effect of sulfide oxidation is the potential to increase the loading of metals, sulfate and acidity in the receiving environment. Oxidation of ferrous Fe and hydrolysis of ferric Fe at pH > 2 provide the additional source of acidity through the reactions:

$$4Fe^{2+}(aq) + 10H_2O + O_2(g) = 4Fe(OH)_3(s) + 8H^+(aq) \quad (2)$$

The study of gossans has evolved to assess macro-textures and mineralogy to determine the significance of gossan colours and mineralogy for mineral prospecting (Emmons 1900, 1917; Locke 1926; Blanchard 1939; Thornber 1984; Thornber and Taylor 1992; Taylor 2011). Blanchard pioneered a rigorous macro-assessment of gossan textures (Blanchard 1968) and geochemistry (Blanchard and Boswell 1934; Blanchard 1944). The appearance of different gossanous materials is shown at hand specimen scale in Fig. 1.

Based on the early fieldwork and identification of gossans, the analysis started to be undertaken in Australia (Zimmerman 1964; Hallberg and Thompson, 1985; Butt 1979; Thornber et al. 1981; Taylor and Scott 1982; Mann 1983; Nickel 1984; Nickel and Daniels 1985; Thornber 1985; Scott 1987; Butt 1989, 2016; McQueen 2009; Gahlan and Ghrefat 2017; Parker et al. 2017), Africa (Webb and Tooms 1959; Tooms and Webb 1961; Cornwall 1961; Garlick 1961; Blain and Andrew 1977; Andrew 1980; Scott 1986; Bowell 1994; Randabel et al. 2018) and in the Americas (Boyle 1957; Greig 1984; Brimhall and Crerar 1987; Leybourne et al. 2002; Boyle 2003; Sillitoe 2009). The aim of much of this work was the recognition of gossans by their characteristic structure, mineralogy and chemical composition. Although gossans refer senso stricto to Fe-rich deposits, they can be related to other supergene base metal deposits associated with weathering of sulfide mineralization (chapter "Supergene Non-Sulfide Base Metal Deposits"). 'Non-sulfide deposits' is a relatively modern term that comprises a series of oxidized base metal minerals (Boni and Mondillo 2015). It has also been used to define a specific deposit type of Zn, Cu and, more rarely, Pb formed by supergene hydromorphic dispersion of metals. Typically, gossans are derived from massive sulfides, such as Volcanogenic Hosted Massive Sulfides (VHMS), Sedimentary Massive Sulfides formed by Exhalation of Sulfides on the seafloor (SEDEX, also referred to as sediment-hosted massive sulfide deposit or SHMS) or from sulfide matrix-supported mineralization in shears or breccias (Hannington et al. 1988; Large 1992).

In terms of supergene deposits, a continuum of geological characteristics can be defined for gossans and similar base metal deposits that show similar origins in terms of sulfide oxidation, element mobility and attenuation. Although no longer targeted as ores themselves, gossans are still sought in modern exploration as a guide to concealed sulfide deposits, whereas non-sulfide deposits are growing as potential sources of Zn and other metals.

As such, the deposits can be:

- Residual products formed by the in situ oxidation and weathering of sulfide mineralization. This accounts for Fe-rich gossans and siliceous, commonly manganiferous, gossans formed from the weathering of Fe-poor sulfide mineralization (e.g., carbonate-hosted Pb-Zn deposits) and some non-sulfide deposits.
- Hydromorphically-transported metals reprecipitated distally from primary mineralization. These are primarily non-sulfide Zn and Pb deposits (chapter "Supergene Non-Sulfide Base Metal Deposits") but can also include some Cu deposits, such as the malachite-rich ores in the Copperbelt (chapter "Supergene Ores of the Central African Copperbelt").

2 Gossan Nomenclature

Several definitions of the weathered portion of sulfide deposits have been published. The

Fig. 1 Hand specimen gossan and ironstone textures (**a**) Copper-bearing gossan showing "spongy goethite boxwork", epimorphs after chalcopyrite and pyrite coated by massive goethite rim. Einsleigh Copper Mine, Northern Queensland Australia. 12 (l) x 7(h) x 4 (w) cm. Kirwin collection, reproduced by permission of the Royal Ontario Museum, Canada; (**b**) Azurite and malachite in goethite boxwork, with the copper minerals representing a solution gossan, Bisbee Mine, Cochise County, Arizona, USA (specimen in the Smithsonian Museum on public display). Approximately 30 cm in length; (**c**) Goethite pseudomorphs after pyrite. Justice Mine, Ashanti, Ghana. The largest pseudomorph is 5 cm long.; (**d**) Erthyrite in a vug of goethite, with remnants of skutterdite and pyrite. Aghbar Mine, Ouarzazate, Morocco. Specimen is 15 (l) x 12 (h) x 8 (w) cm; (**e**) Boxwork texture after tetrahedrite, chalcopyrite, galena and pyrite replaced by cervantite, beudantite, anglesite, stibiconite and malachite. Herodsfoot Mine, Liskeard, Cornwall. Dimensions: 10 (l) x 9 (h) x 7 (w) cm. Epimorphs of the sulfides are clearly preserved in the goethite matrix with other secondary minerals partially infilling the voids. (**f**) "False" gossan or pseudogossan. Ironstone is formed by pedological processes of goethite replacement of Fe-rich silicate matrix, unassociated with sulfide minerals. Approximately 15 cm long. Kirwin collection, reproduced with permission of the Royal Ontario Museum, Canada

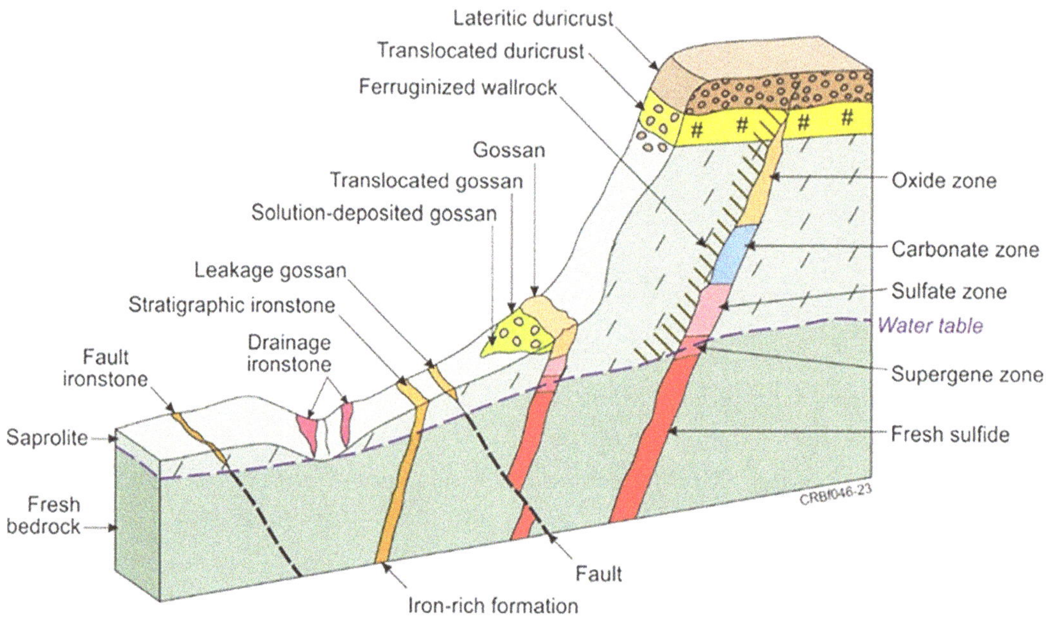

Fig. 2 Types of Gossans and Ironstones (taken from Taylor et al. 1980; Taylor and Thornber 1992; reproduced with permission, CSIRO)

traditional definition has focused on the massive aggregate commonly termed "limonite" that comprises goethite, hematite, quartz and jarosite with variable portions of resistant phases (Locke 1926). Later definitions distinguished gossans from ironstones formed by the weathering of Fe-rich lithologies (Blanchard 1968; Butt 1989; Thornber and Taylor 1992). The identification of various types of gossans and ironstones has been cited as a complication in mineral exploration with the challenge being distinguishing gossans related to sulfide mineralization and ironstones related to pedological concentration (Taylor et al. 1980; Taylor 2011). The most recognized system of gossan characterization (Fig. 2) is that developed by Taylor et al. (1980), based on topographic location and textures, following earlier work by Blanchard and others.

3 Gossan Maturity

3.1 Regolith Setting

Another aspect of gossan description is the maturity of the gossan profile (Blanchard 1968; Taylor et al. 1980; Butt 1989). Mature gossans tend to show a mineralogical stratigraphy or vertical zonation that has a well-developed Fe oxide cap under which there are zones of secondary minerals, namely the Carbonate zone that includes secondary carbonates and other minerals (Fig. 1) and sulfate zone with Fe-rich sulfate minerals. A deeper zone of supergene sulfide minerals may also be present (Boyle 2003). The maturity of a gossan profile can be used in providing an assessment of the degree of weathering (Scott et al. 2001).

Most gossans are the weathered residue of rocks that contain matrix or massive sulfide mineralization. As Fe sulfide minerals generally dominate this style of mineralization, the resulting matrix is rich in Fe oxyhydroxide and oxide minerals. Gossans formed from Fe-poor sulfide mineralization (for example, the Cu-Pb-Zn deposits in carbonate rocks such as those of Tsumeb, Namibia; Bowell 2014) form Fe-poor calcrete gossans or siliceous, commonly Mn-rich gossans. The types of gossans and macroscopically similar ironstones that have been defined (Taylor et al. 1980) include barren gossans formed by weathering of Fe sulfide minerals with no elements of economic significance; Direct or true gossan—formed in situ after base metal-Fe sulfide mineralization (such as shown in Fig. 1a–e); False gossan or pseudo gossan—discrete pedological accumulation of Fe oxide minerals that appear similar to gossans but are not associated with a primary rock sulfide occurrence as shown in Fig. 1f; Solution gossan—hydromorphic transported Fe derived from a sulfide source, can also be termed ironstones.

- Translocated gossan—mechanically moved and commonly recemented.

Within deeply weathered environments, the challenge is to identify the mineralogical and geochemical characteristics of primary sulfide mineralization that persist into the regolith in the form of the modified residual material i.e., gossan. This needs to be distinguished from transported anomalies related to the concentration of Fe in tropical, lateritic regolith (Blain and Andrews 1977; Taylor et al. 1980; Taylor and Thornber 1992; Taylor 2011). By comparison, Fe-poor "non-sulfide base metal deposits" are a relatively modern term that encompasses "calamine zinc", "exotic copper", "carbonate hosted" and non-sulfide hosted deposits of Ag, Pb and V that comprises a series of oxidized base metal mineral deposits (chapter "Supergene Non-Sulfide Base Metal Deposits"). The common feature of these is they are formed by supergene hydromorphic dispersion (Fig. 3).

In terms of supergene deposits, a continuum of geological characteristics can be defined for gossans and similar base metal deposits that show similar origins in terms of sulfide oxidation, element mobility and attenuation. Although no longer a targeted as ores themselves, gossans are still sought in modern exploration as a guide to concealed sulfide deposits, whereas non-sulfide deposits are growing as potential sources of Zn and other metals (Cohen and Bowell 2014; Farooq and Govil 2014; Gahlan and Ghrefat 2017).

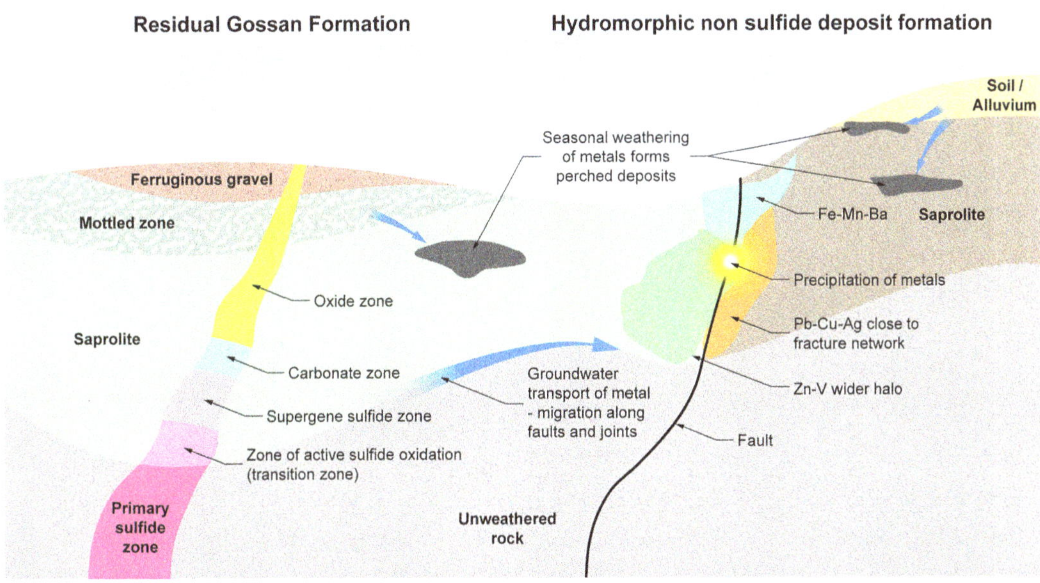

Fig. 3 Conceptual model of the association of gossans and non-sulfide mineral deposits (modified after Butt 1989; Borg 2015)

3.2 Case Study: Iberian Pyrite Belt

Regolith Setting Detailed mapping and mineralogical evaluation of gossans in the Iberian Pyrite Belt by Velasco et al. (2013) have identified three zones, with gradual contacts roughly parallel to the current topography, across deposits in southern Spain (Fig. 4). The zones are: (1) a lower zone dominated by goethite and subordinate jarosite, with significant enrichment in S, As, P, Pb, Sn, Sb, Ag and Au; (2) a principal or main stage dominated by goethite with no jarosite, which is depleted in sulfur and chalcophile elements; and (3) the late stage or upper zone that is near the surface and comprises hematite and quartz with weak anomalies in P, Pb and Sn. The formation of the profile is initiated with an acidic stage of gossan development related to sulfide oxidation that forms a complex Fe-rich oxyhydroxide and sulfate mineral assemblage along with other potential metal sulfate and secondary minerals.

Climate and Topography The Iberian Pyrite Belt is a geographical area that stretches along the south of the Iberian Peninsula, from Portugal to Spain. It is approximately 250 km long and 50 km wide, running NW to SE. Topographically, the region comprises rugged terrain characterized by landforms including mountains, valleys, plateaux and river systems. The belt is known for its mineral-rich deposits, principally sulfide minerals such as pyrite, that historically have driven mining activities.

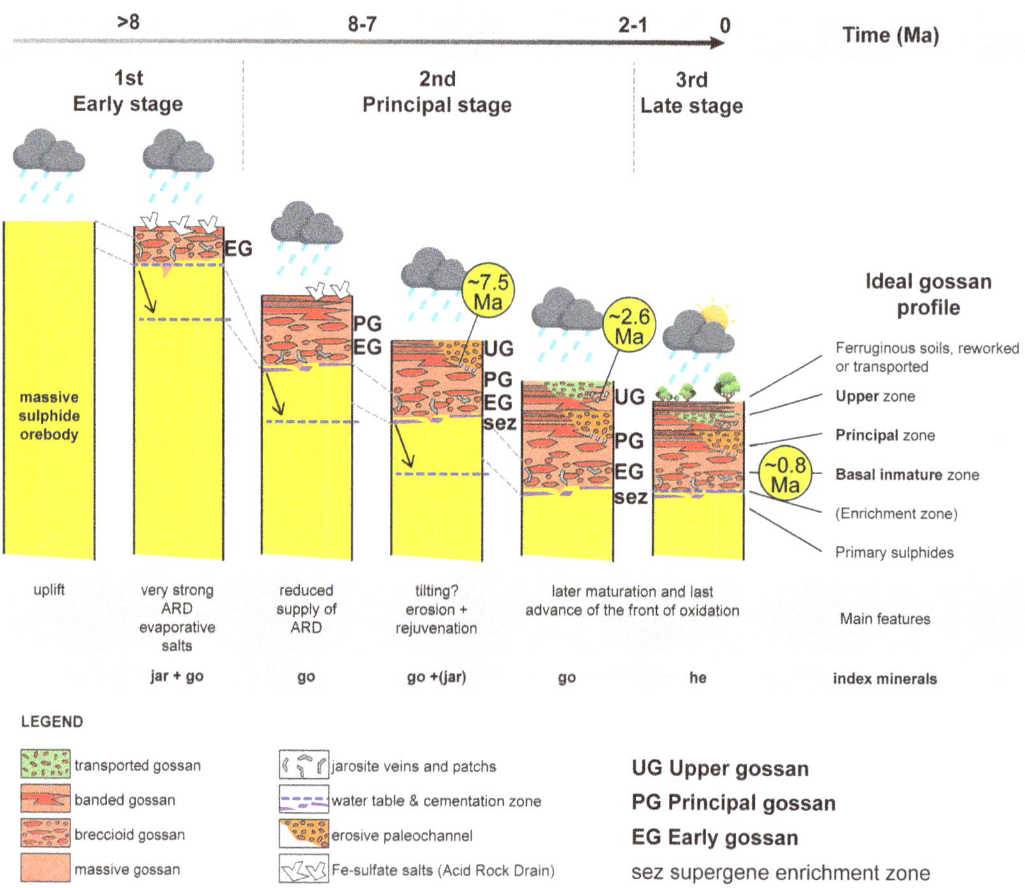

Fig. 4 Hypothetical gossan profile evolution during weathering of the massive sulfide orebodies of the Iberian Pyrite Belt based on an image in Velasco et al. (2013). ARD, acid rock drainage

The climate in the Iberian Pyrite Belt varies depending on factors such as altitude, proximity to the Atlantic Ocean and regional geography. Generally, the coastal areas experience a Mediterranean climate characterized by hot, dry summers and mild, wet winters. Inland regions may have more continental influences, with greater temperature extremes and lower precipitation.

Geology A detailed example of the evolution of mineralogy associated with a buried gossan cap and associated sulfide deposit has been published by Yesares et al. (2017) for the Las Cruces Cu deposit in Spain. Las Cruces is a volcanogenic massive sulfide deposit located north of Huelva in the eastern portion of the Iberian Pyrite Belt. The primary orebody is a polymetallic massive sulfide deposit with an underlying Cu-rich stockwork and a completely preserved overlying supergene profile that includes a Cu-rich secondary ore (initial reserves of 17.6 Mt. @ 6.2% Cu) and a gossan cap (initial reserves of 3.6 Mt. @ 3.3% Pb, 2.5 g/t Au and 56.3 g/t Ag).

The deposit morphology is shown in Fig. 5. The deposit is buried under 140–150 m of post-Alpine carbonate-rich sediments, including marls, overlying a thick sequence of black shales, felsic volcanic and associated clastic rocks. The deposit is orientated approximately E-W and extends for more than 1 km dipping 35°N (Silva et al. 1990). The massive sulfide body is altered such that it comprises gossan, Cu-rich secondary ore, primary massive sulfide mineralization and Cu-rich stockwork beneath.

Gossan Mineralogy and Evolution The gossan has a complex mineral assemblage comprising siderite, calcite, smithsonite and cerussite; Fe-sulfides including pyrite, marcasite, greigite and pyrrhotite; galena, stibnite, tennantite-tetrahedrite phases, cinnabar, bismuthinite, Au-Ag-Hg amalgam and various Pb-Bi-, Fe-Pb-Sb- and Ag-Hg-Sb sulfosalts. The gossan matrix is a hematite and goethite boxwork with secondary siderite associated with secondary galena (Yesares et al. 2017; Scheiber et al. 2018). Based on the mineralogy, the following genesis and evolution of the deposit have been proposed. Sulfide oxidation and alteration of Fe-oxyhydroxide minerals and their relict textures were replaced by carbonate and sulfide minerals indicate that gossan formation occurred under changing physical and chemical conditions (Velasco et al. 2013; Yesares et al. 2017; Scheiber et al. 2018). These replacements varied from the older gossan formed by prolonged tropical weathering from sulfide oxidation, mineral dissolution and redistribution of chemical elements. Later seawater alteration of the gossan and interaction with carbonate-rich, reducing basinal fluids led to carbonatization and sulfidation of the gossan, and thus to the generation of Fe-carbonate and Pb-Sb-sulfide minerals. Changes in redox-pH chemistry in the weathering environment coupled with detailed mineralogy have been used by Yesares et al. (2017) to develop a conceptual paragenesis for the deposit (Fig. 6).

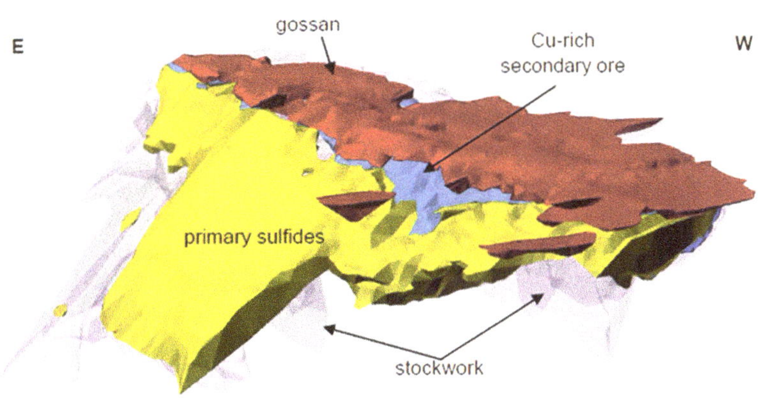

Fig. 5 3D model of the Las Cruces deposit looking south. Deposit scale is approximately 1 km in strike. (From Yesares et al. 2017)

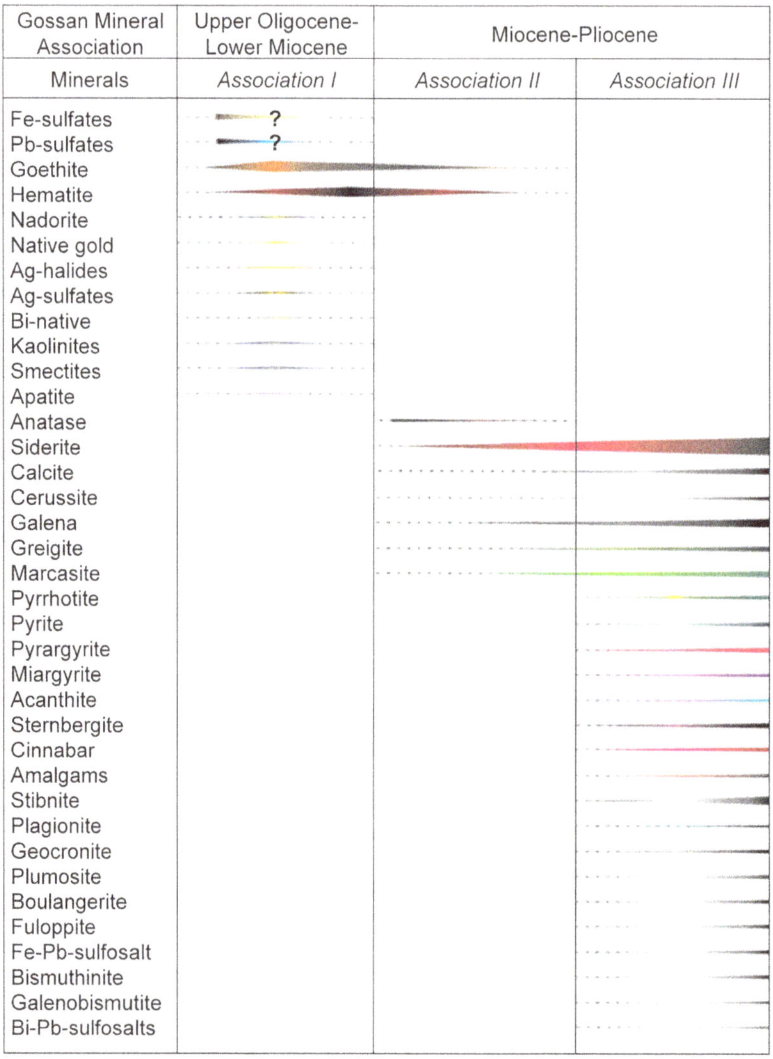

Fig. 6 Mineral paragenetic sequence for the Las Cruces gossan (from Yesares et al. 2017)

The internal structures of the gossan profile at Las Cruces have been assessed using magnetic fabrics and palaeomagnetic dating (Essalhi et al. 2011). The upper parts of the gossans are commonly older than the lower parts, indicating that the alteration proceeded from top to bottom of the profiles. In the upper part, the older age and the horizontal "mature" fabric is interpreted to be a high maturation stage of alteration of massive sulfides, whereas the younger, lower part is considered the immature vertical Hercynian fabric). As noted in other studies of gossan maturity in the Pyrite Belt these different stages reflect changes in paleoclimate, erosion activity and tectonic movement and were not continuous but occurred in episodes (Velasco et al. 2013).

In a detailed study of the galena and secondary siderite formation, Scheiber et al. 2018 proposed reducing groundwater conditions prevailed, based on the presence of H_2S, high sulfur isotope values in sulfate and negative Eh values. The role of decomposing organic matter in generating these conditions is indicated by high concentrations of NH_3, B, I and low $\delta^{13}C$ values in dissolved inorganic carbon of groundwater. Dissolution of goethite-hosting Pb produced an assemblage of galena, siderite and Fe sulfide minerals (mackinawite). Thermodynamic calculations and mineral speciation confirmed siderite formation from waters having low concentrations of Fe (< 10 μg/L) and low $\delta^{13}C$ values, similar to current groundwater. The high $\delta^{34}S$ values in galena indicate sulfide was derived by the reduction of groundwater sulfate, coupled with the reduction of sulfate minerals, as the concentrations required are up to two orders of magnitude higher than could be accounted for by goethite dissolution alone (Scheiber et al. 2018). In calculations using gossan composition, present-day groundwater chemistry and flux, siderite-bearing units were predicted to have formed within 1 Ma, with no additional reactants required (Scheiber et al. 2018).

Initial development of a gossan starts with meteoric water and/or groundwater oxidizing sulfides, mobilizing some metals and residual enrichment of others such as Fe and Pb (Stage 1). Over time, gossans mature (Stage 2), with elements leached progressively downwards through the profile. The ongoing processes lead to alteration, dissolution and reprecipitation of Fe-oxyhydroxide minerals and the loss of the more soluble secondary metal minerals. Components adsorbed to Fe minerals and clays are released as minerals are transformed from poorly crystalline forms to more ordered structures. Local uplift and erosion of gossans and rejuvenation of oxidation can overprint this pattern. The third stage leads to the formation of hematite and quartz. Possible major breaks in this gossan development of the Iberian Pyrite Belt occurred in Messinian times (7–8 Ma) and at the beginning of the Early Quaternary (1–2 Ma?), greatly complicating the stratigraphy due to localized reworking of gossan profiles (Velasco et al. 2013).

4 History of Gossan Exploitation

Mining has motivated human society since the dawn of time. Initially with flints and shaped stone and then through the working of Cu, Au and Fe through to the plethora of metals that aid modern society (Mighall and Chambers 2022). The exploitation of metals and pigments from Fe oxide-rich weathered residual components of sulfide mineralization has been utilized for more than 6000 years (Agricola 1556; Boyle 1957; O'Brien 1996; Graddock 1995; Boyle 2003; Timberlake 2003; Buckley 2005; Topping and Lynott 2005; Mellado et al. 2006; Rosenthal et al. 2013; Mighall and Chambers 2022).

Metal exploitation contributed significantly to the civilization of ancient societies (O'Brien 1996). Ancient societies in this context relate to prehistory up to the end of the medieval period before the onset of the Industrial Revolution (c. 1750 AD). However, the processes involved in metal recovery and metallurgical waste deposits can adversely impact the chemical signatures of soils of former mining and metallurgy sites (Mighall and Chambers 2022). The ancient Egyptians, Phoenicians and others centred in the Middle East and Eastern Mediterranean understood basic metallurgy and this knowledge helped develop metallurgical centres further west such as Lavrion in Greece (Vourlakos and Fitros 2019). Even religious texts refer to ancient metallurgical practices as symbolic of spiritual refinement; for example, in the Bible (New International Version), the book of Ezekiel (chap. 22, verses 17–22) states:

> As silver, copper, iron, lead and tin are gathered into a furnace to be melted with a fiery blast, so will I gather you in my anger and my wrath and put you inside the city and melt you.

The mines of Mineria in Spain have been exploited for 5000 years and much of this involved the processing of gossans and supergene enrichment zones (Topping and Lynott 2005; Mellado et al. 2006; Rosenthal et al. 2013). Archaeological discoveries indicate that mining was conducted in prehistoric times from gossans in situ and at least Cu and gold were recovered. Mining was critical to the development of civilization in the Mediterranean and this was aided by the abundance of exposed gossans over massive sulfide mineralization. Early archaeological evidence for the use of oxidized minerals to produce Cu was observed along the Mediterranean as early as 9500 BC, for example in the Attica Peninsula, at Hallan Cemi in Anatolia, Turkey and from the Mineria area of Andalusia in Spain (Yalçin 2000; Topping and Lynott 2005; Vourlakos and Fitros 2019).

Some of the oldest mines continuously operated include the Skouriotissa Cu mine in Cyprus, the Cu mines of Rio Tinto at Mineria in Spain and the Pb-Ag mines of Laurion in Greece (Fig. 6). Mining in the Skouriotissa area dates back as far as 4000 BC and was described by Pliny in the first century AD and reported by Agricola in 1556 (Agricola 1556; Mellado et al. 2006; Rosenthal et al. 2013). Some operations are still continuing. Similarly, early exploitation of gossans and oxidized ores occurred in other regions. For example, in the Copperbelt of Central Africa working of malachite for Cu is estimated to have occurred several centuries before Europeans arrived (Gunning 1961). Tin and Cu production in Cornwall also benefitted from the presence of Cu-rich gossans associated with rich deposits, and the term "gossan" is derived from old Cornish slang (Pryce 1778; Buckley 2005).

One of the most complete early treatments of mining methods in Europe is by the German scholar Georgius Agricola in his De re Metallica (1556). Within this, he describes the importance of "Eiserner Hut" or "iron hat" as an important source of Cu and Au ore. The influx of this mineral wealth from the New World financed the Renaissance. With the development of deeper sulfide deposits, gossans became less of an ore and more of an indicator of mineralization. Prospecting ideas of developing gossans as a rich source of ore influenced mineral exploration throughout the Americas, Africa, India and eventually Australia (Blanchard 1939, 1944, 1968). Examples of various workings and associated gossans are shown in Fig. 7.

Fig. 7 Examples of gossans in the field: (**a**) Open pit, Tharsis Mines, Tharsis, Alosno, Huelva, Andalusia, Spain showing the gossanous "ore body" showing significant oxidation to a greater depth than the adjacent volcanic host rock; (**b**) Cerro Colorado, Rio Tinto Mines, Minas de Riotinto, Huelva, Andalusia, Spain with a gossan zone developed over the Cu-rich ore and the pyritized wallrock generating a wide gossan expression in the pit; (**c**) Gossan development of massive galena-sphalerite-pyrite mineralization at Rosh Pinah, Namibia where tertiary Zn-Fe sulfate salts have formed due to later alteration of the gossan assemblage. The "vein like zone" is semi-massive smithsonite and calcite; (**d**) Aragonite-flow zone rich in Cu-Zn sulfate salts, (**d**) Kamareza mining area, Agios Konstantinos, Lavreotiki, Attica, Greece

5 Element Mobilization and Mineral Precipitation in Gossan Formation

5.1 Effects of Exposure

Erosion and oxidation of massive or disseminated sulfide mineralization commonly lead to the formation of gossans, which dominate the space between the water table and the surface above the primary sulfide deposit (Emmons 1917; Andrew 1980; Williams 1990; Taylor and Thornber 1992; Boyle 2003; Skarpelis and Ariadne 2009; Velasco et al. 2013; Yesares et al. 2017). Here, surface chemical and mineral alteration of the sulfide minerals leads to the mobilization of metals, with reprecipitation of some to form secondary Fe- and Mn-bearing (hydro)oxides and minerals of the jarosite supergroup (Taylor et al. 1980; Butt et al. 2005; Atapour and Aftabi 2007). Mineralogical and geochemical compositions of gossan are influenced by the composition of primary ore (Taylor and Thornber 1992; Butt et al. 2005; Boyle 2003). However, over time, maturity and different mobilities under varying redox and pH conditions modify the primary geochemical signature so that in highly mature gossans, the concentration and association of chemical elements may no longer reflect the precursor material (e.g., Leybourne et al. 2006). In intense tropical weathering, this can lead to confusion over gossan geochemistry and relation in terms of depth and the spatial extent to the primary ore (Scott et al. 2001; Butt et al. 2005).

5.2 Sulfide Oxidation

The mechanism of sulfide oxidation has been presented in chapter "The Role of Regolith Formation and Evolution in the Development of Supergene Mineral Deposits". Essentially, gossans and associated deposits form through the weathering of sulfide minerals and deposition of metals either in situ or mobilized elsewhere (Sato 1960; Thornber 1975a, b; Thornber and Wildman 1984). Sulfide mineral oxidation reactions occur at the water-solid interface at mineral surfaces or along fractures in the sulfide minerals (Sato and Mooney 1960; Thornber 1975a; Thornber 1983). The reduction-oxidation reactions proceed by a similar mechanism to that of corrosion of metal alloys in which a chemical cell is established with reduction at a cathode and oxidation at the anode and can occur within the massive sulfide and sulfide-gangue matrix (Fig. 8).

The development of the gossan texture (fabric) is controlled by the supply of oxygen in the system and the rate at which oxygen reacts with primary sulfide minerals. The connectivity of sulfide minerals in mineralization will control the rate and the extent to which the galvanic cell can develop. In general, massive sulfides dominated by pyrite, pyrrhotite and chalcopyrite will be able to transfer electrons better than sulfide minerals that have no requirement for oxidation as part of the dissolution mechanism, such as pure sphalerite, galena, covellite or chalcocite (Thornber 1975a, b, 1983). In addition, the presence of insulators such as carbonate and silicate minerals in the matrix in disseminated mineralization will limit the transfer of electrons and thus the extent and rate of oxidation. Gangue minerals are generally weathered by reaction with sulfuric acid and cation hydrolysis. Reaction with carbonate and silicate minerals results in alkaline buffering attenuating metal oxyanion phases (see chapter "The Geochemistry of Supergene Processes"). In terms of gossan development, this buffering leads to the preservation of the gossan fabric by Fe oxyhydroxide and other metal oxyanions forming and the diagnostic boxwork features common to gossans (Fig. 1).

Primary mineral textures textures are not only preserved in macroform but also in the microfabric of the gossan (Fig. 9). Here mineral replacement along joints and fractures is common (Fig. 9a) and initiates replacements that can lead to residual sulfide material in Fe-oxyhydroxide boxworks (Fig. 9b), partial replacement by a complex network of secondary minerals (Fig. 9c) or complete pseudomorphs of primary

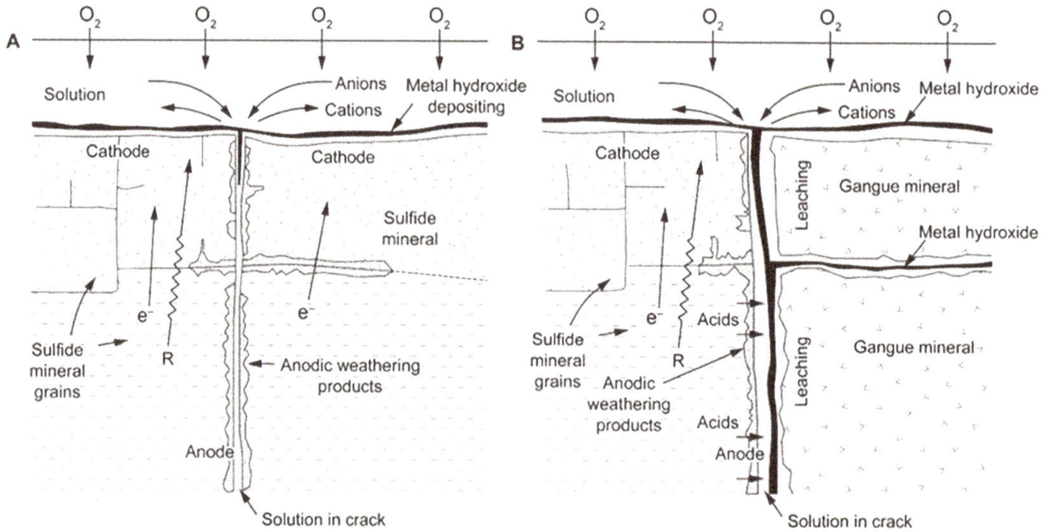

Fig. 8 Schematic diagram reflecting the cathode-anode relationships in (**a**) massive sulfide; and (**b**) sulfide-gangue mineral matrices. (Modified and redrawn from Thornber and Taylor 1992, reproduced with permission, CSIRO)

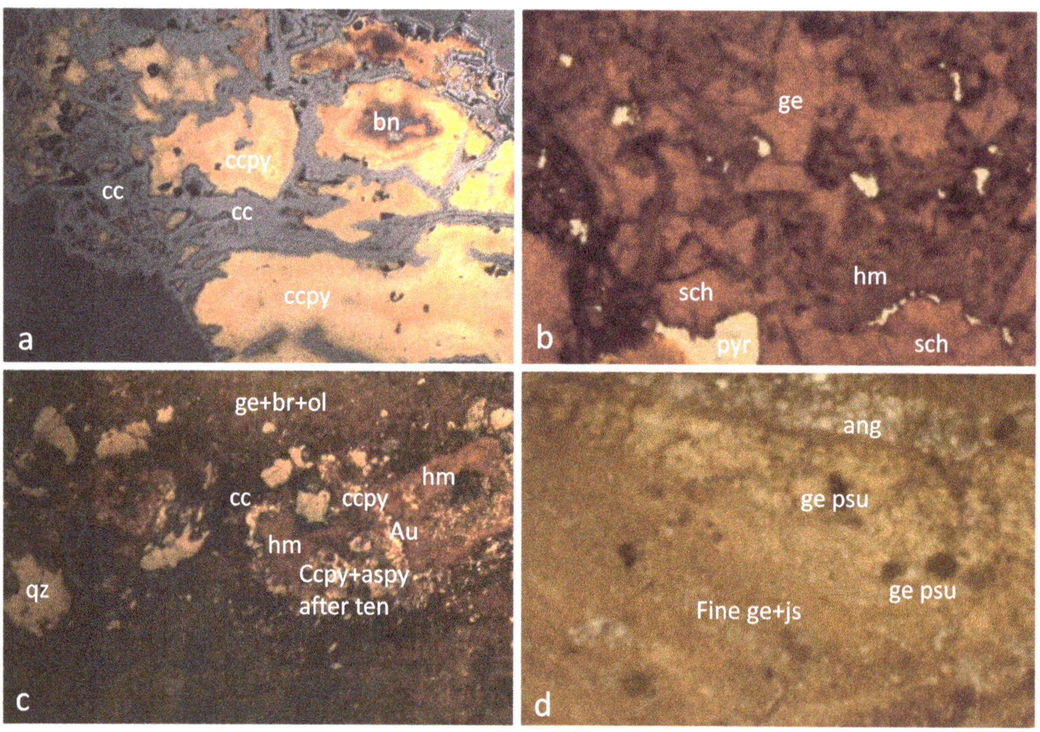

Fig. 9 Photomicrographs of gossan boxwork textures. (**a**) Chalcocite and goethite boxwork. Rosh Pinah Mine, Karas, Namibia. The field of view is 2 mm under reflected light; (**b**) Goethite breccia and replacement of massive pyrite. Yenipazar, Aydın Province, Turkey. The field of view 2 mm under transmitted light; (**c**) Goethite and secondary Cu minerals replacing chalcopyrite. Khandiza VMS deposit, Surxondaryo, Uzbekistan. The field of view is 0.2 mm unde reflected light; (**d**) Goethite pseudomorphs after pyrite associated with Fe-stained matrix. Farlo deposit, Burkina Faso. The field of view is 2 mm under transmitted light. [Key: ang, anglesite; aspy, arsenopyrite; Au, native gold; bn, bornite; br, brochantite; cc, chalcocite; ccpy, chalcopyrite; ge, goethite; hm, hematite; ol, olivenite; psu, pseudomorph; sch, schwertmannite]

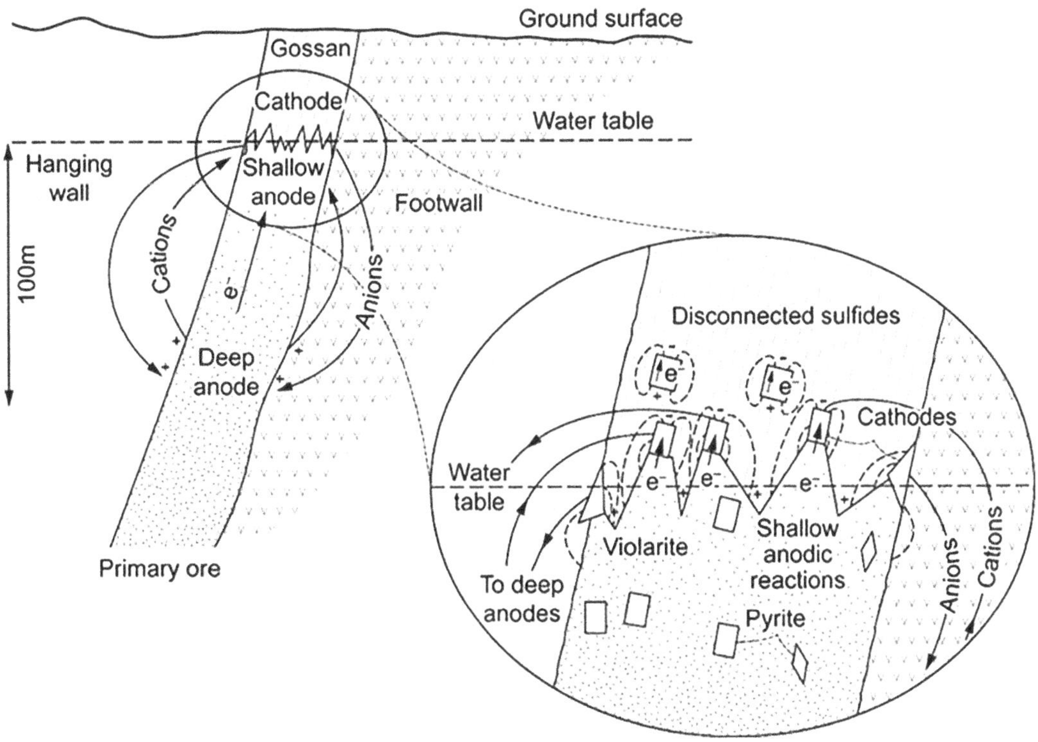

Fig. 10 Schematic representation of the reduction-oxidation cell at the Kambalda Ni sulfide deposit, Western Australia (Nickel et al. 1974; Thornber 1975a, 1983, 1985; Dunn and Howes 1996; reproduced with permission, CSIRO)

mineral retaining elements of the primary texture by secondary minerals (Fig. 9d).

A limitation in the extent of sulfide oxidation is the development of more stable phases that can act as an insulator in the electrochemical cell (Fig. 10). For example with Ni-Fe sulfide minerals at Kambalda, Western Australia, where violarite developed at the expense of pentlandite and pyrrhotite, electron transfer was reduced, limiting the extent of oxidation (Nickel et al. 1974; Thornber 1975a, 1983; Nickel and Daniels 1985). The redox reactions occurred at the water table with the development of deep oxidative weathering below, with violarite formed due to the connection of sulfide minerals (Fig. 10). In the overlying gossan, disconnected sulfide minerals are only partially oxidized as they form "isolated" batteries and therefore the rate of oxidation is slower (Thornber 1975a). This was compared with electrochemical experiments using cyclic voltammetry and intermittent galvanostatic polarization carried out on electrodes of pentlandite and violarite at pH values of 0, 2, 4, 10 and 11 (not shown).

Pentlandite ($(Ni, Fe)_9S_8$) is the principal Ni mineral at Kambalda, with millerite (NiS) being extremely rare (Nickel et al. 1974). Pyrrhotite (Fe_8S_9) and pyrite (FeS_2) are abundant in the ore shoots, together with minor Cu and Co sulfide minerals. Violarite ($(Ni, Fe)_2S_4$) is an important supergene sulfide mineral. Thornber (1983) coupled the mineralogical data to determine the relative thermodynamic stability fields for pentlandite, violarite, millerite, pyrite, pyrrhotite and sulfur as well as sulfate molecules and hydrogen sulfide gas as a function of pH, Eh and the activities of Fe^{2+}, Ni^{2+} and S^{2-} in solution to provide an understanding of their relationships (Fig. 11).

The mechanisms of the cathodic and anodic reactions at the electrolyte/sulfide interface are postulated and the similar behaviour of

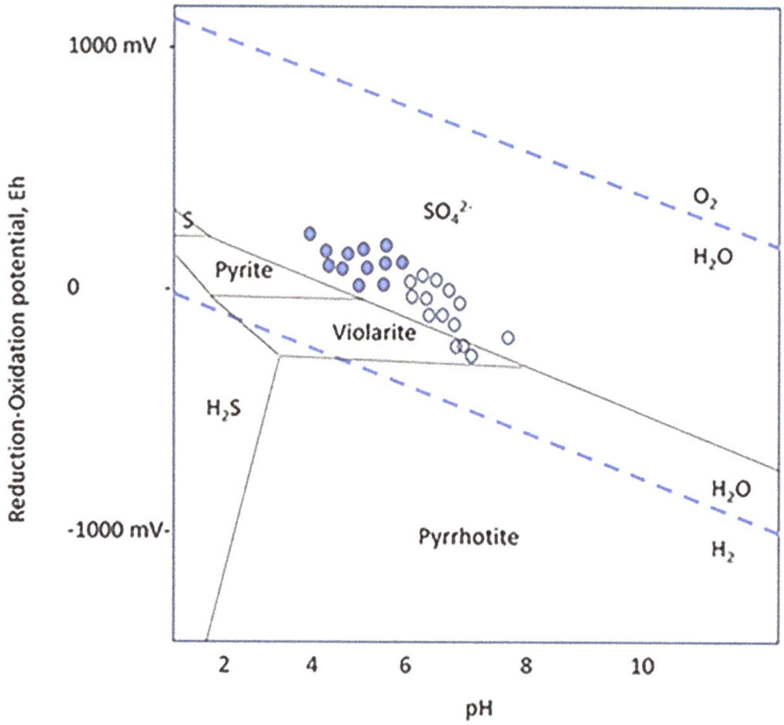

Fig. 11 Eh-pH-Activity diagrams at 298 °C drawn from data listed in Thornber (1983). Activities of Fe SO_4^{2-} and Ni where constant, are taken to be 10^{-6} M and H_2 S activity is assumed to be 10^{-2}. Py = pyrite, Pn = pentlandite, Vi = violarite, Po = pyrrhotite (millerite is not stable under these conditions). Data points are Eh-pH measurements made on pentlandite-violarite assemblages (reproduced with permission, CSIRO)

pentlandite and violarite is demonstrated. Below pH 4, anodic dissolution favours sulfur formation and above pH 4 sulfate is formed. Cathodic dissolution favours H_2S formation over H_2 at low pH.

Electrochemical measurements indicate that, for violarite and pentlandite, the solution-mineral interfaces behave similarly, and these results have been plotted within co-ordinates of Eh and pH so that the measured reactivity of these minerals can be compared with their thermodynamic stability (Fig. 12). This approach demonstrates the potential conditions needed for slow geological weathering of Ni-Fe sulfide ores and the development of the extensive gossans at Kambalda and further demonstrates why the gossan development is deeper relative to deposits with more silicate gangue such as shear zone hosted matrix sulfide-quartz gold deposits (Boyle 1979).

An important control alluded to above in the Kambalda example is the position of the water table as this will influence oxygen ingress to sulfide surfaces. With a rising water table, the rate of sulfide oxidation and thus formation of gossan textures will decline as oxygen accessibility is reduced. A static water table has much the same effect, as once the redox reactions have penetrated to the limits of oxygen ingress then reactions will cease (Williams 1990).

Based on the above mechanism, it is reasonable to assume that massive sulfide ores are good electrical conductors and thus have the potential to form extensive gossans, for example in the Iberian Pyrite Belt or Kambalda in Western Australia, as shown schematically in Fig. 13.

A considerable distance can develop between the cathode or zone of precipitation at the water table and the anodic leaching front in the sulfides (Fig. 10), with the caveat that good conductivity between the two is maintained (Thornber and Taylor 1992). Indeed, the separation aids the development of the leaching reaction by isolating the leaching front from the neutralization reaction at the cathode. As discussed in chapter "The Role of Regolith Formation and Evolution in the Development of Supergene Mineral Deposits", pyrite oxidation liberates additional acidity through the hydrolysis of ferrous Fe and pH can

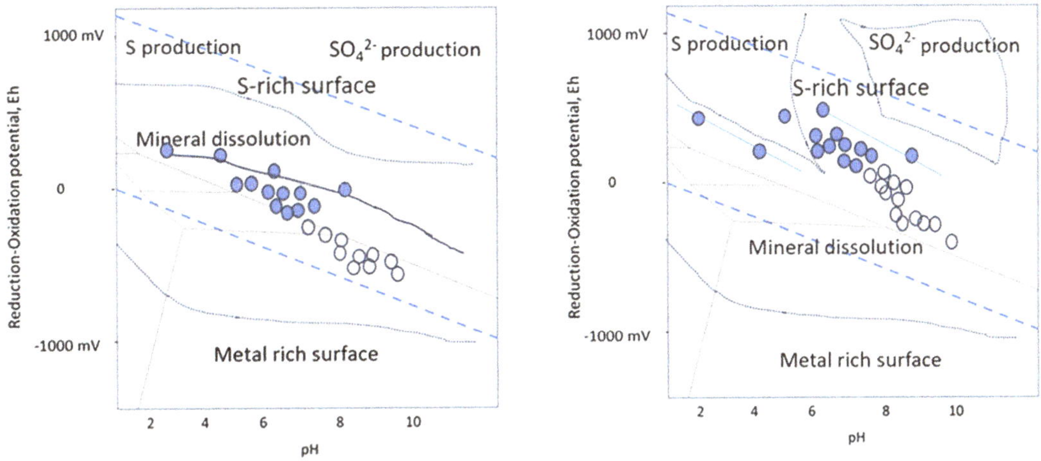

Fig. 12 Potential-pH diagrams for pentlandite and violarite, compiled from the Intermittent galvanostatic polarization and cyclic voltammetry data of Thornber (1975b, 1983; reproduced with permission, CSIRO)

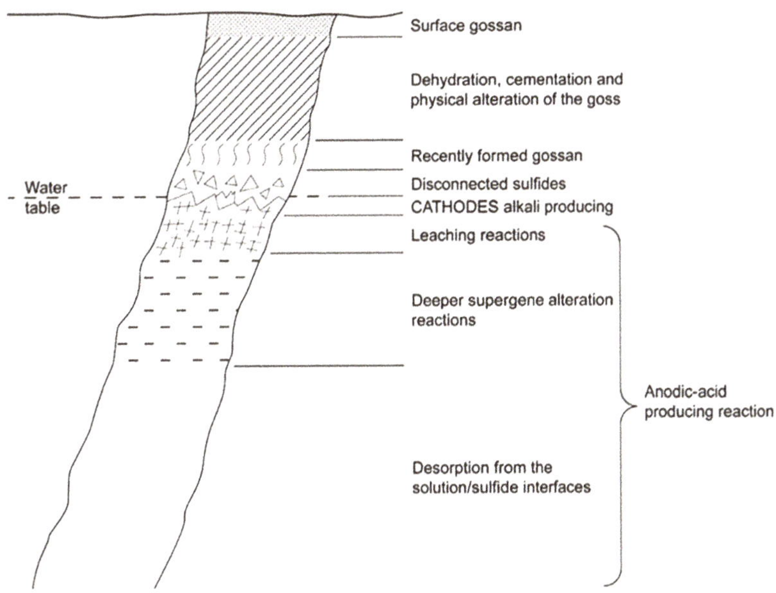

Fig. 13 Schematic representation of the development of gossans showing the connection between cathode and anode zones (Thornber and Taylor 1992; reproduced with permission, CSIRO)

remain extremely low; in extreme examples, pH can be negative, such as at the Iron Mountain massive sulfide mine in California (Nordstrom and Alpers 1999).

Matrix-supported sulfide ores can host considerable gangue but may still maintain electrical contact through interconnected sulfide minerals to allow for the development of boxwork textures. However, due to the proximity of sulfide minerals to insulating gangue minerals, metal mobilization is typically limited so that base metals precipitate close to the site of dissolution, commonly pseudomorphing the precursor sulfide texture (Fig. 8). With disseminated sulfides, each individual sulfide mineral acts as a separate electrical cell and only influences its own oxidation. As such, oxidation is limited to the area of oxygen ingress at the water table and gossan development is minor and generally surficial.

5.3 Supergene Mineralogy

The precipitation or adsorption of metal(loid)s with or by Fe-bearing minerals is an important aspect of retaining metals within the gossan structure, as described earlier in this chapter. Experimental and field studies have demonstrated the importance of the Fe-oxyhydroxide phases in metal retention through co-precipitation and adsorption (Thornber and Wildman 1984; Bowell et al. 2020).

Gossans and similar deposits are composed of secondary minerals with residual primary minerals. Goethite, hematite and jarosite are the most common and generally show an antipathetic relationship to fine-grained secondary quartz. Goethite, together with other Fe oxide, Mn oxide and phyllosilicate minerals are the main phases that attenuate metal(loid)s (Blanchard 1968; Thornber 1975a, 1992; Bowell et al. 2015, 2020). Identification of accessory or residual minerals is important in determining potential primary lithology or mineralization. Significantly, this can include residual sulfide mineral grains that may be armoured from oxidation by insulating gangue minerals (Fig. 4h).

The mineralogy of gossans reflects the primary sulfide assemblage and as such can vary. A schematic summary is given in Table 1 of the common and minor minerals that can be expected in gossan profiles for selected parameters. Secondary minerals such as the jarosite-alunite series and other secondary metal(loid) oxyanion minerals (Fig. 4) can host high concentrations of ore and associated pathfinder elements. The association of these minerals will be a function of Eh, pH, primary ore mineralogy and groundwater chemistry.

The most important precursor sulfide minerals in Cu-Pb-Zn ores are pyrite, chalcopyrite, galena and sphalerite, and in Ni-Co ores, pentlandite, pyrrhotite, cobaltite, chalcopyrite and Co-Ni arsenides. In the presence of oxygen and a catalyst, they react to generate a metal-bearing potentially acidic groundwater (chapter "The Geochemistry of Supergene Processes"). These solutions react with wall rock to form a wide range of supergene minerals as shown in Table 1. In some deposits, such as Lavrion in Greece, due to the potentially high number of minerals that may form involving Cu, Pb, Zn, Ni and As, there are literally hundreds of minerals in the oxide zone (approximately 700 at present, Vourlakos and Fitros 2019). Although oxygen in groundwater is an effective oxidizing agent, it decreases with depth and the ability of water to oxidize is reduced so that the redox potential of the water decreases, which favours the precipitation of the metal load (Baas Becking et al. 1960).

In addition to the usual process of weathering by meteoric water associated with gossan formation, volcanogenic massive sulfides can also be subject to oxidation through interaction with seawater (seafloor weathering, Hannington 1993; Ridley 2012). This process is accompanied by the formation of Cu sulfide minerals and Cu chloride phases such as atacamite (Hannington 1993). Within the transition zone, the mineralogy is mostly confined to chalcocite, bornite, covellite, digenite and enargite (Hannington et al. 1988; Hannington 1993). These phases may replace or be intergrown with primary sulfide minerals such as pyrite, chalcopyrite, sphalerite and galena.

Gold can also be enriched within the transition zone and the gossan assemblage of some deposits such as at Rio Tinto, Spain; Lavrion, Greece; Justice, Ghana; and Golden Grove, Western Australia. In more temperate zones Au has also been reported from gossans such as Murray Brook in Canada (Boyle 1996). The complexity of supergene mineral assemblages in gossans can lead to distinctive zonation, for example, at Lavrion, Greece (Fig. 14).

Initial alteration of the less resistant phases, chalcopyrite, tennantite-tetrahedrite and sphalerite, are replaced by secondary sulfides that include chalcocite, digenite, covellite, secondary galena and wurtzite. The low-temperature Ag sulfide, acanthite, may also form in this environment as can some Ag-Sb-sulfosalts (Sillitoe 2009b). As alteration advances, more stable sulfides such as pyrite may be altered. Dissolution of arsenopyrite and galena tends to be slow but the release of As leads to precipitation of stable arsenates,

Table 1 Comon Mineralogy of Gossans for different metals and depths

Weathering zone	Ore zone	Fe	Mn	Co	Ni	Cu	Zn	Ag	As	Pb
Soil	Oxide Gossan	Goethite Jarosite	Pyrolusite Manganite Hureaulite Purpurite	Heterolite	Honessite Jamborite Morenosite Nickelblödite Retgersite	Cu, cu-sulfates Cuprite	Tarbutite Frapionatite Hopeite Zn-apatite Zn-sulfates	Ag Ag-jarosite	As, trace element with Fe-oxides	Pyromorphite Anglesite Minium
Mottled clay		Goethite Jarosite						Ag Ag-halides		
Saprolite		Goethite Jarosite Hematite				Cu, cu-sulfates Cuprite Cu-chloride Delafossite				
		Goethite Hematite Fe-sulfate salts		Erythrite	Annabergite	Cu-arsenates, phosphates	Smithsonite Adamite Hemimorphite Willemite Rosasite		Scorodite Bukovyskite Kamarizaite Kankite Cu-arsenates Zn-arsenates Pb-arsenates	Pyromorphite Anglesite Leadhillite Linarite Mimetite Wulfenite
	Carbonate	Siderite	Rhodochrosite Pyrolusite Manganite Birnessite	Co-calcite	Nicholsonite	Malachite Azurite	Smithsonite			Cerussite
	Secondary Sulfide Transition	AVS-salts Marcasite	Alabandite Birnessite Hausmannite Hollandite Hübnerite	Co-calcite Cobaltite Clino-safflorite	Violarite	Chalcocite Covelline Digenite Spinkopite Yarorwite	Sphalerite Wurzite	Acanthite Silver Sulfosalts	Arsenolite Karibibite Schneiderhöhnite	Galena Bayldonite Mimetite Cerussite
Unweathered rock	Sulfide	Pyrite Pyrrhotite		Cobaltite Carrollite Ni-co Arsenides	Pentlandite Ni-pyrrhotite Ni-co- Arsenides	Chalcopyrite Bornite Carrollite Tetrahedrite	Sphalerite Tetrahedrite	Silver Argentite Sulfosalts Trace element in sulfides	Arsenian pyrite Arsenopyrite	Galena Pb-sulfosalts

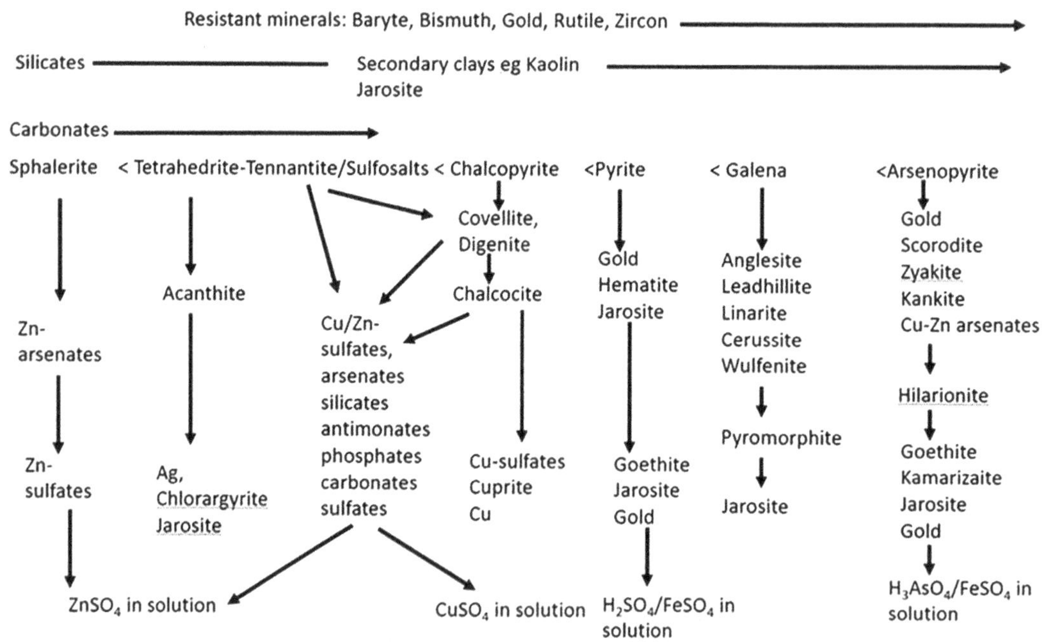

Fig. 14 Schematic paragenetic sequence of supergene alteration, Lavrion mines, Greece based on mineral identification. Mineral groups show weathering sequences for the main metals, sulfate, arsenate, phosphate, halide. silicate and carbonate minerals that may participate in mineral-water interactions but do not necessarily show the same paragenetic sequence as sulfides and weathering products

including scorodite, even at low pH (Majzlan et al. 2014). Coincident with sulfide mineral oxidation is the dissolution of primary carbonate and aluminosilicate minerals that are replaced by clays and silica. Mature gossans host a mineral assemblage of goethite, hematite, quartz, clays, and sulfate and carbonate minerals overlying the transition or supergene enrichment zone.

The basic weathering profile (as shown in Figs. 3 and 13) therefore comprises:

- Leached cap dominated by Fe oxides, quartz and simple clay minerals;
- Oxide zone dominated by metal oxides, with deeper zones enriched in carbonate and sulfate minerals;
- Transition zone dominated by secondary sulfides, particularly in Cu-rich systems.

This profile overlies the primary sulfide zone at depth. More complex zonation can occur, particularly in polymetallic deposits or where As, P and Sb are present (Table 1). The development of the weathering zones is highly variable and increases in development closer to the margins of the sulfide deposit and along structural zones where groundwater may penetrate deeper into the sulfide ore (e.g., Fig. 5c from Rosh Pinah Mine, Namibia).

The stability of the minerals will also be a function of climate, physicochemical environment and groundwater chemistry. Some minerals, such as changoite ($Na_2Zn(SO_4)_2{\cdot}4H_2O$), kröhnkite ($Na_2Cu(SO_4)_2{\cdot}2H_2O$) and salesite ($Cu(IO_3)(OH)$) are only stable in highly arid and/or saline environments. Others such as arsenolite or claudetite (both As_2O_3) are polymorphs that require oxygen-reduced environments.

Where crystalline sulfide minerals are coarse, it is possible that their textures (fabric) or crystallographic features can be preserved as perfect pseudomorphs. Such structures show the replacement of original fabric by secondary minerals through in situ precipitation of ferric oxyhydroxide minerals and the co-precipitation of other metal(loid)s.

Colloform precipitation of solution-deposited oxide minerals may preserve structures such as banding. Pseudomorphic textures after some gangue minerals may also form where precipitates form along grain boundaries and the gangue phase is dissolved resulting in epimorphic boxworks (Fig. 1b).

Such textures can be used to define gossans and distinguish them from other ironstones. However, if the primary sulfide mineralization is fine-grained or the gossan subject to prolonged weathering then such textures may not have formed or have since been destroyed. Overgrowths and replacement textures are indicators of incipient supergene alteration (Fig. 9a). A common physical feature is increased porosity associated with supergene zones. Spongy or porous textures have been noted from most gossans (for example Blanchard 1968). Cellular boxworks after sulfide minerals are a characteristic texture (Fig. 1a, e).

An extension of this increased porosity can lead to solution breccias that form after the dissolution of primary minerals. For example, a cemented Fe oxide matrix is shown in Fig. 9b. Poorly sorted layers of quartz or goethite sand or, in less weathered terrains, pyrite sands, may occur below gossans. It is postulated this form is a result of erosion of the sulfide zones through groundwater advection flow (Boyle 1996). Pseudomorph replacement and mineralization filling of sulfide relic forms may also occur and result in finer-grained material than the precursor sulfide lode as discussed in chapter "The Role of Regolith Formation and Evolution in the Development of Supergene Mineral Deposits".

5.4 Role of Groundwater in Gossan Formation

5.4.1 Groundwater Chemistry

Fluctuations in the water table allow the development of gossan textures at the expense of sulfide mineralization and this can occur seasonally, such as in a humid tropical climate or as a function of climatic or geological change. As discussed in Sect. 3 (above) for the Iberian Pyrite Belt, this can lead to a complex stratigraphy related to changes in paleoclimate, groundwater chemistry and tectonic events (Velasco et al. 2013; Scheiber et al. 2018). Thus, the weathering front moves downwards as the cathodic alkaline oxygen-reducing reactions remain nearer the surface and the anodic leaching and acid-producing reactions migrate (Fig. 13).

Where groundwater can penetrate further, through deep-seated faults or similar structures, then multiple gossans or oxidation zones may form at depth in a deposit, creating a complex sequence of primary, transition and oxide zones. These tend to be better developed in karstic environments where groundwater may penetrate more widely and create solution caverns in which oxide zones can form. The classic example of this is the Tsumeb polymetallic pipe in Namibia (Lombaard et al. 1986; Bowell 2014). Here a complex array of secondary minerals forms as a function of the polymetallic sulfide ore, variation in groundwater chemistry and variable oxidizing and reducing conditions.

As the Tsumeb system is Fe-poor, the matrix is supported not only by Fe-oxyhydroxide minerals but also by Cu-Pb-Zn-Ge oxide, hydroxide, arsenate, carbonate, silicate, sulfate, molybdate and halide minerals. An important control is the change in groundwater chemistry in contact with the deposit (Fig. 15); because the base metals have different oxidation behaviours, these changes lead to zonation of the oxide zone (Fig. 16).

Each of the waters has a different composition) and is associated with different supergene mineral assemblages (Bowell 2014). Throughout all the oxidation zones, metal carbonates are common and occur within sulfide zones as well. Sulfide mineralogy changes from chalcocite to tennantite-dominated with the middle levels showing various levels of sphalerite, galena, chalcopyrite and Ge-sulfide minerals (Fig. 16).

A simpler mineral assemblage, although no less impressive, has been formed by groundwater that penetrated through a thick sequence of Quaternary clay and Carboniferous to Devonian sedimentary units (100–150 m thick) to oxidize the

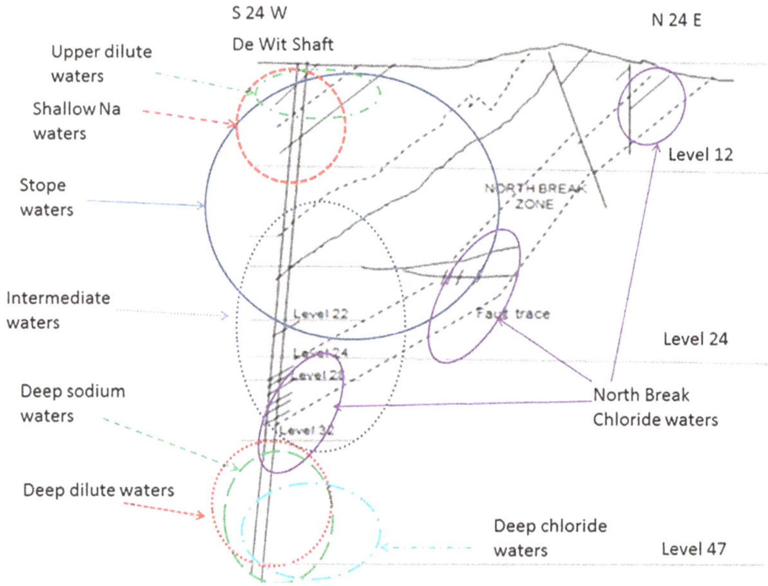

Fig. 15 Zonation of hydrogeochemical water types at Tsumeb (from Bowell 2014). Levels refer to drifts within the mine workings

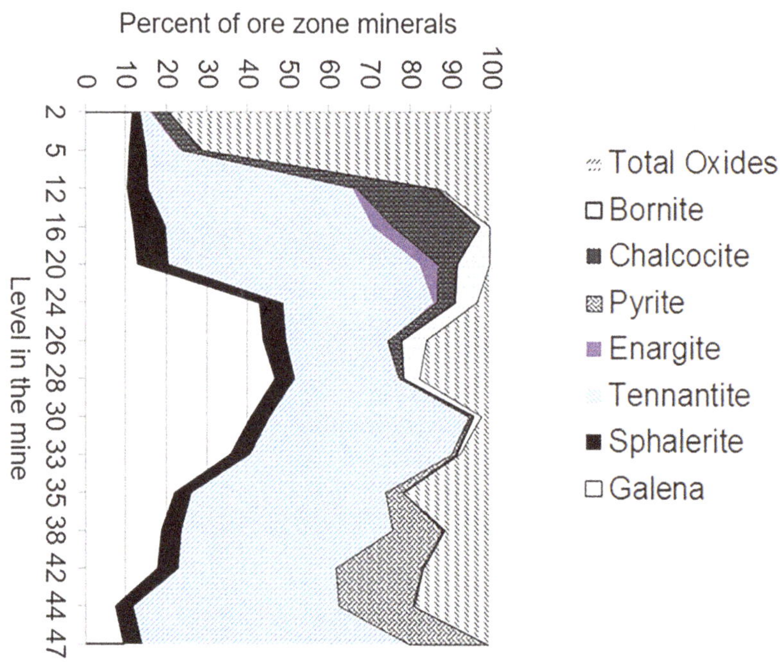

Fig. 16 Approximate proportion of oxide zone minerals at different depths in the Tsumeb Mine (from Bowell 2014)

Rubtsovskoe Cu-Ag deposit, North-West Altai, Russia (Pekov and Lykova 2011). Here, weathering under palaeo-tropical conditions during the Neogene-Pleistocene led to deep circulation of groundwater and penetration to depths of 150 m where descending meteoric water contacted Cu sulfide mineralization. The mineralization contains large nuggets of native Cu, pristine cuprite crystals and a rare assemblage of Ag and Cu iodide minerals (marshite, miersite and iodargyrite, Pekov et al. 2011). The complexity of the supergene zone is also reflected in the mineralization of the Broken Hill orebodies in New South Wales, Australia (Birch 1999).

5.4.2 Case Study Broken Hill, Australia

Introduction The Broken Hill orebodies represent one of the largest known geological resources of base metals comprising more than 280 million tonnes of ore (Stevens 1999; Spry and Teale 2021). The orebodies were formed approximately 1685 million years ago when seawater, heated by magmatism associated with volcanic activity, flowed up through the underlying rocks to the seafloor where it mixed with cold ocean water, forming clouds of black sulfide precipitates (Stevens 1999). These collapsed to form layers of sediment rich in Zn, Cu, Pb and Ag sulfide minerals (Plimer 1999). Erosion over the past 500 million years has exposed these minerals to weathering. Regolith development over this period has led to a complex stratigraphy and mineral assemblages, with more than 1500 minerals reported (Birch 1999).

Physical Setting Broken Hill is in the far west of New South Wales, Australia. The area has a semi-arid climate, with relatively cool winters and highly variable, mostly hot and dry summers. The mean maximum temperature during summer is about 33 °C, with an average of 25% humidity, although storms and cooler weather do occur. The average rainfall is 235 mm pa.

History Charles Rasp staked the first claim on the Broken Hill outcrop on 5 September 1883, believing it represented a cassiterite orebody. With others, he staked the entire outcrop and in 1884, the syndicate reorganized as the Broken Hill Mining Company. Silver was discovered in 1885 and the Broken Hill Proprietary Company Limited was organized. What Rasp had discovered was the gossan of massive Pb-Zn sulfide ores on a feature known as Broken Hill. Rasp reported finding massive galena, sphalerite, cerussite and other oxide minerals, but was most concerned with the galena, a primary source of Pb.

Broken Hill was exploited initially by small prospectors working the gossan for the abundant Ag and galena with the ore rich enough to justify shipping to Wales, UK for processing. Open-pit mining of oxide Ag-rich ores continued this trend until 1898 when underground mining produced Zn-Pb-rich concentrates that were exported. By 1915, a smelter was established, and mining development continued at the surface and underground both to the north and south.

Consolidation of smaller claims eventually resulted in the formation of a single company, Broken Hill Proprietary Company, which ceased mining operations in the area in 1972. The leases were acquired by Minerals Mining & Metallurgy (MMM). Underground mining continued until 1976 and open pit mining concentrated in the Blackwood Open Cut, which commenced in 1973, and the Kintore Open Cut. Pasminco acquired the leases of Zinc Corporation and New Broken Hill Consolidated in 1976 and still produces some two million tonnes of ore pa.

Geology The Broken Hill ore body is hosted within the gneisses of the Willyama Supergroup, a Mesoproterozoic sequence of sillimanite gneisses (Figure 14). The terrane in which Broken Hill is hosted has undergone several metamorphic deformations at amphibolite facies. This has resulted in the elongation of the sulfide mineralization into a "boomerang" morphology with variations in the proportion of Zn- and Pb-sulfide-rich deposits related to the effects of metamorphic remobilization and deformation.

The primary ore consists of massive, commonly recrystallized, sphalerite-rich, galena-sphalerite and galena-rich sulfide lenses in folded horizons. The ore can comprise up to 100% Pb-Zn sulfide minerals (van der Heyden and Edgecombe 1990). The ores have a distinctive mineralogical and chemical composition, with ore boundaries parallel to the bedding planes. Broken Hill is classified as a SEDEX deposit that has been extensively reworked and modified by metamorphism and shearing (Mackenzie and Davies 1990; Plimer 1999; Spry and Teale 2021).

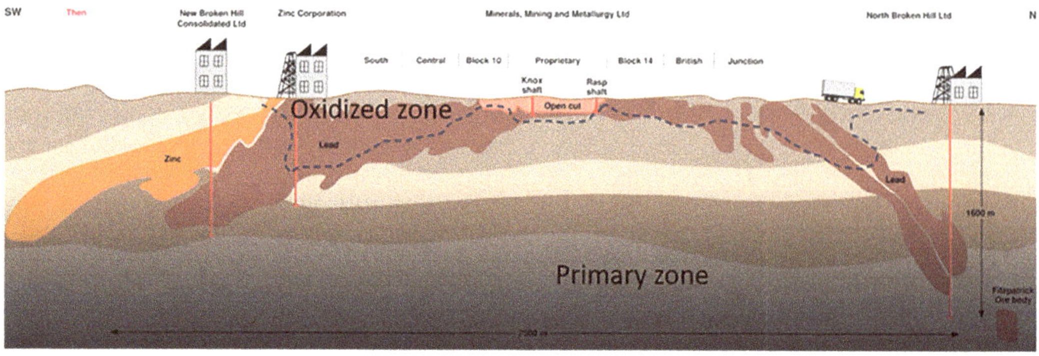

Fig. 17 Schematic cross-section of the Broken Hill deposits (based on work by Plimer 1984, 1999; Birch and van Heyden 1997)

Formation of the Supergene Deposits Erosion over the past 470 million years exposed sulfide minerals to multiple stages of chemical weathering, including the Permian ice age. Global warming was followed by fluvial sediment deposition in a humid tropical environment. Intense lateritic weathering from 60-20 Ma resulted in a thick oxidation zone (Van Moort and Swenson 1981; Plimer 1984). This led to the formation of a complex suite of secondary minerals including smithsonite, cerussite, Ag halides, and more than 260 other minerals (Birch 1999). There is, therefore, a highly varied zone of oxidation in the Broken Hill area with no uniform extensive zone of supergene mineralization (Fig. 17).

Based on the landscape and gossan textures, the age of weathering has been estimated as Tertiary to Quaternary (Van Moort and Swenson 1981; Plimer 1984). Based on mineralogical and geological studies, formation during the Holocene has more recently been proposed (Birch and van der Heyden 1997). Although no direct dating of secondary minerals has yet been undertaken it is likely that due to the complex evolution of the landscape and multiple climatic changes, most preserved secondary minerals are only from the Holocene and that earlier assemblages have largely been reworked.

The secondary zone extends to a depth of 75 m with isolated oxidation channels to 85 m. Oxidation channels are most developed in Pb-rich lodes, where they reach 180 m depth, allowing for deep penetration and lateral flow of oxygenated groundwater (Plimer 1984). Secondary minerals have precipitated in these structures, enclosed by primary sulfide ore. The oxide zone hosts a complex array of carbonate, sulfate, phosphate, arsenate, halide and oxide minerals, together with native metals. The grade varies widely, with the coronadite-quartz gossan containing grades of 10–15% Pb and 900 g/t Ag contrasting with average sulfide ore grades of 4% Pb and 9500 g/t Ag (Plimer 1999). Zonation of the oxide ore (Fig. 18) is based on detailed mapping and mineralogy and includes suites of arsenate, phosphate, and sulfate minerals (Birch and van der Heyden 1997).

The complexity of the different zones can be observed from the mapping and reflects highly localized controls on mineral stability and hydrogeochemistry within the same ore body, both laterally and with depth (Fig. 18). These represent localized episodes of mineral crystallization. Textural variations are equally complex with massive, fracture fill, stalactitic, botryoidal, recemented impregnations and cavern fills. Minerals are overprinted, partially dissolved and psuedomorphed (Birch 1999). Isotopic studies on Mn carbonate minerals indicate that groundwater chemistry has been highly variable over the last 300,000 years (Boettcher et al. 1993).

The lack of a well-preserved, thick, uniform supergene zone rather than a highly variable oxidized zone indicates the palaeo-water table changed over time above the current base of

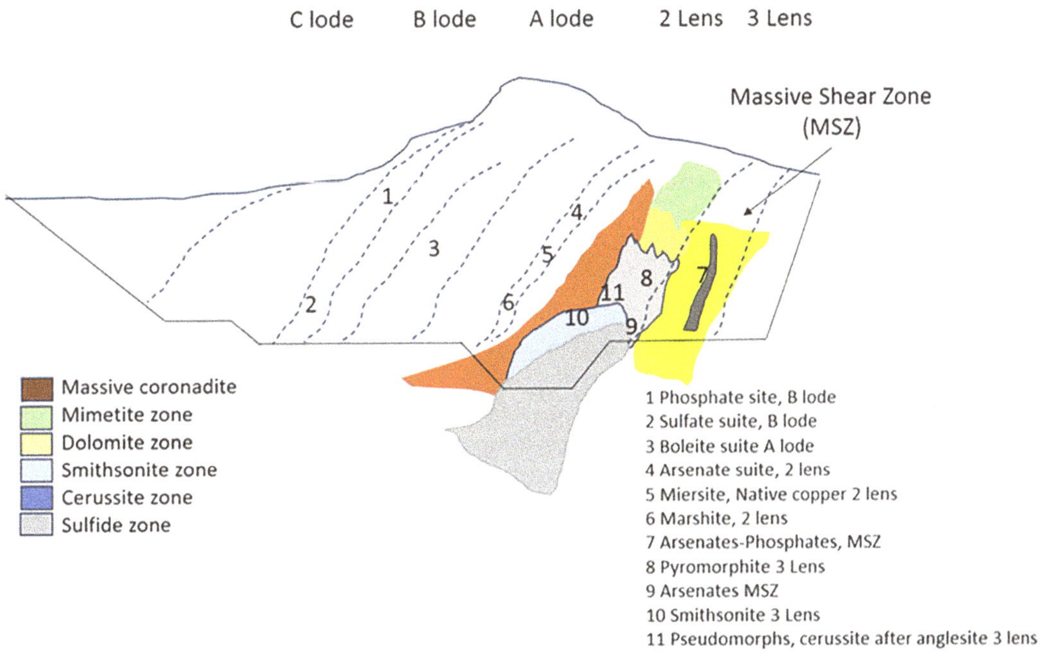

Fig. 18 Cross-section of Block 14 opencut showing projected positions of various mineral assemblages in the oxidized zone of the Broken Hill deposit (based on work by Plimer 1984, 1999; Birch and van Heyden, 1997)

oxidation. The present water table is 100 m below surface and is subject to seasonal and diurnal variation that can be up to 20 m (Plimer 1999).

Ultimately, the mineralogical variation and textural complexity of the Broken Hill oxide zone, the variable oxidation depth and lack of uniformity in the supergene zone as well as geological evidence on uplift and landscape evolution all point to a long history with multiple oxidation events under different climatic conditions.

5.5 Role of Geomorphology and Climate in Gossan Formation

5.5.1 Geomorphological Control on Gossan Formation

In arid environments, the geomorphological control on gossan formation can lead to the preservation of a complex stratigraphic sequence with zones of formation of different mineral phases and strongly enriched secondary zones, for example, Kambalda, Australia (e.g., Thornber 1975a, b). In humid tropical climates, reworking of the oxide zone can overprint or remove the surface and near-surface expression of a gossan leaving only a diffuse trace or none of potential deeper mineralization (Butt 1995). Historically, base metal discoveries have targeted surface expressions such as gossans or enrichment zones to identify mineral deposits. Mapping of gossans has been successfully used in many arid environments for this purpose (Blanchard 1968; Butt 1995; Taylor 2011). By comparison, in tropical regions, the recording and identification of gossans has been considerably less. Most discoveries have required stream sediment surveys followed by geophysics to identify blind targets. Deposits such as Perkoa in Burkina Faso are considered rare by comparison. The reasons for this are many and could be:

- Erosion and intense leaching of near-surface anomalies;

- Failure to recognize ferruginous gossans or detrital products in Fe-rich lateritic terrain;
- Failure in the sampling applied to identify anomalies.

Climate and relief play an important role in the formation and preservation of gossans. In terms of climatic effects, temperature is important in the promotion of sulfide oxidation rates and rainfall will control the rate of leaching. The intensity of leaching and the secondary dispersion of elements from the gossan are influenced by the nature of the relief or geomorphology. Deeply weathered terrains evolve over long periods and have been subject to changes in climate and geomorphology (chapter "The Role of Regolith Formation and Evolution in the Development of Supergene Mineral Deposits"). Gossans appear to be best preserved within arid to mildly temperate environments in areas of low to moderate relief. Sulfide oxidation and element leaching are most intense in rainforest and seasonally humid tropical climates, but gossan preservation appears to be uncommon (Butt 1995). This could be due to gossans being obscured in dense vegetation or considerable reworking, possibly involving organic acids, resulting in dissolution of Fe oxides (Fig. 19). Gossan textures can be lost as Fe is remobilized to form Fe-rich soils.

A common challenge in the evaluation of gossans, or even tropical regolith, is quantifying the extent to which there has been profile reduction, erosion and/or chemical remobilization and alteration. This is particularly true in more humid environments where active lateritization is reworking an existing regolith profile. In southwest Ghana, it has been established that profile reduction has occurred with the removal of the ferruginous duricrust by weathering under a recent more humid tropical climate (Bowell 1993, 1994). Profile reduction has been accompanied by physical erosion of the landscape and, due to the low relief in the area and high vegetation, it is difficult to identify or distinguish residual duricrusts, ironstones or gossans. Much like other regolith development reactions, newly precipitated Fe-oxyhydroxide minerals are initially less crystalline due to excess water and reorganize over time to produce more rigid structures. Reworking of these textures may occur, particularly in seasonally humid climates. Over time the more mobile (adsorbed, soluble) components are completely removed from the gossan profile, for example in the development of the Justice Mine gossan in Ghana (Bowell 1994).

The development of gossans under different climates can be better described by the assessment of contrasting case studies from West Africa. The first, Perkoa in Burkina Faso, formed in an arid environment where the regolith stratigraphy is better preserved and the second, Justice in Ghana, formed where a more tropical humid environment has led to reworking of the surface anomaly.

These examples reflect the preservation of gossans, which tends to occur more in arid environments than in humid ones (Gahlan and Ghrefat 2017). In humid environments, continued lateritic weathering obliterates the primary zone stratigraphy in the near-surface and potentially masks the surface geochemical expression. This reflects the recent nature of soil processes that are ongoing in humid environments compared to more arid or savanna environments where regolith profiles have not been subject to significant reworking (Bowell et al. 1996).

5.5.2 Case Study: Gossan Development at Perkoa, Burkina Faso (Savanna Climate)

Introduction A deep regolith has formed in Burkina Faso over the Perkoa deposit, facilitated through alternating wet-dry seasons over a long period (Séranne 1999). Such climatic variations have resulted in the development of multiple duricrust surfaces indicating successive erosional periods with intermittent periods of regolith development through chemical weathering (Michel 1973; Grandin 1976; Chardon et al. 2006; Burke and Gunnell 2008). Relicts of each palaeo-land surface in the sequence are commonly capped by residual and/or transported ferruginous duricrusts. The Perkoa metamorphosed VHMS deposit in Burkina Faso is the only known example of this type of mineralization in West

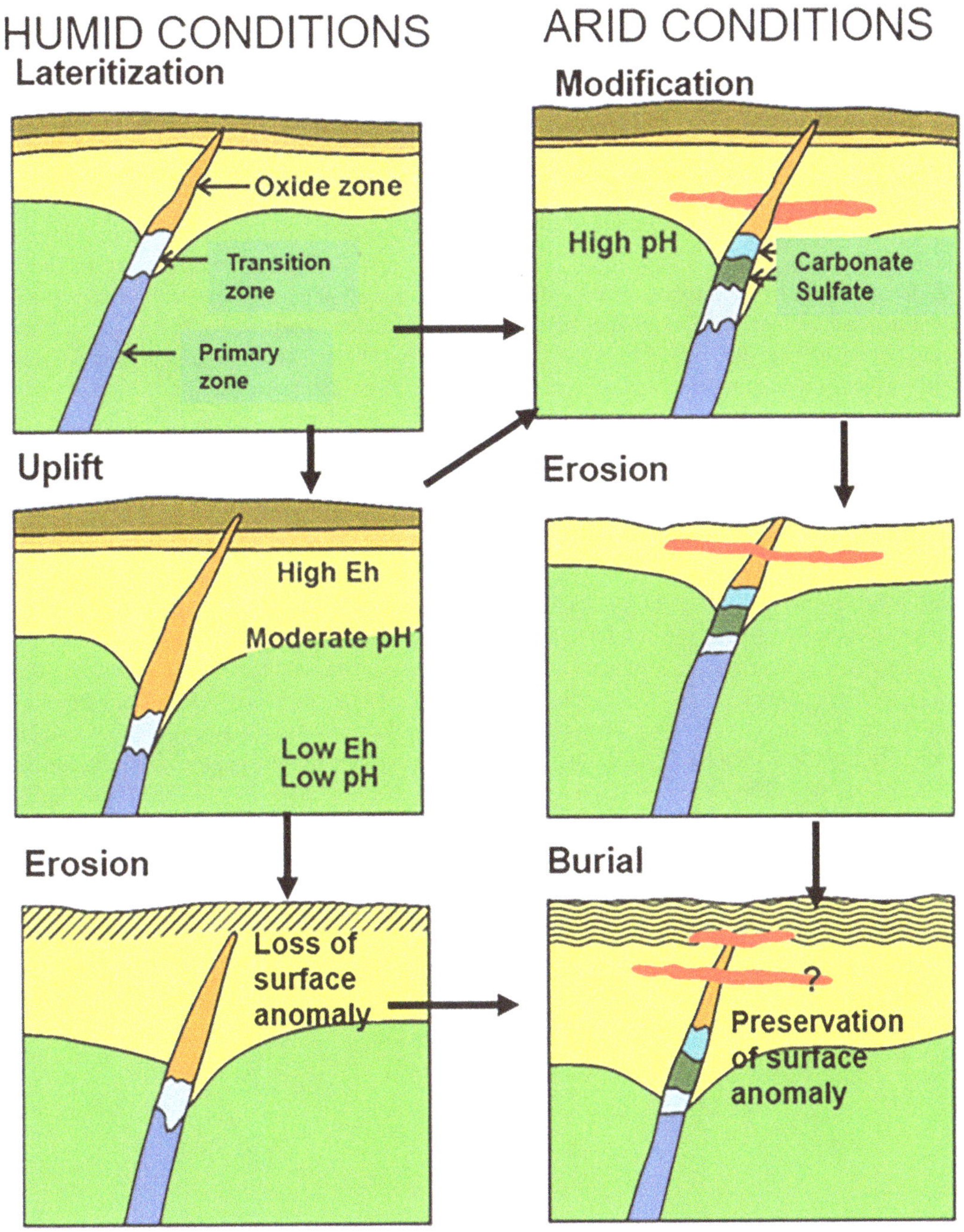

Fig. 19 Gossan formation and evolution under different climate and geomorphological environments (after Butt 1995; reproduced with permission, CSIRO)

Africa. The deposit is unique in West Africa because of the dominance of Zn, and it is capped by a relic gossan (Kříbek et al. 2016). The estimated ore reserves are 6.3 Mt. at 14.5 wt.% Zn, 8.0 wt.% Ba, 0.06 wt.% Pb, 24 ppm Ag and 50 ppm Cu (Schwartz and Melcher 2003).

Geology The Perkoa deposit is in the Palaeoproterozic Baoulé-Mossi Domain of West Africa, hosted by the Lower Birimian greenstone belt (Milési et al. 1992). The sequence comprises andesite, rhyolite and rhyodacite tuffs with intercalated schists with actinolite-clinozoisite and biotite-garnet assemblages. The upper Birimian assemblages of the belt comprises silicified andesites and carbonaceous schists. The deposit is cut by diorite and andesite dykes (Schwartz and Melcher 2003). Sulfide mineralization occurs as two ore bodies, both of which were subject to metamorphism and deformation with peak metamorphic temperatures above 460 °C and at pressures of 1 kbar or more. Temperature and pressure estimates are based on the presence of almandine-rich garnet and andradite, as well as the arsenopyrite-pyrrhotite geothermometer (Schwartz and Melcher 2003). The main ore body is approximately 400 m long and has an average thickness of 10 m, whereas the hanging-wall orebody is thinner with an average thickness of 4 m. Mineralization is dominated by sphalerite with lesser galena, chalcopyrite, arsenopyrite, pyrrhotite, pyrite, cinnabar, acanthite and sulfosalt minerals. Gangue minerals are mainly quartz, calcite and albite, with minor barite, siderite, fluorite and chlorite.

Gossan Characteristics Three separate facies have been described in the Perkoa deposit (Kříbek et al., 2016); massive, banded and brecciated forms of gossans (Fig. 20). The massive gossan shows colloform to cellular textures with well-developed boxworks. A crust of Fe oxyhydroxide minerals fills pore spaces within this unit. The banded facies are limited in distribution and is interpreted to represent the precursor massive sulfide. Brecciated gossans are the most common, particularly at surface. A strong alteration halo flanks the gossan with bleached rocks. Sporadic areas of high Mn most likely reflect reworked hydrothermal alteration and oxidized equivalents including disseminated sulfide minerals that flank the more massive ore bodies.

The mineralogy of the gossan is dominated by hematite and goethite along with clay minerals, brucite and sulfate minerals (corkite, alunite, natroalunite, hinsdalite and hidalgonite). Residual quartz contains inclusions of chalcopyrite, sphalerite and cinnabar (Kříbek et al. 2016). Compared to the primary sulfide ore, the gossan is enriched in As, Ba, Pb, P and Sb and depleted in Ag, Co, Cu, Hg, Mo, Ni and Zn (Schwartz and Melcher 2003; Kříbek et al. 2016).

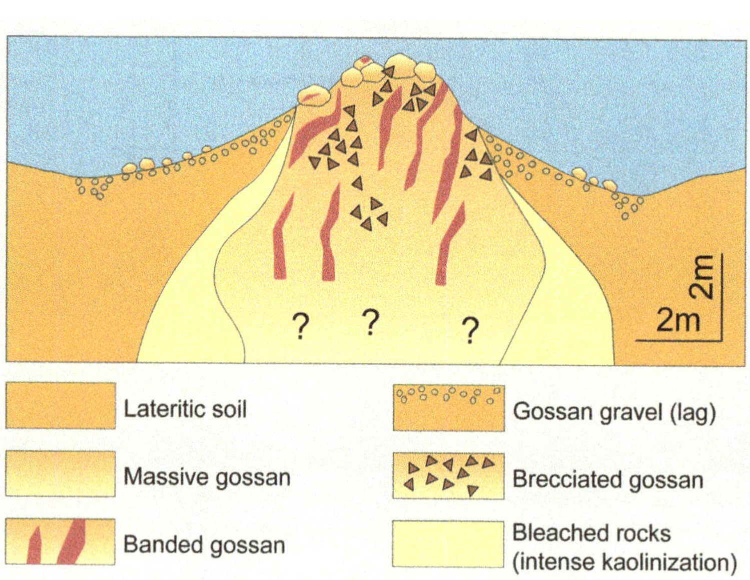

Fig. 20 Schematic illustration depicting the position and structure of the Perkoa gossan and its bleached envelope (from Kříbek et al. 2016)

Iron-rich Soils The near-surface is comprised of Fe-rich crusts and soils derived from the weathering of the deeper horizons, primary or pre-existing regolith or through erosion, transport, and concentration of weathering products. At the surface there is widespread Fe-rich gravel and, in low-lying areas, a surface soil containing humidified organic matter, a bleached zone and more clay-rich mottled soils; this has been classified as a regosol (Kříbek et al. 2016). Where the gossans are compared to the Fe-rich crusts, they are significantly enriched in most ore-related elements, namely As, Pb, Sb and Zn and to a lesser extent Ag, Ba, Cd, Mo, S and Sr.

Geochemical Characteristics The detailed study of gossan geochemistry by Kříbek et al. (2016) shows that the gossan tends to have lower Zn (< 500 ppm) than the primary orebody (> 5000 ppm), but higher concentrations of Pb (up to 3.1 wt.%) and As (up to 872 ppm). This is consistent with enrichment through profile reduction and concentration of residual phases for Pb and As, whereas Zn is mobile and lost from the system. The relative enrichment of elements in the gossan compared to the primary sulfide ore was assessed by normalizing the primary concentrations to Ti and Zr, which are generally immobile in the weathering environment (Fig. 21).

The normalized data reflect enrichment of Pb, As, Sb, Bi and Ba and depletion of Cu, Zn, Ag, Ni, Co, Cd, In, Se, FeO, MgO, MnO, CaO and S in the gossan compared to the primary sulfide ore. Furthermore, the light rare earth elements (LREE; La, Ce, Pr and Nd) are relatively enriched, whereas the rest of the REE are depleted. This

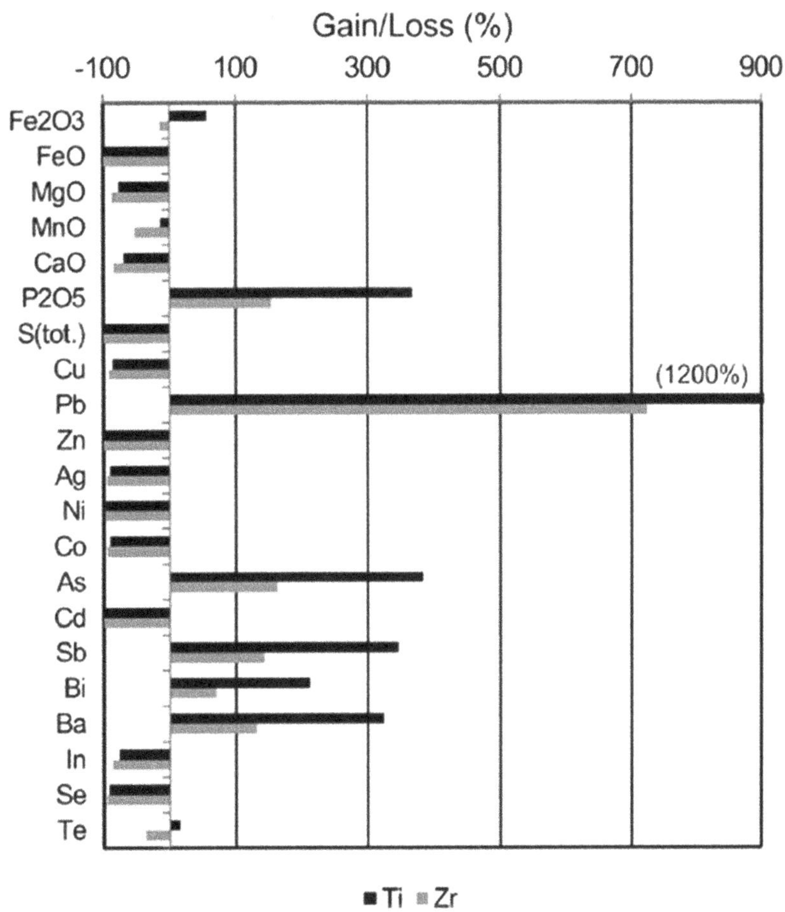

Fig. 21 Changes (in %) in the concentration of some chemical elements/oxides in the Perkoa gossan relative to the primary sulfide ores. The calculation is based on Ti and Zr as immobile elements. Median values of primary ores (n = 7) and gossan (n = 8) were used for the calculation (from Kříbek et al. 2016)

LREE enrichment in the gossan relative to the primary sulfide ore is probably due to a strong affinity of the LREE to Fe-oxyhydroxide mineral surfaces. Compared to the Fe-rich crusts, gossans have higher concentrations of most trace elements, (especially As, Pb, Sb and Zn) by more than an order of magnitude.

Summary The Perkoa VHMS generated gossan consists predominantly of hematite and goethite, with an admixture of hydrous Mn-phases, sulfate minerals and clay minerals. The gossan shows characteristic textures that allow distinction from surrounding Fe-rich soils and particularly from barren Fe-rich soils in the wider area. The gossan has a distinctive composition, with high concentrations of As, Pb, Sb, Zn, Ag, Ba Cd, Mo and Sr compared to the Fe-rich soils. The high mobility of Zn results in little difference in fractionation or isotopic composition of Zn in the different materials compared to Cu and Pb.

From this, it can be interpreted that in the contrasting wet-dry climate in the Perkoa area, following erosion and continued weathering, was largely of an in situ regolith that with little evidence of lateral mobilization of anomalies or transport of material. Conseuqently the surficial Fe duricrust reflects the underlying geochemical characteristics of the precursor material. The exception to this appears to be Zn, which is more mobile than Pb or Cu at neutral pH and has been depleted.

5.5.3 Case Study: Gossan Formation in the Justice Gold-Sulfide Deposit, Ghana (Humid Tropical Climate)

Introduction An atypical example of gossan partially preserved in a humid tropical climate occurs at the Justice Mine, Ashanti, SW Ghana (Talbot et al. 1984). The Mine is situated in the central part of the Ashanti concession and has been worked continuously for 80 years.

Geology Mineralization at Justice occurs in shear zones (3 m wide) containing Au-bearing, matrix-supported sulfide minerals in meta-tholeiites (Junner 1932; Bowell 1994). The shear zones contain up to 50 % sulfide minerals in the lode, mainly pyrite and arsenopyrite (that hosts much of the gold), but also includes chalcopyrite, bornite, galena, pyrrhotite and sphalerite. The Au-bearing shear zones are defined by a multivariate anomaly of Au-Ag-As-Cu-Pb-Zn, whereas the more extensive zones of wallrock alteration are characterized by a Au-As-Cu-Rb-K-S-$C0_2$ halo; importantly, these haloes propagate into the overlying regolith (Bowell 1994).

Gossan Characteristics The gossan is observed in the Justice open pit as an Fe-stained shear zone in which the steep dip (80o–90o W) can be readily observed (Fig. 22). Within the gossan, boxwork and pseudomorph textures after sulfide minerals can be observed, rarely associated with Au or relict sulfide grains encapsulated in goethite

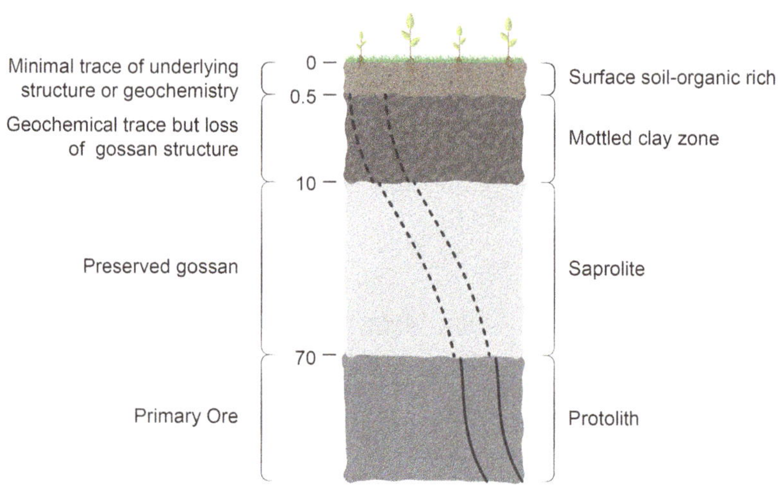

Fig. 22 Regolith Profile, Ashanti region, Ghana (Bowell 1993). Depth scale in metres

and quartz. During sulfide oxidation, arsenopyrite has been replaced by amorphous and crystalline Fe-Mn arsenate minerals, goethite, hematite and arsenolite in box- and ladder-work textures. In the extremely weathered mottled clay zone, goethite, hematite and minor scorodite are present as pseudomorphs of oxidized arsenopyrite.

In the saprolite, sulfide minerals become unstable because of increasing Eh and are replaced by goethite, hematite and secondary minerals that form box- and ladder-work textures. With further oxidation, the "arsenopyrite-dominated" gossans form a porous granular assemblage of loosely aggregated but firmly cemented Fe oxide grains, owing to the partial breakdown of scorodite. In the more common, "mixed-sulfide" gossans, a complex cellular structure is formed, and the gossan consists of a series of weathering-derived products dominated by goethite. In the extremely weathered portion of both gossan types, goethite, scorodite and hematite pseudomorph oxidized arsenopyrite, and these pseudomorphs are used as a visual pathfinder for Au-As mineralization.

Soil In the soil horizons, mineralization can only be delineated geochemically, as all physical features associated with the mineralization have been lost. The profile shows truncation of the typical lateritic regolith profile due to Tertiary uplift and erosion and consequently is relatively simple (Bowell, 1993) with lateritic duricrust generally absent (Fig. 22).

In general, the gossan features are overprinted in the regolith, particularly near surface. Deeper in the saprolite, some remnants of matrix sulfide mineralization can be preserved. In the upper saprolite and organic-rich soils, some nano- to micron-sized Au grains have crystalline habits and are composed of pure Au with no variation between core and rim (Bowell 1993), possibly indicating a supergene formation (Hough et al. 2009), whereas other grains show evidence of Au-enriched rims with Ag varying from 5–10% in the centre to less than 1% in the rims (Bowell 1993; Bowell et al. 1993). Such Ag-depleted zones are typically associated with grain boundaries and on the edge of gold grains or within fractures in the grains. Similar textures have been reported from coarse Au nuggets and are a common weathering effect (Hough et al. 2009).

Elements such as Ca, Na and K show a strong depletion as intensity of weathering increases at both sites. This reflects the instability of primary silicate minerals (Velbel 1989, 1992). Trace elements such as Mn are largely depleted in the profile, although at both sites Mn enrichment is observed in the zone of ferrolysis. Several elements, e.g. Nb, Zr and Ti, appear to be essentially immobile although tending to be enriched in surface soils. Trace metals such as As, Sb, Cu, Pb and Zn show a general depletion as the intensity of weathering increases closer to surface. The enrichment of As and Sb may be due to the adsorption of both elements as oxyanions onto Fe-oxyhydroxide minerals in mildly acidic to near-neutral soils.

To estimate the extent of weathering and erosion of the Justice gossan, a series of mass balance equations were undertaken (see Brimhall et al. 1985; Brimhall and Dietrich 1987; Alpers and Brimhall 1989; Bowell 1994). An assessment of element retention and loss in the weathering profile of the Justice sulfide-rich lode determined that Au mobility is limited in the Ashanti area, with significant enrichment only observed in the lower saprolite-saprock interface (Bowell et al. 1993; Bowell 1994). The concentration of Au and Ag in the gossan and soils shows subtle variation but they are mass-balanced compared to hypogene values. However, Au enrichment at the water table and loss of Au and Ag in the surface soils can be observed. Similar patterns of Au enrichment in surface soils have been recorded elsewhere in Africa (Brimhall et al 1988; Zeegers and Leduc 1991; Colin and Vieillard 1991; Colin et al. 1993; Freyssinet 1994). The truncated nature of the weathering profiles in the Ashanti region distorts this pattern and makes further predictions of the total mass of transferred Au impossible to obtain. The reduction in specific gravity in the regolith is accompanied by leaching and leads to increased porosity. At Ashanti, this can vary from almost zero, due to a large influx of silica, to 40%, where sulfide leaching occurs without concomitant silica or Fe oxide precipitation.

Controls on porosity change influence the textural evolution of the gossan (Blain and Andrew 1977). For example, if all the Fe from sulfide mineral oxidation is precipitated as goethite, then a volume reduction of 80% will occur in the replacement of bornite by goethite and 21% for goethite alteration of arsenopyrite (Blanchard 1968). Much of the precipitated goethite is fine-grained, possibly colloidal, leading to a high surface area and a greater adsorption capacity for leached metal(loid)s. The influx of externally derived (exotic) goethite would complicate this qualitative relationship, leading to a supergene mineralogy markedly different from that of the primary sulfide matrix.

The geochemical trends at Justice indicate that some lateral reworking has occurred in the surface soils. This greatly reduces the connection of surface soil geochemical trends to those in the underlying gossan and primary mineralization. It in turn reflects the truncated nature of the regolith profile in the Ashanti region, which developed during the Mesozoic. More recent surface soils cover the profile that has been subject not only to in situ weathering but also some lateral mobilization.

6 Geological and Geochemical Characteristics of Gossans and Ironstones

Traditional exploration for gossans has emphasized the role of gossan identification based on macroscopic features such as classic boxwork textures, the presence of secondary base metal minerals and field relationships, coupled with follow-up microscopic features (Locke 1926; Blanchard 1968; Blain and Andrews 1977; Thornber and Taylor 1992; Taylor 2011). This form of exploration is largely historic as most regions where gossans are prominent have been explored and readers are referred to these texts for details. However, geochemical evaluation remains relevant as even buried or modified gossans may still preserve some of the chemical features of precursor mineralization (Bowell 2013). In addition, exploration in the Middle East and the Himalayas has identified gossans based on textures, geochemistry and hyperspectral response (Volesky and Stern 2001; Farooq and Govil 2014; Gahlan and Ghrefat 2017).

Geochemical zonation with the development of primary and secondary dispersion haloes is a common feature in many mineral deposits. Some, such as VHMS deposits, have a distinct vertical zonation from Cu- through polymetallic- to Ba-rich zones (Ridley 2012). Both primary and secondary dispersion patterns can be preserved in a gossan but are commonly overprinted by surficial hydromorphic remobilization that can greatly expand the geochemical halo around a source (Fig. 23).

Relative mobilities of geochemical components are dependent on groundwater chemistry, climate and other constituents in solution. The reactivity of elements and molecules will occur at different rates and under different conditions, so they can thus be segregated in the supergene environment.

Ironstones can generally be distinguished from gossans by a lack of relic sulfide boxwork textures, lack of chalcophile elements (such as Cu, see Fig. 23) and absence of diagnostic resistant minerals such as cassiterite (e.g., Boyle 2003). Conversely, some ironstones are enriched in lithophile elements such as Cr and V. Base metal enrichments in ironstones developed on lithological contacts can have geochemical similarities to gossans. For example, Fe and, in places, Mn oxide minerals precipitated along the contacts between dolerite and shale may scavenge Cu and Zn. These outcrop as gossan-like ironstones ("pseudogossans") and can contain up to 1800 ppm Cu or 600 ppm Zn, similar in abundance to gossans over VHMS deposits (Butt 1995). In addition, the lack of a hydromorphic

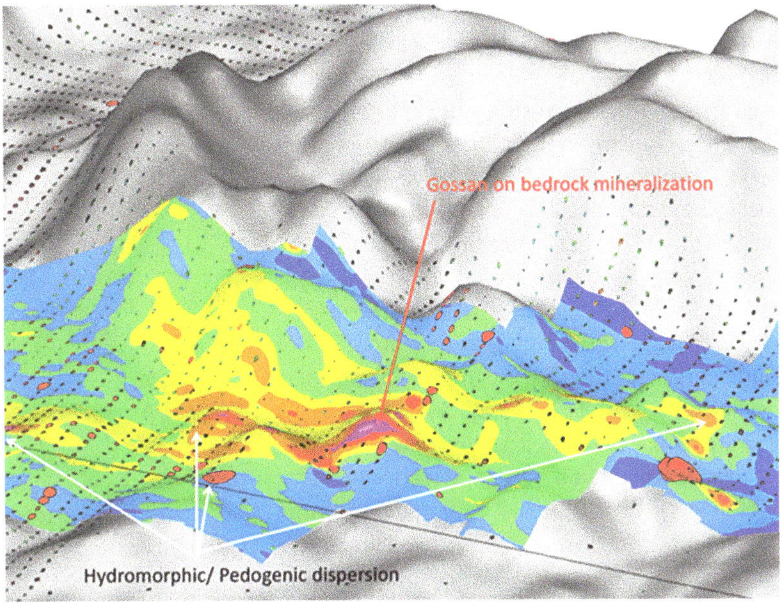

Fig. 23 Schematic Cu dispersion, Solwezi, Zambia

dispersion halo around the ironstones is indicative of pseudogossans but is not a diagnostic criterion.

7 Summary

Sulfide mineralization in deeply weathered terrains has generally been subject to significant chemical weathering and erosion. Gossans and the surface residual structural relics of sulfide mineralization are important exploration indicators of deeper mineralization. The evaluation of such features should involve mapping, mineralogy, structural assessment and geochemistry to define the features better and place them into the context of geomorphology and climate.

The preserved features in a gossan are a function of the receiving environment's physical-chemical conditions and the climate and topography. Except in extremely acidic environments, ferric Fe cannot stay in solution and so precipitates typically within proximity of the liberation as ferrous Fe from sulfide oxidation. As such, ferric Fe can replace or pseudomorph the structure of the primary mineralization and maintain these relic features even in mature gossans.

Chemical weathering of sulfide mineralization produces a series of characteristic mineralogical and geochemical features. In the saprolite, sulfide minerals become unstable because of increasing Eh and are replaced by goethite, hematite and secondary minerals forming box- and ladder-work textures. Intense alteration leads to the formation of porous granular gossans. Higher in the saprolite, these structures begin to break down due to reworking from a fluctuating water table. Truncation of the surface soils and extensive reworking remove most traces of underlying mineralization with only subtle geochemical anomalies observed in the reworked soils. Deeper in the saprolite, the gossan textures and composition are preserved and it remains possible to distinguish between mineralized and unmineralized zones in the deeper parts of the regolith. In more humid environments, due to recent and perhaps current conditions being dynamic and remobilizing chemical elements coupled with profile truncation, the surface soils do not directly reflect the underlying geology. In extremely weathered and reworked regolith zones, recognition of bedrock geology and geochemistry is hampered by the loss of relict structures, such as gossans, and the associated geochemical signatures which are much more subtle and, in some cases, may be lost completely.

The composition of a gossan is a reflection, in part, of the primary geochemical and mineralogical anomalies associated with a particular style of mineralization, any hydrothermal alteration and the physico-chemical conditions of the environment. Despite this, any evaluation should not be focused on the absolute chemical abundances within the gossan but rather on the nature of the element suite and the mineralogy that helps guide knowledge of the primary ore deposit. For example, to distinguish lateritic ironstone from Ni-Cu sulfide mineralization the diagnostic features may be the presence of As or platinum group elements that are absent from the ironstone but occur even in low concentrations in the associated gossan (Thornber et al. 1981). Developments in hyperspectral analysis are likely to become more useful in regional targeting but follow-up surveys will still benefit from the identification of physical textures, mineralogy and geochemical expression related to gossans and the need to use these to distinguish such features from pedological ironstones.

Acknowledgements The authors would like to thank Stephen Phillips for expert drafting in this chapter. The input of authors who kindly approved the use of their published material in this chapter is greatly appreciated. Thank you to Charles Butt, Walid Salama, Gustavo Pires, and an anonymous reviewer who graciously gave their time and knowledge to greatly improve this chapter. Katherine Dunnell at the Royal Ontario Museum, Toronto is thanked for assistance in obtaining images of specimens from the Doug Kirwin collection.

CSIRO is acknowledged for approval to reproduce Figs. 2, 8, 10, 11,12,12, and 19. Elsevier Publishing kindly approved reproduction of Figs. 4,5,6,20,21 and 22. Mineralogical Society of America allowed reproduction of Figs. 15 and 16. Broken Hill Town Council approved reproduction of Fig. 17.

References

Agricola G (1556) TTranslation Hoover, Herbert, 1912, De re metallica, Farlang

Alpers C, Brimhall GH (1989) Plaeohydrologic evolution and geochemical dynamics of cumulative supergene metal enrichment at La Escondida, Atacama Desert, northern Chile. Econ Geol 84:229–255

Andrew RL (1980) Supergene alteration and gossan textures of base-metal ores, South Africa. Miner Sci Eng 12:193–215

Atapour H, Aftabi A (2007) The geochemistry of gossans associated with Sarcheshmeh porphyry copper deposit, Rafsanjan, Kerman, Iran: implications for exploration and the environment. J Geochem Explor 93:47–65

Baas Becking LGM, Kaplan IR, Moore D (1960) Limits of the natural environment in terms of pH and oxidation-reduction potentials. J Geol 68:243–284

Birch WD (ed) (1999) Minerals of broken hill. Broken Hill City Council. 289p

Birch WD, van der Heyden A (1997) Minerals of the Kintore and block 14 open cuts at Broken Hill, New South Wales. Aust J Mineral 3:23–71

Blain CE, Andrew RL (1977) Sulphide weathering and the evaluation of gossans in mineral exploration. Miner Sci Eng 9:119–150

Blanchard R (1939) Interpretation of leached outcrops. J Chem Soc South Africa 39:344–372

Blanchard R (1944) Chemical and mineralogical composition of twenty typical limonites. Am Mineral 29:111–114

Blanchard R (1968) Interpretation of leached outcrops. Carson City, NV. Nevada State Bureau Mines Bull 66, 180 p

Blanchard R, Boswell PE (1934) Additional limonite types of galena and sphalerite derivation. Econ Geol 29:671–690

Boettcher ME, Gehiken RL, Birch WD, Usdowski E, Hoefs J (1993) The rhodochrosite-smithsonite solid solution series from Broken Hill (NSW), Australia: geochemical and infrared spectroscopic investigations. Neues Jahrb Mineral Monatshefte 8:352–362

Boni M, Mondillo N (2015) The "Calamines" and the "Others": the great family of supergene nonsulfide zinc ores. Ore Geol Rev 67:208–233

Borg G (2015) A review of Supergene Nonsulfide Zinc (SNSZ) deposits- the 2014 update. In: SM Archibald and SJ Piercey (eds) Current Perspectives on Zinc Deposits IAEG, 123–147

Bowell RJ (1993) Mineralogical and geochemical characteristics of tropical rain forest soils, Ashanti region, Ghana. Chem Geol 106:345–358

Bowell RJ (1994) Geochemistry of gossans overlying disseminated gold-sulfide deposits, Ashanti, Ghana. Int Geol Rev 36:732–753

Bowell RJ (2013) Minerals of the Kaokoveld District, Kunene region, Namibia. Mineralogical Record 44: 485–504

Bowell RJ (2014) Hydrogeochemistry of the Tsumeb deposit: implications for arsenate mineral stability. Rev Mineral Geochem 79:589–630

Bowell RJ, Foster RP, Gize AP (1993) The mobility of gold in tropical rain forest soils. Econ Geol 88:999–1016

Bowell RJ, Laffoley NA, d'Afreh EO, Hanssen E, Abe S, Yao RK, Pohl D (1996) Gold and base metal distribution and exploration in tropical soils: four contrasting

case studies from West Africa. Exploration and Mining in Africa: towards 2000. (ed: R.P.Foster). Trans Inst Mining Metal 105:B12–B33
Bowell RJ, Luetcke R, Luetcke MR (2015) Mineralogy of the Gallagher vanadium property and Manila Mine, Cochise County, Arizona. Rocks Min 90:338–352
Bowell RJ, Donkervoort L, Griffiths R, Bailey AS, Clarkson B, Prestia A (2020) Adsorption of arsenic and other elements in alluvium: implications for exploration and environmental geochemistry in the great basin. In: vision for discovery. Proceedings GSN Symposium, Reno, Nevada. May 2020. 651–670
Boyle RW (1957) The geology and geochemistry of the silver-lead-zinc deposits of Galena Hill, Yukon Territory. Department of Mines and Technical Surveys, Ottawa. Geological Survey of Canada. Map 4–1957, 140p. and 6 Figures
Boyle RW (1979) Geology, Geochemistry and Genesis of gold deposits. Geological Survey of Canada Ottawa. 400p
Boyle DR (1996) Supergene base metals and precious metals. In: Eckstrand OR, Sinclair WD and Thorpe RI (eds) Geology of Canadian Mineral Deposits. Geological Survey of Canada. Geology of Canada no 8; Geological Society of America, Decade of North American Geology v.P1, 92–108
Boyle DR (2003) Preglacial weathering of massive sulfide deposits in the Bathurst mining camp—economic geology, geochemistry, and exploration applications. In: Goodfellow WD, McCutcheon SR, Peter JM (eds) Massive Sulfide deposits of the Bathurst mining camp, New Brunswick, and Northern Maine, Economic geology monograph, vol 11, pp 689–721
Brimhall GH, Crerar DA (1987) Ore fluids: magmatic to supergene. Rev Mineral Geochem 17:235–321
Brimhall GH, Dietrich WE (1987) Constitutive mass balance relations between chemical composition, volume, density, porosity, and strain in metasomatic hydrochemical systems: results of weathering and pedogenesis. Geochim et Cosmochim Acta 51:567–587
Brimhall GH, Alpers CN, Cunningham AB (1985) Analysis of supergene ore-forming processes and groundwater solute transport using mass balance principles. Econ Geol 80:1227–1256
Brimhall GH, Lewis CJ, Ague JJ, Dietrich WE, Hampel J, Teague T, Rix P (1988) Metal enrichment in bauxites by deposition of chemically mature aeolian dust. Nature 333:819–824
Buckley A (2005) The story of mining in Cornwall. Cornwall Editions Limited, Fowey, Cornwall. 240 pp.
Burke K, Gunnell Y (2008) The African erosion surface: a continental-scale synthesis of geomorphology, tectonics, and environmental change over the past 180 million years. Geol Soc Am Mem 201:80
Butt, C.R.M., 1979. Geochemistry of a pseudo-gossan developed on a black shale—dolerite contact, Killara, Western Australia. J Geochem Explor, 11: 131–156
Butt CRM (1989) Geochemical exploration for base metals in deeply weathered terrain. Recent Dev Base Metal Geol Expl AIG Bull 16:17–31
Butt CRM (1995) Geochemical exploration for base metals in deeply weathered terrain. In: Ho SE, Amann WJ (eds) Recent developments in base metal geology and exploration Bulletin 16. Australian Institute of Geoscientists, Sydney, pp 17–31
Butt CRM (2016) The development of regolith exploration geochemistry in the tropics and sub-tropics. Ore Geol Rev 73:380–393
Butt CRM, Robertson IDM, Scott KM, Cornelius M (Eds.) (2005). Regolith expression of Australian ore systems. CRC LEME, Perth 423 pp
Chardon DV, Chevillotte A, Beauvais A, Grandin G, Boulangé B (2006) Planation, bauxites and epeirogeny: one or two plaeosurfaces on the West African margin? Geomorphology 82:273–282. https://doi.org/10.1016/j.geomorph.2006.05.008
Cohen DR, Bowell RJ (2014) Exploration geochemistry. In: Holland HD, Turekian KK (eds) Treatise on geochemistry, Elsevier, Chapter 13.24, pp 623–650
Colin F, Vieillard P (1991) Behavior of gold in lateritic equatorial environment: weathering and surface dispersion of residual gold particles, at Dondo Mobi. Gabon Applied Geochem 6:279–290
Colin E, Vieillard P, Ambrosi JP (1993) Quantitative approach to physical and chemical gold mobility in equatorial rainforest lateritic environment. Earth Planet Sci Lett 114:269–285
Cornwall FW (1961) Geochemistry chartered exploration. In: Mendelsohn F (ed) The geology of the Northern Rhodesian Copperbelt. Macdonald & Co. (Publishers) Ltd, London, pp 187–207
Dunn JG, Howes VL (1996) The oxidation of violarite. Thermochim Acta 282/283:305–316
Emmons SF (1900) The secondary enrichment of ore deposits. Am Inst Min Metal Petrol Eng Trans 30: 177–217
Emmons WH (1917) The enrichment of ore deposits. Bull United States Geol Sur 625:530p
Essalhi M, Sizaret S, Barbanson L, Chen Y, Lagroix F, Demory F, Nieto JM, Sáez R, Capitán MA (2011) A case study of the internal structures of gossans and weathering processes in the Iberian Pyrite Belt using magnetic fabrics and plaeomagnetic dating. Mineral Deposita 46:981–999
Farooq S, Govil H (2014) Mapping Regolith and gossan for mineral exploration in the eastern Kumaon Himalya, India using hyperion data and object-oriented image classification. Adv Space Res 53:1676–1685
Freyssinet P (1994) Gold mass balance in lateritic profiles from savannah and rain forest zones. Catena 21:159–172
Gahlan H, Ghrefat (2017) Detection of gossan zones in arid regions using Landsat 8 OLI data: implication for mineral exploration in the eastern Arabian shield, Saudi Arabia. Nat Resour Res 27:109–124
Garlick WG (1961) Geochemistry. Rhodesian selection trust. In: Mendelsohn F (ed) The geology of the Northern Rhodesian Copperbelt. Macdonald & Co. (Publishers) Ltd, London, pp 208–213

Graddock PT (1995) Early metal mining and production. Edinburgh University Press Ltd, Edinburgh. 225p
Grandin G (1976) Aplanissements cuirassés et enrichissement des gisements de manga- nèse dans quelques régions d'Afrique de l'Ouest. Mém ORSTOM 82 (276 pp.)
Greig D (1984) Geology of the Teutonic Bore Cu-Zn-Ag massive sulfide deposit., Western Australia. Australian Institute of Mining and Metallurgy
Gunning HC (1961) The early years. In: Mendelsohn F (ed) The geology of the Northern Rhodesian Copperbelt. Macdonald & Co. (Publishers) Ltd, London, pp 3–11
Hallberg JA, Thompson JFH (1985) Geologic setting of the Teutonic bore massive sulfide deposit, Archean Yilgarn Block, Western Australia. Econ Geol 80: 1953–1964
Hannington MD (1993) The formation of atacamite during weathering of sulfides on the modern seafloor. Can Mineral 31:945–956
Hannington MD, Thompson G, Rona PA, Scott SD (1988) Gold and native copper in supergene sulfides from the mid-Atlantic ridge. Nature 333:64–66
Hough RM, Butt CRM, Fischer-Bühner J (2009) The crystallography, metallography and composition of gold. Elements 5:297–230
Junner NR (1932) The geology of the Obuasi goldfield. Gold Coast geological survey. Memoir 2: 43 p
Křibek B, Zachariáš J, Knésl I, Miková J, Mihaljevič M, Veselovský F, Bamba O (2016) Geochemistry, mineralogy, and isotope composition of Pb, Zn, and Cu inprimary ores, gossan and barren ferruginous crust from the Perkoa base metal deposit, Burkina Faso. J Geochem Explor 168:49–64
Large RR (1992) Australian volcanic-hosted massive sulfide deposits: features, styles, and genetic models. Econ Geol 87:471–510
Leybourne MI, Goodfellow WD, Boyle DR (2002) Sulphide oxidation and groundwater transport of base metals at the Halfmile Lake and Restigouche Zn-Pb massive sulfide deposits, Bathurst mining camp, New Brunswick. Geochemist Explorat Environ Anal 2(1): 37–44
Leybourne MI, Peter JM, Layton-Matthews D, Volesky J, Boyle DR (2006) Mobility and fractionation of rare earth elements during supergene weathering and gossan formation and chemical modification of massive sulfide gossan. Geochim Cosmochim Acta 70:1097–1112
Locke A (1926) Leached outcrops as guides to copper ores. Baillere, Tindall and Cox, London. 175p
Lombaard AF, Günzel A, Innes J, Kruger TL (1986) The Tsumeb lead– copper–zinc–silver deposit, South West Africa/Namibia. In: Anhaeusser, C.R., Maske, S. (eds.) Mineral deposits of Southern Africa, Geological Society of South Africa, vol. II, 1761–1787
Mackenzie DH, Davies RH (1990) In: Hughes FE (ed) Broken Hill Lead-Silver-Zinc deposit at Z.C. Mines, in geology of the mineral deposits of Australia and Papua New Guinea. Australasian Institute of Mining & Metallurgy, Melbourne, pp 1079–1084
Majzlan J, Drahota P, Filippi M (2014) Paragenesis and crystal chemistry of arsenic minerals. Rev Mineral Geochem 79:17–184
Mann AW (1983) Hydrogeochemistry and weathering on the Yilgarn block, Western Australia—ferrolysis and heavy metals in continental brines. Geochim Cosmochim Acta 47:181–190
McQueen K (2009) Regolith geochemistry In: Regolith Science, eds: K. Scott and C. Pain. Springer & CSIRO. 73–104
Mellado D, González Clavijo E, Tornos F, Conde C (2006) Geología y estructura de la Mina de Río Tinto (Faja Pirítica Ibérica, España). Geogaceta (in Spanish) sociedadgeologicaes 40:231–234
Michel P (1973) Les bassins des fleuves Sénégal et Gambie: étude géomorphologique. Mém. ORSTOM 63. (383 pp. (in French with English abstract))
Mighall TM, Chambers FM (2022) Early mining and metalworking: its impact on the environment. J Historical Metallurgy 27:71–83
Milési J-P, Ledru P, Feybesse J-L, Dommanget A, Marcoux E (1992) Early Proterozoic ore deposits and tectonics of the Birimian orogenic belt. West Africa Precambrian Res 58:305–344
Nickel EH (1984) The mineralogy and geochemistry of the weathering profile of the Teutonic Bore Cu-Pb-Zn-Ag sulfide deposit. J Geochem Explor 22:239–264
Nickel EH, Daniels JL (1985) Gossans, in Wolf, KH, ed., Handbook of stratabound and stratiform ore deposits. Amsterdam/New York, Elsevier 13:261–390
Nickel EH, Ross JR, Thornber MR (1974) The supergene alteration of pyrrhotite, pentlandite ore at Kambalda, Western Australia. Econ Geol 69:93–107
Nordstrom DK, Alpers CN (1999) Negative pH, efflorescent mineralogy, and consequences for environmental restoration at the Iron Mountain superfund site. California Proc Nat'l Acad Sci 96:3455–3462
O'Brien W (1996) Bronze age copper mining in Britain and Ireland. Shire Publications Ltd, Buckinghamshire. 180p
Parker P, Belford S, Maier R, Lynn S, Stewart W (2017) Teutonic Bore-Jaguar-Bentley volcanogenic massive sulfide field: in Phillips, G.N., (Ed.), 2017 Australian ore deposits. Austral Instit Min Metal Mono 32:167–172
Pekov IV, Lykova IS (2011) Rubtsovskoe deposit (North-West Altai, Russia): mineralogy of the oxidation zone. Mineralogical Almanac 16:93p
Pekov IV, Lyskova IS, Bryzgalov IA, Ksenofontov DA, Zyryanova LA, Litvinov ND (2011) Uniquely high grade iodide mineralization in the oxidation zone of the Rubtsovskoe Base metal deposit, Northwest Altai, Russia. Geol Ore Deposits 8:683–698
Plimer IR (1984) The mineralogical history of the Broken Hill Lode, NSW. Aust J Earth Sci 31:379–402

Plimer IR (1999) The making of the minerals. In: Birch WD (ed) Minerals of Broken Hill. Broken Hill City Council. 70–87

Pryce W (1778) Mineralogia Cornubiensis: Mineralogia Cornubiensis; a treatise on minerals, mines, and mining. Containing the theory and natural history of strata, fissures, and lodes, with the methods of discovering and working of tin, copper, and lead mines, and of cleansing and metalizing their products; shewing each particular process for dressing, assaying and smelting of ores. To which is added, an explanation of the terms and idioms of miners 1778. Pitman Brothers, Bath. 401p

Randabel J, Bowell RJ, Sangare I (2018) Falea: an unconformity-type polymetallic deposit, Mali, West Africa. Proceedings Symposium on Uranium Raw Material for the Nuclear Fuel Cycle:341–354

Ridley WI (2012) Geochemical characteristics in volcanogenic massive sulfide occurrence model. U.-S. Geological Survey Scientific Investigations Report 2010–5070 -C, chap. 14, 18 p

Rosenthal P, Morin D, Herbach R, Photiades A, Delpech S, Jacquemot D, Fadin L (2013) Mining technologies at deep level in antiquity: the Laurion mines (Attica, Greece). In: Anreiter et al. (eds) Mining in European History and its impact on environment and human societies—proceedings for the 2nd Mining in European History Conference of the FZ HiMAT, 7.-10. November 2012, Innsbruck, pp. 89–95

Sato M (1960) Oxidation of sulfide ore bodies II: oxidation mechanism of sulfide minerals. Econ Geol 55:1202–1231

Sato M, Mooney HM (1960) The electrochemical mechanism of sulfide self-potentials. Geophysics 25:226–249

Scheiber L, Ayora C, Vázquez-Suñé E, Soler A (2018) Groundwater-gossan interaction and the genesis of the secondary siderite rock at Las Cruces ore deposit (SW Spain). Ore Geol Rev 102:967–980

Schwartz MO, Melcher F (2003) The Perkoa zinc deposit, Burkina Faso. Econ Geol 98:1463–1485

Scott KM (1986) Elemental partitioning into Mn- and Fe-oxides derived from dolomitic shale-hosted Pb–Zn deposits, Northwest Queensland, Australia. Chem. Geol 57:395–414

Scott KM (1987) The mineralogical distribution of pathfinder elements in gossans derived from dolomitic shale-hosted Pb–Zn deposits, Northwest Queensland, Australia. Chem Geol 64:295–306

Scott KM, Ashley PM, Lawie DC (2001) The geochemistry, mineralogy and maturity of gossans derived from volcanogenic Zn-Pb-Cu deposits of the eastern Lachlan Fold Belt, NSW, Australia. J Geochem Explor 72:169–191

Séranne M (1999) Early Oligocene stratigraphy turnover on the West Africa continental margin: a signature of the tertiary greenhouse-to-icehouse transition? Terra Nova 11:135–140

Sillitoe RH (2009b) Supergene Silver Enrichment Reassessed. In: Titley SR (ed) Supergene environments, processes, and products, Society of Economic Geology, Special Publication 14, pp 25–32

Silva JB, Oliveira JT, Ribeiro A (1990) South Portuguese zone. Structural outline. In: Dallmeyer RD, Martinez Garcia E (eds) Pre-Mesozoic geology of Iberia. Springer-Verlag, Berlin, pp 348–362

Skarpelis N, Ariadne A (2009) Geology and origin of supergene ore at the Lavrion Pb-Ag-Zn deposit, Attica, Greece. Resour Geol 59:1–14

Spry PG, Teale GS (2021) A classification of Broken Hill-type deposits: a critical review. Ore Geol Rev 130: 103935

Stevens B (1999) The geology of Broken Hill. In: Birch WD (ed) Minerals of Broken Hill. Broken Hill City Council. 58–69

Talbot MR, Livingstone D, Palmer PG, Melack JM, Delibrias G, Gulliksen S (1984) Preliminary results of sediment cores from Lake Bosumtwi, Ghana. In: Coetzee JA, Bakker EM (eds) Palaeoecology of Africa. Balkema, Amsterdam, pp 173–192

Taylor R (2011) Gossans and leached caps: field assessment, Springer: 146p

Taylor GF, Scott KM (1982) Evaluation of gossans in relation to Pb-Zn mineralization in the Mount Isa inlier. Queensland Jour Geochem Explor 18:87–110

Taylor GF, Thornber MR (1992) Gossan formation and gossan surveys. In: Butt CRM, Zeegers H (eds) Handbook of exploration geochemistry, Regolith exploration geochemistry in tropical and subtropical terrains, vol 4. Amsterdam/New York, Elsevier, pp 139–201

Taylor GF, Wilmshurst JR, Butt CRM, Smith RE (1980) Ironstones and gossans in: Butt CRM and Smith RE (eds) Conceptual models in exploration geochemistry-Australia. J Geochem Explor 12:118–122

Thornber MR (1975a) Supergene alteration of sulfides, I: A chemical model based on massive nickel sulfide deposits at Kambalda, Western Australia. Chem Geol 15:1–14

Thornber MR (1975b) Supergene alteration of sulfides, II: A chemical study of the Kambalda nickel deposits. Chem Geol 15:117–144

Thornber MR (1983) Mineralogical and electrochemical instability of nickel-iron sulfides—pentlandite and violarite. Jour Applied Electrochem 13:253–267

Thornber MR (1985) Supergene alteration of sulfides, VII: distribution of elements during gossan-forming process. Chem Geol 53:279–301

Thornber MR, Taylor GF (1992) The mechanism of sulfide oxidation and gossan formation. In: Butt CRM, Zeegers H (eds) Handbook of exploration geochemistry, v. 4: Regolith exploration geochemistry in tropical and subtropical terrains. Elsevier, Amsterdam/New York, pp 119–138

Thornber MR, Wildman JE (1984) Supergene alteration of sulfides. VI. The binding of Cu, Ni, Zn, Co and Pb with gossan (iron-bearing) minerals. Chem Geol 44:399–434

Thornber MR, Allchurch PD, Nickel EH (1981) Variations in gossan geochemistry at the perseverance Nickel

Sulfide deposit, Western Australia: a descriptive and experimental study. Econ Geol 76:1764–1774
Timberlake S (2003) Early mining research in Britain: the developments of the last ten years. In: Graddock PT, Lang J (eds) Mining and metal production: through the ages. The Trustees of the British Museum, London, pp 21–42
Tooms JS, Webb JS (1961) Geochemical prospecting investigatins in the Northern Rhodesian copper belt. Econ Geol 56:815–846
Topping P, Lynott M (2005) Miners and mines. In: Topping P and Lynott M (eds) Society for American Archaeology. Meeting (66th, 2001, New Orleans, La.). The cultural landscape of prehistoric mines. Oxford. Oxbow, pp 180–191
van der Heyden A, Edgecombe DR (1990) Silver-Lead-zinc deposit at south mine, Broken Hill. In: Hughes FE (ed) Geology of the mineral deposits of Australia and Papua New Guinea. Australasian Institute of Mining & Metallurgy, Melbourne, pp 1073–1107
Van Moort JC, Swenson C (1981) The oxidized zone of the Broken Hill lode, NSW. In: Amstutz GC (ed) Ore Genesis. Springer, Heidelberg, pp 299–312
Velasco F, Herrero JM, Suarez S, Yusta I, Alvaro A, Tornos F (2013) Supergene features and evolution of gossans capping massive sulfide deposits in the Iberican Pyrite Belt. Ore Geol Rev 53:181–203
Velbel MA (1989) Effect of chemical affinity on feldspar hydrolysis rates in two natural weathering systems. J. Schott, A.C. Lasaga (Eds.), Kinetic geochemistry. Chem Geol 78:245–253
Velbel MA (1992) Geochemical mass balances and weathering rates in forested watersheds of the southern Blue Ridge, III. Cation budgets in an amphibolite watershed, and weathering rates of plagioclase and hornblende. Am J Sci 292:58–78
Volesky JC, Stern RJ (2001) Remote sensing and mineral exploration in the Arabian shield: the Wadi Bidah mining district example. Gondwana Res 4:198–200
Vourlakos NM, Fitros MG (2019) The minerals of Laviron. Library of the scientific Society of Lavreotiki number 13. Lavrion, Greece. 205p
Webb JS, Tooms JS (1959) Geochemical drainage reconnaissance for copper in Northern Rhodesia. Transactions Institute of Mining and Metallurgy, Section B. Appl Earth Sci 68:125–144
Williams PA (1990) Oxide zone geochemistry. Ellis Horwood, London. 286p
Yalçin Ü (2000) Anfänge der Metallverwendung in Anatolien, in: Anatolian Metal, Veröffentlichungen aus dem Deutschen Bergbau-Museum Bochum. Bochum
Yesares L, Sáez R, De Almodóvar GR, Nieto JM, Gómez C, Ovejero G (2017) Mineralogical evolution of the Las Cruces gossan cap (Iberian Pyrite Belt): from subaerial to underground conditions. Ore Geol Rev 80:377–405
Zeegers H, Leduc C (1991) Geochemical exploration for gold in temperate, arid, and tropical rain forest terrains. In: Foster RP (ed) Gold metallogeny and exploration. Blackie and Sons, Glasgow, pp 309–335
Zimmerman DO (1964) Gossans at northern leases, Mount Isa as visible geochemical anomalies. In: Woodcock, JT, Madigan RT and Thomas RC (eds) Proceedings of the Eighth Congress on Mining and Metallurgy. Publication 6, Australasian Institute of Mining and Metallurgy, Melbourne, 359–367

Supergene Non-Sulfide Base Metal Deposits

M. Boni, Rob J. Bowell, N. Sergeev, and R. Burlow

Abstract

Supergene non-sulfide deposits comprise rocks with elevated concentrations of Zn, Cu, Pb and, to a lesser extent, Ag and V. These are formed by the oxidation and mobilization of metals from sulfide mineralization, mobilized in solution and precipitated laterally or above the primary sulfide deposit. In most cases, there is no associated gossan or residual iron cap. Traditionally, Zn-rich non-sulfide deposits were termed "calamine", a common name for an association of smithsonite and hemimorphite, with local cerussite and anglesite. In some cases, calamine deposits are genetically associated with nearby hydrothermal sulfide mineralization, even though their mineralogy and textures are similar to supergene calamine-style deposits.

A relatively recently identified class of non-sulfide Zn deposits are those hosted in siliciclastic rocks, e.g. Yanque in Peru and Skorpion in Namibia. Other metals can also be dominant, e.g. Pb (Magellan/Paroo Station Mine, Australia); Cu (Carlotta, Arizona; Kunene, Namibia); Ag (Sierra Mojada, Mexico; Wonawinta, Australia) or V (Otavi Mountainland, Namibia).

Carbon, oxygen, Zn, Pb and Cu stable isotope studies allow distinction between supergene and hypogene non-sulfide deposits. Periods of paleoclimatic switchovers from seasonally humid/arid to hyperarid climates have been considered the most favourable conditions for the formation and preservation of supergene non-sulfide deposits, even though many of these deposits are currently located in warm humid-weathering climatic zones.

1 Introduction

Non-sulfide deposits (traditionally called "calamine") have until recently been restricted to Zn and Pb ores (Boni and Mondillo 2015), however, other metals can also constitute important by-products or the main commodity such as Cu, Ag and V. Evidence exists that 'calamine'

M. Boni (✉)
Dipartimento Scienze della Terra, dell'Ambiente e delle Risorse (DiSTAR), Università di Napoli "Federico II", Naples, Italy
e-mail: boni@unina.it

R. J. Bowell
SRK Consulting, Churchill House, Cardiff, UK

Department of Geological Sciences and Geological Engineering, Queen's University, Kingston, ON, Canada

N. Sergeev
ERM Australia Consultants Pty Ltd, Perth, WA, Australia

R. Burlow
Rosslyn Hill Mining Pty Ltd, Wiluna, WA, Australia

R. J. Bowell, C. R. M. Butt (eds.), *Geology, Geochemistry and Formation of Supergene Mineral Deposits in Deeply Weathered Terrain*, Mineral Resource Reviews, https://doi.org/10.1007/978-3-031-75733-4_5

deposits were exploited in the Middle East and Central Asia as far back as the third century BC (Kharakwal and Gurjar 2006). Both Greek and Roman cultures used brass alloys made of Cu and Zn and utilized their corresponding ores in Belgium and Sardinia (Pollard and Heron 1996). By the 1600s, several deposits in Poland and Germany were also exploited, and smelting of such ores was described by Agricola (1556). These deposits were again exploited for Zn and Pb in Belgium, Poland, Sardinia, Northern Africa and the Middle East in the late nineteenth century (Large 2001; Hitzman et al. 2003; Borg 2015). They were the basis for the development of the Zn mining and smelting industries. The principal occurrences were: La Calamine (Belgium), Upper Silesia (Poland) and SW Sardinia (Italy) (Boni and Large 2003), as well as Lavrion and Thassos (Greece), Eastern Alps and northern Spain, where non-sulfide ores had been exploited in the past. The non-sulfide Zn (Pb) deposits in the Irish Midlands and Britain represent a minor example of surface oxidation of primary sulfide concentration and preservation under glacial till (Balassone et al. 2008). All these occurrences are now either exhausted, not economically viable, or environmentally dangerous. These older deposits have been described in Boni and Mondillo (2015), and references therein. The exploitation of these ores declined after World War II, and sulfide deposits were favoured instead. However, at the beginning of the twenty-first century, non-sulfide deposits experienced a significant revival, due to new developments in hydrometallurgical acid leaching, solvent extraction and electrowinning techniques. The economic focus has been generally set on Zn non-sulfides, even though most deposits also contain variable amounts of Pb (or consist solely of Pb non-sulfides, e.g. Magellan).

Non-sulfide base metal concentrations can be of hypogene or supergene origin, with the general geological characteristics, genesis and mineralogy providing the classification of each deposit (Hitzman et al. 2003). Hypogene deposits are considered to be hydrothermal and/or associated with metamorphism of primary sulfide ores in high oxygen fugacity fluids, whereas supergene deposits are a result of weathering and oxidation at ambient temperatures. Hypogene deposits are dominated by anhydrous Zn-silicates and oxides, such as willemite, zincite, franklinite, co-existing with some sulfides. The most economically important concentrations of this type are the Vazante deposit (Brazil) in Proterozoic carbonates (Monteiro et al. 2006; Slezak et al. 2014) and the deposits in the Franklin-Sterling mining district (New Jersey, USA), which are associated with high-grade metamorphic processes (Dunn 1995).

In contrast, supergene deposits reflect transport of metals and precipitation as a wide variety of carbonate, silicate and other secondary minerals. These can reflect direct replacement of sulfide minerals within a mineralized zone. Replacement of a portion of the host rock can occur as well as hydromorphically mobilized metals that infill karst structures (Hitzman et al. 2003). All three styles can overlap and grade into each other, although ore replacement deposits tend to contain more Fe and Pb and have been termed “red zinc ores” (Borg 2015). In contrast, wallrock alteration and karst infill deposits have been subjected to geochemical separation and therefore are much purer, commonly with high Zn grades (up to 30%) and low Fe, Cu and Pb. These have been termed “white zinc ores” (Reichert and Borg 2008; Boni and Mondillo 2015; Borg 2015). In one rare case, the Tsumeb mineralization of NE Namibia, all three styles are known in a single deposit, which exhibits complex zonation of multiple supergene and hypogene ores (Bowell 2014). Some of the recently investigated, more significant supergene deposits (though variably interpreted genetically, and of different economic importance) include Skorpion (Namibia), Angouran (Iran), Shaimerden (Kazakhstan), Jabali (Yemen), Accha, Yanque and Bongarà (Peru), and Sierra Mojada (Mexico). Each has been described in great detail (Large 2001; Boni and Large 2003; Borg et al. 2003; Hitzman et al. 2003; Boni 2005; Boni et al. 2007a, 2011a, b; Megaw 2009; Daliran et al. 2013; Borg 2015; Boni and Mondillo 2015).

In this chapter, the major characteristics of supergene deposits are discussed, illustrated by case studies from different environments, principally considering their geology, mineralogy and geochemistry and the controls on ore genesis. A summary of the recently investigated non-sulfide deposits referred to is listed in Table 1, with a map of their occurrence shown in Fig. 1. For purposes of this paper, base metals are extended to include not only Cu, Zn, Ag and Pb, but also V. Table 2 provides formulae and chemical information on the minerals discussed in this chapter.

2 Metal Chemistry of Non-Sulfide Base Metal Deposits

Non-sulfide deposits are dominated by Zn-rich systems (Fig. 2). A striking feature is the separation of metals dependent on the style of mineralization, which reflects their relative chemical mobility in the supergene environment.

Whereas the Zn-rich systems of siliciclastic and carbonate types, and the polymetallic Pb-rich systems, all contain some other base metals, the absence of Zn in Cu-rich system is notable. This is most likely due to the greater supergene mobility of Zn in oxidizing environments compared to Cu and Pb (Sangameshwar and Barnes 1983). In the Kunene district, Namibia, a wide hydrogeochemical anomaly for Zn occurs even though the supergene mineralization contains Cu-Ag±Pb (Bowell et al. 2012).

Most deposits known tend to be associated with carbonate rocks and as such show potential for higher Zn mobility compared to Fe, Cu or Pb. This association is also reflected in the dominant mineralogy of the deposits, an example of this is the supergene mineralization in the Opuwo district, Kunene region, Namibia (Bowell et al. 2012).

3 Mineralogy of Supergene Non-Sulfide Base Metal Deposits

3.1 Zinc Minerals

The major Zn minerals are smithsonite, hydrozincite, hemimorphite and sauconite (a poorly ordered Zn-smectite) (Fig. 3). Deposits in carbonate rocks tend to be dominated by smithsonite and hydrozincite (Sangameshwar and Barnes 1983; McPhail et al. 2003; Reichert and Borg 2008; Reichert 2009). The formation of either hemimorphite or smithsonite is controlled by factors of geology, water rock/ratio and groundwater chemistry. If the solution is dominated by bicarbonate molecules, smithsonite is likely to form, whereas if the host rock for the aquifer water is silica-rich and some silica is present, then hemimorphite or Zn-rich clays are more likely to form. Thus, mineral precipitation at neutral pH favours smithsonite over hydrozincite (Sangameshwar and Barnes 1983). Water-saturated environments and deep-seated supergene alteration will promote higher $CO_2(g)$ fugacity, thus favouring smithsonite formation.

Chlorite and other clay minerals can also host appreciable Zn (Blot et al. 1995; Mondillo et al. 2014b; Choulet et al. 2016). Iron-Mn oxides have also been noted as mineral hosts for high concentrations of Zn at Jabal Dhaylan in Saudi Arabia and at Skorpion, Namibia (Hayes et al. 2000; Gnoinski 2007). More exotic mineralogy has been observed at Skorpion, where the main hosts of Zn are smithsonite and sauconite, but at shallow depths, tarbuttite and hydrozincite are more common, as well as chalcophanite and scholzite (Borg et al. 2003). In addition, phases like smithsonite can have multi-element signatures that include Mg, Fe, Cu and Co. On the peripheral margins of the ores, Zn-dolomite and calcite can occur along with rare mineral phases such as skorpionite (Krause et al. 2008) and minrecordite. At the Jabali deposit, Yemen, up to 20% of Zn occurs within a Zn-rich dolomite phase (Mondillo et al. 2011, 2014a). Similar observations have been carried out also in other deposits by Boni et al. (2011a, 2013).

Willemite occurs in some deposits, such as Bou Arhous in Morocco, generally partly altered to Zn-clays. Moroccan willemite is commonly considered to be partly hypogene in origin (Choulet et al. 2016). Genetically undetermined (i.e., uncertain whether hypogene or supergene) willemite has been found in several Zn deposits. Several willemite samples were first detected in several Belgian deposits (e.g. La Calamine,

Table 1 Main features of non-sulfide base metal deposits

Country	Deposit	Host rock[a]	Deposit type (primary)[b]	Deposit type (supergene)[c]	Major minerals[d]	Non-sulfide resource	References
Australia	Magellan/ Paroo Station	Paleoproterozoic sst, dol	MVT	WR	cer, angl, pyrm	21.4 Mt @ 5.8% Pb	Pirajno et al. (2010)
	Wonawinta	Devonian carbonate rocks	MVT	DR	cer, sm, hem	60 Moz Ag	Boni and Mondillo (2015)
China	Huoshaoyun	Triassic- Jurassic dol-lst	MVT	Pri, DR	sm, sau, hem	18 Mt @23.6% Zn, 5.6% Pb	Chen et al. (2023)
Chile	El Abra	Ter granodior	magmatic	DR	atac, bot, broch, cc, chry, mal, parat	4409 @ 0.436% Cu	Unpublished
Iran	Angouran	Neoproterozoic mica schist-marble	Magmatic/ CRD?	Pri, DR	sm, apy, gal, green, ce, arsenat, hem	17.3 Mt @ 28.1% Zn, 4.4% Pb, 110 g/t Ag	Boni and Mondillo (2015)
Kazakhstan	Shaimerden	Carboniferous lst	MVT	DR, WR	sm, sau, hem	4.3Mt@21% Zn	Boland et al. (2003)
Mexico	Sierra Mojada	Cretaceous carbonate rocks	CRD	DR, WR, karst	sm, acan, Ag, bromarg, chlorarg, hem, tenn	71.2 Mt @ 71.4 g/t Ag	Boni and Mondillo (2015)
Morocco	Atlas	Lower Jurassic lst	MVT	DR, WR, karst	sm, hem, Zn-dol, hem	?	Choulet et al. (2013)
Namibia	Otuani, Kuene	Neoproterozoic carbonate rocks	Tsumeb-type Cu?	DR, karst	di, ch, mal, hm, sh, pla	3 Mt @ 3% Cu	Bowell et al. (2012)
	Berg Aukas	Neoproterozoic carbonate rocks	Berg Aukas-type Zn	karst	sm, des, mott, hem	3.4 Mt @ 15% Zn, 4% Pb, 0.4% V_2O_5	Cairncross (2021)
	Skorpion	Neoproterozoic carbonate rocks/volcanics	VHMS	DR, WR	sau, hem, sm, hydr, tarb, schol	24.6 Mt @ 8% Zn; 0.3% Cu	Borg et al. (2003)
Peru	Bongarà	Upper Triassic-Lower Jurassic dol-lst	MVT	DR, WR, pri	sm, hydr, ce, hem	?	Boni and Mondillo (2015)
	Accha	Mid-upper Cretaceous lst	CRD	DR, WR	sm, hem, hydr, sau	6.6 Mt@ 6.37% Zn, 0.78% Pb	Boni and Mondillo (2015)
	Yanque	Mid-upper Cretaceous siliclastic	CRD	WR	sau, hem, sm, Zn-dol, ce	26.5 Mt @ 2.37% Zn, 2.18% Pb	Boni and Mondillo (2015)

Turkey	Hakkari	Jurassic lst	MVT	DR	sm, hem	10 Mt @15% Zn	Boni and Mondillo (2015)
USA	Carlotta	Ter granodior	magmatic	DR	atac, broch, cc, chry, mal, parat	100 Mt@0.39% Cu	ADMMR (2005)
Yemen	Jabali	Jurassic dol, lst	MVT	DR, WR, HD	sm, Zn-dol, hydr, hem, ce, angl, acan, green	12.6 Mt @ 8.9% Zn, 1.2% Pb, 68 ppm Ag	Boni and Mondillo (2015)

Arizona Department of Mines & Mineral Resources (2005)
Boland et al. (2003)
[a]*dol* dolomite, *granodior* granodiorite, *lst* limestone, *sst* sandstone
[b]*CRD* carbonate replacement deposit, *VHMS* volcanogenic hosted massive sulfide
[c]Deposit types are referred to Hitzman et al. (2003) classification: *DR* direct replacement, *WR* wall rock replacement, *karst* residual and karst filling, *Pri* primary/hydrothermal
[d]Abbreviations: *acan* acantite, *angl* anglesite, *arsenat* arsenates, *apy* Arsenopyrite, *az* azurite, *bor* bornite, *bromarg–chlorarg* bromargirite–chlorargirite, *ce* cerussite, *chalc* chalcocite, *descl* descloizite, *gal* galena, *green* greenockite, *hem* hemimorphite, *hydr* hydrozincite, *mott* mottramite, *psilom* psilomelane, *schol* scholzite, *sm* smithsonite, *tarb* tarbuttite, *tenn* tennantite, *will* willemite, *Zn-dol* zincian dolomite

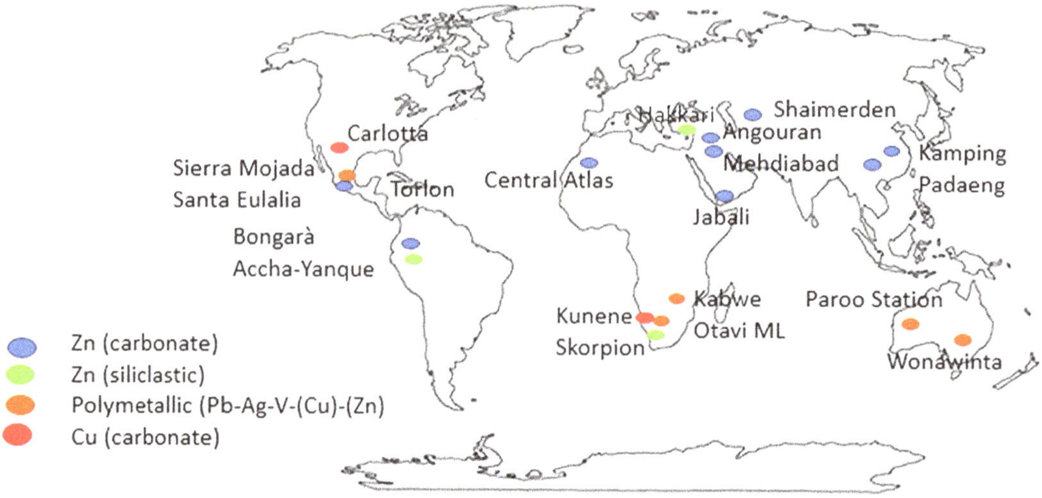

Fig. 1 Occurrence of recently investigated supergene non-sulfide base metal deposits in deeply weathered terrains

Coppola et al. 2008; Choulet et al. 2019). Willemite of several generations, from possibly hypogene to frankly supergene, has been reported in the Otavi Mountainland (Namibia) (Boni et al. 2007b), and the Zambesi metamorphic belt (Zambia) (Boni et al. 2011b). Other Zn-bearing minerals include fraipontite, caryopilite or greenalite (Fransolet and Bourguignin 1975; Guggenheim and Bailey 1990), as well as some micas (Robert and Gaspèrin 1985) and Fe-oxyhydroxides (Boni et al. 2009).

3.2 Lead Minerals

The most common Pb minerals found in association with non-sulfide Zn deposits or in Pb-only ores, such as the Magellan/Paroo Station deposit in Western Australia (Pirajno et al. 2010; see Sect. 7, below) are anglesite and cerussite (Fig. 4). Pyromorphite is also generally available as a late-stage mineral in the paragenesis and occurs commonly pseudomorphing earlier Pb phases (Mason 2004).

3.3 Copper Minerals

In Cu-rich deposits, the mineralogy reflects local geology (Fig. 5). In carbonate-hosted systems, malachite with lesser cuprite, native Cu and chrysocolla are the most common minerals (Munchmeyer 1996; Chavez 2000). In comparison, in siliclastic environments, there is a broader range of Cu minerals that can include dioptase, plancheite, shattuckite, tenorite, brochantite, and in some cases, phosphate-, arsenate-, vanadate-, molybdate- and halide-Cu phases (Williams 1990; Bowell 1992; Crane et al. 2001). In extremely arid environments such as in the Andes, several polymorphs of Cu halides (atacamite, botallackite and paratacamite) are common (Leybourne and Cameron 2008). In some supergene zones, Cu-bearing Mn-Fe oxides ("wad") can also occur (Newberg 1967; Le Mignot et al. 2017).

3.4 Other Minerals

Other minerals occurring in non-sulfide orebodies depend largely on the composition of the host rocks (whether pure carbonate rocks, impure or mixed silicate-carbonate rocks or fully silicate rocks), and on the abundance of Fe sulfides in the protolith (Hitzman et al. 2003). Calcite and dolomite can be a significant Zn host (Fig. 6).

The deposits in rocks mainly consisting of silicates tend to contain hemimorphite- and sauconite-bearing assemblages. Hemimorphite

Table 2 Minerals in non-sulfide deposits quoted in the text

Mineral	Chemical formula	Concentration main element, wt%
Zinc		Zn, wt%
Smithsonite	$ZnCO_3$	52.2
Mg-Smithsonite	$(Zn, Mg)CO_3$	>25
Hydrozincite	$Zn_5(CO_3)_2(OH)_6$	59.55
Hemimorphite	$Zn_4(Si_2O_7)(OH)_2.H_2O$	54.29
Sauconite	$Na_{0.3}Zn_3(Si, Al)_4O_{10}(OH)_2{\cdot}4H_2O$	33.8
Willemite	Zn_2SiO_4	58.68
Zn-dolomite	$Ca(Zn, Mg)(CO_3)_2$	<22
Skorpionite	$Ca_3Zn_2(PO_4)_2(CO_3)(OH)_2.H_2O$	23.2
Goslarite	$ZnSO_4{\cdot}7H_2O$	22.74
Scholzite	$CaZn_2(PO_4)_2{\cdot}2H_2O$	32.96
Tarbuttite	$Zn_2PO_4(OH)$	53.87
Hetaerolite	$ZnMn_2O_4$	27.33
Chalcophanite	$ZnMn^{4+}{}_3O_7{\cdot}3H_2O$	17.09
Lead		Pb, wt%
Anglesite	$PbSO_4$	68.32
Cerussite	$PbCO_3$	77.54
Minium	$Pb^{2+}Pb^{4+}O_4$	90.67
Mimetite	$Pb_5(AsO_4)Cl$	69.61
Plattnerite	PbO_2	86.62
Pyromorphite	$Pb_5(PO_4)Cl$	76.38
Copper		Cu, wt%
Atacamite/Botallackite/Paratacamite	$Cu_2Cl(OH)_3$	59.51
Azurite	$Cu_3(CO_3)_2(OH)_2$	55.31
Brochantite	$Cu_4(SO_4)(OH)_6$	56.2
Chalcocite	Cu_2S	79.85
Chrysocolla	$(Cu, Al)_2H_2Si_2O_5(OH)_4{\bullet}n(H_2O)$	33.86
Copper	Cu	100
Cuprite	Cu_2O	88.82
Dioptase	$CuSiO_2(OH)_2$	40.31
Malachite	$Cu_2(CO_3)(OH)_2$	57.48
Plancheite	$Cu_8Si_8O_{22}(OH)_4{\bullet}(H_2O)$	43.41
Shattuckite	$Cu_5(SiO_3)_4(OH_{)2}$	48.43
Tenorite	CuO	79.89
Silver		
Acanthite	Ag2S	87
Bromoarygrite	AgCl	59.82
Chlorargyrite	AgBr	75.27
Silver	Ag	100
Tennantite	$(Cu, Ag, Zn, Fe)_{12}As_4S_{13}$	<13.65
Vanadium		
Descloizite	$PbZn(VO_4)(OH)$	12.59
Mottramite	$PbCu(VO_4)(OH)$	12.65
Vanadinite	$Pb5(VO_4)_3Cl$	10.79

Anthony et al. (2003)

All mineral formula and metal compositions taken from Anthony et al. (2003) and are as accepted by the International Mineralogical Association

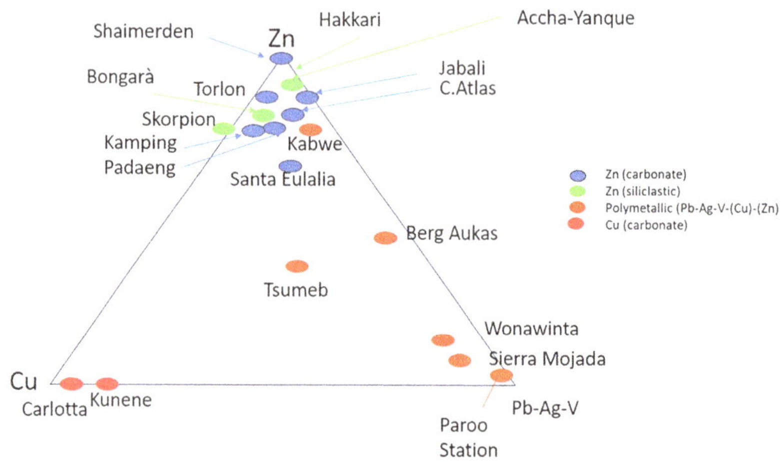

Fig. 2 Schematic representation of metal tenor for various recently investigated non-sulfide base metal deposits

Fig. 3 Typical supergene Zn minerals in non-sulfide deposits; (**a**) smithsonite spherules on matrix. East pit, Skorpion mine, Namibia. (**b**) willemite with minor hydrozincite. Sierra Mojada, Mexico; (**c**) hydrozincite and skorpionite on tarbuttite. Skorpion mine, Namibia. (**d**) hydrozincite on sauconite. Skorpion mine, Namibia. (**e**) hemimorphite replacing smithsonite. Angouran mine, Iran. (**f**) zincian dolomite, Berg Aukas, Namibia

and other non-sulfide minerals as hydrozincite are also present at the late stage of the non-sulfide paragenesis in late concretions and free-growing crystals in cavities. Silver mineralogy is dominated by halide phases (Fig. 6a, c). Following precipitation of carbonate and sulfate phases then clay minerals, hydrosilicates, phosphates, arsenates and vanadates precipitate (Fig. 6b, d).

Fig. 4 Typical Supergene Pb minerals in non-sulfide deposits. (**a**) Anglesite. Abenab mine, Namibia. (**b**) Cerussite. Angouran mine, Iran. (**c**) Pyromorphite after cerussite. Magellan mine, Australia. (**d**) Pyromorphite after cerussite. Ballygown mine, Ireland

Fig. 5 Typical supergene Cu minerals in non-sulfide deposits. (**a**) Dioptase in karstic vug. Okenwasi, Kunene, Namibia. (**b**) Chrysocolla, dioptase, brochantite as cement in river sediment. El Abro, Chile. (**c**) Shattuckite with minor cerussite. Van der Plas mine, Kunene, Namibia. (**d**) Malachite and shattuckite pseudomorph after bornite. Otuani mine, Kunene, Namibia. (**e**) Chrysocolla pseudomorphing chalcopyrite. Otuani mine, Kunene, Namibia. (**f**) Dioptase cementing a cave breccia. Ghuab mine, Tsumeb, Namibia

Fig. 6 "Other" minerals in non-sulfide deposits. (**a**) Bromian chlorargyrite on acanthite (BSE). Fronteria, Sierra Mojada, Mexico. (**b**) Vanadinite and descloizite pseudomorph on galena. Berg Aukas Mine, Namibia. (**c**) Chlorarygrite. Sierra Mojada mine, Mexico. (**d**) Descloizite with arsenian descloizite. Berg Aukas Mine, Namibia

4 Controls on Ore Genesis

4.1 Climate and Geomorphology

The key control in the formation of non-sulfide base metal deposits is the presence of stable climatic conditions for prolonged periods of time. Ideally, the climatic history should have periods of change from humid to arid to hyperarid, together with tectonic stability to preserve the supergene metal concentrations (Reichert and Borg 2008; Choulet et al. 2013). Fluctuating groundwater and increased rainfall will lead to dispersion, particularly for Zn, as discussed in Sect. 2, dispersion can occur even if no suitable structure or trap is encountered (Borg 2009). Most recent supergene non-sulfide deposits currently known occur between 15° and 45° N latitude (Hitzman et al. 2003), thus pointing to a specific weathering climate (warm and humid, with frequent dry spells). Based on field observations, a major difference between supergene non-sulfide Cu and non-sulfide Zn deposits appears to be the climate in which they are preserved. Copper deposits occur almost exclusively in modern arid environments, whereas Zn deposits are present in both arid and humid environments, including temperate regions such as Sardinia, Belgium and Canada (Boni and Mondillo 2015).

Stable isotopic studies of supergene non-sulfide Zn deposits show that the carbonate minerals (smithsonite, cerussite and late calcite) formed at temperatures between 10 and 20 °C (Gilg et al. 2008). This indicates depositional conditions typical of temperate zones, such as those in mid-latitudes, with seasonal contrasts including a distinct dry season. Where such deposits form in karst cavities, these temperatures could be preserved within the karst environment even in arid or tropical climates. By contrast, stable isotope measurements for $\delta^{18}O$ for azurites at Morenci in Arizona indicated requirements for higher than atmospheric partial pressures for P_{CO_2}. Oxygen isotope thermometry indicates that temperatures required are 10°C or more above ambient conditions (Melchiorre and Enders 2003). Such conditions require arid environments to be sustained for prolonged periods.

Modification of hypogene sulfides occurs in many base metal deposits, including precursors of known non-sulfide deposits (Fig. 7). Before oxidation, the primary sulfide ores may also be deformed through compression, folding and

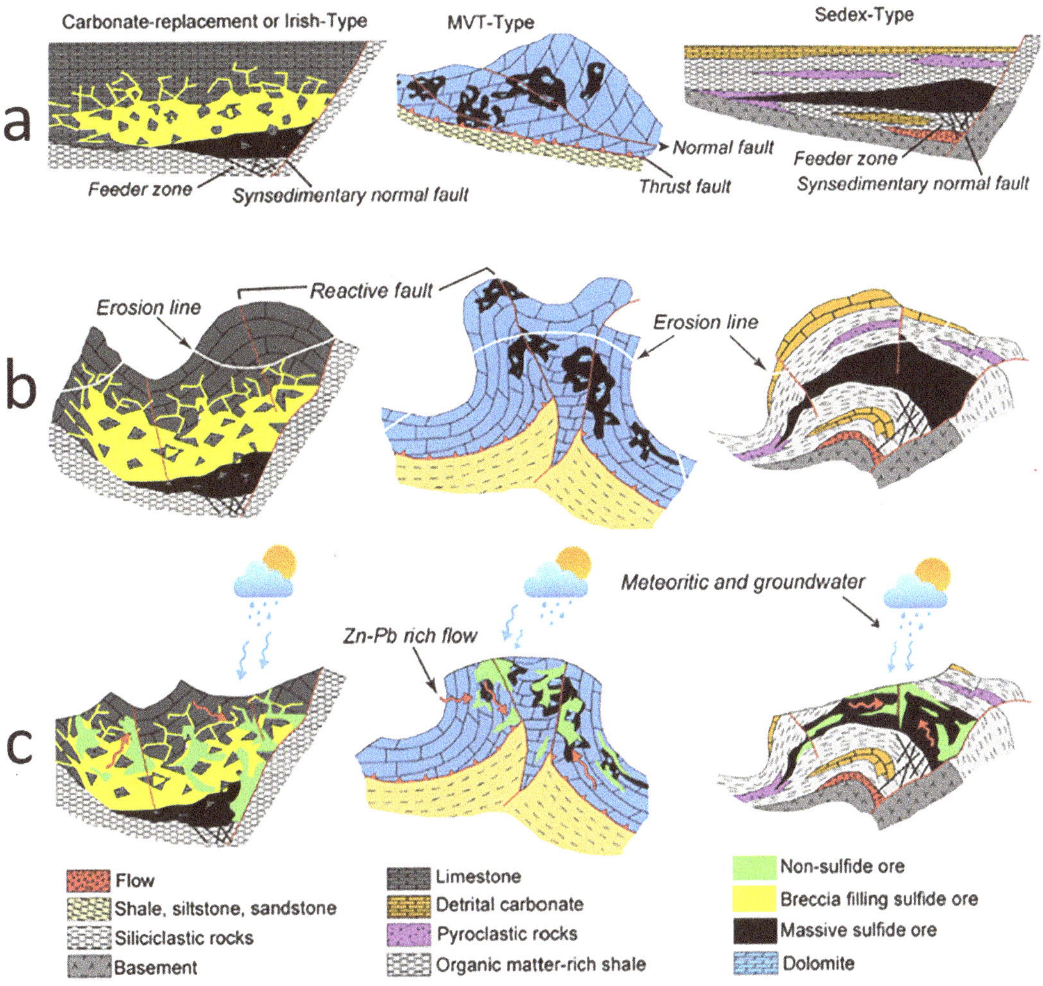

Fig. 7 Evolution of non-sulfide ores from typical Zn-Pb sulfide deposits (after Maghfouri et al. 2018). (**a**) occurrence of sulfide mineral deposits, formed by hypogene processes. (**b**) compression, deformation, uplift and erosion over prolonged period. (**c**) meteoric water interaction with sulfides, supergene oxidation, hydromorphic transport of Zn, formation of karstic structures and infilling by non-sulfide mineralization, post-formation deformation, uplift and erosion

faulting, eventually followed by uplift and erosion. Further interaction of sulfides with meteoric water leads to more supergene oxidation, hydromorphic transport of Zn particularly, modification of karstic structures and infilling by non-sulfide mineralization, post-formation deformation, uplift and erosion. These processes will overlap and develop a complex stratigraphy to the deposit (Fig. 7).

An important aspect of the climate control is the ability to develop prolonged periods of oxidation coupled with fractionation and attenuation of metals in the supergene zone (Borg 2015; Boni and Mondillo 2015). The geomorphological setting is important as it controls the flow of meteoric fluids, particularly those associated with the formation of karstic features and flow along fractures.

High Zn solubility is the main driver behind the formation of enriched Zn-dominant non-sulfide mineralization. In terms of other chemical elements, some deposits can host critical metals such as In, Ga and Ge in the ppm range, e.g. the Tres Marias deposit, Mexico Bou Arhous deposit, Central Atlas, Morocco (Choulet et al. 2017), Kabwe in Zambia (Mondillo et al. 2018a, b) and, in China and Vietnam, deposits including Kamping, Padaeng and Cho Dien (Lien and Pho 2018; Chen et al. 2023). In most of these, deposits, elements such as In, Mo and Cd are typically enriched in goethite, whereas Ga and Ge also occur in some Zn-phases, dependent on temperature and pH of formation, and affinity to the co-precipitating minerals.

Zoned willemite at Tres Marias (Saini-Eidukat et al. 2009) is unusually rich in Ge (ranging up to 4000 ppm, average = 1220 ppm) and Pb (ranging up to 2 wt.%, average = 0.84 wt.% Pb). These are the highest reported Ge contents in willemite. The oscillatory zonation observed reflects trace element incorporation into the willemite structure from the oxidation of primary Ge-bearing sphalerite and galena by siliceous aqueous fluids. At Bou Arhous, willemite is strongly enriched in Ge (up to 1000 ppm) compared to primary Zn sulfides (less than 100 ppm). Minor and trace elements also reveal a strong variability between different willemite types in the old La Calamine deposit (Belgium), although no straightforward relationship to the shape of willemite crystals or their zoning could be demonstrated (Choulet et al. 2019). Among the trace elements, anomalous high contents of P, Cd, As, Pb, Ag and Sb have been measured, as well as significant Ge contents up to 250 ppm. The concentrations measured in willemite are similar to those in sphalerite (averaging 250 ppm) (Choulet et al. 2019).

The degree of metal separation is a function of chemistry but is also influenced by geomorphology in that the flow of water, residence time and contact with reactive wallrock will also play a role in the formation or limitation of a supergene oxidation zone and development of subsurface dissolution. At the Angouran mine in Iran (Boni et al. 2007a), a smithsonite orebody hosted in marble rests on a residual Fe-rich breccia unit with residual galena (Fig. 8). This unit occurs close to surface and indicates that active mineralization has occurred until the Recent, along with physical and chemical weathering. The Angouran area is subject to seasonal rain with infiltration of oxygenated meteoric water that reacts with

Fig. 8 Classic Angouran "red ore" goethite and smithsonite with residual galena in the matrix as "islands" of non-oxidized sulfide. Specimen is 15 cm in length. (After Boni et al. 2007a)

sulfide mineralization, liberating Zn in an acidic leachate that precipitates on contact with marble and neutralizes to form massive smithsonite. Associated spring waters to the karst contain trace levels of Zn and sulfate (Borg 2015).

Under humid savannah climates with pronounced wet seasons, the oxidation of sulfide bodies results in the formation of acidic and oxidized solutions favouring karst development (Bowell 2023). These solutions become progressively metal-rich, particularly for Zn due to its greater mobility. Significant amounts of smithsonite (and lesser hemimorphite, hydrozincite and Zn-clays) may precipitate within karst cavities, thus making karst-hosted non-sulfide deposits most common in this climate setting (Boni and Mondillo 2015).

Seasonal variation in climate is therefore seen as being favourable for the formation of non-sulfide deposits. If rainfall is too frequent, Zn may be dispersed without being concentrated. Vadose zone evaporation and formation of secondary minerals lead to development of thicker supergene zones. Isotopic studies of smithsonite have shown that in arid environments, seasonal variation can lead to precipitation and preservation of a paleoclimatic signature in the carbonate minerals (Gilg et al. 2008). Steep topographic environments, associated with locally deeper groundwater tables adjacent to fault scarps or regional escarpments, can also lead to accumulation and development of thick supergene zones. A typical example is Sierra Mojada, Mexico (Megaw 2009; see Sect. 9).

4.2 Groundwater Hydrogeology and Chemistry

In addition to climate, local geology also influences the intensity of sulfide oxidation and availability of carbonate buffering to limit acidification. Karst development allows fluid flow as well as generating potential trap sites for supergene fluids to pond and saturate. Structural zones allow fluid migration, increase permeability and allow deep circulation of meteoric water that can allow a large supergene system to develop (Borg 2009).

At the Skorpion Zn mine, faults have exerted a strong control on the development of supergene non-sulfide Zn lodes through a combination of increased fracture permeability, deep penetration of high O_2- and CO_2-enriched meteoric waters and flow of Zn-bearing fluids (Borg et al. 2003). Reworking of the mineralization and metal zonation has occurred locally in the surficial zones through seasonal meteoric water infiltration (Fig. 9), developing metal-rich zones in the deposit.

Low pH groundwater generated by sulfide oxidation promotes karst development through

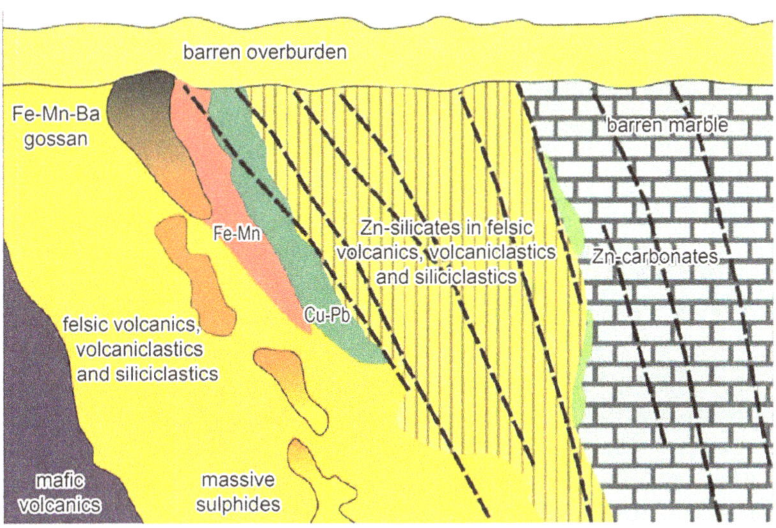

Fig. 9 Faults association with the Zn-orebody, Skorpion mine, Namibia. (After Borg et al. 2003)

reaction with limestone, which acts to neutralize acidic waters and precipitate metals that may pseudomorph minerals or rock structures. Ponding of groundwater can lead to increased salinity. At Tsumeb, Namibia, fluid inclusion data indicate solution salinity of about 25–28 wt % NaCl eq. in supergene minerals (smithsonite, cerussite, Zn-dolomite). In some carbonate-hosted ores, gypsum may form from reaction with acidic solutions and this will effectively shield the limestone from further chemical weathering (Reichert 2009). The presence of goethite and other ferric oxyhydroxide phases that occur as a product of sulfide oxidation is important in the adsorption of metals (Bowell 2023).

4.3 Geological Influence

Geological processes on both a macro and local scale influence the formation of non-sulfide deposits. At a local scale, the nature of the source rocks is important in influencing the nature of the trap zone, mineralogy and morphology of the deposit. Ponding of supergene fluids, especially in karst environments can lead to mass precipitation, particularly of carbonate minerals. In arid environments, this is promoted by high evaporation rates and solute loading in groundwater.

Skorpion, southern Namibia, is the classic example of a siliciclastic- or volcanic-hosted non-sulfide deposit (Borg et al. 2003) in which changes in geology and groundwater chemistry have led to zonation in Zn mineralogy. The protore consists of base metal sulfide ores in late Proterozoic metasedimentary and metavolcanic rocks. Dense faulting and fracturing created a high reactive surface area for the precipitation of Zn from metal-rich supergene fluids. Zonation in the ore has led to development of a proximal gossan-like unit rich in Fe and Mn oxyhydroxides (Fig. 9). A zone of Zn mineralization hosting high Cu concentrations follows along with a Pb-bearing zone dominated by cerussite. The main Zn-orebody which forms almost all of the economic mineralization lies within the metavolcanic and other silicate units and shows lateral and vertical zonation. In siliciclastic rocks and volcanics, Zn-silicates are dominant, whereas, in the marble, Zn-smithsonite prevails. Vertical zonation is shown by the dominance of tarbuttite, sauconite and hydrozincite near the surface, while smithsonite is more common at depth in the marble and hemimorphite in the volcanic rocks (Borg et al. 2003). This zonation also reflects changes in water pH and CO_2 content. In the vadose zone above the seasonal water table, CO_2 is exsolved, lowering CO_2 activity in meteoric water and thereby favouring the formation of phosphate and hydroxide Zn-phases. Below the water table, there is an increased CO_2 activity due to dissolution.

On a more regional scale, a fundamental requirement is tectonic uplift, so that near-surface oxygenated meteoric waters can contact and oxidize sulfide mineralization and transport metals to precipitation sites. As can be observed in Fig. 4, the zone in which metals are soluble is narrow, so even small fluctuations can result in the difference between the dispersion and precipitation of metals.

5 Age of Non-Sulfide Deposits

There are many differences among the possible ages of non-sulfide deposits and the methods used to calculate them. Of course, because some of these deposits can be considered as hypogene, the geochronometers applied to date sulfide ores directly could eventually be used to date some non-sulfides. A typical example is willemite from the Berg Aukas Mine, Namibia (see Sect. 10), which has been considered at least partly suitable to be dated with an Rb–Sr geochronometer (Schneider et al. 2008). Its age has been established at 499 ± 63 Ma and 493 ± 2 Ma, not very dissimilar from the 435 Ma age calculated for willemite at the Beltana-Aroona non-sulfide deposit, Australia by K–Ar dating of coronadite (Groves et al. 2003).

The stage I carbonate ores at Angouran (Iran), which are the most abundant non-sulfides in the mining district, suggests their hypogene formation from low-temperature hydrothermal fluids (Boni et al. 2007a; Daliran et al. 2013). These fluids were probably mobilized during the waning stages of Tertiary–Quaternary volcanic activity when travertine systems were abundant throughout the whole region.

Another example of direct dating of non-sulfide mineralization is the application of the (U-Th)/He thermochronometry to define the possible ages of vanadate ores (descloizite and mottramite) in the Otavi Mountainland, Namibia (Boni et al. 2007b). The (U-Th)/He ages fall into older and younger populations, the older ranging from 26 to 47 Ma and the younger at 0.35 to 1.42 Ma. More reliable ages were reported by Pickford (1993) for the vanadates occurring in some karst fillings, which are paleontologically constrained to the middle to upper Miocene and Pliocene to Pleistocene.

Many other methods to indirectly date supergene non-sulfide ores depend on local geology, depth of primary sulfides and their pyrite contents, nature of the host rocks, tectonic deformation and uplift, and climatic variability.

6 Case Study: Supergene Cu deposit, Otuani Mine, Opuwo District, Kunene Region, Namibia

6.1 Introduction

The Otuani mine is located in the Opuwo district of northern Kunene in NW Namibia. Supergene Cu mineralization occurs in these areas within karstic dolomites as malachite and Cu silicate precipitates and shows elements of ore replacement, wallrock replacement and karstic infill (Fig. 10). The mine has been operational since 2008 and produced more than 30,000 metric tonnes of Cu from oxide zone deposits (Bowell et al. 2012).

Fig. 10 Aspects of supergene non-sulfide copper mineralization, Kunene region, Namibia. (**a**) Wallrock replacement along siltstone-shale lenses within massive dolomite. Mineral assemblage comprises chrysocolla, dioptase, shattuckite, planecheite and malachite. Figure in photograph is approximately 1.6 m. Old German Mine, Kuene. (**b**) Ore zone replacement along mineralized zones with cuprite, chrysocolla, hematite and brochantite replacing chalcocite and pyrite in quartz and calcite veins. Okenwasi pit, Kuene. (**c**) Karstic vug within the mineralized zone with malachite pseudomorph after azurite associated with chrysocolla pseudomorph after cuprite. Okenwasi pit, Kuene. (**d**) Karstic replacement along faults and joints in unmineralized dolomite. Kuene. (**e**) Mottramite, malachite on folded siltstone within massive dolomite associated with hematite alteration of pyrite. Van der Plas mine, Kuene. (**f**) Shattuckite with malachite, within a Karst vug. Okenwasi pit, Kuene. Overall size is 30 cm in length and 16 cm in width (photograph courtesy of W. van der Plas)

6.2 Climate and Topography

The climate in the field area is a hot semi-arid desert with rainfall being less than 400 mm pa mainly falling in the period January to March, with a pronounced dry season from May to November. The temperature range varies from 15–35°C over the year but on a diurnal basis can vary from less than 10°C at night to over 25°C during the day. The topography is a series of undulating hills that rise up to 2000 m amsl and are typically in the region of 1100 m amsl. The regolith varies considerably in thickness and is covered by gravel and silt soils with lateritic duricrust developed in upland areas and, in basins, reddish sandy soils of the Kalahari Formation, in places up to 4 m thick, thickening to the east.

6.3 Geology

The geology of the area comprises sedimentary rocks of the Otavi Group of the Damara System (900–650 Ma). These weakly metamorphosed and dominantly carbonate formations spread out in a wide arc around the western and southern rims of the Etosha Basin, from the Angolan border north of Opuwo in the north and extending eastwards to Tsumeb (Miller 2008). The Damara System, which represents a carbonate platform sequence that extends throughout northern Namibia, consists of regionally deformed and gently folded arenite and calcareous chemical sedimentary rocks that have undergone low-grade metamorphism. The oldest unit in the area is the Nosib (900–750 Ma), which occurs as siltstone and intermittent sandstone in the core of several anticlines. More widely occurring is the overlying Otavi Group (750–650 Ma): this comprises the Abenab Subgroup, a sequence of predominantly carbonate rock types including limestone, dolomite, calcareous arenite, and argillite, with minor oolitic limestone and evaporate. Deformation of the sequence produced a kilometre-scale south plunging folded sequence, with the outer limbs comprised of the Chuos Formation ironstones. These form a prominent V-shaped outcrop pattern with high hills that flank both sides of the licence area.

Later regional metamorphism led to some minor recrystallization to form slate and quartzite. There is evidence of a second, minor episode of deformation with a crenulation cleavage that developed in some of the shale showing NE-SW compression.

6.4 Mineralization

The primary ore comprises mainly chalcocite, minor bornite, galena and pyrite. Tetrahedrite, chalcopyrite and sphalerite are present although rare. Silver in the mineralization appears restricted as a trace element associated with galena. The main gangue minerals are quartz, chlorite, calcite and baryte.

Oxidation processes are developed in the area, with the zone of oxidation to a depth of 50 m observed in workings and drill core (Bowell et al. 2012). Oxidation and weathering of the primary ore deposits have led to supergene Cu enrichment at the base of weathering. The common presence of pseudomorphs and relative in situ nature of the oxidation indicates that Cu dispersion was limited.

6.5 Hydrogeochemistry

As observed in a Pourbaix diagram developed for the groundwater chemistry at Kunene (Fig. 11), the Zn^{2+} ion is predicted to be more stable at higher pH than other metal ions and would be stable within the measured field of groundwater Eh-pH for the natural system. Copper, Pb and Fe under similar conditions are all predicted to have speciation dominated by solid phases. The mineralogy of supergene non-sulfide deposits reflects the nature of the host rocks and of groundwater fluids that migrate to the precipitation site via structural and lithological controls (Borg 2009).

Whereas goethite, malachite, cerussite, cuprite and anglesite are predicted to be stable under the prevailing groundwater chemistry, Zn is

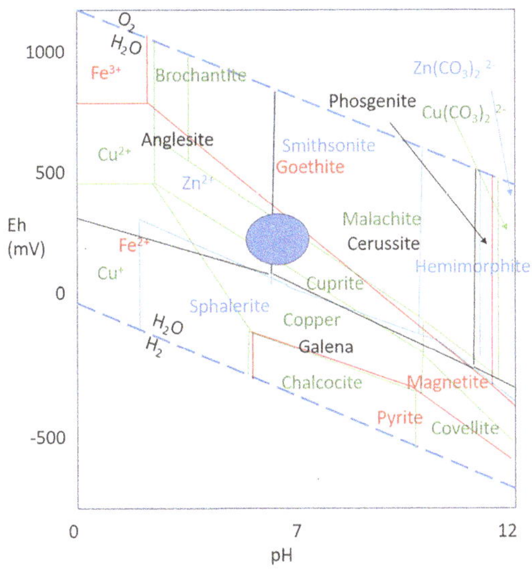

Fig. 11 Pourbaix (Eh-pH) diagrams for groundwater in the Kunene karst environment at 298 K (blue circle indicates typical groundwater Eh-pH conditions)

predicted to be present as Zn^{2+} or, at higher pH, as smithsonite. Thus, Zn is mobilized away from the other metals and forms a wider but lower concentration halo around the mineralized lodes in the regolith and surface sediments (Fig. 12).

Zinc mobility can also lead to the incorporation of Zn into gangue phases such as dolomite (Boni et al. 2011a). In some deposits, Zn-dolomite can form a significant host for Zn in the supergene environment (up to 20–25% at Jabali in Yemen, Mondillo et al. 2011). Aragonite and calcite can also host some Zn but at much lower concentrations.

7 Case Study: Supergene Zn Deposit, Bongarà District, Amazonas, Peru

7.1 Introduction

There are several small but high-grade Zn deposits dominated by smithsonite in the Amazonas region of northern Peru (Boni and Mondillo 2015; Arfè et al. 2018; Chirico et al. 2022). Mineralization consists of mixed sulfide and non-sulfide ores, with the non-sulfide ores being the product of weathering of primary sulfide bodies (Chirico et al. 2022). One example of these is the Cristal prospect in the northern part of the Bongarà district, approximately 740 km north of Lima (Fig. 13).

As discussed above, alternating humid tropical and arid climatic conditions favour intensive sulfide oxidation, regolith development and formation of supergene non-sulfide deposits (Hitzman et al. 2003; Reichert 2009; Boni and Mondillo 2015). Supergene Zn deposits are commonly found in both modern arid and humid tropical environments, as well as in paleo-environments as described above. The geology of the Bongarà area is dominated by carbonate rocks with well-developed karst features that are potential trap rocks for supergene deposits (Arfè et al. 2018; Chirico et al. 2022).

7.2 Geology

The host rocks in the mining district are the early Jurassic carbonates of the Pucarà Group, which in turn overlies a Paleozoic basement metamorphic complex. In the Bongará area, the sedimentary sequences that are representative of a rift phase consist of red sandstones and polymictic continental conglomerates. The Pucarà Group contains platform carbonates and deeper argillaeous limestones (Fig. 14). During regional uplift and erosion karst features developed in the platform carbonate sequence.

Late Andean tectonics affected the Bongará area from the Oligocene to mid-Miocene (20.5 to 12 Ma; Pfiffner and Gonzalez 2013). Further tectonic pulses have been recognized during the late Miocene (~8–7 Ma), Pliocene (~5.4 Ma) and Pleistocene (~2.5 Ma).

7.3 Paleoclimate from Miocene to Recent

A comprehensive study on the Cristal deposit assessed the evolution of the climate from the Oligocene through to the recent and utilized marine paleotemperatures derived from oxygen

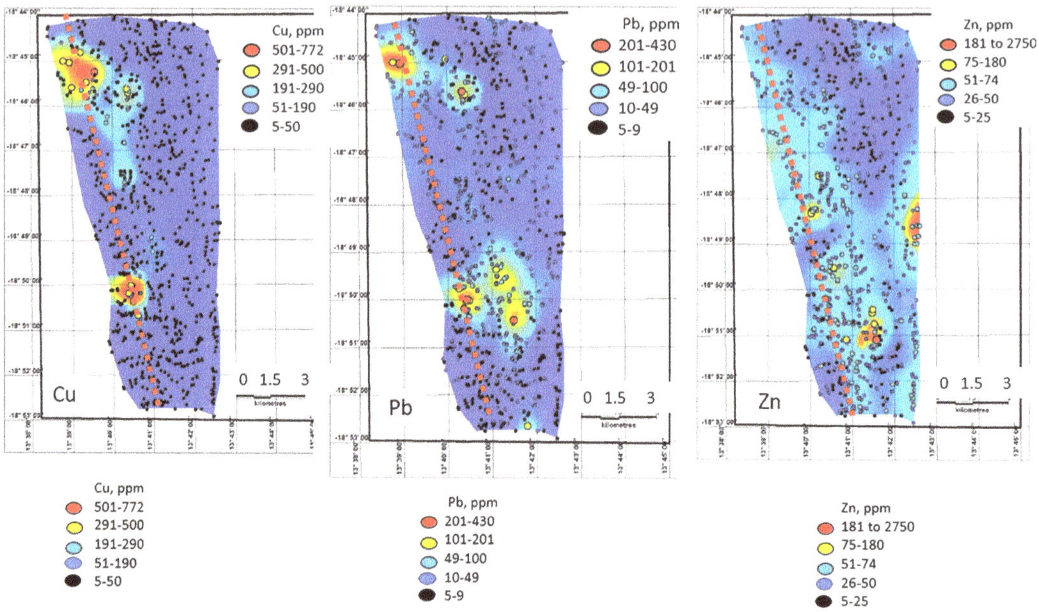

Fig. 12 Concentrations of Cu, Pb and Zn in surface regolith and sediments shown as 'heat maps' over the Kunene Licence (scale bar is 3 km). Mineralization in bedrock is shown by the dashed red line. (Taken from Bowell et al. 2012)

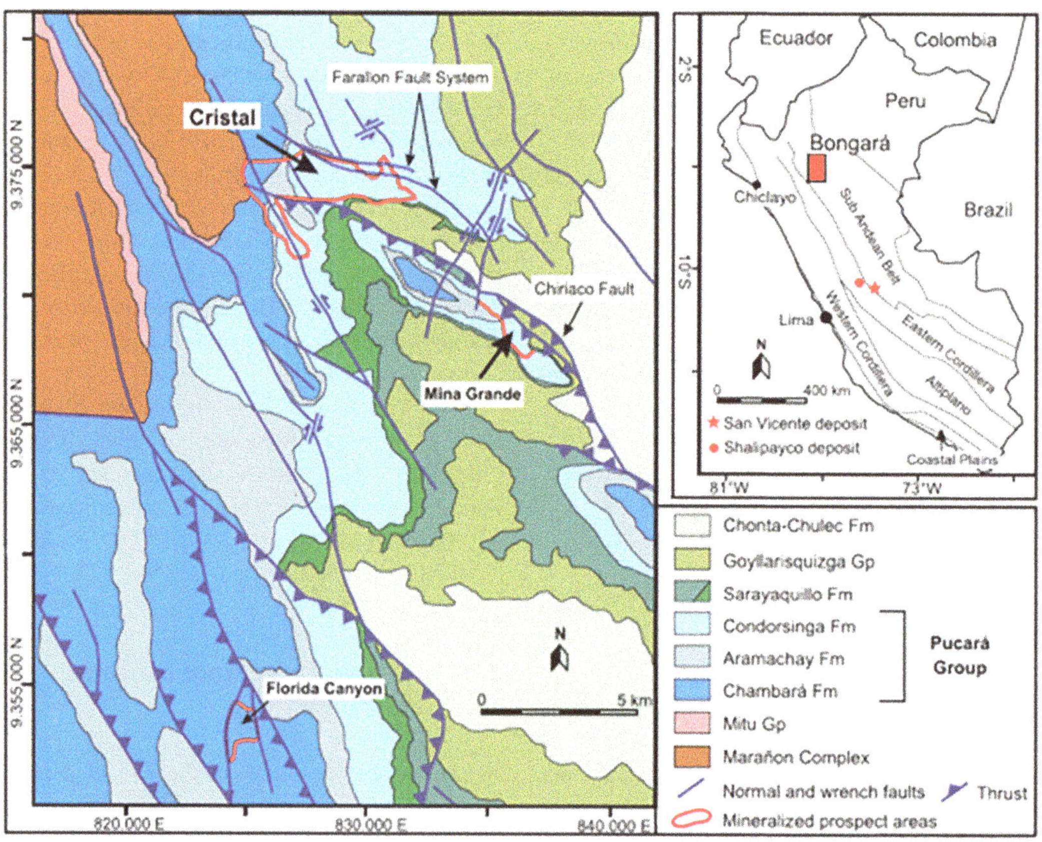

Fig. 13 Location and geology map, Bongarà district, northern Peru. (Modified after Arfè et al. 2018)

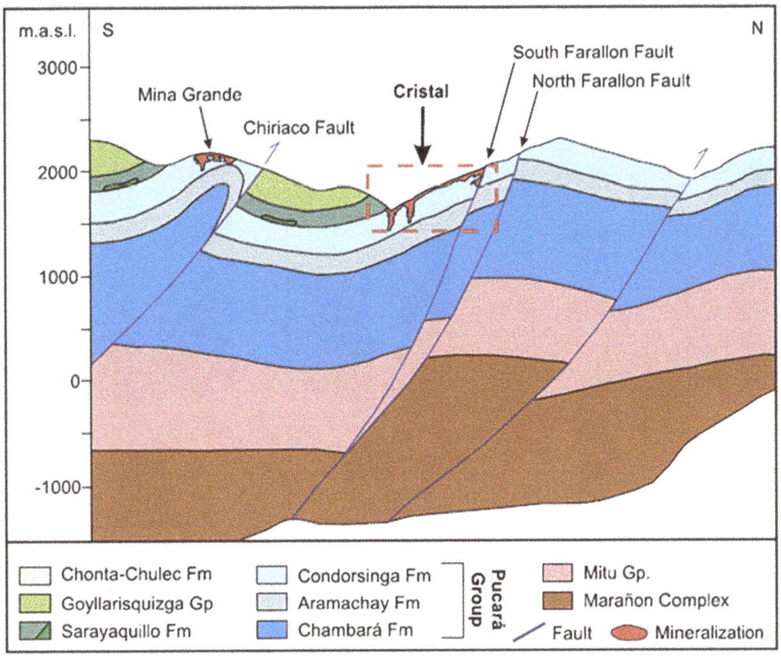

Fig. 14 Schematic section of the Mina Grande and Cristal (dashed red squares) mineralized areas in Peru. (Modified after Arfè et al. 2018)

isotope analyses of Amazonian bivalves, as well as paleontological, palaeobotanical and palynological studies to reconstruct the climatic patterns in the northern Andes (Arfè et al. 2018). Detailed assessment of the Tertiary environment of the Amazon indicates a prolonged humid climate (Hoorn et al. 2010). From the late Oligocene (~ 23 Ma) to the late Miocene (~10 Ma) a large wetland termed "Pebas" occupied the western Amazon basin (Fig. 15a).

The area consisted of shallow lakes and swamps that prograded west from ~12 Ma onward, contemporaneous with the development of the Northern Andes. Subsequently, an uplift at about 10 Ma promoted the onset of the formation of the Amazon River basin (Hoorn et al. 2010), during which a fluvial-tidal-dominated wetland, "Acre" (Fig. 14b), developed. The latter disappeared around 7 Ma and was replaced by the current rainforest ecosystem (Hoorn et al. 2010). The present climate is equatorial, that is, hot and rainy all year round with precipitation averaging around 160 mm per month for most of the year. The Cristal Zn deposit is located between 1600 masl and 2000 masl, and has an average annual temperature range from 23° to 25 °C.

7.4 Non-Sulfide Zinc Mineralization

Supergene non-sulfide Zn deposits derived from weathering of primary sulfides have been described within the Bongará district for Mina Grande (Arfè et al. 2017), Cristal (Arfè et al. 2018), and Florida Canyon (Chirico et al. 2022). These have been described as "ore" (in situ) and wallrock ("hydromorphic replacement") styles of Zn mineralization (using the Hitzman et al. 2003 classification).

The Mina Grande deposit, situated in the hanging wall of a reverse fault, is completely oxidized. The mineralization occurs as several lenses, which are associated with faults and fault-developed karstic structures. Pulses of uplift and erosion at Mina Grande probably led to progressive stages of erosion and exposure of the precursor sulfide orebody (Arfè et al. 2017). At least two stages of oxidation occurred, with early coarse-grained smithsonite and minor hemimorphite and hydrozincite partially replaced and cemented by a later generation of finer-grained hemimorphite, hydrozincite and calcite. The different stages were formed during pulses of uplift and erosion in the Miocene and the Pliocene-Pleistocene. Karst features host Zn-rich

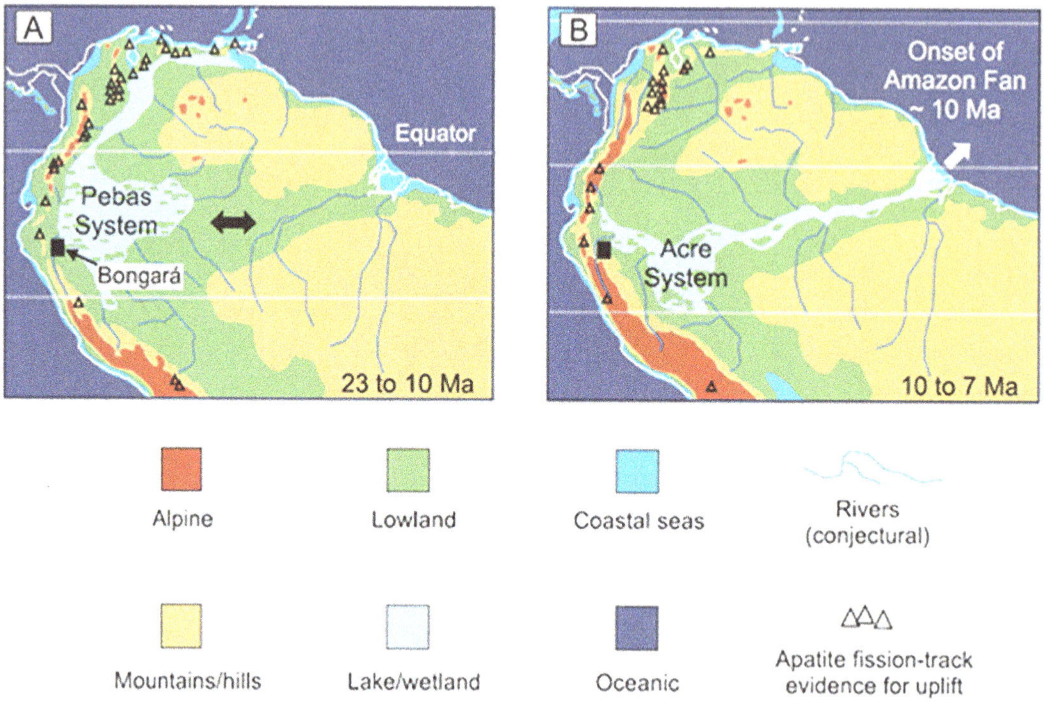

Fig. 15 Paleogeographic maps of the North Andes, showing the transition from "Pebas" to "Acre" mega-wetlands and the location of the Bongará district. (From Hoorn et al. 2010; Arfè et al. 2018)

layers of these minerals that can be more than 100 m thick. Dolomite within the deposit contains appreciable amounts of Zn, similar to other deposits elsewhere in Peru such as Accha (Boni et al. 2009).

The non-sulfide Zn layers have an overall NE-SW strike dominant trend at Cristal, and a dip toward NW at angles between 30° and 60° (Arfè et al. 2018). Zinc grades can vary from about 17 wt % through to 30 wt % in the different lenses, with higher grades associated with direct sulfide replacement (ore replacement) zones compared to wallrock replacement or karst zones. Within the host rock, silicate layers interbedded with dolomite rock have a different Zn mineral assemblage, with smithsonite associated with sauconite, Zn-Mn-bearing micas, goethite and chalcophanite. Later smithsonite occurs in veinlets and as crusts within cavities and on wallrock.

The Florida Canyon deposit (16 km SSE from Cristal; Fig. 12) is hosted in a broad anticlinal structure consisting of dolomitized limestones. Ore mineralization comprises hypogene (sphalerite, galena, pyrite, marcasite) and supergene (mainly smithsonite and hemimorphite) mineral assemblages (Chirico et al. 2022). Smithsonite and hemimorphite occur as in situ replacements of sphalerite, generally preserving the original texture. Cerussite and anglesite are more restricted in occurrence. The oxidized mineralization is a polymetallic replacement deposit. However, its shape is distorted by fault-related karst features and propagated faults that gave rise to a "Christmas-tree" like pattern. The karst activity at Florida Canyon affected the carbonate rocks of the Chambará Formation during pre- and post-ore stages. Karstification controlled a limited remobilization of the supergene ore (Chirico et al. 2022).

The mineralogy of the non-sulfide bodies was determined by (1) uplift rates, and (2) host rock lithology. Uplift was controlled by the activity of local faults, which in the Bongará area allowed the exposure of the sulfide protore at different altitudes during distinctly different time periods.

7.5 Isotopic Data of Supergene Non-Sulfide Mineralization

The $\delta^{18}O$ of sulfide-replacing first-phase smithsonite in the Cristal prospect is between 24.6 and 25.8‰ (Arfè et al. 2018), with an average value of 25.1‰ VSMOW. In contrast, the $\delta^{13}C$ values of the same smithsonite show a very large variation, ranging from −11.7‰ to −0.6 VPDB.

The host rock-replacing second-phase smithsonite has $\delta^{18}O$ values varying between 26.6 and 27.3‰ VSMOW (mean 26.9‰). These values are higher by ~2‰ than those recorded for the first phase of smithsonite but by comparison, the $\delta^{13}C$ values are comparable, with $\delta^{13}C$ of the latter ranging from 0.1 to −11.7‰ VPDB.

The $\delta^{13}C$ and $\delta^{18}O$ compositions of the supergene carbonate minerals vary with the host lithotypes and the presence of texturally different mineral assemblages, but they do not show appreciable variations between the different macrotextures such as cements of breccias or massive bodies (Gilg et al. 2008; Arfè et al. 2017; 2018; Mondillo et al. 2018a, b).

Different $\delta^{13}C$ values of otherwise texturally comparable minerals are associated with different host rocks. The most negative $\delta^{13}C$ values of smithsonite (−11.7‰ VPDB) and calcite (−7.4‰ VPDB) have been measured in siltstones and sandstones, whereas the most positive values (0.7 and 0.1‰ VPDB for calcite and smithsonite, respectively) were obtained from dolomite-hosted layers. The $\delta^{13}C$ variations are not significant among the different mineral generations, but $\delta^{18}O$ values become progressively higher from the early sulfide-replacing smithsonite to the later generations of dolomite and open space fillings (Fig. 16).

Additionally, the sulfide- and dolomite-replacing smithsonite indicate clear variation in $\delta^{18}O$ within the mineralized body with values in the centre of the body higher than those at the boundaries in wallrock, and this correlates to zones of smithsonite development. The final generation of smithsonite and supergene calcite (both open space filling) co-precipitated from a fluid with a $\delta^{18}O$ value around −5.1‰ VSMOW (Table 1).

Studies of the 'non-traditional' isotopes, such as Zn ones, have also been applied to these deposits (Gilg et al. 2008; Mondillo et al. 2018b; Kyser et al. 2020). These isotopes undergo fractionation as a function of changes in redox, supergene weathering and associated mineral transformations, secondary mineral precipitation and adsorption/desorption reactions, among others. Interesting observations have been made for Zn isotope fractionation in secondary Zn minerals (Mondillo et al. 2018b). For Zn, partial sphalerite dissolution is followed by precipitation of initial supergene phase that preferentially incorporates heavier isotopes. The isotopic

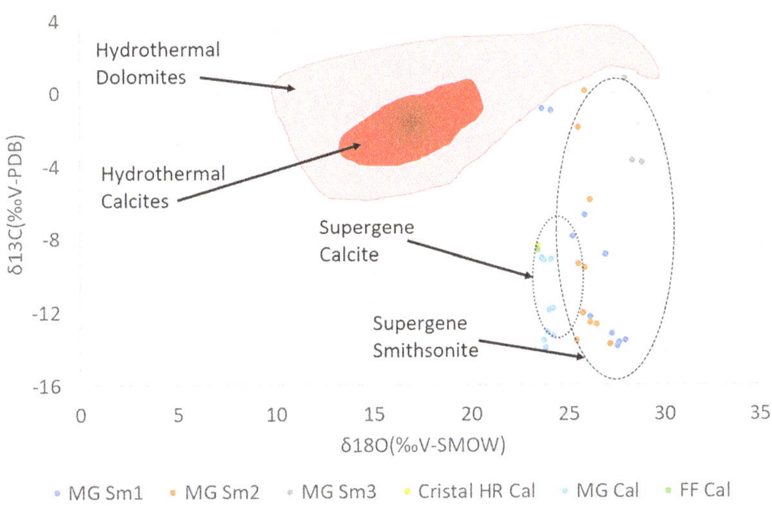

Fig. 16 $\delta^{13}C$ and $\delta^{18}O$ values of supergene carbonate minerals at Cristal, comparison with published isotope values of supergene smithsonites and calcites (supergene smithsonite and calcite fields, Gilg et al. 2008) (after Arfè et al. 2018). SM1 is sulfide-replacing smithsonite; SM2 is dolomite-replacing smithsonite; SM3 is open space filling smithsonite; FFCal is open space calcite. (Data for Arfè et al. 2017, 2018; Mondillo et al. 2018b)

composition does depend not only on the difference between the fractionation factors of the involved phases but also on the amount of the secondary mineral precipitated after the dissolution of primary sulfide (Mondillo et al. 2018b). Progressive precipitation from migrating fluids that form phases enriched in heavy zinc isotopes would lead to a gradual decrease in the $\delta^{66}Zn$ values of such phases, and the fluids involved, in time and space. Strong negative isotopic shifts are almost only observed for late crystallizing phases, such as those in vugs. These are interpreted to reflect precipitation from residual, isotopically light fluids that are the inevitable highly-fractionated product of the above-described process.

7.6 Genesis of the Non-Sulfide Mineralization

The Cristal Zn non-sulfide mineralization post-dates hypogene sulfides in the Bongarà mining district, and occupies different lithotypes, leading to the formation of smithsonite- and hemimorphite-rich assemblages. Mineralization occurs as *mantos*-like lenses. As identified in other studies, faults and karstic structures along faults are important loci for formation of non-sulfide base metal orebodies (Borg 2009; Boni and Mondillo 2015). Smithsonite and hemimorphite occur as paragenetically distinct generations that replace hypogene sulfides in veins (first generations), saddle dolomite in the host rock (second generations), and as late infilling of open spaces cutting the previously precipitated phases (third generations).

The supergene mineralization was initiated by precipitation of siderite as veinlets or as inclusions in smithsonite. This reflects Eh variations from slightly reducing to oxidizing conditions of the supergene solutions during the initial stages of sulfide weathering. Reducing conditions, necessary for siderite precipitation in a supergene environment, can be achieved through microbial activity in subsurface sediments, or by local variations in groundwater, possibly related to seasonal climatic changes or tectonic activity (Mortimer et al. 1997). The presence of greenockite (CdS) as crusts also indicates reducing conditions early in the supergene cycle. In contrast, otavite ($CdCO_3$) has been observed in the Mina Grande deposit (Arfè et al. 2017). The difference between the two deposits is that less oxidizing and alkaline weathering conditions prevailed during the formation of the Cristal deposit (Arfè et al. 2018).

At Cristal, smithsonite occurs as pseudomorphs after sulfides, indicating minimal transport in solution, most likely due to buffering and precipitation from the host dolomite of any generated acidic solution. The $\delta^{13}C$ composition of smithsonite phases clearly indicates a mixed contribution of both isotopically heavy carbon (from carbonate host rocks) and isotopically light carbon (from possible organic matter in the regolith). The $\delta^{13}C$ values of the Cristal supergene carbonates are consistent with sulfide oxidation occurring in an "open system", characterized by a fresh supply of organic and wallrock carbon to groundwater (Gilg et al. 2008). The highest $\delta^{13}C$ values occur in dolomite-hosted smithsonite whereas in siliciclastic sediments Zn-clays and hemimorphite are present. In terms of $\delta^{18}O$ values, an increase in the ratio is observed in smithsonite from rim to the core zone of mineralization and most likely this reflects an evolution of the groundwater system over time. Mineral chemistry changes also occurred as the supergene system evolved with the dissolution of dolomites and ion exchange leading to lower Fe and higher Zn. The leaching of Zn and Mn from micas leads to a common association of Zn minerals with chalcophanite. At the Mina Grande deposit the most abundant Mn-hydroxide is rancíéite ($(Ca, Mn^{2+})_{0.2}(Mn^{4+}, Mn^{3+})O_2\ 0.6H_2O$; Arfè et al. 2017). Its presence, rather than that of chalcophanite, reflects the intensity of weathering (Nimfopoulos et al. 1997). Buffering of acidic solutions by carbonate minerals and concurrent increase in pH would promote neutral groundwater and lead to rancíéite precipitation indicating more prolonged weathering than at Cristal. The common presence of hydrozincite similarly reflects prolonged weathering at Mina Grande. A karst-fed flux of groundwater resulted in the widespread propagation of a supergene weathering front, with smithsonite replaced in the surficial zones by hydrozincite. In near-surface conditions in the vadose zone, where CO_2 fugacity approaches atmospheric values, hydrozincite and hemimorphite have replaced smithsonite (Hitzman et al. 2003; McPhail et al. 2003; Reichert and Borg 2008; Reichert 2009). In comparison, at Cristal, the lack of hydrozincite and the preservation of smithsonite indicates minimal post-oxidation alteration.

8 Case Study: Supergene Pb Deposit, Paroo Station Mine, Western Australia

8.1 Introduction

The Paroo Station Mine Pb deposit (formerly Magellan) is located 30 km W of Wiluna, Western Australia, at approximately 26.52 S latitude and 119.89 E longitude. The project includes six non-sulfide Pb deposits: Magellan, Cano, Pinzon and Gamma in the Magellan Hill area, and Pizarro and Drake in the Finlayson Range area to the SW. The deposit description is mostly derived from Sergeev et al. (2017).

8.2 Physical Setting

The climate is semi-arid, with dry periods of up to nine months, usually commencing before the end of August. Temperatures range from cool during winter, with a mean daily minimum of 5 °C in July to very hot in summer, with a mean daily maximum of 38 °C in January. The average annual rainfall is 249 mm, varying from a high of 712 mm to a low of 49 mm.

There are four main landforms: long ridges, small, isolated mesas, scarcely wooded colluvial slopes and densely vegetated drainages. The Magellan, Cano and Pinzon deposits are situated on a mesa about 5 km by 2.5 km, raised above the surrounding alluvial plain by approximately 25 to 50 m to an elevation between 550 to 570 m amsl. Pizarro and Drake are on low hills. West Creek, part of the Paroo Palaeochannel, is the major local drainage system and aquifer, that flows SE towards Wiluna. The drainage channel is marked by thick vegetation and a surficial valley calcrete.

8.3 Geological Setting

The Paroo Station Mine deposits are located in the southeastern corner of the Paleoproterozoic Yerrida Basin in the sedimentary outlier of the onlapping Earaheedy Basin (Fig. 17).

The Basin comprises coastal to outer shelf, mostly siliciclastic, sedimentary sequence, with a local basal fluvial sandstone and conglomerate (Fig. 18). The formation is interpreted to have deposited at a passive continental margin along the northeastern margin of the Yilgarn Craton (Jones et al. 2000; Pirajno et al. 2009). The mineralization is hosted by a transgressive marine succession of sandstones, shales and carbonates of the Yelma Formation at the base of the Earaheedy Group.

A major NNW- and E-striking fault system transects the Yerrida Basin and the Archaean basement beneath the Paroo Station Mine. Two minor faults crosscut and slightly displaced sediments of the Mooloogool Group. One fault trends NW, parallel the southeastern edge of the Magellan Hill, with the second fault striking W-E on the northern side of the Hill (Etheridge and Henley 1994). The Yelma Formation sediments in the Magellan area are interpreted to have undergone minor structural deformation. The sedimentary packages form a monocline gently dipping to the NE at up to 5°. Dips greater than 50° along the southeastern edge of Magellan Hill are thought to be related to paleo erosion of the Maralou unconformity.

The residual regolith in the deposit area is commonly overlain by thin colluvial soil. The weathered zone varies in thickness from several metres on the plain to up 40–50 m below the mesas. The hilltops are occupied by quartz-clay collapse breccia unit up to 45 m thick (20 m on average). Locally, the breccia, principally the product of weathering of the Yelma dolomite, is underlain by the relic dolomite horizon (Fig. 17).

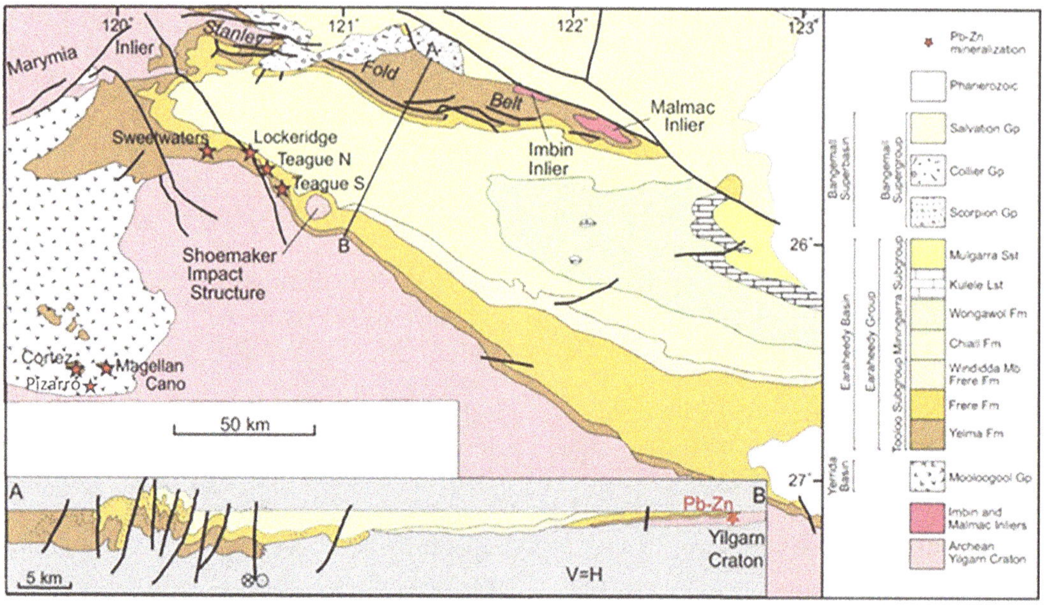

Fig. 17 Generalized geological map and stratigraphy of the Earaheedy Basin. (Modified after Muhling et al. 2012) showing the major Pb–Zn prospects and mines. Inset presents an N–S cross-section of the basin (Muhling et al. 2012) after Bunting et al. (1986)

The breccia is commonly light grey to creamy-brown in colour with occasional mottled or wavy ferruginization textures. This rubble collapse-type breccia mainly comprises chert, drusy and saccharoidal chalcedony and quartz fragments up to 10 cm within the clay and fine-grained quartz matrix. The upper part of the breccia unit is strongly silicified, ranging in texture from loose siliceous gravel to well-cemented rock. Silicification gradually intensifies upwards, commonly with a bed of massive silcrete at the top. The breccia unit is underlain by structureless clay and siltstone saprolite (Fig. 19).

The siltstone unit is generally less than 4 m thick at the Magellan and Pinzon deposits. The sandstone below is altered to saprolite grading saprock, with the basal few metres of the unit being commonly unweathered. At Pinzon, weathering extends deeper, altering the top several metres of the black carbonaceous Maralou shale below into purple to purple-brown saprock.

8.4 Deposit Characteristics

8.4.1 Mineralization

The mineralization is poorly expressed on the surface on hilltops, but Pb anomalies to over 1000 ppm Pb occur on slopes and around the bases of mesas. Cerussite was also visible in sandstones locally outcropping on slopes of the Magellan Hill. Lead mineralization at Magellan Hill occurs as a series of stratiform, generally subhorizontal sheets contained completely within the Yelma Formation. Mineralization at Pizarro and Drake occurs within the Yelma Formation and also along the unconformity between the Juderina Formation (Windplain Group) and small outliers of the Earaheedy Group. The mineralization is Pb-only and consists of supergene minerals; Zn concentrations are low, reaching a maximum of 0.5% at the base of the breccia unit. The distribution of Pb in the deposit indicates higher Pb at the base of the breccia (Fig. 20).

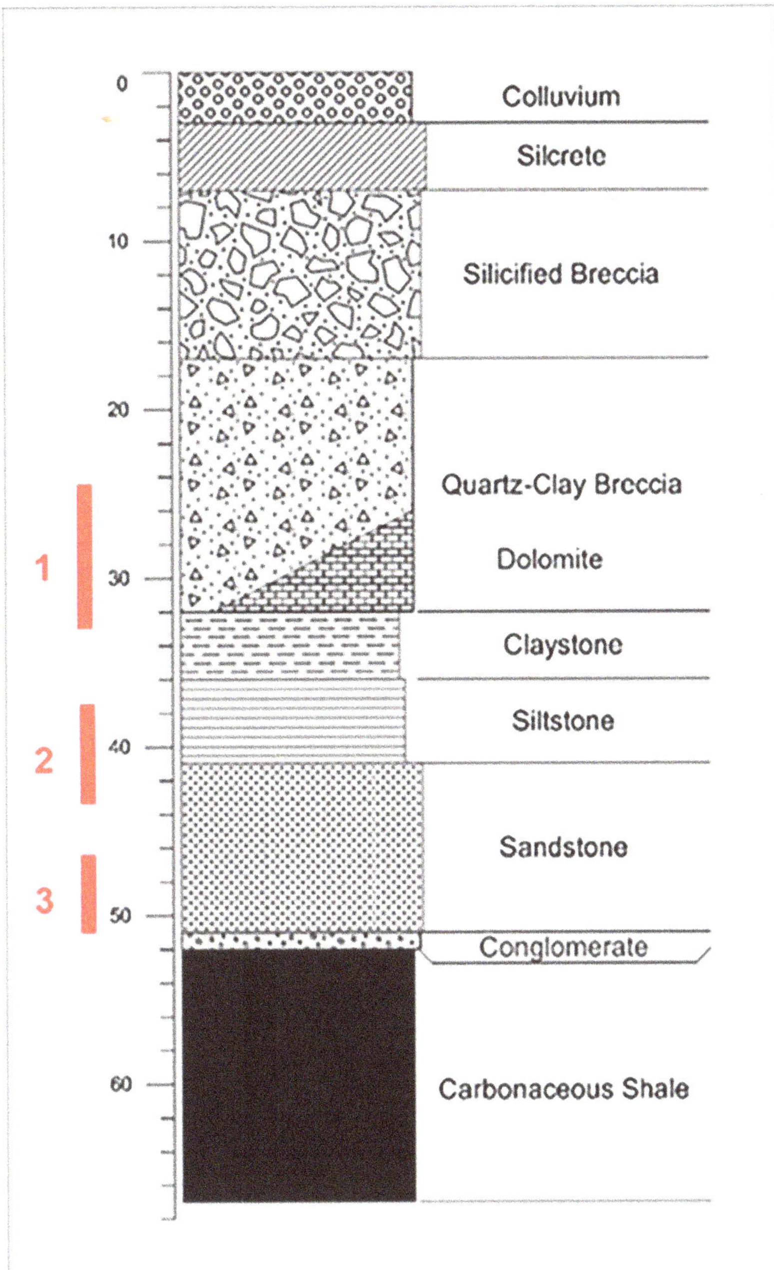

Fig. 18 Generalized stratigraphic column in the Magellan Hill area. Red lines represent the vertical positions of the high-grade mineralization zones 1 to 3. (Modified after Sergeev et al. 2017)

Positions of the high-grade zones generally approximate to NE, NW and N trends, suggesting potential control by a network of vertical structures within the carbonate rocks. However, most of the protore low-grade mineralization was probably stratiform (see Sergeev et al. (2017) for further details). Abundancies of Zn, Cu, As and Ba are low in the deposit, although each element locally reaches elevated levels. Similar to Pb, these elements are depleted in the upper silicified breccia and enriched in the two zones below. Two stratiform enriched horizons range from 4 to 24 m thick (at a 100 ppm cutoff). An upper Zn-Cu-As-Ba enrichment commonly envelops the two upper Pb-mineralized zones although the vertical positions of their peaks can vary by several

Fig. 19 Weathered units of the Yelma Formation, the southern wall of the Magellan A open pit: Bx—Breccia; Dm—Dolomite; Cy—Claystone; St—Siltstone; and Ss—Sandstone (Sergeev et al. 2017)

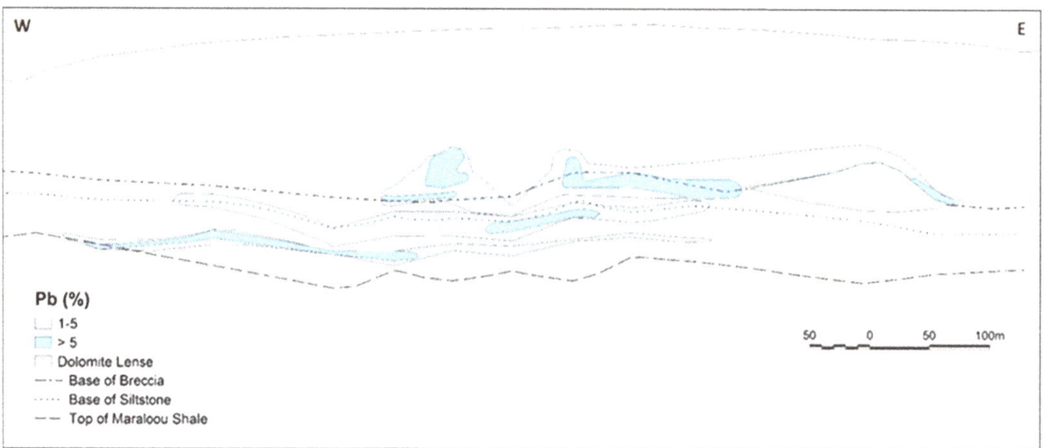

Fig. 20 Lead distribution for the 7,063,550 mN Magellan deposit cross-section, showing positions of the Upper, Middle and Lower Mineralization Zones relative to major lithological boundaries. Vertical exaggeration—4:1

metres. Their concentrations in this zone peak at approximately 5100 ppm Zn, 2600 ppm Cu, 2700 ppm Ba and 1250 ppm As.

8.4.2 Ore Mineralogy and Zoning

The essentially supergene Paroo Station Mine mineralization is relatively simple in the mineral composition: it is dominated by cerussite ($PbCO_3$) and anglesite ($PbSO_4$), with minor pyromorphite ($Pb_5(PO_4)_3Cl$), plumbogummite ($PbAl_3(PO_4)_2(OH)_5{\bullet}(H_2O)$ and coronadite ($PbMn_8O_{16}$) with traces of plattnerite (PbO_2). The only specimen of galena at the deposit (actively mined since 2005) was encountered in 2013. Other authors previously reported trace amounts of galena closely intergrowing with anglesite and cerussite in flotation plant ore samples.

The supergene mineralization is commonly very fine-grained and indistinctive, mainly occurring as disseminations, veinlets and replacement of the host rock, with minor concretionary,

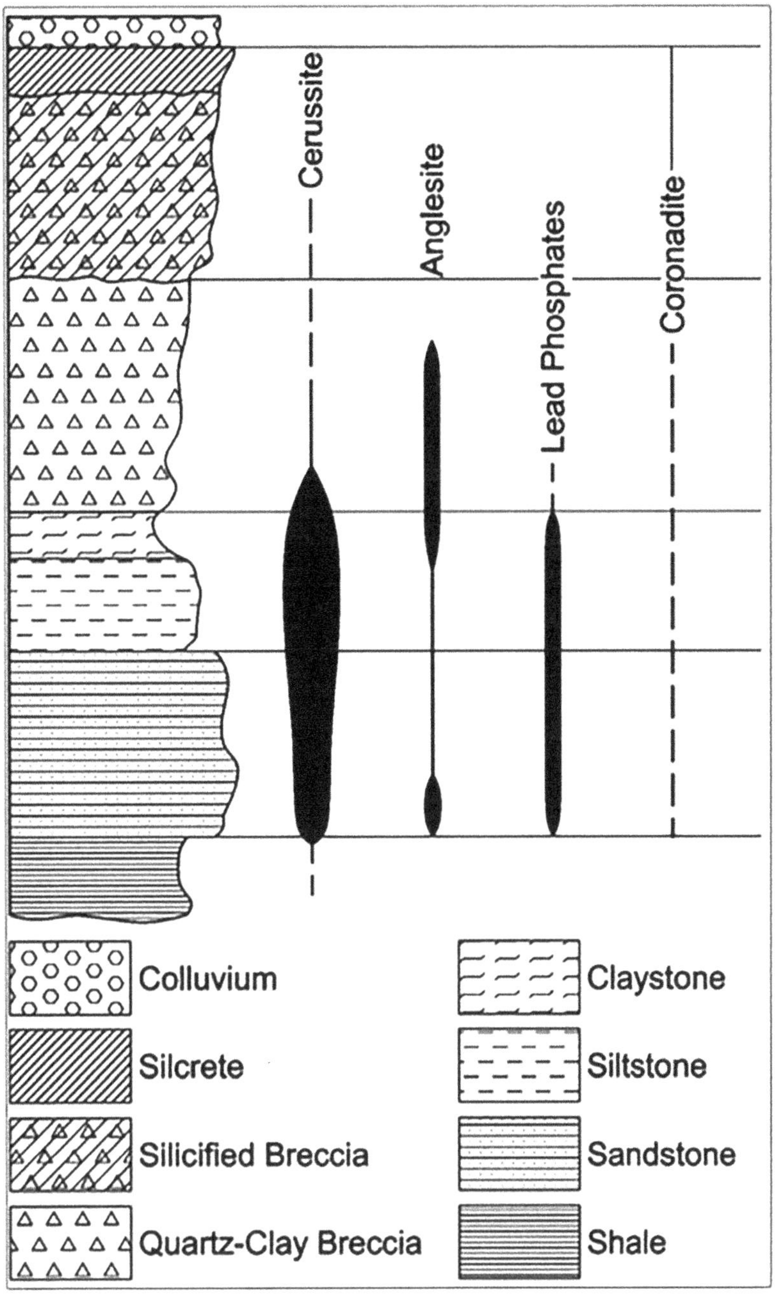

Fig. 21 Generalized vertical mineral zonation, Magellan deposit. The section is approximately 50 m thick (Sergeev et al. 2017)

nodular and coarse crystalline forms. The mineralization exhibits unclear vertical zoning (Fig. 21).

Anglesite is more prevalent than cerussite in the quartz-clay breccia in the upper parts of the Upper Zone and within the remnant high-grade pods above (McQuitty and Pascoe 1998; Sergeev et al. 2017). Cerussite is abundant in the lower parts of the Upper Zone with minor amounts of Pb phosphates and anglesite (Fig. 25). In the clastic sequences below, cerussite is accompanied by Pb phosphates in varying proportions, with the latter prevailing in some intervals. The distributions of plumbogummite and pyromorphite are believed to be controlled by the abundance of phosphorus sourced from detrital and authigenic zircon, monazite and xenotime, which are common in some sandstone beds (Muhling et al. 2012). In the lower parts of the sandstone sequence, near the lower contact with the carbonaceous shale, anglesite again becomes more abundant, although cerussite remains the principal Pb phase. Coronadite is distributed randomly throughout the profile.

8.5 Deposit Formation and Evolution

The large Paroo Station Mine Pb deposit is interpreted to have been formed from relatively low-grade sulfide mineralization through progressive volume reduction of host rocks and redistribution and accumulation of Pb by weathering. The "mushroom-shaped" Pb distribution pattern, comprising the upper depletion zone and enrichment blanket below, is present here, even if not characteristic of supergene non-sulfide Pb deposits. Mineralization is controlled by weathering and hydrodynamic permeability boundaries, with the base of dolomite weathering and the saprolite to clay saprolite contact serving as key barriers for Pb re-precipitation. Oxygen isotopic data indicate formation of cerussite under ambient temperatures of 23–29 °C. These findings strongly support the supergene genesis of the non-sulfide Paroo Station Mine mineralization (Sergeev et al. 2017). The original stratabound sulfide mineralization within the Sweetwaters Well dolomite is likely related to hypogene Mississippi Valley-type (MVT) sulfide deposits. It is believed to be similar in style and synchronous (1.88–1.82 Ga) with the MVT prospects in the Teague area in the Earaheedy Basin to the north (Sergeev et al. 2017). These and many other deposits in this region are linked to the Capricorn Orogeny.

Supergene fluids were moderately acidic during the initial stages of the oxidation of pyrite but were quickly neutralized by the abundant carbonate gangue. Anglesite was the first oxidation product of galena, remaining in situ in residual breccia after the dolomite dissolved. As dolomite weathering progressed, lead and other base metals gradually migrated downwards through a multistage remobilization process, re-precipitating at the lowering dolomite weathering front (Fig. 22). Cerussite became increasingly abundant prevailing the mineralization composition in the lower breccia zone and below.

When the dolomite unit was completely altered and disintegrated leaving behind a residual collapse breccia, the supergene fluids became neutral to slightly alkaline, which increased Pb mobility. Lead migrated deeper and downslope through the most permeable beds of weathered sandstone, following groundwater gradients. Lateral Pb transport may have been enhanced during the later stages of weathering under arid conditions (since the Miocene) due to increased activities of Pb halide complexes in groundwater.

8.6 Summary

The Paroo Station Mine deposit is unusual in several aspects, providing some guidance for further exploration of this type of deposits in the Australian environment. Firstly, the mineralization does not contain any sulfides, being mostly composed of supergene cerussite and anglesite with minor Pb phosphates. The deposit also exhibits a high degree of supergene Pb mobility, leading to the near-complete obliteration of the primary Pb distribution pattern and formation of the supergene-type pattern, characterized by an upper depletion zone coupled with a subhorizontal enrichment blanket beneath.

9 Case Study: Supergene Ag Deposit, Sierra Mojada, Mexico

9.1 Introduction

Supergene Ag enrichment can generally occur in the oxidation zone of Ag-rich primary deposits, as magmatic epithermal Au–Ag mineralization, but also in MVT, SEDEX and especially carbonate-hosted polymetallic ore types. Silver tends to be

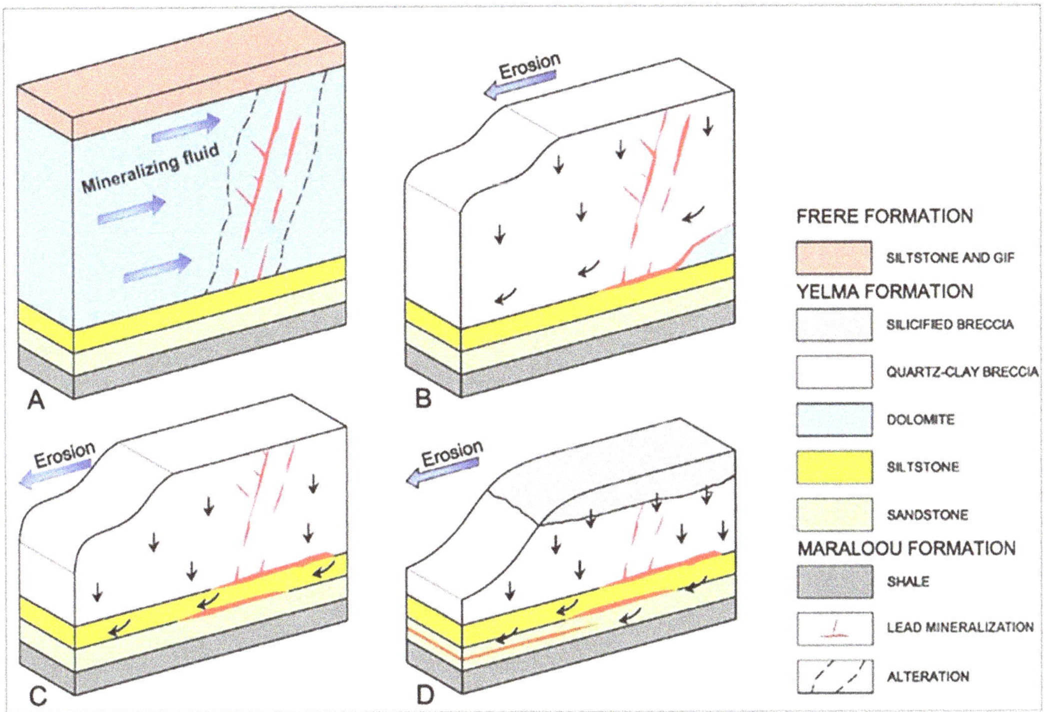

Fig. 22 Genetic model of formation of the Paroo Station Mine Pb deposits; see description in text. (After Sergeev et al. 2017)

retained in oxidized zones, with little metal remaining available in solution to generate secondary sulfides in cementation zones.

Silver occurs mostly as Ag halides, which are extremely insoluble, but also as secondary acanthite, native Ag and as trace element in Fe sulfates and Mn hydroxides. In the carbonate-hosted deposits, large Ag amounts (71.2 Mt at an average grade of 71.4 g/t Ag) can be contained within the "calamines", as in the case of the Sierra Mojada (Mexico) Ag-Zn-Pb belt (Ahn 2010) with its historical high-grade Ag, Pb and Zn concentrations discovered in 1879. This mining district is located in the northwestern part of Coahuila State, Mexico, close to the border with Chihuahua State (Fig. 23).

9.2 Climate and Topography

Sierra Mojada is located west of Sierra Madre Oriental on the Mexican Plateau. The terrain is generally flat, with prominent relief formations of up to 1500, with the crest of the Sierra Mojada range at more than 2400 m above sea level. Most of the mineral concessions are located at the base of the cliffs where there is moderate relief with numerous gullies that erode the surface alluvium. The area is high desert covered by scrub vegetation. The present climate is semi-arid with an average rainfall less than 550 mm pa.

9.3 Geology

The Sierra Mojada deposit is located in the Eastern Zone, one of the three principal geologic zones of Mexico defined by age, tectonics and lithologies (Megaw 2009; Gryger 2010). The other two zones are the Western Zone and the Trans-Mexican Volcanic belt. The Eastern Zone represents a passive plate margin relative to the Western Zone, which abuts a convergent plate margin, and is composed of three major

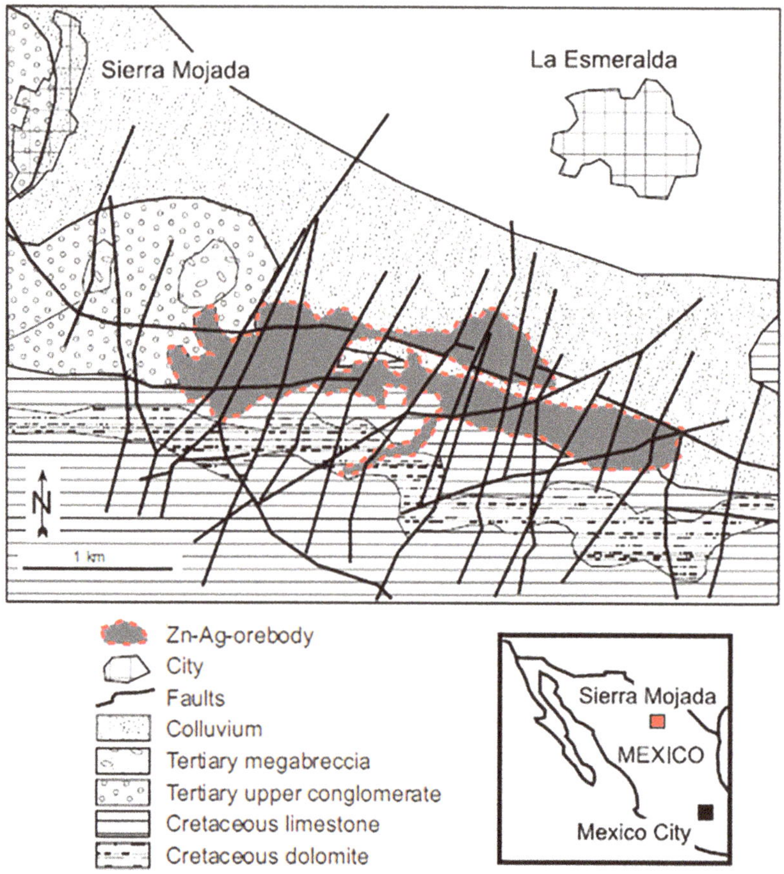

Fig. 23 Geological map of the Sierra Mojada project. (After Boni and Mondillo 2015)

lithostratigraphic terrains: the Coahuila, Maya and Sierra Madre. The boundary between the Eastern and Western terrains is in Chihuahua, just west of the Sierra Mojada project area. Within the Eastern Zone, the deposit is located in the Coahuila terrain. Basement geology comprises moderately metamorphosed andesite volcanics cut by Permian to Triassic granitoid intrusive rocks. The deposit is hosted in Cretaceous to Tertiary sediments within the Sabinas basin, which was formed during Jurassic rifting and the opening of the Mexican Gulf.

The San Marcos fault component of this system exhibits a minimum of four recorded movements and begins with an early normal movement with later left-lateral strike-slip reverse movements beginning in the early Tertiary (Gryger 2010). The initial movement along the San Marcos fault has been attributed to deformation along the Torreón-Monterrey lineament and the Mojave-Sonora megashear together with subsequent isostatic adjustment due to crustal thickening during the Jurassic. The thrust component of the San Marcos fault is locally referred to as the Sierra Mojada thrust. Igneous intrusions associated with the Seveir-Laramide orogeny are believed to be the source of the primary mineralization in the Sierra Mojada district.

9.4 Mineralization

Sulfide mineralization at Sierra Mojada is hosted by the Cretaceous marine carbonate sequence south of the San Marcos fault (Megaw et al. 1988; Megaw 2009). Approximately 80% of the district mineralization is hosted in dolomite and the remainder is in limestone. The district shows a complex history of hypogene sulfide mineralization, followed by oxidation and supergene alteration. The hydrothermal alteration shows a paragenesis of dolomitization, carbonate, sericite and silica alteration. Supergene oxidation events followed with precipitation of carbonates, silica, clay minerals and iron oxides.

The youngest structures in the district are normal high-angle structures varying from 0 to 20° strike, 90 to 55° dip, down-faulted to the E and W, forming a series of horst and graben across the district. These structures are probably related to basin-and-range movements and typically show offsets of 5 to 25 m. The N-S structures are important at Sierra Mojada as they are a major inheritor of remobilized supergene and oxide mineralization.

There are two important and diverse mineralization types, accentuated by a locally dense structural architecture. A major Carbonate Replacement Pb-Zn-Ag-(Cu) Deposit (CRD) is distal to the source intrusion. This kind of deposit was then followed by supergene enrichment, associated with the oxidation of the original sulfides, leading to the current expression of mineralization. There are essentially two overlapping mineralized sections in the Sierra Mojada district. These are the Shallow Silver Zone (SSZ), and the Base Metal Mantos (BMM). The SSZ consists of a supergene Ag ore body located near the surface: it is estimated to contain more than 163 Moz of Ag with an average grade of 70 g/t Ag (at a 25 g/t cutoff) over a 3.8 km strike (Megaw 2009; Gryger 2010). The mineralization in the SSZ is dominated by acanthite, "embolite" (solid solution of bromargyrite–chlorargyrite) and Ag-bearing tennantite. Silver occurs in early to late structures, karst breccias, low-angle fault breccias, mantos and as disseminated replacements in porous hydrothermally altered dolomites.

The mineralization within the BMM begins with the Lead Zone in the highest stratigraphic position, followed by the Red Zinc Zone, and the White Zinc Zone (see Borg 2015). The mineralization dips towards the east at 10 degrees and is controlled by dolomite and subordinate limestone host rocks.

The Zinc Zone is stratigraphically constrained and consists of several ore bodies containing oxidized Zn minerals, situated directly below and adjacent to the Shallow Silver Zone (Megaw 2009, Gryger 2010). Historically Zn hydro-silicate and Zn carbonate minerals ("Zinc *Manto* Zone", with ~20 Mt @ 10% Zn) were discovered underlying the Ag–Pb mineralized horizon and exploited in the early twentieth century. Based on historical records, it has been estimated that more than one million metric tonnes of Ag, Zn, Pb and Cu ores have been mined from the site. Two non-sulfide Zn zones have been identified in the mineral deposit (JDS 2013).

The Red Zinc Zone (dominated by Fe oxide) is a continuous *manto* some 2500 m along strike, up to 200 m wide, and up to 160 m thick. It averages about 80 m thick and about 130 m in width. The mineralization follows reactive dolomite host rocks and karst fill breccia (Ahn 2010). The mineralization consists of massive hemimorphite with subordinate amounts of smithsonite and minor hydrozincite (Borg 2015, Fig. 8, p. 129). Massive red Zn manto mineralization is surrounded by a halo of fault and fracture-controlled red Zn as a result of supergene processes, primarily but not restricted to the footwall. The mineralization is vuggy and shows replacement of zebra textures as well as laminated cave-floor and soft-sediment deformation. Relict pyrite, galena and sphalerite have been noted although the overall level of oxidation is strongly pervasive. Anglesite, cerussite and plattnerite are present, especially in the footwall faults and Ag values are approximately 45–70 g/t and Zn around 20–30% (Ahn 2010).

The White Zinc Zone (smithsonite manto; Borg 2015, Fig. 9, p. 130) lies underneath the Red Zinc Zone and forms a series of mantos, chimneys and filled fault and joint structures. The zone consists of two bodies approximately 100–200 m across each and up to 70 m in thickness. The two bodies of mineralization are separated by the Campamento fault, which has down-thrown the eastern body relative to the western. The thickest section of the Red Zinc Zone directly overlies the White Zinc Zone where total supergene Zn mineralization is over 200 m thick. The mineralization follows reactive limestone and dolomite host rocks and karst fill breccia. Mineralization shows classic karst cave-floor accumulation and soft-sediment deformation and has a strong structural component occupying steeply dipping faults in the zone. Massive White Zinc Zone mineralization grades approximately 25 to 40% Zn with negligible Pb and Ag.

9.5 Genesis

Hypogene Pb-Zn-Ag-Cu sulfide mineral mineralization in the Sierra Mojada district underwent intense oxidation, followed by supergene enrichment and a second oxidation event (Megaw 2009; Ahn 2010; Kyle et al. 2018). The Late Tertiary to Quaternary events were accelerated by the tectonic structure of the area during a period of rapid climate change as the region went from a savanna climate in the Pliocene to cool-wet climates in the Pleistocene and to the present highly arid climate. The non-sulfide base metal mineralization at Sierra Mojada would classify as about 70% ore replacement and 30% wallrock replacement, primarily in structures (Hitzman et al. (2003) classification scheme).

Carbon isotope values from smithsonite are consistent with other supergene smithsonites (Gilg et al. 2008; Ahn 2010). Unusually low and variable $\delta^{18}O$ values were also reported (Ahn 2010). The oxygen isotope compositions indicate that smithsonite at Sierra Mojada was formed at slightly elevated temperature, at least during some mineralization stages and/or that oxidizing waters were isotopically much lighter than in the other non-sulfide districts (Boni and Mondillo 2015). Although local paleometeoric water composition is unavailable for the district, adjacent modern groundwater composition can be used as an estimate of the $\delta^{18}O$ water value. The $\delta^{18}O$ value of groundwater is low (−8‰V-SMOW), translating into estimated precipitation temperatures of about 33 °C for smithsonite.

Under oxidizing conditions in limestone-dolomite host rocks, sphalerite is readily oxidized to its carbonate equivalent, smithsonite, under high partial pressure of CO_2. Upon relaxation of the partial pressures of CO_2, smithsonite altered to hydrozincite prior to the addition of silica, leading to the formation of hemimorphite. Silver mineralization in the deposits forms acanthite in the supergene environment and in places oxidizes to Ag halides and rarely native Ag (Sillitoe 2009).

10 Case Study: Supergene Vanadium Deposit, Berg Aukas, Namibia

10.1 Introduction

In the Otavi Mountain Land of northern Namibia, Zn-Cu-Pb vanadate mineralization at Berg Aukas occurs in collapse breccias and solution cavities in karstic structures associated with Mesozoic paleoland surfaces. The deposits is approximately 500 km NE of the capital Windhoek (Fig. 24).

10.2 Climate and Topography

The rugged topography of the area is formed by weathered carbonate rocks, with the highest mountain at 1677 m amsl. The valleys are characterized by a cover of calcrete (caliche) overgrown by dense thorn bush and scrubby vegetation. The mountain slopes contain even more resilient species of shrubs and cacti, the latter attesting to annual mean temperatures that commonly exceed 30 °C. The rainfall in the Otavi Mountain Land is the highest in Namibia, with an average of 520 mm per annum.

10.3 Geology

The Berg Aukas Mine is situated on the northern limb of the Berg Aukas Syncline, which consists of dolostones, limestones and shales of the Berg Aukas Formation (at the base of Abenab Subgroup/Otavi Group). The Berg Aukas Formation is part of the Neoproterozoic sedimentation on the Otavi Platform (Miller 2008). The Otavi Group is the main lithostratigraphic unit in the area, comprising approximately 8500 m of predominantly

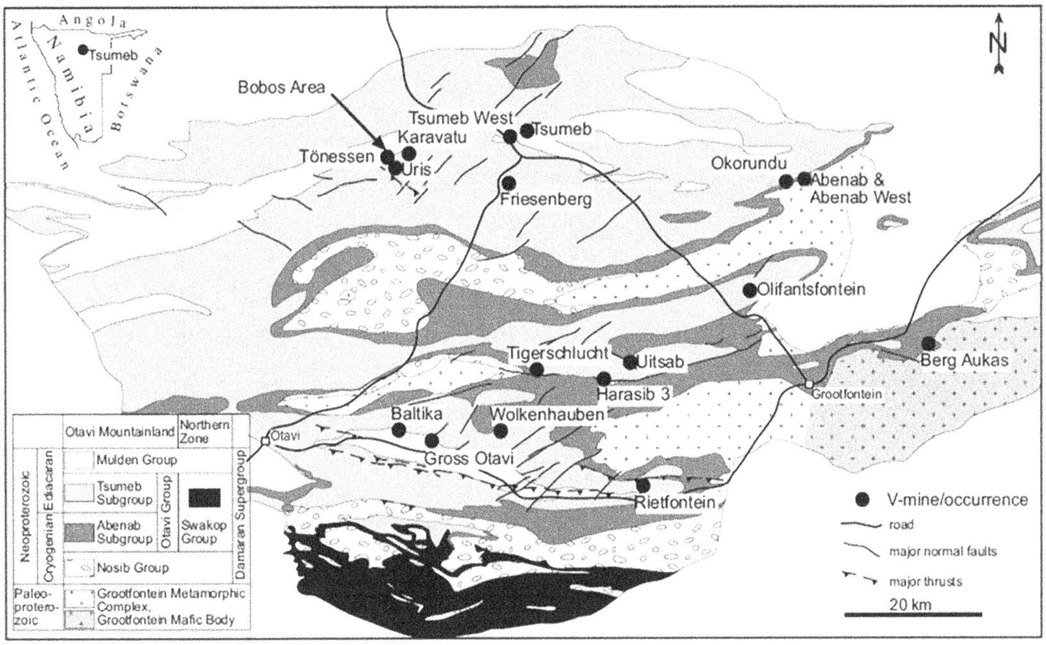

Fig. 24 Geological map of the Otavi Mountain Land (Namibia) with the most important localities where V ores have been exploited. (From Boni et al. 2007b)

dolostone with subordinate clastic shale, siltstone, sandstone, diamictite and banded iron formation (Hoffman and Halverson 2008).

10.4 Mineralization

Mineralization at Berg Aukas is hosted in the Gauss Formation of the Abenab Subgroup (Misiewicz 1988). It is concentrated along the northern, steeply dipping section of the mineralized structure, and where this levels out at depth, mineralization ceases (Misiewicz 1988). The deposit is structurally complex, with the northern and southern limbs of the Berg Aukas syncline undergoing later faulting. The Berg Aukas sulfide mineralization was therefore both structurally and stratigraphically controlled, the latter due to karst features in the dolostones coupled with permeability contrasts that allowed the hydrothermal fluids to deposit ore minerals at the contact between two texturally different carbonate units (Misiewicz 1988).

Supergene concentration of V and base metal minerals was due to the leaching of V from the older country rocks, and of Zn and Pb from the pre-existing primary sulfides. The supergene ores occur in collapse breccias and solution cavities related to the karst network in the deposit. Vanadium mineralogy is dominated by descloizite and less abundant mottramite; both overprinting earlier sulfide-willemite orebodies (Fig. 25). A supergene, post-Damaran origin has been generally accepted: however, the timing and genesis of this mineralization style remain still controversial (Boni et al. 2007b).

Based on the faunal assemblage contained within the descloizite-rich breccias, the age of the Otavi V ore genesis had been estimated as late-Middle Miocene (13 ± 1 Ma) and early-Upper Miocene (9 ± 1 Ma) (Pickford 1993). This age is only partly consistent with radiogenic age dating by Boni et al. (2007b), that estimated an age of 24 to 33 Ma by [U–Th]/He thermochronometry.

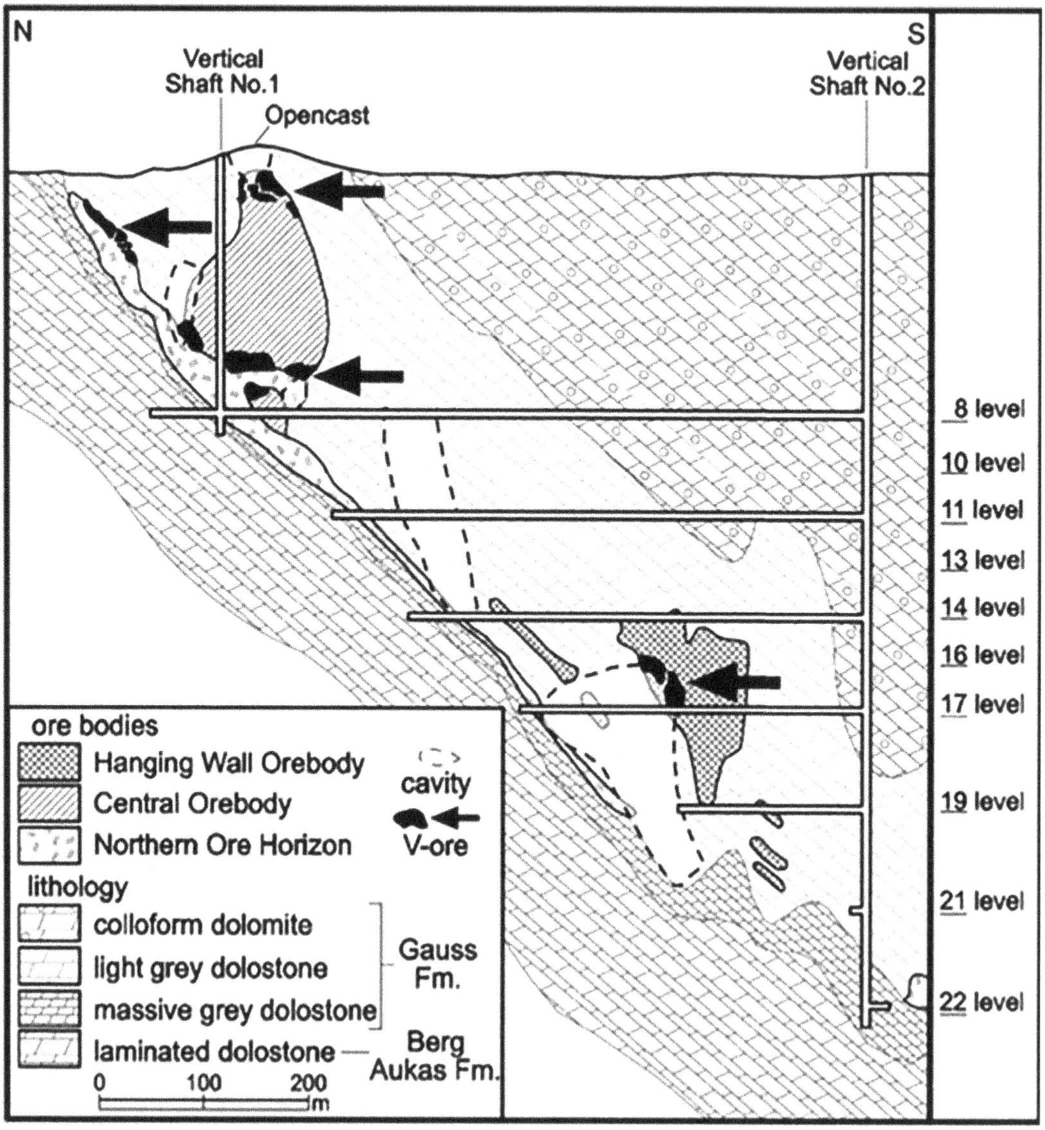

Fig. 25 Location of the vanadate ore in the Berg Aukas Mine associated with oxidized Zn and sulfide Zn ores in karstic features. (Modified from Misiewicz 1988)

10.5 Genesis

The V deposits in the Otavi Mountainland represent an unusual medium-temperature, weathering-related, non-sulfide ore type, that has been found in several areas of southern Africa. Calcite and dolomite gangue are paragenetically coeval with the vanadates and occur in breccia-related ores, and as cement in the deeper parts of the recent karst infills. Acidic conditions of sulfide oxidation coupled with controlled redox conditions related to decomposition of soil organic matter have promoted the formation of vanadate mineralization (Fig. 26) (Boni et al. 2007b).

Results of bulk geochemistry and isotope data indicate that vanadate minerals were formed at elevated temperatures of ~40° to 50 °C by meteoric waters interacting with Neoproterozoic shales, carbonates, and primary sulfide ores

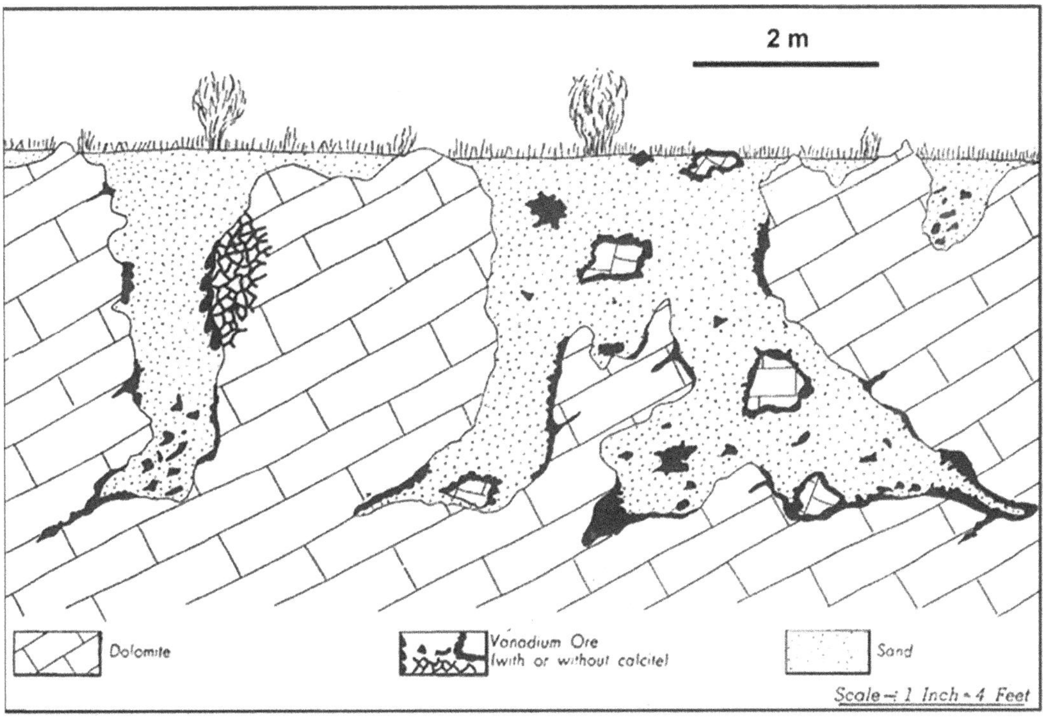

Fig. 26 Field sketch by Schwellnus (1945) illustrating the nature of the karst features in the Otavi Mountainland dolomite and subsequent infilling by vanadiferous sediment. These are referred to as "sand-sack" ores and were the main style of mineralization exploited in the Berg Aukas Central Orebody open pit. (Modified after Boni et al. 2007b)

(Boni et al. 2007b). These reactions occurred over multiple events and led to stacking of the mineralization. The primary source of V in the deposit is cited as being from shales within the Otavi Group and/or from possible Paleoproterozoic mafic volcanics.

11 Conclusions

Supergene non-sulfide base metal deposits have been historically important sources for Zn and to a lesser extent for Cu, Ag, V and Pb. They form in near-surface to surface environments through the oxidation of sulfide mineral deposits and replacement by supergene minerals or through mobilization of metals due to permeability or structural controlled zones to a trap site where the metal load is precipitated.

These deposits form through sulfide oxidation and buffering of acidic solutions by dolomite and calcite rich wallrock that leads to mineralization occurring within the precursor sulfide zone, or at the margins in wallrock.

In the case of Zn, the metal can migrate some distance before being trapped in structural zones such as folds or faults or within karstic structures. In a few cases (Angouran, Iran; Otavi Mountain Land, Namibia) the supergene solutions have been partially mixed with low-temperature hydrothermal fluids, thus changing their mineralogical and chemical compositions.

Lateral metal dispersion is of high exploration significance, particularly where fluids are trapped. Metal transport through its remobilization and "leaking" from the mineralization in groundwaters through the porous zones or units, followed by precipitation on hill slopes and in discharge areas, provides much larger exploration target and vector to the mineralization. This improves the opportunity of detecting such deposits in regional exploration programmes.

Acknowledgements The authors would like to thank Professor Gregor Borg and another reviewer for all their constructive comments on the paper. Mr Stephen Phillips is thanked for drafting the figures. Elsevier are thanked for approval to reproduce Figs. 7, 13, 14, 15, 18, 19, 21, 22 and 23 from their publications. The Society of Economic Geology and Economic Geology Publishing Company are thanked for allowing publication of Figs. 9, 16, 24, 25 and 26. Canadian Institute of Mining and Metallurgy is thanked for approval to reproduce Fig. 12. The authors of all these figures also kindly provided approval to reproduce their work and reference to each is given in the figure caption.

References

Agricola G (1556) De Re Metallica. The 1950 Edition, translated by Hoover H.C. and Hoover L.H. Dover Publication, 672 pp

Ahn HI (2010) Mineralogy and geochemistry of the non-sulfide Zn deposits in the Sierra Mojada District, Coahuila, Mexico. Unpublished Ph.D. thesis, Austin, USA, University of Texas, 179 pp

Anthony JW, Bideaux RA, Bladh KW, Nichols MC (eds) (2003) Handbook of mineralogy. Mineralogical Society of America, Chantilly. http://www.handbookofmineralogy.org/

Arfè G, Mondillo N, Boni M, Balassone G, Joachimski M, Mormone A, Di Palma T (2017) The karst hosted Mina Grande nonsulfide zinc deposit, Bongará district (Amazonas region, Peru). Econ Geol 112:1089–1110

Arfé G, Mondillo N, Boni M, Joachimski M, Balassone G, Mormone A, Santoro L, Medrano EC (2018) The Cristal Zn prospect (Amazonas region, Northern Peru). Part II: An example of supergene enrichments in tropical areas. Ore Geol Rev 95:1076–1105

Arizona Department of Mines & Mineral Resources (2005) Arizona Mineral Resource, #40, 4p

Balassone G, Rossi M, Boni M, Stanley G, McDermott P (2008) Mineralogical and geochemical characterization of nonsulfide Zn–Pb mineralization at Silvermines and Galmoy (Irish Midlands). Ore Geol Rev 33:168–186

Blot A, Barros de Oliveira SM, Magat P (1995) La chloritisation supergene zincifère des phlogopites de Canoas (PR Brèsil). CR Acad Sci Paris 321 (IIa):651–658

Boland MB, Kelly JG, Schaffalitzky C (2003) The Shaimerden supergene zinc deposit, Kazakhstan. Econ Geol 98:787–779

Boni M (2005) The geology and mineralogy of non-sulfide zinc ore deposits. Proceedings LEAD and ZINC '05, Kyoto 17–19 October 2005, 1299–1314

Boni M, Large D (2003) Non-sulfide mineralization in Europe: an overview. Econ Geol 98:715–729

Boni M, Mondillo N (2015) The "Calamines" and the "Others": the great family of supergene non-sulfide zinc ores. Ore Geol Rev 67:208–233

Boni M, Gilg HA, Balassone G, Schneider J, Allen CR, Moore F (2007a) Hypogene Zn carbonate ores in the Angouran deposit, NW Iran. Miner Deposita 42: 799–820

Boni M, Terracciano R, Evans N, Laukamp C, Schneider J, Bechstädt T (2007b) Genesis of vanadium ores in the Otavi Mountainland (Namibia). Econ Geol 102:441–469

Boni M, Balassone G, Arseneau V, Schmidt P (2009) The non-sulfide zinc deposit at Accha (Southern Peru): geological and mineralogical characterization. Econ Geol 104:267–289

Boni M, Mondillo N, Balassone G (2011a) Zincian dolomite: a peculiar dedolomitization case? Geology 39: 183–186

Boni M, Terracciano R, Balassone G, Gleeson SA, Matthews AH (2011b) The carbonate-hosted willemite prospects of the Zambezi Metamorphic Belt (Zambia). Miner Deposita 46:707–729

Boni M, Mondillo N, Balassone G, Joachimski M, Colella A (2013) Zincian dolomite related to supergene alteration in the Iglesias mining district (SW Sardinia). Int J Earth Sci (Geol Rundsch) 102:61–71

Borg G (2009) The role of fault structures and deep oxidation in supergene base metal deposits. In: Titley SR (ed) Supergene environments, processes and products, Economic Geologists Special Publication, vol 14, pp 121–132

Borg G (2015) A Review of Supergene Nonsulfide Zinc (SNSZ) deposits- the 2014 update. In: Archibald SM, Piercey SJ (eds) Current perspectives on zinc deposits, vol 2015. IAEG, pp 123–147

Borg G, Kärner K, Buxton M, Armstrong R, van der Merwe SW (2003) Geology of the Skorpion non-sulfide deposit, southern Namibia. Econ Geol 98: 749–771

Bowell R (1992) Supergene copper mineralassemblages at Botallack, St Just, Cornwall. J Russel Soc 4(2):45–53

Bowell RJ (2014) Hydrogeochemistry of the Tsumeb deposit: implications for arsenate mineral stability. Rev Mineral Geochem 79:589–630

Bowell RJ (2023) Natural attenuation in the vadose zone: nature's gift to mine closure. In: Parshley J, Fourie A, Tibbet M (eds) Mine closure 2023—Behrooz Abbasi. Australian Centre for Geomechanics, Perth, pp 1–14. https://papers.acg.uwa.edu.au/c/mc2023

Bowell RJ, Egorova L, Gurevich D, Sedov V, Holwell DA, Mikhailov A, Kordi P (2012) Grassroots exploration in remote areas: an example from the Kuene region, Namibia. CIM 3:3–12

Bunting JA (1986) Geology of the eastern part of the Nabberu Basin. Geological Survey of Western Australia, Bulletin 131, 130 pp

Cairncross B (2021) Minerals of Berg Aukas, Otavi Mountainland, Namibia. Rocks Minerals 96:110–145

Chavez WX (2000) Supergene oxidation of copper deposits: zoning and distribution of copper oxide minerals. SEG Newsl Soc Econ Geol 41:1–13

Chen X, Duan D, Zhang Y, Zhou F, Yuan X, Wu Y (2023) Genesis of the giant Huoshaoyun non-sulfide zinc-lead deposit in Karakoram Xinjiang: constraints from mineralogy and trace element geochemistry. Minerals 13: 842–853

Chirico R, Mondillo N, Boni M, Joachimski M, Ambrosino M, Buret Y, Mormone A, Leigh LENB, Flores WH, Balassone G (2022) Genesis of the Florida Canyon non-sulfide Zn ores (Northern Peru): new insights into the supergene mineralizing events of the Bongarà District. Econ Geol 117:1339–1366

Choulet F, Charles N, Barbanson L, Branquet Y, Sizaret S, Ennaciri A, Badra L, Chen Y (2013) Non-sulfide zinc deposits of the Moroccan High Atlas: multi-scale characterization and origin. Ore Geol Rev 56:115–140

Choulet F, Buatier M, Barbanson L, Guégan R, Ennaciri A (2016) Zinc-rich clays in supergene non-sulfide zinc deposits. Miner Deposita 51:467–490

Choulet F, Barbanson L, Buatier M, Richard J, Vennemann T, Ennaciri A, Zouhair M (2017) Characterization and origin of low-T willemite (Zn2SIO4) mineralization: the case of the Bou Arhous deposit (High Atlas, Morocco). Miner Deposita 52:1085–1102

Choulet F, Richard J, Boiron MC, Dekoninck A, Yans J (2019) Distribution of trace elements in willemite from the Belgium non-sulfide deposits. Eur J Mineral 31(5–6):983–997

Coppola V, Boni M, Gilg HA, Balassone G, Dejonghe L (2008) The "calamine" nonsulfide Zn–Pb deposits of Belgium: Petrographical, mineralogical and geochemical characterization. Ore Geol Rev 33:187–210

Crane MJ, Sharpe JL, Williams PA (2001) Formation of chrysocolla and secondary copper phosphates in the highly weathered supergene zones of some Australian deposits. Rec Aust Musc 53:49–56

Daliran F, Pride K, Walther J, Berner ZA, Bakker RJ (2013) The Angouran Zn (Pb) deposit, NW Iran: evidence for a two stage, hypogene zinc sulfide–zinc carbonate mineralization. Ore Geol Rev 53:373–402

Dunn PJ (1995) Franklin and Sterling Hill, New Jersey: the world's most magnificent mineral deposits. 2 vol. Excalibur Mineral Company, 1000 N, Division Rd, Peekskill, NY, 877 pp

Etheridge and Henley Geoscience Consultants (1994) Basinal framework analysis of the southern Glengarry Basin as a basis for exploration targeting. Report for RGC Exploration Pty Ltd., June 1994, 59 pp

Fransolet AM, Bourguignin P (1975) Donnès nouvelles sur la frapontite de Moresnet (Belgique). Bull Soc Fr Miner Cristallogr 98:235–244

Gilg HA, Boni M, Hochleitner R, Struck U (2008) Stable isotope geochemistry of carbonate minerals in supergene oxidation zones of Zn–Pb deposits. Ore Geol Rev 33:117–133

Gnoinski J (2007) Skorpion Zinc: optimization and innovation. J S Afr Inst Min Metall 107:657–662

Groves IM, Carman CE, Dunlap WJ (2003) Geology of the Beltana willemite deposit, Flinders Ranges, South Australia. Econ Geol 98:797–818

Gryger SM (2010) Geologic framework of the Sierra Mojada mining district, Coahuila, Mexico; An integrative study of a Mesozoic platform-basin margin, MSc. thesis, University of Texas-Austin, December 2010, 376p

Guggenheim S, Bailey SW (1990) Baumite discredited. Am Min 75:705

Hayes TS, Kadi K, Balkhiyour M, Siddiqui A, Beshir Z (2000) Phanerozoic sediment hosted base-metal mineralizing systems in Saudi Arabia. Saudi Geological Survey Technical Report USGS-TR-00-3 (IR-964), 29 pp

Hitzman MH, Reynolds NA, Sangster DF, Allen CR, Carman CE (2003) Classification, genesis, and exploration guides for non-sulfide zinc deposits. Econ Geol 98:685–714

Hoffman PF, Halverson GP (2008) Otavi Group of the western Northern Platform, the Eastern Kaoko Zone and the western Northern Margin Zone. In: Miller RMG (ed) The geology of Namibia, Neoproterozoic to Lower Palaeozoic, vol 13. Ministry of Mines and Energy, Geological Survey, Windhoek, pp 69–136

Hoorn C, Wesselingh FP, Hovikoski J, Guerrero J (2010) The development of the Amazonian mega-wetland (Miocene; Brazil, Colombia, Peru, Bolivia). In: Hoorn C, Wesselingh FP (eds) Amazonia, landscape and species evolution. Wiley, New York, pp 123–142

JDS Energy & Mining Inc. (2013) Silver Bull Resources, Sierra Mojada Silver–Zinc Project, Coahuila (Mexico), NI 43-101. Technical Report 345 pp

Jones JA, Pirajno F, Hocking RM, Grey K (2000) Revised stratigraphy for the Earaheedy Group: implications for the tectonic evolution and mineral potential of the Earaheedy Basin, Geological Survey of Western Australia, Annual Review 1999–2000, 57–64

Kharakwal JS, Gurjar LK (2006) Zinc and brass in archaeological perspective. Ancient Asia 1:139–159. https://doi.org/10.5334/aa.06112

Krause W, Effenberger H, Bernhardt H-J, Medenbach O (2008) Skorpionite, $Ca_3Zn_2(PO_4)_2CO_3(OH)_2{\cdot}H_2O$, a new mineral from Namibia: description and crystal structure. Eur J Mineral 20(271–280):104935

Kyle JR, Ahn H, Gilg, HA (2018) Nature and origin of the nonsulfide zinc deposits in the Sierra Mojada District, Coahuila, Mexico: constraints from regional geology, petrography, and isotope analyses. Miner Deposita 53:1095–1116

Kyser TK, Leybourne MI, Layton-Matthews D (2020) Advances in the use of isotopes in geochemical exploration: instrumentation and applications in understanding geochemical processes. Geochem Exp Environ Anal 20:199–204

Large D (2001) The geology of non-sulfide zinc deposits—an overview. Erzmetall 54:264–276

Le Mignot E, Scebenaller L, Beziat D, Andre-Mayer A-S, Reisberg L, Salvi S, Velasquez G, Zimmerman C, Nare A, Francheschi G (2017) Palaeoproterozoic copper deposits in Burkina Faso. Econ Geol 112:88–112

Leybourne MI, Cameron EM (2008) Source, transport, and fate of rhenium, selenium, molybdenum, arsenic, and copper in groundwater associated with porphyry-Cu deposits, Atacama Desert, Chile. Chem Geol 247: 208–228

Lien NT, Pho NV (2018) Formation of secondary non-sulfide zinc ore in Cho Dien Pb-Zn deposits. Vietnam J Earth Sci 40:228–239

Maghfouri S, Hosseinzadeh MR, Rajabi A, Choulet F (2018) A review of major non-sulfide zinc deposits in Iran. Geosci Front 9:249–272

Mason J (2004) Supergene lead mineralization in Central Wales. UK J Mines Min 24:35–46

McPhail DC, Summerhayes E, Welch S, Brugger J (2003) The geochemistry of zinc in the Regolith. In: Roach IC (ed) Advances in Regolith. CRC for landscape environments and mineral exploration, pp 287–291

McQuitty BM, Pascoe DJ (1998) Magellan lead deposit. In: Berkman DA, Mackenzie DH (eds) Geology of Australian and Papua New Guinean mineral deposits, pp 293–296

Megaw PKM (2009) Evaluation of oxidized Pb–Zn–Ag-carbonate replacement deposits of Mexico in light of supergene zinc and residual lead enrichment processes. Society of Economic Geologists, Spec. Publ. No.14, 51–58

Megaw PKM, Ruiz J, Titley SR (1988) High-temperature, carbonate-hosted Ag–Pb–Zn(Cu) deposits of northern Mexico. Econ Geol 83:1856–1885

Melchiorre EB, Enders S (2003) Stable isotope geochemistry of copper carbonates at the northwest extension deposit, Morenci District, Arizona: implications for conditions of supergene oxidation and related mineralization. Econ Geol 98:6-7-621

Miller RMG (2008) The geology of Namibia, vol 1–3. Ministry of Mines and Energy, Geological Survey of Namibia, Windhoek

Misiewicz JE (1988) The geology and metallogeny of the Otavi Mountain Land, Damara orogen, SWA/Namibia, with particular reference to the Berg Aukas Zn-Pb-V deposit—a model for ore genesis. M.S Thesis (unpublished), Rhodes University, Grahamstown, South Africa

Mondillo N, Boni M, Balassone G, Grist B (2011) In search of the lost zinc: a lesson from the Jabali (Yemen) non-sulfide zinc deposit. J Geochem Explor 108:209–219

Mondillo N, Boni M, Balassone G, Joachimski M, Mormone A (2014a) The Jabali non-sulfide Zn–Pb–Ag deposit, western Yemen. Ore Geol Rev 61:248–267

Mondillo N, Boni M, Balassone G, Villa IM (2014b) The Yanque prospect (Peru): from polymetallic Zn–Pb mineralization to a non-sulfide deposit. Econ Geol 109:1735–1762

Mondillo N, Herrington R, Boyce AJ, Wilkinson J, Santoro L, Rumsey M (2018a) Critical elements in non-sulfide Zn deposits: a reanalysis of the Kabwe Zn-Pb ores (central Zambia). Min Mag 82:89–114

Mondillo N, Wilkinson JJ, Boni M, Weiss DJ, Mathur R (2018b) Global assessment of Zn isotope fractionation in secondary Zn minerals from sulfide and non-sulfide ore deposits and model for fractionation control. Chem Geol 500:182–193

Monteiro LVS, Bettencourt JS, Juliani C, De Olivera TF (2006) Geology, petrography, and mineral chemistry of the Vazante non-sulfide and Ambrosia and Fagundes sulfide-rich carbonate hosted Zn–(Pb) deposits, Minas Gerais, Brazil. Ore Geol Rev 28:201–234

Mortimer RJG, Coleman ML, Rae JE (1997) Effect of bacteria on the elemental composition of early diagenetic siderite: implications for paleoenvironmental interpretations. Sedimentology 44:759–765

Muhling JR, Fletcher IR, Rasmussen B (2012) Dating fluid flow and Mississippi Valley type base-metal mineralization in the Paleoproterozoic Earaheedy Basin, Western Australia. Precambrian Res 212–213:75–90

Munchmeyer C (1996) Exotic deposits – products of lateral migration of supergene solutions from copper deposits. In: Camus F, Sillitoe RH, Petersen R (eds) Andean copper deposits: new discoveries, mineralisation, styles and metallogeny. Soc Econ Geologists Spec Publ No. 5:43–58

Newberg DW (1967) Geochemical implications of chrysocolla-bearing alluvial gravels. Econ Geol 62: 932–956

Nimfopoulos K, Michailidis KM, Christofides G (1997) Zincian rancieite from the Kato Nevrokopi manganese deposits, Macedonia, northern Greece. Geol Soc London Spec Publ 119:339–347

Pfiffner OA, Gonzalez L (2013) Mesozoic-Cenozoic evolution of the Western Margin of South America: case study of the Peruvian Andes. Geosciences 3:262–310

Pickford M (1993) Age of supergene orebodies at Berg Aukas and Harasib 3a, Namibia. Commun Geol Surv Namibia 8:157–160

Pirajno F, Hocking RM, Reddy SM, Jones JA (2009) A review of the geology and geodynamic evolution of the Palaeoproterozoic Earaheedy Basin, Western Australia. Earth-Science Reviews 94:39–77

Pirajno F, Burlow R, Huston D (2010) The Magellan Pb deposit, Western Australia; a new category within the class of supergene non-sulfide mineral systems. Ore Geology Reviews 37:101–113

Pollard M, Heron C (1996) Archaeological chemistry, 2nd edn. Royal Society of Chemistry Publishing, 375 pp

Reichert J (2009) A geochemical model of supergene carbonate-hosted nonsulfide zinc deposits. Soc Econ Geol Spec Publ 14:69–76

Reichert J, Borg G (2008) Numerical simulation and a geochemical model of supergene carbonate-hosted non-sulfide zinc deposits. Ore Geol Rev 33:134–151

Robert J-L, Gaspèrin M (1985) Crystal structure refinement of hendricksite, a Zn- and Mn- rich trioctahedral potassium mica: a contribution to te crystal chemistry of zinc-bearing minerals. Mineral Petrol 34:1–14

Saini-Eidukat B, Melcher F, Lodziak J (2009) Zinc–germanium ores of the Tres Marias Mine, Chihuahua. Mexico Miner Deposita 44:363–370

Sangameshwar SR, Barnes HL (1983) Supergene processes in zinc–lead–silver sulfide ores in carbonate rocks. Econ Geol 78:1379–1397

Schneider J, Boni M, Laukamp C, Bechstädt T, Petzel V (2008) Willemite (Zn_2SiO_4) as a possible Rb–Sr geochronometer for dating non-sulfide Zn–Pb mineralization: examples from the Otavi Mountainland (Namibia). Ore Geol Rev 33:152–167

Schwellnus CM (1945) Vanadium deposits in the Otavi Mountains, southwest Africa. Trans Geol Soc S Afr 48:49–73

Sergeev N, Burlow R, Tessalina S (2017) The Paroo Station Mine supergene lead deposits, Western Australia: geological and geochemical constraints. Ore Geol Rev 80: 564–593

Sillitoe RH (2009) Supergene silver enrichment reassessed. Soc Econ Geol Spec Publ 14:15–32

Slezak PR, Olivo GR, Diniz Oliveira G, Marcel A, Dardenne MA (2014) Geology, mineralogy, and geochemistry of the Vazante Northern Extension zinc silicate deposit, Minas Gerais, Brazil. Ore Geol Rev 56: 234–257

Williams PA (1990) Oxide zone geochemistry. Ellis Horwood Series in Organic Chemistry, 556 pp

Supergene Gold Deposits in Intensely Weathered Regolith

R. R. Anand, C. R. M. Butt, W. S. Salama, and C. G. Porto

Abstract

Supergene Au deposits are commonly found in highly weathered landscapes in humid sub-tropical and tropical climates, including savannas, equatorial rainforests and some adjacent Mediterranean areas. They are also present in climatic zones, such as semi-arid belts, where humid tropical climates prevailed in the past. Their formation involves physical accumulation and chemical and biological redistribution of Au during ore-body weathering and sedimentary processes. Based on regolith-landform evolution and Au distribution in the weathered profile, three types of Au enrichments occur in the regolith. These include (i) *ferruginous gravel and duricrust (FGD) deposits*, in which enrichment is in residual nodules and pisoliths (lateritic residuum) and/or ferruginized sediments (ferricrete) overlying upper parts of residual profiles (ii) *saprolitic deposits,* where enrichment is in the lower part (saprolite, saprock) of residual profile overlain by FGD or Au-poor saprolite and/or mottled zone and (iii) *paleochannel deposits,* in which the enrichment is either in the saprock or saprolite below-transported paleochannel sand or clays or within the basal clays. Short descriptions of the Igarapé Bahia, Brazil and Moolart Well, Australia, deposits illustrate different styles of supergene Au mineralization in contrasting climatic environments.

1 Introduction

Prior to the 1980s, many earth scientists considered Au to be chemically inert and that all surficial deposits were solely due to the physical accumulation of particulate primary Au. Those in soil and other regolith were thought to be due to residual enrichment, whereas placer deposits formed by erosion of regolith and basement, followed by mechanical transport and concentration in alluvial sediments (e.g., Hails 1976). This is now found to be incorrect, with Au being chemically mobile in many surficial environments. Indeed, nearly a century before, Liversidge (1893a, b) had made observations and conducted experiments that demonstrated Au mobility in the weathering environment. Since the mid-1980s, Liversidge's conclusions have been confirmed by numerous workers in

R. R. Anand (✉) · C. R. M. Butt · W. S. Salama
CSIRO Mineral Resources, ARRC, Kensington, WA, Australia
e-mail: Ravi.Anand@csiro.au

C. G. Porto
Programa de Pós-graduação em Geologia, Instituto de Geociências, Universidade Federal do Rio de Janeiro, Rio de Janeiro, RJ, Brazil

R. J. Bowell, C. R. M. Butt (eds.), *Geology, Geochemistry and Formation of Supergene Mineral Deposits in Deeply Weathered Terrain*, Mineral Resource Reviews, https://doi.org/10.1007/978-3-031-75733-4_6

many weathered terrains around the world, leading to an understanding of the role of weathering of primary mineralization on the mobility and physical occurrence of Au in some surface environments. These processes have economic importance, by liberating Au from encapsulating quartz and sulfides, thereby increasing its recovery (Butt 1998). Furthermore, exposure of metallic Au to reactive ligands enables hydromorphic dispersion and secondary enrichment. Determining which factors control the remobilization of primary Au is essential to (i) determining the weathering and landscape environments in which supergene deposits have formed and (ii) devising effective exploration for primary ore bodies.

Supergene Au deposits are commonly found in highly weathered landscapes in the sub-tropics to humid tropics, including savannas, rainforests and some adjacent Mediterranean zones, as well as in other climatic environments where similar climates prevailed in the past. Gold deposits hosted by ferruginous gravels, duricrusts and saprolite have been major contributors to the establishment of an Au industry in these terrains. The deposits have formed by combinations of physical accumulation and chemical and biological mobilization of Au during the weathering of primary mineralization. The deposits are low grade (<1–5 g/t Au), mostly small (<500,000 ounces, Table 1) and exploited by open pit mining. They may represent (i) the only mineable reserves overlying otherwise uneconomic primary mineralization; (ii) a minor proportion of total reserves of major deposits but yielding an early cash flow; (iii) a group of small supergene deposits close to larger operations; or (iv) combined as an economic cluster, such as in paleochannel systems (Butt 1998; Anand et al. 2019).

A variety of regolith materials is present in deeply weathered landscapes, but Au enrichments generally occur in three types of materials. These include (i) *ferruginous gravel and duricrust (FGD) deposits*, where enrichment is in residual nodules and pisoliths (lateritic residuum) or similar ferruginized sediments (ferricrete) overlying barren sediments or mottled saprolite; (ii) *saprolitic deposits* in which enrichment is in mid to lower saprolite and saprock in residual profiles, overlain by FGD or saprolite depleted in Au; and (iii) *paleochannel deposits* where the enrichment is either in the basal sands and clays of the paleochannel or in underlying saprolite or saprock (Fig. 1). Short descriptions of the Igarapé Bahia, Brazil and Moolart Well, Australia, deposits (Sects. 6 and 7 below) illustrate different styles of FGD mineralization in contrasting climatic environments.

The term 'ferruginous gravel and duricrust' (FGD) here is used as a general term to describe the upper part of a weathered ferruginous profile cemented by Fe oxides, irrespective of its origin. The process of Fe oxide precipitation, continued dissolution of clay, dehydration, nodule and pisolith development have commonly been repeated several times under changing environments. They have developed in a variety of materials in various topographic settings such that there is a continuum from essentially residual deposits (as lateritic residuum) to entirely transported (as ferricrete; Anand et al. 2019). The correct genetic classification of these materials is commonly difficult to determine.

2 Deposit Characteristics

2.1 Ferruginous Gravel and Duricrust (FGD) Deposits

Supergene FGD Au deposits are sub-horizontal units, 1–10 m thick, that occur at or near the land-surface, or buried beneath shallow (<20 m) transported cover (Figs. 1, 2, 3 and 4). Gold enrichment in these deposits may extend several kilometres along strike, but less than a few hundred metres across (Costa 1993; Freyssinet and Itard 1997; Anand 2001; Anand et al. 2019).

Deposits in Lateritic Residuum are products of the relative concentration of Au in FGD. In places, the upper Au-rich ferruginous materials may be partly transported and be considered as 'semi-residual', but nevertheless be overlying identifiable primary mineralization. Such deposits are common and occur in many locations worldwide (e.g., Boddington and Calista, Western

Table 1 Examples of supergene Au deposits in deeply weathered terrains

Deposit	Location	Type	Gold
Mt Gibson	WA	FGD	350,000 oz FGD
Moolart Well	WA	FGD/SP	493,881 oz FGD; 227,316 oz SP
Boddington	WA	FGD/B/SP	7,780,640 oz
Calista	WA	FGD/SP	280,000 oz
Hannan South	WA	SP	34,840 oz
Sunrise Dam	WA	SP	360,000 oz @ 4.2 g/t
Ity	Côte d'Ivoire	FGD/SP	1,000,000 oz
Igarapé Bahia	Brazil	FGD/SP	3,096,474 *oz*
Mavikou	Gabon	FGD	70,500 oz
Yatela	Mali	FGD/SP	1,000,000 oz
Syama	Mali	FGD/SP	80,000 oz
Challenger-Swordsman	WA	PS	250,000 oz
Rose	WA	PS	70,970 oz
Kalana	Mali	PS	140,000 oz
Rosebel	Suriname	FGD/SP	310,000 oz
Sadiola	Mali	SP	2,722.000 oz @ 2.2 g/t

FGD ferruginous gravels and duricrust, *SP* saprolite, *B* bauxite zone, *PS* paleochannel sediments, *WA* Western Australia, Australia

Australia; Igarapé Bahia, Brazil (Sect. 6); Rosebel, Suriname; Syama and Yatela, Mali; Table 1).

Boddington is situated in deeply weathered undulating terrain in a Mediterranean climate. It has an almost continuous 2–4 m thick Fe-rich lateritic residuum overlying a 3–6 m bauxite zone and 25–70 m saprolite. Primary Au mineralization is hosted by felsic to intermediate volcanic rocks; in the overlying regolith, supergene Au mineralization is concentrated in the lower- to mid-saprolite, bauxite zone and lower fragmental duricrust (Anand 1998, 2001). Mineable lateritic reserves of over 700,000 ounces at 1–2 g/t occurred over 4–5 km^2.

In contrast, Calista is in a semi-arid region, with saprolite and bedrock exposed on topographically higher areas, surrounded by plains dominated by transported cover overlying a thick residual regolith. The Calista FGD deposit is buried below 15–20 m of transported overburden, and underlain by 40–50 m saprolite (Anand 2001). The sulfide-rich primary mineralization is hosted in overturned, steep, westerly-dipping volcano-sedimentary rocks. In detail, the FGD is composed of pisolitic lateritic residuum. Gold enrichment is stronger and more consistent in the lower unit than the upper, probably semi-residual unit, although the Au distribution is erratic (11–6860 ppb; Fig. 2).

Rosebel (Sirois et al. 2002) is situated in the equatorial rainforest in intensely weathered, undulating terrain. Several small occurrences exploited in the area are typical of FGD and associated saprolitic deposits. They developed from orogenic shear zone mineralization in early Proterozoic felsic volcanic and sedimentary rocks (Fig. 3; Sirois et al. 2002). Pisolitic gravels and massive clays form a residual to semi-residual surficial FGD unit up to 4 m thick, mantling hills and upper slopes, locally grading 1–2 g/t Au. Igarapé Bahia, see Sect. 6 below, is also in tropical rainforest, with an FGD deposit similar to that at Boddington.

Deposits in Ferricrete are FGD Au deposits hosted by transported cover. Ferricrete Au deposits occur on flanks of paleochannels beneath less than 20 m of colluvium-alluvium. Source mineralization may be in bedrock or saprolite immediately underlying or upslope of the deposit and in flanking valleys, but is generally sub-economic (Anand et al. 2019). Such deposits are especially common in the Yilgarn Craton of Western Australia (e.g., Moolart Well (Sect. 7),

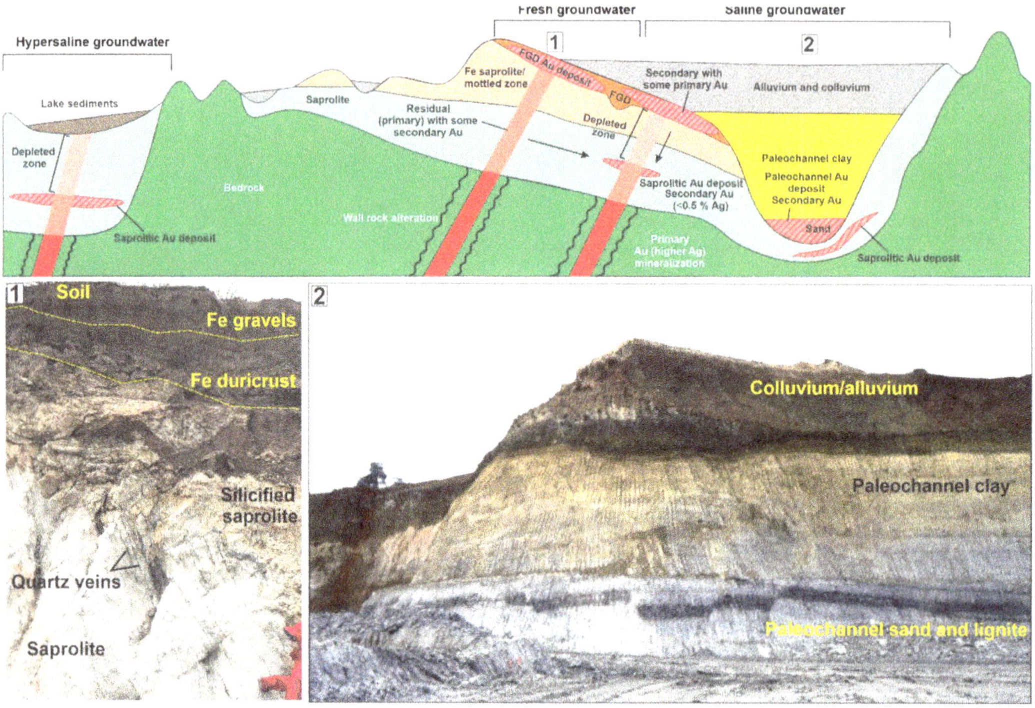

Fig. 1 An idealized diagram showing regolith-landform relationships and the types of supergene Au deposits. In Ferruginous gravels and duricrust (FGD) deposits, Au occurs as a large, flat lying zone, a mixture of largely secondary Au (Ag < 0.5 wt%) with some primary Au (mainly Ag >5 wt%). In saprolite deposits, Au itself is mostly residual primary and some secondary with little or no dispersion. Under certain arid conditions, with oxidizing saline groundwaters, Au is mobilized and precipitates as secondary Au in saprolite. In some hypersaline environments, Au enrichment may extend laterally over weakly mineralized alteration zones. In paleochannels, flat-lying Au enrichment may occur in basal sands or in saprolite/saprock below the unconformity. **1**: ferruginous gravels and duricrust (FGD) overlying saprolite (approximately 3.8 m vertical profile). **2**: paleochannel sediments (approximately 11 m vertical profile) overlying saprolite

Mt Gibson). At Mt Gibson, Au-rich ferricrete, 2–4 m thick occurs as loose or cemented nodules and pisoliths (Smith and Anand 1992; Anand et al. 2019). Prior to mining, some Au-rich ferricrete and lateritic residuum occurred close to the surface; elsewhere, both were overlain by several metres of silicified colluvium and alluvium. Gold-multi-element anomalies extended for about 4 km along strike, with a width of up to 1 km (Fig. 4). The greatest Au concentrations, to 13 and 6.5 ppm, occur in loose and cemented ferricretes, respectively, whereas Au concentrations are low (median 75 ppb) in underlying clay-rich saprolite, except over quartz-hematite veins. There is insufficient Au in bedrock and quartz-hematite veins to be the major sources of the regional Au halo, suggesting that at least some Au was sourced from primary mineralization in uplands to the west (Anand et al. 2019).

Gold (1–3 g/t Au) in FGD deposits is hosted by both magnetic and non-magnetic nodules and pisoliths. There can be a wide range of Au abundance within and between individual structures in a deposit. For example, at Moolart Well, pisoliths developed in local sediments may contain 45 ppm to 40 ppb Au within a single sample. Both primary (3->20% Ag) and secondary (<0.5% Ag) particles occur in these deposits. Thus, the purity and grade of Au in these deposits are higher than in the underlying saprolite and bedrock sources—depending upon the relative

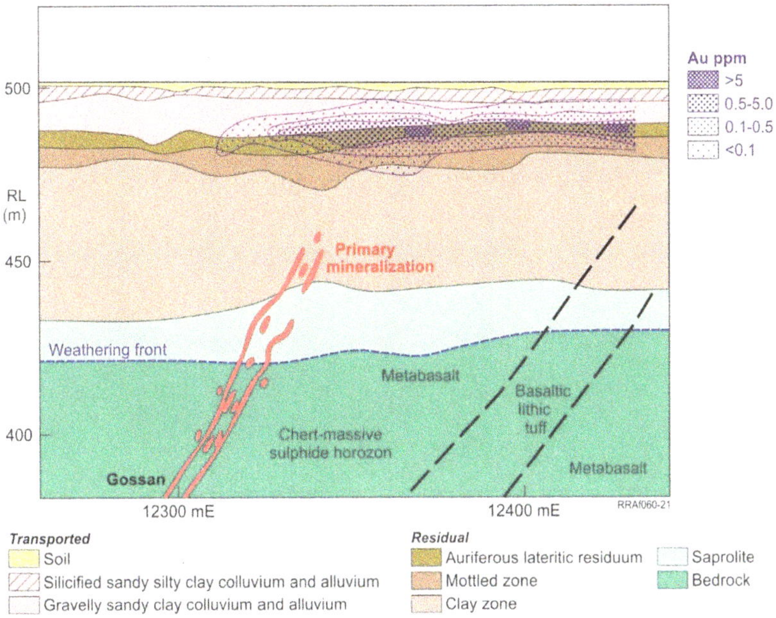

Fig. 2 A residual ferruginous gravels and duricrust (FGD) Au deposit buried beneath 15-20 m of colluvial-alluvial transported cover, Calista, Yilgarn Craton. RL: elevation above mean sea level. (Modified after Anand 2001)

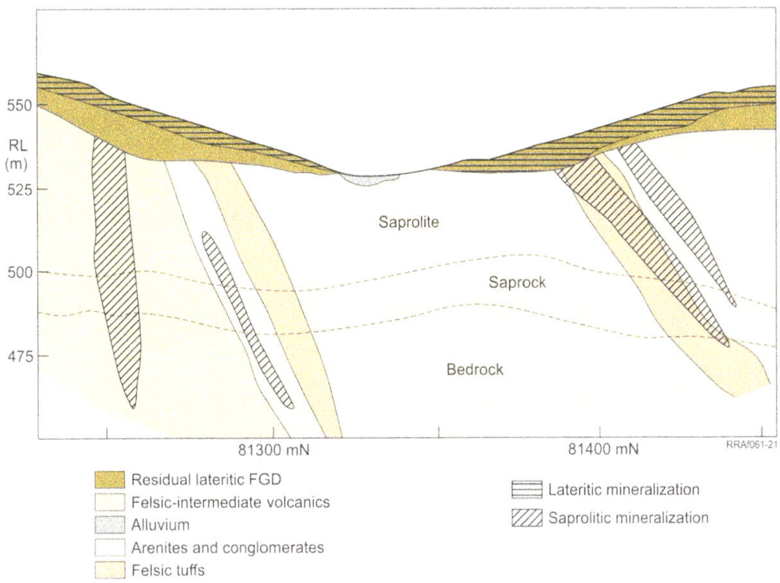

Fig. 3 Residual to semi-residual FGD and saprolitic Au mineralization, Royal Hill deposit, Rosebel, Suriname. (After Sirois et al. 2002)

proportions of primary and secondary Au (Freyssinet et al. 2005).

The style of mineralization and the weathering environment control the proportions of secondary and primary Au in the FGD. Gold trapped in vein quartz is protected from weathering. Thus, it is preserved more readily in FGD than that occurring freely or hosted by more easily weathered sulfide mineralization in hypogene ore. Primary Au grains in FGD deposits vary from pristine and angular to being partly or wholly rounded by weathering or transport, commonly with an outer rim depleted in Ag. Some grains have a xenomorphic morphology, usually after quartz or pyrite. Rare nuggets (2 mm->10 cm) found in FGD (and in soil derived from saprolite and in

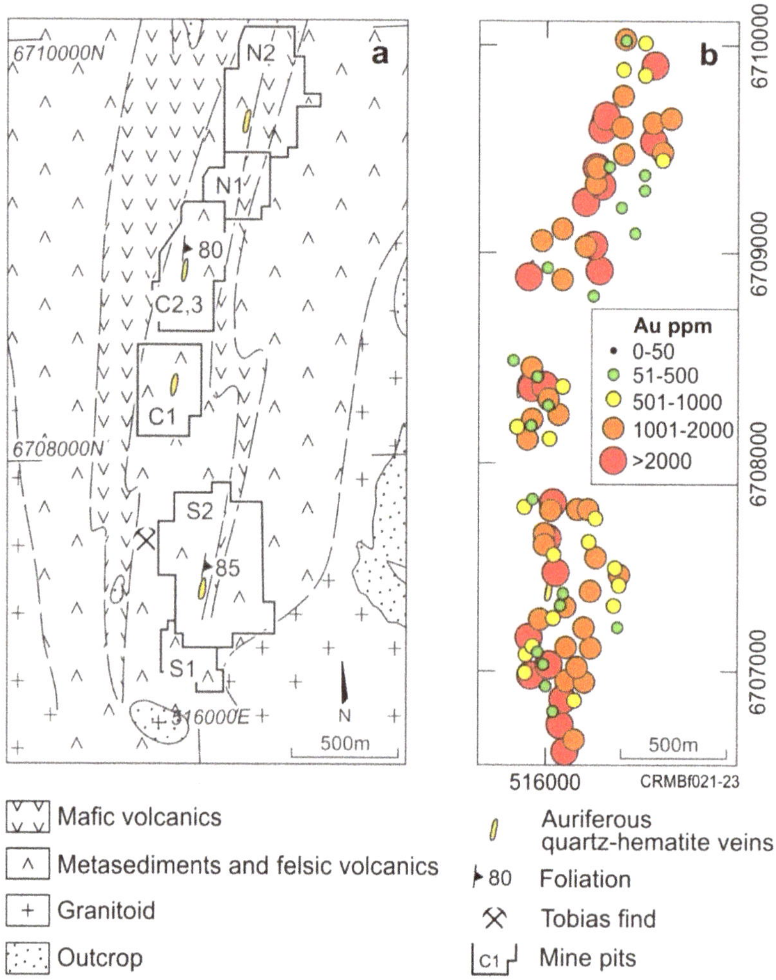

Fig. 4 Mt. Gibson Au deposit, W. Australia. (**a**) Geology; (**b**) Distribution of ore-grade Au in FGD, Mt Gibson The Au-bearing quartz veins are also shown. (After Smith and Anand 1992; Anand et al. 2019)

shallow sediments) are hypogene, with Ag depletion on external surfaces and along internal crystal boundaries (Butt et al. 2020). At Rosebel, coarse Au, presumably hypogene, occurs as free grains attached to goethite or hematite, or within cubic xenomorphic voids, after pyrite in quartz. Nuggets up to 8 mm in diameter have been found lying on the surface after heavy rain (Voicu 2010).

There are a variety of morphologies of secondary Au on organic and inorganic substrates, including clusters, wires, spheres, framboids, laths, chains, irregular anhedral forms, dendrites and hexagonal and triangular sheets (Hough et al. 2008; Anand et al. 2019). In FGD, these occur largely in goethite in cortices, cracks and voids (Fig. 5), as well as cracks and voids of cores of pisolith and nodules, and within organic carbon (Moolart Well). However, a high proportion of secondary gold also occurs as very fine-grained (<20 μm) particles. A higher proportion of secondary gold is associated with environments where complexing ligands (organic matter, halides and thiosulfate) are abundant or have been in the past.

2.2 Saprolitic Deposits

Saprolitic deposits are enrichments in the regolith between the mottled zone and the weathering front. Primary structures in the wall rocks and

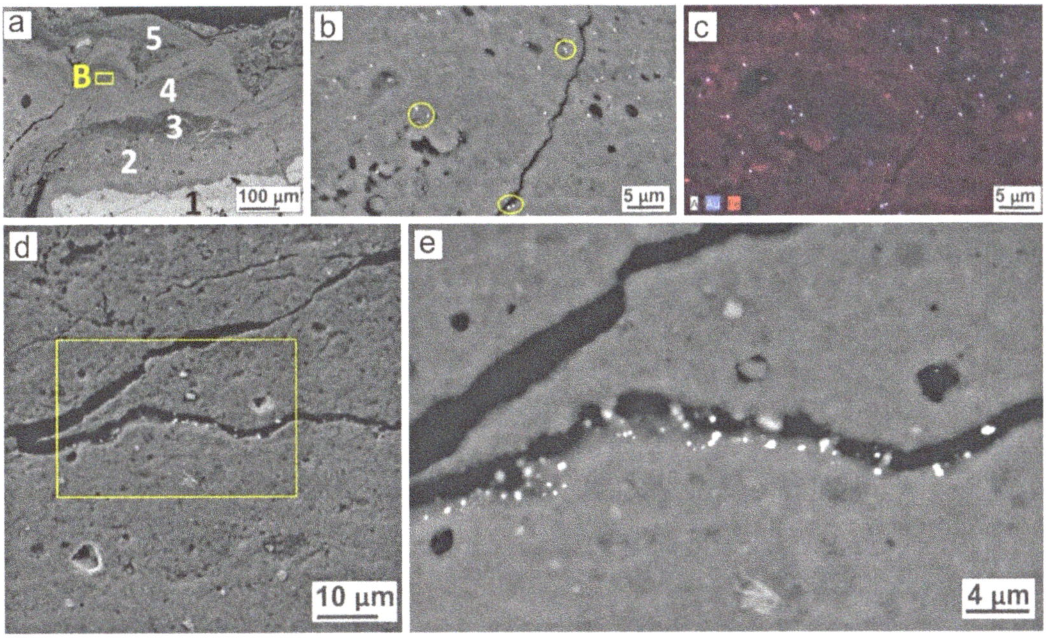

Fig. 5 SEM-BSE images of sectioned pisolith from FGD Au, Mt Gibson, W. Australia. (**a**) hematite-rich core (1), goethite cortex (2), kaolinite-goethite cortex (3), goethite cortex (4) and goethite-kaolinite cortex (5). Boxed area in **a** is shown in **b**. (**b**) Numerous Au grains in goethite cortex and cavities, some grains are highlighted by yellow circles. (**c**) SEM-EDX element map of Au and Fe. (**d**) Au grains in cavities of core. Boxed area in **d** is shown in **e**. (**e**) High magnification image of Au grains in cavity. (After Anand et al. 2019)

mineralization are preserved, with Au grades generally similar to those in unweathered ore. Minor leaching of Au is commonly offset by relative enrichment due to mass loss (Gray et al. 1992; Butt 1998; Freyssinet and Itard 1997; Costa 1993; Freyssinet et al. 2005; Anand et al. 2017). Gold is mostly residual primary grains (<50–1000 µm; Fig. 6); secondary (<100 µm) grains are rare and there is minor or no lateral dispersion (Figs. 2 and 3). Stronger leaching under tropical rainforest conditions has led to greater residual enrichment in saprolite at Rosebel (Fig. 3) and Igarapé Bahia. However, where groundwater is acidic and saline in arid environments—such as in the south of the Yilgarn and Gawler Cratons in Australia—Au leached from the mottled zone and upper saprolite has precipitated at greater depths, close to the water table or related redox fronts, and at porosity barriers. The resultant Au enrichments occur as patchy sub-horizontal to horizontal horizons commonly 10–40 m thick in the saprolite and saprock over primary mineralization. There is commonly a sharp boundary between this supergene enrichment and an overlying zone strongly depleted in Au (Fig. 7). Where profiles are complete, FGD and saprolite deposits are generally separated by a 'deplete zone' 5–30 m thick, barren of Au (e.g., Mt Gibson; Fig. 7). Where the FGD has been eroded, the depleted zone generally occurs at the surface or immediately below-transported cover. Gold is most strongly enriched in the upper 5 m of the supergene zone, with a gradual decrease downwards. There is generally little (<100 m) or no broadening relative to primary mineralization and associated alteration zone (Gray et al. 2008). Some sites (e.g., Hannan South and Kanowna Belle in the southern Yilgarn) show well-developed broadening of the saprolitic Au enrichment, but ore grades only extend a few tens of metres from the oxidized primary source (Fig. 8; Lawrance 2001). Gold is typically secondary and mostly as much smaller grains (<50 µm) than free hypogene particles, and in many cubic crystal morphologies including flat

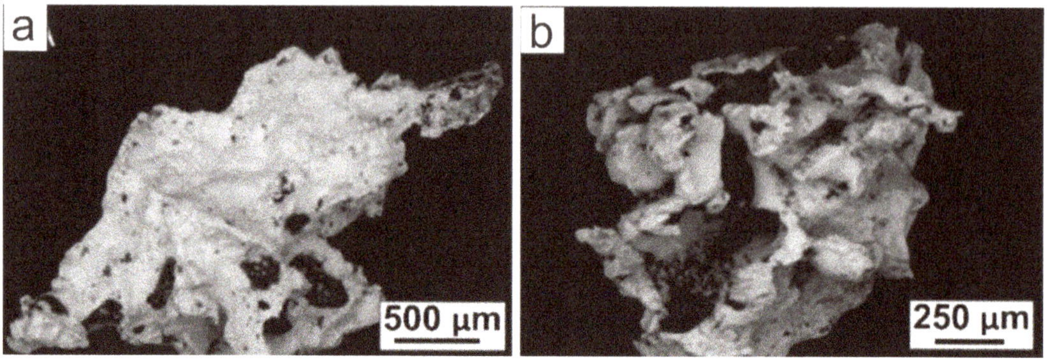

Fig. 6 SEM-BSE images of irregular primary Au grains panned from saprolite, Bulchina Au deposit, W. Australia. Both grains have high (8–10%) Ag contents. Grain **b** has dissolution cavities. (After Anand et al. 2019)

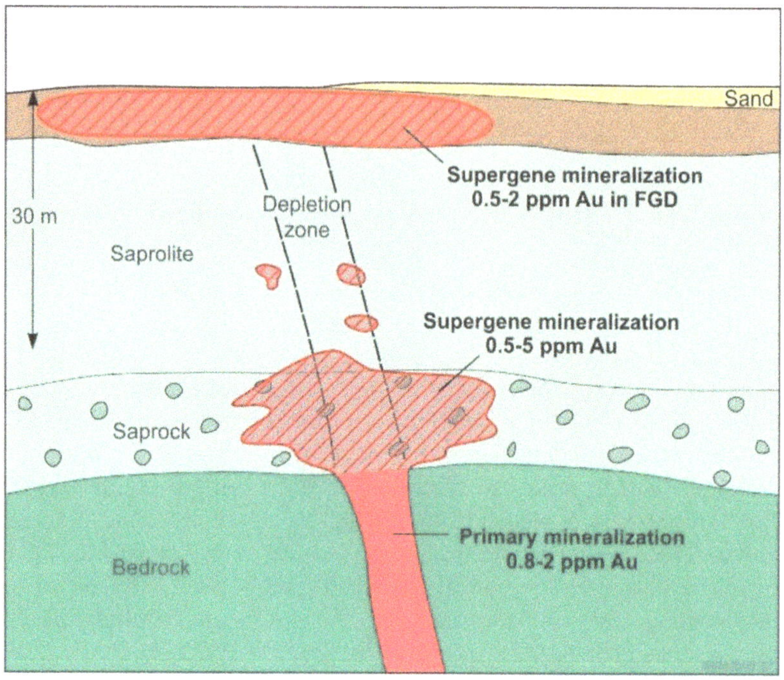

Fig. 7 Schematic depiction of ore-grade supergene Au mineralization in FGD (ferruginous gravels and duricrust) and saprock in a saline groundwater environment, Mt Gibson, W. Australia. A depletion zone lies between the mineralized zones

triangles and hexagons, octahedra and rhombohedra. There is, however, no evidence for the formation of secondary nuggets (Butt et al. 2020). Examples of differing primary and secondary Au grain morphologies in a single profile at Bardoc, Western Australia are shown in Freyssinet and Butt (1988)—see also Butt and Zeegers (1992, Figure III.3–9).

No saprolitic enrichment or depletion zones similar to those in the Yilgarn have been reported from the humid tropics. Minor increases in Au and Ag contents occur at the water table and saprock-saprolite contact in the gossan profile at the Justice matrix sulfide deposit at Ashanti, Ghana (Bowell et al. 1993; chapter "Gossans: In Situ Supergene Base Metal Deposits"). Significant supergene deposits have, however, formed in saprolite on calcareous rocks at Ity, Cote d'Ivoire (Fig. 9) and Sadiola, Mali, over otherwise sub-economic resources. The Proterozoic orogenic hydrothermal primary mineralization at Ity is unusual in that it occurs as skarns at the contact

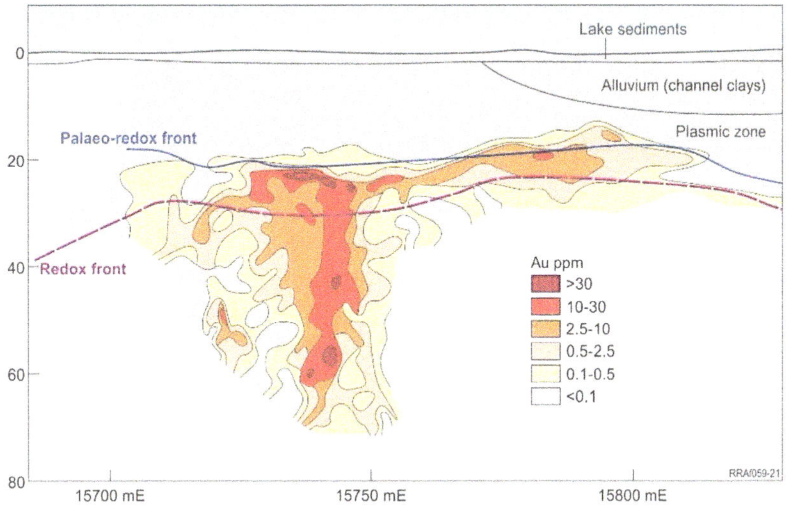

Fig. 8 Saprolitic Au deposit with a well developed supergene mineralization in a hypersaline environments, Hannan South, W. Australia. (After Lawrance 1991)

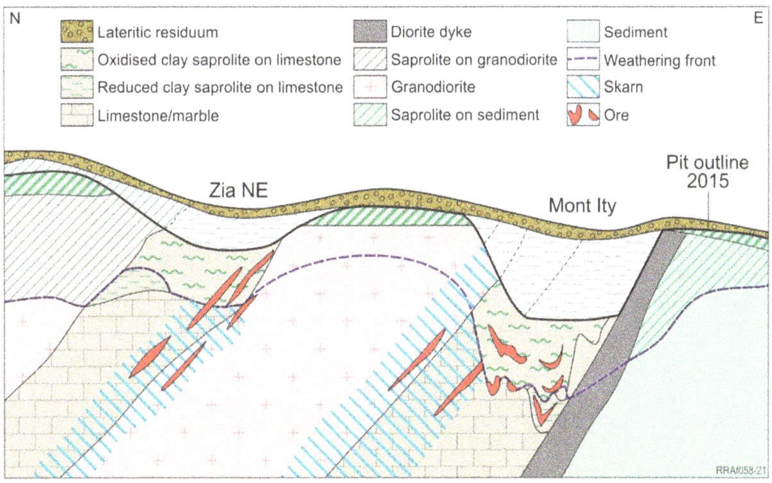

Fig. 9 Cross section through the Ity Au deposit, Côte d'Ivoire, showing auriferous lateritic residuum and saprolite. The mineralized skarn underlying the supergene Au deposit is sub-economic. (Modified after Beziat et al. 2016)

of limestones and carbonate-rich sediments with granodiorite-tonalite intrusions. The skarns consist of garnet, diopside, tremolite and phlogopite in a calcite matrix, with Au associated with a range of sulfides, tellurides and selenides. The whole area has been intensely weathered, with a nearly continuous lateritic residuum (FGD), commonly 10–15 m thick, overlying clay-rich saprolite to depths of 50–120 m (Body et al. 2015; Beziat et al. 2016). Saprolite includes highly contorted and brecciated zones caused by lateral flow and collapse due to the dissolution of limestones and the formation of karst cavities during weathering; in places, FGD components are incorporated in the saprolite. Economic mineralization is largely confined to the (semi-) residual FGD, grading 2–7 g/t Au, now mined out, and more localized saprolitic ore developed above the mineralized skarns, with grades of 2->10 g/t Au. These grades are higher and over greater widths than the hypogene mineralization, with enrichment partly residual, due to mass loss (SG 2.7 in fresh limestone, 1.5 in saprolite and FGD), and partly to Au remobilization during weathering.

Intense weathering, to depths of 220 m of the carbonate-rich host rocks at Sadiola, has similarly enhanced Au grades in the low-density saprolitic ore (SG 1.7) which plays a significant role in the formation of economic ore (Robertson 2004).

2.3 Paleochannel Deposits

Paleochannels are common in weathered terrains and are a significant source of Au and water resources. They are generally U-shaped and symmetrical, having moderately broad valley floors and straight, gently inclined sides, incised into a deeply weathered landscape. Subsequently, they were filled by erosion products and commonly have a sequence of high-energy basal sands overlain by clay-rich lacustrine to estuarine sediments. In Australia, these sediments are commonly Late Middle to Late Eocene respectively, overlain by younger alluvial and colluvial deposits. The channel sediments have undergone post-depositional weathering, during which Fe has been mobilized and redeposited as Fe oxide--rich mottles, nodules and pisoliths. Gold deposits occur predominantly within the basal sands but can also occur in oxidized clay. However, in some deposits, mineralization is almost entirely contained within saprolite and saprock below the sand (e.g., Freddo Au deposit; Fig. 10). The deposits tend to be flat-lying and sheet-like to lenticular in cross-section. These deposits are individually small, but they may be semi-continuous along a particular drainage system, thereby forming a significant resource, such as at Lady Bountiful, Challenge-Swordsman and Higginsville in the Yilgarn Craton (Devlin and Crimeen 1990; Mazzuccheli 2005). Bonanza grades are not uncommon, particularly in conglomerate lenses. Gold occurs as secondary particles and the enrichment zones may transgress sedimentary structures, including the basal unconformity of the channel. Gold in the saprolite/saprock is a flat-lying enrichment below the unconformity. The auriferous zones appear to be similar in geometry and origin to the saprolitic deposits described above but can be laterally more extensive.

Although sometimes described as 'deep leads', these paleochannel-hosted deposits differ from those in Victoria, Australia. The Victorian deposits are alluvial placers, consisting dominantly of free particles of primary Au, including large nuggets, and Au in vein quartz, deposited in basal sands in drainage channels. Most of the deposits are overlain by Paleocene-Pliocene basalt flows and later sediments following the same channels—many of which have subsequently been intensely weathered. The deep

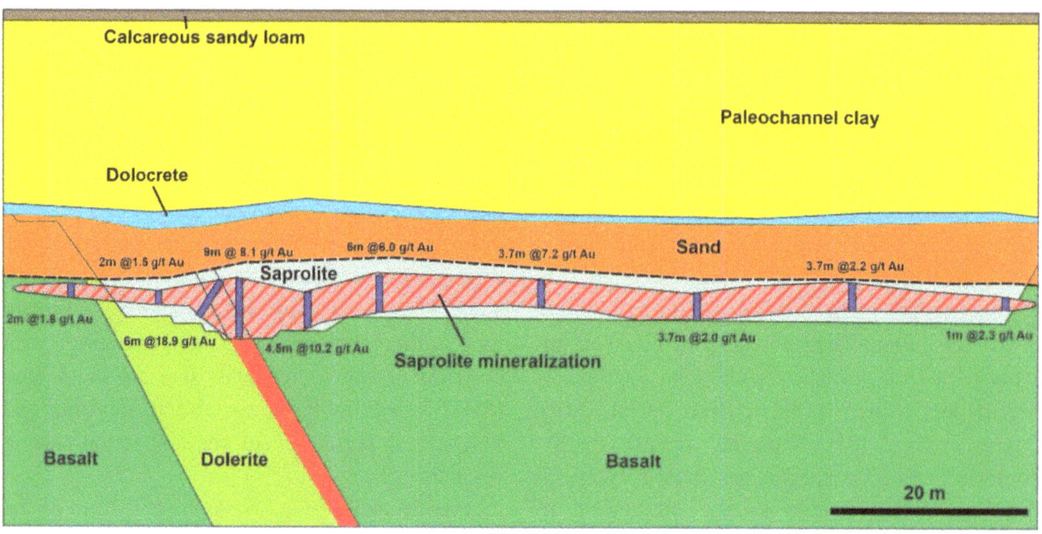

Fig. 10 Location of supergene Au enrichment in saprolite, underlying paleochannel basal sands, Freddo Au deposit, Yilgarn Craton, Western Australia

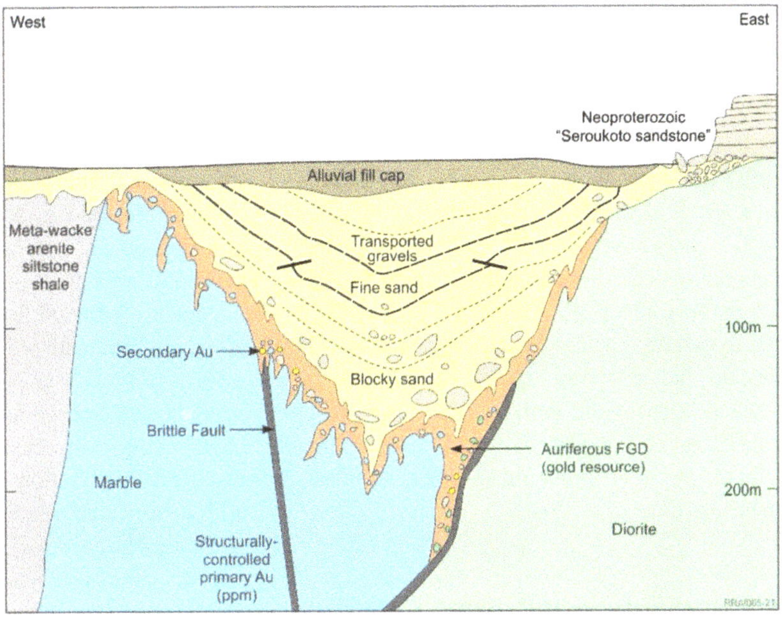

Fig. 11 Schematic geological cross section through the Yatela deposit, Mali showing idealized karst infill stratigraphy. (Modified from Hanssen et al. 2004)

leads yielded over 11% of the Au mined in Victoria prior to 1988 (Canavan 1988).

At Yatela, Mali, an Au-rich ferruginous residue derived from dissolution of the marbles hosting primary mineralization is overlain by sediments up to 150 m thick (Masurel et al. 2016). These are dominantly coarse and fine sands, and ferruginous lateritic debris, infilling karst-like troughs along a fault zone, but seem not to be true paleochannels (Fig. 11).

3 Global Distribution

Economic or near-economic supergene concentrations of Au occur in lateritic regoliths in most climatic zones including seasonally humid tropics (savannas), humid tropical rainforests and semi-arid to arid environments. Examples of supergene Au deposits in FGD are from Australia (Butt 1989, 1998; Anand 2000; Anand et al. 2019), West and Central Africa, South America and India (Michel 1987; Ellison and Baylock 1990; Santosh et al. 1990; Andrade et al. 1991; Costa 1993; Bowell et al. 1993; Freyssinet et al. 2005; Porto 2016; Table 1). Mining of these deposits in Australia has been largely carried out in regions having Mediterranean to semi-arid climates, such as the Yilgarn Craton (e.g., Boddington, Mount Gibson, Moolart Well). There are similar deposits in Africa in Mali (Syama, Yatela), Cote d'Ivoire (Ity), Gabon (Mavikou) and South America in the Guianas (Rosebel, Yuaou), Brazil (Igarapé Bahia) and Venezuela. Saprolitic enrichments are mined mainly in dry savanna to semi-arid environments principally in Australia, but some of the largest (1 million oz) are in tropical rainforest-humid savanna at Ity, and semi-arid conditions at Sadiola and Yatela.

4 Factors Influencing Occurrences

4.1 Influence of Primary Mineralization on Supergene Au Dispersion

Supergene Au mobility is influenced by the style of bedrock mineralization through (i) controlling the access of weathering solutions to sulfide host minerals and free primary Au; (ii) determining the physical state of the primary Au, and (iii) influencing the composition of supergene solutions. Primary Au in vein quartz is protected from weathering and is residually concentrated at or near the surface; there is thus only minor depletion in the upper regolith and little or no dispersion in saprolite and saprock (Freyssinet et al. 2005; Gray et al. 2008). In contrast, Au in sulfides is more mobile. The sub-micron particle size, easy access by weathering solutions and high concentrations of various oxyanions generating corrosive conditions favour Au dissolution. At Calista, W. Australia (Fig. 2), this has led to strong Au remobilization in the regolith, despite the low salinity of the groundwater (Anand 2015). Gold mobilization, depletion and supergene enrichment in the southern Yilgarn Craton is more pronounced over sulfide-rich deposits (Gray et al. 2008).

4.2 Influence of Bedrock Geology on Supergene Au Dispersion

The degree of supergene Au mobility is also affected by bedrock type, faults and fractures. Ultramafic and, to a lesser degree, mafic rocks buffer groundwaters more than more felsic lithologies, leading to reduced Au depletion (Anand 2000; Gray et al. 2008). Two adjacent deposits at Gourdis, in the northern Yilgarn, have contrasting lithologies and regolith Au distributions. At the 81400 deposit, hosted by sheared quartz-feldspar porphyry and basalt, the upper regolith is strongly depleted in Au, whereas at the 81800 deposit (400 m away), hosted by sheared mafic rocks, the upper regolith is enriched in Au. This probably reflects the importance of mafic rocks in enhancing the hydrothermal precipitation of Au, so that system was initially richer in Au than the felsic-hosted mineralization. Strong fracturing and faulting, and calcareous host rocks have a significant influence on the extensive enrichment in saprolite at Ity, Côte d'Ivoire and Sadiola, Mali.

4.3 Weathering History and Climate

Continental landmasses between 35°N & 35°S have significant similarities in regolith development and landscape evolution (chapter "The Role of Regolith Formation and Evolution in the Development of Supergene Mineral Deposits"). They have all been tectonically stable, with important climatic and weathering episodes in common. In particular, they have all been subjected to seasonally humid tropical climates similar to those in contemporary humid savannas, resulting in the development of an intensely weathered, leached and commonly lateritic regolith and planation of the landscapes. Such conditions were geographically widespread from the late Cretaceous to the mid-Cenozoic, but have since contracted to their present near-equatorial distribution, to be succeeded by semi-arid to arid and Mediterranean climates in the mid-latitudes. Similar histories affected the terrain of associated island arcs in the Caribbean, SE Asia and Oceania as they emerged during this period.

The shield landscapes were not affected by glaciation throughout the Cenozoic. Their landscapes are developed by physical processes related to changes in erosional base-levels due to epeirogenic uplift and downwarping, and ongoing chemical weathering under changing climates. In consequence, uplifted paleolandscapes were isolated as relics overlooking younger active landforms. Slow uneven erosion exhumes deeper regolith, but the erosion materials were partially retained in the landscape, deposited in colluvial-alluvial plains. Successive weathering episodes (e.g., ferruginization, silicification) affect both exhumed residual and transported material. In places, relief inversion has resulted in ferricretes occurring in elevated areas. These situations are especially common in weathered terrains in West Africa, where the underlying basin and swell structure controls topography, and the Brazilian Shield. These differing histories of regolith and landscape evolution have had a significant bearing on the development of supergene Au deposits.

Deep saprolites form due to result of in situ weathering under humid conditions in areas of low relief, where the downward rate of chemical weathering exceeds that of surface erosion. Relative enrichment of primary Au, with only minor secondary remobilization, is common over mineralized bedrock and may result in exploitable supergene reserves, amenable to low-cost open-cut mining, even where the bedrock resource is uneconomic. Further supergene enrichment in saprolite (and within and beneath paleochannel sediments) characteristically only occurs where the climate has changed to semi-arid or arid, and groundwaters have become highly saline, acidic and oxidizing. In contrast, the FGD units overlying saprolite are heterogeneous and structurally complex, usually products of multiple stages of mechanical, biological and chemical processes. Despite this, FGD have a high potential for hosting supergene ore. Depending on their nature (lateritic residuum to ferricrete), the Au source may be either (or both) underlying or upslope or now missing altogether due to erosion. In addition, alluvial cover on pediments and in paleochannels may host placers derived from bedrock mineralization upslope.

5 Deposit Formation and Evolution

5.1 Gold Mobility During Weathering

In bedrock mineralization, Au occurs as tellurides, solid solutions within sulfides or as particulate metal. In most low-temperature weathering environments, Au is chemically inert. However, its mobility increases in the presence of certain complexing ligands in groundwater and soil (Gray et al. 1992), depending on the weathering environment, geomorphological setting and climatic conditions (Butt 1998). Gold can form complexes with organic, thiosulfate, halide and hydroxide ions, with each complex occurring under specific dissolution conditions, and precipitating as pure Au or Au-Ag alloys where conditions change

Table 2 Conditions of Au mobility in regolith environments

Species	Possible origin	Solubility conditions	Product
$AuCl_2^-$ / $AuCl_4^-$	Oxidative dissolution of gold under acidic saline conditions $Au + 2Cl^- \rightarrow AuCl_2^-$	Oxidized/saline/acidic; pH <4.0, Cl>35,000 mg/L	High fineness gold
AuI_2^-	Dissolution of gold under moderately oxidative conditions $Au + 2I^- \rightarrow AuI_2^-$	Enhanced where I^- is released by decomposition of organic matter	High fineness gold
$Au(SH)_2^-$	Dissolution of gold by reduced waters during early supergene alteration or by biologically generated reducing conditions $Au + 2SH^- \rightarrow Au(SH)_2^-$	Reduced/neutral	Medium-high fineness gold
$Au\ (S_2O_3)_2^{3-}$	Weathering of gold and/or pyrite in neutral to alkaline solution $Au + 2S_2O_3^{2-} \rightarrow Au\ (S_2O_3)_2^{3-}$	Alkaline to weakly acidic (metastable)	Medium fineness gold
$AuCN_2^-$	Interaction of cyanide with gold $Au + 2CN^- \rightarrow AuCN_2^-$	Cyanide present, e.g., derived from decomposition of organic matter	Low fineness gold
Au-organic matter	Interaction of organic matter with gold under oxidizing conditions $Au + OM^- \rightarrow Au^+ - OM$?	Not certain, may depend on source of material	High fineness gold fine-grained
Colloidal gold	Possibly formed during reduction of Au by organic matter $Au^+ + OM^- \rightarrow Au°$		High fineness gold fine grained
$AuOH(H_2O)°$	Dehydrated soil system $Au + H_2O + \frac{1}{4}O_2 \rightarrow AuOH(H_2O)°$	Neutral to alkaline solutions only confirmed under experimental conditions	Possible high fineness gold

After Gray et al. (1992), Freyssinet et al. (2005)

(Table 2). Several workers (Andrade et al. 1991; Reith et al. 2012, Anand et al. 2019) have pointed out the role and importance of organic complexes in Au mobility, but the nature of the ligands and the conditions under which they are significant are largely unresolved. Biological activity and vegetative cycling also contribute to dissolution and precipitation of Au in near-surface environments. Nanoparticulate Au and organic-Au complexes may also be mobilized as colloids.

During FGD and saprolite formation, chemical mobilization and dispersion of Au has resulted in a net loss of mass but conversely leads to enrichment as supergene deposits. Lateritic regolith provides two types of chemical environments. The bioturbated Fe- or Al-rich FGD and mottled zones (the pedolith) and the underlying saprock and saprolite (the saprolith) may behave as independent systems with different environmental controls. For example, in FGD, biological processes have played major roles in transport and precipitation of Fe and Au (Anand et al. 2019) but only marginally affect the deep underlying saprolite (Monteiro et al. 2014). Furthermore, the mechanical transport of nodules and pisoliths is an important process in FGD formation. Thus, fine Au-bearing clasts and micro- to nanoparticulate Au may be laterally transported in FGD, whereas in saprolite, vertical collapse is the only physical movement and lateral transport is minimal. These contrasting environments provide different complexes appropriate for Au transport and different mineral hosts for Au precipitation. Dominant reactions have varied with time, responding to successive climatic and tectonic changes. Two major climatic regimes have controlled the mobility and distribution of Au in the regolith. Firstly, humid conditions prevailed over most of the mid-to-lower latitudes prior to the late Miocene when saprolite and FGD formed and, secondly, in higher latitudes, the subsequent temperate to arid periods that physically and chemically modified the saprolite.

5.2 Gold Mobility Under Humid Tropical Conditions

During intense weathering in seasonally humid, warm to tropical climates, oxidation at the weathering front deep below the water table occurs under neutral to acidic conditions, with low pH favoured by felsic rocks and high sulfide contents. Gold associated with tellurides or held in the lattices of the oxidizing sulfides and other minerals may be released, but the free metal remains immobile due to the absence of suitable complexing ligands (Butt 1998). For thiosulfate ions to form complexes with Au and Ag (Mann 1984), sulfide oxidation takes place in neutral to alkaline conditions in which concentrations of organic matter and groundwater chloride ions are very low. This may provide a favourable environment and mechanism for the formation of saprolite and FGD deposits from carbonate-rich rocks (e.g., Wau, Papua New Guinea, Webster and Mann 1984; Sadiola, Mali, Robertson 2004; Yatela, Mali, Masurel et al. 2016). However, at Ity, Côte d'Ivoire, Body et al. (2015) proposed that oxidation of sulfide-rich skarn assemblages was acidic and led to dissolution of carbonate host rocks and the formation of karst cavities. Impurities in the carbonate and the calc-silicate zones hosting the hypogene mineralization left a clay-rich saprolite that collapsed into the cavities, with relative enrichment in Au grades caused by mass loss. Thiosulfate production is rare. Thus, apart from some minor etching and reduction of grain size, there is minimal supergene enrichment or lateral dispersion of Au into saprolitic wall rocks. Consequently, primary Ag-rich Au grains persist through the saprolite and into unconsolidated upper regolith and may form an Au-rich blanket in lateritic residuum.

A combination of mechanisms acting over long periods is responsible for Au enrichment in FGD. *Relative enrichment* of Au during land-surface reduction, minor Au mobility in solution or as particulates, and local colluvial movement of unconsolidated upper regolith may form an Au-rich blanket in lateritic residuum (Fig. 12a). Conversely, an *absolute enrichment* may form by lateral dispersion of primary Au followed by reprecipitation of secondary Au, in places tens to hundreds of metres from primary mineralization. Such dispersion is typical of the humid tropics, indicating that Au has undergone considerable physical and chemical transport (e.g., Igarapé Bahia—Sect. 6; see also Colin et al. 1993; Costa 1993; Freyssinet and Itard 1997;

Freyssinet et al. 2005). The proportions of relative and absolute enrichment are controlled by the nature of the ferruginous material (lateritic residuum vs ferricrete), with ferricrete developed in sediments having no relative concentration of Au.

Rain enhances greater surface biomass in humid, tropical regions, providing a high concentration of organic matter to complex and mobilize Au in solution or as colloids, as noted at Igarapé Bahia. Plant roots, in association with microbial activity, produce organic acids, resulting in precipitation and cementation of Fe oxides in the rhizosphere. Evidence for this process with FGD is observed as ferruginized paleorhizospheres, root channels, organic carbon and microbial fossils, suggesting significant biological activity (Genise et al. 2011; Anand et al. 2019). Thus, the greater biomass in these climates leads to the formation of organic and cyanide complexes in the soil and FGD. The role of biologically generated ligands (e.g., humic acids, CN^{-}) in Au mobility, especially in FGD, have been considered by numerous authors (see Gray et al. 1992; Freyssinet et al. 2005; Anand et al. 2019). These organic ligands dissolve Au in surface horizons. Subsequent reduction of Au complexes leads to the formation of fine-grained Au structures in goethite in the cortices, voids and cracks in pisoliths and nodules. The association of Au and Fe oxides can also be explained by the preferential adsorption of colloidal Au or Au complexes on Fe oxides (Greffié et al. 1996). Several authors (e.g., Greffié et al. 1996, Freyssinet et al. 2005) report pure Au spherules associated with Fe oxides in Mali, Brazil and Ghana. Such mechanisms (organic, mechanical and chemical) can together form flat-lying Au enrichments in FGD, containing secondary and primary Au, even over thin horizon of mineralized saprolite.

Gold deposits in paleochannel sediments can be products of both humid and arid weathering regimes. Enrichments in sediments may be alluvial (derived from upstream) or supergene (derived from underlying bedrock). The sediments have generally undergone post-depositional weathering; for example, at Garden Well, in the northern Yilgarn W. Australia, Au in ferruginous nodules formed in the basal sandy unit is largely secondary (Anand et al. 2021). The abundances of the lithological- and ore-related elements Cr, Au, As, Cu (Fig. 13), Bi, Sb, Se, In and S are much greater in the nodules formed in basal sediments than in those in overlying pisolitic clay. Gold dispersion is at least three times greater in units characterized by ferruginous nodules than in the underlying saprolite. The Au-enriched zones at the base of the paleochannel are broadly coincident with weathered primary mineralization in the underlying saprolite, suggesting that Au and associated pathfinder elements were physically emplaced, with clasts of eroded auriferous saprolite being mixed with fluvial sand at the base of the channel. Components of mineralized saprolite also migrated into the overlying unit during the slow deposition of clay. Quartz-goethite nodules formed during later weathering of the channel sediments, with chemically remobilized Au and other elements concentrating in goethite. Dating (U-Th/He) of goethite in nodules and pisoliths suggests that this occurred during humid periods in the mid-Miocene (11.5–18.6 Ma; Anand et al. 2021).

Flat-lying Au enrichments in saprolite beneath the channels are commonly associated with an underlying primary source. The enrichment is largely due to mobilization in acidic saline groundwaters in an arid climate.

5.3 Gold Mobility During Profile Modification Under Rainforest and Mediterranean Climate

The effects of a change to rainforest environments are shown in Fig. 12b–d. The FGD in present-day rainforests probably formed in seasonally humid tropical savannas and the associated supergene Au deposits resemble those in savanna regions. Mineralization in FGD is mushroom-shaped, following the topography, with saprolitic ores generally confined to the original mineralized structures (Fig. 12). Examples include Ashanti (Bowell et al. 1993), and the Ire prospect (Bowell et al. 1996) in Ghana, Ovala in Gabon (Edou-

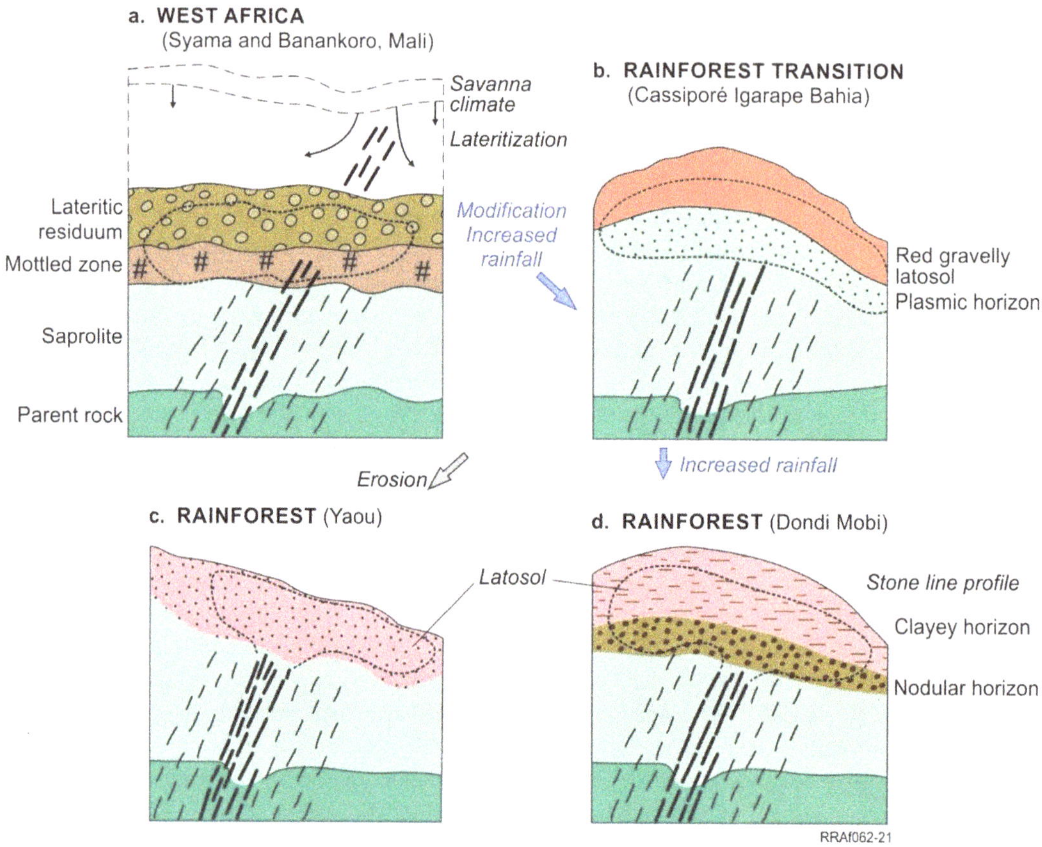

Fig. 12 Model illustrating the main types of Au supergene deposits and their evolution due to climatic changes from savanna to rainforest. (Modified after Freyssinet et al. 2005). See text for explanation

Minko et al. 1992), Ity in Côte d'Ivoire (Lecomte and Zeegers 1992; Body et al. 2015; Beziat et al. 2016), Cassiporé, (Costa et al. 1993) and Igarapé Bahia (Andrade et al. 1991; Sect. 6) in Brazil; Yaou, (Freyssinet 1994) in French Guiana and Rosebel (Sirois et al. 2002), Suriname. The saprolitic ore has similar features to that in the savannas but, in the upper part of the profile, the FGD units are composed of coarse, hard ferruginous nodules and blocks in an ochre or red clay-rich matrix formed by chemical disaggregation (Lecomte and Zeegers 1992; Freyssinet et al. 2005; Fig. 12b–d). Overlying red gravelly latosols are depleted in Au in the upper horizons, but enriched beneath (Costa et al. 1993). Where the FGD has largely been re-weathered, the upper horizon consists of clayey latosols richer in Au than their underlying parent saprolite, as at Igarapé Bahia.

In very humid regions, a stone line can form above the saprolite (Fig. 12d). Examples of mineralized stone-line profiles come mostly from Gabon (e.g., Dondo Mobi, Lecomte and Colin 1989). The Au distribution in saprolite is again similar to that in savanna regions. It is widely dispersed throughout both the lower clay-rich horizon and the nodular horizon (Fig. 12d), which can be compared to the FGD of less humid regions. Where the ferruginous horizon has been destroyed by erosion, the stone line consists mainly of vein quartz, with supergene mineralization is only in saprolite.

Ferruginous gravels and duricrusts at Boddington, W. Australia, are actively weathering under the current Mediterranean climate and display near-surface Au depletion similar to that in tropical rainforests (Fig. 14a, b; Anand and Butt 2010).

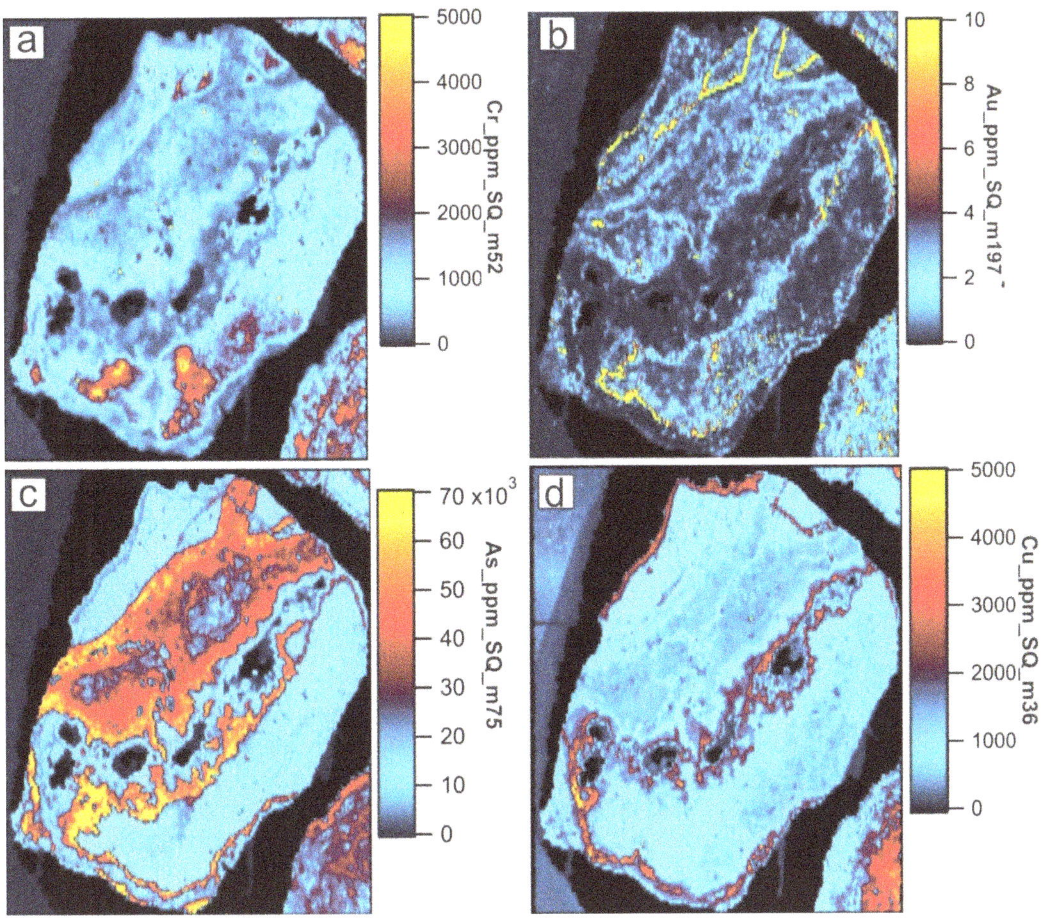

Fig. 13 Laser ablation ICP-MS elemental maps showing Cr (**a**), Au (**b**), As (**c**) and Cu (**d**) for a ferruginous nodule (approximately 10 mm in diameter) formed in paleochannel basal sand, Garden Well Au deposit. (After Anand et al. 2021)

5.4 Gold Mobility During Profile Modification Under an Arid Climate

The effects of climate change on more arid conditions since the late Miocene, coupled with minor tectonic uplift, are shown in Fig. 15. Similar changes have affected Au deposits in the African Sahel and much of inland Australia. Semi-arid and arid landscapes in Australia mainly have low relief due to limited mechanical erosion, so that thick (30->80 m), highly weathered regolith has been preserved. There has been some erosion of exposed FGD units, with duricrusts degraded to friable ferruginous gravels, but locally re-cemented by amorphous silica, hematite and/or goethite. Elsewhere, the residual regolith units are covered by colluvium and alluvium (Anand 2001). Lowering of the water table, coupled with decreased leaching and, in the south, increasing groundwater salinity has affected saprolite. The slow erosion has continued, with a variety of aeolian, colluvial, alluvial and lacustrine sediments deposited on lower slopes and valley floors. These environments have, in effect, formed closed systems, so that alkali and alkaline earth elements, sourced largely from rainfall or slow weathering, respectively, are retained in the

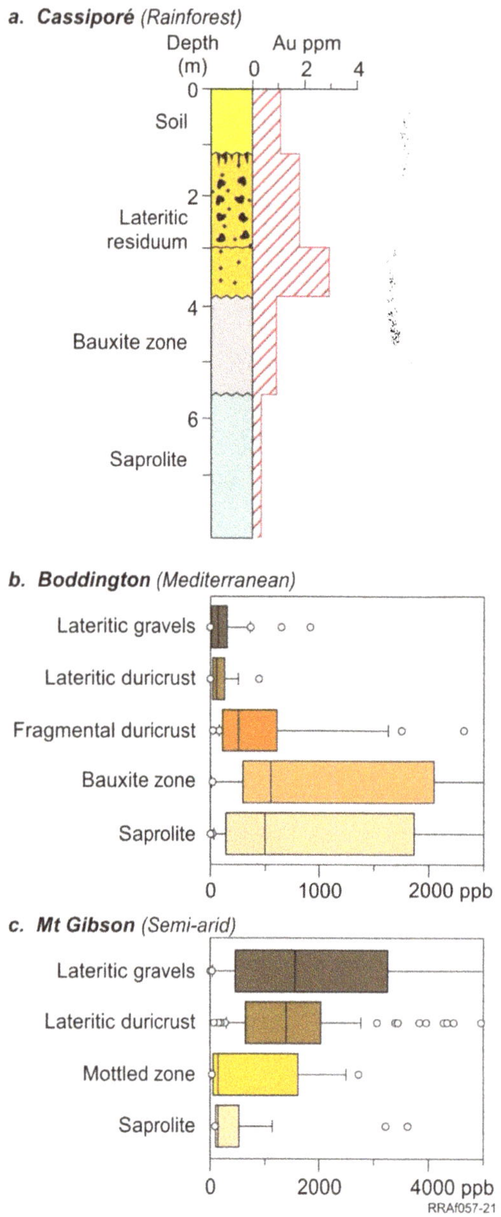

Fig. 14 Comparison of Au distributions in FGD between deposits in different climatic zones. (**a**) rainforest (Cassiporé, after Costa 1993); (**b**) Mediterranean (Boddington; modified after Anand 2001) and (**c**) arid climate (Mt Gibson; after Anand and Butt 2010) Au deposits

landscape. Gold deposits in FGD formed under pre-existing savanna conditions remain largely unchanged, retaining their "mushroom" distribution, even where buried (e.g., Calista, Fig. 2). The main difference compared to deposits in savannas is that in arid areas, there is no leaching of Au from surficial duricrusts and nodular and pisolitic gravels (Mt. Gibson, W. Australia, Fig. 14c).

Gold distribution in saprolite has been largely unaffected by aridity, except where groundwaters have become highly acidic and saline (Fig. 15). To date, this has only been observed in Australia, south of about latitude 30°S. Here, the main effects are (i) development of depletion of Au immediately beneath the FGD (or from upper saprolite where FGD is never formed or has been eroded); (ii) supergene enrichment in the saprolith, and (iii) in places, enrichment in pedogenic carbonates. Strong leaching of the upper saprolite by saline, acid groundwaters has led to marked mobilization and depletion of Au (Butt 1989; Lawrance 2001; Gray et al. 1992) as halide (generally chloride) complexes. During stillstands in the water table decline, the mobilized Au has accumulated as sub-horizontal enrichment zones of pure secondary Au within the saprolite/saprock. Gray et al. (2008) suggested that such Au enrichment is formed by (i) relative concentration of Au due to loss of rock mass during weathering, (ii) strong overlying Au depletion and (iii) chemogenic accumulation of Au in the upper few metres of the enrichment zone, derived by downward migration of Au leached from FGD and upper saprolite. Dispersion in the saprolite is usually confined to the upward projection of primary mineralization defined by the hydrothermal alteration zone and associated shearing. Dispersion is much less than in FGD units as it is entirely hydromorphic and further constrained by the limited stability of halide complexes, impermeability of the clay matrix of weathered wall rocks and low groundwater flow rates under arid conditions. Significant lateral dispersion can occur in highly saline environments, especially in paleochannels, which commonly follow the same faults and shears associated with primary mineralization.

Mobilization of Au by plants under present-day conditions (Anand et al. 2014) may also cause depletion, but is also implicated in the association of Au with pedogenic carbonates (Lintern 2001, 2015). In the small FGD-saprolite deposit at Callion, the highest grades are in calcrete precipitated in the FGD

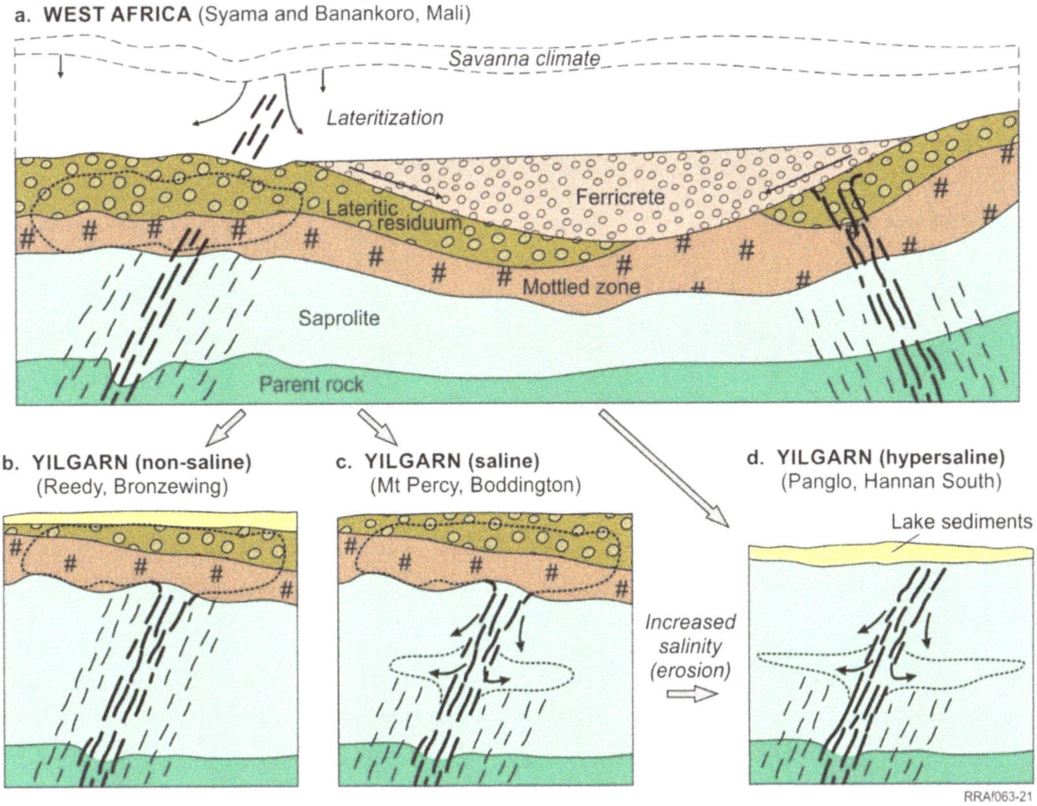

Fig. 15 Model illustrating the main types of Au supergene deposits and their evolution due to climatic changes from savanna to semi-arid. (Modified after Freyssinet et al. 2005). See text for explanation

and locally replacing ferruginous nodules (Glasson et al. 1988). The Au-carbonate association is an important phenomenon in the southern Yilgarn and Gawler Cratons; calcareous soils are important as exploration sample media.

6 Igarapé Bahia Gold Deposit, Carajás Mineral Province, Brazil

6.1 Location and Geological Setting

The Igarapé Bahia Cu–Au deposit, located in the deeply weathered Carajás mineral province, Brazil, was discovered by following stream-sediment Cu anomalies detected in regional surveys. Although geochemical signatures of the mineralization have been obliterated on the lateritic plateau that covers the deposit, Cu–Au soil anomalies on the slopes led to its discovery. The bedrock sulfide Cu–Au resource is estimated at 219 Mt @ 1.4% Cu and 0.86 g/t Au; supergene reserves, identified only for Au, were originally estimated at 30 Mt @ 3.2 g Au/t. This description is derived largely from Porto (2016).

Igarapé Bahia is an IOCG type Cu–Au deposit within a Late Archaean volcano-sedimentary sequence (Fig. 16). The primary ore is hosted by a hydrothermal chlorite breccia on the contact between a lower iron formation and mafic volcanic unit and an upper metasedimentary unit. The ore is locally massive and composed of chalcopyrite, magnetite, gold, siderite and monazite, with associated U–Mo–Pb–Sn–Ag–Bi–Sb–W–P–REE (Dreher et al. 2005; Tallarico et al. 2005).

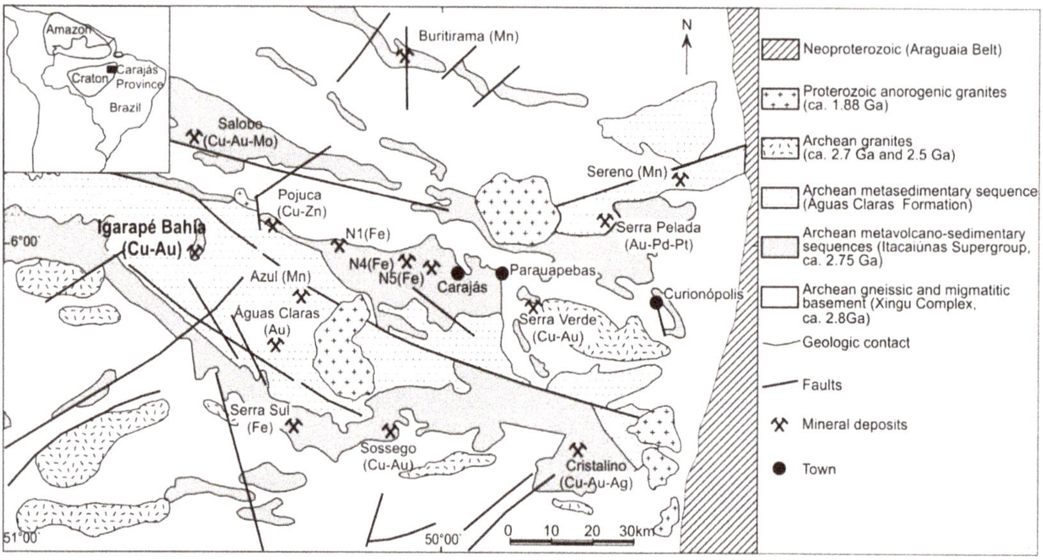

Fig. 16 Geology of the Carajás Mineral Province with location of Igarapé-Bahia and other important deposits (Tallarico et al. 2005)

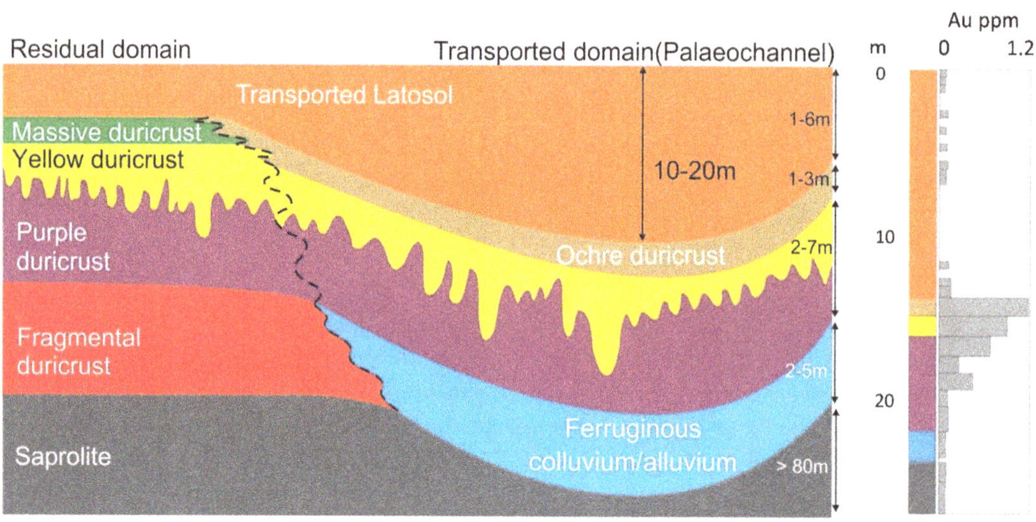

Fig. 17 Regolith stratigraphy of the residual and palaeochannel domains, Igarapé Bahia. Gold distribution in the paleochannel domain (right) is from a representative drill hole

The Carajás region is dominated by an extensive lateritic plateau, 750–900 m high, surrounded by lowland areas at 250–350 m. The plateaux are capped by ferruginous duricrust overlain by a latosol (Ker 1997). In the deposit area, the regolith is at least 100 m thick, subdivided into a residual domain and a subordinate paleochannel domain (Fig. 17). Although the saprock contains abundant secondary Cu minerals, most of the supergene Au ore is hosted by the overlying 80 m thick saprolite. In the residual domain, the saprolite grades upwards into a collapsed fragmental duricrust overlain by ferruginous duricrust. These consist of a dominant lower purple duricrust, overlain in turn by yellow pisolitic duricrust resulting from intense fluid percolation coupled with biogenic action, and a massive, strongly indurated, pisolitic duricrust. In the paleochannel domain, the saprolite has an erosional contact with ferruginous colluvium/alluvium, which is overlain by the purple and yellow duricrusts and an upper friable and highly leached ochre duricrust. The duricrust units are around 10 m thick. The whole area is covered by a transported clay-rich latosol, generally from 1 to 6 m thick, but up to 20 m in the paleochannel domain. The present climate is typical of tropical rainforest with annual precipitation between 1500 and 2000 mm.

6.2 Regolith Mineralogy and Fabric

Kaolinite is the dominant mineral in the saprolite, occurring as "booklets" intermixed with Fe oxide pseudomorphs of chlorite. These features are partly preserved in all ferruginous duricrusts. Gibbsite is present in the matrix in the residual domain but is more abundant in the paleochannel domain, where it occurs in pores and fractures in the matrix and within ferruginous nodules (Fig. 18a). The latosol contains disseminated ferruginous nodules including well-rounded pisoliths and irregularly shaped concretions, some of which are magnetic. Relict kaolinite "booklets" (Fig. 18b) are present in all types of nodules, except pisoliths, over residual and paleochannel domains. Pisoliths have homogeneous texture and composition (Fig. 18c) and are interpreted as authigenic (Thorne et al. 2014; Anand and Verrall 2011).

6.3 Gold Distribution in the Residual and Paleochannel Domains

The Au distribution in the residual terrain over the mineralization and lateral dispersion zones is shown in the multiple box plots of Fig. 19. There are significant Au enrichments in the top 10 to 20 m of the saprolite (median 2.45 to 1.79 g/t Au) and fragmental duricrust (median of 3.24 g/t). These enrichments may be relative due to chemical wasting and mass loss of saprolite and collapse of the fragmental duricrust but may have been augmented by Au leached from the overlying ferruginous duricrust (median 1.34 g/t), similar to that at the Boddington Au deposit, Australia (Anand 1998). Where the protolith and saprolite are virtually barren, the near-surface dispersion zone shows a sharp increase in Au grade from saprolite (median 0.16 g/t Au) to fragmental (median 0.65 g/t Au) and ferruginous (median 0.69 g/t Au) duricrusts. This enrichment is greater than in the ore zone and is probably due to lateral input of Au. In the fragmental duricrust, the dispersion is probably mostly chemical, but in the ferruginous duricrust, some contribution of colluvial transport from the ore zone is likely. In the latosol, the Au grades are invariably low due to its transported nature. In the paleochannel domain (Fig. 20) Au is enriched in the ochre, yellow and upper purple duricrusts, and the coarse gibbsitic nodular fraction of the duricrusts.

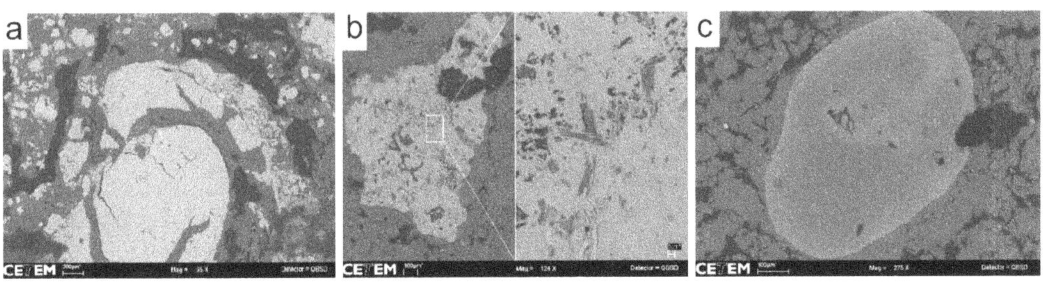

Fig. 18 SEM-BSE image of Fe oxide concretions, Igarapé Bahia. (**a**) gibbsite filling a fractured ferruginous nodule and in the matrix. (**b**) irregular concretion showing relicts of kaolinite "booklets". (**c**) smooth pisolith in latosol matrix

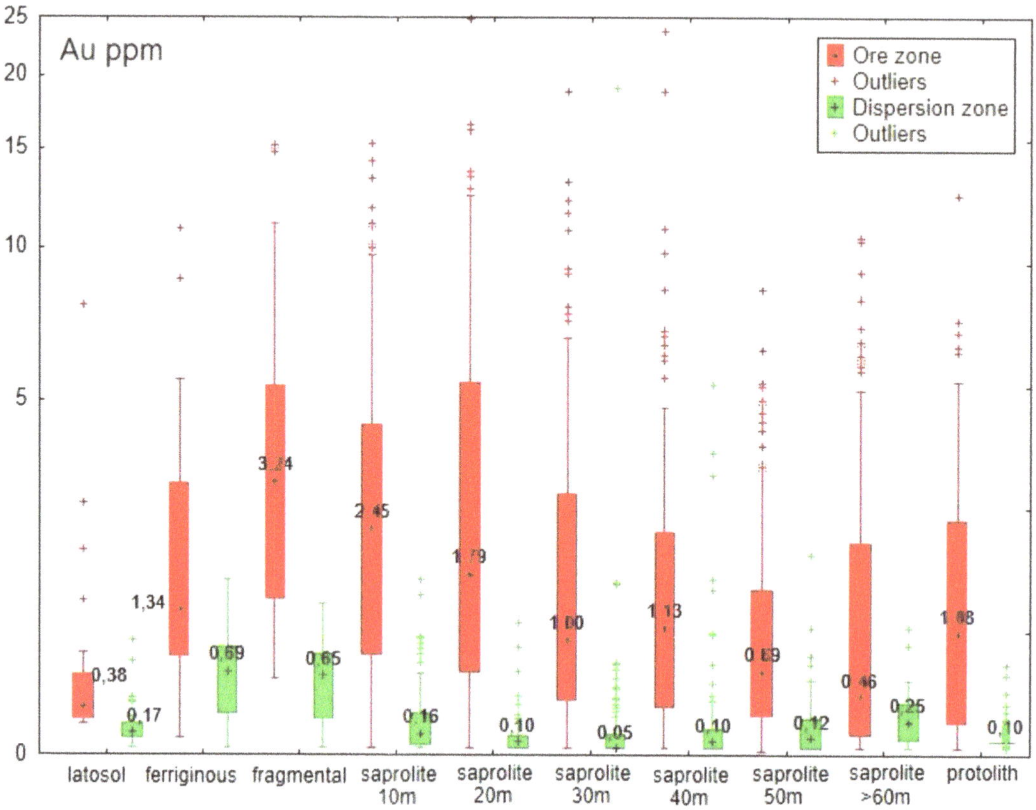

Fig. 19 Box-plots of Au grades (cut at 25 ppm Au) in the regolith over the ore and dispersion zones, Igarapé Bahia

6.4 Gold Distribution in Latosol Components

A comparison of the three main types of nodules and the fine fraction (<53 μm) of the latosol shows that irregularly shaped concretionary nodules (Fig. 18b) are richest in Au. However, there is no meaningful contrast in Au abundance in nodules between paleochannel and residual domains over either primary ore or the dispersion zone, implying that the nodules have been spread over these domains during latosol deposition. However, Au grades tend to be higher in the deeper parts of the latosol, closer to the underlying Au-enriched duricrusts. The Au-rich nodules that lack gibbsite but show relic saprolite fabric are derived from fragmental or ferruginous duricrusts originally present upslope, closer to the ore zone. Magnetic nodules and authigenic pisoliths generally have low Au contents; the former are rich in U, Mo, Sn and W and are probably derived from magnetite-rich units associated with the primary ore, whereas the latter do not contain any metals associated with Au mineralization. The fine fractions of the latosol are invariably Au-poor.

6.5 Deposit Formation and Evolution

Gold in the saprolite ore and related hydrothermal alteration products is residual whereas, in the lateritic unit, Au has been partly to significantly mobilized and dispersed both physically and chemically. Early in the development of the lateritic units, channels were incised away from the main outcropping gossan-rich and resistant ore zones. These channels were filled with material eroded from the upper saprolite by colluvial/alluvial mass flow. Colluvial transport was dominant in the residual domain and strongly contributed to lateral Au dispersion.

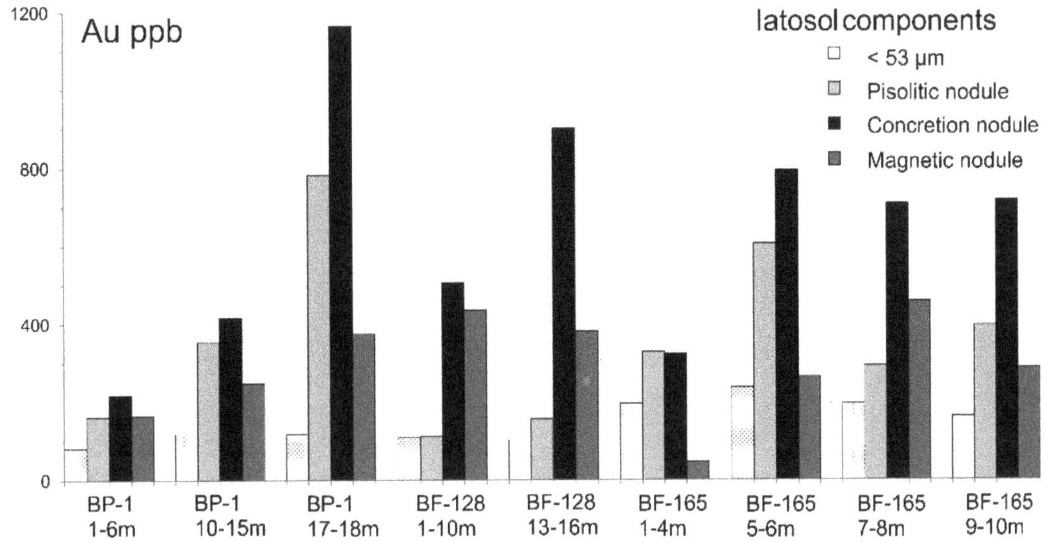

Fig. 20 Distribution of Au in profile BP-1 over the paleochannel, Igarapé-Bahia

With the onset of morphoclimatic conditions conducive to lateritization, the purple hematitic duricrust, subsequently overprinted by the yellow goethitic duricrust, formed over a collapsed and ferruginous saprolite and over sediments in the paleochannels. The massive duricrust in upland positions probably formed in response to dehydration and hardening. Gold was leached from the upper ferruginous duricrusts and reprecipitated in the lower fragmental duricrust, resulting in Au enrichment and lateral dispersion similar to that observed by Freyssinet (1993) in West Africa. Despite leaching, Au displays wide dispersion haloes in the ferruginous duricrusts. Further climatic change towards more humid equatorial conditions resulted in the bauxitization of all duricrusts. This phenomenon has been identified elsewhere in the Carajás region (Costa 1997) and is particularly strong over the paleochannels where increased groundwater flow culminated in the formation of the ochre duricrust. There are indications that the Au enrichment of ferruginous duricrusts in the paleochannel domain may have an important chemical contribution, possibly assisted by complexation by organic matter. Gold-organic complexes may readily migrate within the paleochannel, and precipitate Au where conditions change, such as becoming more oxidizing. As noted above, Au mobilization and bauxitization are also important features in the Boddington deposit, Western Australia (Anand 1998). Finally, the deposition of the sediment precursor to the homogeneous latosol smoothed the landscape over the region. The latosol incorporates Au-bearing concretions derived from the erosion of duricrusts located closer to mineralization and may be an effective geochemical indicator of the presence of Au mineralization at depth.

7 MoolartWell Gold Deposit, Yilgarn Craton, Western Australia

7.1 Introduction

Moolart Well is a prime example of an FGD deposit in which the ferruginous duricrust and gravels form a ferricrete developed in sediments. Moolart Well is 400 km NE of Kalgoorlie in the Yilgarn Craton, Western Australia. The deposit was discovered in 2001 by stratigraphic RAB drilling, and mining began in 2010, with an expected ten-year life span. Much of this description of the deposit is derived from Anand et al. (2017, 2019). The main economic deposits are supergene and restricted to (i) an upper, flat-lying, zone of goethite-coated, transported ferruginous nodules and pisoliths (FGD) within the paleochannel cover (FGD supergene deposit, 1.20 g/t Au; 493,880 ounces Au) and (ii) an underlying saprolite (saprolitic supergene deposit, 0.70 g/t Au; 227,300 ounces Au), sourced from weathering of primary mineralization. Primary mineralization (0.3–0.5 g/t Au; 100,000 ounces Au) beneath the channel is uneconomic; it is hosted predominantly by intermediate diorite intrusive rocks within an N-trending shear zone with quartz lodes and 2–5% disseminated sulfides (Balkau et al. 2017).

7.2 Geological, Regolith and Physical Setting

A sequence of high-Mg basalts with interbedded chert and sulphidic sediments strikes N-S and dips E at a moderate angle (Fig. 21). The basalts are overlain by an ultramafic komatiitic unit,

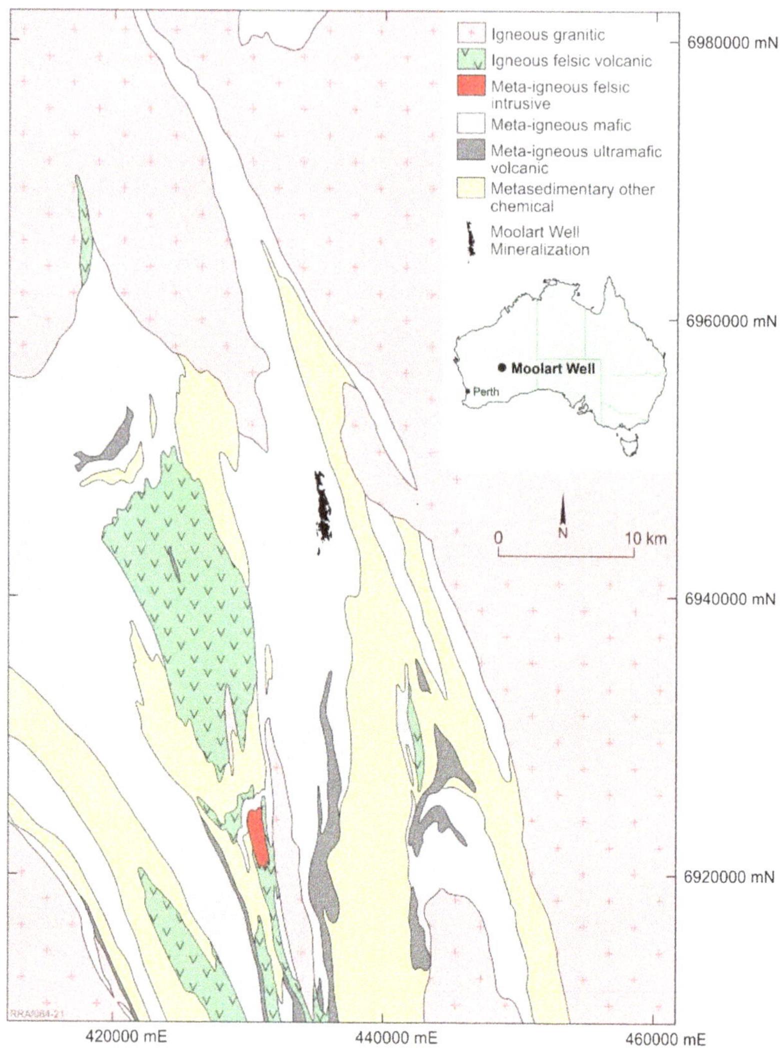

Fig. 21 Geology of the Moolart Well Au deposit and environs. (After Geological Survey of Western Australia 2011)

50–100 m thick, that forms the hanging wall of the mine sequence. The footwall contact of the ultramafic unit is possibly a regional dextral strike-slip fault (Balkau et al. 2017). The shear zones with biotite-chlorite-carbonate and sulfide alteration are the structures carrying Au in the primary mineralization. Arsenopyrite, pyrite and traces of chalcopyrite, galena and sphalerite dominate the disseminated sulfide assemblage. Arsenopyrite is a host to some Au but there is none in pyrite (Balkau et al. 2017).

The deposit is located in a broad N-trending drainage in a paleovalley (1–2 km wide) flanked by low hills of saprolite and saprock. The aeromagnetic data delineates the paleochannel due to magnetic nodules and pisoliths in the sediments (Fig. 23). The regolith is comprised of in situ (saprolite) and *transported* units (older ferricrete, younger silicified colluvium and alluvium and soil) of various ages. The ferricrete consists of poorly-sorted, yellow, red and black magnetic (hematite-maghemite-rich) and non-magnetic (hematite-goethite-rich) nodules and pisoliths (2–30 mm) in ferruginous kaolinitic matrix. The pisoliths and nodules have polymictic cores formed fragments of pisoliths, saprolite and matrix material, overprinted by ferruginization, suggesting several stages of reworking. A yellow to pale brown 0.5 to 3 mm-thick layer of goethite, kaolinite and quartz coats many nodules and pisoliths. Younger transported cover (silicified colluvium and alluvium) is 1–3 m thick east of the ferricrete Au deposit, trending to over 10 m on the west. Gold ore zones are present a few metres below the alluvium-saprolite unconformity at the eastern end to as deep as 25 m below the unconformity at the western end. Saprolite is 20–70 m thick.

Moolart Well has a semi-arid climate (mean annual rainfall of 220 mm with evapotranspiration rates >3300 mm). The water table is ~30 m below surface. The groundwater is fresh and contains no detectable Au (<5 ppt). *Acacia aneura* is the dominant vegetation.

7.3 Regolith Expression

7.3.1 Saprolite Ore

The saprolitic ore is largely confined within the weathered hydrothermal alteration halo, with little lateral Au dispersion into the wall rocks. Due to a lack of chloride and thiosulfate complexing ligands, Au released from sulfides and quartz has remained largely in situ. Gold grains are angular, consisting of a mixture of Ag-rich (3–5% Ag), residual primary grains (median size 545 μm) and Ag-poor secondary grains (median size 580 μm). Many Ag-rich Au grains remain embedded in quartz, consistent with a primary Au origin. The presence of secondary grains suggests some Au solubilization but little or no dispersion.

7.3.2 Ferricrete Ore

The ferricrete deposit is over 5 km N-S and up to 1.5 km E-W and has a mean thickness of 4 m (Figs. 22 and 23). The high-grade zones in the ferricrete appear to be closely related to the structures and Au mineralization of the underlying saprolite zone (Balkau et al. 2017).

Over four hundred 1 kg samples of silicified colluvium, alluvium, ferricrete, soil and saprolite were analysed from eight profiles at 0.25 m vertical intervals to depths of 12 to 15 m on a section across the deposit. The ferricrete overlying saprolite is 5–8 m thick. Gold content is very variable within the ferricrete profile (Fig. 24). The greatest Au concentrations (maximum 10360 ppb Au) in yellowish brown ferricrete coincides with the mineralized saprolite (maximum 9030 ppb). However high concentrations (to 6650 ppb) in ferricrete also occur over non-mineralized saprolite. Here, the overall enrichment ratio relative to the saprolite is a factor of 10 or more. By contrast, As concentrations in ferricrete over mineralized saprolite are significant (>500 ppm), but are reduced (50–100 ppm) where saprolite is unmineralized. Colluvium and alluvium also have significant concentrations of Au, especially towards the base (up to 500 ppb). Gold concentration in soil is very poor (<20 ppb).

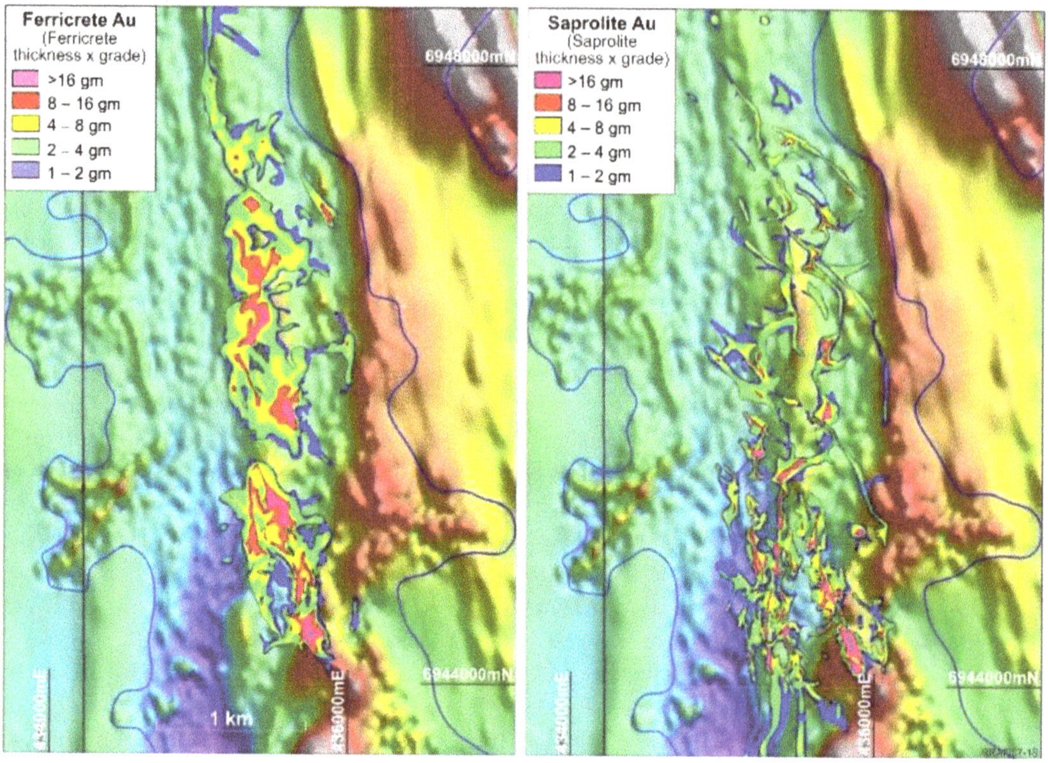

Fig. 22 Gold concentrations in ferricrete and saprolite overlain on a magnetic image, Moolart Well Au deposit. A paleochannel (1–2 km wide) bounded by magnetic gravels is shown by the blue line. (Modified after Anand et al. 2017, 2019)

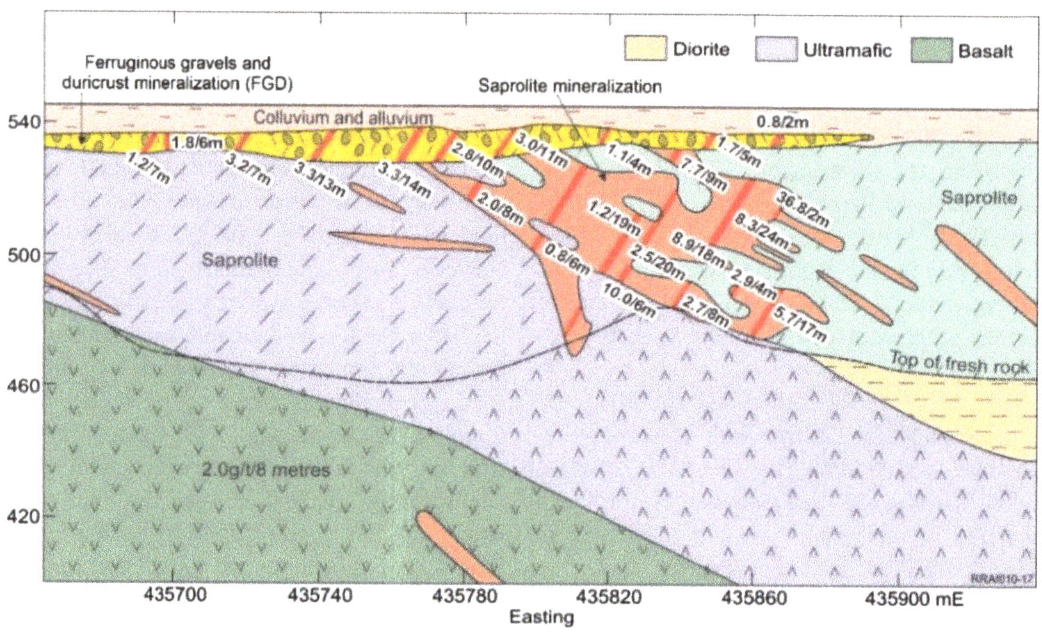

Fig. 23 Gold grades (g/t) over the expected thickness (metres) in ferruginous gravels and duricrust, Moolart Well. There is little or no Au dispersion in saprolite. (After Anand et al. 2017)

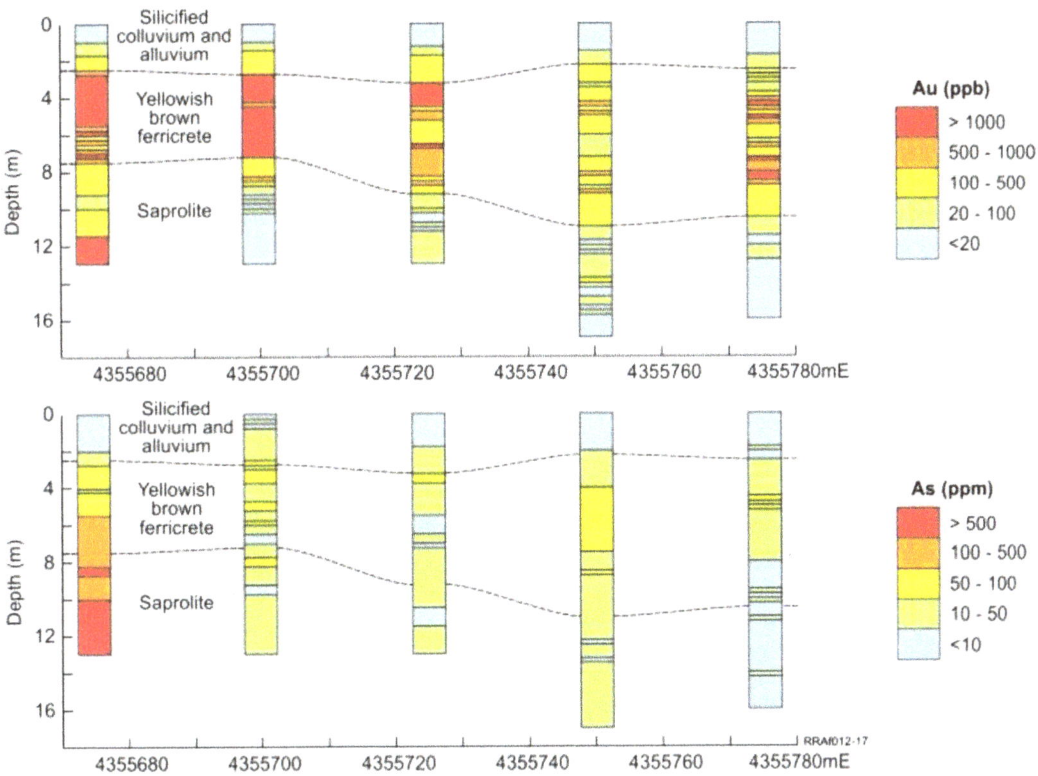

Fig. 24 E-W cross section showing Au and As abundances in silicified colluvium, ferricrete and saprolite, on the eastern side of the Moolart Well. Gold in ferricrete is highly variable. Samples taken at 25 cm intervals from the surface. (Modified after Anand et al. 2017, 2019)

Gold occurs in all fine and coarse size fractions, but the highest concentrations (1210–6290 ppb) are in the goethite-coated nodules and pisoliths dominated >2000 μm fraction in yellowish brown ferricrete. The fine-grained goethite-kaolinite-quartz matrix (<2 μm and 2–53 μm fractions) also contains Au (700–1900 ppb), but comprises under 10 wt% of the ferricrete, suggesting that pisoliths are the main host of Au. Pisoliths (7–35 mm in diameter) separated from an Au-rich sample (bulk content 6200 ppb Au) demonstrate a wide range of Au concentrations from 40 to 45,300 ppb (Fig. 25a), preferentially enriched in those with cortices (Fig. 25b). There are significant differences in Au abundance between the magnetic and non-magnetic pisoliths. Magnetic pisoliths are 2–3 times more enriched in Au than adjacent non-magnetic pisoliths, probably due to their formation at surface by heating of goethite and hematite to maghemite and interaction with organic material. Gold grains in ferricrete are small (<3 μm) and Ag-poor compared to saprolite and thus interpreted as secondary. However, there are also a few coarse (100 μm) angular detrital primary grains with variable Ag contents in ferricrete; some are similar to primary grains in saprolite (3–5% Ag), whereas others are richer in Ag (12%), suggesting a mixed provenance. A variety of Au morphologies occur on both inorganic and organic media. The most prominent examples of secondary Au are clumps and larger clusters in organic-rich cortices and cracks in pisoliths (Fig. 26). Nanoparticulate spherical Au (<300 nm) is common on organic substrates or on nanoparticulate goethite grown on an organic substrate.

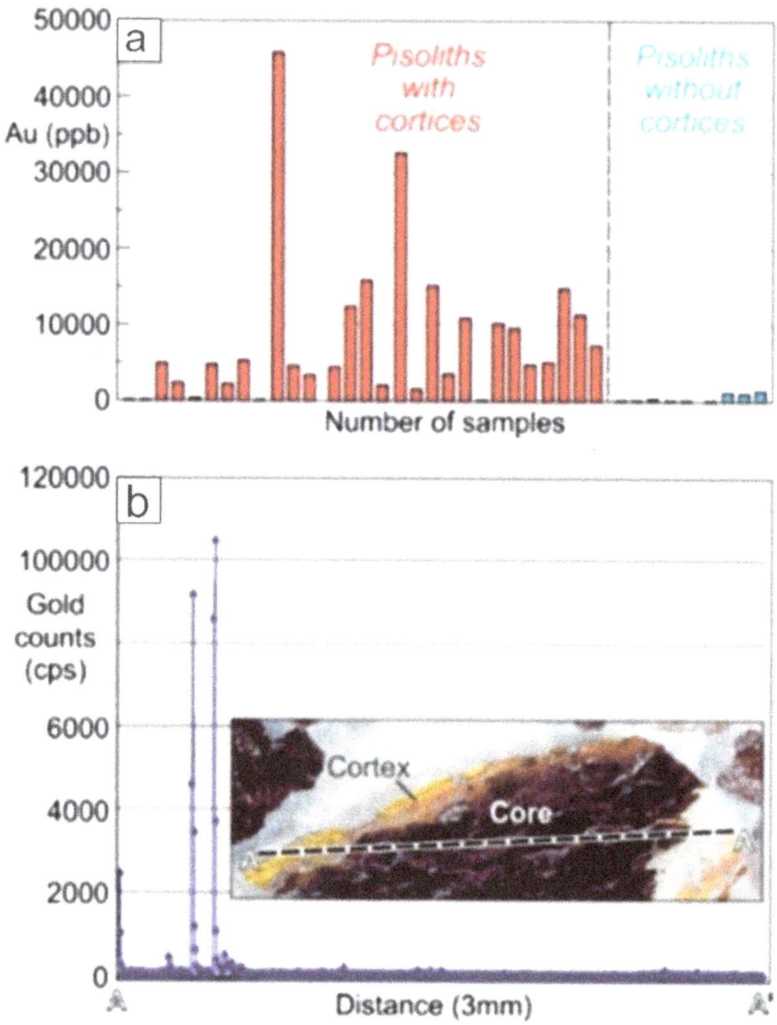

Fig. 25 Gold distribution in ferruginous pisoliths, Moolart Well. (**a**) Analyses of individual pisoliths showing Au is enriched in pisoliths with cortices. (**b**) Laser ablation traverse showing Au hosted by a goethite-kaolinite cortex. (Modified after Anand et al. 2017, 2019)

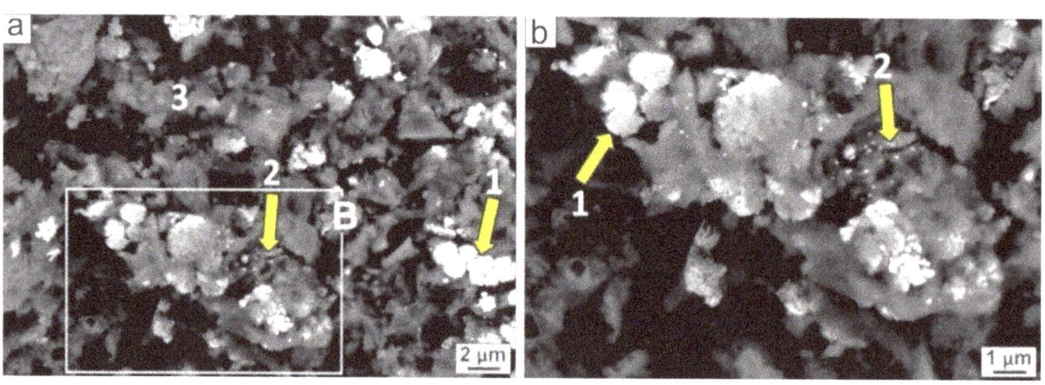

Fig. 26 SEM-BSE images of broken surfaces of pisoliths, Moolart Well, showing granules of Au (1) and nanoparticulate Au spheres (2) in a matrix rich in organic carbon (3). Boxed area in (**a**) is shown in (**b**). (Modified after Anand et al. 2017)

7.3.3 Gold in Vegetation, Termite Mounds, Silicified Colluvium and Soil

Mulga *(Acacia aneura)* leaves are anomalous in Au and directly delineate the deposit. The Au is probably sourced from ferricrete and/or saprolite rather than groundwater, which is at 30 m depth and Au-poor. The location of mineralization is also shown by sampling termite mounds, which have anomalous Au contents (4–5 ppb Au) over mineralization, compared with background concentrations <2 ppb in both <250 and > 250 μm fractions (Stewart et al. 2012). The highest Au values (>5000 ppb) occur in >2000 μm fractions in termite nests over mineralization. This is considered to be largely due to termites bringing Au-rich nodules and pisoliths vertically from 1–4 m depth. Colluvium, sampled from 1 to 1.5 m depth, is anomalous in Au (2 to 350 ppb) and shows a broad halo over the ferricrete deposit. Although metal-rich leaf litter is deposited on the ground surface, the abundance of Au in soil is very low. This is probably due to the slow decomposition of the litter in the arid climate and its periodic removal by fluvial and wind erosion and by bush fires.

7.4 Deposit Formation, Evolution and Implications for Exploration

Because of the lack of a genetic relationship between the ferricrete and the underlying saprolite, Au precipitation in each of these materials has occurred by different mechanisms and in different environments. Saprolite was formed by weathering of underlying mineralized bedrock under an earlier humid environment (Paleocene-mid-Miocene). Lateral dispersion of Au was limited but large angular to euhedral grans of Au accumulated residually, with only minor secondary (<0.5%) Au. In contrast, the formation of the ferricrete Au deposits was complex, with several stages of mechanical, chemical and biological transport and reprecipitation of Au in the paleochannel environment. This resulted in relatively high concentrations of Au and extensive dispersion haloes compared to the saprolite. The Au-rich ferricrete broadly overlies Au-mineralized structures in saprolite, suggesting much of the Au in the ferricrete was physically and chemically mobilized from the saprolite and deposited within the younger precipitates (goethite cortex and voids) of ferricrete. However, the presence in ferricrete of detrital primary Au grains with variable Ag contents suggests a mixed provenance. For example, particulate Au may have been derived from erosion of small Au occurrences on low hills above the plains.

Most of the Au in ferricrete is secondary, of high purity, and hosted in goethite, organic carbon and a mixture of goethite and kaolinite. Abundant vegetation growing over the paleochannel would have provided abundant organic carbon, an important sink for Au in this environment. This could precipitate ionic Au from waters containing thiosulfate, halide or organic complexes (Bowell et al. 1993), and/or trap particulate colloidal Au (Hough et al. 2011). Gold may also have been transported as colloidal Au^0 sorbed or trapped in Fe oxides. A combination of organic, mechanical and chemical mechanisms could form the broad flat-lying Au enrichments of secondary and primary Au in the ferricrete, compared to the relatively narrow mineralized zone in saprolite. The Au in vegetation, termite mounds and silicified colluvium indicate active dispersion even in an arid climate. On a broad scale, sampling of ferricrete can be safely used to detect the general area where Au deposits may occur. Follow-up will require sampling saprolite, saprock and bedrock.

8 Summary and Conclusions

Supergene Au deposits formed in FGD, saprolite and paleochannel sediments have played a very important role in the development of the Au industry in lateritic terrains. The deposits are formed by a combination of physical accumulation and chemical and biological mobilization and redistribution of Au during weathering and sedimentary processes. The FGD and saprolite are separate systems with different environmental controls and hence different conditions for Au mobilization. Physical lateral transport of Au-bearing ferruginous particles or colloidal Au on slopes, together with the mobilization of Au as organic complexes during weathering under humid tropical climates, formed wide dispersion haloes in the FGD. In contrast, relative enrichment in saprolite due to mass loss is largely offset by leaching, and lateral dispersion is minimal. The Au distribution generally remains unchanged even where the surface environment has experienced vertical tectonism or climatic change. Such tectonism may cause erosion or burial. In humid regions, including Mediterranean climates, Au is leached from soil and upper FGD horizons and reprecipitated in the deeper gravels and duricrusts. In most arid regions, these supergene Au deposits remain largely unaffected. However, where aridity has caused salinization of groundwater, the Au distribution can be strongly affected by (i) strong leaching and depletion of Au beneath the FGD, (ii) absolute enrichment deeper in the saprolite, characterized by secondary Au of high purity, and (iii) in some environments, enrichment in pedogenic carbonates. In general, the Au distribution delineated by enrichment in saprolite is almost exclusively residual and broadly corresponds to that of the primary mineralization, including the alteration zone. Laterally extensive supergene enrichment only occurs within and beneath paleochannel sediments, typically in the axes of saline drainages.

Acknowledgements This chapter is a summary of many years of work by the research groups at CSIRO and CRC LEME in Australia, the Federal University of Rio de Janeiro, and by numerous other individuals in Australia, Brazil, France and elsewhere with whom the authors have had the privilege of collaborating. Dr R.P. Foster is thanked for his thoughtful and constructive review of the manuscript. Many of the Figures were drafted by Angelo Vartesi.

References

Anand RR (1998) Regolith-landform evolution and geochemical dispersion from the Boddington gold deposit, Western Australia. Open File Report 3, CRC LEME CSIRO, Perth, Australia. CSIRO, pp 156. https://crcleme.org.au/Pubs/OFRSindex.html

Anand RR (2000) Regolith and geochemical synthesis of the Yandal greenstone belt. In: Phillips GN, Anand RR (eds) Yandal greenstone belt, regolith, geology and mineralisation, vol 31. AIG Bulletin, pp 79–112

Anand RR (2001) Evolution, classification and use of ferruginous regolith materials in exploration, Yilgarn Crato. Geochem Explor Environ Anal 1:221–236

Anand RR (2015) The importance of 3D regolith-landform control in areas of transported cover implications to exploration. Geochem Explor Environ Anal 16:14–26

Anand RR, Butt CRM (2010) A guide for mineral exploration through the regolith in the Yilgarn Craton, Western Australia. Aust J Earth Sci 57:1015–1114

Anand RR, Verrall M (2011) Biological origin of minerals in pisoliths in the Darling Range of Western Australia. Aust J Earth Sci 58:823–833

Anand R, Lintern M, Noble R, Aspandiar M, Macfarlane C, Hough R, Stewart A, Wakelin S, Townley B, Reid N (2014) Geochemical dispersion through transported cover in regolith-dominated terrains—toward an understanding of process. SEG Special Publ 18:97–126

Anand R, Lintern M, Hough R, Noble R, Verrall M, Salama W, Balkau J, Radford N (2017) The dynamics of gold in regolith change with differing environmental conditions over time. Geology 45:127–130

Anand R, Hough R, Salama W, Aspandiar M, Butt C, González-Álvarez I, Metelka V (2019) Gold and pathfinder elements in ferricrete gold deposits of the Yilgarn Craton of Western Australia: a review with new concepts. Ore Geol Rev 104:294–355

Anand R, Wells M, Lintern M, Schoenveld L, Danišík M, Salama W, Noble R, Metelka V, Reid N (2021) The (U-Th)/He chronology and geochemistry of ferruginous nodules and pisoliths formed in the paleochannel environments at the Garden Well gold deposit, Yilgarn Craton of Western Australia: implications for landscape evolution and geochemical exploration. Minerals 11(7):679. https://doi.org/10.3390/min11070679

Andrade WO, Machesky ML, Rose AW (1991) Gold distribution and mobility in the surficial environment, Carajas Region, Brazil. J Geochem Explor 40:95–114

Balkau J, French T, Ridges T (2017) Moolart Well, Garden Well and Rosemont gold deposits. In: Phillips GN (ed) Aust Ore Deposits. The Australasian Institute of Mining and Metallurgy, pp 261–266

Beziat D, Siebenaller L, Salvi S, Chevalier P (2016) A weathered skarn-type mineralization in Ivory Coast. Ore Geol Rev 78:724–730

Body KJ, Mudau M, Cunningham GI, Bosc R, Perez, P., Baker J, Gauthier D, Larochelle P, Sangam HP (2015) Technical report for the Ity gold mine, Côte d'Ivoire. Coffey Mining South Africa, for Endeavour Mining, 256 pp (unpublished). https://s21.q4cdn.com/954147562/files/doc_downloads/technical_report/Ian-Hamilton-technical-report-ity.pdf

Bowell RJ, Foster RP, Gize AP (1993) The mobility of gold in tropical rain forest soils. Econ Geol 88:999–1016
Bowell RJ, Afreh EO, Laffoley NA, Hanssen E, Abe S, Yao RK, Pohl D (1996) Geochemical exploration for gold in tropical soils—four contrasting case studies from West Africa: Institution of Mining and Metallurgy. Transactions 105:B12–B33
Butt CRM (1989) Genesis of supergene gold deposits in the lateritic regolith of the Yilgarn Block, Western Australia. In: Keays RR, Ramsay WRH, Groves DI (eds) The geology of gold deposits: the perspective in 1988, Economic geology monograph, vol 8, pp 460–470
Butt CRM (1998) Supergene gold deposits. AGSO J Geol Geophys 17:89–96
Butt CRM, Zeegers H (1992) Regolith exploration geochemistry in tropical and subtropical terrains, Handbook of exploration geochemistry, vol 4. Elsevier, Amsterdam, p 607
Butt CRM, Hough RM, Verrall M (2020) Gold nuggets: the inside story. Ore Energy Resour Geol 4–5. https://doi.org/10.1016/j.oreoa.2020.100009
Canavan F (1988) Deep lead gold deposits of Victoria. Bulletin 62, Geological Survey of Victoria, Melbourne, 101 pp
Colin F, Vieillard P, Ambrosi JP (1993) Quantitative approach to physical and chemical gold mobility in equatorial rainforest lateritic environment. Earth Planet Sci Lett 114:269–285
Costa ML (1993) Gold distribution in lateritic profiles in South Ameria, Africa, and Australia: application to geochemical exploration in tropical regions. J Geochem Explor 47:143–163
Costa ML (1997) Lateritization as a major process of ore deposit formation in the Amazon region. Explor Mining Geol 6:79–104
Costa ML, Vieira-Costa JA, Angelica RS (1993) Gold-bearing bauxitic laterite in a tropical rain forest climate, Cassipore, Amapa, Brazil. Chronique de la Recherche Minière 510:41–51
Devlin SP, Crimeen JD (1990) Lady Bountiful Extended gold deposit. In: Hughes FE (ed) Geology of the mineral deposits of Australia and Papua New Guinea. The Australasian Institute of Mining and Metallurgy, Melbourne, pp 405–410
Dreher AM, Xavier RP, Martini SL (2005) Fragmental rocks of the Igarapé Bahia Cu–Au deposit, Carajás Mineral Province, Brazil. Revista Brasileira de Geografia 35:358–368
Edou-Minko A, Colin F, Lecomte P, Trescases JJ (1992) Altération latéritique du gîte aurifère d'Ovala (Gabon) et formation d'une anomalie superficielle de dispersion. Mineral Deposita 27:90–100
Ellison D, Baylock G (1990) The Syama gold mine project. Mali Min Mag 162:346–349
Freyssinet P (1993) Gold dispersion related to ferricrete pedogenesis in south Mali: application to geochemical exploration. Chronique de la Recherche Minière 510: 24–40
Freyssinet P (1994) Gold mass balance in lateritic profiles from savanna and rain forest zones. Catena 21:159–172
Freyssinet P, Butt CRM (1988) Morphology and geochemistry of gold in a laterite profile, Bardoc Mine, Western Australia. Open File Report 5, CRC LEME, Perth, 28 pp. http://crcleme.org.au/Pubs/OFRSindex.html
Freyssinet P, Itard Y (1997) Geochemical mass balance of gold under various tropical weathering conditions: application to exploration. In: Gubins AG (ed) Proceedings of exploration, Fourth Decennial Conference on Mineral Exploration, vol 97, Toronto, pp 347–354
Freyssinet P, Butt CRM, Morris RC, Piantone P (2005) Ore-forming processes related to lateritic weathering. In: Hedenquist JW, Thomson JFH, Goldfarb RJ, Richards JP (eds) Economic geology 100th anniversary volume. Economic Geology Publishing Company, New Haven, pp 681–722. https://doi.org/10.5382/AV100.21
Genise JF, Bellosi ES, Verde M, González MG (2011) Large ferruginized palaeorhizospheres from a Paleogene lateritic profile of Uruguay. Sediment Geol 240: 85–96
Geological Survey of Western Australia (2011) East Yilgarn Western Australia 1:100 000 Geological Information Series. Geological Survey of Western Australia, Department of Primary Industry and Resources. http://www.dmp.wa.gov.au. Accessed Dec 2021
Glasson MJ, Lehne RW, Wellmer FW (1988) Gold exploration in the Callion area, Eastern Goldfields, Western Australia. J Geochem Explor 31:1–19
Gray DJ, Butt CRM, Lawrance L (1992) The geochemistry of gold in lateritic terrains. In: Butt CRM, Zeegers H (eds) Handbook of exploration geochemistry, Regolith exploration geochemistry in tropical and sub-tropical terranes, vol 4. Elsevier, Amsterdam, pp 461–481
Gray DJ, Sergeev NB, Britt AF, Porto CG (2008) Supergene mobilization of gold and other elements in the Yilgarn Craton-Final Report. Open File Report 228, CRC LEME CSIRO, Perth, Australia. CSIRO, 79 pp https://crcleme.org.au/Pubs/OFRSindex.html
Greffié C, Benedetti MF, Parron C, Amouric M (1996) Gold and iron oxide associations under supergene conditions: an experimental approach. Geochemica Cosmochimica Acta 60:1531–1542
Hails JR (1976) Placer deposits. In: Wolf KH (ed) Handbook of strata-bound and stratiform ore deposits, vol 3. Elsevier, Amsterdam, pp 213–244
Hanssen E, Kaisin J, Tessougue S (2004) The Yetela gold mine, Western Mali: Supergene enrichment of lower Proterozoic gold mineralization in a surficial karst. In: Conference Proceedings-CIM Annual Meeting, Edmonton

Hough R, Noble R, Hitchen G, Hart G, Reddy S, Saunders M, Clode P, Vaughan D, Lowe J, Gray D, Butt CRM, Verrall M (2008) Naturally occurring Au nanoparticles and nanoplates. Geology 36:571–574. http://crcleme.org.au/Pubs/OFRSindex.html

Hough R, Noble R, Reich M (2011) Natural gold nanoparticles. Ore Geol Rev 42:55–61

Ker JC (1997) Latossolos do Brasil: Uma Revisão. Geonomos 5:17–40. https://doi.org/10.18285/geonomos.v5i1.187

Lawrance LM (1991) Distribution of gold and ore-associated elements within lateritic weathering profiles of the Yilgarn Block, Western Australia. PhD thesis, The University of Western Australia, 191 pp (unpublished)

Lawrance LM (2001) Multi-element dispersion within salt-lake environments: case study of the buried Hannan South gold deposit, Western Australia. Geochem Explor Environ Anal 1:323–339. https://doi.org/10.1144/geochem.1.4.323

Lecomte P, Colin F (1989) Gold dispersion in a tropical rain forest weathering profile at Dondo Mobi, Gabon. J Geochem Explor 34:285–301

Lecomte P, Zeegers H (1992) Exploration in areas of low to moderate relief, Humid tropical terrains (rainforests). In: Butt CRM, Zeegers H (eds) Handbook of Exploration Geochemistry: Regolith exploration geochemistry in tropical and subtropical terrains, vol 4. Elsevier, Amsterdam, pp 241–294

Lintern MJ (2001) Exploration of gold using calcrete—lessons from the Yilgarn Craton, Western Australia. Geochem Explor Environ Anal 1:237–252

Lintern MJ (2015) The association of gold with calcrete. Ore Geol Rev 66:132–199

Liversidge A (1893a) On the origin of gold nuggets. J Proc Royal Soc New South Wales 27:303–343

Liversidge A (1893b) On the crystallization of gold in hexagonal forms. J Proc Royal Soc New South Wales 27:343–346

Mann AW (1984) Mobility of gold and silver in lateritic weathering profiles, some observations from Western Australia. Econ Geol 79:38–49

Masurel Q, Miller J, Hein K, Hanssen E, Thebaud N, Ulrich S, Kaisin J, Tessougue S (2016) The Yatela gold deposit in Mali, West Africa: the final product of a long-lived history of hydrothermal alteration and weathering. Afr J Earth Sci 113:73–87

Mazzuccheli RH (2005) Paleochannel gold deposits, Mount Pleasant district. In: Butt CRM, Robertson IDM, Scott KM, Cornelius M (eds) Regolith expression of Australian ore systems. CRC LEME, Perth, pp 302–303

Michel D (1987) Concentration of gold in *in situ* laterites from Mato Grosso. Mineral Deposita 22:185–189

Monteiro HS, Vasconcelos PM, Farley KA, Spier C, Mello CL (2014) (U–Th)/He geochronology of goethite and the origin and evolution of cangas. Geochimica Cosmochimica Acta 131:267–289. https://doi.org/10.1016/j.gca.2014.01.036

Porto CG (2016) Geochemical exploration challenges in the regolith dominated Igarapé Bahia gold deposit Carajás, Brazil. Ore Geol Rev 73:432–450

Reith F, Stewart L, Wakelin SA (2012) Supergene gold transformation: secondary and nano-particulate gold from southern New Zealand. Chem Geol 320:32–45

Robertson D (2004) A technical report on the Sadiola gold mine, Mali. Iamgold Corporation, Canada (unpublished). www.sec.gov/Archives/edgar/data/1203464/000127956904000149/courtesycopy.pdf

Santosh M, Omana PK, Yoshida M (1990) Gold grains in laterite weathering profiles of Nilambur: South India and a model for the genesis of supergene gold deposits. J Mineral Petrol Econ Geol 85:416–423

Sirois R, Croal A, Cimon D, Menard R, Runnels DF, Johnson PM (2002) Rosebel Project, Suriname, South America. Technical Report. Cambior Inc., 142 pp (unpublished)

Smith RE, Anand R (1992) Mount Gibson gold deposit, Western Australia. In: Butt CRM, Zeegers H (eds) Regolith exploration geochemistry in tropical and sub-tropical terrains, vol 4. Elsevier, Amsterdam, pp 313–318

Stewart AD, Anand RR, Balkau J (2012) Source of anomalous gold concentrations in termite nests, Moolart Well, Western Australia. Geochem Explor Environ Anal 12:327–337

Tallarico FHB, Figueiredo BR, Groves DI, Kositcin N, McNaughton NJ, Fletcher IR, Rego JL (2005) Geology and Shrimp U–Pb geochronology of the Igarapé Bahia deposit, Carajás copper–gold belt, Brasil: an Archean (2.57 Ga) example of iron oxide Cu–Au(U–REE) mineralization. Econ Geol 100:7–28

Thorne R, Anand RR, Suvorova A (2014) The formation of fluvio-lacustrine ferruginous pisoliths in the extensive palaeochannels of the Yilgarn Craton, Western Australia. Sediment Geol 313:32–44

Voicu G (2010) Rosebel Mine, Suriname. Technical Report NI 43-101, IAMGOLD Corporation. https://www.sec.gov/Archives/edgar/data/1203464/000119312510071064/dex991.htm, 147 pp (unpublished)

Webster JG, Mann AW (1984) The influence of climate, geomorphology and primary geology on the supergene migration of Au and Ag. J Geochem Explor 22:21–42

Carbonatite Hosted Residual REE Deposits

Pete R. Siegfried, Frances Wall, and Philip L. Verplanck

Abstract

Rare earth elements (REEs) occur in magmatic rocks but are especially enriched in carbonatite and alkaline silicates. If these rocks are chemically weathered, then the REEs may become further enriched within the regolith developed from these rocks. Primary magmatic REE minerals, as well as the various carbonate minerals and apatite, provide the REEs which, under pervasive chemical weathering, are incorporated within low-temperature REE minerals forming within the regolith. Many of these minerals, as well as their textures, are characteristic of this mode of formation. Lateritic conditions of weathering are instrumental in producing a thick, weathered, or regolith, profile, and the roles of sulfide oxidation, fluctuating groundwater tables, and downward mass wasting due to carbonate dissolution are identified as the most important controls on REE enrichment in the regolith.

1 Introduction

Rare earth elements (REEs) are a group of elements necessary to enable much of the decarbonation of the world, and the demand for them is rising steadily, driven by their use in magnets for electric vehicles and wind turbines. The REEs include the lanthanoids (La-Lu) plus scandium and yttrium, and as a group, are considered "critical metals" (Wall 2014). The individual REEs are used in a range of different technologies of varying economic importance, and the calculation of their economic scarcity potential demonstrates the contemporary importance of Dy, Nd, Pr, Tb, and Gd compared to the rest (Pell et al. 2019).

The most important REE deposits are those associated with magmatic rocks, particularly alkaline igneous rocks and carbonatites (Chakhmouradian and Zaitsev 2012). Figures 1 and 2 show the location of major REE deposits and the distribution of noteworthy REE projects being explored and mined today. There is a marked split between light REE (LREE) and heavy REE (HREE; Tb through to Lu + Y) deposits, and this is crystallographically defined by the filling of the 4f shell (Weng et al. 2015). In

P. R. Siegfried (✉)
Camborne School of Mines, University of Exeter, Penryn, Cornwall, UK

GeoAfrica Prospecting Services, Windhoek, Namibia

F. Wall
Camborne School of Mines, University of Exeter, Penryn, Cornwall, UK

P. L. Verplanck
U.S. Geological Survey, Denver Federal Center, Denver, CO, USA

R. J. Bowell, C. R. M. Butt (eds.), *Geology, Geochemistry and Formation of Supergene Mineral Deposits in Deeply Weathered Terrain*, Mineral Resource Reviews, https://doi.org/10.1007/978-3-031-75733-4_7

Fig. 1 Global distribution of important REE-enriched fresh and weathered carbonatites

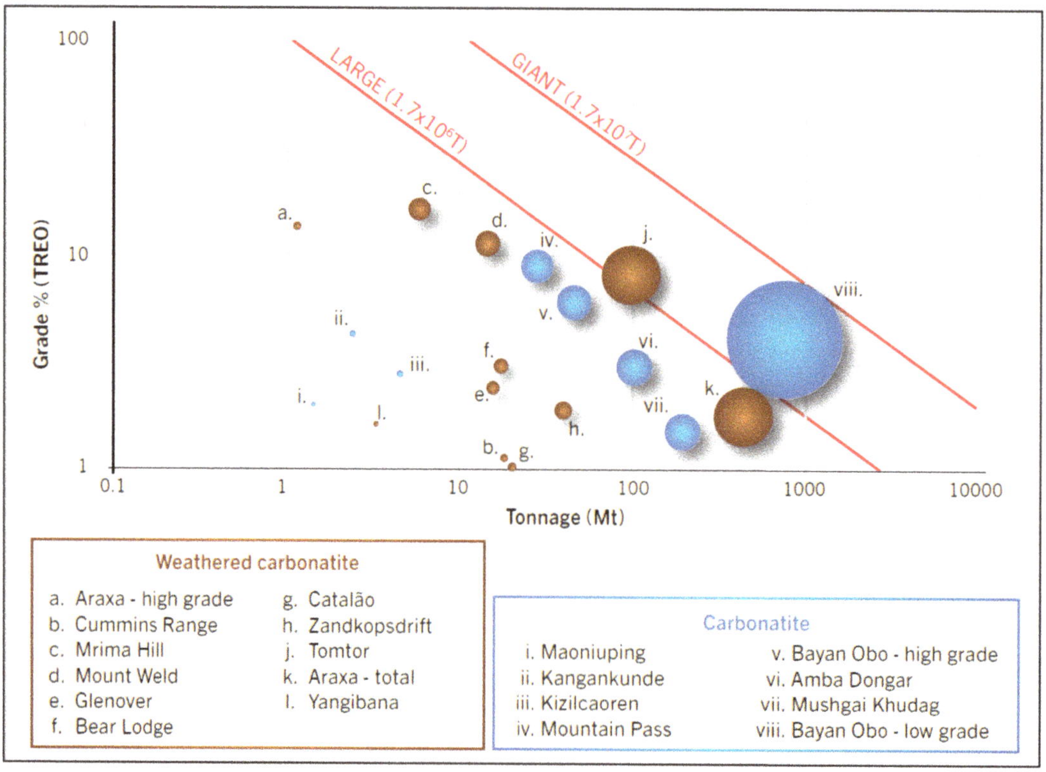

Fig. 2 Grade-Tonnage distribution diagram of some of the various carbonatite-hosted REE deposits showing both weathered carbonatite and carbonatite-related fresh rock examples. The size of the bubble denotes the total REE contained. (Adapted from Smith et al. 2016; Harper et al. 2011; Dahlberg et al. 2014)

the case of the latter, the HREE mineralization is dominantly hosted by the mineral xenotime (Ni et al. 1995) and REE ion adsorbed clays (Sanematsu and Watanabe 2016), and in the former, by monazite and apatite, or bastnäsite and other REE fluorocarbonates, or a mixture of both.

Two weathered REE deposit types, ion adsorption clay deposits (see chapter "Ion Adsorption Mineralisation in Regolith-Hosted REE Deposits") and residual deposits above carbonatites, are significant current REE producers and targets of active exploration. For example, Mount Weld in Australia—a classic residual supergene deposit developed in the regolith above a carbonatite—produces over 12% of the world's REEs (Lynas 2021). In 2021, over 60% of all the world's REEs were mined in China (U.S. Geological Survey 2021), in consequence, exploration for REE deposits outside of China is of major interest for industrial REE users. This review provides an overview of the carbonatite-hosted REE deposits formed by supergene processes, including mineralogy, considerations for processing and production, and some case studies.

2 Fundamentals of Carbonatites

Carbonatites are uncommon igneous rocks defined by the International Union of Geological Sciences (IUGS) system of igneous rock classification as having more than 50 modal percent primary carbonate minerals (such as calcite, dolomite, and ankerite) and less than 20% SiO_2 (Le Maître 2002). On average, they contain the highest concentration of REEs of any igneous rock. Many elements that are usually only encountered in trace amounts in other igneous rocks (e.g., REEs, Sr, P, Nb, and Ba) can be major elements in carbonatites, which in turn is reflected in their mineralogy. These mineral assemblages further impact the nature of weathering and the resultant lateritic regolith. Carbonatites occur worldwide (Verplanck et al. 2014) and, although commonly encountered in areas of extension or rifting, have been identified within other terranes, with noteworthy examples being hosted within transpressional terranes such as the Mianning–Dechang REE belt, southwestern China (Liu and Guo 2019). The common factor appears to be the presence of thin continental crust or major structures that allow the carbonatite melts to pass (Pell 1994) through thicker upper crust and cratonic areas.

Where undeformed, carbonatite intrusions are generally pipe-like, with a diameter between hundreds of meters and several kilometers. Alkali silicate rocks, such as nepheline syenites, are commonly associated with these intrusions (Beard et al. 2022). For a more comprehensive overview of carbonatite intrusive rocks, see Mitchell (2005), Woolley (1982), Woolley and Kempe (1989), Woolley and Kjarsgaard (2008), and Le Bas (1987). A particularly informative recent review is provided by Yaxley et al. (2022). Unlike most silicate rocks, common rock-forming minerals in carbonatites are calcite, dolomite and ankerite instead of feldspars, quartz, pyroxenes and amphiboles. The mineralogy of carbonatites is quite variable, and they commonly contain minor silicates (olivine, pyroxenes, or micas), oxides (magnetite, ilmenite, or rutile), phosphates (apatite and monazite), sulfides (e.g., pyrite, galena.) and sulfates (barite, celestite). Common REE minerals include apatite, monazite, and fluorocarbonate minerals (bastnäsite, parisite and synchysite).

Another important difference compared to silicate magmas is that carbonatite magmas can contain two to three times the amount of water at a given pressure and temperature (Keppler 2003), so that as carbonatite magmas evolve, these fluids typically exsolve and lead to hydrothermal alteration. The resulting alkali alteration, termed "fenitization," is characterized by reactions with Na and K (Elliott et al. 2018) and the precipitation of Na and K-rich minerals such as aegirine, albite, and arfvedsonite.

Most known carbonatites are unweathered, with only a few having developed a thick regolith above them (Cocker 2014). Observations made in the weathered zones of the few carbonatites currently being mined provide a "snapshot" into the internal parts of the regolith and provide information in understanding the processes by which these rocks have been weathered in what appear to be almost exclusively hot, humid climates. These excavations have also allowed us to

Fig. 3 Typical lateritic regolith profile observed during mining operations at Catalão-1, Brazil. The exposure of the pit wall shows the different zones and where apatite enrichment occurs. The saprolite is more deeply weathered over light-colored areas, where carbonatite dykes are more common

illustrate the important and fundamental controls of dissolution and reprecipitation of all constituent elements that may form deposits within the supergene zone itself. An instructive example of the up to 150 m thick regolith developed over the Catalão-1 carbonatite is presented in Fig. 3.

Because most of the rock-forming minerals in carbonatites are susceptible to chemical breakdown, the survival of carbonatite minerals under conditions of intense chemical weathering is poor. Under the action and chemical control of most active groundwater regimes, carbonate minerals are rapidly destroyed, leading to not only a substantial mass decrease but a significant upgrading of the immobile elements in more resistant minerals. This is shown in the many weathered carbonatite deposits that are being mined today, not only for REEs but other important critical elements such as Nb and P (Table 1).

3 Weathering Processes, Water Table, and Groundwater Chemistry

To understand the processes of rock weathering better, it is best to start with the actual precursor protolith. The individual minerals in the fresh rock will exert ultimate control on the type and selection of secondary minerals precipitated in the regolith. Unfortunately, in the real world, this transition is often only possible to observe with the drilling of deep diamond core holes.

An essential component of the processing system at work during weathering is the position of the rest water level (RWL). It is here at the water table where fundamental changes in oxidation, pH, and electrolytic state take place. The level at which these processes are at their most active is at the interface between the overlying vadose zone (or zone of desiccation and saturation and usually dominated by an oxidative process) and the phreatic zone, which is saturated by groundwater (Fig. 4). Groundwater flows under the action of forced head or hydrostatic load, as well as being controlled by the surface topography. Along with these factors, climate introduces and may augment the groundwater through the introduction of significant amounts of meteoric water. Active recharge will also affect the rate at which ions exchange between the groundwater and those contained within the carbonatite itself. The vadose zone becomes saturated and desiccated periodically as the groundwater moves up and down within the regolith, with the RWL being, therefore, largely determined by seasonal changes with the active groundwater in this zone referred to as "phreatic."

Table 1 List of residual and weathered carbonatite hosted REE deposits

Deposit (mines in bold)	Country	Commodity	Primary igneous host rock	REE minerals in primary rock	Active weathering	Depth of weathering	REE minerals in weathered rock	Age
Mabounié	Gabon	REE, Nb, P	Calciocarbonatite	Ap, Pcl	Yes	60 m	Pcl, crandallite, Mnz?	660 Ma
Zandkopsdrift	South Africa	REE, Mn, P	Magnesiocarbonatite	Mnz, Ap	No	30–70 m	Mnz, crandallite	57 Ma
Glenover	South Africa	REE, P, Sc	Magnesiocarbonatite>calciocarbonatite	Ap, Syn	No	20–100 m	Mnz, Ap	1249 Ma
Ngualla	Tanzania	REE, P, CaF2	Calciocarbonatite>Fe-magnesiocarbonatite	Bst, Syn, Ap	No	70–80 m, up to 150 m	Bst, Mnz	1040 Ma
Longonjo	Angola	REE	Carbonatite, fenite	Ap, Mnz, Bst	No	30–50 m	Mnz, Rhb	120 Ma
Kieshöhe	Namibia	REE	Magnesiocarbonatite	Mnz, Bur	No	50 m?	Mnz, parisite	
Mrima Hill	Kenya	REE, Nb	Calciocarbonatite>magnesiocarbonatite	Ap, Pcl	Yes	120–140 m	Mnz	120 Ma
Buru Hill	Kenya	REE	Calciocarbonatite		Yes	40 m	Bst (Mnz not reported)	
Lueshe	DRC	Nb, REE	Calciocarbonatite	Pcl, Ap, Anc	Yes	40 m	Pcl, Mnz, Rhb	516 Ma
Xiluvo	Mozambique	REE, Nb, CO_3	Calciocarbonatite	Ap, Pcl	No	30 m	Mnz?	
Fen	Norway	REE,	Magnesiocarbonatite	Bst	No		Bst, Mnz	540 Ma
Tomtor	Russia	REE, P, Nb	Calciocarbonatite	Pcl, Ap	No	300 m, top 100 m Permian cover	Mnz-Ce, Cra	600 Ma
Storkwitz	Germany	REE	Calciocarbonatite	Ap, Pcl	No	100 m?	Mnz, Syn	71–75 Ma
Sokli	Finland	Nb, P, REE	Calciocarbonatite	Pcl, Ap	No	150 m with 20 m till cover	Ap, Mnz, Pcl	600 Ma
Chuktukon	Russia	REE, Sc, Nb	Calciocarbonatite>>magnesiocarbonatite	Ap, Pcl, Syn	No		Mnz, florencite, churchite	252 Ma
Nam Xe, Phong Tho	Vietnam	REE	Calciocarbonatite>ferrocarbonatite	Ap	Yes	40 m		50–55 Ma
Maoniuping	China	REE	Calciocarbonatite	Bst	No	Unknown, laterite not mentioned	Bst	30–40 Ma
Mount Weld	Australia	REE, Nb, P	Calciocarbonatite>magnesiocarbonatite	Ap	No	70–130 m	Mnz, churchite, (cheralite)	2056 Ma
Cummins Range	Australia	REE, P	Calciocarbonatite>magnesiocarbonatite	Ap, Bst, Prs	No	100 m	Mnz, Bst, Cra	905 Ma
Cargill	Canada	P, Nb, REE	Calciocarbonatite, magnesiocarbonatite	Ap	No	100–200 m	Ap	1740 Ma

(continued)

Table 1 (continued)

Deposit (mines in bold)	Country	Commodity	Primary igneous host rock	REE minerals in primary rock	Active weathering	Depth of weathering	REE minerals in weathered rock	Age
St-Honoré	Canada	Nb, REE	Magnesiocarbonatite, Ferrocarbonatite	Bst, Pcl	No	? Covered by shale	Mnz	630–650 Ma
Oka REE	Canada	REE, P	Calciocarbonatite	Pcl, Ap	No		Mnz	130 Ma
Bear Lodge	USA	REE	Calciocarbonatite	Bur	No	150 m	Carbocernaite, Mnz	51 Ma
Araxá	Brazil	Nb, P, REE	Magnesiocarbonatite>calciocarbonatite	Pcl, Ap	Yes	300 m	Mnz	77 Ma
Catalão-1	Brazil	Nb, P, REE	Glimmerite>calciocarbonatite	Ap, Pcl	Yes	150 m	Mnz	84 Ma
Salitre	Brazil	Ti, P, Nb, REE	Pyroxenite>>carbonatite	Ap	Yes			
Tapira	Brazil	Ti, P, Nb, REE	Pyroxenite>>carbonatite	Prv, Ap	Yes	Mainly 30 m but 200 m in REE, Nb	Anatase, pseudorutile, Mnz	70 Ma
Seis Lagos	Brazil	Nb, REE	Ferrocarbonatite	Ap, Pcl, Prv?	Yes	320 m	Mnz, florencite, cerianite	1300 Ma

Data from various sources including Brown (2020), Chebotarev (2017), Harmer and Nex (2016), Harper et al. (2011), Hutchinson (2016), Kasay et al. (2021), Lynas (2021), Niegisch et al. (2020), Onuonga and Bowden (2000), Smith et al. (2016), Witt et al. (2019)

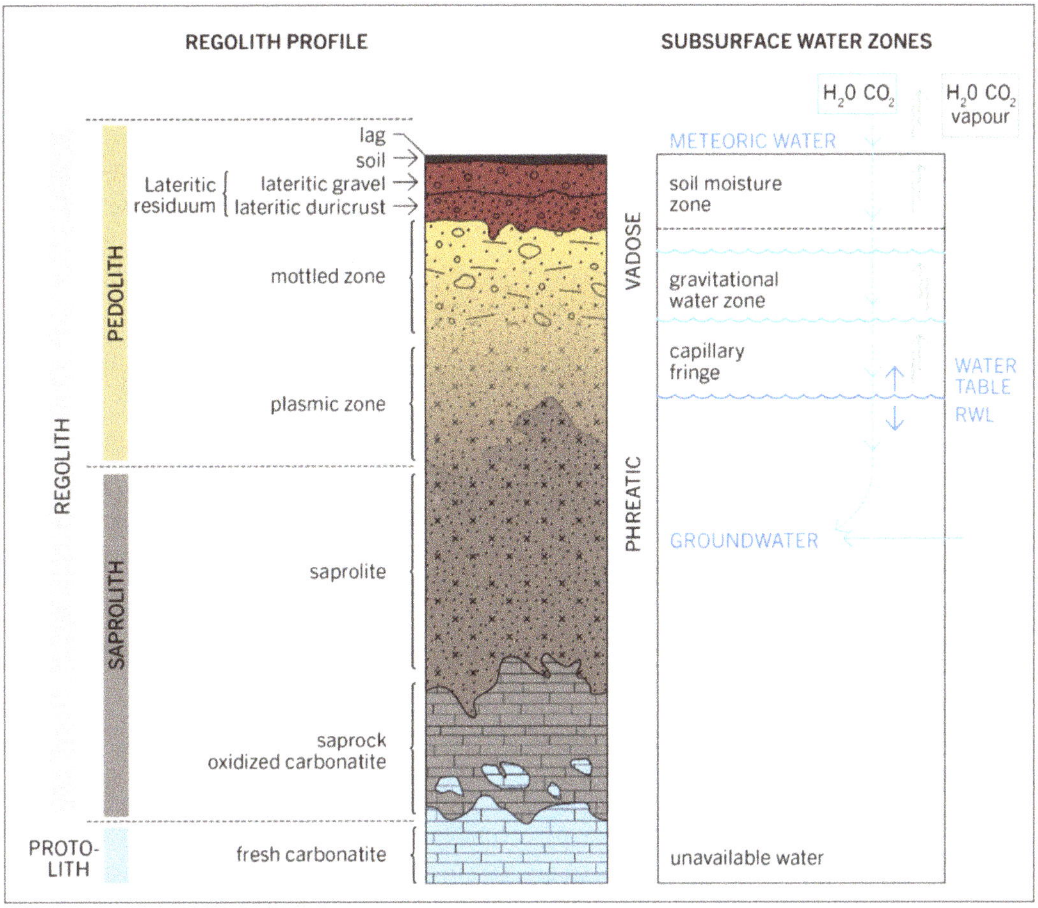

Fig. 4 Regolith and groundwater nomenclature applied to a typical lateritic profile formed over a weathered carbonatite. (Modified after Anand and Paine 2002; Taylor and Butt 1998). RWL rest water level

Therefore, the role of a rising and falling water table is fundamental to this enrichment process, much in line with classic supergene sulfide deposits. Furthermore, these processes affect the composition of the groundwater itself, resulting in a plethora of buffering and reactivity process routes taking place as the regolith evolves. As reactivity controls in the environment above the water table are largely oxidizing, the presence of yellow goethite (senso lato "limonite"), which is a characteristic product in all oxidized zones, can be used to define the position of both active as well as fossil water tables.

It is important to note that these reactions occur at low temperatures (<50°C) and with no or only minor confining hydrostatic pressure. The ability of soils to allow further interaction of the atmosphere with the groundwater means that the role of common atmospheric gases such as O_2, H_2, H_2O, CO_2, and N_2 must be considered. As mentioned, the low temperature and negligible pressure will affect the rates of reaction and potentially will slow them down considerably. A weathering series showing this relatively low-temperature saturated reactivity may be drawn up, which will run from sulfide to carbonate to silicate to oxide minerals, with the oxide minerals being the least amenable to dissolution through chemical weathering.

The formation of supergene-hosted REE-enriched deposits is largely driven by this action of groundwater, and therefore, the presence of an active groundwater system is paramount. Within carbonatites, it is considered that, unlike most carbonate rocks where the most common weathering style is solution carbonation weathering, dissolution of carbonate is important, but oxidation of iron-bearing sulfides is observed as an early reaction in the weathering of carbonatites. In fact, as the most abundant sulfide in most carbonatites is pyrite, which occurs in often accessory amounts (commonly 1% to 2% of the rock, with often local concentrations), the presence of this mineral will importantly define the initial rate of weathering.

Oxidation is usually the dominant chemical weathering process in sulfide deposits (Blain and Andrew 1977; Andreu et al. 2015), and as pyrite and other sulfides are common accessory minerals in carbonatites, they may exhibit significant control on the weathering process. Near-surface oxidization of pyrite results in the formation of Fe minerals such as goethite [FeO (OH)]. This can commonly be seen as an orange discoloration of the carbonatite denoting initial oxidation; with time, these become brown and ever darker. Residual weathering will further indurate the Fe-rich (and generally Mn-rich) material at the surface to form a ferruginous zone that commonly becomes cemented as a lateritic duricrust (Anand and Paine 2002). This Fe-rich layer is commonly crucial for the preservation of the underlying, decomposed, and easily eroded regolith.

The process of hydrolysis—arguably the most important control on the weathering transformation of minerals—is dependent upon the presence of groundwater. By this process, ions are removed from minerals and replaced with OH^- which in part mimics the Bowen reaction series in reverse, so that minerals that are generally stable at the highest temperature, such as olivine and pyroxene, are the first to react and start to break down under the action of chemical weathering (Bowen 1956). Once the rock is saturated by groundwater, ionization is possible with the addition of more H_2O as well as CO_2. Hydrolysis itself involves the incorporation of OH^- and H^+ into pre-existing minerals.

Importantly, hydrolysis reactions can also attack carbonate minerals, and trace elements, including the REEs, will be released. These cations may either be mobile or immobile, depending on the chemistry of the groundwater itself. On the whole, hydrolysis reactions are generally slow, as H^+ is rather uncommon, although it increases significantly with a decrease in pH. pH itself is defined as an inverse and logarithmic measure of the amount of H^+ present, and so is an effective measure of the potential of hydrolytic reactions.

For instance, the following reaction will take place as an example of early-stage hydrolysis:

$$Fe_2SiO_4 + 4H_2O \rightarrow 2Fe^{2+} + 4OH^- + H_4SiO_4$$

This Fe will rapidly be oxidized if it is in the vadose zone, with studies showing that bacteria may play a fundamental role in this oxidative process (Xu et al. 2019).

If the conditions are moist and acid and extra water is added, the extreme reaction is:

$$Al_2Si_2O_5(OH)_4 + 5H_2O \rightarrow 2Al(OH)_3 + 2H_4SiO_4$$

Furthermore, the general lack of H^+ needed to complete hydrolysis is augmented by the breakdown of carbonic acid derived from rainwater through the following reaction:

$$H_2O + CO_2 \rightarrow H_2CO_3 \rightarrow H^+ + HCO^{3-}$$

The continued supply of carbonic acid will also dissolve carbonate minerals as the following reaction relates.

$$CaCO_3 + H_2CO_3 \rightarrow Ca(HCO_3)_2$$

This daughter phase is quite soluble and is then easily removed with similar reactions noted for Mg as below:

$$(Ca, Mg)CO_3 + H_2CO_3 \rightarrow Ca, Mg(HCO_3)_2$$

In contrast to silicate rocks and the addition of water through hydrolysis of silicate minerals, carbonatites are first oxidized, and only then does the process of dissolution become the dominant mode of chemical weathering, with Mg and

Ca being progressively removed as soluble phases.

4 Structural Control and Geomorphology

Faults are an important part of the carbonatite system and include the fundamental, deep-seated fractures allowing the mantle-derived melts to be entrained and emplaced within the upper crust or even erupt (Mitchell 2005) to post emplacement, reactivation tectonics (Giebel et al. 2019). Most studies of carbonatites are focused on their mineralogy and geochemistry, such that little attention has been on documenting structures within them. During weathering, structures will allow substantial migration of groundwater, controlling both chemical oxidation as well as dissolution, with many of the dissolved ions being removed. As carbonatites are located along structures, re-activation may occur. Furthermore, deep weathering along these fundamental structures may cause significant enrichment of the REEs, particularly if episodic movement has allowed deep groundwater circulation. A good example of this occurs along the Thuniberg fault in the Kaiserstuhl carbonatite in Germany (Wimmenauer 2003; Walter et al. 2018), where the most highly weathered material is the most REE-enriched. Similarly, the position of the most intense weathering is partly structurally controlled in the deep regolith at Ngualla (Witt et al. 2019).

Carbonatites generally weather as features that stand proud of the surrounding landscape, with noteworthy examples such as Salpeterkop in South Africa, Tchivira in Angola, and Dikke Willem in Namibia. However, under sustained intense tropical weathering, this may be reversed, with deep internal drainage developed. The carbonatite plug is commonly surrounded by a resistant circular range of hills usually caused by fenitization (alkali metasomatism) of the immediate country rocks, forming a much more resistant lithology. In many examples, it can be noted that centripetal drainage may be developed over the intrusion itself – for example, the Kruidfontein complex in South Africa (Woolley and Kjarsgaard 2008). Fenite aureoles are commonly resistant rocks; in the Goiás region in central Brazil, the country rocks are largely quartz sandstones which, when fenitized, become very hard and resistant to weathering. Notable examples include Tapira, Araxá, Catalão-1, Catalão-2 and Salitre. Lacustrine conditions are commonly developed within these hills and over the carbonatite and have been identified overlying carbonatites such as Mount Weld, Catalão-1, and Seis Lagos (Giovannini et al. 2017, 2021).

During lateritization (Retallack 2010), downward mass wasting within the regolith is important. Progressive leaching of the carbonate minerals accelerates this effect, and with time, the residual land surface descends (Fig. 5). In some examples, such as Glenover (Verwoerd 1967) and Cummins Range (Jaques et al. 2016), solution collapse breccias are developed on the regolith/hard rock interface, with the underlying carbonatite being oxidized, dissolved and the remaining semi-weathered more resistant minerals falling into the voids generated, which then become cemented by carbonate, Fe oxide, silica or even apatite. The presence of a recognizable regolith-driven depression in the landscape, combined with an internally directed drainage or major diversion of existing regional drainages, may assist in recognizing the presence of significant weathering, which may not be apparent at the surface.

5 Geochemistry, Mineralogy, Mineral Textures

The bulk of carbonatite-hosted weathered deposits, including the REE deposits, are a function of existing enrichment in certain elements within the carbonatite itself. The general bulk carbonatite composition, although hosting highly anomalous amounts of REEs, is rarely enriched to economic grades. It is clear, therefore, that an enriched protolith is vital and that the pre-existing mineralogy will also control the minerals that form in the regolith itself.

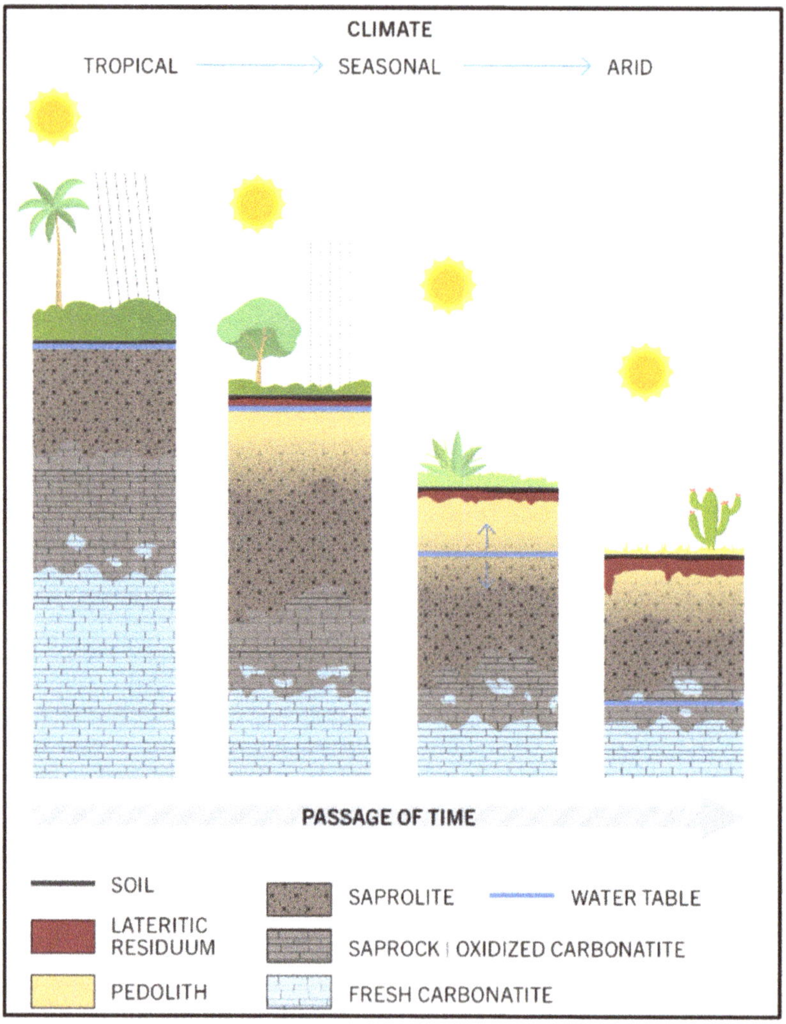

Fig. 5 Diagram showing downwards directed mass wasting and evolution of a lateritic profile through time and climatic change. (Modified after Anand and Paine 2002)

5.1 Geochemistry

An important step during the initial review and analysis of REE mineralization is to look at the chondrite-normalized distribution curves of the elements themselves. A carbonatite source may be distinguished from an alkali granite source through the Eu anomaly or lack thereof (Fig. 6). This plot further illustrates the general distribution curves of carbonatites demonstrated by Mount Weld, Zandkopsdrift and Wigu Hill, as well as the typical alkaline silicate curve with a pronounced Eu anomaly represented by Hoidas Lake.

Chondrite-normalized plots not only allow rapid review of the systematics of possible fractionation trends and source origins but also allow identification of discrepancies in the analytical results. As these distribution patterns follow fairly simple and predictable curves, the presence of anomalous or deficient amounts of one or more of the REEs allows potential errors in the analytical process to be identified. In contrast to almost all trivalent REEs, Ce can be oxidized and exists in a 4+ state. Ce^{4+} is less soluble than Ce^{3+}, and as the other REEs are mobilized through groundwater dissolution, on reprecipitation, they do so with a negative Ce anomaly in the resultant rocks (Sanematsu and Watanabe 2016). Similar negative Ce anomalies are seen in Ni laterites (Braun et al. 1990).

The role of carbonate dissolution is explicit owing to the importance of carbonates as the

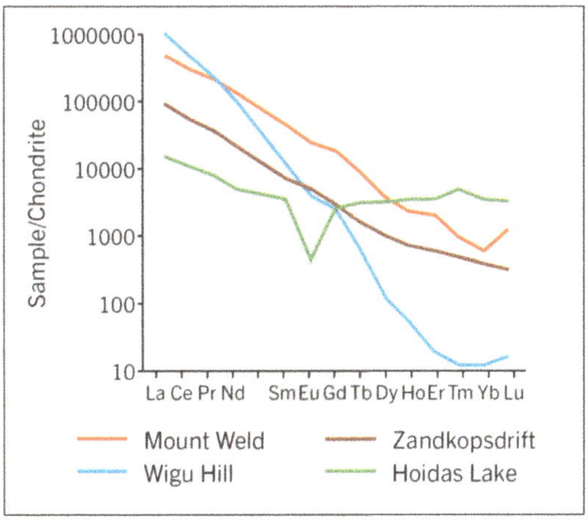

Fig. 6 Chondrite-normalized REE distribution curves of select REE projects. Mount Weld and Zandkopsdrift are both supergene examples. (Data from AMIS (2009), Harmer and Nex (2016), Harper et al. (2011), Great Western Minerals Group (2008), chondrite normalized values after Anders and Ebihara (1982))

major rock-forming minerals in carbonatites. It is intriguing that the study of many of the most REE-enriched carbonatites shows that they are dominated by both varied and uncommon carbonate species (Mitchell 2005), many of which are relatively Fe-rich. The presence of Fe as a 2+ state is the norm in fresh carbonatite, be it included in magnetite or included within the carbonate species. There is a common magmatic evolution of primary carbonate from calcic to more dolomitic to ankeritic, concomitant with progressively stronger oxidizing environments. These carbonate minerals are admixtures of Mn, Fe, Ba, Sr, Mg and Ca. There may be enrichment in REEs within these carbonate species (Chakhmouradian et al. 2016), commonly up to a few 1000 ppm in calcite and dolomite. During weathering, these ions will be released from the carbonate lattice. The various cations (Ca, Mg, Mn, Fe, Ba, Sr) making up carbonate minerals all have different solubilities. They, therefore, will be expected to be leached and removed from the carbonate lattice at different rates. As cation solubility is largely a function of pH, the buffering of reactions through the presence of carbonate minerals will result in high variability in pH and, subsequently, the respective cation solubility. Secondary "daughter" or transition carbonate minerals will form as leaching is determined by the differential solubilities of the constituent elements, which, as they are removed, will quickly produce a wide range of "daughter" compositions. Magnesium appears to be the most mobile of the cations, being rapidly removed from the lattice. A progressively-weathered dolomite will become increasingly more Mn- and Fe-rich until it approaches the composition of ankerite and then siderite. Transition carbonate phases that are Ba, Sr and Mn enriched should, therefore, also be considered and expected when studying these rocks.

On a large scale, the regolith generated from weathered carbonatites is also dependent upon the presence or absence of related silicate rocks. These may take the form of associated discrete intrusive or extrusive alkaline silicate rocks such as nepheline syenite, phonolite, nephelinite, trachyte or similar, and the amount and composition of fenite material contained as xenoliths and that may be accounted for. In several carbonatites, there is commonly over 50% of this included fenitized material, all of which must also be weathered. These associated silicate rock types substantially increase the amounts of Al, Si and K, which may be expected – elements that are less common in carbonatite rocks—and will lead to further speciation of other secondary minerals that will now be present in the regolith.

The precipitation of secondary Ti, Mn and Fe oxides and hydroxides is a further consideration, and these minerals assist greatly in defining the dissolution trends of the original protolith. Manganese and Fe become practically immobile in the regolith once they are oxidized to Fe^{3+} or Mn^{4+}, initially precipitating as amorphous oxides before crystallizing to goethite and hematite – distinguished by their respective yellow-brown or red streaks. Goethite is further generated during oxidative dehydration and the removal of constituent water. It has been demonstrated that goethite grains in lateritic regolith profiles in Australia adsorb the Sc ion onto their surface and, during the dry season, are dehydrated to form hematite, within which some Sc becomes locked (Chassé et al. 2016). A similar situation is suggested to be occurring with the REEs and is expanded upon below.

5.2 Mineralogy

The main REE minerals in weathered carbonatites are monazite, minerals of the crandallite group and, in some deposits, bastnäsite (Tables 1 and 2). Apatite is also an important REE host mineral. Each of these minerals contains the LREEs, and all minerals of economic interest are the (Ce)-predominant species. They occur as residual minerals from the fresh rock concentrated in the saprolite and pedolith, as well as newly formed minerals precipitated in the profile. As a general rule, the LREE-enriched nature of the protolith is retained in the regolith.

Many of the REE-hosting minerals in the supergene zone are hydrated, for instance, the mineral rhabdophane is almost only found within the regolith of weathered carbonatite intrusions. As most of these secondary minerals are derived from the release of ions through the destruction of existing anhydrous minerals such as carbonates and feldspars, the formation reactions are generally seen to be driven by hydrolysis, as described earlier.

Monazite is a LREE phosphate ((La, Ce, Pr, Nd) PO_4) and a ubiquitous REE accessory mineral in almost all rock types, forming in geological environments ranging from igneous and metamorphic to hydrothermal and weathering. Carbonatites form at the end stages of magma crystallization in calcite, dolomite and ferroan dolomite carbonatites, in hydrothermal and deuteric alteration, and during supergene weathering. Monazite is perhaps the most important REE mineral in weathered carbonatites (Fig. 7). It persists through the profile if present in the fresh carbonatite but also precipitates as a consequence of apatite dissolution, which removes Ca and releases P, which combines with available REEs.

All monazite of economic interest is of the Ce-predominant variety, monazite-(Ce). Monazite-(Nd), with a higher proportion of Nd compared to La and Ce, has been reported from Mount Weld (Willett et al. 1989). Thorium contents in monazite tend to be lower in weathered carbonatites than in fresh rock, and all carbonatite monazite (fresh and weathered) tends to have lower Th than monazite in rocks such as granite. Secondary monazite crystallization within the regolith of the South Australian craton is presented in studies by Collins (2014). Tiny seed grains approximately 20 μm in size were noted within the regolith, having formed during times of a fluctuating, probably brine-rich groundwater table (Benison and Bowen 2015). Monazite from the Matamulas REE deposit in Spain (Vergara and Burkhalter 2019), as well as Glenover (Broom-Fendley et al. 2021), formed within the sediment or regolith indicating low temperatures ($<50^{\circ}C$) of formation.

Owing to its high resistance to weathering, monazite also occurs in eluvial deposits near carbonatites and in colluvial deposits derived from carbonatites (Witt et al. 2019).

Rhabdophane ((La,Ce, Pr,Nd)PO_4 $0.6H_2O$) forms in weathered horizons. It is almost always fine-grained and difficult to identify and study. Occasionally, it can contain significant Th. The most common species is rhabdophane-(Ce), but rhabdophane-(La) is also reported. Significant amounts are documented at Longonjo in Angola (Lapido-Loureiro 1973; Brown 2020).

Xenotime (YPO_4) is the main HREE phosphate in fresh and weathered carbonatites. However, owing to the LREE characteristic of carbonatites, it is usually a rare accessory mineral. A notable exception in fresh rock occurs at Lofdal, Namibia (Wall et al. 2008; Swinden and Siegfried 2011), but there is no weathered carbonatite with predominant xenotime known. Xenotime has been recorded at Mount Weld (Willett et al. 1989).

Churchite ($YPO_4.H_2O$), the HREE-hydrated phosphate, is very rare in fresh and weathered carbonatite. It is, however, known from Mount Weld (Lottermoser 1990) as well as Bear Lodge (Andersen et al. 2017).

Bastnäsite ((La, Ce, Pr, Nd)CO_3F) is the most important of the REE fluorocarbonates in weathered carbonatites. It is the predominant REE mineral at Ngualla (Witt et al. 2019) and occurs as a minor mineral, especially in the saprolith horizons in various other lateritic carbonatites, including Bear Lodge, where bastnäsite with a negative Ce anomaly is regarded to point to the involvement of low-temperature meteoric water

Table 2 List of rock forming minerals and supergene equivalents

Mineral	Protolith: Fresh rock (magmatic, hydrothermal, metamorphic)	Saprolith: Saprock	Saprolith: Saprolite 'apatite zone'	Pedolith: 'crandallite zone'	Pedolith: Oxide-rich uppermost zone
REE and main REE-bearing minerals					
Ancylite	F	R?	–	–	–
Apatite	FF	R	R F	(R)	–
Bastnäsite	F	R F?	R F	(R?)	–
Cerianite	–	–	F?	FF	F?
Churchite-(Y)	–	?	?	F	?
Crandallite	–	?	F	FFF	?
Florencite	(F)	F	F	F	?
Gorceixte	–	F?	F	FF	?–
Goyazite	(F)	F?	F	F	?
Monazite-(Ce)	F	R F?	R F	R F	R?
Monazite-(Nd, La)	–	–	–	F	F
Parisite	F	R	?	?	?
Rhabdophane	-	?	F	R F	FF
Synchysite	F	R	(R)	F	–
Xenotime-(Y)	F	R	R?	F	?
Other minerals					
Aluminium oxides	–	?	F	F	F
Alunite			?	F	F
Anatase	–	?	F	?	?
Ankerite	FFF	FF	R	–	–
Baddeleyite	F	R	R	R?	R
Barite	FF	RF?	RF?	F	FF
Biotite-phlogopite	FF	(R)	–	–	–
Chalcedony	–	F	F	–	F
Calcite	FFF	R?	–	–	–
Chlorite	F	?	–	–	–
Dolomite	FFF	R	–	–	–
Ferroan dolomite	FF	R	–	–	–
Fluorite	F	–	–	F	F
Goethite/Fe hydroxides	?	F	FF	FFF	FFF
Hematite	F	R	F	F	FF
Hollandite	–	?	FF	F	F
Ilmenite	F	R	R	–	–
Kaolinite	–	?	F	F?	–
K-feldspar (orthoclase)	FF	R	R?	R	(R)
Maghemite	–	–	F	?	?
Magnetite	FF	R	R	–	–
Martite		F	FF	FFF	R
Olivine	F	(R)	–	–	–
Pyrolusite	–	?	?	?	?
Pyrite (and other sulfides, chalcopyrite, galena, sphalerite)	F	(R)	–	–	–
Pyrochlore	F	R*	R*	(R*)	–

(continued)

Table 2 (continued)

Mineral	Protolith Fresh rock (magmatic, hydrothermal, metamorphic)	Saprolith Saprock	Saprolite 'apatite zone'	Pedolith 'crandallite zone'	Oxide-rich uppermost zone
Pyroxene (aegirine, augite, diopside)	F	(R)	–	–	–
Pyrrhotite	F	–	–	–	–
Quartz	F	?	?	?	?
Riebeckite (tremolite, other amphibole group)	F	R	(R)	–	–
Romanèchite	–	–	F	F	FF
Rutile (niobian rutile)	F	?	R	R	R
Strontianite	F	R	(R)	–	–
Vermiculite	–	F	F	FF	–
Zircon	F	R	R	R	?

R residual, *F* formed in this horizon with number of letters e.g. F or FFF signifying abundance. All minerals assumed –(Ce) varieties unless specifically mentioned

R*—Pyrochlore is residual but exchanges cations and so changes composition through the profile, becoming progressively more deficient in it's A site cations while the Nb, Ta, Ti-bearing B site remains stable. Of the REE, Ce in particular can be absorbed into pyrochlore in laterite. At Mount Weld, pyrochlore breaks down in the supergene regolith (Willett et al. 1989)

(R)—mineral reported in this horizon as undergoing dissolution

Fig. 7 Typical monazite—crandallite aggregates (light-colored material) hosted in largely goethite groundmass exposed in the trench at Zandkopsdrift. This material will contain between 4 and 6% TREO

(Andersen et al. 2017). It is a minor component of the pedolith at Araxá (Neumann and Medeiros 2015). It may become residual and concentrated by weathering as the carbonates are removed but is also reported as a replacement of the Ca-bearing REE fluorocarbonate, synchysite (Witt et al. 2019). In both these cases, it retains the strong LREE-enrichments of the fresh carbonatite.

Synchysite (Ca(La, Ce,Pr,Nd)$(CO_3)_2$F) is also regarded as residual at Ngualla (Witt et al. 2019) and is described as being replaced by bastnäsite during weathering. No reports of newly formed synchysite in the carbonatite regolith are known, although unspecified REE fluorocarbonates are reported as forming towards the base of some bauxite profiles (Maksimovic and Panto 1996).

Parisite (Ca(La,Ce,Pr,Nd)$_2(CO_3)_3F_2$) is common in hydrothermally-formed REE assemblages associated with carbonatites but only as a minor phase in weathered carbonatites (Table 2). It has been recorded to be of supergene origin at Bear Lodge (Andersen et al. 2017).

Apatite ($Ca_5(PO_4)_3$(F, OH, CO_3)) is a rock-forming mineral common in many carbonatites and a key control on REE during weathering. In fresh rock, fluorapatite and F-rich hydroxylapatite are the most common species, containing up to about 4.5 wt% LREE oxide (Chakhmouradian et al. 2017). During lateritic weathering, apatite first survives as a residual mineral in the saprolite and can accumulate as calcite and dolomite are removed, as noted at the base of the Mount Weld weathered profile (Duncan and Willett 1990) or as observed at Catalão-1 (Toledo et al. 2004). Apatite is entirely dissolved in the more intensely weathered pedolith. The Ca and P components migrate downwards through the profile and reprecipitate as apatite in the saprolite. This weathered apatite commonly coats subhedral primary apatite laths with an anhedral mass of secondary apatite. The reprecipitation of apatite can form indurated rock, with loss of obvious igneous textures in the saprolite.

Evidence of one of many reaction routes is the generation of francolite (carbonate fluorapatite) from original high-temperature magmatic apatite, where secondary supergene apatite is then transformed into low-temperature apatite or francolite (Broom-Fendley et al. 2021). The origin of some francolite, such as that at the Kovdor phoscorite carbonatite complex in Russia, as having either a low-temperature hydrothermal or weathering origin is a subject of discussion, but much of it now seems the result of weathering (Broom-Fendley et al. 2021).

Crandallite Group (plumbogummite group), including florence, goyazite, gorceixite and crandallite, are hydrated phosphates that contain Al and, importantly, large ions of REEs, Ba, Sr and Th. Although commonly incorporating a substantial proportion of the REE hosted in the regolith, the crandallite group of minerals is at present not economically important due to the difficulty of concentrating them by commercial mineral processing techniques (Neumann and Medeiros 2015) and even less potential to recover the contained REEs due to their complex composition, and importantly, their lower REE contents compared to monazite and bastnäsite. They are, however, a vital part of understanding what happens during the weathering transition and the release of Al, P and REEs through the weathering of primary apatite, potassium feldspar and biotite/phlogopite. They are considered as a group here because there is a substantial solid solution between the end members. True end-member species are rarely formed. They precipitate within the regolith developed over carbonatites, are unknown as products of magmatic crystallization, and are rare in hydrothermal environments (Table 2). The formation of crandallite minerals in the regolith depends largely on the co-availability of Al and P, therefore linking the original protolith composition of the carbonatite and later hydration with the incorporation of Ca to form crandallite, Sr, to form goyazite, Ba to form gorceixite and REE, to form florencite (Hussak and Prior 1899). Florencite may be Ce, La, Nd and Sm dominated with the REE suffix used to name the mineral. Generally, Ba-rich carbonatites have up to 2000 ppm Ba included within carbonate minerals such as calcite or dolomite. These carbonates can release a substantial amount of Ba during weathering and will produce gorceixite-dominated assemblages, which, importantly, will also incorporate some REEs and Sr if available (e.g., at Zandkopsdrift). The presence of Ba-enriched carbonate minerals such as norsethite, baryto-calcite, or kutnahorite is uncommon in carbonatites, but as discussed, the dolomitic phases can vary with enrichment in Ba and other elements such as Fe, Sr, Mn and REEs.

Cerianite (CeO_2) is rare in fresh carbonatite but forms in some parts of weathered profiles, especially in the pedolith (Table 2). Cerianite forms in the ferruginous duricrust at Ngualla (Witt et al. 2019) and Gakara (Ntiharirizwa et al. 2018), is present in parts of the Mount Weld regolith (Lottermoser 1990) and is a minor component at Araxá (Neumann and Medeiros 2015). It is normally strongly Ce-enriched, with Ce oxidized to Ce^{4+}. As far as the economic recovery of REEs is concerned, cerianite is similar to the crandallite group in its processing difficulty.

Other Minerals The rock-forming carbonates calcite, dolomite, ferroan dolomite and ankerite host trace element levels of REEs (Hornig-Kjarsgaard 1998; Chakhmouradian et al. 2016; Dawson and Hinton 2003) that are released as the carbonates are dissolved in the weathering profile. Given the abundance of these minerals (by definition, >50% of carbonatite composition), they may contribute a significant proportion of REEs in the weathered profile.

Pyrochlore ($A_2Nb_2(O, OH)_6Z$, A is Na, Ca, Sr, Fe, Pb^{2+}, Y, U^{4+} and Th. Z is OH, F, O, H_2O or a vacant site) is the principal host and ore mineral for Nb. It is formed mainly by magmatic processes in carbonatites (also occasionally by hydrothermal activity such as in fenite) but then survives through stages of hydrothermal alteration and weathering until it is broken down in the most extreme alteration in the upper parts of the pedolith (Table 2). Although the pyrochlore is residual, it is the Nb cations that are stable, whereas the A sites cations are leached and exchanged. Pyrochlore hosts minor REEs and can incorporate REEs within the weathering profile, including the uptake of Ce, as described at Lueshe, DRC (Wall et al. 1996).

Sulfides are common minor constituents of carbonatites. Pyrite, pyrrhotite, chalcopyrite and sphalerite are reported. They do not survive in the regolith (Table 2) but have an important role in the geochemistry of weathering, as discussed.

Manganese minerals, especially hollandite and romanèchite (hydrated form), both contain Ba as well as Mn; small amounts of Th, REEs and other 3+ valence ions are common, especially within the ferruginous upper horizons.

5.3 Mineral Fabrics

Fabric (or the rock microstructure) is defined by the shape and orientation of the carbonate minerals, entrained fenite clasts, crystals, and phenocrysts, matrix composition, the configuration and organization of these features within the rock as a whole, along with any structural overprints which may have occurred. The effect of porosity is fundamental in controlling the migration of ions through the progressively weathering carbonatite rock itself.

Characteristic fabrics are both preserved and developed in the rocks of the altered carbonatite, and those textures characteristic of the latter include botryoidal, atoll, concretionary and pisolitic mineral habits. These are noted in soils and rocks driven in their formation by the progressive migration of seasonal or longer-term groundwater cycling. These growth rings are commonly inward-directed with parallel trends in evolving mineral composition, indicating the infilling of cavities that may have been generated through the extent dissolution of carbonate minerals. Surrounded "seed" grains of more resistant minerals are commonly observed. These characteristic botryoidal growth layers occur in many parts of the regolith, with layer parallel horizons often noted—the result and involvement of a pre-existing water table. The most characteristic is the precipitation of secondary botryoidal Mn oxides within the supergene environment. In fact, many of these residual REE deposits have significant resources of Mn associated with them. Zandkopsdrift, for example, was first explored for its Mn potential. Intensely hardened and indurated layers of Mn oxide and Fe oxide form both as a surface residuum (lateritic duricrust) overlying the weathered carbonatites and deeper within the regolith itself. In the former case, these ferruginous enrichment layers are important as they preserve much of the underlying weathered rock from being eroded and/or reworked.

6 Examples of REE Enriched Residual Carbonatites

In the examples presented below, common themes of mineralogy, alteration, and formation are presented. It is important to consider that the regolith may be many 100 s of metres thick, as in Seis Lagos, Araxá or Tomtor (Kravchenko and Pokrovsky 1995). The similarities between, for example, those of Araxá (Verplanck et al. 2016) in Brazil and Mount Weld in Australia (Duncan and Willett 1990; Lottermoser 1990) are substantial even though they are separated in time, place, and present climate.

In order to understand these deposits better, they should be seen in the context of lateritic weathering and the processes associated with their formation. Laterites are the focus of extensive literature, especially in countries such as India, Australia and Brazil, where they form an important part of the landscape. The reader is referred to Anand and Paine (2002), Freyssinet et al. (2005) and chapter "The Role of Regolith Formation and Evolution in the Development of Supergene Mineral Deposits", for detailed reviews of the formation and interpretation of laterites.

An important control on the development of these weathered profiles is the role of the water table, which in turn is largely controlled by alternating dry and humid tropical seasons. The Araxá and Catalão-1 deposits in Brazil fall within a warm and high rainfall savanna climate, where weathering is clearly active and taking place in the regolith today, whereas Mount Weld in the western desert of Australia, and Tomtor in the Siberian tundra in Russia, are evidence of tropical, hot and moist climates in earlier times (Table 1). If there was less climatic seasonality, there would only be limited movement of the water table, thus restricting the extensive, intense weathering required for lateritization. The position and vertical fluctuation of the water table are of fundamental importance in facilitating REE enrichment in the following examples.

6.1 Mount Weld, Western Australia

The Mount Weld carbonatite, located in Western Australia, is Paleoproterozoic in age (2025 +/- 10 Ma; Graham et al. 2004). Detailed studies of the deposit are presented by Duncan and Willett (1990), Lottermoser (1990), Willett et al. (1989), and Zhukova et al. (2021). The intrusion is chiefly composed of dolomite with lesser calcite, ferroan carbonates, magnetite with apatite, and, in places, significant REE enrichment largely as monazite and lesser amounts of the fluorocarbonate mineral, parisite. The intrusion is circular in plan and has a diameter of over 4 km. It was first defined by the strong magnetic response identified during an airborne survey of the area in early 1966. It was subsequently drilled, and the REE mineralization within the residual regolith was recognized beneath ~60 m of largely Tertiary – Quaternary age overburden. These terrestrial sediments include paleo-lake sediments and a top cover of 20 m of unconsolidated sand and mud.

The regolith developed from the fresh carbonatite is about 50 m thick, with an irregular, karstified lower contact. As the deposit is so deeply weathered, only now, through the recent mining activities, has it been possible to gain a proper three-dimensional understanding of the mineralization trends in the paleo-regolith.? A well-defined ferruginous duricrust caps the deposit and probably prevented the erosion of the REE-enriched material before the deposition of Tertiary-age sediments. The mine at Mount Weld produces about 12% of the world's REEs (U.S. Geological Survey 2021), most of which are light REEs. During the 2021 fiscal year, over 14,000 t of REO were produced, of which 5460 tonnes were of a mixed NdPr concentrate (Lynas 2021).

Within the regolith, there are various zones of enrichment noted, which include anomalous P, Nb, Ba, and the REEs (Duncan and Willett 1990). Some of the REE grades reported are exceptional (>45% TREO), making it probably the richest, largest REE deposit on the planet (Zhukova et al. 2021). A persistent feature of the extreme REE enrichment at Mount Weld is the precipitation of REE minerals from REE-enriched groundwater, proposed by Duncan (2016). A prominent dolerite dyke crosses the center of the intrusion. This dyke is also weathered but, importantly, is more resistant than the decomposed carbonatite, which has resulted in focusing migrating groundwaters (phreatic zone) through the regolith. This "pregnant" groundwater is alkaline, reducing, and enriched in REEs and P after having leached these elements from the carbonatite. The slow migration of this groundwater is interrupted by the dyke, which forces it upwards, resulting in a mixing with meteoric (rain) and oxidizing waters, which percolate downwards. This interaction caused REE phosphate minerals such as monazite to crystallize (Fig. 8).

The most REE-enriched parts of the regolith appear to be the result of terrestrial plant-related organic fixing (Zhukova et al. 2021), which effectively brackets an inferred maximum age of the Carboniferous for the formation of this regolith. Importantly, Ce is depleted in this zone and in the overlying ferruginous duricrust, indicating the secondary source of Ce^{3+} (and the other REEs) in solution compared to Ce^{4+}.

6.2 Longonjo, Angola

This carbonatite forms a volcanic centre located just outside the town of Longonjo, Huambo Province, central Angola. The intrusion is about 4 km across, with prominent hills forming the northern crater rim of the intrusion. Longonjo has not been dated, but its intimate relationship with other carbonatites within the Lucapa corridor is clear

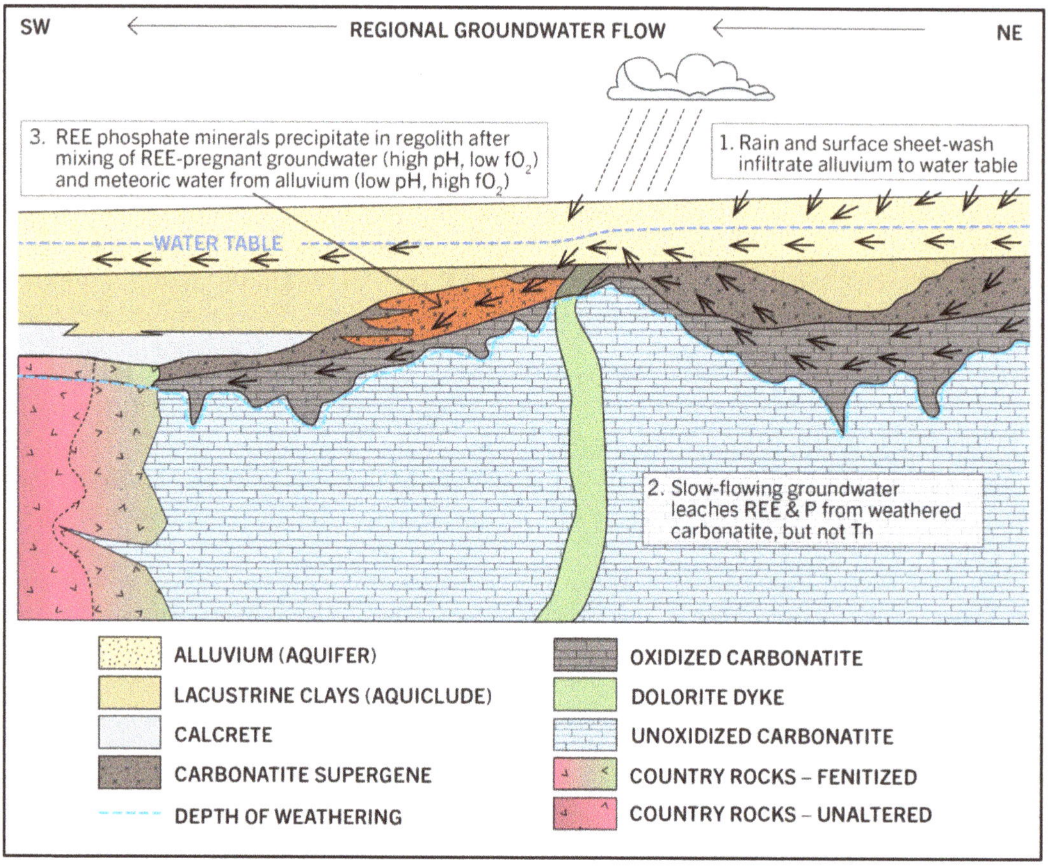

Fig. 8 Schematic groundwater section of Mount Weld showing the possible origin of high-grade mineralization zones. (Adapted from Duncan 2016; Duncan and Willett 1990)

(Machado 1958), suggesting an early Aptian age of about 120 Ma, similar to the intrusions of Brazil and Namibia (Marsh 1973). The carbonatite is chiefly dolomite carbonatite but has extensive amounts of included fenite material. The fenite is composed of biotite/phlogopite and potassium feldspar, and in some rare instances, original granite remnants are recognized.

Weathering is active today, as shown by the largely hydrated mineral assemblages that have been identified (Lapido-Loureiro 1973; Brown 2020). There is a strong seasonality in the climate, which is mirrored in the annual change in the water table. Large parts of the mineralized regolith are dominated by minute grains of rhabdophane as well as monazite and apatite (Calvo et al. 2011). Although the academic study of the deposit has been limited (Lapido-Loureiro 1973; Machado 1958), within the last four years, a detailed exploration program has been completed (Brown 2020), leading to the announcement of a new REE mine project. Of note was the effective use of ground-penetrating radar to map the thickness of the regolith (Fig. 9).

6.3 Zandkopsdrift, South Africa

This intrusion occurs in the western part of South Africa and is regarded to be part of early Tertiary kimberlite and melilite activity in this region (Moore and Verwoerd 1985). The approximately 800 m diameter pipe—which today is expressed as dark coloured hill—is dominated by aillikite breccias that form the majority of the rocks comprising the intrusion. They erupted over

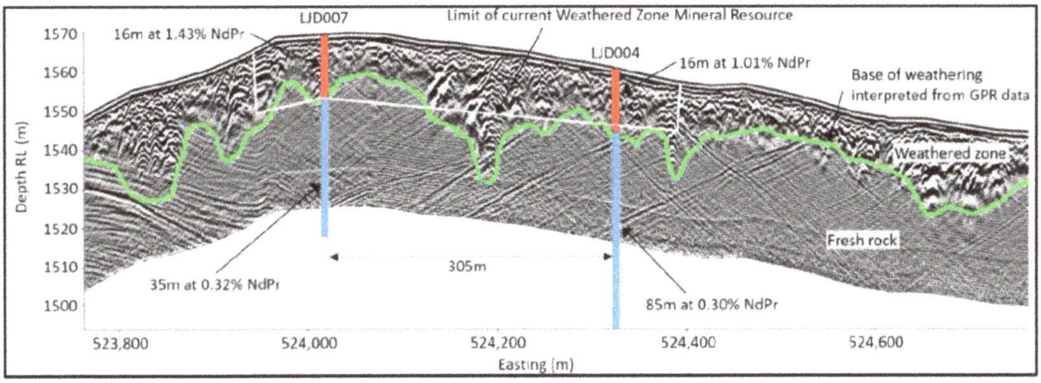

Fig. 9 Ground Penetrating Radar (GPR) section through Longonjo showing strong regolith control of mineralization (Brown 2020)

only a short time interval, and a plethora of different rock types have been noted (Ogungbuyi et al. 2016, 2022). These rocks consist largely of phlogopite but have a significant amount of apatite and magnetite present (commonly 5% of the rock) and are correctly aillikites. These aillikite breccias are cross-cut by narrow dykes of Mn-dolomite and Fe-dolomite, with the dykes, in turn, enriched in monazite and apatite. The dykes contain 2–3% pyrite, barite and strontianite.

Weathering was intense, and the regolith was developed down to a depth of 70 m. Significant Mn enrichment is noted, and botryoidal masses and horizons of pyrolusite, psilomelane and romanèchite are common, especially as surficial layers. The Mn is derived from the weathering of the Mn-dolomites. In contrast to the deeply weathered zone of carbonatite dykes, the surrounding aillikite is weathered to a lesser degree, and the surrounding gneisses are hardly affected at all. This suggests that the carbonatite regolith has been preserved by the protection provided by the overlying Mn and Fe-rich duricrusts (Fig. 10).

The REE mineralization within the laterite is hosted by monazite, with more than 95% of the REE reporting in this mineral (Harper et al. 2011). The REEs also occur in the two crandallite minerals gorceixite and goyazite, although neither make up an appreciable component of the regolith, the majority being goethite, hematite and Mn oxides. The regolith is very porous and friable, with a density of 1.2. This is an important consideration for both mineral resource studies and for the formation and preservation of porosity over such depths.

Fig. 10 Secondary Mn oxides (black/dark brown) forming a well-developed duricrust above weathered carbonatite exposed at Zandkopsdrift. Note the yellow colors indicative of goethite

Deep, humid tropical weathering took place in this part of southwestern Africa during the Paleocene – Eocene thermal maximum (PETM), which started at 55.8 Ma (Pickford 2016) and was further enhanced given the low elevation of the intrusion. Preservation resulted from the progressive aridification of the region in the Miocene, leading to both a steady lowering of the water table and cementation of the residual ferruginous capping.

6.4 Bear Lodge, USA

The Bear Lodge alkaline complex, Wyoming, USA, hosts the Bear Lodge REE deposit and is an advanced exploration project primarily focused on the oxidized zone above a suite of carbonatite dykes, veins and stockwork. Identified zones within the carbonatite resource include an unoxidized zone (Stage I and II), a transitional zone (Stage III), an oxide-carbonate zone (Stage IV), and an oxidized zone (Stage V), all loosely correlated within the regolith nomenclature presented in Fig. 4. The deepest zone is characterized by unoxidized, hydrothermally-altered carbonatite with mineralized pseudomorphs and veins, which contain a REE mineral assemblage of predominantly ancylite and carbocernaite (Andersen et al. 2017).

The carbonatite dykes contain a variety of minerals, including calcite, dolomite, ankerite and strontianite, with subordinate biotite, phlogopite, potassium feldspar, pyroxene, barite, apatite, sulfides and oxides (Hutchinson 2016; Van Rythoven et al. 2020). The narrow transition zone, or Stage III of Hutchinson et al. (2022), is generally 6 m thick, occurs at the top of the unoxidized zone, and represents the initial stages of sulfide oxidation and replacement of ancylite by Ca-REE fluorocarbonate minerals (Dahlberg et al. 2014). The oxide-carbonate zone (saprolite) is characterized by nearly total oxidation of sulfides, partial dissolution of groundmass calcite, and substantial replacement of ancylite by fluorocarbonates. The upper part of the deposit (pedolith) primarily consists of intensely oxidized dykes and veins termed FMR dykes (Fe oxide-Mn oxide-REE). In this zone (up to 150 m thick), sulfides are completely oxidized, and the groundmass carbonate is completely leached. The REE mineral assemblage consists of fluorocarbonates (bastnäsite and parisite), with variable but subordinate phosphates and cerianite. Gangue minerals include relict potassium feldspar and biotite with iron oxides, manganese oxides and clays (Fig. 11).

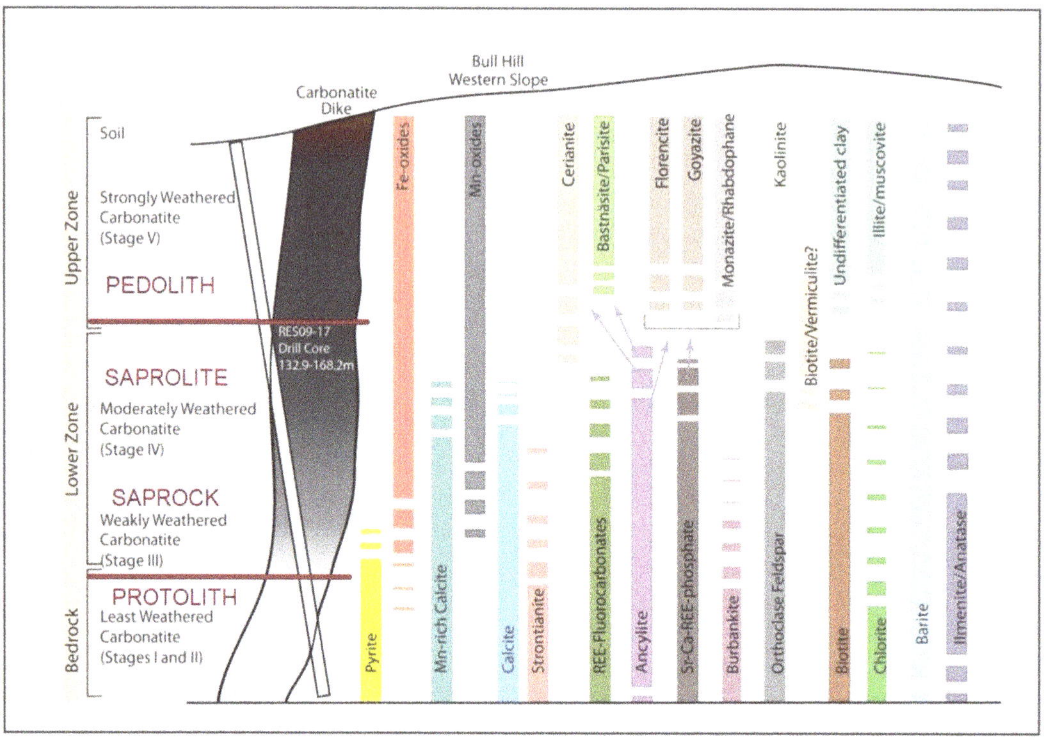

Fig. 11 Cross section of the Bear Lodge deposit showing mineral transformation and regolith divisions. (Modified after Hutchinson et al. 2022)

To evaluate chemical changes resulting from weathering processes, a suite of carbonatite samples collected from the drill core were compared to near-surface samples collected from an exploration trench that was dug above the carbonatite. The trench contains FMR dykes and weathered host rocks. The chemical composition data for these samples are published in Verplanck et al. (2020) and summarized in Fig. 12. Figure 12a displays variations of Ca vs Fe and highlights the loss of Ca during weathering due to the dissolution of carbonate minerals. Although Fe is variable in the carbonatite samples (1.64 to 15.2 wt %), concentrations in the weathered zone display an approximately two-fold increase (3.01 to 30.4 wt %). Similar to Ca, Sr is also lost during weathering (Fig. 12b).

Hutchinson (2016) noted a 40% increase in porosity in the weathered zone compared to the unweathered carbonatite, consistent with the loss of carbonate minerals. With the loss in mass, conservative estimates of element concentrations should be roughly doubled in the weathered zone. Although Fe and Mn have a wide range in concentrations in the carbonatite, the highest concentrations are in the weathered zone (Fig. 12c). Similarly, the highest concentrations of Th and Y also occur in the weathered zone (Fig. 12d). Although the highest Ba and La concentrations do occur in the weathered zone, the plot of La vs. Ba does not show a clear trend of increasing concentration with weathering (Fig. 12e). Two populations for both Ba and La are seen in both the deeper and shallower suites. Overall, the results presented here for the Bear Lodge deposit are broadly similar to other weathered carbonatites, but the complexity arises from heterogeneity in the mineralogy of the parent rocks and multiple processes at work during weathering.

7 Mineral Processing

The production of REEs from mined ore usually consists of three steps: the separation of a concentrate consisting of REE minerals such as monazite, the cracking of the REE minerals to release the REE into an aqueous solution, and the separation of the REEs into their individual elements. Succinct summaries of REE mineral processing are provided by Verbaan et al. (2015) and Beard et al. (2022). Currently, extractive processing of REEs from supergene carbonatite ores is limited to the Lynas Rare Earths facility in Malaysia. This treats the high-grade mined and concentrated ore from Mount Weld (Lynas 2021). Flotation at the mine site in Western Australia upgrades the original ore to a concentrate suitable for shipping. This then undergoes further beneficiation, including cracking and REE separation at the Malaysian operation. The beneficiation of ore from the Mount Weld deposit was a significant milestone because it established that fine-grained lateritic material from a weathered carbonatite can be economically mined and processed.

A review of REE processing by Jordens et al. (2013) introduced some new and novel techniques that had not yet been employed in REE recovery, including pyrometallurgical processing, such as the calcination of the lateritic ore to convert REE carbonates to oxides. Generally, however, processing studies conducted on lateritized material, for example, Catalão-1 (Tassinari et al. 2001) and Araxá (Neumann and Medeiros 2015), have clearly demonstrated the problems in attempting any physical processing and upgrading.

A litany of challenges face the processing of these ores, notably:

1. Fine grain size
2. Tightly intergrown polymineralic mineral intergrowths
3. Diversity of minerals
4. Ionic effects caused by goethite resulting in a collapse of froth flotation methods
5. The presence of REE-bearing minerals which are not a commercial source of these elements, such as the REE-bearing crandallite and florencite-(Ce)
6. The presence of excessive Fe, P, Mn, Ba, Sr and Th that affect the quality of the concentrate

Standard ore processing includes initial physical concentration through the application of recovery methods most suited to the select

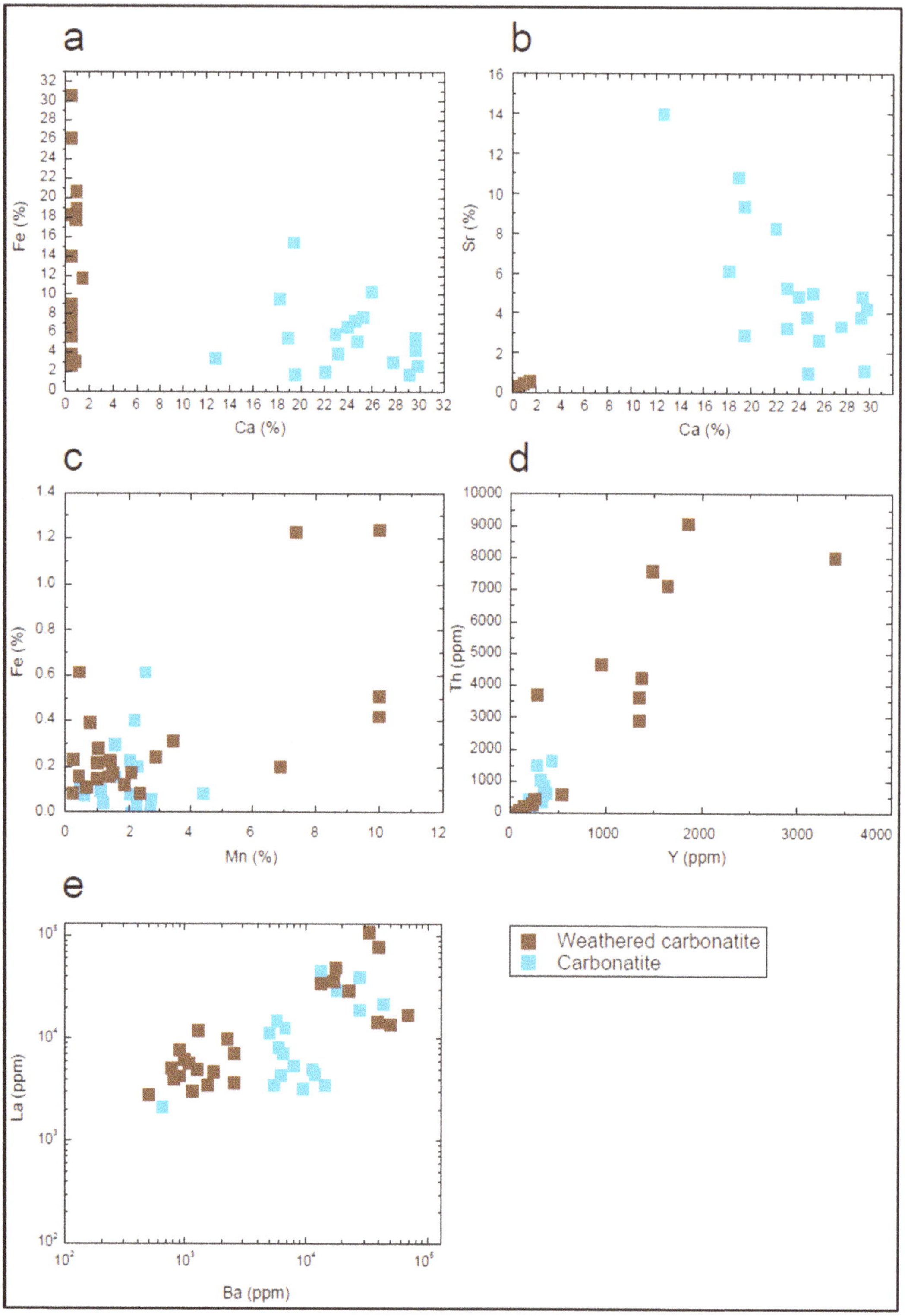

Fig. 12 Plots of carbonatite samples from Bear Lodge collected from fresh drill core compared with near surface and weathered samples recovered from above the carbonatite. (Data from Verplanck et al. 2020)

characteristics of the individual target minerals. These include magnetic separation (both wet and dry), density separation (most commonly employed where there are large contrasts between the ore minerals and the gangue, and including gravity spirals and air classification), froth flotation, and, lastly, spectral-, radiometric- or fluorescence-based sorting such as employed by the company Tomra. The field of REE processing has progressed significantly in the last decade, and the use of selective collectors and microbubble froth floatation is an effective way of pre-processing (Ross et al. 2019). Unfortunately, very few of these techniques are successful in beneficiation of lateritic material that hosts REE mineralization. Indeed, leaching of the entire run of mine ore is anticipated to be the usual way of processing this material (Zhang and Edwards 2013).

Transitional laterite ores, such as the examples of Ngualla (Witt et al. 2019) and Longonjo (Brown 2020), both include an important consideration given to froth flotation to concentrate the monazite. These ores include primary minerals such as apatite, monazite and bastnäsite, as well as others of supergene origin such as rhabdophane, crandallite group and synchysite.

Detailed metallurgical test work was completed on a range of weathered carbonatites from Brazil (Antoniassi et al. 2020). These included samples from four complexes, including Araxá, Salitre, Morro de Ferro and Catalão-1. A comprehensive study showed that a full understanding of the mineralogy, mineral intergrowths and textures is necessary to optimize recovery methods. It was concluded that “the low degree of liberation and occurrence in ultrafine-grained REE-bearing minerals are a challenge for the development of mineral processing routes since existing technologies are limited for those features.” Petrographic studies are recommended early on in any exploration project. SEM study has been highlighted as of the greatest importance for the study of these microcrystalline aggregates (Tassinari et al. 2001) as crystal grains are commonly in the order of a few microns in diameter. These studies are relatively low cost and important early on, as failing to identify “fatal flaws” that will impact metallurgical processing later are crucial in any initial resource studies.

A number of projects in development have presented summaries of their envisaged processing metallurgy, and these include several innovative ways to deal with these inherently complex ores. An acid bake (generally at 400°C) utilizing sulfuric acid is proposed for both Araxá (Clay and Ackroyd 2013) and Zandkopsdrift (Harper et al. 2011). Advances in Continuous Ion Exchange and Continuous Ion Chromatography (CIX/CIC) methods as alternatives to solvent extraction are proposed for the treatment of “cracked” REE minerals. Cracking is the process by which the REE minerals are dissolved, and the REE ions are put into a solution so that they may be recovered. Details of these methods, as well as some of the issues that go hand in hand with consideration of Th, are dealt with at length by García et al. (2020).

The mining of a limited REE-enriched mantle overlying the fresh rock means it will be ultimately exhausted, and it is, therefore, important to consider both fresh and weathered rock types for early-stage processing tests. A cautionary tale is told of Jacupiranga, where, although today’s operations only mine the hard rock resource, the majority of the early production was of apatite-enriched weathered material, which was mined out in the 1980s. An entire reconfiguration of the plant was necessary in order to process the fresh carbonatite, as all flow sheets up to that time had been defined solely on the regolith material.

8 Summary of Weathering Process and Conclusions

The primary protolith composition of the carbonatite complex determines the composition of the overlying regolith. This includes the composition of the majority of carbonate minerals as well as the accessory minerals, as all have different responses to weathering. In contrast to sedimentary carbonate rocks, there is a wider range of minerals in carbonatites, which weather and release Fe oxides. These include pyrite, which itself is an important driver of the chemical weathering of carbonatites. The S-rich groundwaters or supergene fluids developed during oxidation of these sulfides are importantly buffered substantially by carbonates in the carbonatite. Nonetheless, it is within this environment that the release of ions takes place during weathering, and the crystallization of low-temperature "neo" or secondary minerals such as monazite, various Mn oxides, crandallite minerals and rhabdophane occurs.

Leaching of the carbonate minerals is progressed through groundwater rich in dissolved oxygen and carbon dioxide. Downward migration of meteoric water results in heightened oxidation, and additional sulfur is then released from the sulfide minerals in the rocks to form sulfuric acid. As the reaction with the carbonatite continues, Cl^- and F^- are released, especially from apatite, progressing weathering further. At the water table, there is an important change in the redox conditions with the environment changing from an oxidizing setting to a reducing one and, with that, a distinct change in cation solubility. Iron is an important consideration in reducing conditions; it rapidly goes into solution as $Fe^{2+,}$ and similar activity is expressed by a number of 3^+ and 2^+ valence cations—the most important in the carbonatite environment being P, Nb, Ti, Fe, Mn, Sc and REEs.

Residual REE-bearing apatite and monazite are progressively dissolved, releasing both P and REEs, which may be sequestered in newly formed secondary, sometimes hydrated minerals such as monazite, churchite and rhabdophane.

The more soluble ions percolate downwards under the action of downward migrating groundwater, whereas less soluble ions such as Fe, Ba, Mn and Ce^{4+} remain in the oxidized zone where they form secondary minerals. These minerals are characteristic of the oxidation zone, notably the crandallite group, particularly florencite. Secondary apatite, monazite, barite, strontianite and the Mn oxides hollandite and romanèchite are also typical. Botryoidal, atoll and open space-filling textures are characteristic of the appearance of these minerals. Microcrystalline aggregates of commonly bladed crystals a few microns in diameter and which define colloform textures are typical and diagnostic of this mode of formation.

Below the indurated Fe oxide capping, a leached zone is usually developed. In many lateritic regoliths, either a bleached (previously called the pallid) zone or mottled zone is commonly described as evidence of the leaching of significant amounts of Fe, causing a paler appearance. In some places, this is due to active leaching on erosional slopes in dissected lateritic terrain and may not represent the original process that occurred in the regolith itself. However, REE content does appear to be reduced within this zone.

The net effect of these supergene processes is to move metal ions, such as the REEs, from the leached zone to the enriched zone, increasing the concentration of REEs to levels higher than in the unmodified primary zone or protolith.

It is essential that basic mass balance calculations are made between that of the fresh and the weathered material, as these answers can be used to better understand the amount of chemical weathering that has taken place.

Acknowledgments Much of this research has been supported by the European Union's Horizon 2020 research and innovation program under grant No 689909 (HiTech AlkCarb) as well as the Natural Environment Research Council [grant number 118637] "Weathering of carbonatite REE deposits" or WREED. Illustrations have been drafted by Carl Lamprecht of ThoughtSculptor, who is thanked for his efforts. Comprehensive reviews by Erin Marsh, Ed Loye, and Charles Butt were much appreciated and greatly improved this chapter.

Any use of trade, firm, or product names is for descriptive purposes only and does not imply endorsement by the U.S. Government.

References

AMIS (2009) AMIS0185. Certified Reference Material, Rare Earth Elements, Wigu Carbonatite Complex, Tanzania. Certificate of Analysis. www.amis.co.za

Anand RR, Paine M (2002) Regolith geology of the Yilgarn Craton, Western Australia: implications for exploration. Austr J Earth Sci 49:3–162. https://doi.org/10.1046/j.1440-0952.2002.00912

Anders E, Ebihara M (1982) Solar-system abundances of the elements. Geochimica et Cosmochimica Acta 46: 2363–2380

Andersen AK, Clark JG, Larson PB, Donovan JJ (2017) REE fractionation, mineral speciation, and supergene enrichment of the Bear Lodge carbonatites, Wyoming, USA. Ore Geol Rev 89:780–807

Andreu E, Torró L, Proenza JA, Domenech C, Garcia-Casco A, de Benavent CV, Chavez C, Espaillat J, Lewis JF (2015) Weathering profile of the Cerro de Maimón VMS deposit (Dominican Republic): textures, mineralogy, gossan evolution and mobility of gold and silver. Ore Geol Rev 65:165–179

Antoniassi JL, Uliana D, Contessotto R, Kahn H, Ulsen C (2020) Process mineralogy of rare earths from deeply weathered alkali-carbonatite deposits in Brazil. J Mater Res Technol 9(4):8842–8853

Beard CD, Goodenough KM, Borst AM, Wall F, Siegfried PR, Deady EA, Pohl C, Hutchison W, Finch AA, Walter BF, Elliott HAL, Brauch K (2022) Alkaline-silicate REE + HFSE systems. Econ Geol. (in press)

Benison K, Bowen B (2015) The evolution of end member continental waters: the origin of acidity in southern Western Australia. GSA Today 25(6):4–10

Blain CF, Andrew RL (1977) Sulfide weathering and the evaluation of gossans in mineral exploration. Miner Sci Eng 9:119–149

Bowen NL (1956) The Evolution of the Igneous Rocks. Dover, Canada, p 332

Braun J-J, Pagel M, Muller J-P, Bilong P, Michard A, Guillet B (1990) Cerium anomalies in lateritic profiles. Geochimica Cosmochimica Acta 54:781–795

Broom-Fendley S, Siegfried PR, Wall F, O'Neill M, Brooker RA, Fallon EK, Pickles JR, Banks DA (2021) The origin and composition of carbonatite-derived carbonate-bearing fluorapatite deposits. Mineral Deposita 56:863–884

Brown R (2020) Competent persons report on the geology and mineral resources of the Longonjo Project, Angola. Report prepared for Pensana by SRK Consulting Australasia. 86 pp, AIM listing document

Calvo AC, Draper JC, Abella PA, Bambi ACJM, Gonçalves AO (2011) Nb and REE mobilization at the Longonjo carbonatite, Angola. Proceedings of the Eleventh Biennial SGA Meeting, Antofagasta, Chile, September 2011, pp 681–683

Chakhmouradian AR, Zaitsev AN (2012) Rare earth mineralization in igneous rocks: sources and processes. Elements 8:347–353

Chakhmouradian AR, Reguir EP, Couëslan C, Yang P (2016) Calcite and dolomite in intrusive carbonatites. II. Trace-element variations. Mineral Petrol 110:361–377

Chakhmouradian AR, Reguir EP, Zaitsev AN, Couëslan C, Xu C, Kynický J, Mumin H, Yang P (2017) Apatite in carbonatitic rocks: compositional variation, zoning, element partitioning and petrogenetic significance. Lithos 274–275:188–213

Chassé M, Griffin WL, O'Reilly SY, Calas G (2016) Scandium speciation in a world-class lateritic deposit. Geochem Persp Lett Eur Assoc Geochem 3:105–114

Chebotarev DA, Doroshkevich AG, Klemd R, Karmanov N (2017) Evolution of Nb-mineralization in the Chuktukon carbonatite massif, Chadobets upland (Krasnoyarsk Territory, Russia). Periodico di Mineralogia 86:99–118

Clay AN, Ackroyd B (2013) Preliminary Economic Assessment in the form of an independent Technical Report on MBAC Fertilizer Corp. Araxá project located in Minas Gerais state, Brazil. Prepared by VENMYN Rand PTY. Ltd, 111 pp

Cocker MD (2014) Lateritic, supergene rare earth element (REE) deposits. In: Conway FM (ed) Proceedings of the 48th Annual Forum on the Geology of Industrial Minerals, Phoenix, Arizona, May 2012. Arizona Geological Survey Special Paper #9, Chapter 4, pp 1–18

Collins T (2014) Understanding the geochemistry and mineralogy of regolith hosted REE mineralization. Honours Thesis, University of Western Australia, 67 pp

Dahlberg PS, Noble AC, Pickarts T, Rose WL, Jaacks JA (2014) Technical Report on the Reserves and Development of the Bull Hill Mine, Wyoming. Rare Element Resources, Ltd., Bear Lodge Project Canadian NI 43-101 Pre-Feasibility Technical Report, Oct. 9, 2014

Dawson JB, Hinton RW (2003) Trace-element content and partitioning in calcite, dolomite and apatite in carbonatite, Phalaborwa, South Africa. Mineral Mag 67:921–930

de Machado FJS (1958) The volcanic belt of Angola and its carbonatites. In. Publication Commission for Technical Cooperation in Africa South of the Sahara. Joint Meeting Leopoldville 44:309–317

Duncan R (2016) Mount Weld Poster. In: Fenite and geophysics as exploration tools. HiTech AlkCarb Expert Council Meeting. Zomba, Malawi

Duncan RK, Willett GC (1990) Mount weld carbonatite. Geol Miner Depos Aust Papua N Guin 1:591–597

Elliott HAL, Wall F, Chakhmouradian AR, Siegfried PR, Dahlgren S, Weatherley S, Finch AA, Marks MAW, Dowman E, Deady E (2018) Fenites associated with carbonatite complexes: a review. Ore Geol Rev 93:38–59

Freyssinet P, Butt CRM, Morris RC, Piantone P (2005) Ore-forming processes related to lateritic weathering. Econ Geol 100th Anniversary Volume:681–722

García AC, Latifi M, Amini A, Chaouki J (2020) Separation of radioactive elements from rare earth element-bearing minerals. Metals 10:1524. https://doi.org/10.3390/met10111524

Giebel RJ, Parsapoor A, Walter BF, Braunger S, Marks MAW, Wenzel T, Markl G (2019) Evidence for Magma–Wall Rock Interaction in Carbonatites from the Kaiserstuhl Volcanic Complex (Southwest Germany). J Petrol 60:1–31

Giovannini AL, Bastos AC, Porto CG, Pereira VP, Takehara L, Barbanson L, Bastos PHS (2017) Mineralogy and Geochemistry of Laterites from the Morro dos Seis Lagos Nb (Ti, REE) Deposit (Amazonas, Brazil). Ore Geol Rev 88:461–480

Giovannini AL, Bastos Neto AC, Porto CG, Takehara L, Pereira VP, Bidone MH (2021) REE mineralization (primary, supergene and sedimentary) associated to the Morro dos Seis Lagos Nb (REE, Ti) deposit (Amazonas, Brazil). Ore Geol Rev 137:104308

Graham S, Lambert D, Shee S (2004) The petrogenesis of carbonatite, melnoite and kimberlite from the Eastern Goldfields Province, Yilgarn Craton. Lithos 76:519–533

Great Western Minerals Group (2008) The integrated Rare Earths producer. Presentation at 4th International Rare Earths Conference, Kowloon, Hong Kong

Harmer R, Nex PAM (2016) Rare earth deposits of Africa. Episodes 39:381–406

Harper F, Venter M, Hall M, Siegfried PR, Brown J, Stripp G, Jager C, Scheepers J, Wiid G, Vivier K, Grobler M (2011) Independent technical report on the results of a preliminary economic assessment of Frontier Rare Earths Rare Earths Project, located in the Northern Cape Province of South Africa. Venmyn Rand NI 43-101, report to Frontier Rare Earths (available from www.sedar.com)

Hornig-Kjarsgaard I (1998) Rare earth elements in sövitic carbonatites and their mineral phases. J Petrol 39: 2105–2121

Hussak E, Prior GT (1899) A new Brazilian mineral, florencite. Nature 61:118–120

Hutchinson MB (2016) REE enrichment in weathered carbonatite, Bull Hill: Bear Lodge Mountains, Wyoming. Doctoral thesis, Colorado School of Mines (unpublished)

Hutchinson MB, Slezak P, Wendlandt R, Hitzman M (2022) Rare Earth Element Enrichment in the Weathering Profile of the Bull Hill Carbonatite at Bear Lodge, Wyoming, USA. Econ Geol 117: 813–831

Jaques AL, Verrall M, Sweetapple MT, Fletcher IR, McNaughton NJ, Rasmussen B (2016) Hydrothermal alteration textures and REE mineralization in the Cummins Range Carbonatite Complex, Kimberley region, W.A. Poster W. Australia Museum. https://doi.org/10.13140/RG.2.2.13124.94084

Jordens A, Cheng YP, Waters KE (2013) A review of the beneficiation of rare earth element bearing minerals. Min Eng 41:97–114

Kasay GM, Bolarinwa AT, Aromolaran OK, Nzolang C, Mambo VS (2021) A review of the geological settings, ages and economic potentials of carbonatites in the Democratic Republic of Congo. Appl Earth Sci 130(3):143–160. https://doi.org/10.1080/25726838.2021.1911585

Keppler H (2003) Water solubility in carbonatite melts. Am Mineral 88:1822–1824

Kravchenko SM, Pokrovsky BG (1995) The Tomtor alkaline ultrabasic massif and related REE-Nb deposits, Northern Siberia. Econ Geol 90:676–689

Lapido-Loureiro FE (1973) Carbonatitos de Angola. Memorias e Trabalhos do Instituto de Investigacao Cientifica de Angola. 11

Le Bas MJ (1987) Nephelinites and carbonatites. Geological Society, London, Special Publications 30:53–83

Le Maître RW (ed) (2002) Igneous rocks: a classification and glossary of terms. Cambridge University Press, Cambridge, UK, p 193

Liu Y, Guo D (2019) Occurrence and geochemistry of bastnäsite in carbonatite-related REE deposits, Mianning–Dechang REE belt, Sichuan Province, SW China. Ore Geol Rev 107:266–282

Lottermoser BG (1990) Rare-earth element mineralization within the Mt. Weld carbonatite laterite, Western Australia. Lithos 24:151–167

Lynas (2021) Financial Report for the Year 2020–2021, Lynas Corporation, 99 pp

Maksimovic ZJ, Panto G (1996) Authigenic rare earth minerals in karst-bauxites and karstic nickel deposits. In: Jones AP, Wall F, Williams CT (eds) Rare Earth Minerals: chemistry, origin and ore deposits, Mineralogical Society Series 7. Chapman and Hall, London, UK, pp 257–279

Marsh JS (1973) Relationships between transform directions and alkaline igneous rock lineaments in Africa and South America. Earth Planet Sci Lett 18:317–323

Mitchell RH (2005) Carbonatites and carbonatites and carbonatites. Can Mineral 43:2049–2068. https://doi.org/10.2113/gscanmin.43.6.2049

Moore AE, Verwoerd WJ (1985) The olivine melilitite—"kimberlite"—carbonatite suite of Namaqualand and Bushmanland, South Africa. South Afr J Geol 88:281–294

Neumann R, Medeiros EB (2015) Comprehensive mineralogical and technological characterisation of the Araxá (SE Brazil) complex REE (Nb-P) ore, and the fate of its processing. Int J Min Process 144:1–10

Ni Y, Hughes JM, Mariano AN (1995) Crystal chemistry of the monazite and xenotime structures. Am Min 80:21–26

Niegisch M, Kamradt A, Borg G (2020) The upper zone of the Storkwitz Carbonatite, Geochemical and mineralogical characterization of the REE-mineralization in the upper zone of the Storkwitz Carbonatite Complex Geoprofil, Heft 15 from drill core SES-1/2012, 91 pp

Ntiharirizwa S, Boulvais P, Poujol M, Branquet Y, Morelli C, Ntungwanayo J, Midende G (2018) Geology and U-Th-Pb Dating of the Gakara REE Deposit, Burundi. Minerals 8:394. https://doi.org/10.3390/min8090394

Ogungbuyi PI, Janney PE, Harris C (2016) Radiogenic and stable isotope constraints on the source regions of Phanerozoic carbonatites and associated igneous rocks from western South Africa and southern Namibia. Poster presentation at International Geological Congress 35 Cape Town, South Africa

Ogungbuyi PI, Janney PE, Harris C (2022) Carbonatite, aillikite and olivine melilitite from Zandkopsdrift, Namaqualand, South Africa: constraints on the origin of an unusual lamprophyre-dominated carbonatite complex and the nature of its mantle source. Lithos, pp 418–419

Onuonga IO, Bowden P (2000) Hot-spring and supergene lanthanide mineralization at the Buru carbonatite centre, Western Kenya. Mineral Mag 64:663–673

Pell J (1994) Carbonatites, nepheline syenites, kimberlites and related rocks in British Columbia; British Columbia Ministry of Energy, Mines and Petroleum Resources, Bulletin 88, 133 pp
Pell RS, Wall F, Yan X, Bailey G (2019) Applying and advancing the economic resource scarcity potential (ESP) method for rare earth elements. Resour Policy 62:472–481
Pickford M (2016) Cenozoic geology of the Northern Sperrgebiet, Namibia. Commun Geol Surv Namibia:10–104
Retallack GJ (2010) Lateritization and bauxitization events. Econ Geol 105:655–667
Ross V, Singh A, Pillay K (2019) Improved flotation of PGM tailings with a high-shear hydrodynamic cavitation device. Min Eng 137:133–139
Sanematsu K, Watanabe Y (2016) Characteristics and genesis of ion adsorption-type rare earth element deposits. In: Verplanck PL, Hitzman MW (eds) Society of economic geologists reviews in economic geology, vol 18, pp 55–79
Smith MP, Moore K, Kavecsánszki D, Finch AA, Kynicky J, Wall F (2016) From mantle to critical zone: a review of large and giant-sized deposits of the rare earth elements. Geosci Front 7:315–334
Swinden HS, Siegfried P (2011) Amended 43-101 technical report on the rare earth element occurrences in the Lofdal carbonatite complex, Kunene region, Khorixas district, Namibia, Technical Report for Namibia Rare Earths Ltd, 2011 (available from www.sedar.com)
Tassinari MML, Kahn H, Ratti G (2001) Process mineralogy studies of Corrego do Garimpo REE Ore, Catalao-1 Alkaline Complex, Goias, Brazil. Min Eng 14:1609–1617
Taylor G, Butt CRM (1998) The Australian regolith and mineral exploration. AGSO J Aust Geol Geophys 17(4):55–67
Toledo MCM, Lenharos SLR, Ferrari VC, Fontan F, de Parseval P (2004) The compositional evolution of apatite in the weathering profile of the Catelao 1 Alkaline carbonatitic complex, Goias, Brazil. Can Mineral 42: 1139–1158
U.S. Geological Survey (2021) Mineral Commodity Summaries 2021. https://doi.org/10.3133/mcs2021
Van Rythoven AD, Pfaff K, Clark JG (2020) Use of QEMSCAN ®to characterize oxidized REE ore from the Bear Lodge carbonatite, Wyoming, USA. Ore and Energy Resource Geology, 2–3
Verbaan N, Bradley K, Brown J, Mackie, S (2015) A review of hydrometallurgical flowsheets considered in current REE projects. In: Simandl GJ, Neetz M (eds) Symposium on Strategic and Critical Materials Proceedings, November 13–14, 2015, Victoria, British Columbia. British Columbia Ministry of Energy and Mines, British Columbia Geological Survey Paper: 147–162
Vergara R, Burkhalter E (2019) Matamulas Rare Earth Project in Spain. Poster presentation at Second European Rare Earth Resource Conference, Santorini
Verplanck PL, Van Gosen BS, Seal RR, McCafferty AE (2014) A Deposit Model for Carbonatite and Peralkaline Intrusion-Related Rare Earth Element Deposits; U.S. Geological Survey Scientific Investigations Report 2010–5070-J. U.S. Geological Survey, Reston, 58 pp
Verplanck PL, Mariano AN, Mariano AN Jr (2016) Rare earth element ore geology of carbonatites. In: Verplanck PL, Hitzman MW (eds) Society of economic geologists reviews in economic geology, vol 18, pp 5–32
Verplanck PL, Andersen AK, Van Gosen BS (2020) Whole-rock geochemical data for the Bear Lodge alkaline complex, Wyoming: 2009–2013: U.S. Geological Survey data release. https://doi.org/10.5066/P9RCPAMC
Verwoerd WJ (1967) The carbonatites of South Africa and South-West Africa. Handbook 6. Department of Mines and Geological Survey of South Africa, 452 pp
Wall F (2014) Rare earth elements. In: Gunn AG (ed) Critical metals handbook. Wiley, pp 312–339
Wall F, Williams CT, Woolley AR, Nasraoui M (1996) Pyrochlore from Weathered Carbonatite at Lueshe, Zaire. Mineral Mag 60:731–750
Wall F, Niku-Paavola VN, Storey,C, Müller A and Jeffries T (2008) Xenotime-(Y) from carbonatite dykes at Lofdal, Namibia: unusually low LREE: HREE ratio in carbonatite, and the first dating of xenotime overgrowths on zircon. The Canadian Mineralogist 46:861–877
Walter BF, Parsapoora A, Braunger S, Marks MAW, Wenzel T, Martin M, Markl G (2018) Pyrochlore as a monitor for magmatic and hydrothermal processes in carbonatites from the Kaiserstuhl volcanic complex (SW Germany). Chem Geol 498:1–16
Weng Z, Jowitt SM, Mudd GM, Haque N (2015) A detailed assessment of global rare earth element resources: opportunities and challenges. Econ Geol 110:1925–1952
Willett GC, Duncan RK, Rankin RA (1989) Geology and economic evaluation of the Mount Weld carbonatite, Laverton, Western Australia. In: Ross J (ed) Kimberlites and Related Rocks Volume 2. Geological Society of Australia Special Publication 14, pp 1215–1235
Wimmenauer W (2003) Kaiserstuhl. In: Geologische Karte Von Baden-Württemberg 1: 25.000 Mit Erlauterungen. Landesamt für Geologie, Rohstoffe und Bergbau Baden-Württemberg, Freiburg
Witt WK, Hammond DP, Hughes M (2019) Geology of the Ngualla carbonatite complex, Tanzania and the origin of the weathered bastnäsite zone REE ore. Ore Geol Rev 105:28–54
Woolley AR (1982) A discussion of carbonatite evolution and nomenclature, and the generation of sodic and potassic fenites. Mineral Mag 46:13–17
Woolley AR, Kempe DRC (1989) Carbonatites: nomenclature, average chemical compositions, and element distribution. In: Bell K (ed) Carbonatites: genesis and evolution. Unwin Hyman Ltd, pp 1–14

Woolley AR, Kjarsgaard BA (2008) Paragenetic types of carbonatite as indicated by the diversity and relative abundances of associated silicate rocks—evidence from global database. Can Mineral 46: 741–752

Xu F, You X, Li Q, Liu Y (2019) Can primary Ferroan Dolomite and Ankerite be precipitated? Its implications for formation of submarine Methane-Derived Authigenic Carbonate (MDAC) chimney. Minerals 9:413. https://doi.org/10.3390/min9070413

Yaxley GM, Anenburg M, Tappe S, Decree S, Guzmics T (2022) Carbonatites: classification, sources, evolution, and emplacement. Annual Rev Earth Planet Sci 50:261–293

Zhang J, Edwards C (2013) Mineral decomposition and leaching processes for treating rare earth ore concentrates. Can Metal Quart 52:243–248. https://doi.org/10.1179/1879139513Y.0000000084

Zhukova IA, Stepanov AS, Jiang S-Y, Murphy D, Mavrogenes J, Allen C, Chen W, Bottrill R (2021) Complex REE systematics of carbonatites and weathering products from uniquely rich Mount Weld REE deposit, Western Australia. Ore Geol Rev 139. https://doi.org/10.1016/j.oregeorev.2021.104539

Regolith Hosted Niobium Deposits in Carbonatite Complexes in Brazil

Claudio G. Porto, Artur C. Bastos Neto, and João P. Proença Bento

Abstract

The total supergene niobium resource associated with carbonatite complexes in Brazil is close to 880 Mt, with most coming from five deposits in the Alto Paranaiba Igneous Province (APIP) in Central Brazil. This resource might be increased considerably if the speculated 3.8 Bt resource of the Seis Lagos deposit in the Amazon is included. The carbonatites are all covered by a lateritic regolith that may be 50 to 250 m thick. The carbonatites of the APIP are of Late-Cretaceous age and are rimmed by fenitized quartzite and schists with a marked topographic expression that protects the lateritic regolith from erosion. Niobium mineralization is mainly in the saprolite as secondary Ba-pyrochlore or residually enriched primary pyrochlore. The upper ferruginous parts of the regolith are generally barren. The Seis Lagos carbonatite complex is of a probable Precambrian age and consists of siderite carbonatite. The host gneissic rocks are not preserved in the landscape, but the more resistant Fe duricrusts that contain the Nb deposit form a prominent hill. Despite intense alteration of pyrochlore in the siderite carbonatite, Nb concentrations remain high in most of the Fe duricrusts due to its scavenging by secondary oxides, in particular goethite, brookite and rutile.

1 Introduction

Brazil holds 88% of the world's Nb resources, with 94.5% of global Nb production derived from the Igneous Alkaline Province of Alto Paranaíba (APIP). The main occurrences of alkaline carbonatite complexes in Brazil are shown in Fig. 1, highlighting the APIP. In the northwest, the Seis Lagos carbonatite complex is overlain by a thick regolith cover that hosts the world's largest Nb resource.

The Brazilian Nb production derives from three mines in the APIP where the Nb ore is extracted from residual accumulations in the regolith over the Araxá, Catalão I and Catalão II carbonatite complexes. Two other carbonatite complexes in this Province, Tapira and Salitre-Serra Negra, host non-producing regolith Nb deposits. The Nb resources and other geological

C. G. Porto (✉) · J. P. Proença Bento
Programa de Pós-graduação em Geologia, Instituto de Geociências, Universidade Federal do Rio de Janeiro, Rio de Janeiro, RJ, Brazil
e-mail: porto@geologia.ufrj.br

A. C. Bastos Neto
Programa de Pós-graduação em Geociências, Instituto de Geociências, Universidade Federal do Rio Grande do Sul, Porto Alegre, RS, Brazil

R. J. Bowell, C. R. M. Butt (eds.), *Geology, Geochemistry and Formation of Supergene Mineral Deposits in Deeply Weathered Terrain*, Mineral Resource Reviews, https://doi.org/10.1007/978-3-031-75733-4_8

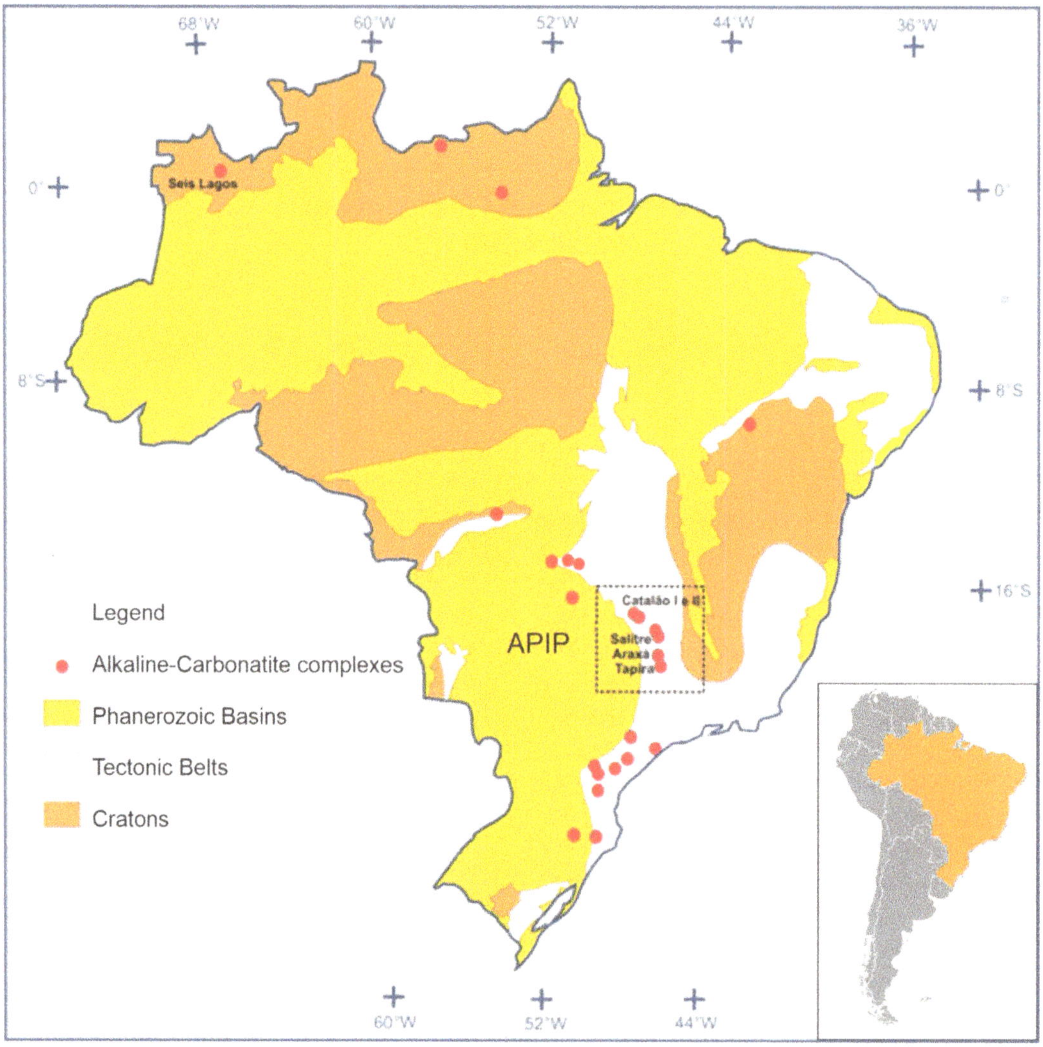

Fig. 1 Main occurrences of Alkaline-Carbonatite Complexes in Brazil highlighting the ones mentioned in this study as part of the Alto Paranaiba Igneous Province (APIP) and the Seis Lagos carbonatite Complex in the extreme North. (After Ribeiro et al. 2014; Seer and Moraes 2018)

characteristics of these deposits are listed in Table 1. The Seis Lagos carbonatite complex has a very diverse nature, and the most recent research work conducted in the area is presented here. However, further exploration and mining are not allowed because it is located within an Indigenous Territory, within a National Park, and within a Biological Reserve. It has never been exploited. The National Park and the Biological Reserve were created after the discovery of the deposit in the 1970s. In this paper, the main geological features of the APIP and Seis Lagos Nb deposits are presented, including their regolith setting and Nb ore mineralogy and a discussion of their contrasting characteristics.

2 The Niobium Deposits of the APIP

2.1 Geological Setting

The APIP is a NW trending province of Late-Cretaceous age composed of alkaline igneous

Table 1 Niobium deposits hosted in the regolith over carbonatite complexes in Brazil

Deposit	Resources	Size of complex	Regolith thickness	Associated resources	Main carbonatite rock
Seis Lagos (a)	10.5–28.3 Mt@ 2.97% Nb_2O_5 1.4–3.8 Bt @ 1.17 wt% Nb_2O_5	11 km^2	>250 m	Ti; REE	Siderite carbonatite
Araxá (b, c,d)	462 Mt@2.48% Nb_2O_5 Primary: 940 Mt@1.6% Nb_2O_5 (j)	16 km^2	~200 m	Phosphate; barite; REE	Beforsites, Sovites
Tapira (b,e)	166 Mt @0.73% Nb_2O_5	36 km^2	>200 m	Phosphate; Ti	Sovites
Salitre (f)	196 Mt @0.48% Nb_2O_5	100 km^2	130 m	Phosphate; Ti	Sovites
Catalão I (g,h)	19 Mt@1.08% Nb_2O_5 Primary:29 Mt@1.22% Nb_2O_5	36 km^2	80 m	Ti; REE; vermiculite	Sovites
Catalão II (i)	3.5 Mt 1.67% Nb_2O_5	14 km^2	50 m		Sovites

(a) Bento et al. (2022); (b) Ribeiro et al. (2014); (c) Schobbenhaus (1986); (d) Murilo and Gaspar (1995); (e) Melo et al. (1997); (f) Biondi (2005); (g) Cordeiro et al. (2011); (h) Hirano et al. (1990); Rodrigues and Lima (1984); (i) Rocha et al. (2001); (j) Issa Filho et al. (1984); Silva (1986)

intrusive rocks associated with carbonatite complexes (Fig. 1). The primary igneous rocks comprising these complexes are mostly alkaline-ultramafic rocks cut by multiple generations of veins and larger masses of carbonatitic rocks. These complexes generally occur in shapes varying from circular to oval and elongated. They are rimmed by fenitized quartzite and schists, having a strong topographic expression. The most important associated commodities in the regolith include phosphate, titanium, rare earth, vermiculite, barite and magnetite. Thus, the APIP is of great economic interest and can provide key information for mineral exploration of carbonatite-related deposits.

2.2 Regolith

The regolith that covers these complexes is invariably thick, generally from 50 m to more than 200 m, and protected from erosion by the fenitized quartzite and schists that rim the complexes. The regolith surface is dominantly flat but tends to be lower towards the centre of the complex and may be dominated by transported materials such as lake sediments or alluvium-colluvium. The regolith generally constitutes a highly leached and barren lateritic upper zone composed of ferruginous duricrust and soil that varies in thickness from a couple of metres to a few tens of metres. This is underlain by a transition to the saprolite, which dominates the regolith with a thickness that may exceed 100 m in places. This is the zone in which most of the Nb and phosphate ore minerals are residually enriched and where carbonatite vein structures are preserved. The transition to fresh rock constitutes a zone dominated by fractures and incipiently weathered rock that may extend several tens of metres into fresh rock.

2.3 Niobium Mineralization

Niobium ore minerals are invariably derived from the alteration of primary Na-Ca-pyrochlore generally associated with phoscorite, dolomite, and calcic carbonatite (sovites). The residual accumulation of these minerals is the main cause of Nb enrichment in the saprolite. The most common Nb ore mineral is supergene Ba-pyrochlore, which is more resistant to weathering than its primary counterpart. This mineral, formerly known as pandaíta $(Ba, Sr)_2(Nb, Ti)_2(O, OH)_7$, defined by Hogarth (1977), is today considered a variety of pyrochlore that is rich in Ba and Sr and has been renamed as "Ba-enriched" kenopyrochlore according to the nomenclature of Atencio et al. (2010).

3 The Niobium Deposit of the Araxá Carbonatite Complex

3.1 Geology and Resources

This is the best-known complex of the APIP. Phosphate ore was identified in the 1940s, and Nb in 1953. The Barreiro mine produces 78.3% of the Nb output in Brazil. The resources are estimated at 462 Mt @ 2.5% Nb_2O_5 (Ribeiro et al. 2014; Paraiso and Fucio 1982). Primary Nb ore is of the order of 940 Mt@1.6% Nb_2O_5 (Issa Filho et al. 1984; Hirano et al. 1990; Silva 1986) but has not been mined. Other mineral resources associated with the regolith include phosphate resource of 460 Mt @15.07% P_2O_5; barite as residual hydrothermal veins, disseminations, and concretions are estimated at 463 Mt @ 20.67% $BaSO_4$; potential REE resources associated with silexite veins and disseminations of monazite (Seer and Moraes 2018).

The domed elongate shape of the Complex occupies a surface area of 16 km^2 and has a regolith thickness of about 200 m (Fig. 2). The central plug is dominated by dolomite

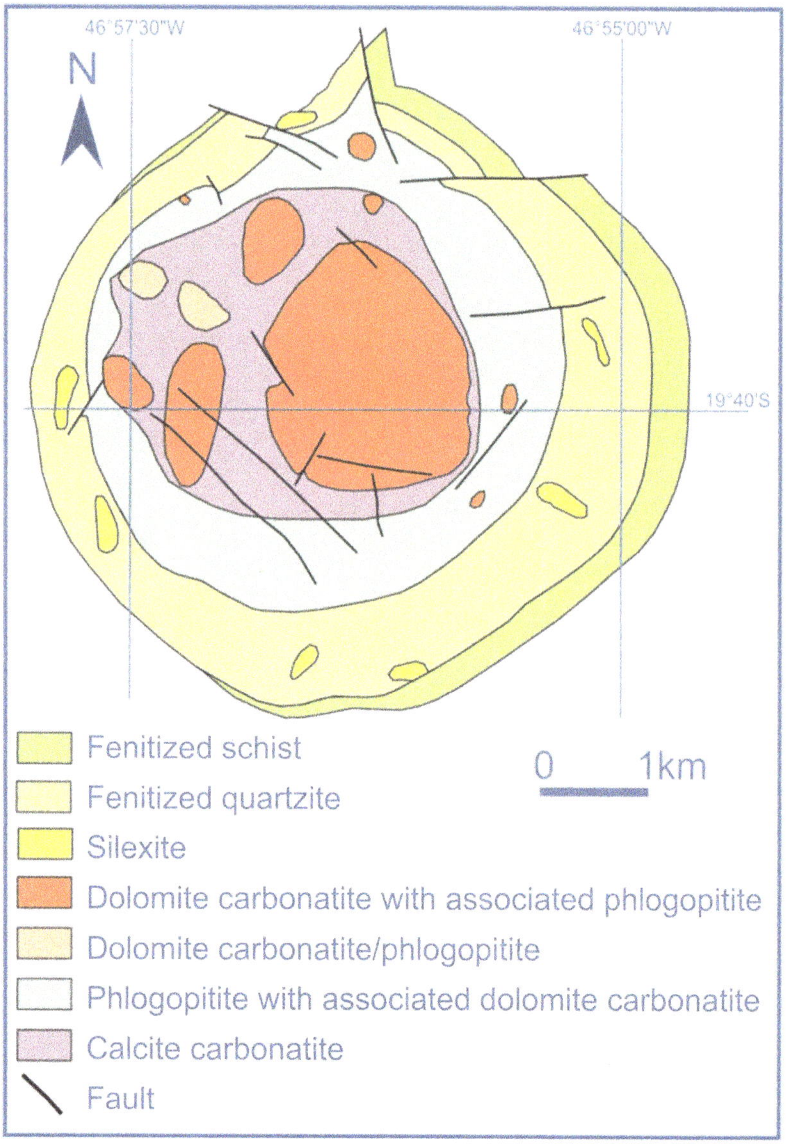

Fig. 2 Geological map of the Araxá Carbonatite Complex (Silva et al. 1979; Seer and Moraes 2018)

carbonatites, phoscorites and sovites (Silva 1986). An aureole of metasomatized ultramafic rock (glimmerites), cut by carbonatites and phoscorites, occurs between this nucleus and the fenitized host rock. Metasomatic phlogopitite veins are volumetrically significant. The Nb and phosphate ores are concentrated residually over the carbonatites and phoscorite, where primary concentrations of apatite and pyrochlore are high and where weathering has been most intense.

3.2 Regolith

In the Nb mine, the regolith profile is described by Silva (1986) from top to bottom as: (i) Top soil: reddish brown magnetitic clays, 0.5 to 2 m thick; (ii) Transition zone: the soil grades downwards into a red to yellow clay-rich and magnetitic material with an average thickness of 25 m; (iii) Main ore zone—Saprolite: the Ba-pyrochlore mineralized layer with an average thickness of 100 m. It is a ferruginous brown to yellow, Mn-bearing saprolite containing mica flakes and magnetite concentrated in veins generally related to silexite masses that may reach a few 10 s of metres, sometimes crosscut by barite veinlets, in abrupt contact with (iv) Saprock: a 10 to 40 m thick green-grey zone containing fragments of fresh carbonatite rock, silexite, and mineral aggregates surrounded by a dark grayish clay-rich matrix. It is fractured and contains numerous cavities left by carbonate dissolution, with a sharp basal contact to (v) Fresh carbonatite.

Apatite is unstable and releases P to reprecipitate as neoformed apatite at the base of the regolith, or it may remain in the upper parts of the regolith as Al-phosphates. Gierth and Baecker (1986) describe Ba-p*yrochlore* preserved in the lateritic soil as fine grains enveloped by magnetite. The phosphate ore in the regolith occurs in a lower 120 m-thick zone of the saprolite (Torres and Gaspar 1995). This zone comprises phlogopite with apatite and magnetite veins, in which textures are generally better preserved than the saprolite matrix, especially in its middle and lower parts. In the upper saprolite, apatite is substituted by secondary Al-phosphates, and phlogopite transforms into vermiculite and then kaolinite at the top.

3.3 Niobium Mineralization

The principal primary Nb ore mineral is Ca-Na-pyrochlore (Ribeiro et al. 2014). Upon weathering, it transforms into Ba-pyrochlore with little change in the Nb_2O_5 content. The resultant Ba-pyrochlore is a neoformed resistate mineral and is, therefore, efficiently accumulated residually in the saprolite. As well as Ba-pyrochlore, the ore paragenesis in the regolith includes Fe oxy-hydroxides, barite, magnetite, gorceixite, monazite, ilmenite and quartz (Issa Filho et al. 1984).

4 The Niobium Deposit of the Catalão I Carbonatite Complex

4.1 Geology and Resources

The Catalão I complex intruded Late Proterozoic fenitized quartzites and schists. It has a circular shape and occurs as a vertical pipe with a surface area of 36 km^2 and a domelike structure. Cordeiro et al. (2011) divide the complex into an outer zone dominated by phlogopitite and an inner zone composed mostly of dolomite carbonatites and phoscorite-series rocks (Fig. 3). The pipe-like Nb orebodies consist of dyke swarms of late-stage phoscorite-series rocks (apatite ilmenite-rich nelsonites) that cut previous phlogopitite and carbonatite. Carvalho and Bressan (1997) describe carbonatite veins up to 20 m thick and rich in phlogopite. The regolith is about 80 m thick and includes TiO_2, REE, phosphate and vermiculite as associated commodities. The complex is capped by a 10 to 70 m thick barren cover. Mining of the regolith-hosted Catalão I Nb deposit started in 1976 with a reserve of 19 Mt at 1.08 wt.% Nb_2O_5. Resources in the fresh rock are 29Mt@1.22% Nb_2O_5.

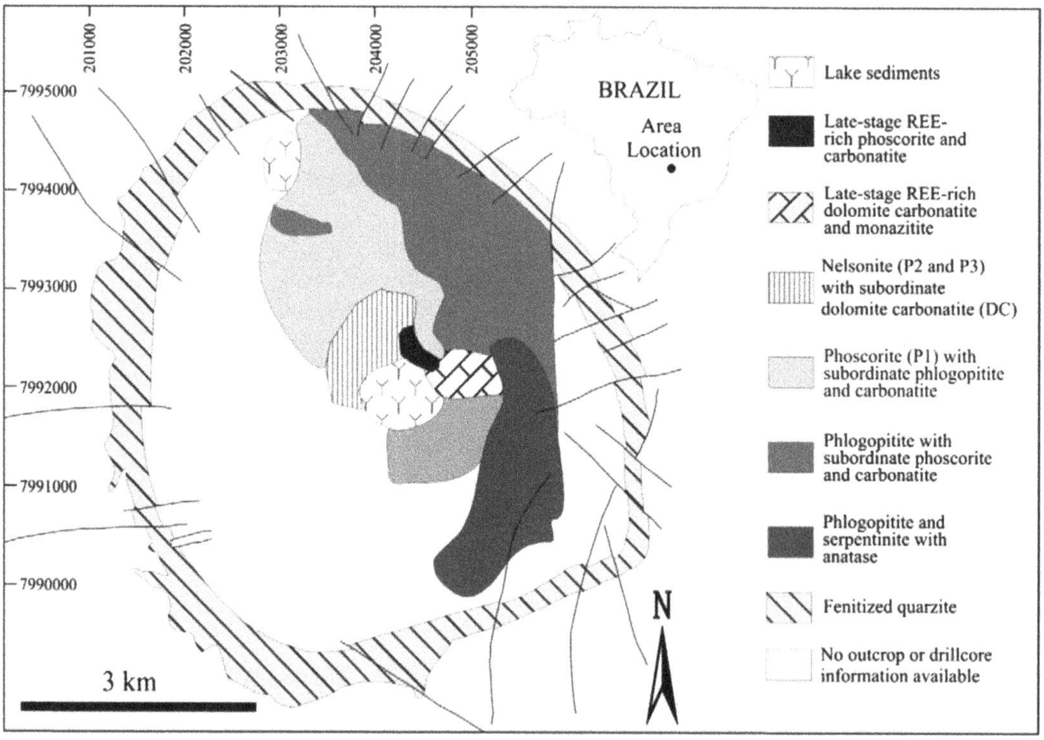

Fig. 3 Geological map of the Catalão I Complex. The fresh rock niobium deposit is in the centre of the complex comprising the nelsonite unit. (Adapted from Brod et al. 2004)

4.2 Regolith

The regolith in the phosphate mine is described by Toledo et al. (2004) and is schematically represented in Fig. 4. The thickness and terminology follow the conventions of Ribeiro et al. (2014). The fresh rock is found generally below 100 m depth, although weathering features can be found even deeper than 100 m. It consists of carbonatite and phlogopitite in variable proportions, with subordinate phoscorite. It is overlain by saprock, in which the phosphate ore constitutes siliceous carbonated fresh blocks within friable material. This grades up into the saprolite, which hosts most of the phosphate and Nb ore. This comprises micaceous ore where micas are better preserved than the saprolite clay-rich matrix. Oliveira and Imbernon (1998) recognized a lower portion where a less weathered grey saprolite contains large amounts of phlogopite, a middle portion where phlogopite is partially preserved, and an upper clay-rich portion where primary textures are no longer preserved and where apatite is substituted by secondary Al-phosphates and phlogopite transforms into vermiculite and then kaolinite at the top (Torres and Gaspar 1995). Above is a ferruginous zone with a more homogeneous composition in which the phosphate ore is oxidized. It is composed of a loose reddish-yellow sandy material with occasional relict silicified carbonatitic veins. There is a sharp contact with an overlying barren reddish argillaceous overburden. At the top is a red soil with minor relicts of Ba-pyrochlore preserved as fine grains enveloped by magnetite (Gierth and Baecker 1986).

4.3 Niobium Mineralization

Fava (2001) investigated the mineralogy of pyrochlore and concluded that weathering induced Ba substitution in the A site of residual primary pyrochlore, resulting in Ba-pyrochlore in

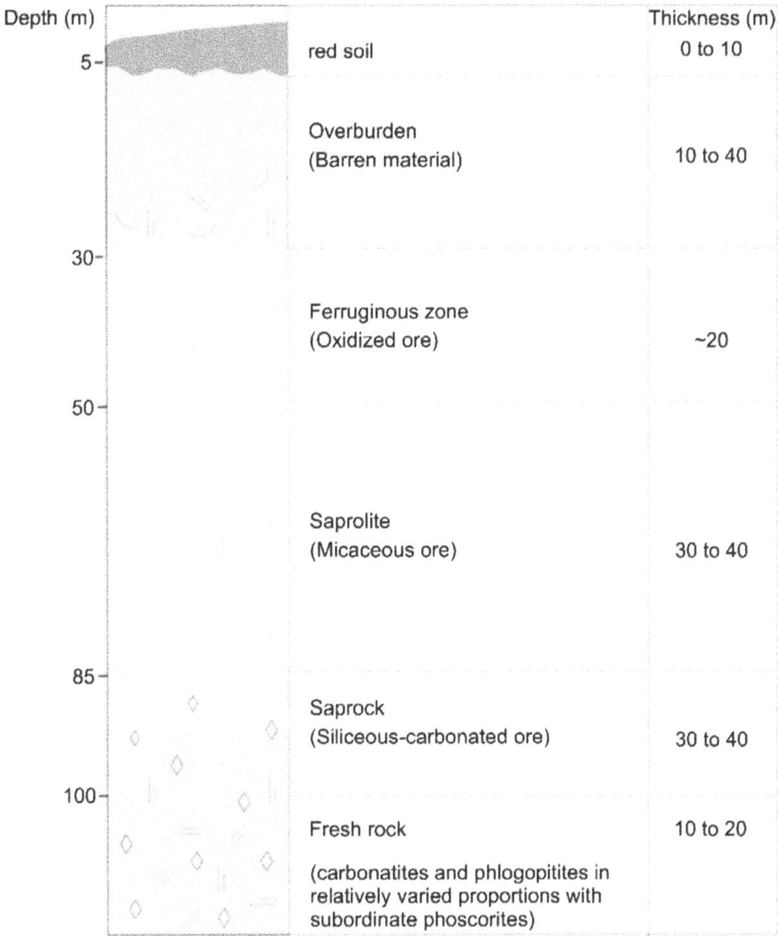

Fig. 4 Regolith profile of the Catalão I Complex. (Modified from Toledo et al. 2004)

the regolith. He also found Ba-enriched kenopyrochlore in the primary carbonatite, but these show a negative Sr–Ca correlation that leads toward Sr-enriched kenopyrochlore. This correlation has also been observed in a few regolith pyrochlore crystal cores, but the majority of pyrochlore crystals in the regolith lack the negative Sr–Ca correlation, suggesting that the processes responsible for the formation of primary kenopyrochlore differ from those in the regolith. Kenopyrochlore in the primary rock formed by interaction with Sr-bearing hydrothermal fluids. This interpretation was supported by oxygen isotope studies of the associated carbonates. The residual Ba-enriched kenopyrochlore formed by depleting Ca and Na from primary pyrochlore. As weathering progressed, Ba was eventually leached from pyrochlore, leading to its destruction and subsequent formation of secondary Nb-enriched minerals in the soil (Wall et al. 1999).

5 The Niobium Deposit of the Tapira Carbonatite Complex

5.1 Geology and Resources

The Tapira carbonatite Complex has an elliptical shape, with an area of 36 km^2 and a regolith thickness of about 200 m (Melo 1997). It has been mined for phosphate since the 1970s, and although it contains TiO_2 resources in the upper parts of the regolith with grades of up to 15%, these cannot be recovered commercially (Seer and Moraes 2018). The Nb resources are estimated at 166 Mt @ 0.73% Nb_2O_5 (Ribeiro et al. 2014).

Primary rock is mostly comprised of ultramafic intrusions cut by carbonate veins in the north part of the complex and carbonatite bodies in the central southern part. Carbonatites represent approximately 10% of all primary rocks and are mostly sovites (Melo et al. 1997). Studies by Brod (1999) in fresh rock did not detect the presence of nelsonites with primary Nb mineralization as in Catalão I and II. A fenitization aureole occurs restricted to the contact area with the hosting quartzites. The contact with the ultramafic host rocks produced metasomatic phlogopitites (phlogopite + magnetite + dolomite). Dolomitic carbonatites are restricted to late veins.

5.2 Regolith

Melo (1997) and Seer and Moraes (2018) described the distribution of the ore zones within the regolith (Fig. 5). A barren surface zone composed of reddish-light brown clays and Fe oxy-hydroxides overlies the greenish Ti-rich mineralized zone. There is a transition between this and the phosphate ore below. The Nb mineralization is more restricted to the central part of the complex and tends to occur between the Ti-rich zone and the underlying phosphate ore.

5.3 Niobium Mineralization

Pyrochlore in the unweathered carbonatite in the central part of the complex occurs associated with La and Ce and, to a minor extent, Ti and P (Melo et al. 1997). Niobium grades may reach 1.2% Nb_2O_5. In the regolith, Nb is residually enriched, and the grades may reach 5% Nb_2O_5.

6 The Niobium Deposit of the Catalão II Carbonatite Complex

6.1 Geology and Resources

The Nb resources are estimated to be about 3.5 Mt @ 1.67% Nb_2O_5 (Rocha et al. 2001). The complex has an elongated elliptical shape and is divided into two pipe-like structures about 2 km apart. A wide variety of carbonatitic rocks are present in a complicated pattern of cross-cutting veins and dyke-like structures within the pipes. This contrasts with the comparatively simple dome-

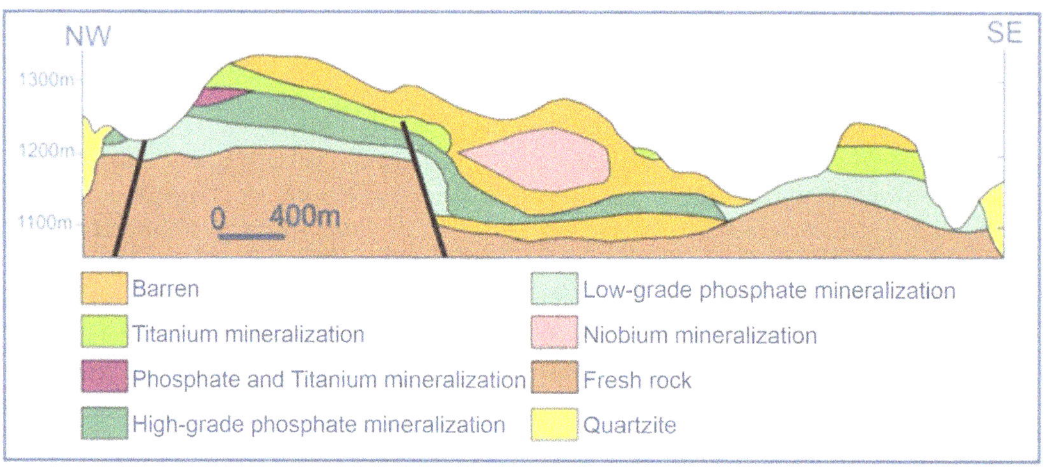

Fig. 5 Geological section of the mineralized zones in the regolith over the Tapira Carbonatite Complex (Melo 1997; Seer and Moraes 2018)

like structure of the neighboring and better-known Catalao I complex. Rocha et al. (2001) identified five types of carbonatite veins with a thickness varying from centimeters to meters. Their composition varies from sovite, silico-sovite, dolomitic, ankeritic, and a sovitic-phoscoritic dyke that is coarse-grained and distinctly rich in pyrochlore which can account for as much as 20% of the rock volume. It is associated with apatite, which may reach nearly 50% of the rock volume, and magnetite, ilmenite, and phlogopite.

6.2 Regolith

There is a general lack of information about the geology of the regolith, except that it is relatively thinner, about 50 m, than the other complexes described here. However, Nb mineralization in the regolith is apparently similar, with the main Nb ore mineral being Ba-pyrochlore.

6.3 Niobium Mineralization

Crystallochemical analysis by Rocha et al. (2001) shows that the weathering of primary pyrochlores resulted in major loss of Na, Ca and F. The loss of A-site Ca and Na is partly compensated by Ba and Ce introduction. The increased vacancies are accompanied by hydration. The compositional change from primary to weathered pyrochlore is considerable and occurs between a depth of 30–40 m, corresponding to a horizon of pedogenic silcrete. The precipitation of the silcrete is considered to represent a paleo-water-table level, reflecting drainage conditions radically different than the present.

The exchange reactions between pyrochlore and groundwater are generally consistent with relatively low pH, low activities of Na^+, Ca^{2+}, F^- and elevated activities of Ba^{2+} and/or Ce^{3+}. The availability of Ba in weathering solutions may have been derived from barite dissolution, whereas the Ce might come either from dissolved REE-carbonate minerals or from the pyrochlore itself through selective immobilization of Ce as Ce^{4+}.

7 The Niobium Deposit of the Salitre Carbonatite Complex

The Salitre Complex is relatively less studied than the other Complexes described here. Salitre is close to the neighbouring Serra Negra Complex. Mapping by Melo et al. (1997) shows that both Complexes are circular, with a total area of about 100 km^2. The primary carbonatite of the Salitre Complex is mostly sovite, hosted by ultramafic intrusive rocks. According to Barbosa (2009), nelsonites are rare, and pyrochlore is disseminated in the primary carbonatite but residually concentrated in the regolith. The regolith thickness is close to 130 m, and its resources are similarly distributed to those of Tapira, where the Nb pyrochlore ore is below the Ti-rich zone and above the phosphate ore. Reported Nb resources in the regolith are of the order of 196 Mt @ 0.48% Nb_2O_5 (Biondi 2005).

8 The Niobium Deposit of the Morro dos Seis Lagos Carbonatite Complex

8.1 Geology and Resources

The Seis Lagos Carbonatite Complex consists of three circular bodies discovered in the 1970s by the RADAM Project (Pinheiro et al. 1976). An exploration program conducted by the Brazilian Geological Survey (Viegas Filho and Bonow 1976) in the major body, the Morro dos Seis Lagos (Fig. 6), demonstrated that the hill capped by Fe duricrusts corresponds to the Morro dos Seis Lagos Nb deposit (MSLD). It stands out among the nineteen key examples of Nb deposits associated with carbonatites selected by Chakhmouradian et al. (2015) as being the world's largest Nb deposit and for being the only one associated with siderite carbonatite.

The Nb resource associated with the Fe duricrusts of the MSLD was estimated to be 2.89 Bt @ 2.81 wt% Nb_2O_5 (Justo and Souza 1984, 1986). Bento et al. (2022), based on geological

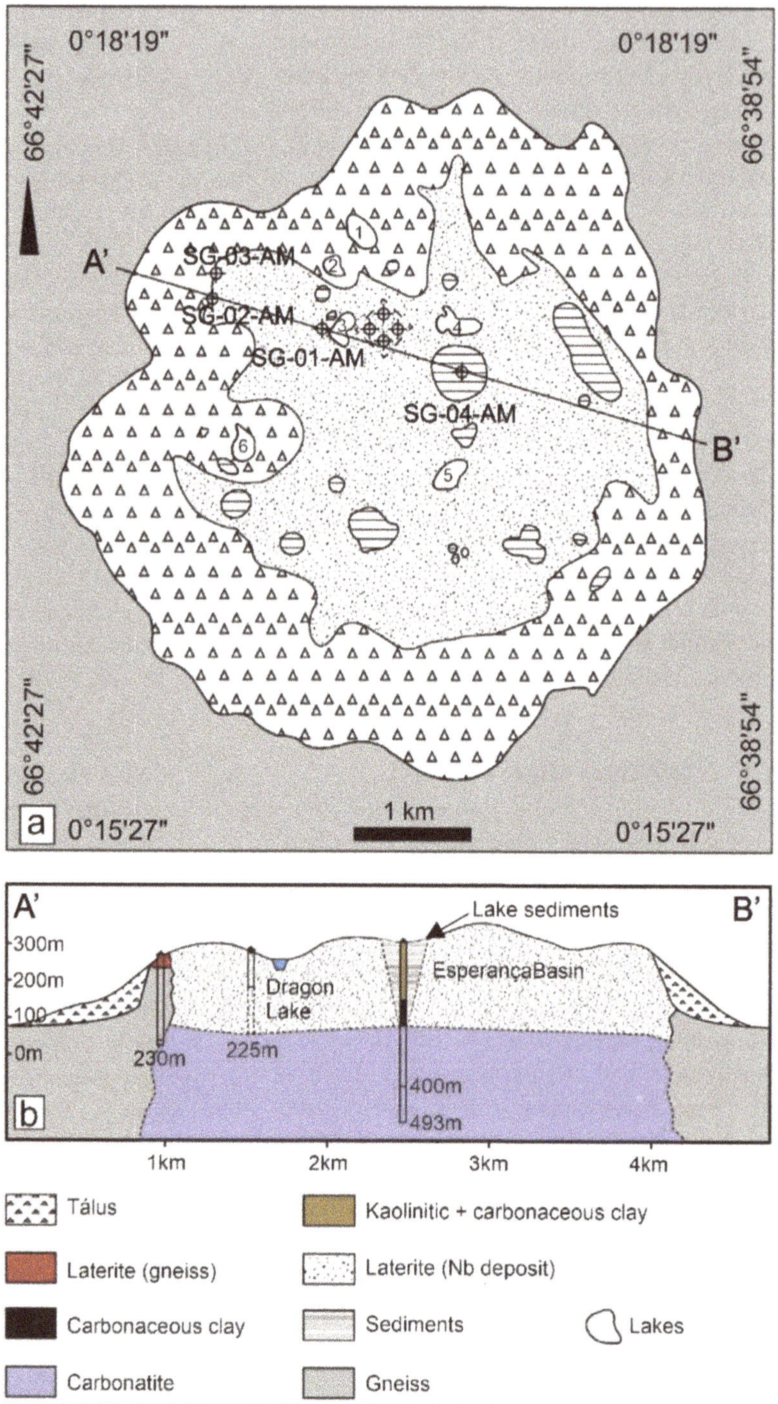

Fig. 6 (**a**) Geological map and (**b**) vertical section of the Morro dos Seis Lagos carbonatite body and associated Nb-deposit. (Modified from Santos et al. 2006; Viegas Filho and Bonow 1976; Giovannini et al. 2020)

modeling of the area and supported by new geochemical analyses of the drill cores that divided the MSLD into two domains: in Domain 1, it has been estimated between 10.5–28.3 Mt @ 2.97 wt % Nb_2O_5 and 0.71 wt% REE_2O_3, and in Domain 2, between 1.40–3.78 Bt @ 1.17 wt% Nb_2O_5 and 0.46 wt% REE_2O_3. High grades of Ti, Sn and W were also found in the lateritic crust. Recent lake-filled sediments are also present in the Seis Lagos Hill. In one of them, the sediments of the

Esperança Basin (Fig. 6b) contain 11.8 Mt @ 1.76 wt% REE_2O_3 and 2.36 wt% P_2O_5, whereas the siderite carbonatite shows high grades of Nb, REE, Sc and Zn, further increasing the area's mineral potential.

The MSLD is located in Guyana Shield, in the Amazonas Craton, in the southeastern part of the Rio Negro Province. The wall rock of the carbonatite body is a ~1.8Ga monzogranitic augen gneiss (Almeida et al. 2007; Santos et al. 2000, 2003, 2006). The age of the emplacement of the carbonatite is uncertain. The U-Pb age of zircons from the carbonatite provided two populations of ages 1.8 Ga and 1.3 Ga (Rossoni et al. 2017). The age of 1.3 Ga suggests an evolution related to the K'Mudku belt. The $^{87}Sr/^{86}Sr$ (0.70411–0.70573) and $^{144}Nd/^{143}Nd$ (0.512663–0.512715) isotopic data suggest the carbonatite is mantle-derived with essentially no crustal contamination and has an age of 200 Ma. The carbonatite complex represents the uppermost parts of a differentiated carbonatite magmatic system, whereas the siderite carbonatite is related to late-magmatic-to-carbo-hydrothermal processes (Giovannini et al. 2020).

The bedrock is known from only two drill holes (Fig. 6b), and in both cases, it is a siderite carbonatite and related saprolite. Three types of siderite carbonatite are recognized: (1) a brecciated and oxidized core siderite carbonatite consisting of up to 95 vol% siderite together with hematite, pyrochlore, Nb-brookite, Ti-maghemite and thorobastnäsite; (2) a REE- and P-rich variety of the core siderite carbonatite consisting of siderite (up to 95 vol%), hematite, minor pyrochlore, monazite and bastnäsite; (3) a border hydrothermal siderite carbonatite with ~70 vol% siderite, barite (~15 vol%), gorceixite (~7 vol%) and minor rhabdophane and pyrochlore.

8.2 Physical Setting

The Seis Lagos hill (elevation approximately 400 m) stands above fairly flat terrain with an average elevation of approximately 90 m. The six lakes occur on Fe duricrusts or, at intermediate altitudes, in talus zones (Fig. 6a). The circular shape of the Morro dos Seis Lagos body is outlined by the base of the talus deposit. However, the shape at a higher elevation resembles a polygon (Fig. 7), suggesting that the intrusion was controlled by brittle structures. The relief on

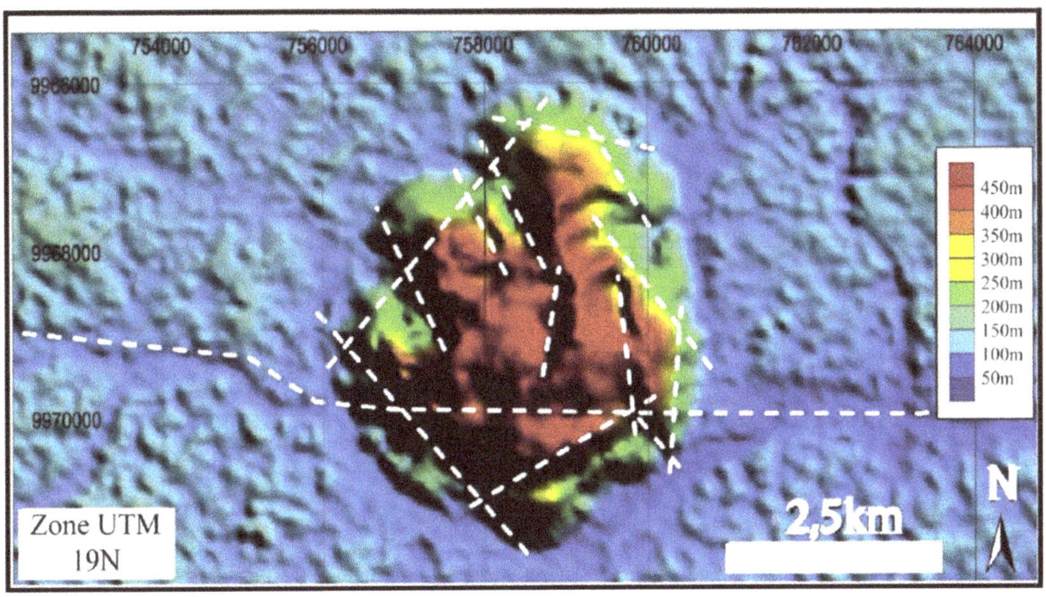

Fig. 7 Digital terrain model of the Morro dos Seis Lagos Hill. N-S, NW and NE showing the faults that control the relief of the deposit. The E-W structure controlled the positioning of the carbonatite body; it was reactivated and affected the southern part of the deposit, where it controls a series of depressions filled by sediments (see Fig. 6b). (Modified from Rossoni et al. 2016; Giovannini et al. 2017)

the hill is controlled by NNE-SSW structures in the NE part, by an NNE-SSW structure in the central part, and by an E-W structure in the southern part (Fig. 7). All the lakes and a number of depressions filled with clays (assigned as sediments in Fig. 6b) are associated with these structures (Rossoni et al. 2016).

8.3 Regolith

The lateritic profile is dominated by residual Fe duricrusts, sub-divided (Giovannini et al. 2017) into six textural and compositional types, which from the top to the base of the sequence are (Fig. 8): (1) pisolitic Fe duricrust (2 and 4)

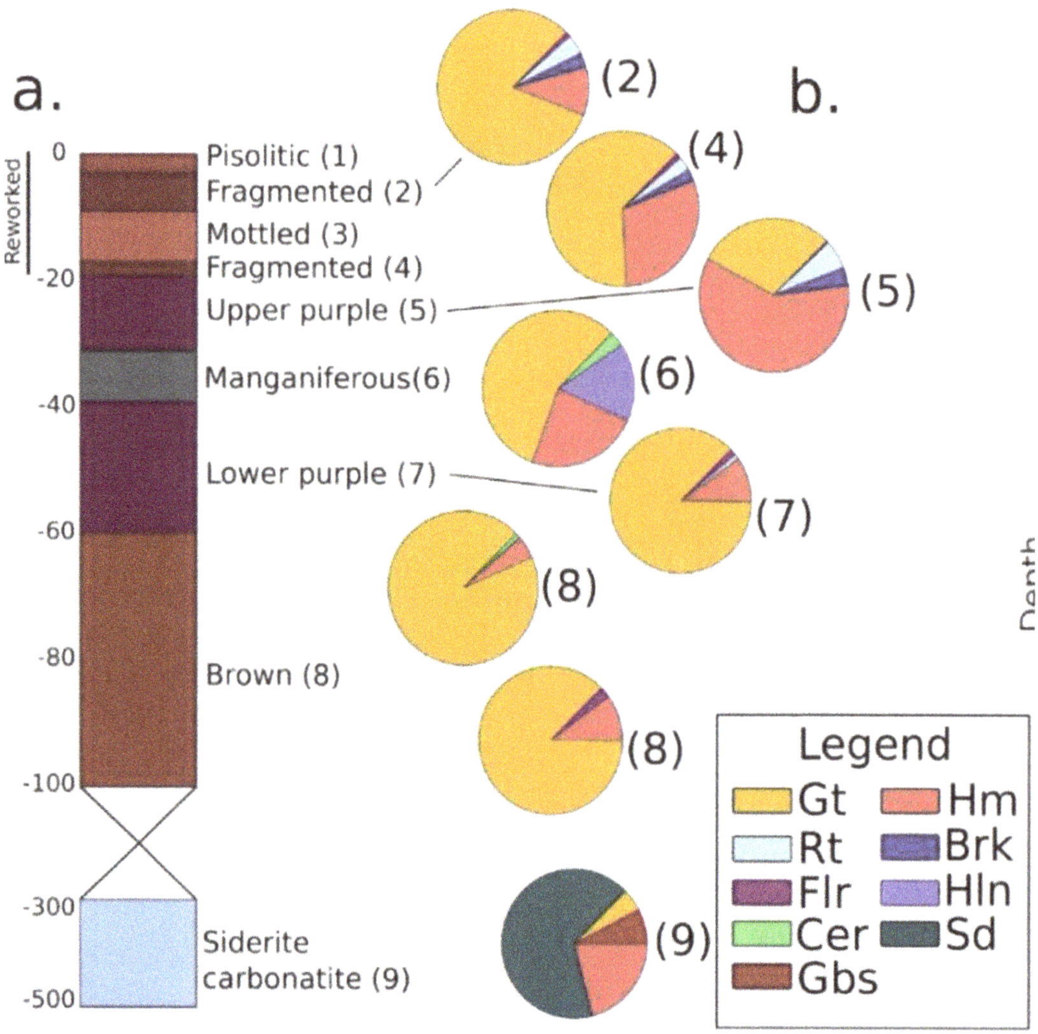

Fig. 8 Mineralogical and geochemical evolution of the lateritic profile of the Morro dos Seis Lagos deposit. (**a**) Schematic evolution of the lateritic profile indicating the names of the different Fe duricrust horizons identified petrologically and their thickness. Horizons (1), (2), (3), (4) were subject to physical and chemical reworking. (**b**) Mineralogical proportions obtained by Rietveld refinement of HR-XRD patterns (Gt: goethite, Hm: hematite, Rt: rutile, Brk: brookite, Flr: florencite, Hln: hollandite, Cer: cerianite, Sd: siderite, Gbs: gibbsite). (Modified from Bollaert et al. 2023)

fragmented Fe duricrust, (3) mottled Fe duricrust, (5 and 7) purple Fe duricrust, (6) manganiferous Fe duricrust and (8) brown Fe duricrust. All Fe duricrusts (Fig. 8) are composed mainly of goethite (predominant in the lower and upper varieties) and hematite (more common in the intermediate types, formed from goethite dehydroxylation). The upper Fe duricrusts have been reworked, resulting in goethite formation. In the manganiferous Fe duricrust (10 m thick), the Mn oxides (mainly hollandite, with associated cerianite) occur as veins or irregular masses, formed in a late event during the development of the Fe duricrusts, precipitated from a solution with higher oxidation potential than that for Fe oxides, closer to the water table. Representative examples of the siderite carbonatite and the main iron crust textures are presented in Fig. 9.

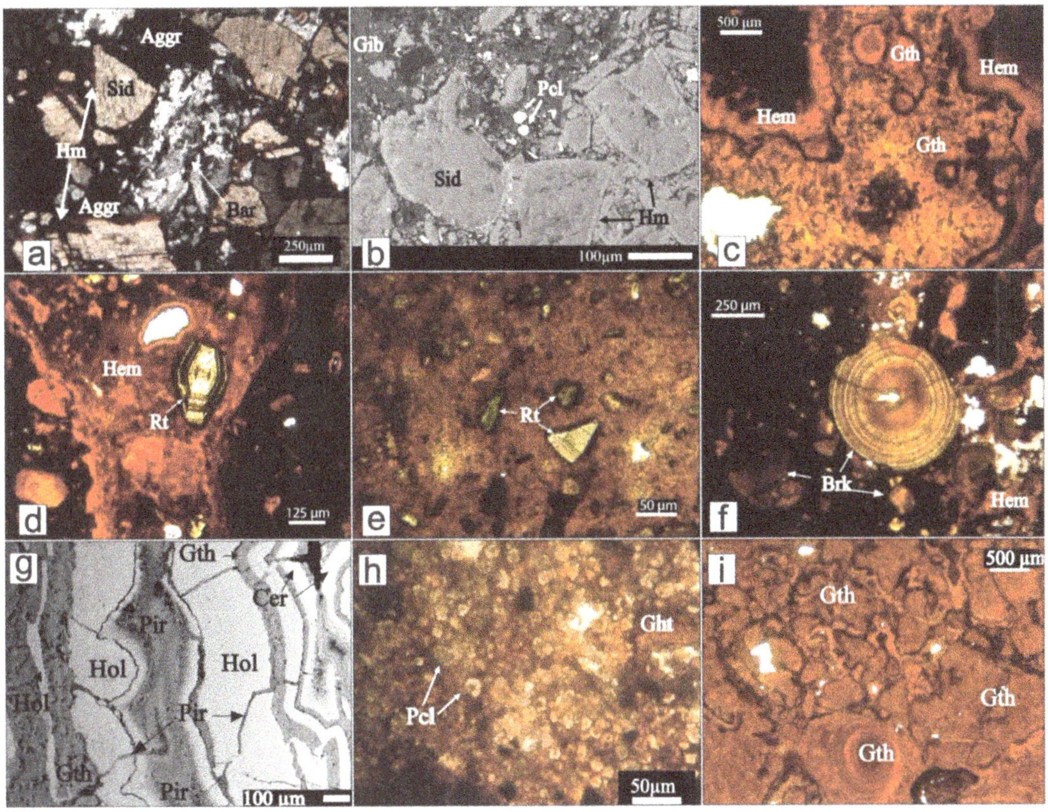

Fig. 9 Microscopic features, SEM images and photomicrographs of the carbonatite and Fe duricrusts of the Morro dos Seis Lagos deposit. (**a**) Typical border siderite carbonatite formed by siderite crystals cut by hematite and the matrix formed of aggregates of barite crystals and aggregates (dark zones) of gorceixite, monazite and pyrochlore (PL). (**b**) Core siderite carbonatite with siderite, pyrochlore euhedral crystals and gibbsite veinlets (BSE image). (**c**) Fibrous goethite filling tubular cavities and as pisolites, darker red Fe-bands are more hematitic (pisolitic Fe duricrust). (**d**) Zoned Nb-rich rutile crystal (fragmented Fe duricrust). (**e**) Zoned and fractured Nb-rich rutile crystals set in a ferruginous cavity (mottled Fe duricrust). (**f**) Spherules of Nb-rich brookite (purple Fe duricrust). (**g**) Banded Mn-rich veins constituted of hollandite, pyrolusite, cerianite and Mn-bearing goethite (manganiferous Fe duricrust). (**h**) Pyrochlore small crystals with core transformed to goethite (brown Fe duricrust). (**i**) Typical microscopic texture formed by botryoidal goethite evenly distributed (brown Fe duricrust). Abbreviations: Sid (siderite), Bar (barite), Mon (monazite), Gor (gorceixite), Aggr. (aggregates), Hmt (hematite), Gth (goethite), Rt (Nb-rich rutile), Brk (Nb-rich brookite), Hol (holandite), Cer (cerianite), Pir (pyrolusite), Pcl (pyrochlore). (Giovannini et al. 2017)

8.4 Niobium Mineralization

The relative abundances of the minerals hosting Nb in the weathering profile (Fig. 10) determined by SEM and EPMA vary depending on the horizon (Bollaert et al. 2023). Pyrochlore is dominant in the core carbonatite with rarer Nb-bearing Ti oxides; the presence of Nb-bearing goethite is inferred from EPMA analyses. Iron oxides are ubiquitous hosts for Nb in all the Fe duricrust horizons. In the brown Fe duricrust, pyrochlore and Nb-bearing Ce oxides are readily detected, whereas Nb-bearing Ti oxides are rarer. The lower purple horizon is characterized by the common occurrence of pyrochlore, but Nb-bearing Ti and Ce oxides are scarce. In the manganiferous Fe duricrust, only hollandite and Ce oxides contain significant amounts of Nb. In the upper purple horizon, Nb-bearing phases are like that of the lower purple horizon, except for abundant Nb-bearing Ti oxides. In the fragmented Fe duricrust horizon, pyrochlore and Ce oxides are rare, whereas Nb-bearing Ti oxides are abundant.

The main Nb host, the Nb-Ti oxides brookite and rutile, occur with varied textures and can be zoned or not (Fig. 11). The formation of Nb-Ti and Nb-Fe oxides results from the alteration of pyrochlore (Fig. 9). Giovannini et al. (2017) interpreted the Pb-Ba-pyrochlore (border siderite carbonatite) as an early product of primary pyrochlore weathering, followed by Ce-Ba-pyrochlore (core siderite carbonatite, strongly weathered due to the karstic process that formed the Esperança Basin) and then by Ce-pyrochlore (lower part of the brown Fe duricrust). Bollaert et al. (2023) were able to follow the evolution in the composition of pyrochlore throughout the lateritic profile (Fig. 10).

The Nb-Ti oxides occur in all Fe duricrusts, with the crystals showing omnipresent oscillatory zoning with respect to their Ti, Nb and Fe contents. The zoning is evident in optical or BSE images (Fig. 12). Representative compositions and structural formulae are provided in Table 2. The main Nb-Ti oxides have the following ranges: TiO_2

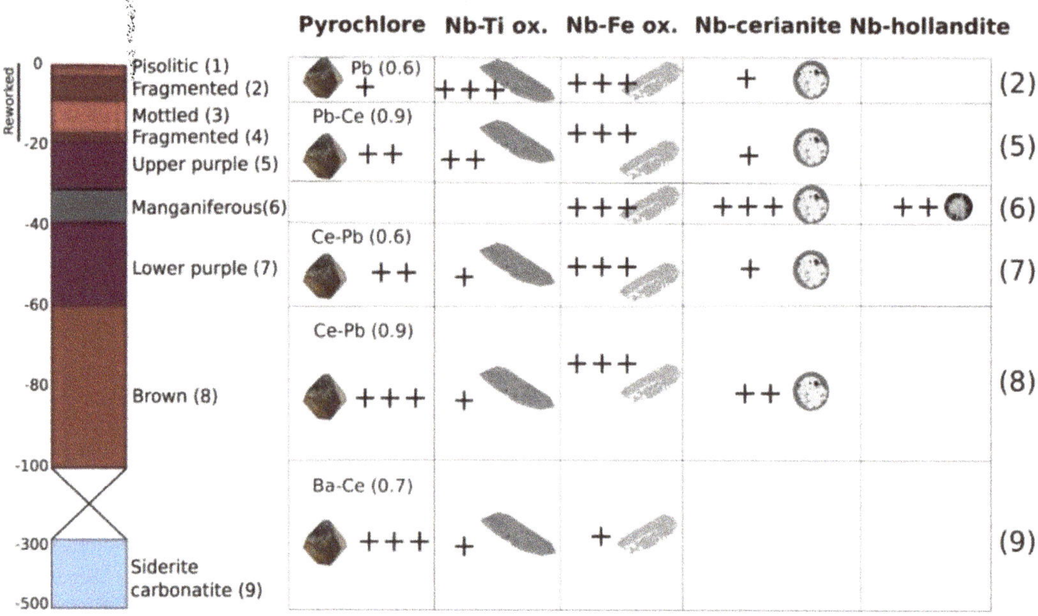

Fig. 10 Morro dos Seis Lagos deposit./Qualitative relative abundance of Nb carriers as obtained by *SEM* and *EPMA* analyses. Relative abundances range from not observed to very abundant (+++) phases. The crystal-chemistry and the A-site occupancy of pyrochlore is shown in brackets (Bollaert et al. 2023)

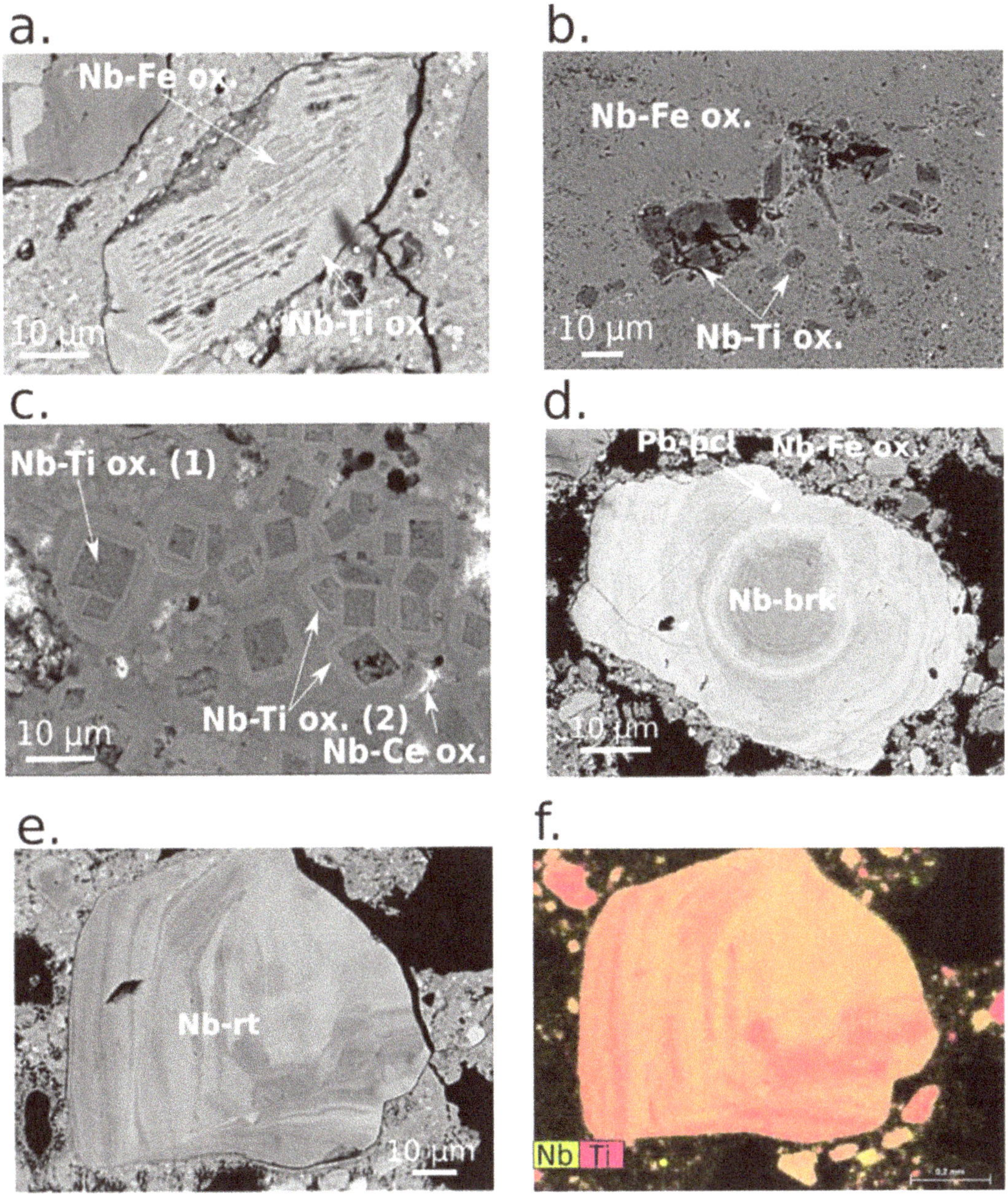

Fig. 11 Niobium-bearing Ti oxides in Fe duricrusts at the Morro dos Seis Lagos deposit. (**a**) Euhedral Nb-bearing Ti oxide partially altered into Nb-bearing Fe oxide. (**b**) Relics of primary Nb-bearing Ti oxides. (**c**) Porous square sections of Nb-bearing Ti oxide (Nb-Ti ox. 1) overgrown by successive generations of Nb-bearing Ti oxide (Nb-Ti ox. 2). (**d**) Zoned Nb-bearing brookite (Nb-brk) with inclusions of Pb-bearing pyrochlore (Pb-Pcl). (**e**) Neoformed Nb-bearing rutile (Nb-rt) displaying Nb and Ti zoning as evidenced by EDS chemical map (**f**). (Bollaert et al. 2023)

(57.84–80.92 wt%), Nb_2O_5 (11.26–22.23 wt%), Fe_2O_3 (8.96–14.95 wt%), up to 1.27 wt% WO_3, and up to 1.09 wt% SiO_2. The average structural formula is (Ti 0.75, Fe 0.13, Nb 0.11, Si 0.01, W 0.01)O_2. A distinctive feature is the absence of Ta, corroborated by whole rock analyses (Ta < 2.3 ppm in carbonatite and Ta < 0.70 ppm in the Fe duricrusts).

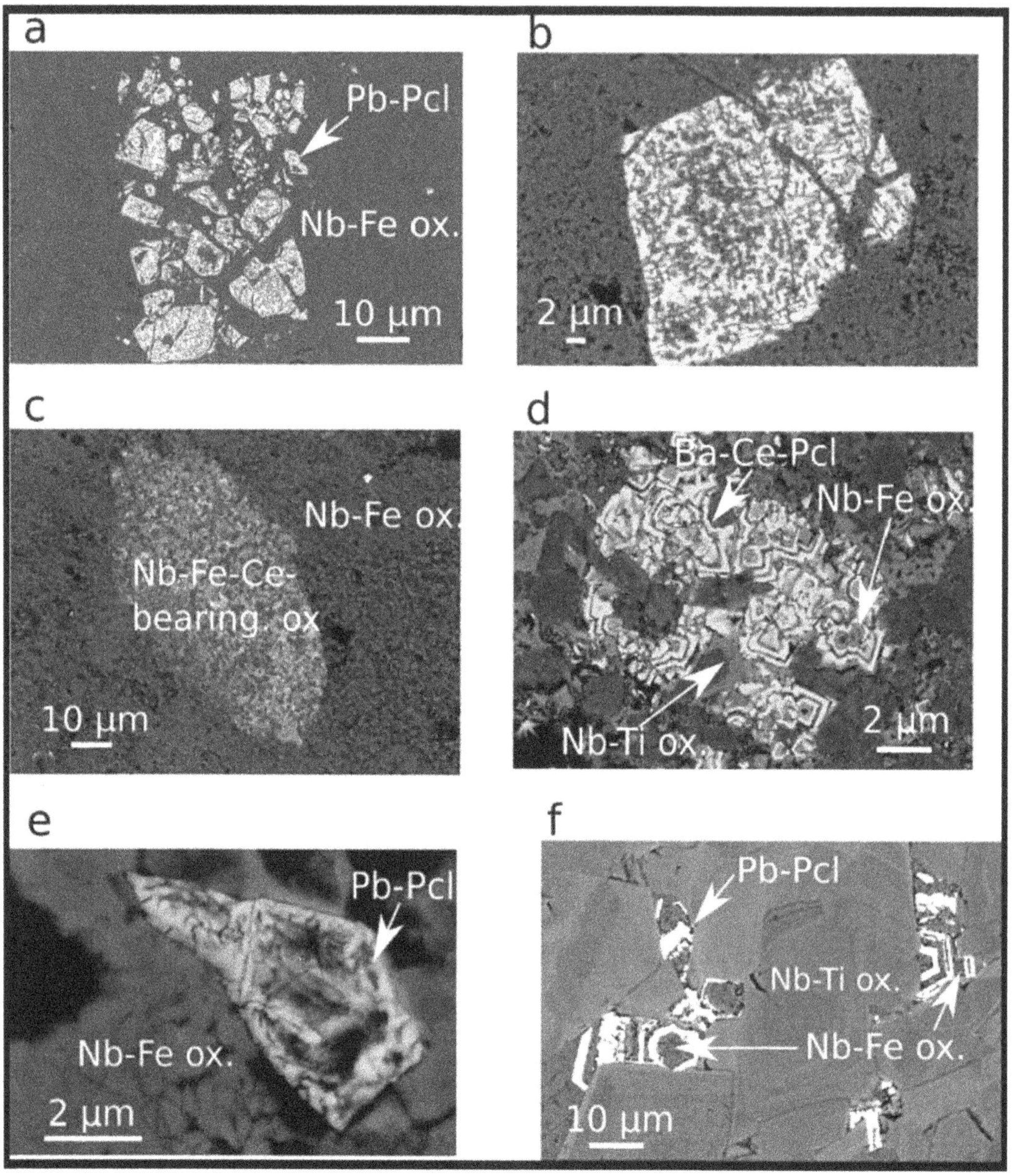

Fig. 12 Pyrochlore textures in the Fe duricrust, Morro dos Seis Lagos deposit. (**a**) Agglomerates of micrometric crystals of Pb-enriched pyrochlore (Pb-Pcl) hollowed by Nb-bearing Fe oxide (brown Fe duricrust). (**b**) Close-up of a weathered Pb-enriched pyrochlore crystal (brown Fe duricrust). (**c**) Pseudomorph of a mixed Nb-bearing Fe and Ce oxides after an unidentified mineral in an Nb-bearing Fe oxide-rich matrix (brown Fe duricrust). (**d**) Alteration of Ba-Ce-enriched pyrochlore (Ba-Ce-pcl) in concentric zones into Nb-bearing Ti and Fe oxides (purple Fe duricrust). (**e**) Skeletal crystal of weathered Pb-enriched pyrochlore (fragmented Fe duricrust). (**f**) Replacement of Pb-enriched pyrochlore by zoned Nb-bearing Ti oxide (fragmented Fe duricrust). (BSE images, Bollaert et al. 2023)

Table 2 Representative compositions and structural formulae of Nb-Ti oxides from the borehole SG-01-AM from the Morro dos Seis Lagos deposit: fragmented Fe duricrust (sample 4), mottled Fe duricrust (5 crystals in sample 6) and upper purple Fe duricrust (sample 12) and brown Fe duricrust (5 crystals in sample 34). Giovannini et al. (2017)

Nb-rich rutile								Nb-rich rutile (brown crust)				
	4.1	6.1	6.2	6.3	6.4	6.5	12.1	34.1	34.2	34.3	34.4	34.5
Fe_2O_3	8.96	12.18	13.94	11.12	14.95	10.52	17.78	14.92	11.64	11.67	11.18	12.43
SiO_2	0.00	1.09	0.45	0.63	0.62	0.52	1.00	0.33	0.27	0.34	0.30	0.28
TiO_2	80.92	70.73	68.92	72.45	66.26	74.08	57.84	58.26	67.34	68.66	66.92	57.86
Nb_2O_5	11.26	16.59	19.08	16.71	18.89	14.89	22.23	25.46	19.16	17.47	20.28	27.61
WO_3	0.16	0.35	0.31	0.70	0.33	0.64	1.27	0.20	0.33	0.19	0.85	0.62
Total	101.30	100.94	102.70	101.61	101.05	100.65	100.12	99.17	98.74	98.33	99.53	98.80
Structural Formulae												
Fe^{3+}	0.093	0.130	0.148	0.118	0.162	0.112	0.198	0.168	0.128	0.128	0.123	0.140
Si	0.000	0.015	0.006	0.009	0.009	0.007	0.015	0.005	0.004	0.005	0.004	0.004
Ti	0.841	0.753	0.729	0.766	0.715	0.786	0.643	0.653	0.740	0.754	0.732	0.653
Nb	0.070	0.106	0.122	0.106	0.123	0.095	0.149	0.172	0.127	0.115	0.134	0.187
W	0.001	0.001	0.001	0.003	0.001	0.002	0.005	0.001	0.001	0.001	0.003	0.002
Cations	1.005	1.005	1.006	1.002	1.010	1.002	1.010	0.999	1.000	1.003	0.996	0.986

Atoms per formula unit are calculated on the basis of two oxygens

9 Final Considerations

The carbonatite complexes of the Alto Paranaíba Alkaline Province (APIP) have experienced similar evolution since their emplacements in 80 to 90 million years ago. They were subjected to intense lateritic weathering from the onset of the Sul-Americano geomorphological cycle in the Oligocene, which produced a thick regolith cover varying from 50 m to more than 250 m. Their Fe-rich (ultramafic) and carbonate-rich protoliths promoted the development of ferruginous regolith in which deposits of Nb, Ti, P and REE formed by residual enrichment. Their invariably near-circular shape, protected by resistant aureoles of fenitized host rocks, helped preserve the regolith and avoid intensive flushing of the weathering solutions, thereby promoting reprecipitation of leached metals in the lower portions of the regolith or their immobilization as altered ore minerals or neoformed phases.

In contrast, the Morro dos Seis Lagos deposit (MSLD) lacks the protective resistant rim, probably because it is older and has consequently been exposed to erosion for longer. However, the richer Fe content of the protolith (siderite carbonatite) promoted the development of the thick, resistant Fe duricrust, which was able to maintain its elevated topographic expression, thereby preserving the Nb deposit.

The main difference between MSLD and the APIP deposits is that the Nb-enriched Fe duricrust at MSLD formed independently of the residual accumulation of pyrochlore, as is the case in the world-class Nb deposits of Araxá and Catalão-I. The exceptional Nb concentration in MSLD (on average 2.97 wt% Nb_2O_5) resulted from the combination of the Nb enrichment of the siderite carbonatite inherited from late-carbo-magmatic-to-hydrothermal processes (Giovannini et al. 2020) and Nb scavenging by secondary oxides during weathering (Giovannini et al. 2017; Bollaert et al. 2023).

Despite unexpectedly intense alteration of pyrochlore, the Nb concentration remains high in most of the profile due to its scavenging by secondary oxides, in particular goethite, brookite and rutile. This model of Nb enrichment depends on the capacity of Nb^{5+} to substitute for cations released from the primary Nb minerals in secondary phases formed during alteration. Preservation of primary enrichment will thus depend on the composition of the parent rock, which must provide Fe and Ti for secondary oxide precipitation, and on the physicochemical conditions of formation of the Fe duricrusts, which must favour the precipitation of goethite and Ti oxides (Bollaert et al. 2023). The extremely intense weathering and the type of carbonatite (sideritic carbonatite) performed fundamental roles in the formation of this deposit.

References

Almeida ME, Macambira MJB, Oliveira EC (2007) Geochemistry and zircon geochronology of the I-type high-K calc-alkaline and S-type granitoid rocks from southeastern Roraima, Brazil: Orosirian collisional magmatism evidence (1.97–1.96 Ga) in central portion of Guyana Shield. Precambr Res 155:69–97

Atencio D, Andrade MB, Christy AG, Gieré R, Kartashov PM (2010) The pyrochlore supergroup of minerals: nomenclature. Can Mineral 48:673–698

Barbosa ESR (2009) Mineralogia e Petrologia do Complexo Carbonatítico-Foscorítico de Salitre, MG, Unpublished Ph.D. thesis, University of Brasília, Brazil, 432 pp

Bento JPP, Porto CG, Takehara L, da Silva FJ, Neto ACB, Machado ML, Duarte AC (2022) Mineral potential re-evaluation of the Seis Lagos Carbonatite Complex. Amazon. Brazil. Brazilian J Geol 52(1)

Biondi JC (2005) Brazilian mineral deposits associated with alkaline and alkaline–carbonatite complexes. In: Comin-Chiaramonti P, Gomes CB (eds) Mesozoic to Cenozoic alkaline magmatism in the Brazilian Platform. Editora da Universidade de Sao Paulo. Fapesp, São Paulo, pp 707–750

Bollaert Q, Chasse M, Bastos Neto AC, Baptiste B, Courtain A, Galoisy L, Mathon O, Quantin C, Vantelon D, Calas G (2023) Mechanisms leading to exceptional niobium concentration during lateritic weathering: the key role of secondary oxide. Chem Geol 641:1–20

Brod JA (1999) Petrology and geochemistry of the Tapira alkaline complex, Minas Gerais State, Brazil: Unpublished Ph.D. thesis, University of Durham, U.K., 486 pp

Brod JA, Ribeiro CC, Gaspar JC, Junqueira-Brod TC, Barbosa ESR, Riffel BF, Silva JF, Chaban N, Ferrari AJD (2004) Geologia e Mineralizações dos Complexos Alcalino-Carbonatíticos da Província Ígnea do Alto Paranaíba. 42° Congresso Brasileiro de Geologia, Araxá, Minas Gerais, Excursão, p 1, 29 pp.

Carvalho WT, Bressan SR (1997) Depósitos de fosfato, nióbio, titânio e terras raras e vermiculita de catalão I, Goiás. In: Shobbenhaus C, Queiroz ET, Coelho CES (eds) Principais depósitos minerais do Brasil. Vol. IV parte C, Chapter VI. DNPM, pp 69–94

Chakhmouradian AR, Reguir EP, Kressall RD, Crozier J, Pisiak LK, Sidhu R, Yang P (2015) Carbonatite-hosted niobium deposit at Aley, northern British Columbia (Canada): mineralogy, geochemistry and petrogenesis. Ore Geol Rev 64:642–666

da Silva AB (1986) Jazida de nióbio de Araxá, Minas Gerais. In: Shobbenhaus C, da Coelho CES (eds) Principais depósitos minerais do Brasil, vol II. DNPM, pp 435–453

de Cordeiro OCG, Brod JA, Palmieri M, de Oliveira CG, Barbosa ESR, Santos RV, Gaspar JC, Assis LC (2011) The Catalão I niobium deposit, central Brazil: resources, geology and pyrochlore chemistry. Ore Geol Rev 41:112–121

Fava N (2001) O manto de intemperismo e a química do pirocloro de Catalão I (GO):Um estudo preliminar: Unpublished M.Sc. thesis, University of Brasília, Brazil, 124 pp

Gierth E, Baecker ML (1986) A mineralização de Nb e as rochas alcalinas associadas no Complexo Catalão I, Goiás. In: Shobbenhaus C, Coelho CES (eds) Principais depósitos minerais do Brasil. Vol. II, Chapter 36. DNPM, pp 455–462

Giovannini AL, Bastos Neto AC, Porto CG, Pereira VP, Takehara L, Barbanson L, Bastos PHS (2017) Mineralogy and geochemistry of laterites from the Morro dos Seis Lagos Nb (Ti, REE) deposit (Amazonas, Brazil). Ore Geol Rev 88:461–480

Giovannini AL, Mitchell R, Bastos H Neto AC, Moura CAV, Pereira VP, Porto CG (2020) Mineralogy and geochemistry of the Morro dos Seis Lagos siderite carbonatite, Amazonas, Brazil, Lithos, Volumes 360–361

Hirano H, Kamitani M, Sato T, Sudo S (1990) Niobium mineralization of Catalão I carbonatite complex, Goiás, Brazil. Bull Geol Surv Japan 41:619–626

Hogarth DD (1977) Classification and nomenclature of the pyrochlore group. Am Mineral 62:403–410

Issa Filho A, Lima PRAS, Souza OM (1984) Aspectos da geologia do complexo carbonatítico do Barreiro, Araxá, Minas Gerais, Brasil. In: Rodrigues CS, Lima PRAS (eds) Companhia Brasileira de Metalurgia e Mineração, pp 20–44

Justo LJEC, Souza MM (1984) Jazida de nióbio do Morro dos Seis Lagos, Amazonas. In: Simpósio Amazonense 2, Manaus, pp 467–472

Justo LJEC, Souza MM (1986) Jazida de ni'obio do Morro dos Seis Lagos, Amazonas. In: Schobbenhaus C, Coelho CES (eds) Principais Dep'ositos Minerais do Brasil—Ferro e Metais da Indústria do Aço. Departamento Nacional da Produção Mineral, vol 2, pp 463–468

Melo MTV (1997) Depósitos de fosfato, titânio e nióbio de Tapira. Minas Gerais. In: Schobbenhaus C, Queiroz ET, Coelho CES (eds) Principais depósitos minerais do Brasil, 2. MME/DNPM, Brasília, pp 41–56

Melo MTV, Chaban N, Grossi Sad JH (1997) depósitos de fosfato, titânio e nióbio de salitre, Minas Gerais. In: Shobbenhaus C, Queiroz ET, Coelho CES (eds) Principais depósitos minerais do Brasil. Vol. IV parte C, Chapter V. DNPM, pp 57–68

Murilo GT, Gaspar JJ (1995) Geoquímica do manto de intemperismo da mina de fosfato do complexo alcalino-carbonatítico do barreiro Araxá, M.G. V

Cong. Bras. Geoq. / III Cong. Geoq Países de Língua Portuguesa, Niterói RJ, Exp. Abst.

Oliveira SMB, Imbernon RAL (1998) Weathering alteration and related REE concentration in the Catalao I carbonatite complex, central Brazil. J South Am Earth Sci 11:379–388

Paraiso OS, Fucio J (1982) Araxá Nb mine. Mining Mag 146(2)

Pinheiro SS, Fernandes PECA, Pereira EP, Vasconcelos EG, Pinto AC, Montalvão RMG, Issler RS, Dallagnoll R, Teixeira W, Fernandes CAC (1976) Folha NA.19 Pico da Neblina; geologia, geomorfologia, pedologia, vegetação e uso potencial da terra. Rio de Janeiro, Cap. 2:19–138

Ribeiro CC, Brod JA, Junqueira-Brod TC, Gaspar JC, Palmieri M, Cordeiro PFO, Torres MG, Grasso CB, Barbosa ESR, Barbosa PAR, Ferrari AJD, Gomide CS (2014) Potencial e Controles Metalogenéticos de ETR, Ti e Nb em Províncias Alcalino-carbonatíticas brasileiras. In: Silva MG, Rocha Neto MB, Jost H, Kuyumjan RM (eds) Metalogênese das províncias tectônicas brasileiras. CPRM, Belo Horizonte, pp 559–589

Rocha E, Nasraoui M, Soubiès F, Bilal E, De Parseval P (2001) Évolution géochimique du pyrochlore au cours de l'altération météorique du gisement de Catalao II (Goiás, Brésil). Comptes Rendus l'Academie Sci. – Ser. IIa Sci. la Terre des Planetes 332:91–98

Rodrigues CS, Lima PRA (1984) Carbonatite Complexes of Brazil: Geology. Companhia Brasileira de Metalurgia e Mineração, São Paulo (44 pp)

Rossoni MB, Bastos Neto AC, Saldanha DL, Souza VS, Giovannini AL, Porto CG (2016) Aplicação de técnicas de sensoriamento remoto na investigação do controle do posicionamento do Complexo Carbonatítico Seis Lagos e no estudo do depósito (Nb) laterítico associado (Amazonas, Brasil). Rev Pesquisas 43:111–125

Rossoni MB, Bastos Neto AC, Souza VS, Marques JC, Dantas E, Botelho NF, Giovannini AL, Pereira VP (2017) U-Pb zircon geochronologycal investigation on the Morro dos Seis Lagos Carbonatite Complex and associated Nb deposit (Amazonas, Brazil)'. J South Am Earth Sci 80:1–17

Santos JOS, Hartmann LA, Gaudette HE, Groves DI, Mcnaughton NJ, Fletcher IR (2000) A new understanding of the provinces of the Amazon craton based on integration of field mapping and U-Pb and Sm-Nd geochronology. Gondwana Res 3:453–488

Santos JOS, Potter PE, Reis NJ, Hartmann LA, Fletcher IR, McNaughton NJ (2003) Age, source and regional stratigraphy of the roraima supergroup and Roraima-like sequences in Northern South America, based on U-Pb. Geochronol Geol Soc Am Bull 115(3): 331–348

Santos JOS, Hartmann LA, Faria MSG, Riker SR, Souza MM, Almeida ME, McNaughton NJ (2006) A compartimentação do Cráton Amazonas em províncias: avanços ocorridos no período 2000–2006. SBG-NO, 9o Simpósio de Geologia da Amazonia CD-ROM, Belém

Schobbenhaus C (1986) Geologia do Nb e Ta. In: Shobbenhaus C, Coelho CES (eds) Principais depósitos minerais do Brasil, vol II. DNPM, pp 429–433

Seer HJ, Moraes LC (2018) Nióbio. In: Recursos Minerais de Minas Gerais. http://recursomineralmg.codemge.com.br/substancias-minerais/niobio/

Silva AB, Marchetto M, Souza OM (1979) Geologia do complexo carbonatítico de Araxá (Barreiro), MG. Mineração e Metalurgia 415:14–18

Toledo MCM, Lenharo SLR, Ferrari VC, François F, Parseval P, Leroy G (2004) The compositional evolution of apatite in the weathering profile of the Catalão I Alkaline-Carbonatitic Complex, Goiás, Brazil. Can Mineral 42:1139–1158

Torres MG, Gaspar JC (1995) Geoquímica do manto de intemperismo da mina de fosfato do complexo alcalino-carbonatítico do barreiro Araxá, M.G. V Congresso Brasileiro de Geoquímica e III Congresso de Geoquímica dos países de Língua Portuguesa, Niterói/RJ, 1995. Exp. Abst

Viegas Filho JR, Bonow CW (1976) Projeto Seis Lagos (Internal Report). Companhia de Pesquisa de Recursos Minerais CPRM, Manaus

Wall F, Williams CT, Woolley AR (1999) Pyrochlore in niobium ore deposits. In: Stanley CJ et al (eds) Mineral deposits: processes to processing, Proceedings of the 5th Biennial SGA meeting and 10th Quadrennial IAGOD Symposium London, U.K. Balkema, Rotterdam, pp 687–690

Supergene Ores of the Central African Copperbelt

Rob J. Bowell and C. R. M. Butt

Abstract

The Copper Belt of central Africa hosts the world's largest sedimentary-hosted Cu resource. Primary ore consists of sulfide minerals, chiefly chalcopyrite, bornite, chalcocite, and carrollite hosted in siliclastic and carbonate host rocks. Multiple stages of weathering and erosion from the Miocene onwards have resulted in extensive oxidation of the primary deposits, the intensity of which reflects depth, brecciation, tectonics, and the presence of karstic features that have acted as conduits for meteoric fluids. Much of the ore mined to date has been of high-grade supergene mineralization formed during this period. Extensive development of malachite, chrysocolla and other secondary Cu minerals has occurred through groundwater flow, whereas heterogenite formed by residual enrichment and profile denudation has been responsible for high-grade Co and Mn concentrations in surface zones.

R. J. Bowell (✉)
SRK Consulting, Churchill House, Cardiff, UK

School of Earth Sciences, Queens University, Kingston, ON, Canada
e-mail: rbowell@srk.co.uk

C. R. M. Butt
CSIRO Mineral Resources, ARRC, Kensington, WA, Australia

1 Introduction

The Central African Copperbelt straddles the border between the North West Province of Zambia and the Katanga region of the Democratic Republic of Congo (Fig. 1). Traditionally, the term Copperbelt covered the mining regions of the Zambian Copperbelt Province (notably the towns of Ndola, Kitwe, Chingola, Luanshya and Mufulira) and the Haut-Katanga and Lualaba provinces of the Congo (notably Lubumbashi, Kolwezi and Likasi). It reflects the distribution of the Neoproterozoic sedimentary rocks of the Katanga Supergroup that host the mineralization (Hitzman et al. 2005). The copper-cobalt ores of the region contain more than 200 Mt Cu, equivalent to 15% of known global Cu resources (Boyle et al. 1989; Selley et al. 2005) and the world's largest reserve of Co (Hitzman et al. 2012). In addition, regionally important deposits of Pb, Zn and Ge are known from Kipushi and Kamwe (Selley et al. 2005). The deposits, discovered from their supergene expression in the 1930s, have been continuously exploited for over one hundred years (Mendelsohn 1961; Hitzman et al. 2005, 2012; Pirard 2011; Selley et al. 2005). Several case studies illustrate the characteristics and genesis of these deposits.

R. J. Bowell, C. R. M. Butt (eds.), *Geology, Geochemistry and Formation of Supergene Mineral Deposits in Deeply Weathered Terrain*, Mineral Resource Reviews, https://doi.org/10.1007/978-3-031-75733-4_9

2 The Geology and Formation of the Copperbelt

2.1 Primary Mineralization

Copper-cobalt mineralization in the Copperbelt is hosted by Neoproterozoic metasedimentary rocks of the Katanga Supergroup (600–900 Ma) deposited in a series of intracontinental rift-associated basins and alongside abundant evaporate deposits (Bateman 1930; Annels 1974; McGowan et al. 2003; Hitzman et al. 2005; Master et al. 2005; Rainaud et al. 2005; Broughton and Rogers 2010). Early rift-stage continental rocks were overlain by a sequence of mixed carbonate-evaporite and clastic rocks (Mwashya in the west and the Lower Roan in Zambia). This was followed by renewed rifting deposition of clastic rocks with mafic intrusions (the Mashya gabbros). Subsequently, the Rodinian glacial event modified the stratigraphy, which was then covered by postglacial deposits (diamictite). Subsequent deposition of basin-fill clastic and carbonate rocks preceded the inversion of the basin in the Pan African orogenic event (~560 Ma). The host geological settings of the Copperbelt deposits are varied, with mineralization occurring over a wide part of the stratigraphic sequence (Fig. 2).

There are also differences in the mineralogy, styles and textures of the mineralization, and zoning of alteration. Deposits can occur as disseminated fine-grained sulfides or coarser-grained vein-hosted sulfides in quartz-carbonate gangue. However, most deposits occur at redox-associated stratigraphic or structural boundaries and along reactive carbonate-evaporate horizons (Hitzman et al. 2005; Selley et al. 2005). Despite redox being an important control in the deposition of Cu mineralization, there is no correlation of grade to organic carbon in shale. Zonation of Cu mineralogy occurs on different scales, with the complexity of zonation reflecting the complexity of the host rock (McGowan et al. 2003; Broughton and Rogers 2010; Hitzman et al. 2012). At a macro scale, extensional faults are a major control on fluid flow, influenced by variation in the extent of diagenetic changes to protolith carbon content. In the Katangan sequence in the DRC, the penetration of evaporate diapirs that created anticlinal salt-dome-type structures also played a role in the control of fluid flow (Selley et al. 2005). Regional scale alteration showing potassic, magnesian, silicic, sodic and argillic alteration is controlled partly by precursor lithology and partly by the proximity and composition of basin brines (Hitzman 2010).

Mineralization occurred over a long period of time, spanning diagenesis, basin inversion and metamorphic events that led to the circulation of brines within the Katangan basin (Cailteux 1994; McGowan et al. 2003; Selley et al. 2005; Broughton and Rogers 2010; Hitzman et al. 2012).

3 Supergene Enrichment

3.1 Profile and Mineralogy

Orebodies in the Copperbelt have undergone supergene enrichment, and much of the ore mined to date has been high-grade secondary mineralization. Variable levels of oxidation have resulted in marked changes in mineralogy and increases in the Cu and Co grades of the deposits (Fig. 3). The Primary (bornite-chalcopyrite) and Transition (chalcocite) Zones generally have lower grades than the Oxide Zone, although at NChanga, the Co grade is higher in the Primary Zone than the Transition Zone due to the presence of carrollite ($CuCo_2S_4$).

The secondary minerals in the Oxidized and Transition zones formed during the post-orogenic period throughout the Copperbelt. The Katanga area was on the southern margin of a major sedimentary basin during the Mesozoic, but episodic uplift exposed the metal-rich substratum to the action of meteoric fluids. These utilized vertical faults propagated from the protore to the surface and, to a lesser degree, lithological boundaries and changes in porosity, possibly through mechanisms such as seismic pumping (Cameron

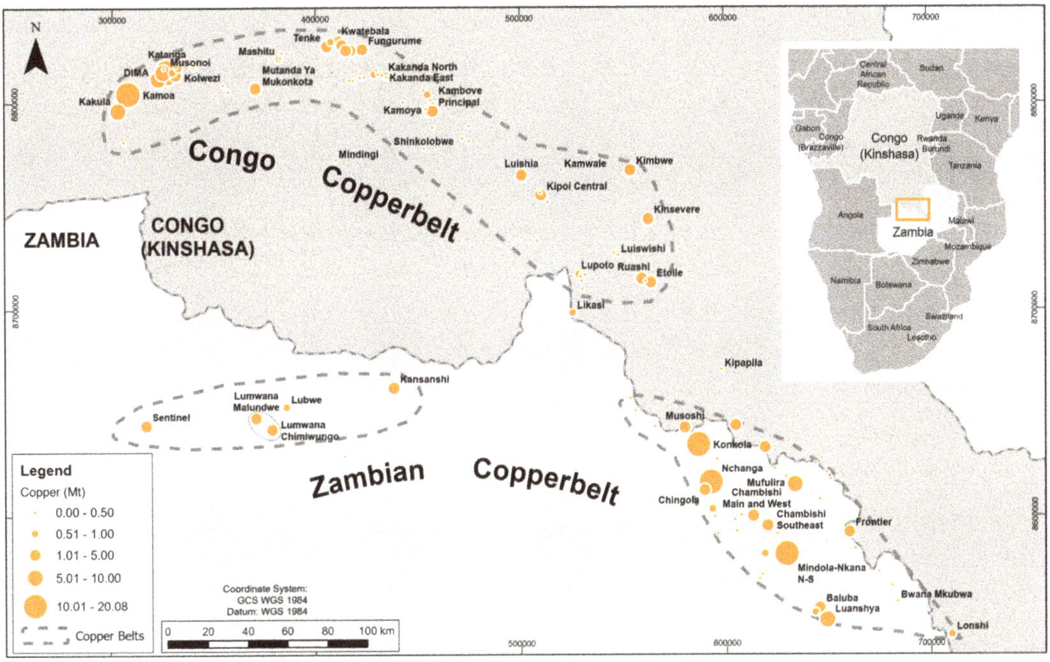

Fig. 1 Location of the Central African Copperbelt and the main known mineral deposits. (From Hitzman et al. 2005)

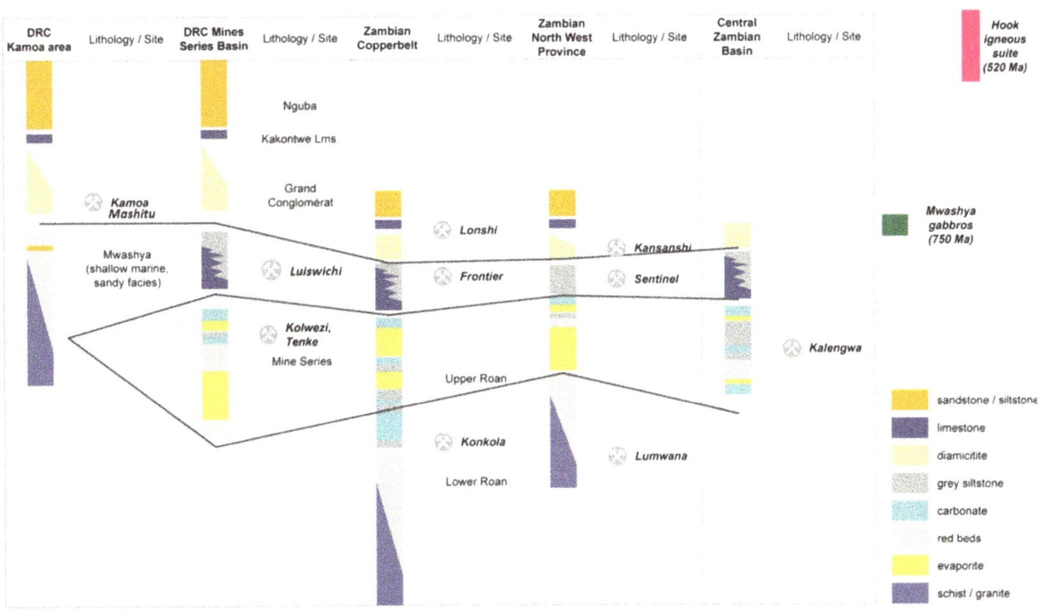

Fig. 2 Schematic representation of stratigraphic location of deposits. (After Hitzman et al. 2012)

and Leybourne 2005). The ensuing weathering led to the formation of rich supergene ores during the Miocene and Pliocene (Decrée et al. 2010; De Putter et al. 2015). The current thickness of the regolith hosting the supergene ores is approximately 10–50 m in the southern part of the Zambian Copperbelt and increases to more than 150 m at Tenke-Fungurume and Masoni in the

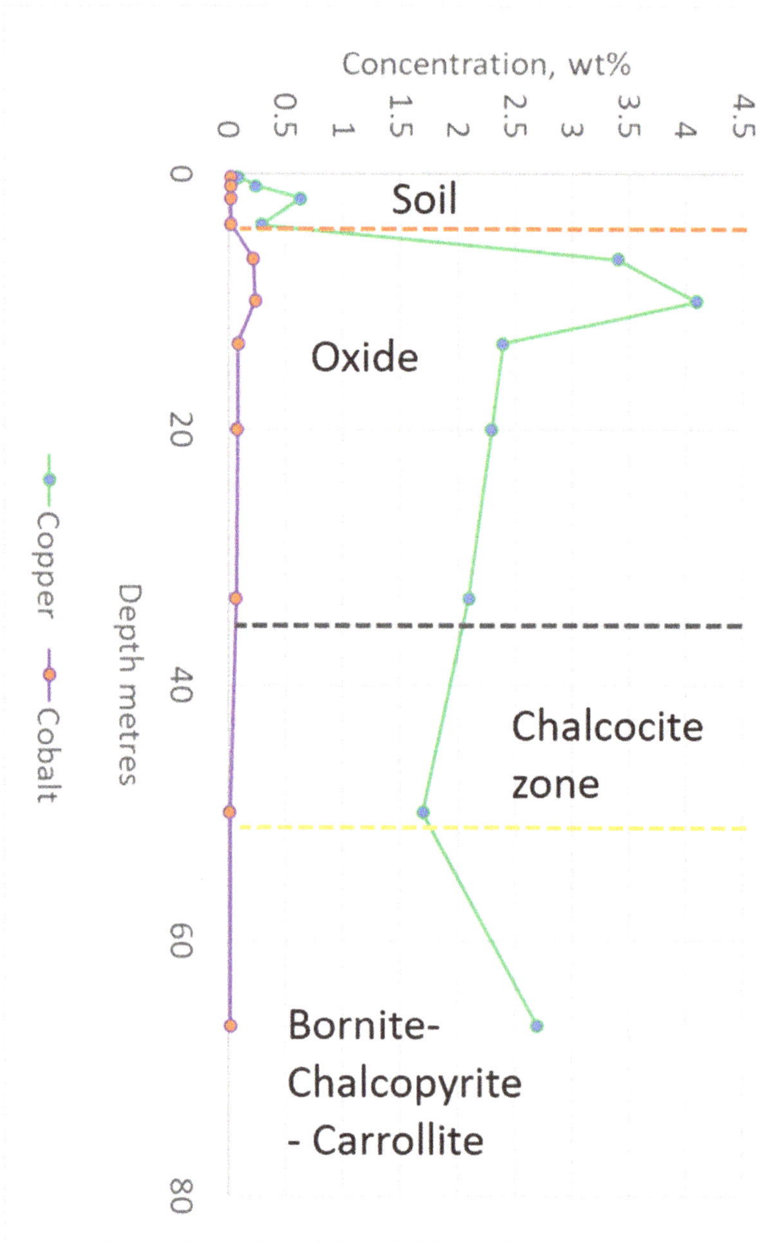

Fig. 3 Variation in Cu and Co grades, NChanga deposit, Zambia. (After McKinnon and Smit 1961). The Transition zone equates to the "Chalcocite zone" and the Primary zone equates to the "Bornite-Chalcopyrite-Carrollite" zone. From 0 to 5 m, the soil is generally depleted in both elements. The Oxide Zone is dominated by malachite, chrysocolla, cuprite and copper. The Transition zone is chalcocite, covellite and bornite, and the Primary Zone is chalcopyrite, bornite and carrollite

North Katanga Belt (Mendelsohn 1961; Selley et al. 2005; Schuh et al. 2012). Due to the action of migrating groundwater, however, supergene native Cu and malachite have been found on fracture planes at 900 m depth at Tenke-Fungurume (R. North, written communication 2023).

The supergene mineral assemblage comprises a goethite-illite-kaolinite-dolomite matrix with a wide range of secondary minerals including chalcocite, chrysocolla, malachite, brochantite, azurite, libethenite, cuprite, native Cu, torbernite, turquoise, heterogenite, sphaeocobaltite, Co-calcite, Co-smithsonite, Mn oxides including cryptomelane, pyrolusite, lithiophorite, manganite and "Mn wad" (variably enriched in Co and Cu), and a complex suite of U minerals (Audeoud

Fig. 4 Examples of supergene ore assemblage from the Copperbelt mines (red bar = 1 cm). (**a**) Cobalt-calcite, heterogenite and kolwezite, Musonoi mine. (**b**) Malachite stalagmites, Kolwezi district. (**c**) Cuprosklodowskite, torbernite, malachite and francevillite, Musonoi mine. (**d**) Azurite, NChanga Mine (open pit). (**e**) Malachite, Bwana M'Kubwa Mine, Ndola. (**f**) Native gold, torbernite, pitchblende, kasolite, Musonoi mine

1982; Gautier et al. 1980; Master et al 2005; De Putter et al. 2015; Sośnicka et al. 2019; Fontaine et al. 2020). Some of these are illustrated in Fig. 4. Near-surface grades can be 3–6% Cu, declining to 2–3% Cu at depth. Cobalt is less mobile than Cu and commonly forms high-grade near-surface zones (Fay and Barton 2012).

3.2 Luiswichi Cu-Co Deposit, Katanga, DRC

The deposit is located ~20 km to the NNW of Lubumbashi. The hypogene protore is composed predominantly of chalcopyrite, carrollite and pyrite, with lesser bornite (Lerouge et al. 2005). It occurs mainly as disseminations or clusters in thin beds, nodules or narrow fracture-filling veins in fine-grained stratified dolostone and silicified-stromatolitic dolomites, alternating with chloritic silty dolomitic units (Cailteux and De Putter 2019). The general stratigraphy is shown in Fig. 5. The supergene mineralization is comprised of malachite, chrysocolla, azurite and heterogenite in a Fe oxide-rich matrix. Supergene Cu-Co oxide minerals are mainly concentrated along bedding planes, in cavities and cracks, and in fracture zones associated with faults, and decrease in abundance with depth. In a study of the supergene mineralogy, Fontaine et al. (2020) recognized the series of zones typical of the Copperbelt, similar to that illustrated in Fig. 3:

- Primary ore zone (Protore), containing sulfides and massive dolomite.
- Transition ore zone, a mixed supergene cementation zone with co-existing oxides and sulfides.
- Oxidized zone within the saprolite comprised only of oxidized minerals.

The Transition zone, which continues to about 100 m depth, was formed by the oxidation of the primary sulfides. Its location coincides with a surface dome feature marked by dolomite and Mg-calcite in the host rocks. As acidity was released, it was rapidly neutralized by dolomite and Mg-chlorite in the host rock. This neutralization reaction led to saturation and precipitation of a wide range of oxidized mineral phases. Copper silicates such as chrysocolla are commonly poorly crystallized; indeed, Fontaine et al. (2020) considered most to be amorphous. A

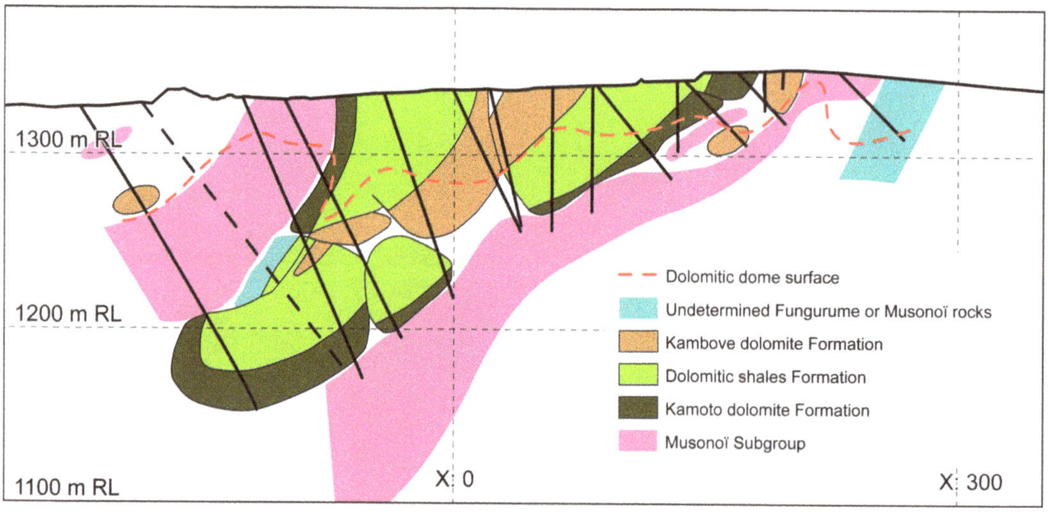

Fig. 5 Cross-section through the Luiswishi ore deposit (Fontaine et al. 2020)

schematic representation of the mineralogy in the different parts of the Luiswishi deposit is shown in Fig. 6. In places, a cemented supergene zone with dominant chalcocite has been identified that may have been derived from low-temperature fluids, such as those described below at the Mashitu South deposit (Sośnicka et al. 2019).

In the primary sulfides, Cu is concentrated within chalcopyrite and, to a lesser extent, carrollite and bornite, which typically contain less than ~15% Cu. Cobalt is hosted within carrollite and cobaltoan pyrite. The mixed ores in the Transition zone contain partially weathered primary sulfides and residual goethite, with secondary sulfides (Cu: S $\approx$ 2 minerals, such as chalcocite) and goethite typically replacing chalcopyrite. As oxidation proceeds, chalcopyrite is entirely replaced by goethite higher in the profile. Carrollite and cobaltoan pyrite are not weathered in the Transition zone (Fontaine et al. 2020). Bornite occurs associated with the transition from a pyrite-chalcopyrite-dominated system to a chalcocite-goethite system and, from mineralogical evidence, may well be both primary and supergene in origin.

The Transition zone represents the initial step of the oxidation of the sulfides in the primary ore zone. Groundwater commonly penetrates more deeply in the sulfide-rich mineralized units than in the wall rocks, where higher pH and lower redox potential result in less aggressive fluids (Thornber and Taylor 1992). The preferential weathering of chalcopyrite over pyrite has been observed in several studies and attributed to galvanic reactions associated with different sulfides (Sato 1992; Thornber and Taylor 1992; Bowell 1994).

The distribution of REEs in the weathering profile at Luiswishi can be used to track weathering processes, with REE mobilized during weathering, then fractionating and concentrating in the different zones of the profile (Jin et al. 2017). At Luiswishi, there is an overall HREE enrichment in the upper Oxidized zone due to preferential leaching of LREE and abundance of host phases for HREE elements. The Transition zone has less intense LREE enrichment than the Primary zone. The mobility of the REE in the Transition and Oxide zones is probably influenced by a higher oxidizing environment. The LREE that migrates downwards is attenuated to some degree in the relatively clay-rich Transition zone. The application of REE patterns in understanding weathering processes is hindered by the lack of discrete REE minerals in the regolith.

3.3 Mashitu South Cu-Co Deposits, Katanga, DRC

The Mashitu deposits are situated about 50 km E of Kolwezi. Hypogene and supergene mineralization are hosted by Neoproterozoic rocks of the Mines and Red "Roches Argilo-Talqueuses"

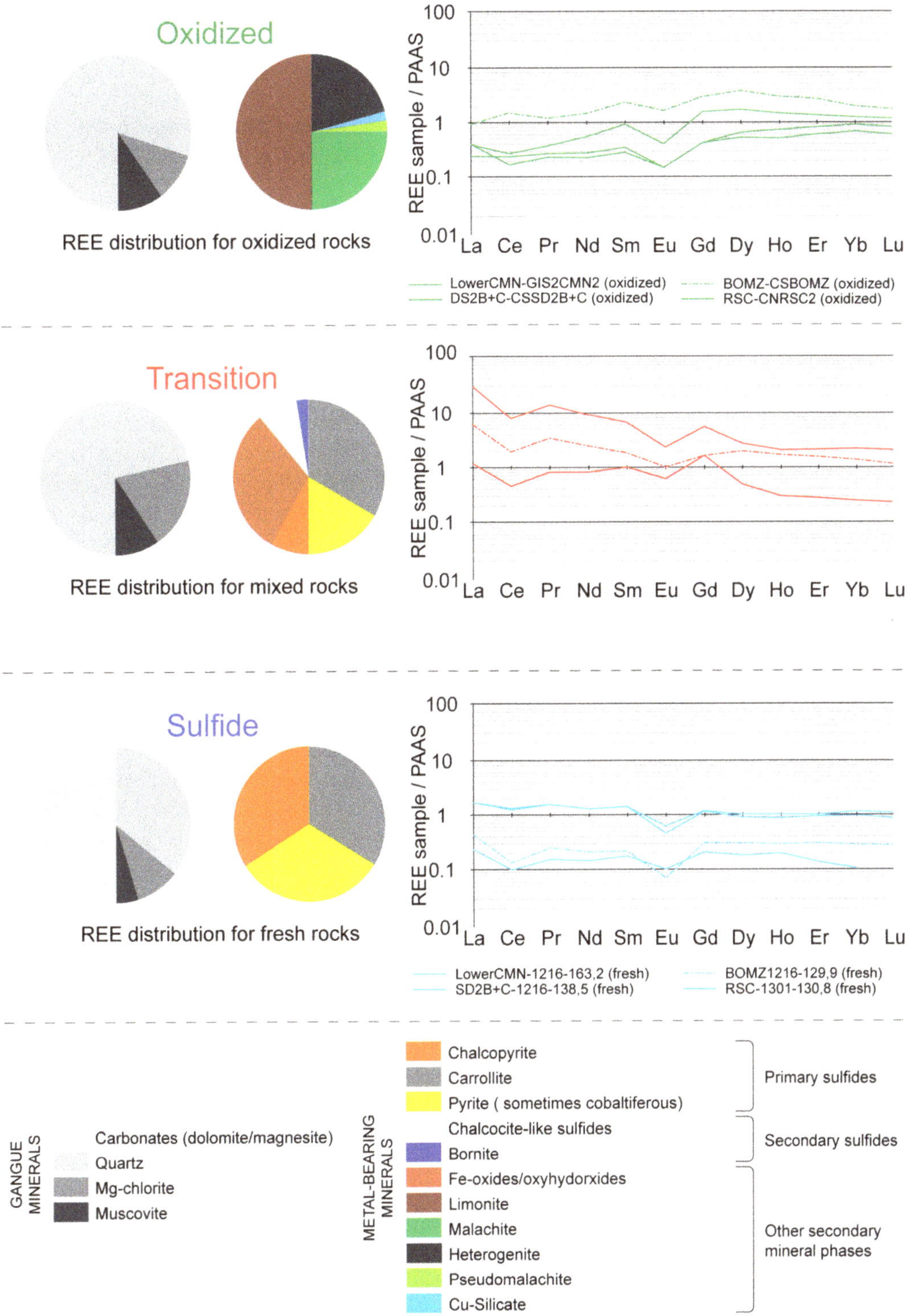

Fig. 6 Overview of the mineralogy and of the REE geochemical behavior in the three zones of the supergene ore of the Luiswishi deposit (Fontaine et al. 2020)

(R.A.T.) subgroups of the Katangan Supergroup. Copper and Co sulfide mineralization at the Mashitu South deposit is distributed throughout the stratigraphy, but high-grade mineralization is mostly restricted to stratified silty dolomites, finely laminated siliceous dolomites dolomitic shales and black carbonaceous dolomitic shales. Petrographic, fluid inclusion and stable isotope data have provided constraints for fluid geochemistry and ore-forming processes that indicate large-scale migration of metal-rich high-salinity brines, sulfate-rich and meteoric fluids across the Katangan Copperbelt (Sośnicka et al. 2019). This suggests that the strata bound, hypogene, hydrothermal Cu-Co ore veins precipitated at high temperatures (~247–316 °C) from mixed high-salinity Mg-rich/Na-K-Ca-Mg-Cl brines and Ba-K-Na-Ca ± Cl-SO_4 fluids with sulfide formation connected to the activity of thermochemical sulfate reduction, as indicated by negative $\delta^{13}C_{V\text{-}PDB}$ ratios (down to −7.4‰). A number of these veins were subjected to dynamic recrystallization under the influence of low salinity/meteoric fluids during active deformation at the onset of the Lufilian Orogeny (Sośnicka et al. 2019). Isotopic compositions from calcite-dolomite veins that host coarse carrollite, chalcopyrite, bornite and chalcocite crystals from the Kamoya South II deposit overlap with isotope data from Mashitu South hydrothermal veins, suggesting similar parent fluids were basin-wide. Metalliferous (Cu, Zn) Na-K-Cl brines, which provided metals in the post-tectonic ore veins from the Kamoya South II deposit (about 80 km ESE of Mashitu), are postulated to have originated from evaporation of seawater (Sośnicka et al. 2019).

Other essential precursors to the formation of the supergene ores in the Copperbelt include late deformation brecciation, associated deuteric alteration and dissolution/re-precipitation processes. At Mashitu, primary mono-phase aqueous inclusions and isotopic signatures of C and O from free-growing supergene aragonite, cobaltoan dolomite and bladed dolomite-malachite-hematite record temperature < 50 °C, implying precipitation from predominantly low salinity waters, with local pulses of saline fluids during the formation of supergene Cu and Co mineralization (Sośnicka et al. 2019). A summary of the various processes at the Mashitu South deposit is shown in Fig. 7.

3.4 Tenke-Fungurume Co-Cu Deposits, Katanga, DRC

Tenke-Fungurume is a cluster of sediment-hosted deposits, ~75 to 100 km E of Kolwezi, along a potentially mineralized ~120 km strike length (Porter 2014). The deposits have been mined for Cu for at least 500 years (Porter 2014), with modern exploration commencing in 1919. However, whilst most of the main deposits were outlined over the next 50 years, significant mining development only commenced in 2006 due to political risk. The bulk of the primary mineralization occurs within a 20–45 m thick sequence in the reduced lower Mines Subgroup immediately above the regional lithogeochemical contact with the red, oxidized, hematitic, Roches Argilo-Talqueuses (R.A.T.) sediments. The mineralization consists of disseminated chalcocite, bornite, carrollite, chalcopyrite, and minor pyrite, and as blebby grains and sulfide nodules, the latter with quartz-dolomite interiors. The mineable resource, at least initially, has been dominantly supergene and extends to 80–150 m depth, although it may locally be much deeper (Fig. 8). A significant sulfide resource also exists within the Dipeta syncline (R. North, written communication 2023). The principal supergene minerals are malachite, azurite, cobaltoan dolomite, Cu phosphates, Cu silicates, cuprite and heterogenite, together with goethite and hematite. The highest-grade oxide ores are commonly at depths of 15 to 30 m. The underlying Transition zone occurs at depths of 50 to 200 m, comprising cobaltoan dolomite, chrysocolla and supergene chalcocite with minor covellite and digenite, replacing earlier sulfide phases. The base of the Transition zone is highly variable, from 50 to >400 m depth. Strongly leached zones are generally not well developed due to carbonate buffering. However, where there are acid-leachable, carbonate-poor and hydrologically favourable units, leaching may extend to 50–80 m depth down-dip, below which exceptionally rich oxide ore may be developed.

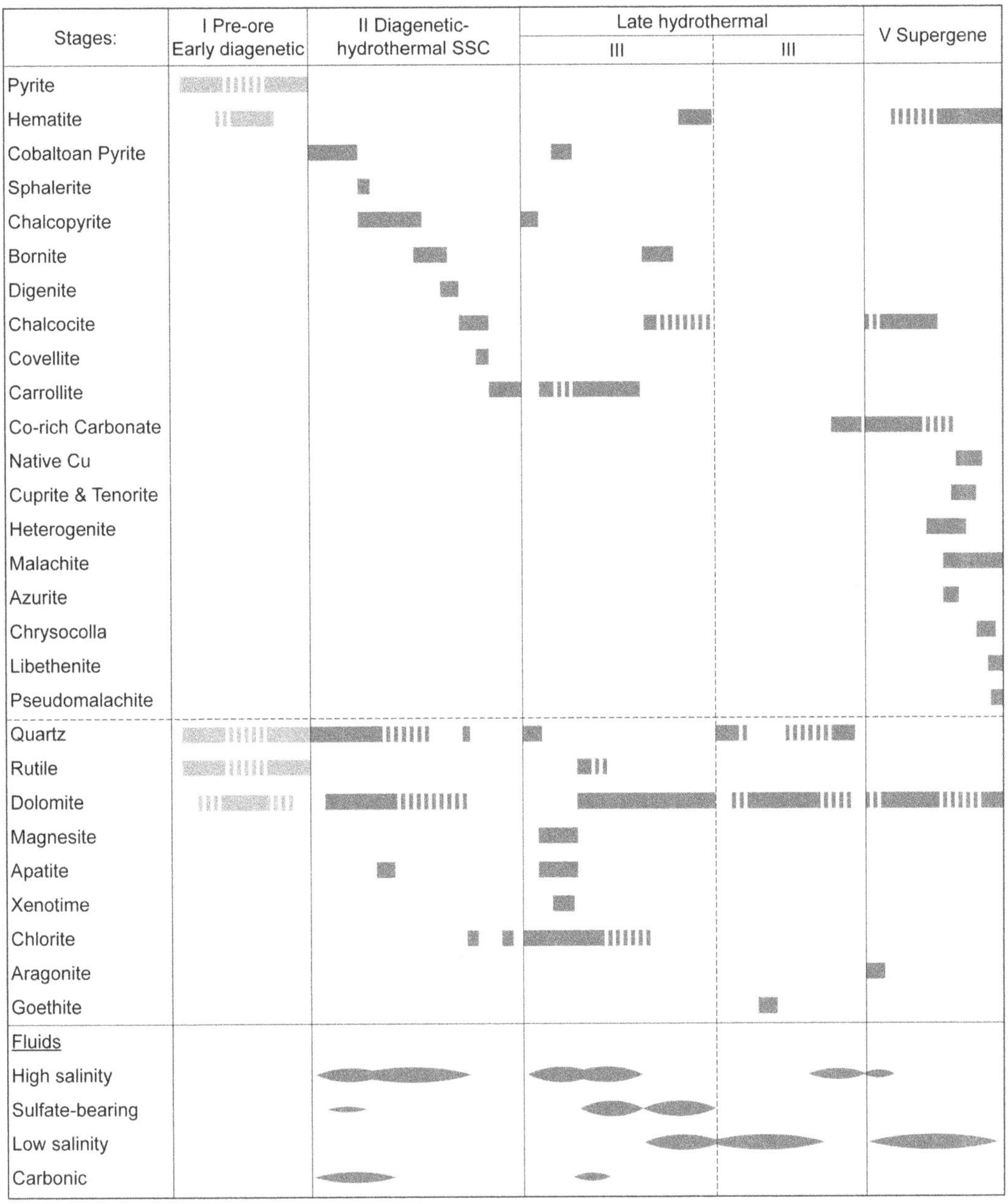

Fig. 7 Mineral paragenetic sequence, fluid types and evolution stages distinguished in the Mashitu South deposit (Sośnicka et al. 2019). SSC = Sediment-hosted Stratiform Copper

3.5 Mn-Co Cap rocks, DRC

Manganese oxide-rich surficial "caps" (duricrusts) about 50 m thick are present within the oxidized zone across the peneplain that extends for over 250 km westward from Katanga. They overlie primary Mn-rich dolomites and typically occur as E-W elongated hills. The elevation of the peneplain declines from east to west, from about 1500 m near Kolwesi to about 1100 m at Kisenge, with the Mn oxide-capped hills rising a further 30–50 m higher (De Putter et al. 2015). Kawama Hill, about 30 km E of Kolwezi, for example, has remnants of a Co-rich Mn cap

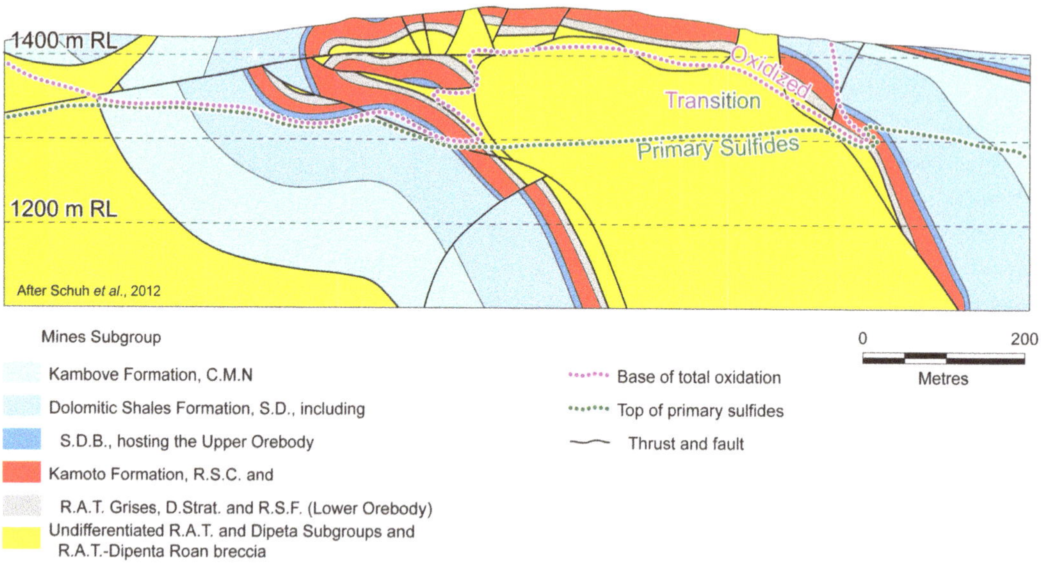

Fig. 8 Geological section through the Kwatabala deposit, Tenke-Fungurume district, Katanga, DRC, showing the principal zones of primary and supergene mineralization. (After Schuh et al. 2012; Porter 2014)

(Fig. 9), exploited by artisanal Co mining, that formed the upper 20 m of the hill (Decrée et al. 2010). The cap developed from supergene silicification of the carbonate rocks from the lower part of the Roan Group. Typically, the carbonate content is reduced to less than 15% (based on field estimates). Within the Co-rich cap, heterogenite occurs as impregnations and is mixed with quartz and complex Fe–Mn–Cu oxides as earthy fillings in fractures and large cavities. In the upper part of the hill, heterogenite is laterally associated with pale pink, kaolinitic clay lenses and with minor Fe oxides. In other mines, such as in the Kolwezi district, heterogenite also occurs as concretions of nearly pure (44–52 wt% Co) botryoidal ore, with varied Cu, Ni, and Mn contents (0–10 wt%.)

Manganese-Co caps also occur in the uppermost part of the regolith in other mines (e.g., Etoile, Luiswishi and Ruashi: Mambwe et al. 2022). However, not all deposits have such Mn caps overlying Cu-rich supergene zones. This is possible because they only formed as strongly resistant duricrusts over the most Mn-rich units, whereas those on protores with lower Mn grades were less well developed and have been eroded. Elsewhere, over carbonate-poor and siliceous rocks, Fe-rich lateritic duricrusts developed instead.

A broadly similar setting, with Mn-rich cap on hilltops, occurs ~250 km W of Kolwesi at Kisenge. Here, supergene enrichment of Mn has formed a world-class Mn deposit from a Late Paleoproterozoic rhodochrosite-spessartine-bearing deposit (ca 50% MnO) associated with interlayered mudstones and shales (De Putter et al. 2015). The enrichment has resulted in a thick supergene Mn oxide deposit outcropping along a series of low hills, rising to 35 m above the plain over a strike length of over 6 km.

4 Discussion: Formation of Supergene Cu-Co-Mn Mineralization in the Copperbelt

Several studies (McGowan et al. 2003; Hitzman et al. 2005, 2012; Selley et al. 2005; Sośnicka et al. 2019) indicate some supergene ores formed from late-stage Na-Ca-Cl-SO_4 fluids (<12.7 mass % NaCl equiv.) whereas others developed through pervasive interaction with low salinity

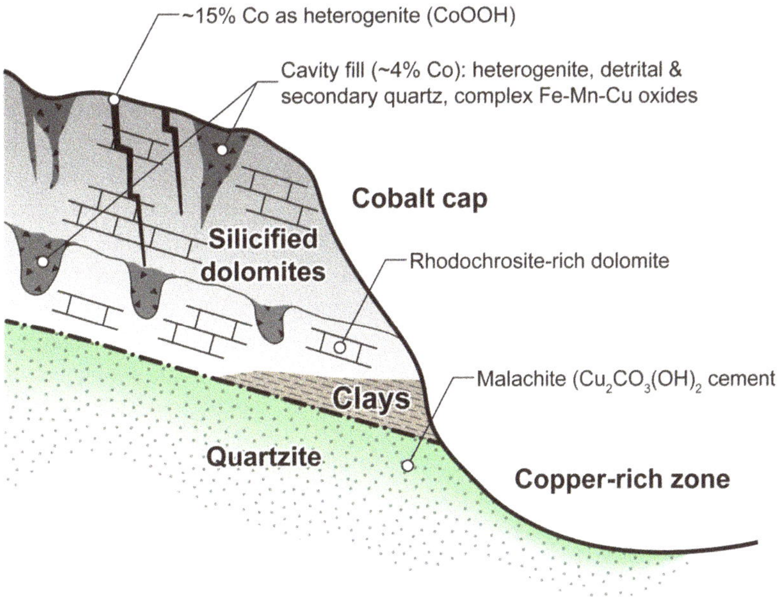

Fig. 9 Remnant cobalt cap, Kawama Hill, Katanga. (After Decrée et al. 2010)

meteoric fluids within formation waters stored in hematitic siliciclastic rocks.

Detailed studies on the Mashitu deposit involved the determination of the isotope signature of freely growing acicular aragonite crystals. These are estimated to fall within the range of Neoproterozoic dolomites observed in other studies and indicate a marine origin (δ18OV-PDB = −6.3 to −5.6‰, δ13CVPDB = −4.6 to −2.6‰). Bladed dolomite underwent replacement in the development of late supergene fibrous massive malachite mineralization (Fig. 4e). The dolomite underwent full replacement by the supergene Cu carbonates accompanied by Fe release from the dolomite and precipitation of associated botryoidal hematite. Maximum temperatures of about 40–50 °C were inferred from fluid inclusions; applying the dolomite-H_2O oxygen isotope fractionation equation (for T = 0–130 °C), Chacko and Deines (2008) determined that the isotopic compositions (δ18OSMOW) of the fluid were in a range between 2.8 and 4.6‰ (Sośnicka et al. 2019). These ratios indicate heavy oxygen isotope enrichment relative to the Neoproterozoic seawater (δ18OSMOW = −2 ± 1‰, Hudson and Anderson 1989) reflecting freshwater addition. Local involvement of high salinity fluids during remobilization and precipitation of supergene minerals is supported by the high salinity of fluid inclusions enclosed within Co carbonates (Sośnicka et al. 2019). Fluid circulation through the siliclastic rocks from freshwater sources and possibly marine waters would have interacted and leached the hematite-altered host rocks and led to the dilution of the highly saline formation waters of diagenesis. Deeper vein sulfide mineralization in the Mines Subgroup developed as a consequence of heating of the recirculating fluids at temperatures in the range of 25–320 °C (McGowan et al. 2003; Sośnicka et al. 2019). Alteration of these zones was limited, but some occurred due to the reactivation of faults as conduits triggered by the onset of the Lufilian Orogeny

Malachite [$Cu_2(OH)_2(CO_3)$] is a common supergene mineral (Fig. 4) in the Central African Copperbelt and is exploited both industrially and by artisanal miners. It is important in explaining the formation of the supergene ores (De Putter et al. 2010). It is commonly present in the fine-grained siliciclastic host rocks as scattered impregnations parallel to the bedding. These may have formed either by absolute supergene enrichment or by in situ replacement of sulfides. Within voids of various types, including dissolution cavities in carbonate host rocks, malachite has precipitated in distinctive forms, such as thick laminated crusts and even speleothems up to several tens of centimetres in length. The malachite ores have precipitated from groundwater through in situ recombination of carbonate ions derived from dissolution of host-rock carbonates (particularly from bladed dolomite, as discussed above) and downward percolating Cu^{2+} leached from oxidizing Cu sulfides during weathering (Fig. 10).

The formation of malachite requires three basic conditions:

1. The existence of voids of any size within the host rock,
2. The presence of mobile Cu^{2+} aq, and
3. The availability of CO_2, either in the form of atmospheric gas dissolved in fluids or dissolved from carbonate minerals as bi-carbonate or carbonate ions (Fig. 11). The speciation and form are dictated by fluid conditions (Rose 1989).

The trap sites for the formation of malachite and other supergene minerals may occur in micro-scale evaporite solution cavities through to the formation of breccias and large karstic features. Most large-scale voids have formed during recent erosion and weathering by dissolution mechanisms such as karst formation. The migration of Cu could be linked to tectonic or climatic processes, including evaporation, as observed in arid climates (see Cameron and Leybourne 2005; Bowell and Davies 2017). Cobalt is less mobile than Cu and has concentrated in surface zones as heterogenite-rich Co-Mn duricrusts ('Co caps'). Cobalt released from carrollite and other sulfides under acid oxidizing conditions is less mobile than Cu and precipitates with Fe $\pm$ Mn oxides as residual lateritic accumulations where the initially acidic weathering fluids are neutralized by reaction with dolomite (Decrée et al. 2010, 2015). Absolute enrichment of Co and Mn also occurs where heterogenite and Mn oxides precipitate from carbonate-rich fluids, together with malachite, deeper in the profile. The relative supergene enrichment of Co, therefore, tends to be greater than Cu. For example, at NChanga, the Cu/Co ratio in primary ores is 20–30, it decreases to 10–15 in the supergene ore (Fig. 3). Trace elements associated with malachite include Ni, Zn and Cd, which can also be strongly concentrated in heterogenite.

Where silica is present (with high groundwater pH, typically above 8.5), then chrysocolla or amorphous Cu silicates may also coexist with malachite (Fig. 11). Karstic features, common in the Katangan Copperbelt, constitute highly favourable sites for the accumulation of large amounts of secondary minerals and may form economic orebodies, such as at Tenke Fungurume (Schuh et al. 2012).

The REE geochemistry is consistent with this mechanism, with the mineralogical transformations in the weathered profile corresponding with geochemical variations, including leaching of LREE and enrichment of HREE. The leached LREE accumulates in the cementation zone, where both secondary chalcocite-type sulfides and residual and replacement goethite form by oxidation within the rims and cracks in chalcopyrite (Fontaine et al. 2020). A similar enrichment of LREE in deeper supergene zones, relative to Co-Mn-rich surface duricrusts, was noted by Decrée et al. 2015, where heterogenite is closely admixed with REE-rich carbonate.

5 Chronology of Supergene Enrichment

The secondary ores were formed by sulfide oxidation under humid tropical weathering conditions (Cailteux et al. 2005a, b; De Putter et al. 2010). The weathering fluids were probably mildly acidic (pH ~ 6) and oxidizing on a regional scale, with the acidity of these fluids enhanced locally by the extensive oxidation of primary sulfides, which released high quantities of sulfuric acid as evidenced by widespread sulfide oxidation. These acid fluids were progressively neutralized by the reactive gangue minerals such as carbonates (magnesite and dolomite) and Mg-chlorite, with the dissolution of carbonates proceeding faster than for Mg-chlorite (Sverdrup 1990). This led to the development of the supergene ores in the near-surface environment, as

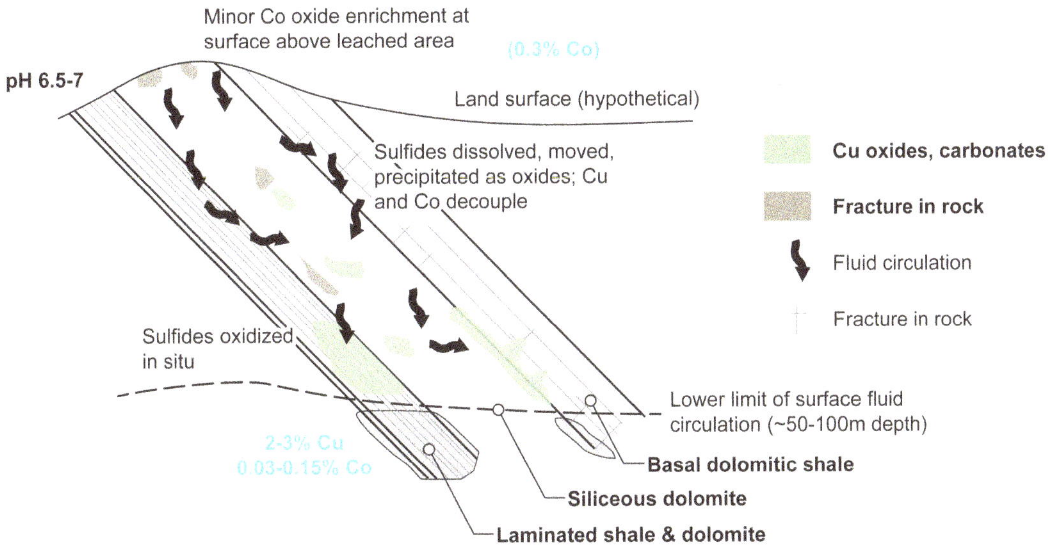

Fig. 10 Schematic representation of fluid circulation and Cu solution, mobilization and precipitation at Tenke-Fungurume. (After Fay and Barton 2012)

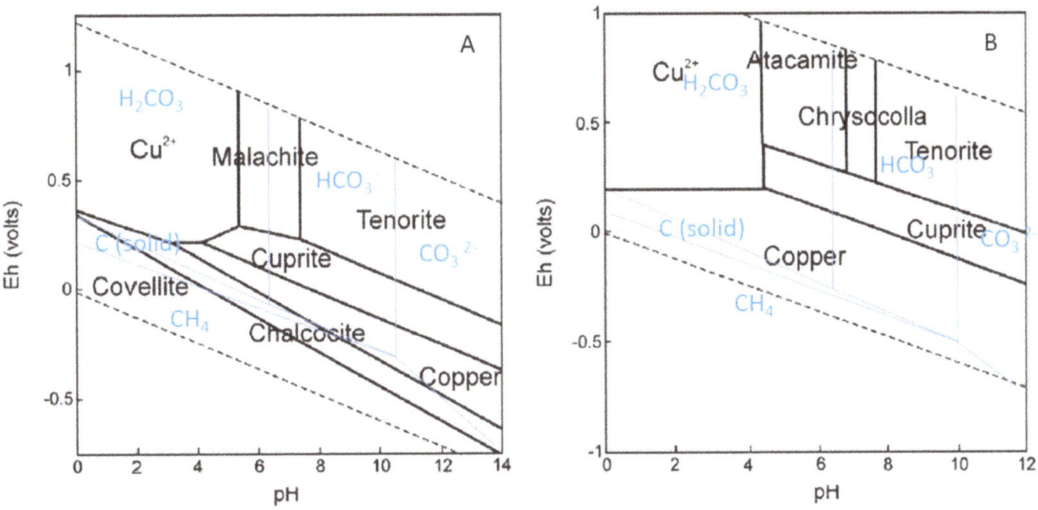

Fig. 11 Eh–pH diagrams for Cu phases and species: common parameters to all scenarios are $aCu^{2+} = 10^{-4}$ M, T = 25 °C, and P = 1,013 bar. (**a**) Groundwater chemistry of present-day Copperbelt water ($PCO_2 = 10^{-2}$ and $[SO_4^{2-}] = 200$ mg/L). (**b**) Field of chrysocolla and atacamite by addition of silica and chloride. (Modified after De Putter et al. 2010)

described above. The high concentrations of metals hosted by the secondary mineral phases are due to residual enrichment from the tenor of the primary sulfide mineralization by the dissolution of silicates and carbonates during weathering (Cailteux et al. 2005a, b; De Putter et al. 2010), supplemented by absolute enrichment, especially of Cu, leached from overlying horizons of the regolith.

The age of supergene ore formation in the Katanga region has been estimated from ^{39}Ar-^{40}Ar dating of cryptomelane from the western part of the Katanga, and U–Pb SIMS dating of heterogenite from the east, which provide evidence for multiple phases of weathering (Decrée et al. 2010; De Putter et al. 2010, 2015, 2016; De Putter and Ruffet 2020). The dating record covers a period of approximately 80 Ma starting in the Late Cretaceous (c. 76 Ma) and showing significant events in multiple pulses from the mid-Eocene (c. 45 Ma and c. 35 Ma), the Oligo-Miocene (c. 23.8 Ma) Miocene (10.5 Ma) and Pliocene (3.6 and 2.4 Ma). The regolith hosting the supergene ores is considered to be equivalent to the lateritic regolith developed over the largely unmineralized siliclastic country rocks of the region (Decrée et al. 2010; Alexandre 2002; Selley et al. 2005), which have ferruginous near-surface horizons overlying kaolinitic saprolite.

Although no direct dating of malachite is possible, if the conceptual formation indicated in Fig. 10 is correct, then it is contemporaneous with surface heterogenite deposits, which are Pliocene in age (Decrée et al. 2010).

6 Summary

The Central African Copperbelt (CAC) spans across Zambia and the Democratic Republic of Congo (DRC). The CAC contains approximately 15% of the world's known Cu and more than half of the world's known Co resources. In addition, it hosts substantial resources of other metals, including Au, U, Zn, Pb and Ge. Mining has been focused on high-grade supergene deposits formed by oxidation of the primary Cu and Co sulfide minerals, principally chalcopyrite ($CuFeS_2$), bornite (Cu_5FeS_4), chalcocite (Cu_2S) and carrollite ($CuCoS_2$), hosted in Neoproterozoic fine-grained siliciclastic and carbonate sedimentary rocks. The Oxidized and Transition zones occur down to about 100–150 m in depth. Rapid neutralization of the released acidity by carbonate minerals (mainly dolomite) and Mg-chlorite in the host rock, and concomitant precipitation of a wide range of oxidized mineral phases: goethite, malachite, heterogenite, pseudomalachite and amorphous Cu silicates initially results in only minor supergene enrichment. However, significant enrichment with recurring mineralogical successions, including heterogenite–malachite or malachite-Cu-silicate aggregates, develop under the influence of multiple weathering events and associated groundwater flow. Fluid inclusion data suggest that supergene minerals precipitate at temperatures below <50 °C from predominantly low salinity waters, with local pulses of saline fluids during the formation of decoupled supergene Cu and Co mineralization.

The age of supergene ore formation in the Katanga region has been estimated from ^{39}Ar-^{40}Ar dating of cryptomelane from the western part of the Katanga and U–Pb SIMS dating of heterogenite from the east, which provide evidence for multiple phases of weathering. The dating record covers a period of approximately 80 Ma starting in the Late Cretaceous (c. 76 Ma) and showing significant events in multiple pulses from the mid-Eocene (c. 45 Ma and c. 35 Ma), the Oligo-Miocene (c. 23.8 Ma) Miocene (10.5 Ma) and Pliocene (3.6 and 2.4 Ma). The regolith hosting the supergene ores is equivalent to the lateritic regolith developed over the largely unmineralized siliclastic country rocks of the region, which have ferruginous near-surface horizons overlying kaolinitic saprolite. Although no direct dating of malachite is possible, the contemporaneous occurrence of surface heterogenite deposits in many areas indicates Pliocene.

Acknowledgments Michael Whitbread and Robert North are thanked for their comments on a draft of the manuscript, including helpful insights into the mineralogy of the deposits. Society of Economic Geology Publishing Company is thanked for approval to reproduce Figs. 1, 2 and 8. Elsevier Publishing is thanked for approval to reproduce Figs. 5, 6, 7, and 9. Springer Publishing is thanked for approval to reproduce Fig. 10.

References

Alexandre J (2002) Les cuirasses latéritiques et autres formations ferrugineuses tropicales. Exemple du Haut Katanga meridional. Annales du Musée Royal de l'Afrique Centrale, Sciences Géologiques 107: 118

Annels AE (1974) Some aspects of the stratiform ore deposits of the Zambia Copperbelt. In: Bartholomè P (ed) Gisement stratiformes et provinces cuprifères Société Géologique de Belgique, pp 235–254

Audeoud D (1982) Les mineralisations uraniferes et leur environment a Kamato, Kambove et Shinkolobwe, Shaba, Zaire Centre de Recherches sur la Geologie de l'Uranium, Nancy, France and the Universite Claude-Bernard, Lyon, France. These de troisieme cycle, 204 p

Bateman AM (1930) Ores of the Northern Rhodesian Copperbelt. Econ Geol 25:365–418

Bowell RJ (1994) Sulfide oxidation and production of gossans, Ashanti Mine, Ghana. Int Geol Rev 36:732–752

Boyle RW, Brown AC, Jefferson CW, Jowett EC, Kirkham RV (eds) (1989) Sediment-hosted Stratiform copper deposits, vol 36. Geological Association of Canada, Special Paper, pp 97–110

Bowell RJ, Davies AA (2017) Assessment of supergene uranium-vanadium anomalies, Meob Bay deposit, Namibia. Geochem Explor Environ Anal 17:101–112

Broughton DW, Rogers T (2010) Discovery of the Kamoa copper deposit, Central African Copperbelt, DRC. Econ Geol Special Publications 15:287–297

Cailteux J (1994) Lithostratigraphy of the Neoproterozoic Shaba-type (Zaire) Roan Supergroup and metallogenesis of associated stratiform mineralization. J Afr Earth Sci 19:279–301

Cailteux J, De Putter T (2019) The Neoproterozoic Katanga Supergroup (D.R. Congo): state-of-the-art and revisions of the lithostratigraphy, sedimentary basin and geodynamic evolution. J Afr Earth Sci 150: 522–531

Cailteux JLH, Kampunzu ABH, Batumike MJ (2005a) Lithostratigraphic position and petrographic characteristics of R.A.T. ("Roches Argilo-Talqueuses") Subgroup, Neoproterozoic katangan belt (Congo). J Afr Earth Sci 42:82–94

Cailteux JLH, Kampunzu AB, Lerouge C, Kaputo AK, Milesi JP (2005b) Genesis of sediment-hosted stratiform copper-cobalt deposits, central African Copperbelt. J Afr Earth Sci 42:134–158

Cameron EM, Leybourne MI (2005) Relationship between groundwater chemistry and soil geochemical anomalies at the Spence copper porphyry deposit, Chile. Geochem Explor Environ Anal 5:135–145

Chacko T, Deines P (2008) Theoretical calculation of oxygen isotope fractionation factors in carbonate systems. Geochimica Cosmochimica Acta 72:3642–3636

De Putter T, Ruffet G (2020) Supergene manganese ore records 75 Myr-long Campanian to Pleistocene geodynamic evolution and weathering history of the Central African Great Lakes Region—Tectonics drives, climate assists. Gondwana Res 83:96–117. https://doi.org/10.1016/j.gr.2020.01.021

De Putter T, Mees F, Decrée S, Dewaele S (2010) Malachite, an indicator of a major Pliocene Cu remobilization in karstic environment (Katanga, Democratic Republic of Congo). Ore Geol Rev 38:90–100

De Putter T, Ruffet G, Yans J, Mees F (2015) The age of supergene manganese deposits in Katanga and its implications for the Neogene evolution of the African Great Lakes Region. Ore Geol Rev 71:350–362

De Putter T, Bayon G, Mees F, Ruffet G, Smith T (2016) Cenozoic sedimentation history of the Congo Basin revisited. In: Programme and abstracts of the SGF workshop source to sink: a long term perspective of sediment budgets and sources characterization. Rennes, France, pp 51–52

Decrée S, Deloule E, Ruffet J, Dewaele S, Mees F, Marignac C, Yans J, De Putter T (2010) Geodynamic and climate controls in the formation of Mio-Pliocene worldclass cobalt and manganese oxidized ores in the Katanga (DRCongo). Mineral Deposita 45:621–629

Decrée S, Pourret O, Baele J-M (2015) Rare earth element fractionation in heterogenite (CoOOH): Implication for cobalt oxidized ore in the Katanga Copperbelt (Democratic Republic of Congo). J Geochem Explor 159: 290–301

Fay I, Barton MD (2012) Alteration and ore distribution in the Proterozoic Mines Series, Tenke-Fungurume Cu–Co district, Democratic Republic of Congo. Min Deposita 47:501–519

Fontaine L, De Putter T, Bernard A, Decrée S, Cailteux J, Wouters J, Yans J (2020) Complex mineralogical-geochemical sequences and weathering events in the supergene ore of the Cu–Co Luiswishi deposit (Katanga, D.R. Congo). J Afr Earth Sci 161:1–16

Gautier G, Francois A, Deliens M, Piret P (1980) Famous Mineral Localities: the uranium deposits of the Shaba region, Zaire. Mineral Rec 20:265–288

Hitzman M (2010) Formation of sedimentary rock-hosted stratiform copper deposits through earth history. Econ Geol 105:627–639

Hitzman MH, Thorson J, Broughton DW, Kirkham RV, Selley D (2005) The sediment-hosted stratiform copper ore system. Econ Geol 100th Anniversary Volume:609–642

Hitzman MW, Broughton D, Selley D, Woodhead J, Wood D, Bull S (2012) The Central African Copperbelt: diverse stratigraphic, structural, and temporal settings in the World's Largest Sedimentary Copper District. In: Hedenquist JW, Harris M, Camus F (eds) Geology and genesis of major copper deposits and districts of the world: a tribute to Richard H. Sillitoe. Society of Economic Geologists, Inc. Special Publication, pp 487–514

Hudson JD, Anderson TF (1989) Ocean temperatures and isotopic compositions through time. Earth Environ Sci Trans R Soc Edinburgh 80:183–192

Jin L, Ma L, Dere A, White T, Mathur R, Brantley SL (2017) REE mobility and fractionation during shale weathering along a climate gradient. Chem Geol 466: 352–379

Lerouge C, Cailteux J, Kampunzu AB, Milesi JP, Flehoc C (2005) Sulphur isotope constraints on formation conditions of the Luiswishi ore deposit, Democratic Republic of Congo (DRC). J Afr Earth Sci 42:173–182

Mambwe P, Shengo M, Kidyanyama T, Muchez P, Chabu M (2022) Geometallurgy of Cobalt Black Ores in the

Katanga Copperbelt (Ruashi Cu-Co Deposit): a new proposal for enhancing cobalt recovery. Minerals 12: 295. https://doi.org/10.3390/min12030295

Master S, Rainaud C, Armstrong RA, Phillips D, Robb LJ (2005) Provenance ages of the Neoproterozoic Katanga Supergroup (Central African Copperbelt), with implications for basin evolution. J Afr Earth Sci 42:41–60

McGowan RR, Roberts S, Foster RP, Boyce AJ, Coller D (2003) Origin of the copper-cobalt deposits of the Zambian Copperbelt; an epigenetic view from Nchanga. Geology 31:497–500

McKinnon D, Smit NJ (1961) NChanga. In: Mendelsohn F (ed) The geology of the Northern Rhodesian Copperbelt. Macdonald & Co. (Publishers) Ltd, London, pp 234–275

Mendelsohn F (ed) (1961) The geology of the Northern Rhodesian Copperbelt. Macdonald, London, 523 p

Pirard E (2011) A history of early copper exploration in Katanga (D.R. Congo). In: Ortiz JE, Puche O, Rábano I, Mazadiego LF (eds) History of research in mineral resources, Madrid, pp 271–280

Porter M (2014) Tenke-Fungurume-Kwatebala, Goma, Fwaulu Katanga, Dem. Rep. Congo. https://portergeo.com.au/database/mineinfo.asp?mineid=mn873

Rainaud C, Master S, Armstrong RA, Robb LJ (2005) Geochronology and nature of the Palaeoproterozoic basement in the Central African Copperbelt (Zambia and the Democratic Republic of Congo), with regional implications. J Afr Earth Sci 42:1–31

Rose AW (1989) Mobility of copper and other heavy metals in sedimentary environments. In: Boyle RW, Brown AC, Jefferson CW, Jowett EC, Kirkham RV (eds) Sediment-hosted Stratiform copper deposits, vol 36. Geological Association of Canada, Special Paper, pp 97–110

Sato M (1992) Persistency-field Eh-pH diagrams for sulfide and their application to supergene oxidation and enrichment of sulfide ore bodies. Geochimica Cosmochimica Acta 56:3133–3156

Schuh W, Leveille RA, Fay I, North R (2012) Geology of the Tenke-Fungurume Sediment-Hosted Strat-Bound Copper-Cobalt District, Katanga, Democratic Republic of Congo. In: Hedenquist JW, Harris M, Camus F (eds) Geology and genesis of major copper deposits and districts of the world: a tribute to Richard H. Sillitoe. Society of Economic Geologists, Inc. Special Publication, pp 269–302

Selley D, Broughton D, Scott RJ, Hitzman M, Bull SW, Large RR, McGoldrick PJ, Croaker M, Pollington NL, Barra F (2005) A new look at the geology of the Zambian Copper belt. Econ Geol 100th Anniversary Volume:965–1000

Sośnicka M, Gigler GM, Kinnaird JA, Przybyłowicz WJ, Chimviga J, Master S, Plessen B (2019) Mineralizing fluids of the supergene-enriched Mashitu South Cu-Co deposit, Katanga copperbelt, DRC. Ore Geol Rev 109: 201–228

Sverdrup HU (1990) The kinetics of base cation release due to chemical weathering: Lund University Press, Lund, 246 p

Thornber MR, Taylor GF (1992) The mechanism of sulfide oxidation and gossan formation. In: Butt CRM, Zeegers H (eds) Handbook of exploration geochemistry 4: Regolith exploration geochemistry in tropical and subtropical terrains. Elsevier, Amsterdam, pp 119–202

Lateritic Bauxite Deposits

Maxim Seredkin and Graham Jeffress

Abstract

The majority of modern lateritic bauxite deposits are found in a small number of provinces located in the present-day tropical and subtropical belts. Ancient bauxites formed under similar climatic conditions are located at higher latitudes.

Bauxites form in hot, humid climates with pronounced wet and dry seasons in well-drained environments. Under similar but drier conditions, ferruginous lateritic regoliths form instead. Where climates change to very wet, rainforest conditions, such as in Amazonia, bauxites are replaced by clay materials. Surface zones are overburdened sedimentary clays and, with depth, metasomatic kaolinite, then bauxites.

The weathering processes that result in the formation of bauxitic lateritic regoliths are typically the result of acid leaching by infiltration of rainwater, which has a pH of 5.0–5.5, mainly due to dissolved atmospheric CO_2. A Transition zone at the base of the bauxite horizon is controlled by seasonal fluctuations of the groundwater level due to the direct penetration of water from the surface by open fissures during the wet season.

Bedrock composition and geomorphology are critical controls in the formation of bauxite deposits. In order to form bauxites, bedrock should be enriched in alumina, have good filtration coefficients, and ideally have sub-horizontal bedding for stratal formations. The distribution and quality of bauxites are influenced by geomorphological factors. Many of the examples are drawn from the Fouta Djallon Bauxite sub-province, Guinea.

1 Introduction

'Bauxite is a naturally occurring, heterogeneous material composed primarily of one or more aluminium hydroxide minerals, plus various mixtures of silica, iron oxide, titania, aluminosilicate, and other impurities in minor or trace amounts. The principal aluminium hydroxide minerals found in varying proportions with bauxites are gibbsite and the polymorphs boehmite and diaspore' (USGS 2021a). Bauxites are named after type deposits at Les Baux in Provence, France (Freyssinet et al. 2005).

Aluminium is primarily sourced from bauxite and refined to alumina (using the Bayer Process) before being smelted (via the Hall-Heroult

M. Seredkin (✉) · G. Jeffress
Environmental Resource Management (ERM), West Perth, WA, Australia
e-mail: Maxim.Seredkin@erm.com; maximseredkin75@gmail.com; graham.jeffress@erm.com

R. J. Bowell, C. R. M. Butt (eds.), *Geology, Geochemistry and Formation of Supergene Mineral Deposits in Deeply Weathered Terrain*, Mineral Resource Reviews, https://doi.org/10.1007/978-3-031-75733-4_10

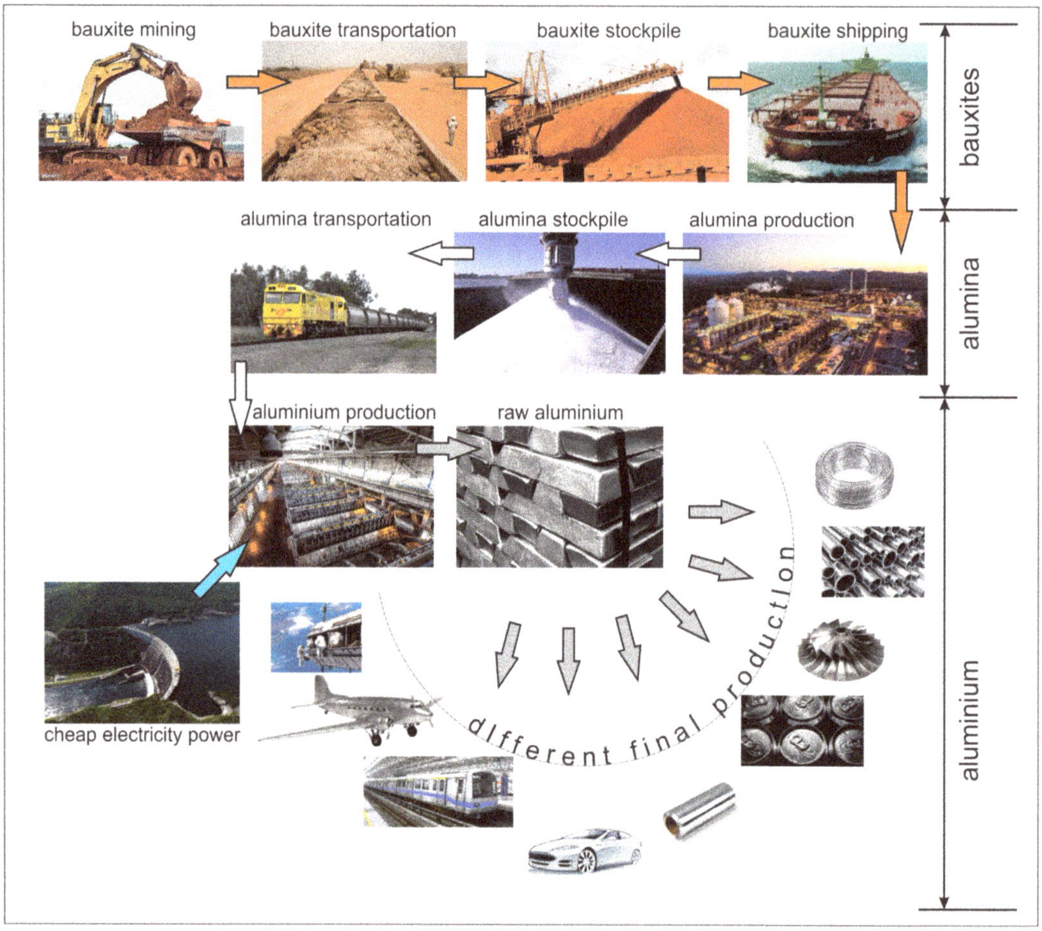

Fig. 1 Aluminium production cycle. (Prepared using images from open sources)

process). The process of transformation of bauxite to aluminium is summarised in Fig. 1. Other natural sources of aluminium, such as nepheline rocks and clays, are very minor.

The common classification of bauxites is based on their genesis and the rocks from which they are derived (Bardossy 1982; Bardossy and Aleva 1990; Tardy and Roquin 1998). Two main types of bauxite deposits are identified (Freyssinet et al. 2005):

- *Lateritic bauxite deposits*, developed from the intense weathering of aluminosilicate rocks under subtropical to tropical conditions, comprise approximately 90% of the global bauxite resources and,
- *Karst bauxite deposits*, which are Al oxide accumulations caused by the dissolution of carbonates and the concentration of Al derived from the weathering of associated aluminosilicates, e.g., in interbedded volcanic units.

Most karst deposits formed under similar conditions and at the same periods as lateritic bauxite deposits (Freyssinet et al. 2005) and may be regarded as a special case of lateritic bauxites formed in cavernous carbonate-rich rocks, usually with well-developed drainage.

The majority of lateritic bauxite deposits are found in a small number of Provinces. Bauxites of

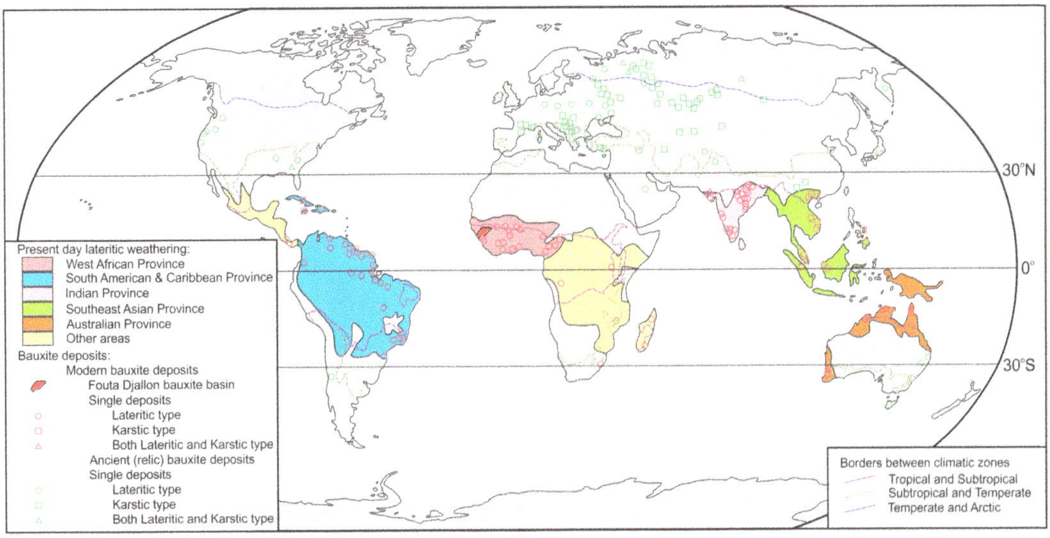

Fig. 2 Location of the main lateritic bauxite deposits and present-day lateritic weathering. (After Bardossy and Aleva 1990; Tardy and Roquin 1998; USGS 2021b)

Tertiary to Recent origin are only located in current-day tropical and subtropical belts (Fig. 2), whereas ancient bauxites formed during the Palaeozoic to Mesozoic Periods are mostly located in higher latitude areas. These older bauxite deposits are considered to have formed under similar conditions as modern bauxites, but with later resilicification or limonitisation with degradation due to tectonics and global climate changes (Tardy and Roquin 1998).

Five major provinces for lateritic bauxites are recognised (Freyssinet et al. 2005):

- *West African Province*: predominantly Guinea, with some deposits in Guinea-Bissau, Côte d'Ivoire, Burkina Faso, Sierra Leone, Ghana and Mali;
- *South American and Caribbean Province*: the Guiana Shield (Guyana, French Guiana, Suriname), central Brazilian Shield, Jamaica and other smaller deposits;
- *Indian Province*: on the plateaux inland from the western and eastern coasts;
- *Southeast Asian Province*: principally China and Vietnam, but also Laos, Cambodia and Malaysia; and,
- *Australian Province*: comprising regions in the southwest and northeast of the continent.

Some minor provinces with small- and medium-sized deposits are also found in SE Africa, the southern United States, the Urals and Timan regions of Russia, Kazakhstan, and SE Europe (Freyssinet et al. 2005).

World bauxite production is concentrated in the major bauxite provinces (Fig. 3) (Williams 2019). The largest and most productive region, the Fouta Djallon sub-province, is part of the large West African bauxite province (Fig. 2) and hosts bauxite resources of more than 42 billion tonnes at grades ranging from 40% to 65–70% Al_2O_3—representing around 45% of the world's known resources (Fig. 4) (Mamedov et al. 2010).

There is extensive literature on bauxite occurrence and formation in most of these Provinces. Rather than summarise this literature, the chapter concentrates on bauxites in the Fouta Djallon region of Guinea for the following reasons:

- It is the world's largest and most productive bauxite region
- It has examples of most of the known different types of bauxite

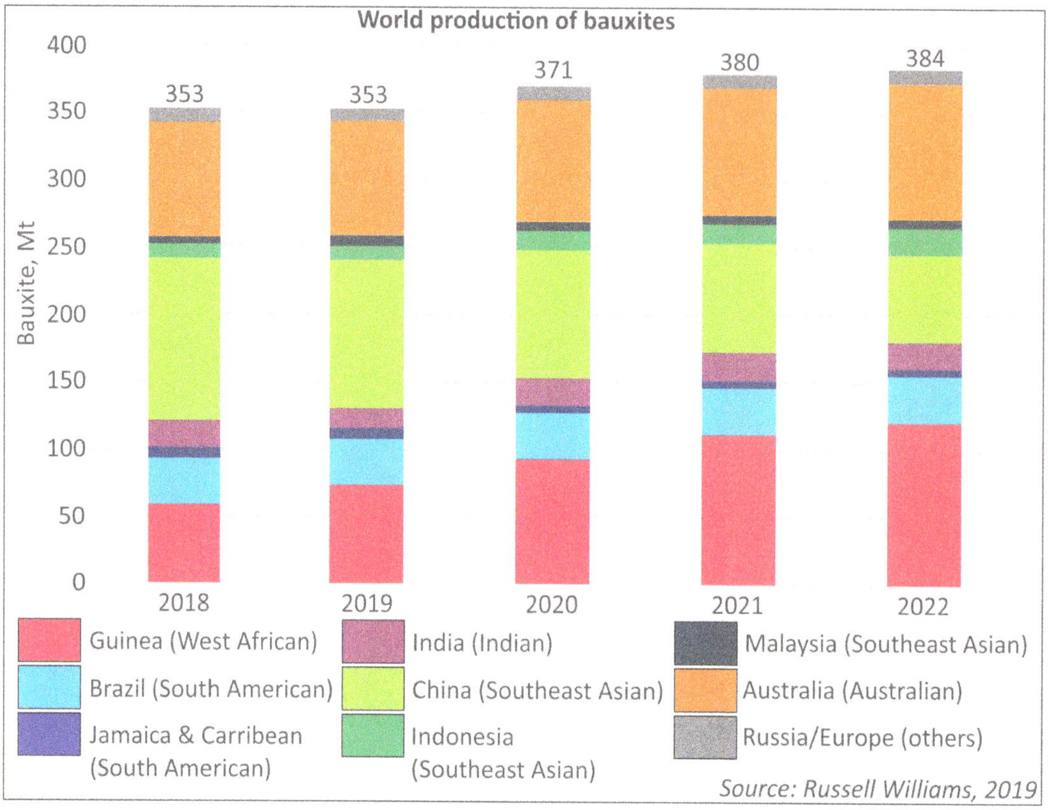

Fig. 3 World bauxite production estimates. (After Williams 2019)

- It is poorly represented in the mainstream (English language) literature
- The chapter contains previously unpublished data and interpretations
- The author worked as a chief geologist in the lead exploration company in this sub-province for five years and later as a consultant for many clients in this region.

Bauxite mining in the Fouta Djallon sub-province commenced in the 1950s on the Los Islands, and over a period of almost 50 years to 2015, three mines with a total capacity of 15–20 mtpa were in operation. Over the next seven years, production increased by 500% (Fig. 5), and this is not a final limit of potential production.

There are several different types of bauxite in the Fouta Djallon sub-province province, which, together with those in the S. American province, provide a model for bauxite genesis.

2 Deposit Characteristics

Lateritic bauxites can develop over a variety of aluminium-rich rocks—regardless of the igneous, sedimentary or metamorphic origin—under humid, tropical conditions. The structure of the bauxite-bearing weathering profile is analogous to ferruginous or nickeliferous lateritic profiles (Nahon 1991; Tardy 1997). Most studies of the economic geology of lateritic bauxite deposits were undertaken during the 1970s to 1990s (Valeton 1972; Jacob 1984; Bardossy and Aleva 1990).

The general zonation of a classical bauxite-bearing regolith profile is shown in Table 1.

A bauxitic regolith may vary in thickness from 1–2 m to 50–100 m depending on local geological, hydrogeological and geomorphological conditions. It may have the full sequence of

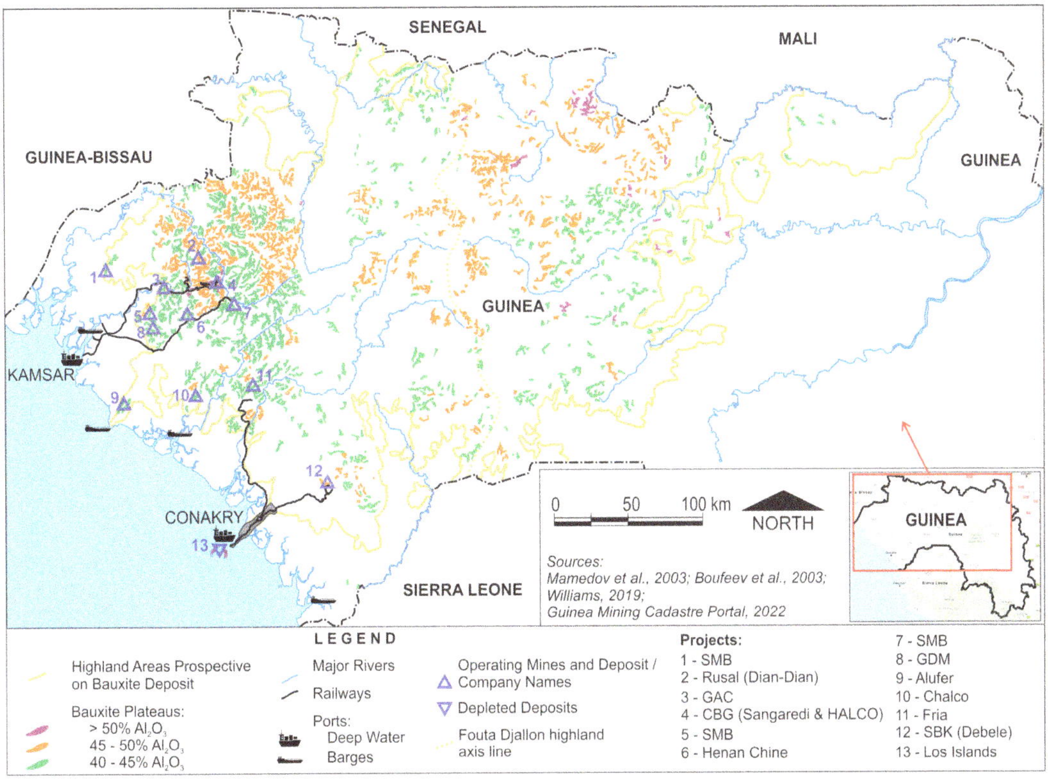

Fig. 4 Infrastructure, bauxite distribution by grade and bauxite producers in Guinea (the principal part of the Fouta Djallon sub-province). (Prepared by the authors based on Mamedov et al. 2010; Boufeev et al. 2003; Williams 2019; Guinea Mining Cadastre Portal 2022)

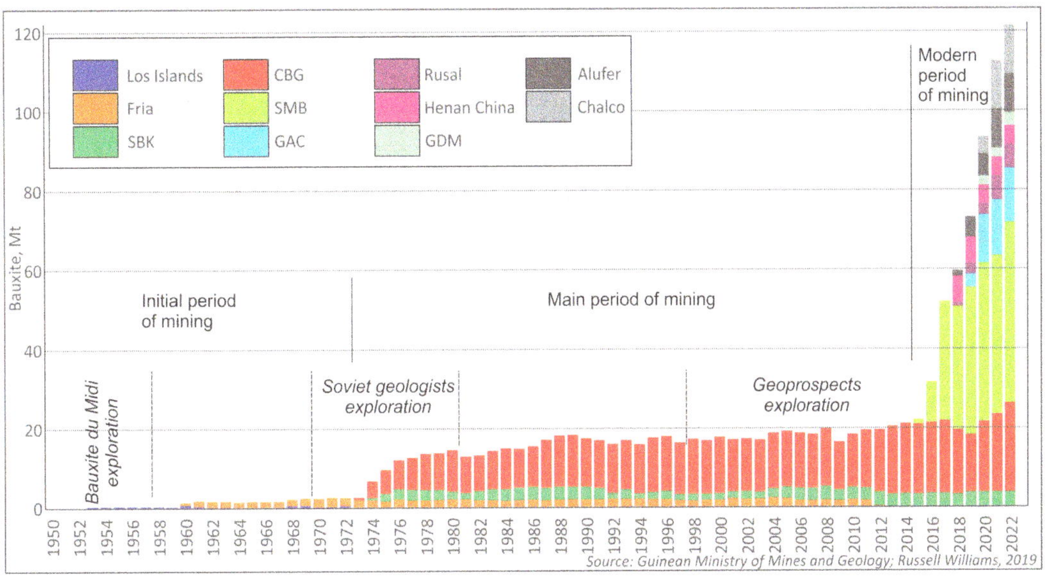

Fig. 5 Historical bauxite production in Guinea (1950s–2022). (After Guinean Ministry of Mines and Geology 2022; Williams 2019)

Table 1 Idealised zonation of bauxite bearing lateritic regolith, detailed mineralogy is discussed below. (After Mamedov et al. 2010)

Regolith	Zone	Main elements	Common major minerals
Pedolith	Soil	All leached, transported/ sedimentary material	Kaolinite, quartz, organic matter
	Ferruginous cap	Fe	Goethite (± hematite)
	Bauxite zone	Al, Fe	Gibbsite (± boehmite), goethite (± alumogoethite, hematite)
	Upper transition zone	Fe, Al	Goethite (± alumogoethite, hematite), gibbsite (± boehmite)
Saprolith	Lower transition zone	Fe, Si	Goethite (± hematite), kaolinite, quartz
	Saprolite	Si, Al, Fe, Mg, Ca	Kaolinite, quartz, hydromica, Illlite, chlorite, halloysite, montmorillonite
	Saprock	Si, Al, Fe, Mg, Ca, Na, K	(Disintegrated and weakly weathered zone)
Protolith	Bedrock	(Primary composition)	(Primary rock minerals)

zones, or one or more may be absent due to various factors such as erosion.

Three types of lateritic bauxite are distinguished according to the classification developed by Tardy (1997):

- *Orthobauxites* are considered the classical bauxite profile and are formed in a single weathering stage without significant later modifications. The bauxite-bearing horizon is located above the saprolite/clay zone and is composed of gibbsite, goethite, and hematite. Iron content increases from the bottom to the top of the profile ('direct' zonation).
- *Metabauxites* are lateritic bauxites that show a similar succession of horizons to orthobauxites but are more aluminous and less ferruginous, they also contain relatively greater proportions of boehmite. Metabauxites are considered by Tardy (1997) to result from orthobauxite transformation due to a palaeoclimatic change to less humid conditions; the transformations are due to dehydration that progresses downwards from the surface, creating conditions that modify the solubility of Fe and Al oxyhydroxides.
- *Cryptobauxites* differ from orthobauxites and metabauxites by the presence of a kaolinite horizon in the top part of the profile (thickness can vary from a few metres to 70 m). This kaolin zone develops upon a gibbsite horizon from top to bottom. The gibbsite horizon consists of a soft matrix composed of goethite, gibbsite and minor kaolinite. Hematite nodules occur in the middle part of this horizon. The most probable explanation for the development of the upper kaolinite horizon is due to climate change (Lucas 1989; Negrão et al. 2018). The profile formed initially as a classic lateritic profile, with Fe accumulation in a ferruginous duricrust. Later formation of kaolin requires an external source of silica, and some may have formed due to the development of a soil horizon enriched in silica and organic matter or by termite activity. Also, some of these kaolin zones may be (at least partly) sedimentary in origin (Negrão et al. 2018).

A schematic comparison of 'idealised' lateritic weathering crust profiles for orthobauxites, metabauxites and cryptobauxites is shown in Fig. 6.

3 Bedrock Geology

Bedrock composition and location are important for the formation of bauxite deposits. In order to form bauxites, bedrock should be enriched in alumina to have good filtration coefficients (permeability), and ideally, stratiform rocks should have sub-horizontal bedding. Lateritic bauxites are developed on nepheline syenite (Pocos de

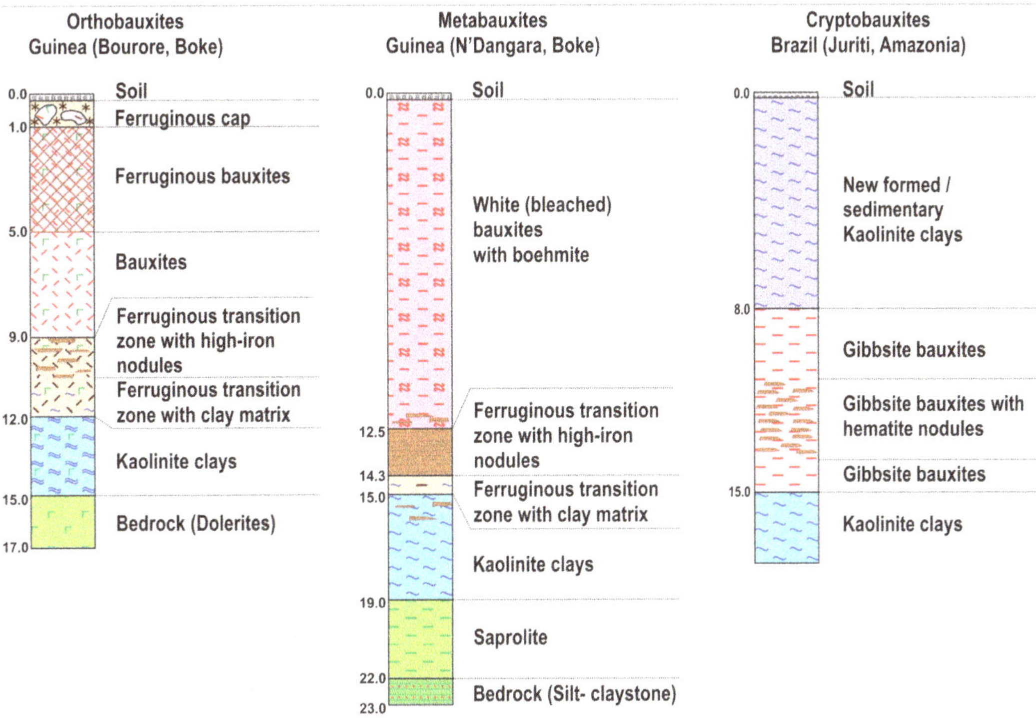

Fig. 6 Types of bauxite-bearing profiles. (Based on the author's observations and Mamedov et al. 2010; Freyssinet et al. 2005; Lucas 1989; Negrão et al. 2018)

Caldas in Brazil, Los Islands in Guinea), granitic rocks (Darling Range in Australia, some deposits in Brazil, Suriname and Guinea), basalt (Deccan plateau in India, many deposits in Guinea) and on clastic sediments (Weipa, Australia, most deposits in Guinea). Karst bauxites have been formed from silicate substrates such as tuffs in or overlying carbonate bedrock such as those in SE Europe, Russia and the Caribbean region (Jamaica, Dominican Republic).

The Fouta Djallon sub-province, located in the southwestern margin of the West African Craton, provides some of the best examples that demonstrate the importance of bedrock composition for the genesis of bauxite deposits. The Archaean-Proterozoic basement is covered by sedimentary rocks from Neoproterozoic to Palaeozoic age in three basins (Mamedov et al. 2010) (Fig. 7). A multi-layered doleritic sill system and ultramafic and alkaline complexes have been intruded into both the basement and cover sediments.

The most favourable bedrock is nepheline syenite in the Los Islands, 5 km SW of Conakry, where high-quality bauxites (>50% Al_2O_3) have formed in less than a million years (Mamedov et al. 2010). Most of the moderate-high quality (>43–45% Al_2O_3) bauxites at Fouta Djallon were formed in sub-horizontal clastic sediments with quite high permeability in silt- and clay-stones of the Faro Series in the Bowe Basin, and sediments in the Tougue Basin (Fig. 7). Bauxites formed in dolerite are usually ferruginous, with 38–43% Al_2O_3. Some bauxite deposits have also formed upon basement rocks of granite-diorite composition in gneissic and magmatic complexes (Fig. 7). Bauxites at Fouta Djallon are typically orthobauxites without bleaching (see Sect. 4 below).

Sediments of the Télimélé Series have favourable compositions for bauxite formation, but their permeability is low due to thin layers of claystone, so fewer bauxite deposits have formed than on the Faro Series. The filtration

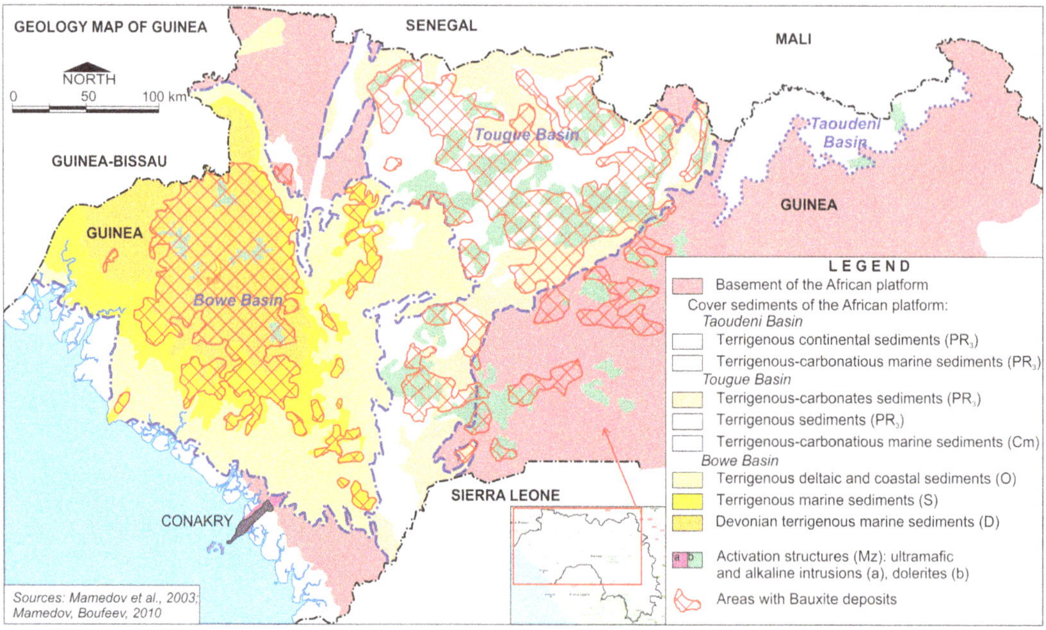

Fig. 7 Bedrock geology and location of bauxite deposits in Guinea (which comprises most of the Fouta Djallon Bauxite sub-province). (Mamedov et al. 2003, 2010)

coefficient/permeability and bedding planes of rocks significantly influence bauxite formation. In particular, the water table rises in the rainy season within local synformal structures if the permeability is low, leading to the development of highly ferruginous lateritic rocks but not bauxite. Steeply dipping rocks have highly variable thicknesses of regolith/bauxite due to uneven infiltration of groundwater and are not usually favourable for the formation of large- bauxite deposits.

The Pita Series is generally an unfavourable host for bauxite deposits as it is composed of quartz sandstones. Bauxites have only formed where the Pita Series overlies dolerites (Fig. 7). Ultramafic rocks are also not favourable for bauxite formation due to their very low Al contents; deposits of Fe ore (e.g., Kaloum Peninsula) and Ni-Co laterite (e.g., Kalia deposit) are common on these bedrocks.

4 Regolith Geology

The structure of lateritic regoliths is generally independent of bedrock composition and is usually comprised of two units (Mamedov et al. 2010) (Table 1, Fig. 8):

- Saprolith—lower unit, mainly composed of silicate minerals.
- Pedolith—upper unit, mainly composed of hydroxide/oxide minerals (in bauxite: ferric iron and Al oxides and hydroxides).

The saprolith, 2 to 25 m thick, is composed of clay-rich weathered rock retaining the original bedrock fabric as pseudomorphs (Fig. 8):

- The lowest horizon is saprock, in which the physical and initial chemical disintegration of fresh rock takes place.
- Saprolite consists of hydromica, illite, quartz and/or chlorites or montmorillonite clay. The zone is characterised by low concentrations of Fe and high concentrations of silica and Mg. The Fe, silica and Mg concentrations depend on the lithology and composition of the bedrock.
- The upper part of the saprolith consists of kaolinitic clays, pale in the lower part and multi-coloured in the upper part, with red colours due to ferruginous components (usually goethite).

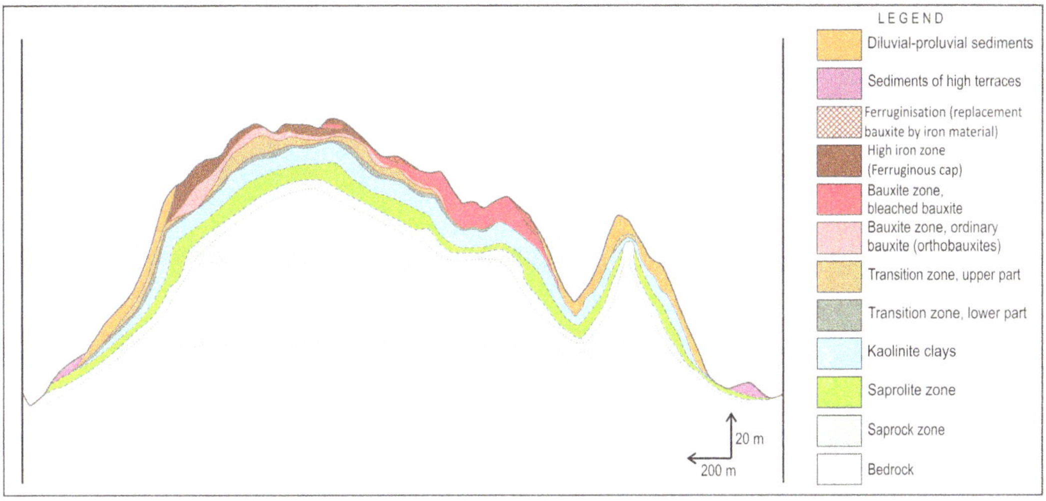

Fig. 8 Typical weathering crust profile. Telimele group of bauxite deposits

The pedolith, 2 to 25 m, is composed mainly of Al and Fe hydroxides and oxides (Fig. 8):

- The lowest part of the pedolith is a Transition zone (Figs. 8 and 9f–k) and is commonly unconsolidated. It is characterised by decreasing silica, with Fe concentrations increasing from the base to the middle part of the zone and then decreasing as the alumina concentration increases upwards. Iron-rich concretions, known locally as 'ferriplantites' (Fig. 9h–i), are typical for the middle or upper parts of the Transition zone, which are close to the groundwater table during the rainy season.
- The Bauxite Zone forms in the middle of the pedolith (Figs. 8 and 9b–e). It is characterised by a relatively sharp increase in Al and a corresponding decrease in Fe immediately above the base of the Transition zone. The middle and upper parts of the Bauxite Zone are characterised by decreasing Al and increasing Fe concentrations due to Fe oxides precipitating in fissures and gradually replacing Al minerals. Illuviation of soil material in fissures from the surface results in higher silica contents. Bauxite does not occur everywhere, only in favourable environments.
- The upper part of the pedolith, the lateritic duricrust, is known locally as the Ferruginous Cap (Figs. 8 and 9a). This zone is characterised by decreasing Al and increasing Fe contents from bottom to top, continuing features of the Bauxite Zone. Bauxite occurs as relict blocks within the ferruginous lateritic duricrust. Soil material with quartz and kaolinite is common within fissures and cavities.

Soil, 0.1–2 m thick, covers the lateritic regolith. Soils are enriched by siliceous material (quartz, kaolinite) and contain relict bauxite and ferruginous lateritic blocks.

The typical bauxitic regoliths described above (Figs. 8 and 9) are examples of orthobauxites (see Fig. 6, above; Tardy 1997) or the '*in situ* laterite type' (Mamedov et al. 2010). Bedrock fabrics and structures are visible in the Transition zone and bauxites (Figs. 9 and 10), indicating they formed by residual weathering.

However, there are many other types of bauxite and regolith profiles due to differences in parent lithologies and weathering environments:

- Lateritic regolith formed from recent colluvium on slopes of plateaux (Figs. 8 and 11a) consists of ferruginous and bauxitic blocks and fragments eroded from the plateaux, cemented by a ferruginous matrix. This colluvial material rarely contains economic bauxite deposits.
- Bauxites formed from continental terrigenous sediments of the Miocene Sangaredi Series in Fouta Djallon (Figs. 11b–e and 12) are

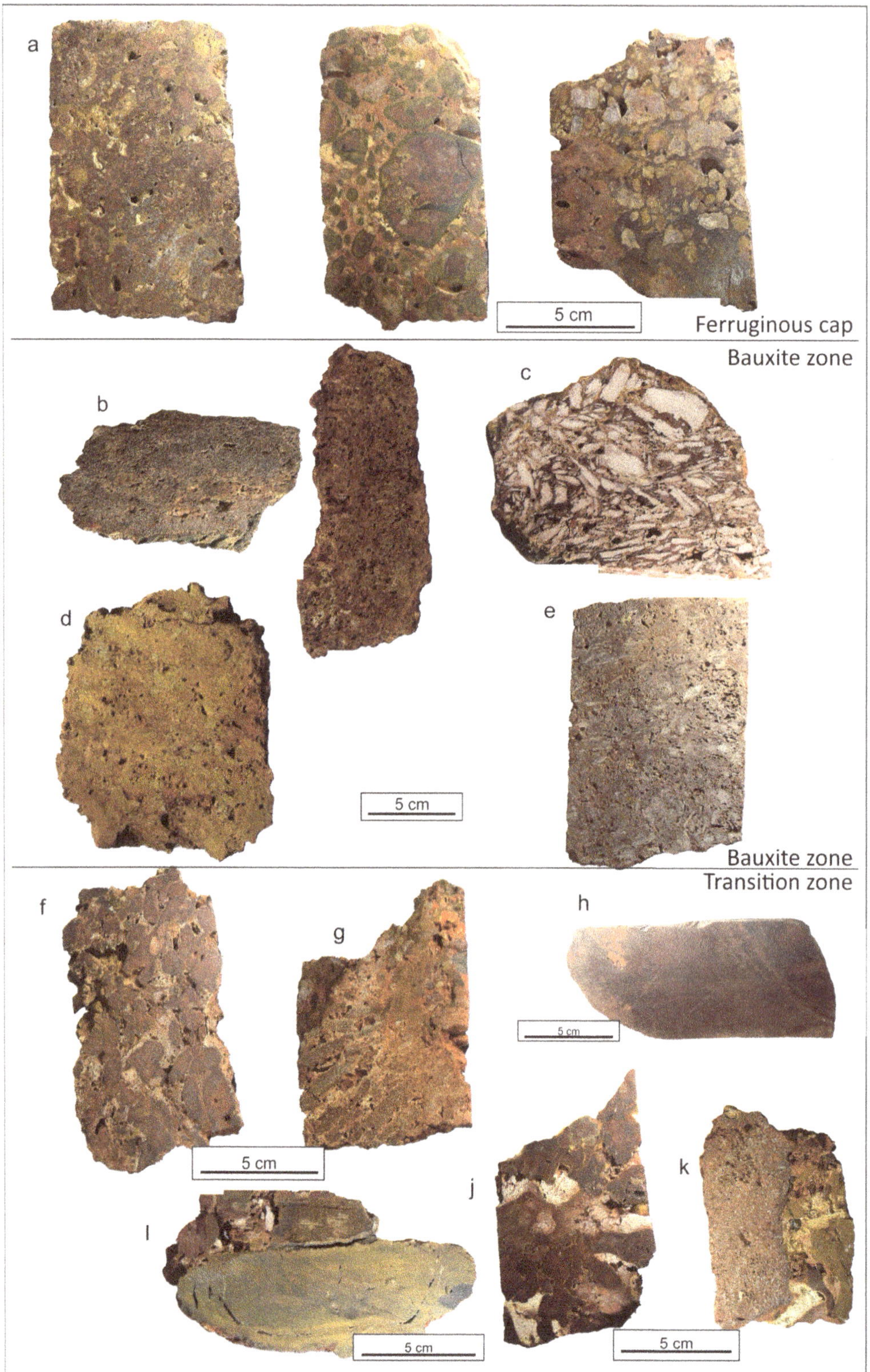

Fig. 9 Samples from a lateritic orthobauxite profile with N'Dangara, Paravi and Bourore deposits underlying the Sangaredi group, Boke basin (**a**) Ferruginous cap with goethite and clay-soil material in cement. (**b**) Red bauxite from upper part of a bauxite horizon formed on dolerite. (**c**) Red bauxite from the upper part of a bauxite horizon

distributed very locally—mostly at Sangaredi and adjacent plateaux in the Kogon and Tinguilinta River basin, as small bodies on the Kavelin plateau in Tomine River basin and probably others near Labe city. These are major bauxite deposits (hundred million tonnes, now mostly depleted) of extraordinarily high quality (52–60% Al_2O_3 and higher), with high grades of boehmite (monohydrate) (Fig. 12). The Sangaredi Series bauxites are generally located on slopes of paleo-valleys as pebble-gravel facies (Figs. 11d and 13), whereas some relict bodies occur on the upper surfaces of the Sangaredi and adjacent plateaux (Fig. 13) as gravel-, sand- and even clay-stone facies (Fig. 11b–c).

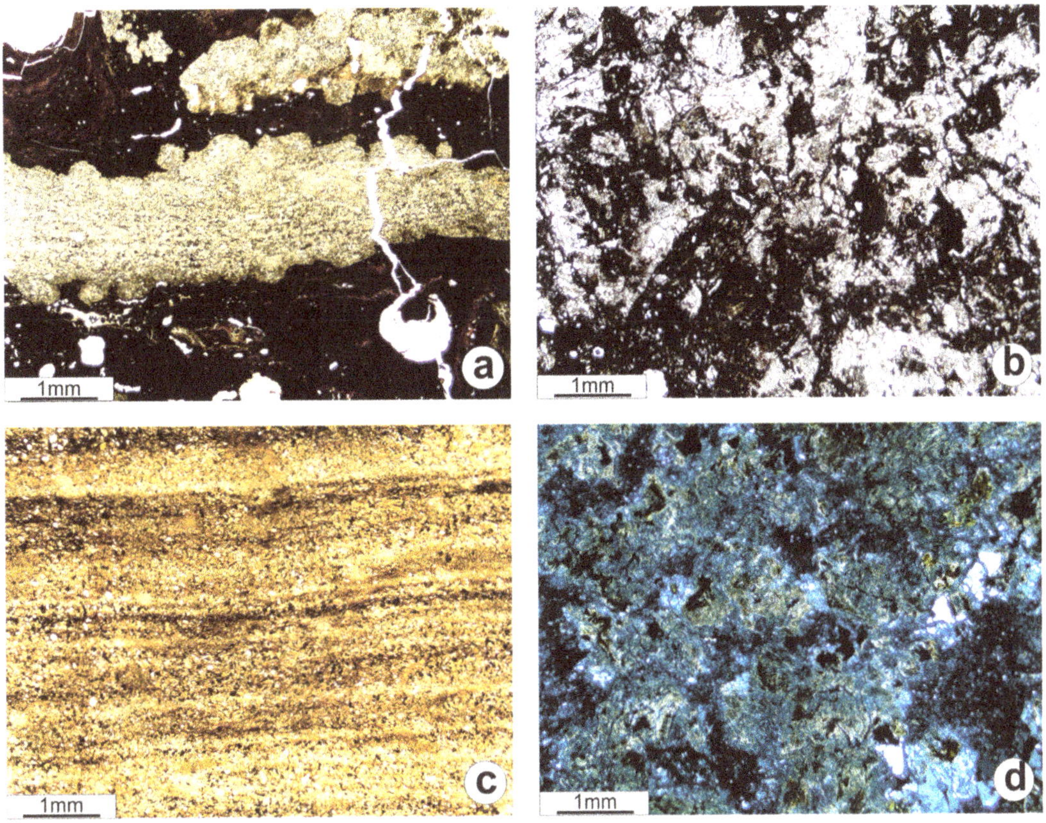

Fig. 10 Thin sections, Bourore deposit (Boke basin). (Photographs: M. Seredkin): (**a**) bauxites formed in siltstones (plane polarised light), (**b**) bauxites formed in dolerites (plane polarised light), (**c**) kaolinite clays formed in siltstones (plane polarised light), (**d**) kaolinite-montmorillonite clays formed in dolerites (crossed polarised light)

Fig. 9 (continued) formed on siltstone (brecciated collapsed saprolite). (**d**) Yellow bauxite from the lower bauxite horizon formed on dolerite. (**e**) Pale-cream bauxite from the lower bauxite horizon formed on siltstone (brecciated collapsed saprolite). (**f**) Ferruginous Transition zone with Al hydroxide cement (upper part) formed on siltstone. (**g**) Ferruginous Transition zone with Al hydroxide cement (upper part) formed on dolerites. (**h**) Iron-rich duricrust ('ferriplantite' after V. Mamedov), with hematite predominant. (**i**) Iron-rich duricrust ('ferriplantites' after V. Mamedov) with goethite predominant. (**j**) Ferruginous clay-rich Transition zone (lower part) formed on siltstone. (**k**) Ferruginous clay-rich Transition zone (lower part) formed on dolerite. (Photographs: M. Seredkin)

Fig. 11 Bauxites/ferruginous rocks in diluvial-colluvial-alluvial-lacustrine formations at the N'Dangara, Sangaredi and Paravi deposits, Boke basin. (Photographs: M. Seredkin): (**a**) formed in recent diluvial-colluvial formations with bauxites and ferriplantites in ferruginous cement, (**b**–**e**) formed in Sangaredi Series sediments: (**b**) Sandstone facies, (**c**) Gravelstone facies, (**d**) Pebble-gravelstone facies, (**e**) Breccia-gravelstone facies

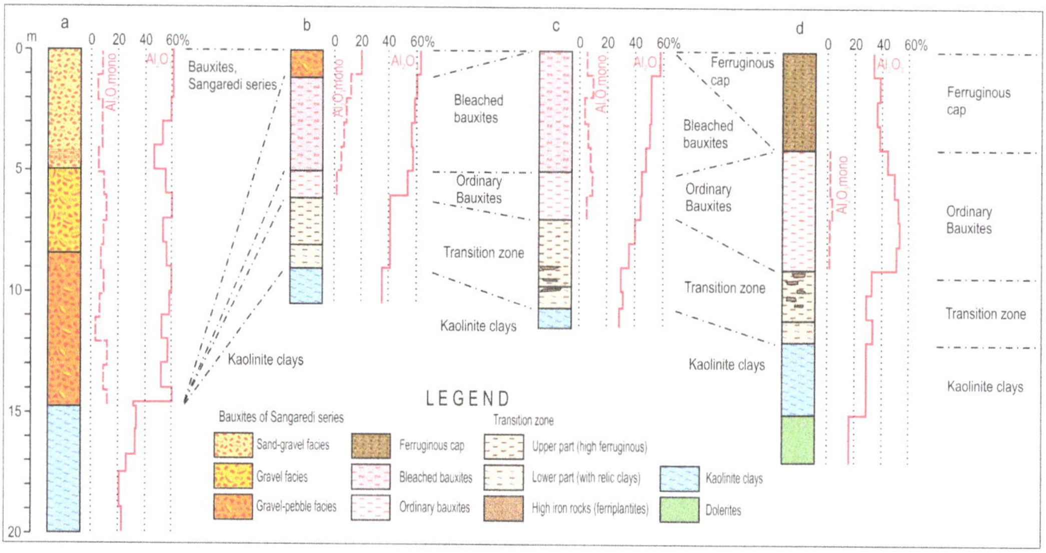

Fig. 12 Lateritic bauxite profiles of the Fouta Djallon sub-province. (After Mamedov et al. 2010). (**a**) Bauxite profile of the Sangaredi series—a form of Cryptobauxite. (**b**) Regolith profile with bleached bauxites overlain by bauxites of the Sangaredi series. (**c**) Bleached bauxite profile with (**d**) Orthobauxite profile

The Sangaredi bauxites are a form of cryptobauxite (Tardy 1997), as demonstrated below.

- Bleached bauxites (Fig. 14) are more common than Sangaredi bauxites and, over favourable bedrock, usually have high grades (>48%

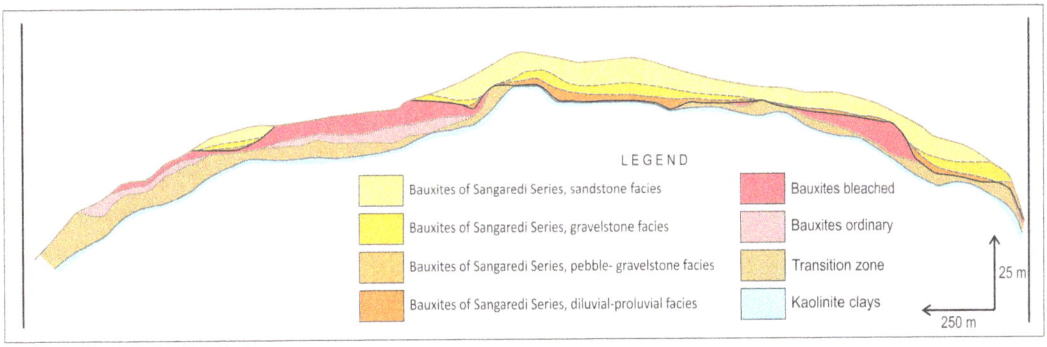

Fig. 13 Regolith section of the N'Dangara deposit, Boke basin

Fig. 14 Bleached bauxites formed from different bedrock lithologies, N'Dangara and Boundou-Waade deposits, Boke basin. (Photographs: M. Seredkin). (**a**) Aphanite bauxites. (**b**) Microoolitic bauxites formed from siltstones. (**c**) Oolitic bauxites. (**d**) Bleached bauxites formed from siltstones. (**e**) Bleached bauxites formed from dolerites

Al_2O_3) with a patchy distribution of monohydrate alumina. Bleached bauxites are probably types of metabauxite (Tardy 1997). Bleached laterite profiles are characterised by an inverse grade pattern, with increasing Al and decreasing Fe grades from bottom to top in the Bauxite zone (Fig. 12). This type of pattern is atypical for classical lateritic bauxites (Mamedov et al. 2010) and is considered to be due to initial bleaching of saprolite or ferruginous horizons under reducing gley conditions in the Middle Miocene, in a similar manner to bauxites below the Belterra Clay in Brazil (Negrão et al. 2018) (Fig. 15d–e). Ferruginous lateritic units commonly occur at the top of these profiles. Detailed mapping of the plateaux has shown that bleached bauxites are commonly spatially and genetically connected with the Sangaredi Series (Figs. 12 and 17).

Not all types of bauxite and bauxite profiles are present at Fouta Djallon. Two important types found elsewhere are karst bauxites and cryptobauxites:

- Karst bauxites are usually only represented by the upper horizons of lateritic profiles (Fig. 15). These have formed from silicate sediments, commonly of volcanic origin, overlying well-drained carbonate rocks. The bauxites rest directly on the limestone or dolomitic bedrock, and as a result, there is no underlying clay-saprolite.
- Bauxitic regoliths overlain by clays (Fig. 16), exemplified by lateritic bauxites overlain by the Belterra Clay, Brazil (Negrão et al. 2018). Iron has been leached from the Belterra Clays and upper zones of the lateritic profile and redeposited at depth. Ferruginous material and bauxites below the clays (Fig. 16) result in the formation of high-quality bauxites

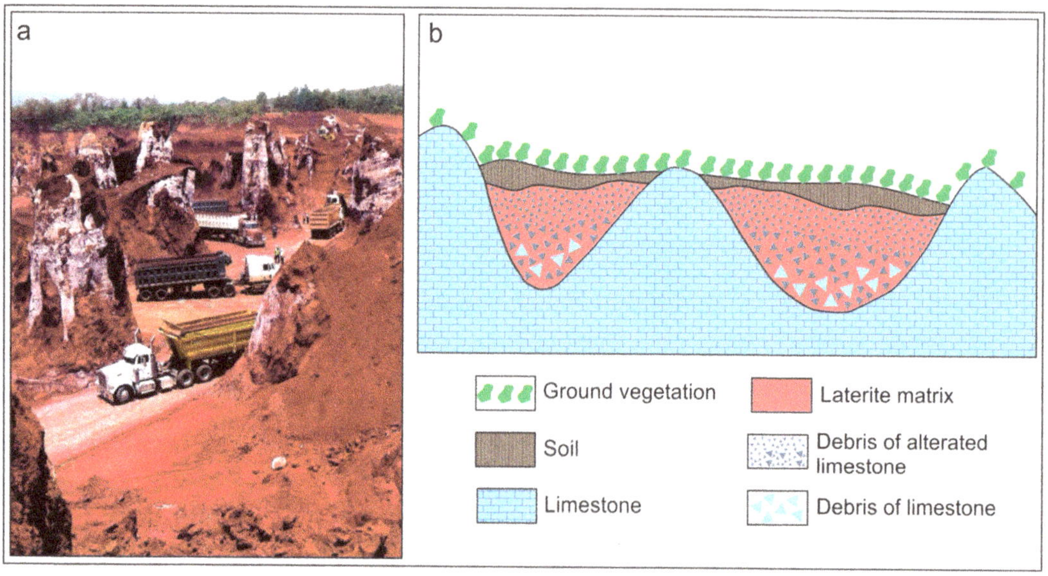

Fig. 15 Lateritic profiles in the Caribbean (Torró et al. 2017). (**a**) Karst bauxite deposit in Dominican Republic (red–bauxites, white–carbonate rocks). (**b**) Diagrammatic sketch of karst bauxites

($Al_2O_3 > 50\%$). These bauxites are classified as cryptobauxites (Tardy 1997).

5 Geomorphological Control

Geomorphological factors influence the distribution and quality of bauxites and are well illustrated in the Fouta Djallon sub-province. A continuous bauxitic lateritic regolith occurs at all elevations across the sub-province except in valleys, where it is partially or wholly eroded and covered or replaced by colluvial and alluvial sediments (Figs. 8 and 17). The regional geomorphology is represented by a 'stair' of plateau surfaces (Mamedov et al. 2010) (Figs. 17 and 18):

- The central part of the Fouta Djallon range has the highest geomorphological surfaces:
  - Late Jurassic—Early Cretaceous level (J_3–K_1) (Post-Gondwanan) is the highest surface level, with a single plateau remnant preserved on the north of the range.
  - Cenomanian—Campanian level (K_2^1–K_2^5) occurs as plateau remnants.
  - Maastrichtian—Middle Eocene level (K_2^6–E_2^2) (African surface): an erosional-denudation surface, with some isolated remnants preserved in other areas.
- Two large groups of bauxite-bearing plateaux located in the Boke and Tougue basins represent mid-level surfaces:
  - Late Eocene—Late Oligocene level (E_2^3–E_3^2) occurs as preserved plateau remnants and as the highest erosional-denudation surface.
  - Miocene level (N_1) is a prevailing erosion-denudation and accumulation surface and is the most productive for high-quality bauxites.
  - Pliocene—Early Pleistocene level (N_2–Q_I) forms the lowest erosion-denudation and accumulation surfaces.
- The youngest geomorphological surfaces occur in river valleys and along the ocean coast:
  - Middle Pleistocene level (Q_{II}): erosional or accumulation surfaces as high and middle river and ocean terraces.
  - Late Pleistocene—Holocene level ($Q_{III\text{-}IV}$): erosional and accumulation surfaces as lower terraces, floodplains, river beds and coastal plains.

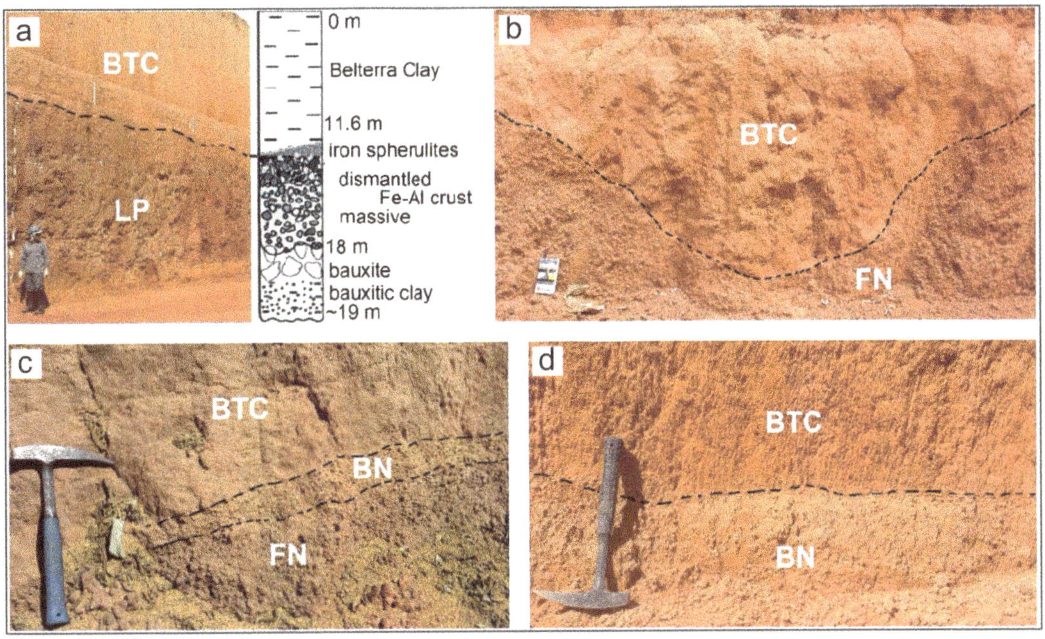

Fig. 16 Lateritic profiles in the South American Province (Negrão et al. 2018). (**a**–**d**) Lateritic regolith profile covered by upper Belterra Clays (BTC), Amazonia sub-province, Brazil. (**a**) General view of the Branco bauxite pilot mine showing the lateritic profile (LP) with its profile (right) overlain by thick Belterra Clay. (**b**) Erosive paleochannel developed in the ferruginous nodules (FN) filled by Belterra Clay, Ciríaco mine; (**c**) sharp contact between BTC and the underlying bauxitic nodule (BN) horizon and ferruginous nodule (FN) horizons, Décio mine; (**d**) contact between the BTC and the BN horizon, Branco mine

Three surfaces can be identified quite reliably—the modern surface, the Middle Miocene surface by the location of bauxites formed on the Sangaredi Series and the African surface, which is common across wide areas in Africa. Other surfaces located between these surfaces are estimated from their relative positions.

Weathering of the Sangaredi Series has produced bauxites of extremely high quality, and its distribution mostly controls the occurrence of high-grade bleached bauxites. It is important for the discovery of similar deposits in Guinea.

In some areas of the Fouta Djallon sub-province, the Miocene and Pliocene–Early Pleistocene levels were depositional surfaces for marine transgression in the Miocene and filling of valleys by debris (Mamedov et al. 2010, 2011). The total thickness of the Sangaredi sediments reached a maximum of 48 m (average 22 m) over a vertical range from 135 to 245 m asl (Mamedov et al. 2011). Relicts of the bauxites formed on the Sangaredi Series are common on Miocene surfaces, although deep palaeo-valleys also occur on Pliocene—Early Pleistocene levels (Fig. 19). The Sangaredi sediments occur as lateritised alluvial sediments in palaeo-valleys and as lacustrine-marsh sediments at higher palaeo-levels. Three possible palaeo-basins with Miocene Sangaredi sediments are shown in Fig. 17.

Bleached metabauxites are common on Miocene and Pliocene–Early Pleistocene surfaces, where they are spatially and genetically connected to the Sangaredi sediments (Fig. 19). Orthobauxites are common on erosional Miocene and Pliocene—Early Pleistocene surfaces (Fig. 19).

The Sangaredi Series is not present on Late Eocene—Late Oligocene or higher levels, but bleached bauxites are found in many plateaux amongst orthobauxites (Fig. 19). The occurrence of bleached bauxites may be due to the presence

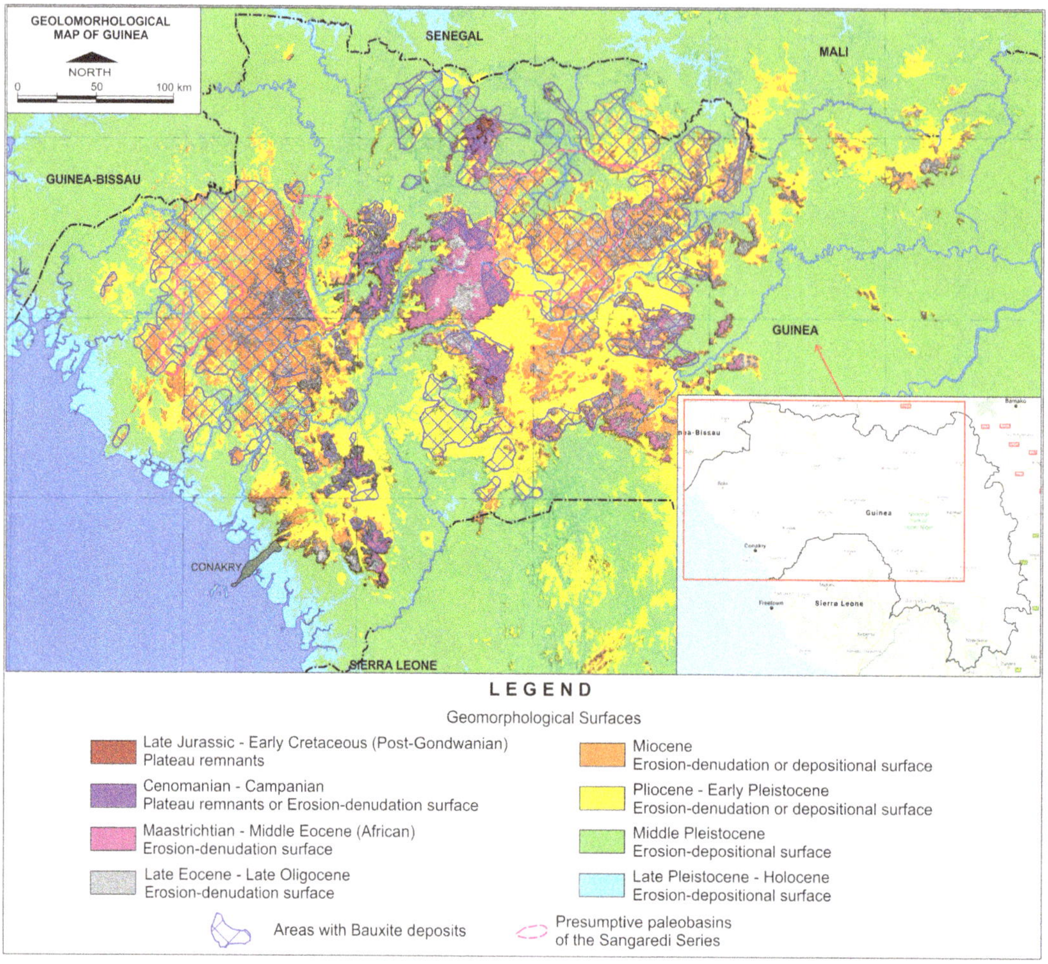

Fig. 17 Geomorphological map of the Fouta Djallon sub-province

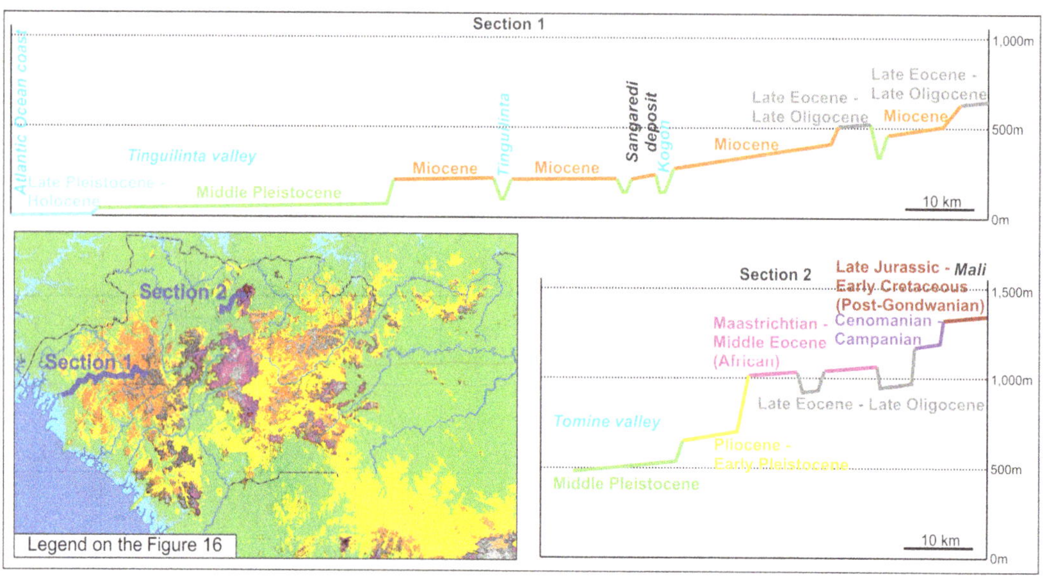

Fig. 18 Geomorphological levels of the Fouta Djallon sub-province

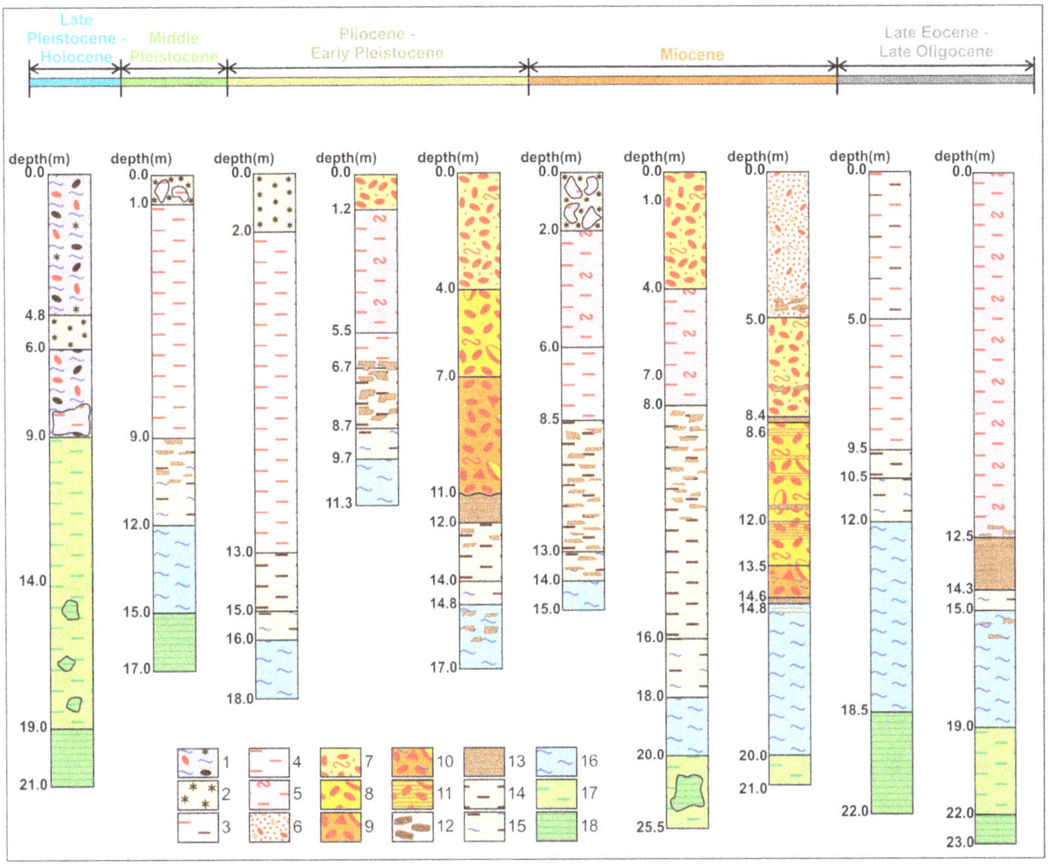

Fig. 19 Lithology and zonation of regolith profiles on different geomorphological surfaces, Fouta Djallon sub-province, Boke basin, observed in diamond drill core 1—alluvial loam with gravel, 2—ferruginous cap, 3—low-quality orthobauxites(<40–42% Al_2O_3), 4—moderate/high-quality orthobauxites (Al_2O_3 > 42–45%), 5—bleached bauxites, 6–11—bauxites formed in Sangaredi Series: 6—sandstone facies, 7—gravelstone facies, 8—pebble- gravelstone facies, 9—breccia - gravelstone facies, 10—with clay-rich cement, 11—with ferruginous cement, 12—Fe-rich concretions ('ferriplantites'), 13—sub-horizontal Fe-rich duricrust ('ferriplantites'), 14—ferruginous upper Transition zone with alumina cement, 15—ferruginous lower transition zone with clay-rich cement, 16—kaolinite clays, 17—saprolite, 18—bedrock

of local small palaeo-lakes and swamps at higher levels in the Miocene, similar to the palaeo-lakes and swamps in the main -basins.

Lateritic regoliths on Middle Pleistocene surfaces occur as classical in situ profiles with low- medium grade bauxites (Fig. 19). The regolith on Late Pleistocene—Holocene surfaces is ferruginous and clay-rich in places with lenses of low-quality bauxites (Fig. 19). These profiles may be regarded as ferruginous analogues of the cryptobauxite profiles in Brazil, having only poor grade or no bauxite.

6 Hydrogeology and Gas Regime

The base of the bauxite horizon corresponds to the maximum height of the groundwater table (Fig. 20) (Mamedov et al. 2010), with the Transition zone situated in the seasonal range of water table fluctuation (Fig. 21) caused by the infiltration of water from the surface through open fissures. This water is oxidised, resulting in the ferruginisation of the Transitional zone. Other water percolates through the porous matrix and

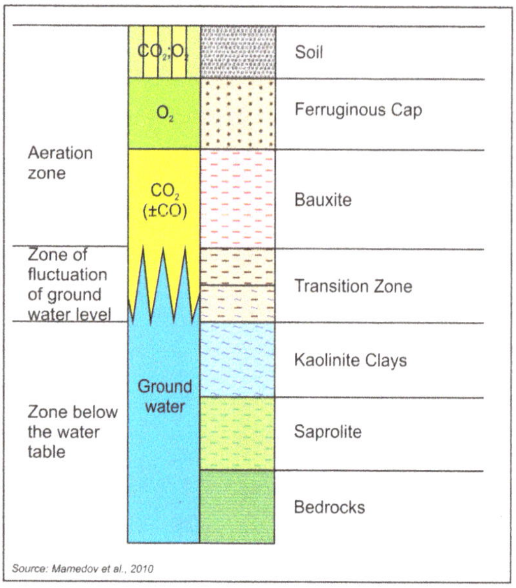

Fig. 20 Principal scheme of bauxite-bearing regolith profile formation. (After Mamedov et al. 2010)

gives rise to zonation of the regolith (Fig. 20). There is a general correlation between the thickness of bauxites and the depth of the groundwater table (Fig. 21).

7 Mineralogy

The most typical minerals of the upper horizon of lateritic bauxite profile are gibbsite, goethite, alumogoethite, hematite and, in places, boehmite and alumohematite. The proportion of these minerals defines the bauxite or ferruginous composition of the pedolith (Fig. 22). Quartz and kaolinite are mainly found in fissures, but in places, they occur as relicts in bauxite. Soil material is enriched in quartz by accession of aeolian material brought by dust storms.

Alumogoethite is a solid solution between goethite and diaspore, containing up to 35% diaspore. The highest levels of the diaspore component in alumogoethite are common in the bleached bauxites and bauxites of Sangaredi. Similarly, high diaspore contents occur in alumogoethite in bauxites beneath the Belterra Clays in Brazil (Negrão et al. 2018), where bleaching of bauxites is comparable to that occurring at the base of the Miocene Sangaredi Series sediments. Diaspore as a separate phase is typical in metamorphosed bauxites, for example, in Greece and Turkey.

Small quantities of corundum are known in bauxites of the Darling Range (Western Australia) (Anand and Gilkes 1987; Anand et al. 1991; Hickman et al. 1992). Alumohematite (a solid solution between hematite and corundum) was also identified in these bauxites, with the grade of the corundum component reaching 10–15% (Gilkes 1998).

The lower Transition zone consists predominantly of kaolinite and goethite, with goethite, alumogoethite, hematite, gibbsite and, in places, boehmite increasing in abundance in the upper Transition zone (Fig. 22).

Gibbsite, goethite (alumogoethite) and hematite are the most common minerals in orthobauxites, with minor kaolinite, quartz and Ti oxides/hydroxides (Fig. 22). Quartz is usually derived from the overlying soil and corroded by kaolinite.

Boehmite is common in the bleached bauxites and bauxites of the Sangaredi Series (Fig. 22); the clastic components of these bauxites are composed mostly of gibbsite with hematite/goethite, although cemented by boehmite material and minor hematite (Fig. 23). Under present conditions, boehmite is not in equilibrium (metastable) and commonly partly or fully re-crystallised to fine-grained gibbsite aggregates (Fig. 23).

Some highly ferruginous basal units of the Sangaredi Series have a slightly altered cement and clastic material (Fig. 24). Other units have higher silica contents due to clay-rich interbedded units in the parent sediments (Fig. 19).

The following relationships between minerals are common in different horizons of bauxitic regoliths:

- In goethite developing in upper horizons of clays.
- In the lower Transition zone, kaolinite is commonly replaced by goethite, with relict kaolinite present as lenses.
- Ferruginous phases (goethite, hematite) are strongly concentrated In the middle Transition

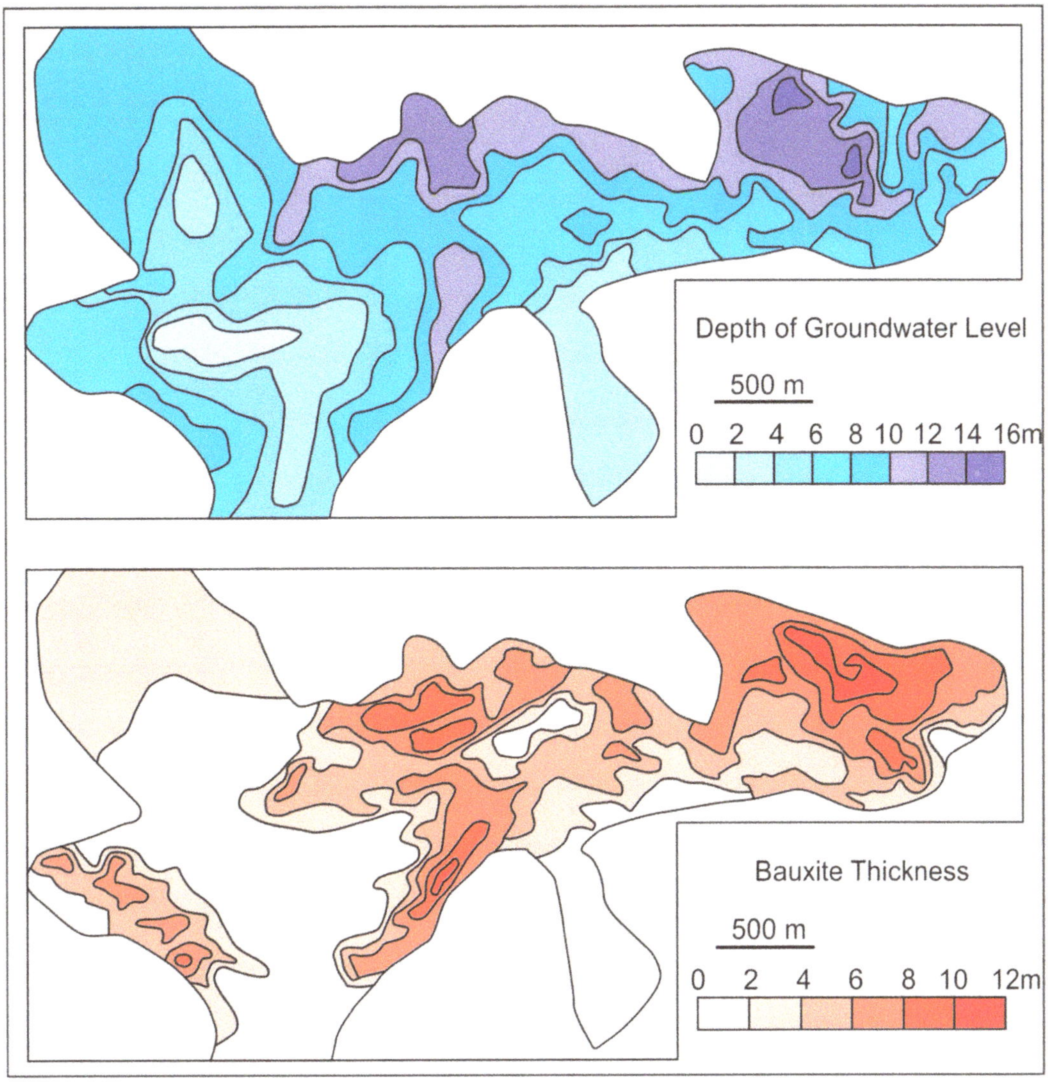

Fig. 21 Comparison of the depth of water table (in rainy season) and bauxite thickness, Debele deposit, Kindia. (After Mamedov et al. 2010)

zone as Fe-rich ferriplantites. Some ferriplantites have formed in white Sangaredi bauxites in zones of marked groundwater table fluctuation (Fig. 24).

- In the upper Transition zone, goethite and hematite are cross-cut and replaced by gibbsite, or gibbsite-boehmite, making a gradual transition to bauxites (Fig. 23e–f); in addition, the Al content of alumogoethite increases.
- In the upper bauxite horizon or along cavities and open fissures, gibbsite and boehmite are replaced/covered by Fe oxides, predominantly goethite (Fig. 23g–h).

8 Deposit Formation and Evolution

Classical bauxite lateritic deposits are located in humid tropical and adjacent subtropical regions (Fig. 2). Bauxites in the tropics are forming now, whereas those in the many subtropics are most probably relicts, due largely to the much lower

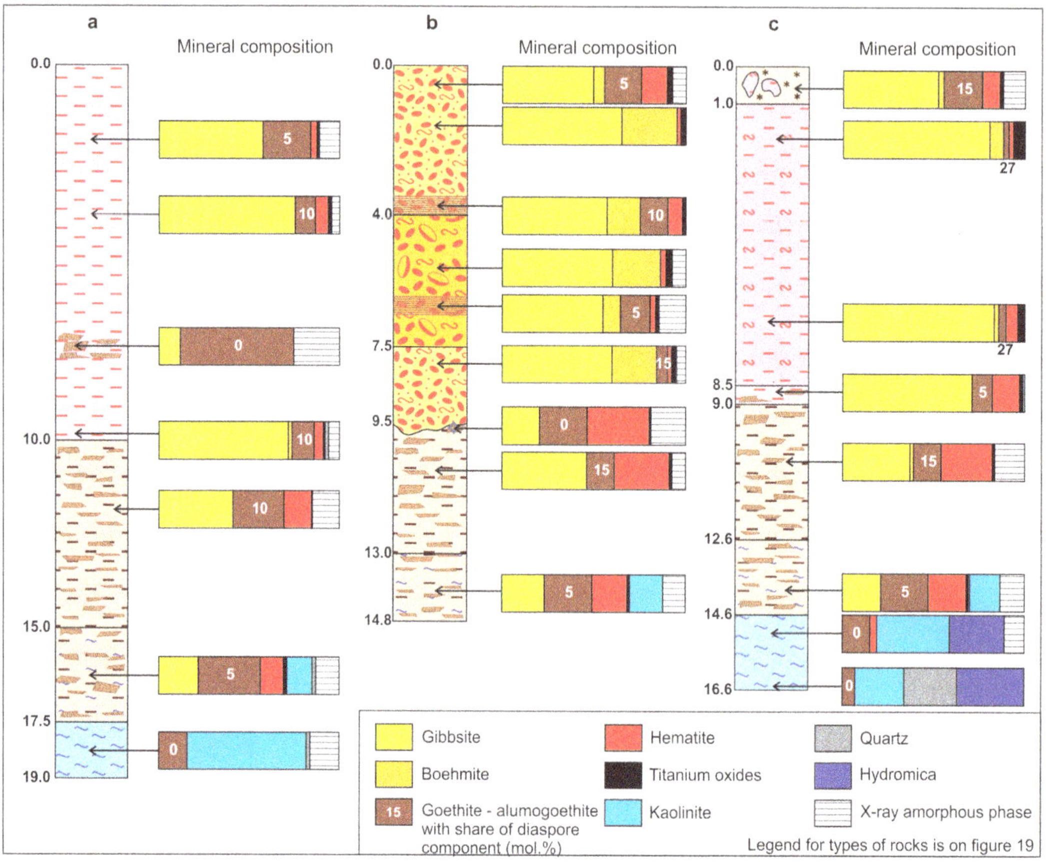

Fig. 22 Mineralogical compositions of regolith profile, N'Dangara and Bourore deposits (Boke basin). (**a**) Orthobauxites, (**b**) bauxites developed from Sangaredi Series sediments, (**c**) bleached bauxites

rainfall in these regions. Much rarer types of bauxites are represented by relict ancient weathering crust in temperate and even arctic zones, such as those in Kazakhstan and Russia (Timano-Pechora basin) (Fig. 2). Some bauxites are formed in karst cavities in tropical and adjacent subtropical areas, too. Classical deposits of this type are located in Jamaica. Relict karst deposits are located in southern and eastern Europe (France, Greece, Turkey) and Russia (Northern Urals bauxite basin).

Bauxite formation requires a mostly hot, seasonally wet climate, with marked dry and rainy seasons (humid savannas). For example, in Fouta Djallon, the average annual temperature is 20.5–27.2 °C (Table 2), reaching 35–37 °C daily in the dry season and less in the rainy season (about 20–25 °C). Night temperatures are much lower than daily temperatures. Annual precipitation is from 1366 to 4352 mm in the summer months (Table 2),

Tropical seasonal climates with lower rainfall (dry savannas) lead to the formation of ferruginous lateritic regoliths instead of bauxitic regoliths. In very wet climates (e.g., Amazonia, Brazil), bauxites are replaced by clays (Lucas 1989; Negrão et al. 2018).

The weathering processes that form lateritic regoliths are typical of acid leaching by infiltration of rainwater, which has a pH of 5.0–5.5, mainly due to dissolved atmospheric CO_2. They result in the 'ideal' zonation of a bauxitic regolith (Table 1), modified by the reactions occurring in the Transition zone. In addition to the infiltration reactions between groundwater and rock minerals, the direct flow of oxidised rainwater

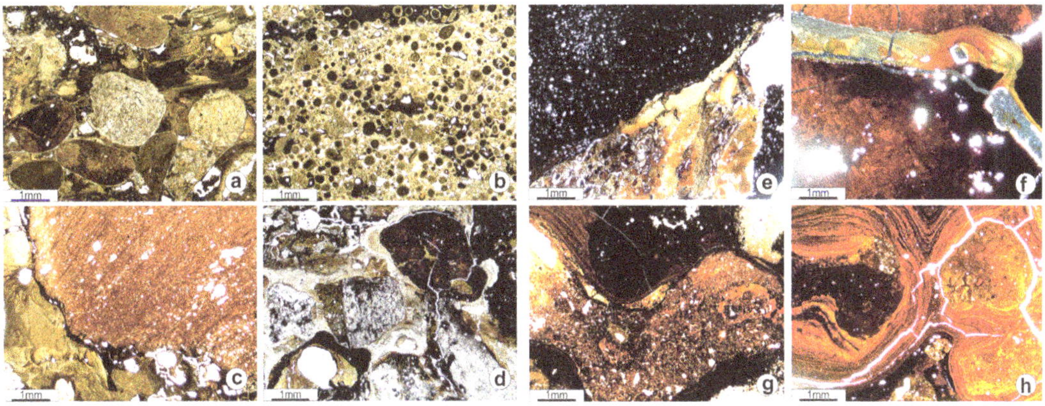

Fig. 23 Relationships between minerals in bauxites, transition zone and ferruginous cap. N'Dangara and Sangaredi deposits, Boke basin. Photomicrograph of thin sections (plane polarised light) by M. Seredkin. (**a**–**d**) Bauxites with boehmite, Sangaredi Series. (**a**) Gravelstone facies, gibbsite-goethite/hematite gravels with dominantly boehmite cement; (**b**) Oolitic bauxites, dominantly boehmite; (**c**) Gravelstone facies, gibbsite-goethite/hematite gravels, with dominantly boehmite cement partly re-crystallised to gibbsite; (**d**) Gravelstone facies, gibbsite-goethite/hematite with dominantly boehmite cement almost fully re-crystallised to gibbsite. (**e**–**h**) Interactions in bauxite-bearing regolith profile: (**e**, **f**) Gibbsite and gibbsite-boehmite 'veins' in high ferruginous Transition zone; (**g**, **h**) Goethite in bauxites, upper bauxite horizon

Fig. 24 Ferruginous bauxites, Sangaredi Series, N'Dangara deposit, Boke basin. (Photograph: M. Seredkin). (**a**) Fe-rich clastic material. (**b**) Sub-horizontal Fe-rich 'ferriplantite' zone

through open fissures to the Transition zone results in the formation of an Oxidised zone within the weathering column (Fig. 20). The preservation of primary bedrock fabrics in saprolite is evidence that weathering is initially a type of metasomatic alteration.

Bauxite-bearing profiles form in climates that have alternating wet and dry seasons, and it is possible that in the dry season, water may be considered to be a thermodynamically inert component—atypical for geological processes. Each thermodynamically inert component forms its own equilibrium phase, as shown in Fig. 25 for Guinean bauxites.

Gibbsite–goethite (alumogoethite) and gibbsite–goethite (alumogoethite)–hematite are

Table 2 Climatic regime of the Fouta Djallon sub-province (Mamedov et al. 2010)

Meteorological station	Elevation, m	Average temperature, °C			Precipitation, mm			Rainy days per year
		August (minimum)	April (maximum)	Annual	January (minimum)	August (maximum)	Annual	
Boke	69	25.8	29.7	27.2	0	822	2830	
Conakry	26	24.7	27.8	26.3	1	1327	4352	170
Koubia	790						1366	87
Mali	1441	18.4	23.7	20.5	0	492	1807	
Labe	1025	20.7	25.1	22.5	2	368	1712	122
Mamou	782	22.1	26.2	23.7	5	437	1976	142

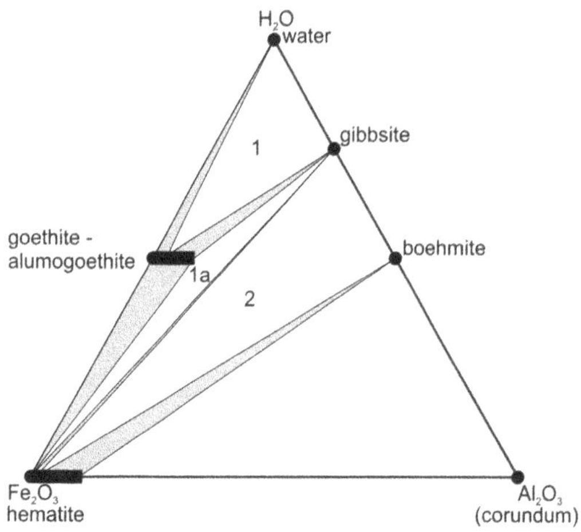

Fig. 25 Composition—mineral paragenesis diagram for bauxites with thermodynamically inert water. Mineral parageneses 1 and 1a are stable in present-day conditions in Guinea, whereas paragenesis 2 is metastable in current conditions in Guinea

the most common mineral associations in orthobauxites. Boehmite is present in bauxites that have been enriched in Al by thermodynamically inert water to form metabauxites. Boehmite is the metastable phase in modern conditions in Guinea. This diagram may also be applied to analyse the paragenesis of bauxitic regoliths in other Provinces.

Lateritic profiles with high-quality bauxites and inverse zonation are common in Guinea and make this region unique. Inverse zonation is mineralisation in which higher grades of alumina occur in the upper parts of the profiles; this is typical for most bauxites, which are overlain by a Fe-rich horizon. These deposits commonly have bleached white bauxite with a high grade of boehmite (metabauxites). As described above, there is a spatial and genetic association between these bauxites and those of the Sangaredi Series formed from Middle Miocene alluvial and lacustrine sediments (Mamedov et al. 2011). Similar processes are occurring under present-day conditions in Brazil, where bleaching of bauxites is occurring beneath the Belterra Clay.

The principal scheme of formation of the bauxites of the Sangaredi Series and bleached bauxites with high grades of alumina is shown in Fig. 26:

- Stage 1: Formation of orthobauxite profile;
- Stage 2: A rise in the base of erosion, with depressions filled by alluvial, lacustrine-alluvial, and swamp-lacustrine sediments under very humid climates in the Middle Miocene. Formation of a clay-rich horizon in the upper part of the profile and bleached bauxites beneath the clays (cryptobauxites).
- Stage 3: Formation of Fe-poor clay-rich sediments in the floor of a swampy lacustrine peneplain (no bauxite formation).
- Stage 4: Formation of lateritic regoliths on the clay-rich Sangaredi sediments and bleaching of underlying rocks, forming high-quality white bauxites (metabauxites).
- Stage 5: further deep erosion with bauxite formation below the bleached bauxites—a typical situation in the Fouta Djallon sub-province(metabauxites).
- Stage 1 again: Deep erosion removing all bleached bauxites, with the profile reverting to an orthobauxite.

This genetic scheme can explain most features of bauxites and the alteration of their composition due to changes in climate. Consequently,

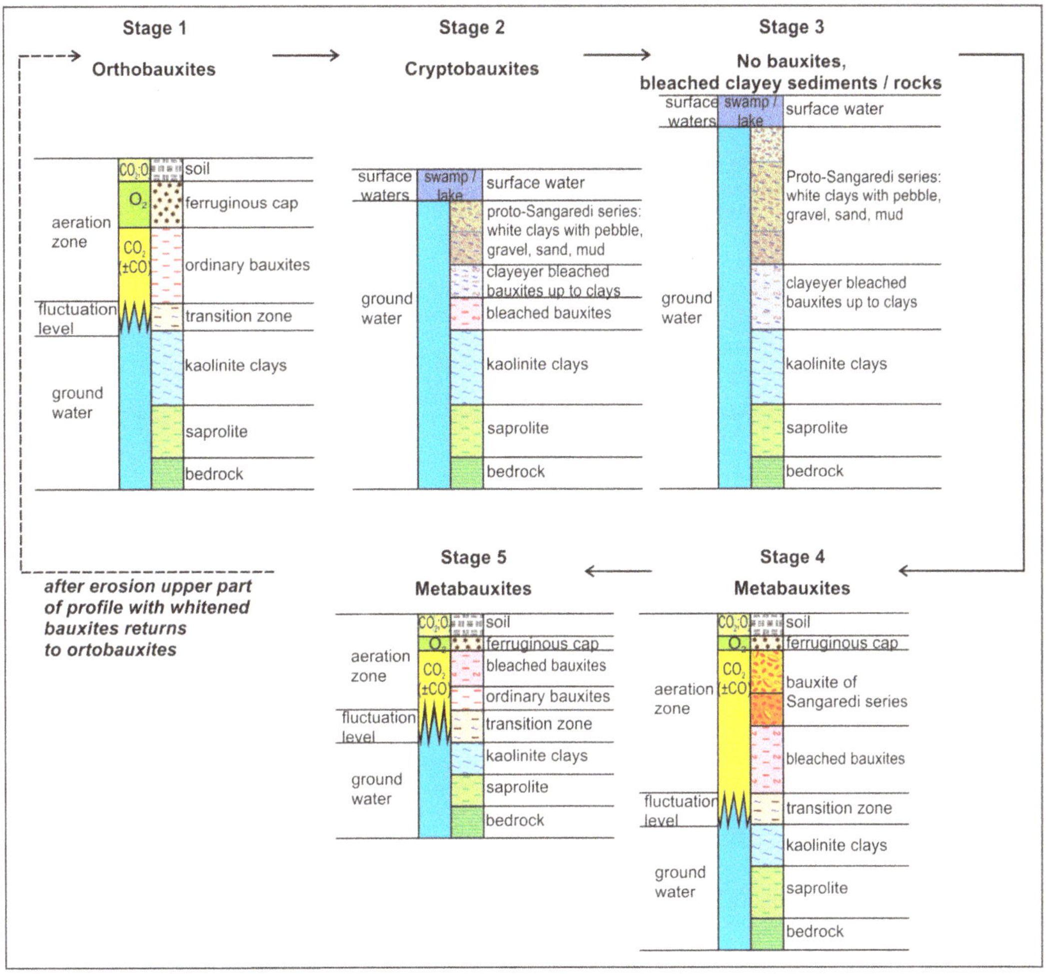

Fig. 26 Stages of formation of orthobauxites, bauxites of the Sangaredi Series and bleached bauxites. (Interpreted from Mamedov et al. 2011)

orthobauxites, metabauxites and cryptobauxites may interchange their characteristics depending on climatic conditions.

9 Implications for Exploration and Exploitation

Investigations of bauxite composition and features of profile zonation are important for the exploration and exploitation of bauxite deposits.

Bauxites are generally divided into two geometallurgical types:

- For low-temperature Bayer Process (140–150 °C)—mainly gibbsite bauxites
- For high-temperature Bayer Process (230–250 ° C)—boehmite (and diaspore) bauxites

The relationship between mineralogical composition and geometallurgical parameters is shown in Fig. 27. Modern technologies allow surface miners to select higher-quality bauxites for the high-temperature Bayer Process and ordinary lower-grade bauxite for selective mining (Fig. 28). High-quality bauxites are commonly located at the top of bauxite profiles, so they can

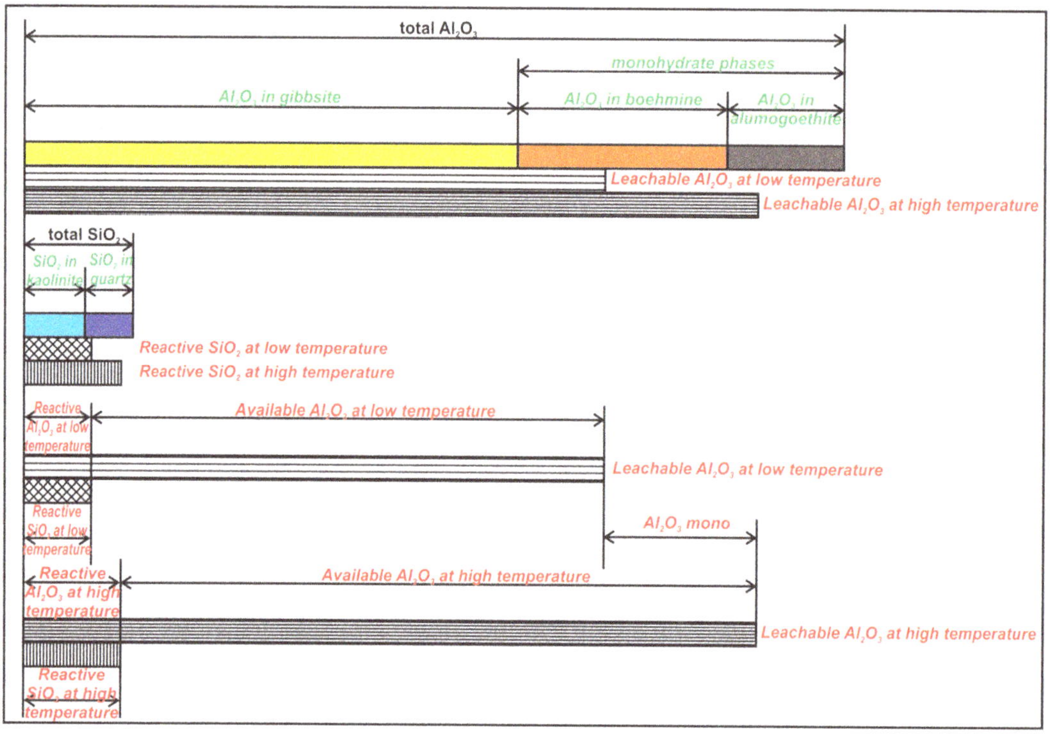

Fig. 27 Relationship between mineralogical composition and geometallurgical parameters of bauxites

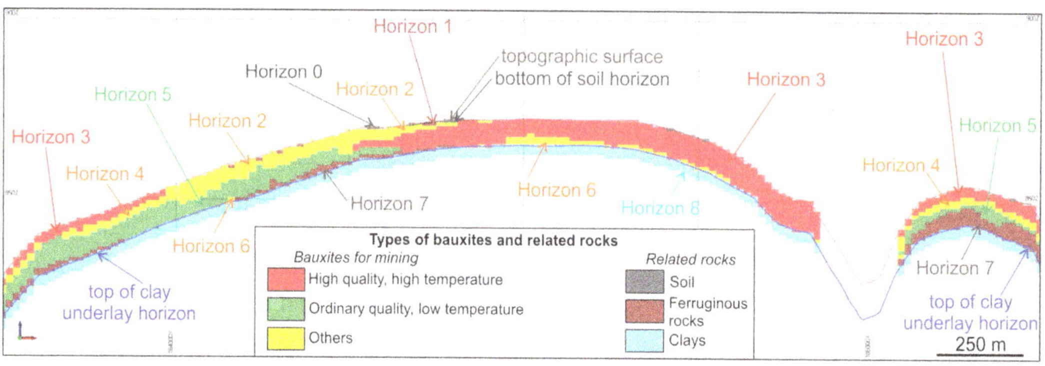

Fig. 28 Example of bauxite selection for surface mining

be accessed during the initial stages of mine development.

10 Summary and Conclusions

Aluminium is primarily sourced from bauxite and then refined to alumina.

The majority of lateritic bauxite deposits are found in a small number of provinces. Bauxites of Neogene to Recent origin are only located in current-day tropical and subtropical belts (Fig. 2), whereas many older bauxites, formed during the Palaeozoic to early Cenozoic, are located at higher latitudes. These older bauxite deposits are considered to have formed under similar conditions as modern bauxites but with later alterations due to changes in tectonics and global climate.

Bauxites form in hot, humid climates with pronounced wet and dry seasons. Under similar but drier conditions, ferruginous lateritic regoliths form instead. Where climates change to very wet, rainforest climates, as in Amazonia, bauxites are replaced by clay materials.

Laterite regoliths form as a result of two processes: classical infiltration and metasomatism, and direct flow of oxidised rainwater through open fissures to form an oxidised Transition zone in the middle to upper part of the metasomatic column.

Ordinary lateritic bauxites (orthobauxites) form as hypergenic 'metasomatites'. These are typical infiltration profiles from saprolite and clay zones at depth to a strongly leached near-surface bauxite, the result of acid leaching by clean rainwater that has an acid pH between 5–5.5 derived from anions in the atmosphere and almost without cations.

The Transition zone is located at the base of the bauxite horizon and is controlled by the changing levels of the groundwater due to its seasonal fluctuations resulting from direct penetration of water from the surface through open fissures. As the water is oxidised, the Transitional zone is usually ferruginous. The water then percolates and filtrates through the porous matrix, resulting in the metasomatic zonation of the weathering crust.

Bauxite-bearing profiles form in climates with alternating rain and dry seasons. In the author's opinion, in the dry season, water may be a thermodynamically inert component, atypical for geological processes. Each thermodynamically inert component forms its own equilibrium phase. Gibbsite—goethite (alumogoethite) and gibbsite—goethite (alumogoethite)—hematite are the most common mineral associations in orthobauxites. Boehmite occurs in Al-rich bauxites with thermodynamically inert water (metabauxites). Boehmite is a metastable phase in modern orthobauxites.

Bedrock composition and location are important for the formation of bauxite deposits. In order to form bauxites, bedrock should be enriched in alumina, have good filtration coefficients, and, if sedimentary, ideally have sub-horizontal bedding. Geomorphological factors influence the distribution and quality of bauxites.

The principal scheme of formation of orthobauxites, metabauxites and cryptobauxites may be explained by changing climate conditions:

- Stage 1: Formation of ordinary bauxite-bearing profile (orthobauxites).
- Stage 2: A rise in the base level of erosion, with depressions filled by alluvial, lacustrine-alluvial, and swamp-lacustrine sediment and a very humid climate—resulting in a clay-rich upper horizon, overlying bleached bauxites (cryptobauxites).
- Stage 3: Formation of clay-rich sediments with low Fe contents in the floors of swamp-lacustrine peneplain (no bauxite).
- Stage 4: Formation of lateritic crusts on clay-rich sediments and underlying bleached rocks, producing high-quality white bauxites (metabauxites).
- Stage 5: Further deep erosion, forming orthobauxites beneath bleached bauxites (metabauxites).
- Stage 1 again: Deep erosion with removing all bleached bauxites—reverting to orthobauxite.

Under dry climate conditions, bauxites degrade to ferruginous duricrusts.

Acknowledgements The authors thank CSA Global/Environmental Resource Management (ERM) for permitting the publication of this paper, Charles Butt for the invitation to contribute to the monograph and for the detailed review and editing of the text.

References

Anand RR, Gilkes RJ (1987) The association of corundum and maghemite in Darling Range laterites, Western Australia. Aust J Soil Res 35:303–311

Anand RR, Gilkes RJ, Roach GID (1991) Geochemical and mineralogical characteristic of bauxites, Darling Range. Appl Geochem 6:233–248

Bardossy G (1982) Karst bauxites. Bauxite deposit on carbonate rocks, Developments in economic geology 14. Elsevier, Amsterdam, 441 p

Bardossy G, Aleva GJJ (1990) Lateritic bauxites, Developments in economic geology 27. Elsevier, Amsterdam, 624 p

Boufeev Y, Diawara S, Mamedov V (2003) Catalogue of bauxite deposits for estimation potential exploration target of Republic of Guinea. (in French)

Freyssinet P, Butt CRM, Morris RC, Piantone P (2005) Ore-forming processes related to lateritic weathering. Econ Geol 100th anniversary volume:681–722

Gilkes RJ (1998) Mineralogy of Darling Range bauxite. Minerals and Energy Research Institute of Western Australia (MERIWA) report no. 38, Perth

Guinea Mining Cadastre Portal, 2022. https://guinee.cadastreminier.org/en/

Guinean Ministry of Mines and Geology, 2022. https://mines.gov.gn/

Hickman AH, Smurthwaite AJ, Brown IM, Davy R (1992) Bauxite mineralisation in the Darling Range, Western Australia. Geological Survey of Western Australia. Report 33, Perth

Jacob L (1984) Bauxite. In: American Institute of Mining, Metallurgical, and Petroleum Engineers, Bauxite symposium, Los Angeles, Proceedings, 1984. Society of Mining Engineers of American Institute of Mining, Metallurgical, and Petroleum Engineers, New York, 142 p

Lucas Y (1989) Systèmes pédologiques en Amazonie brésilienne: Equilibres, Déséquilibres et Transformations: Unpublished Ph.D. thesis. University of Poitiers, 153 p (in French)

Mamedov V, Boufeev Y, Makstenik I (2003) Map of potential bauxite distribution in Republic of Guinea, (in French)

Mamedov V, Boufeev Y, Nikitine Y (2010) Geologie de la Republique de Guinee. Ministère des Mines et de la Géologie de la République de Guinée, Moscow, (in French)

Mamedov VI, Chausov AA, Kanishev AI (2011) Formation stages of the unique sangaredi bauxite-bearing group, Fouta Jallon-Mandingo province, West Africa. Geol Ore Deposits 53:177–201

Nahon D (1991) Introduction to the petrology of soils and chemical weathering. Wiley, New York, 313 p

Negrão LBA, Lima da Costa M, Pöllmann H (2018) The Belterra clay on the bauxite deposits of Rondon do Pará, eastern Amazon. Braz J Geol 48:473–484, (in Portuguese)

Tardy Y (1997) Petrology of laterites and tropical soils. Balkema, Rotterdam, 408 p

Tardy Y, Roquin C (1998) Dérive des continents, latérites et paléoclimats tropicaux. BRGM Editions, France, 472 p, (in French)

Torró L, Proenza JA, Aiglsperger T, Bover-Arnal T, Villanova-de-Benavent C, Rodríguez-García D, Ramírez A, Rodríguez J, Mosquea LA, Salas R (2017) Geological, geochemical and mineralogical characteristics of REE-bearing Las Mercedes bauxite deposit, Dominican Republic. Ore Geol Rev 89: 114–131

USGS (2021a) Bauxite and alumina statistics and information. https://www.usgs.gov/centers/nmic/bauxite-and-alumina-statistics-and-information?qt-science_support_page_related_con=0#qt-science_support_page_related_con

USGS (2021b) Bauxite deposit world map, 2014. https://www.usgs.gov/media/images/bauxite-deposit-world-map

Valeton I (1972) Bauxites, Developments in soil science, 1. Elsevier, Amsterdam, pp 106–109

Williams R (2019) Global bauxite resources and game changers. AZ-China. AZ Global Consulting, Hong Kong, 58 pp

Nickel-Cobalt Laterite Deposits

C. R. M. Butt, M. Elias, A. Kadarusman, and N. W. Brand

Abstract

Nickel laterite ores supply nearly 70% of global Ni production. They are regolith units formed by intense weathering of mostly serpentinized olivine-rich ultramafic rocks in which Ni and, commonly, Co have been concentrated to over 1.0 wt% and 0.3 wt%, respectively. A few are also residually enriched in Sc and PGM. There are three main styles of deposit, defined by the principal host minerals, namely *oxide* (Ni and Co hosted by Fe and Mn oxyhydroxides), *hydrous silicate* (Ni in serpentine, "garnierite," nepouite) and *clay silicate* (Ni in smectitic phyllosilicates). The genesis, mineral composition and grades of the ores are controlled by the interaction of bedrock lithology, tectonic and climatic history, and geomorphological setting. Many Ni laterites have a multi-stage development, evolving as their climate and/or topographic settings change over time. The richest deposits (>3 wt% Ni) occur in tectonically active regions, especially island arc terranes in SE Asia-Oceania and the Antilles. Here, successive episodes of faulting and epeirogeny have uplifted the landscape, resulting in Ni being leached from the upper, oxide-rich horizons and re-concentrated in actively forming hydrous silicates in the saprolite and saprock. Brief descriptions of the Lapaopao oxide-hydrous silicate deposit, Sulawesi, and the Sunrise oxide deposit, NSW, Australia, illustrate some key features of Ni-Co laterite ores.

1 Introduction

Nickel laterites are intensely weathered regoliths in which one or more horizons contain exploitable reserves of Ni and, commonly, Co. They have developed on olivine-rich ultramafic rocks, mostly partially or wholly serpentinized, situated in continental and island arc setting between about 35°S and 60°N (Fig. 1). Nickel laterites comprise about 55% of global Ni reserves and resources (Mudd and Jowitt 2022). Although laterites have been mined since the 1870s and were the dominant source from 1880 to 1905, in 2000, they accounted for only 25–40% of total Ni production, with the remainder sourced from sulfide deposits. The low

C. R. M. Butt (✉)
CSIRO Mineral Resources, ARRC,
Kensington, WA, Australia
e-mail: Charles.Butt@csiro.au

M. Elias
ERM Australia Consultants Pty Ltd, West Perth,
WA, Australia

A. Kadarusman
Department of Geology and Geophysics, University
of Indonesia, Jakarta, Indonesia

N. W. Brand
Geochemical Services Pty Ltd, West Perth, WA, Australia

R. J. Bowell, C. R. M. Butt (eds.), *Geology, Geochemistry and Formation of Supergene Mineral Deposits in Deeply Weathered Terrain*, Mineral Resource Reviews, https://doi.org/10.1007/978-3-031-75733-4_11

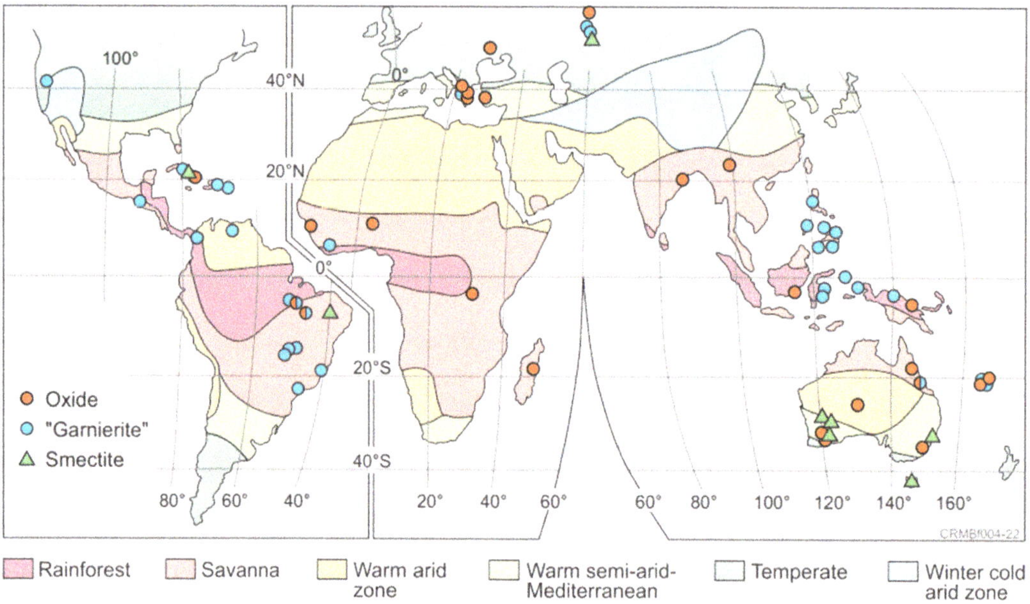

Fig. 1 Global distribution of Ni laterite deposits, classified by the dominant style of mineralization

grades, complex processing, and high energy costs, coupled with distance from centres of industrial demand and appropriate infrastructure, slowed their development. Historically, most production from laterites came from the garnierite-rich deposits of New Caledonia, which have been mined since 1873, and lower-grade deposits in Russia, Kazakhstan, Greece, Cuba and the United States (Oregon, California). Discoveries of Ni laterites have since been made in many other regions, and with increased demand, improved processing and reducing sulfide resources, production has increased. By 2010, the proportion of total Ni production from laterite ores was over 50% in 2010, at about 60% since 2014 (Mudd and Jowitt 2022; USGS 2022) but nearing 70% in 2024. The principal producers of lateritic Ni are Indonesia, Australia, the Philippines, Brazil, Cuba, New Caledonia and Russia. Nickel laterites also host some 36% of total terrestrial Co resources and contributed about 15% of Co production in 2011 (Slack et al. 2017). Nickel laterites are potential sources of Sc and PGM as by-products, but the development of processing facilities for these metals has been slow.

A few Ni laterite deposits, such as those in Sulawesi, have formed in essentially unchanged rainforest conditions since their ophiolitic protores were exposed to weathering in the Miocene. The majority, however, have formed and evolved under changing climatic and/or tectonic conditions over long periods in places since the Palaeozoic, resulting in modified or new ore types. Such multiphase development is typical of lateritic regoliths, whether in cratonic environments exposed to weathering throughout much of the Phanerozoic or in island arc settings subjected to periodic tectonic uplift during the Cenozoic (chapter "The Role of Regolith Formation and Evolution in the Development of Supergene Mineral Deposits"). This Chapter summarizes the characteristics of these deposits and the relationship between their chemical and mineralogical compositions and their genesis. It draws on comprehensive studies by Golightly (1979, 1981), reviews by Butt and Cluzel (2013), Golightly (2010), Freyssinet et al. (2005), Gleeson et al. (2003), Elias (2002) and many other individual studies, exemplified by brief descriptions of two deposits.

2 Deposit Characteristics

2.1 Classification

There are three broad ore types, *oxide*, *hydrous Mg silicate* and *clay silicate*, based on the principal host minerals (Gleeson et al. 2003; Brand et al. 1998) (Fig. 2, Table 1). This subdivision has important economic implications for mining and processing. Most deposits with a complete lateritic profile have two ore types: (i) oxide ore and (ii) either an underlying hydrous silicate or clay silicate ore (Berger et al. 2011). The different mineral hosts have different processing requirements; hence, most mines exploit only one style of mineralization. Exceptions are Goro, New Caledonia, where oxide and hydrous Mg silicate resources have both been exploited (Golightly 2010; Freyssinet et al. 2005), and at Lapaopao, Sulawesi where it is planned to both ores (see Sect. 7, below). Mining at Wingellina, Western Australia, may exploit both the oxide (~70% of total resource) and the clay (~30%) components (Putzolu et al. 2020a, b). Loma Ortega, part of the Falcondo deposit in the Dominican Republic, is unusual in that it has a "hybrid" silicate ore consisting of both hydrous and clay silicate components (Tauler et al. 2017).

2.1.1 Oxide Deposits

Oxide deposits, commonly termed "limonite" in commercial literature, consist dominantly of Fe oxyhydroxides as the mid to upper ferruginous saprolite, extending to the overlying plasmic zone (Figs. 2 and 3). Nickel is mainly hosted by goethite (α-FeOOH), either adsorbed or substituting for Fe. Poorly crystalline Mn oxides, enriched in both Co and Ni, are commonly abundant in these

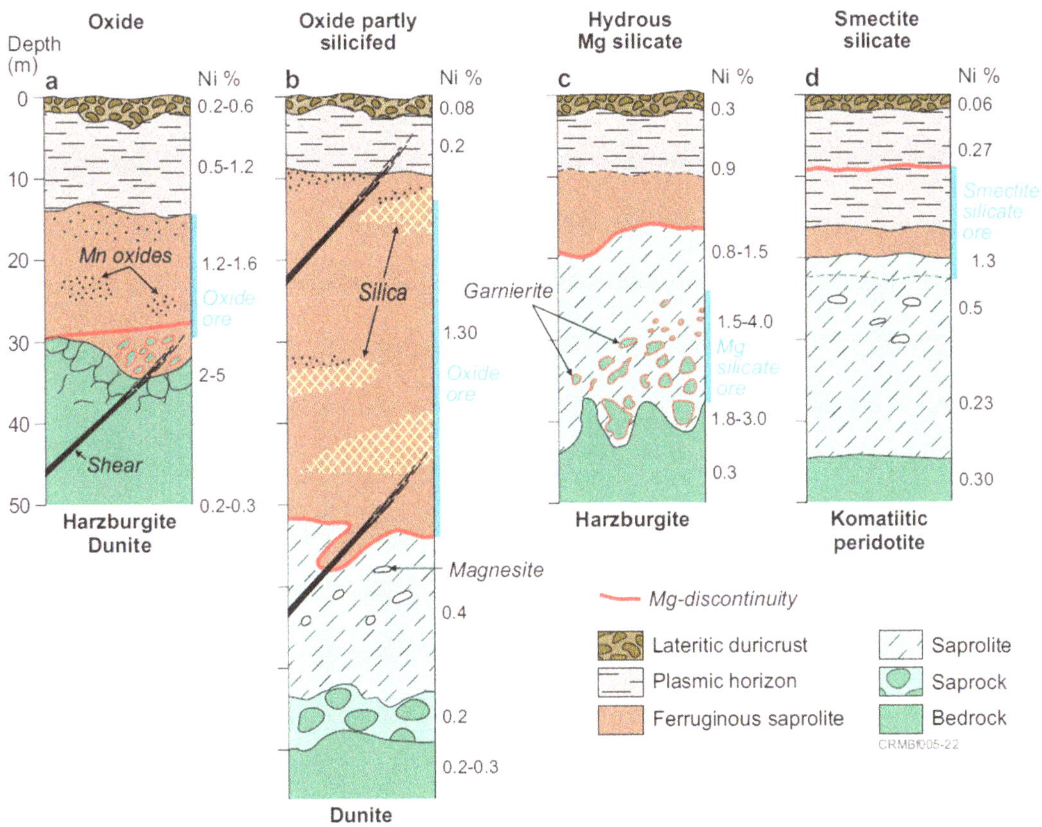

Fig. 2 Regolith profiles of lateritic nickel deposits develop on serpentinized olivine-rich ultramafic rocks showing the principal Ni laterite ores

Table 1 Principal minerals in nickel laterites. Most minerals are poorly crystalline and variable in composition. (Modified after Freyssinet et al. 2005. Comprehensive lists of names and properties of approved and discredited Ni phyllosilicates in hydrous silicate deposits are given in Villanova-de-Benavent (2015))

Mineral	Ideal formula	Group	Ni %
Hydrous Mg silicate deposits			
Lizardite	$Mg_3Si_2O_5(OH)_4$	Serpentine	0.15
Nickeloan lizardite	$(Mg,Ni)_3Si_2O_5(OH)_4$	Serpentine	6.1
Népouite (Ni analogue of lizardite)	$(Ni, Mg)_3Si_2O_5(OH)_4$	Serpentine	32.8
Clinochrysotile	$Mg_3Si_2O_5(OH)_4$	Serpentine	
Pecoraite (Ni analogue of clinochrysotile)	$Ni_3Si_2O_5(OH)_4$	Serpentine	40.5
Antigorite	$(Mg,Fe)_3Si_2O_5(OH)_4$	Serpentine	0.1
Nickeloan antigorite	$(Mg,Ni)_3Si_2O_5(OH)_4$	Serpentine	4.9
Amesite	$Mg_2Al(AlSiO_5)(OH)_4$	Serpentine	
Berthierine	$(Fe^{2+},Fe^{3+},Al)_3(Si, Al)_2O_5(OH)_4$	Serpentine	2.8
Brindleyite (Ni analogue of berthierine)	$(Ni,Al)_3(Si,Al)_2O_5(OH)_4$	Serpentine	22.6
Talc	$Mg_3Si_4O_{10}(OH)_2$	Talc	0.3
Willemseite (kerolite) (Ni analogue of talc)	$(Ni,Mg)_3Si_4O_{10}(OH)_2$	Talc	27.1
Nickel-kerolite (formerly pimelite; variety of willemseite)	$(Mg,Ni)_3Si_4O_{10}(OH)_2$ Poorly crystalline; extreme variation	Talc	15.7
Clinochlore	$(Mg,Fe^{2+})_5Al(Si_3Al)O_{10}(OH)_8$	Chlorite	0.3
Nickeloan chlorite	$(Mg,Fe^{2+},Ni)_5Al(Si_3Al)O_{10}(OH)_8$	Chlorite	7.2
Nimite (Ni analogue of clinochlore)	$(Ni,Mg,Al)_6(Si, Al)_4O_{10}(OH)_8$	Chlorite	16.9
Sepiolite	$Mg_4Si_6O_{15}(OH)_{26}H_2O$	Sepiolite	0.4
Nickeloan sepiolite	$(Mg,Ni)_4Si_6O_{15}(OH)_{26}H_2O$	Sepiolite	2.9
Falcondoite (Ni analogue of sepiolite)	$(Ni,Mg)_4Si_6O_{15}(OH)_{26}H_2O$	Sepiolite	24
7 Å garnierite	Poorly crystalline; extreme variation	Serpentine	15.1
10 Å garnierite	Poorly crystalline; extreme variation	Talc	19.9
14 Å garnierite	Poorly crystalline; extreme variation	Chlorite	3.3
Unclassified garnierite	Poorly crystalline; extreme variation	Serpentine, talc, chlorite, sepiolite, vermiculite	17.6
Oxide deposits			
Goethite	α-$Fe^{3+}O(OH)$ or FeO.(OH)	Diaspore	4%
Hematite	Fe_2O_3	Hematite	1–2%
Asbolane	$(Ni,Co)_{2-x}Mn^{4+}(O,OH)_4\ nH_2O$; or $Mn^{4+}(O,OH)_2{\cdot}(Co,Ni,Mg,Ca)_x(OH)_{2x}\ nH_2O$	Undefined	To 15% Ni and/or Co
Lithiophorite	$(Al,Li)Mn^{4+}O_2(OH)_2$		1% Ni to 6% Co
Heterogenite	CoO.OH		4%Ni, 55% Co

(continued)

Table 1 (continued)

Mineral	Ideal formula	Group	Ni %
Clay silicate deposits			
Nontronite	$Na_{0.3}Fe_2^{3+}(Si,Al)_4O_{10}(OH)_2.nH_2O$	Smectite	To ~3% Ni
Beidellite	$(Na,Ca)_{0.3}Al_2(Si,Al)_4O_{10}(OH)_2.nH_2O$	Smectite	To ~3% Ni
Montmorillonite	$(Na,Ca)_{0.3}(Al,Mg)_2Si_4O_{10}(OH)_2.nH_2O$	Smectite	To ~3% Ni
Saponite	$(Ca,Na)_{0.3}(Mg,Fe^{2+})_3(Si,Al)_4O_{10}(OH)_2.4H_2O$	Smectite	To ~3% Ni

Fig. 3 Oxide Ni laterite deposits. (**a**) Goro area, New Caledonia. (Photo C.R.M. Butt). (**b**) Çaldag, Turkey. (Photo R.L. Thorne). Oxide ore overlying pale-coloured low-grade silicate saprolite

ores. Oxide deposits generally have mean grades of 1.0–1.6 wt% Ni and comprise about 60% of total Ni laterite resources. Oxide deposits are exemplified by the Lapaopao (Sulawesi) and Sunrise (NSW, Australia) deposits, Sects. 7 and 8.

The uppermost plasmic zone and vesicular and/or nodular ferruginous duricrust that overlies oxide ores have much lower Ni contents; the duricrust rarely contains more than 0.03% Ni. Below the Ni-Co-rich ferruginous saprolite, the transition to lower saprolite and saprock is marked by a sharp increase in MgO abundance, from $<2\%$ to $> \sim 20\%$ (Mg-discontinuity) and an increasing presence of silicates, mainly hydrous silicates such as altered primary serpentine and smectites, and remnant primary minerals. The regolith hosting the deposits may vary from dominantly residual and moderately drained in upland sites to having colluvial and alluvial material in the upper horizons in poorly drained to swampy environments in valleys. The transported material may be derived from eroded oxide ore or lower-grade detritus from similarly weathered country rocks. Near Goro, New Caledonia, the oxide horizons may be partly under reducing conditions and contain organic matter such as peat and roots in the sediments or ferruginous rhizomorphs in residuum.

Oxide ores are the main sources of Co in Ni laterites. Asbolane, lithiophorite and other poorly crystalline Mn oxide phases are the principal hosts, typically intimately intergrown with Fe oxides, precipitated as veins, void fillings and coatings on surfaces, or as discrete masses and concretions. The Mn oxides have a wide range of Co and Ni content, from less than 1% to maxima exceeding 10% Co and 15% Ni (e.g., see Sunrise, below; Llorca 1993). Manganese oxides are usually sparse in the upper oxide zone, precipitating at lower levels, close to the water table, or at sites of temporary ponding and saturation above porosity barriers such as the Mg discontinuity and zones of silicification. The vertical distribution of the Co-Mn minerals may reflect Eh and pH conditions within the profile. Thus, in New Caledonia, Llorca (1993) found that from saprolite (pH ~8) to the oxide zone (pH ~6), the Mn oxide mineralogy generally follows the sequence: heterogenite (CoO.OH), Ni-rich asbolane, Co-rich asbolane, asbolane-lithiophorite, lithiophorite.

Many oxide deposits contain abundant secondary silica, mainly as chalcedony and quartz (e.g., Cawse and Ravensthorpe, Western Australia; Onça and Puma, Brazil; Sunrise, New South Wales, Australia), diluting the bulk Ni content. Silicification is typical of lateritically weathered ultramafic rocks, especially dunites and serpentinized dunites, as the low Al content has restricted the formation of clays. Removal of the silica by crushing and screening increases the effective grade and the available resources. The depth of weathering and thickness of ore horizons in oxide deposits varies; at Goro, Cawse and Wingellina, profiles may reach depths over 80 m, with 20–>40 m of oxide ore; Cawse formed from dunitic protore, with the mined resource including upgradeable silica-oxide ore. In comparison, at Moa Bay and E. Pinares, Cuba, the complete profiles are only 20–25 m deep, including about 10 m of oxide ore (Elias 2002; Fig. 3b).

2.1.2 Hydrous Mg Silicate Deposits

Most hydrous Mg silicate deposits are formed on serpentinized ophiolitic harzburgites. The majority occur in tectonically active island arc terranes of high relief in the Antilles and mid-latitudes around the Pacific. The silicate ores, in zones 10–15 m thick, are hosted by saprolite and saprock, with Ni concentrated in nickeloan varieties of serpentine, talc, chlorite and sepiolite; many of these are bright green, poorly-defined mineral phases known informally as "garnierite" (Figs. 2 and 4). Similar assemblages of hydrous silicate minerals occur in some deposits in Greece and the Urals (Glazkovsky et al. 1977), where serpentinites have weathered in karst limestone settings, but it is uncertain whether they can be correctly classified as Ni laterites.

Partly-weathered primary lizardite, in which Mg has been leached and exchanged by Ni, is a significant host mineral in many deposits (Pelletier 1996; Golightly 1979), but the highest grades are associated with garnierite silicate minerals, some of which have very high Ni

a:

Fig. 4 Hydrous silicate Ni laterite deposits. (**a**) HGL pit, New Caledonia. (Photo C.R.M. Butt). (**b**) "Garnierite" vein in fissure through saprolite Lapaopao, Sulawesi. (Photo M. Elias)

contents (3–40% Ni; Table 1). Hydrous silicate ores accordingly form the highest-grade Ni laterite deposits (locally 2–>5% Ni), with accumulations of garnierite as coatings on core stones and as veins in faults and fractures. Historically, the majority of Ni laterite ores were of this type but only comprise about 30% of present-day resources, with a mean grade of 1.45 wt%.

The regolith is rarely thicker than 40 m due to rapid erosion rates. Where preserved, the upper zones of the lateritic profile are similar to those of oxide deposits, with comparable grades and resources. However, due to different processing requirements, they have to be separated and stockpiled, as proposed for Lapaopao (Sect. 7).

Spheroidal weathering is characteristic of many hydrous silicate deposits, especially those on poorly serpentinized harzburgite. The saprolite ("*boulder ore*") consists of blocks of a few millimetres to tens of centimetres, with cores of fresh or partly weathered rock becoming more intensely weathered toward the margin. Garnierite minerals are typically precipitated as veins and coatings on the saprolite blocks and as boxworks and void fillings in relic structures such as shears, joints and grain boundaries. Garnierite and associated minerals such as secondary silica also occur in faults and fissures that extend deep in the saprock and into otherwise unweathered ophiolite bedrock in many locations in New Caledonia (e.g., Koniambo) and Dominican Republic (e.g., Falcondo). Slickensides and striations on these fissure fillings suggest that precipitation was syntectonic, occurring during deformation and uplift associated with ongoing tectonism (Cathelineau et al. 2016, 2017; Fritsch et al. 2016; Villanova-de-Benavent et al. 2014). In some situations, this might be regarded as hydrothermal rather than due to weathering, although the distinction is subtle.

2.1.3 Clay Silicate Deposits

Clay silicate Ni laterite deposits have only recently been recognized and exploited as an ore type, made economical by bulk mining and newly developed processing technology. Nickel-bearing clays have, however, long been known in lateritic regoliths on serpentinized ultramafic rocks. The principal ore minerals are Ni-rich smectites and saponite in the mid to upper saprolite and pedolith (Figs. 2d and 5). Ore grades are generally 1.0–1.5 wt% Ni and known deposits represent less than 10% of total Ni laterite resources.

Known clay silicate deposits are areas of relatively low relief, mainly in shield areas in Australia (Murrin Murrin; Bulong, Wingellina), southern Urals (Buruktal, Russia; Kempirsai, Kazakhstan), Burundi (Musongati) and Brazil (Niquelandia), but also in Cuba (San Felipe, Tauler et al. 2023). Most deposits are developed on serpentinized peridotite, although Niquelandia is formed from pyroxenite (Colin et al. 1990), and the clay silicate

Fig. 5 Clay-silicate Ni laterite deposits, Western Australia. (**a**) Criterion pit, Bulong. (Photo. M. Elias). (**b**) Murrin Murrin North. (Photo: C.R.M. Butt). Deep solution hollows, up to 30 m deep, now filled with oxidized clay and colluvial detritus, are features of both deposits

resource at Wingellina from gabbroic units associated with serpentinized peridotite (Putzolu et al. 2020b). The protores at Murrin Murrin and Bulong are serpentinized olivine ortho- and meso-cumulates that have been weathered to 40–60 m depth. The dominant minerals in the saprock and lower saprolite include primary and weathered serpentine, chlorite and saponite, locally with abundant secondary silica and magnesite (Wells and Butt 2017, 2000). The overlying saprolite and plasmic zones are smectite-rich and form the ore. At Murrin Murrin, Gaudin et al. 2005 reported that the ore is composed of phases intermediate between Al-montmorillonite, Al-beidellite, Fe-montmorillonite and Fe-nontronite, with Ni sited in the octahedral layer. The upper boundary of the ore approximates the Mg-discontinuity, with a sharp transition to a goethitic ferruginous zone, equivalent to oxide-type mineralization, with a lowermost ferruginous smectite unit. An overlying hematitic duricrust is present locally. The total Fe oxide content increases from about 5–10 wt% in saprolite to 50–70 wt% across the Mg-discontinuity (Figs. 2 and 5).

3 Global Distribution

The global distribution of Ni laterite deposits classified according to principal deposit type (Fig. 1) shows that, with the exception of Australia, most Ni laterite resources occur within 25° of the equator in the present humid tropics. This includes all the important producers in SE Asia and Oceania (Indonesia, Philippines, New Caledonia), the Antilles and South America (Cuba, Dominican Republic, Colombia, Brazil). There are, however, numerous deposits at higher latitudes, including those in semi-arid to arid central and SW Australia and humid Mediterranean to temperate regions in SE Australia, the western United States (Oregon and California), Greece, the Balkans, and in the Ural Mountains and plains of central continental Russia and Kazakhstan. All of these regions, however, had warm, humid climates in the past (chapter "The Role of Regolith Formation and Evolution in the Devel opment of Supergene Mineral Deposits"). There does not appear to be any clear relationship between deposit type and past or present climate. Most hydrous Mg silicate deposits are in the present tropics, but similar mineral assemblages are found in temperate regions. Oxide and clay silicate deposits occur in all climatic regimes.

4 Factors Influencing the Distribution and Formation of Nickel Laterite

4.1 Bedrock Geology

Lithology The vast majority of Ni laterite deposits are on serpentinized olivine-rich ultramafic rocks, which provide protores containing 0.2 to 0.4% Ni. A few small deposits in Greece developed by re-weathering sediments containing lateritic debris derived from serpentinized peridotite (Valeton et al. 1987). Some smectitic clay silicate deposits are formed from pyroxenites (Niquelandia; Colin et al. 1990) and gabbros (Wingellina; Putzolu et al. 2020a), probably due to fluid mixing during weathering of these and adjacent peridotites. The lithology of the ultramafic protore influences the mineralogy and, hence, the dominant ore type. Olivine-pyroxene-bearing peridotites weather to oxide and either hydrous Mg silicate or clay silicate assemblages, whereas dunites (>90% olivine) predominantly form oxide deposits, mostly containing abundant secondary silica that may act as a diluent. In peridotites, the degree of serpentinization influences the abundance and mineralogy of hydrous Mg silicates that form in free-draining environments (Pelletier 1996; Golightly 1979). Unaltered rocks, with negligible serpentinization, give rise to oxide-rich ores, with only minor silicate mineralization in saprolite (e.g., West Soroako), whereas in rocks that are weakly to moderately serpentinized, the silicate ore zone is thicker, with abundant garnierite in veins, fracture-fillings and coatings, with nickeloan Fe-Mg smectites formed from olivine. However, altered primary lizardite also hosts Ni, substituting for Mg in octahedral sites (Manceau and Callas 1985). In highly serpentinized rocks, such Ni-rich altered lizardite is the principal ore mineral (Pelletier 1996. Golightly 1979).

Tectonic Setting The *accretionary island arc terranes* of the Circum-Pacific belt, Antilles and the Balkans host approximately 85% of Ni laterite resources, derived mainly from serpentinized ophiolites (harzburgite peridotites and dunites). Tectonic uplift in these terranes has resulted not only in erosion and exposure of the ophiolites but also given rise to a mostly free-draining regolith, promoting the development of oxide and hydrous Mg silicate deposits. Conversely, the high erosion rates limit the thickness of the regolith and the preservation of the upper horizons. The remaining deposits occur in *cratonic terranes* in South America, Africa and Australia, derived from mostly serpentinized protores. These include deposits at Murrin Murrin, Goongarrie, Cawse and Ravensthorpe in the Archaean Yilgarn Craton, Western Australia on komatiitic dunite and peridotite protores, and Niquelandia and Barro Alto in Brazil, Musongati (Burundi) and Wingellina (central Australia) developed on basal peridotites and dunites of Archaean to Palaeozoic layered intrusions. Due to their stability and mostly low relief, these cratons have been subjected to long periods of weathering under a variety of climatic conditions, resulting in the formation of oxide and clay silicate deposits preserved in deep lateritic regoliths. Deposits Russia and Kazakhstan have characteristics of both tectonic settings (Glazkovsky et al. 1977). Intense folding and faulting in the Urals have produced "linear" deposits, some with abundant hydrous silicates (e.g., Cheremshan, Russia), whereas more stable platform environments host "areal" deposits with mainly oxide and clay silicate ores (Buruktal, Russia; Kempirsai, Kazakhstan).

Structure Faults, shear zones and fractures in bedrock and regolith can strongly influence the thickness, grade and, in places, the mineralogy of Ni laterite deposits. Where such structures are pre-existing, their influence is mostly passive, affecting drainage characteristics. Faults and shears may either present barriers to water flow or, more commonly, increase permeability and promote deeper weathering. This can lead to preferential concentration of Ni in shear zones (e.g., Cawse: Brand et al. 1996; Murrin Murrin: Wells & Butt 2017), both in W. Australia. In regions where weathering is contemporaneous with tectonic activity, continuing activity along existing faults or in new low-angle shears in the regolith due to slope failure may focus the concentrations Ni minerals. For example, in New Caledonia, garnieritic hydrous Mg silicates, quartz and other neo-formed minerals have precipitated as overgrowths and coatings in boxworks and fault breccias, as veins and in tension cracks. Differing mineral habits and assemblages and overprinted striations suggest that supergene mineralization and tectonic activity were synchronous, though possibly under slightly different weathering conditions (Cluzel and Vigier 2008). Similar observations have been reported at Falcondo, Dominican Republic (Villanova-de-Benavent et al. 2014). The structural settings of some hydrous silicate veins at (or below) the base of the regolith in the Koniambo deposit, New Caledonia, have been speculatively interpreted as the product of hydrothermal activity rather than being supergene. Fractures and breccia sets in saprock, cemented by successive precipitation of lizardite, Ni-Mg kerolite and quartz, are considered to typify hydraulic fracture-and-seal processes (Cathelineau et al. 2017), with precipitation of these minerals occurring at 50–95 °C. However, the faulting and uplift also change the drainage regime of the regolith, increasing groundwater flow and permitting the transport of dissolved Ni, Mg, Si and other elements not only vertically, deep into the saprolite and the underlying fractured protolith, but also laterally into regolith and unweathered bedrock downslope, distant from the source. Whether minerals precipitated from such groundwaters are considered to be supergene or hydrothermal is a matter of definition, since weathering is itself hydrothermal and meteoric hydrothermal fluids are heated groundwater—as depicted, for example, in the hydrothermal cell hypothesized by Cathelineau et al. (2017). Possibly, to be regarded as "truly" hydrothermal, the critical dissolved chemical components (e.g., Ni, Mg, Si) would have to be derived from reactions at depth, beneath the weathering zone, rather than from circulating groundwater, or by mixing of cool descending surface water containing these elements with warm ascending meteoric water without them.

At La Gloria and Exmibal, Guatemala, Golightly (2010) suggested that high Ni enrichments in saprolite are associated with shears related to downfaulted terrace blocks that had been detached from the lateritic plateau on the margin of a graben.

4.2 Climate

As noted above and shown in Fig. 1, the majority of Ni laterite deposits are in present-day humid tropical regions. Some are situated in equatorial rainforests, with rainfall >1800 mm pa and dry seasons of less than two months. Examples include Soroako, Weda Bay and Lapaopao (Indonesia), Sipolou, Conakry (West Africa), and Onca, Puma, Vermelho, Cerro Matoso (S. America). Most, however, are situated in seasonally humid wet savannas (summer rainfall 900–1800 mm, 2–5-month winter dry season). These include the mines and prospects in the Philippines, New Caledonia, NE Australia, Antilles, Burundi and many in Brazil. Thorne et al. (2012) calculated that Ni laterites develop where rainfall exceeds 1000 mm pa and mean monthly temperatures range between 22–31 °C (summer) and 15–27 °C (winter). Although this could include areas with Mediterranean climates, it is generally the higher temperatures and rainfall of the humid tropics, coupled with high biogenic activity, that promotes the intense chemical weathering necessary for lateritization.

Numerous oxide and clay silicate deposits are located in warm semi-arid to arid regions in central and western Australia and in humid Mediterranean to temperate regions in SE Australia, USA (Oregon and California), the Balkans, Greece, Turkey, Russia and Kazakhstan. All of these regions are considered to have had more humid and warmer, possibly tropical, climates in the past (Scotese 2000; Thorne et al. 2012) typical of lateritization, despite now being situated at high latitudes. Present-day southern Australia, for example, was at latitude 60–65°S in the late Cretaceous to mid-Eocene, with intensely weathered regoliths; Murrin Murrin and other Ni laterites were formed during and after this period (Pillans 2005). Palynological and palaeontological evidence suggests warm to tropical humid conditions that supported rainforest in South Australia prevailed at this time, with similar conditions returning in the late Oligocene and mid-Miocene (McGowran and Hill 2015; Mao and Retallack 2019). The rest of the continent had broadly similar climates, becoming warmer to the north, although no part of the continent was above about 30°S until the Early Eocene (50 Ma). Modification of the deposits under the later arid climates has generally been minor, such as by precipitation of magnesite and silica (e.g., at Murrin Murrin, Wells and Butt 2017), although Putzolu et al. (2020b) report late-stage precipitation of interstratified Ni-clays (chlorite-smectite and talc-smectite) in the clay-silicate resource at Wingellina as neo-formed minerals and replacements of phyllosilicates, following the onset of arid conditions.

The global distribution of deposits (Fig. 1) shows no clear relationship between ore type, grade or size and the present climate. The apparent relationships between clay silicate mineralization and semi-arid climates and hydrous Mg silicates with the humid tropics are due largely to their tectonic, structural and geomorphological settings, which influence drainage status and vulnerability to erosion.

4.3 Age of Weathering

The landscapes hosting Ni laterite deposits formed and evolved over long time periods under a succession of different weathering environments. Accordingly, it is only possible to estimate the periods of most intense weathering rather than more specific dates. Direct dating of regolith has only been undertaken in a few regions and rarely from Ni laterite deposits themselves (see also chapter "The Role of Regolith Formation and Evolution in the Development of Supergene Mineral Deposits").

The oldest deposits occur on cratons, in regions that may have been eroded and exposed for much of the Phanerozoic. The Yilgarn Craton of Western Australia is now mostly semi-arid, but the widespread mantle of deep lateritic regolith hosts several Ni laterite mines and prospects. Palaeomagnetic dating at Murrin Murrin and sites near Cawse and Bulong gives Palaeozoic ages, indicating a very early onset of weathering (Pillans 2005; Anand and Paine 2002). Other dates, however, point to major later phases of humid tropical weathering in the mid-late Cretaceous, Palaeocene and late Eocene-Oligocene. Similarly, episodic weathering occurred in northern Australia and the central South American and West African cratons, but in each of these, humid tropical conditions continued throughout the Cenozoic (Scotese 2000). Deposits in tectonically active terranes are younger, with the most accurate dates estimated for those buried by later sediments of known stratigraphic age. For example, in the Urals, weathering of upper Devonian serpentinites commenced in the late Triassic, with some Ni laterites buried by the mid-Cretaceous (Glazkovsky et al. 1977). Similarly, in Greece, Ni laterites developed on upper Jurassic to Lower Cretaceous ophiolites, and sediments with ophiolitic detritus were buried by mid to upper Cretaceous sediments (Valeton et al. 1987). However, the nearby Çaldağ deposit in Turkey probably formed in the Palaeocene to mid-Miocene (Tavlan et al. 2011). Weathering in islands of the western Pacific (Philippines, Indonesia, New Caledonia) and the Antilles also dates from the Eocene/Oligocene (Sevin et al. 2012). Some deposits may have formed very rapidly: Miocene ophiolites on Sulawesi (e.g., Lapaopao) have probably been exposed for less than 10 Myr.

4.4 Geomorphology

Nickel laterites are typically found in regions with a deep, intensely weathered regolith. These terrains have (1) sufficient tectonic stability and low relief such that the rate of weathering exceeds that of erosion and (2) a low relief so that the regolith is protected from erosion by either burial or armouring by ferruginous or siliceous duricrusts during later tectonic or climatic changes (see chapter "The Role of Regolith Formation and Evolution in the Development of Supergene Mineral Deposits"). These conditions prevail in cratons but are less sustainable in accretionary terranes, where most deposits occur in uplifted remnants of a previously more extensive regolith cover. These terranes are subject to active erosion, and deposits are well preserved only beneath duricrust on crests, upper slopes and terraces.

The topographic setting affects the drainage of the landscape and the regolith, and hence the nature and grade of deposits, especially on serpentinized olivine-pyroxene peridotites. The regolith on stable cratons and locally elsewhere is poorly drained due to low relief and a high water table. This results in reduction in the intensity of leaching and rate of removal of weathering solutions. As a result, Ni concentration is largely residual, with little absolute accumulation. The peridotites weather to form low-grade clay silicate deposits dominated by smectite in the mid to upper saprolite and overlying plasmic zone (e.g., Murrin Murrin, W. Australia; San Felipe, Cuba; Kempirsai, Kazakhstan). The deposits generally have poorly developed oxide zones. Wingellina, W. Australia, has both a smectite clay deposit and a larger oxide deposit with associated silica, but these appear to have developed over different bedrock lithologies (Putzolu et al. 2018, 2020b).

In accretionary terranes or on hills in cratons where there is high topographic relief, the water table is commonly deep, and the regolith is freely draining. The rate of leaching and movement of groundwater is maximized, resulting in absolute enrichment deep in the saprolite, generally expressed by high-grade hydrous Mg silicates and residual concentration of Ni, a well-developed oxide zone. In comparison, dunites tend to form oxide deposits irrespective of drainage conditions (Tiebaghi and much of Goro, New Caledonia). If drainage is impeded, silica accumulation and grade dilution may be greater (e.g., Cawse, Ravensthorpe, and locally at Wingellina, Australia, and possibly Onca and Puma, Brazil).

5 Nickel Laterite Formation and Evolution

5.1 Direct Formation

Many oxide deposits have formed during essentially continuous weathering in humid savanna climates. Such conditions are typical of lateritization on stable cratons and tectonically less active accretionary terranes. In such environments, water tables are high, although fluctuating seasonally, and erosion rates low, thereby favouring weathering to depths of 50–80 m (Fig. 6a). Weathering rates are estimated to be about 8 to 20 m/Myr (Nahon and Tardy 1992; Freyssinet and Farah 2000); hence, a full profile requires at least 2 to 10 Myr to form. In more freely drained near-surface zones, Ni released by the hydrolysis of olivine or serpentinized olivine is largely retained, hosted by goethite, either adsorbed or incorporated within the crystal lattice. Silica and Mg are leached so that porosity increases and density decreases, with total mass losses of up to 70%. In the ferruginous oxide zone above the Mg discontinuity, this alone can residually concentrate Ni to 0.6–1%, mainly associated with goethite. Higher grades, to over 1.5% Ni, are due to

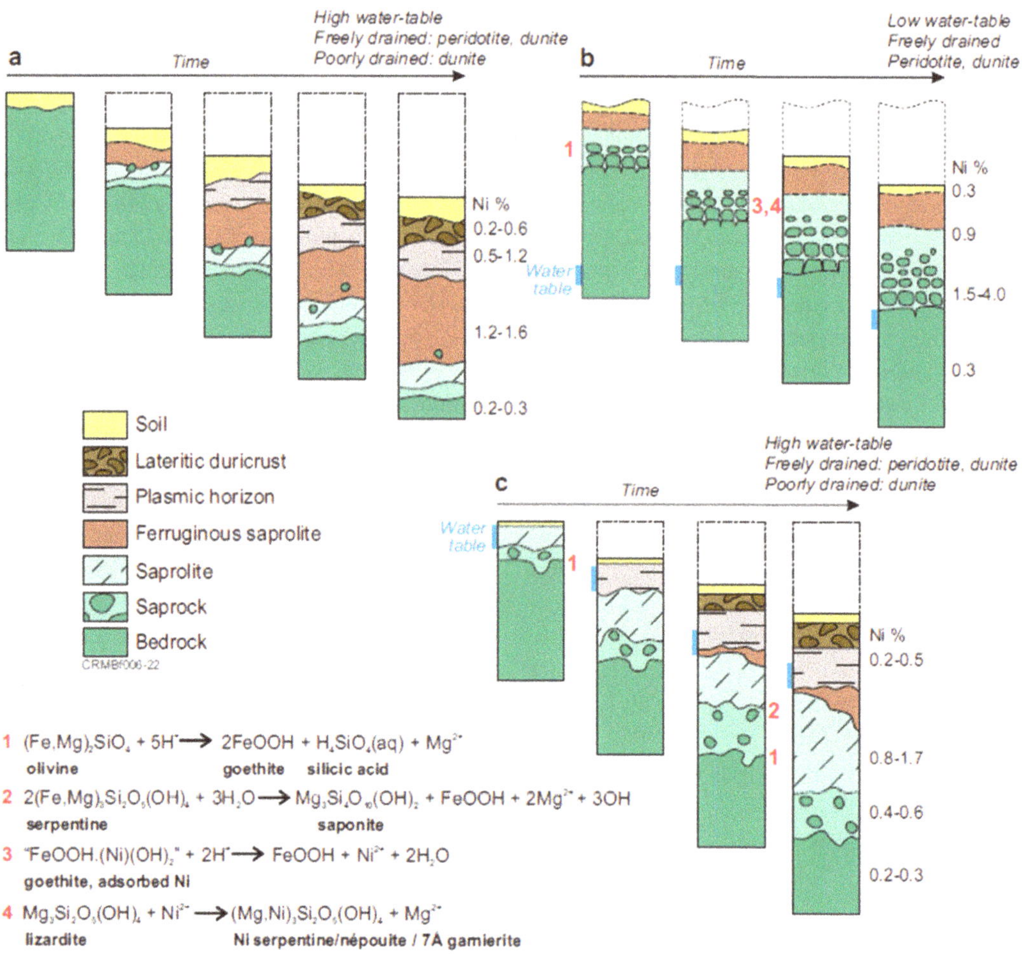

Fig. 6 Direct formation of Ni laterite deposit in humid savanna to rainforest transition climates. (**a**) Oxide. (**b**) Hydrous silicate. (**c**) Clay-silicate. (After Freyssinet et al. 2005; Butt and Cluzel 2013)

absolute enrichment by Ni leached from surface horizons or upslope. Nickel is released as goethite "ages" and coarsens by successive dissolution and recrystallization (Dublet et al. 2015) and is adsorbed to, or co-precipitated with, goethite in the lower oxide zone, including ferruginous saprolite. Manganese oxides, typically lithiophorite and asbolane, also precipitate in the lower oxide zone, concentrating Co and some Ni. Below the Mg discontinuity, there is rarely any significant Ni enrichment in the mid-to-lower saprolite except where faults and shear zones permit increased groundwater movement and formation of hydrous silicates (Fig. 6b). Similar oxide zones develop over serpentinized dunite even in poorly drained environments, but as silica may not be fully leached, it is commonly precipitated as quartz in the lower oxide zone, upper Mg-bearing saprolite and deeper fault zones. Serpentinized peridotites generally have a lower porosity and more stable mineralogy, so they are less strongly leached. Most Mg, Si and Ni are retained in the mid-to-upper saprolite where serpentine has weathered to smectite and saponite, in places forming clay silicate Ni laterite deposits (Fig. 6c).

In juvenile accretionary terranes, deposits such as Soroako (Golightly 1979) and Lapaopao, Sulawesi, have developed under continuing rainforest climates since exposure in the late Miocene. Topographic relief is higher than generally associated lateritization, so the regolith is thinner and less mature because of rapid surface erosion. Due to a lower water table and free-draining conditions, Ni adsorbed to goethite in the oxide zone is also partially leached and translocated deeper in the profile and laterally downslope. Here, the Ni forms absolute enrichments by being adsorbed or co-precipitated with goethite, exchanged for Mg in serpentine, or by reaction with Mg and silica released from primary minerals to form secondary hydrous Mg silicates (Fig. 6b).

Although described and depicted as largely vertical processes within a regolith profile, lateral mobility is a key to leaching, mobilization and precipitation. This may account for local enrichments within a topographic sequence and add uncertainty to simple mass balance calculations.

5.2 Multi-stage Formation

Tectonic activity is important in the formation of some deposits by rejuvenating the topography and lowering previously high water tables in areas of originally low relief. In the humid tropics, this has typically led to the modification of pre-existing low-grade oxide deposits to form higher-grade hydrous silicate deposits. Nickel is leached from the oxide zone by solution and recrystallization, "aging" goethite (Dublet et al. 2015), and initially forms pervasive enrichments in the saprolite by incorporation in serpentines through exchange reactions for Mg (Pelletier 1996). Subsequently, the excess Ni precipitates as hydrous Mg silicates (garnierite) on open surfaces on spheroidally weathering boulders, joint fractures, and faults (Figs. 4 and 7), possibly in response to the changed Eh/pH conditions of the voids. Many high-grade deposits on uplifted and partly eroded plateaux and terraces have formed by such mechanisms and account for locally spectacular occurrences of garnierite, such as those in New Caledonia and the Dominican Republic. There appear to be paragenetic successions within the silicate precipitates, which become progressively enriched in Ni and Si due to increased silica activity over time, with a final assemblage of Ni-poor minerals and quartz (Fritsch et al. 2016; Villanova-de-Benavent 2015; Galí et al. 2012; Golightly 2010). In New Caledonia, garnierites are dominantly lizardite-népouite serpentine phases, whereas at Falcondo, Dominican Republic, they are mainly talc-like, with the apparent precipitation sequence: Fe-bearing Ni-enriched serpentine > Fe-free mixtures of serpentine and kerolite–pimelite > kerolite-pimelite > sepiolite–falcondoite and finally Ni-free sepiolite and quartz (Villanova-de-Benavent 2014; Galí et al. 2012).

The stepped land surface of New Caledonia indicates that lateritic weathering initially occurred on a landscape of low to moderate relief

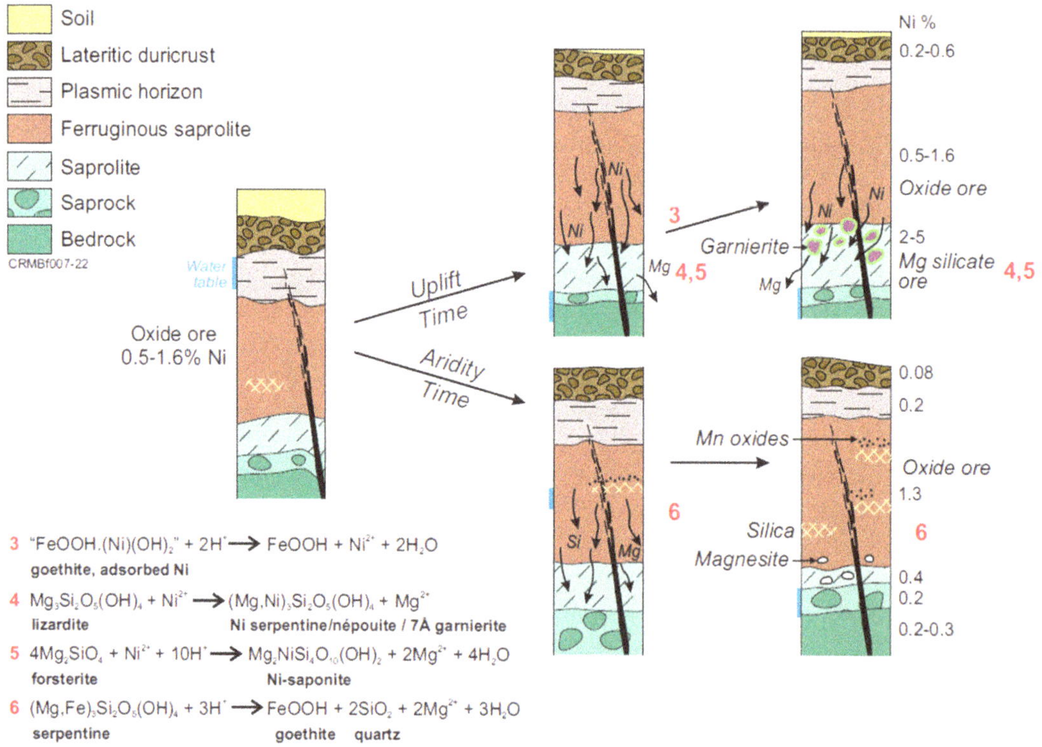

Fig. 7 Multi-stage formation and evolution of oxide and hydrous Mg-silicate deposits. (**a**) Well-differentiated lateritic oxide deposits formed under seasonally humid tropical savanna and rainforest climates in tectonically stable areas (see Fig. 6a). (**b**) With tectonic uplift under similar climates, Ni is leached from the oxide zone (reaction 3) and exchanges with Mg to form Ni-rich serpentine (4; e.g., nickeloan lizardite) and precipitates with Mg and Si to form garnieritic hydrous silicates (5; e.g., nepouite, saponite). (**c**) With change to a more arid climate, ongoing slow weathering causes only minor modifications to the profile, except for precipitation of silica and magnesite (6). (After Freyssinet et al. 2005; Butt and Cluzel 2013)

in the Lower Oligocene and continued whilst these surfaces were uplifted and partially dissected (Sevin et al. 2012; Chevillotte et al. 2006). Similarly, in much of the Antilles, including the Dominican Republic, ophiolites emplaced from the Cretaceous to the Oligocene or as late as the early Miocene were exposed to at least four erosion and weathering cycles, triggered by punctuated uplift due to transpressional strike-slip movement along major faults, commencing in the Miocene. In Cuba, however, uplift during this period was by gradual epeirogenic movement (Lewis et al. 2006), so that groundwater flow and leaching are reduced, favouring the formation of clay silicates in saprolite (e.g., San Felipe, Tauler et al. 2023).

A similar mechanism seems responsible for the origin of some hydrous Mg silicate deposits in the humid tropics situated in more stable areas of generally low relief. Due to relief inversion, there are isolated hills of free-draining, weathered ultramafic rocks protected by ferruginous duricrust. In these, Ni leached from the oxide zone has concentrated in the underlying saprolite. Cerro Matoso, Colombia, is an example. It is on a low hill capped by ferruginous duricrust with a regolith about 40 m thick, including a 5–10 m oxide zone. The principal ore horizon is 'green saprolite' characterized by Ni serpentines, Ni sepiolite and other hydrous silicates including pimelite (Ni kerolite); in the lower saprolite and saprock, fault-and joint fractures contain garnierite veins, dominantly pimelite, similar to hydrous silicate deposits in the Antilles, New Caledonia and Indonesia (Gleeson et al. 2004). The presence of quartz and Si-rich Mg silicates in these veins is ascribed to precipitation during dry seasons. The faults and fractures in the lower saprolite, saprock and protolith are the focus of groundwater flow and ensure that the regolith is freely drained - a situation similar to those in more tectonically active island arc settings.

Many deposits are situated in regions that have been subject to climatic changes since the lateritic host regolith was formed. In much of continental Australia, semi-arid climates have prevailed since the mid-Miocene, resulting in greatly reduced weathering rates, the accumulation of magnesite and further silicification over serpentinized dunite protores. There is little other change to the regolith profile nor to the distribution or abundance of Ni in either smectite silicate or oxide deposits (Figs. 7 and 8) except for dilution by silica in the latter. In Brazil, oxide ore occurs beneath a protective silica cap rock at the Onca and Puma deposits. Golightly (2010) considered that silicification occurred in an earlier phase of weathering under arid conditions, suggesting that these deposits represent a separate genetic model. However, silicification may have been the product of poor drainage during an earlier lateritization event, so these deposits may be further examples of continued weathering after relief inversion without significant climate change.

5.3 Reactive Transport Modelling

Domènech et al. (2017) constructed simple 1D reactive transport models to investigate whether the Punta Gorda oxide deposit at Moa Bay, Cuba, might have formed directly by such a mechanism over a time period of about 10 Myr. The deposit is developed on partially serpentinized ophiolitic harzburgite and has hydrous Mg-silicates in a thin saprolite (~5–10 m) overlain by a thick goethite-rich oxide horizon (30–50 m). The modelling was based on the premise that infiltrating rainwater flows vertically downwards, initially through unweathered rock, to form oxide and saprolite horizons by the dissolution of primary forsterite, enstatite and serpentine, and precipitation of goethite and secondary serpentine. Experimental conditions were based on field observations at Punta Gorda and other Ni laterite deposits reported in the literature, and published mineral reaction kinetics and processes of mineral dissolution/precipitation accounting for chemical

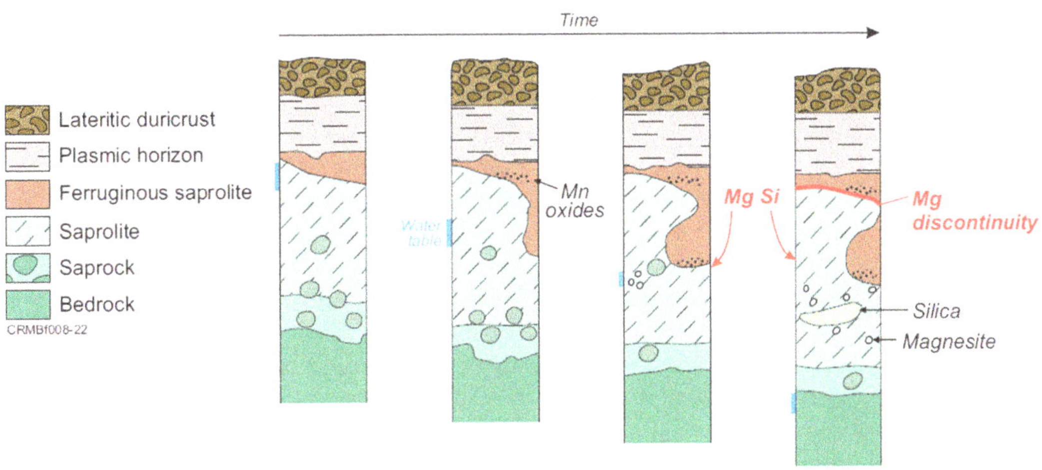

Fig. 8 Evolution of a clay silicate deposit formed under seasonally humid tropical savanna climate and subjected to a change to an arid climate. Ongoing slow weathering causes only minor modifications to the profile, except for precipitation of silica and magnesite

and physical transformations during weathering. The resultant models simulate the formation of the different horizons in the profile. The results also suggest that sorption of Ni onto the goethite surfaces at pH >7 can explain the concentration of Ni in goethite in saprolite and the lower oxide zone, with abundances decreasing upwards due to leaching by acidic rainwater. The model does not, however, consider Ni incorporated in the goethite lattice.

Similar modeling of the formation of oxide-hydrous silicate deposits in New Caledonia was undertaken by Myagkiy et al. (2017, 2019), considering both direct and multi-stage formation models. These studies suggest that increasing thickness of the oxide zone and element mobility, especially of Ni, are controlled by the descent of a pH front. Adsorption of Ni to goethite occurs in the high pH environments associated with silicate weathering in saprock and saprolite, reaching a maximum at pH 8.5, with competition by Mg occurring at higher pH. Gradual acidification of surface horizons caused by rainfall releases the adsorbed Ni (and Mg); some of the Ni enters the crystal lattice during associated dissolution and reprecipitation of goethite, with the remainder being leached deeper in the profile, becoming adsorbed to neo-formed goethite or exchanging for Mg as primary silicates weather to secondary hydrous silicates. Thus, as in the direct model, as the profile matures over a period of several million years, the hydrous silicate component becomes dominant at the expense of the oxide component. In a more comprehensive 3D model, Myagkiy et al. (2019) simulated the effects of uplift on the changed drainage status on the hydrodynamics of groundwater flow and, hence, the dispersion of elements released during ongoing weathering. The simulations match existing empirical suggestions that, under similar seasonally humid tropical climatic conditions, uplift enhances the rate and effectiveness of weathering and leads to higher-grade hydrous silicate deposits deeper in the saprolite and downslope. These effects are magnified in situations where uplift associated with tectonic activity causes faulting and fracturing at a range of scales, permitting penetration of groundwater through the rock mass as well as between larger fault blocks. The downward and lateral progress of the pH front is more rapid, so the silicate deposits form faster than in areas of lower relief, although this is offset by the concomitant increase in erosion.

6 Scandium

Historically, Sc supplies have been recovered as by-products from the processing of ores of Fe, REE, Ti, and Zr (China), U (Ukraine, Kazakhstan) and apatite (Russia), in which initial concentrations are less than 100 ppm Sc (Kaiser Research 2022). Recent attention has turned to tailings and ores of high tonnage supergene resources in regoliths over alkaline igneous rocks, ion-adsorption REE deposits and, especially, in bauxite and Ni laterite. Commercial extraction of Sc from the intermediate mixed Ni-Co sulfide product from high-pressure acid leach (HPAL) commenced at Taganito HPAL Nickel Corporation operations at Surigao del Norte, Philippines, in 2018 (Sumitomo Metal Mining Co., Ltd. 2018). Interest in Ni laterites as a source of Sc has increased with the recognition that some Australian deposits have associated enrichments of 300–600 ppm (see Sect. 8).

Ultramafic bedrock contains about 3–20 ppm Sc, hosted by pyroxenes and amphiboles, which it substitutes for ferrous Fe due to similarities in ionic radius (Williams-Jones and Vasyukova 2018). Overall abundances are determined by the relative proportions of olivine and (clino)-pyroxene. In New Caledonia, Teitler et al. (2019) noted a range from 3–5 ppm Sc in dunite, 6–8 ppm Sc in harzburgite, 10–25 ppm Sc in lherzolite and 130 ppm Sc in hornblende-rich amphibolite. Scandium is residually concentrated in lateritic regoliths over ultramafic rocks. Mean abundances of 80–120 ppm (maximum ~160 ppm) were noted in ferruginous saprolite and lateritic residuum over pyroxenite in the layered Ora Banda Sill in Western Australia—but only 20 ppm over adjacent peridotite (Butt et al. 1992). Scandium is similarly enriched in both saprolitic hydrous silicate (30–40 ppm) and oxide ores (70–120 ppm) of Ni laterite deposits (e.g., Sulawesi, Ito et al. 2021; Cuba, Dominican Republic and New Caledonia (Aiglsperger et al. 2016). In each of these deposits, Sc concentrations progressively increase upwards, reaching a maximum in the lowermost "yellow limonite" of the oxide zone and then decreasing in overlying "red limonite" and duricrust. This distribution is the result of Sc released from pyroxenes being incorporated in goethite, which gradually accumulates to form the oxide zone of the lateritic profile. Subsequent remobilization of Sc occurs as the uppermost horizons are leached and goethite dissolves and recrystallizes, or is partially replaced by hematite. Released Sc is transported downwards and is incorporated in neo-formed goethite in the lower oxide zone, where it may be adsorbed to the surface or possibly substituted for ferric iron in the goethite lattice (Qin et al. 2020). The highest Sc contents in the goethite-rich oxide zones generally reflect increasing modal abundances of clinopyroxene in bedrock (Ito et al. 2021; Teitler et al. 2019).

These modest enrichments (70–130 ppm Sc) in oxide ore are a potential source of Sc as a by-product of HPAL treatment, as noted above, but none forms a significant resource. In contrast, several Ni laterite deposits in eastern Australia have oxide horizons with mean grades of 250–430 ppm Sc, developed on Alaskan-type ultramafic complexes. Examples include the neighbouring Sunrise and Owendale deposits near Fifield and the Gilgai deposits at Nyngan, each of which is enriched in Ni, Co, Pt and Sc. The higher-grade Sc mineralization is formed mainly from the pyroxenites. For example, the dunite and clino-pyroxenite protoliths at Sunrise contain <10 ppm and 70–80 ppm Sc respectively, hence although the grades of oxide ore developed over dunite are similar to deposits elsewhere (50–110 ppm Sc), higher grades are mainly associated with the clino-pyroxenites (see Sunrise, Sect. 8; Chassé et al. 2017, 2019; Williams-Jones and Vasyukova 2018). At Sunrise, Chassé et al. 2017 found that Sc released from clinopyroxene is mainly hosted by ferrosaponite, suggesting weathering under reducing conditions in which Sc is immobile, implying concentrations up to 300 ppm Sc in saprolite are due to relative enrichment. Subsequent intense weathering of ferrosaponite and other Mg-silicates in the upper saprolite and the formation of the lateritic oxide zones above the Mg discontinuity releases Sc, which is then mostly scavenged by adsorption on goethite. Thereafter, successive dissolution–reprecipitation reactions from different generations of Fe oxides with varying compositions progressively concentrate Sc in the lower oxide horizons of the lateritic cover. These are similar to the goethite-aging reactions (Dublet et al. 2015) that remobilize Ni in these horizons. Dehydration of goethite results in up to 20% of the Sc being hosted in hematite. The Sc enrichment in the upper oxide, semi-residual and transported units at Sunrise occurs where there are high abundances of allochthonous and neo-formed kaolinite and gibbsite, whereas Sc concentrated in bauxite is associated with goethite, not Al minerals (Gantzmann et al. 2021).

7 Lapaopao Nickel Laterite Deposit, Sulawesi, Indonesia

7.1 Introduction

The Lapaopao Ni-Co deposit is typical of lateritic mineralization developed on ultramafic bedrock in tropical and subtropical climates. It has a significant thickness of saprolite comprising Ni-enriched serpentine with veins of garnierite, overlain by a well-developed ferruginous oxide ("limonite") zone. Unusually, both the hydrous silicate and oxide zones at Lapaopao contain mineable resources. Lapaopao is located on the coast of the Gulf of Bone, Sulawesi, Indonesia, at latitude 3.86°S and longitude 121.27°E. It is about 130 km W of Kendari and about 40 km NW of Kolaka.

The mineral concession at Lapoapoa covers an area of 6785 Ha. PT INCO commenced exploration in 1969, with an initial phase of reconnaissance auger drilling and test pit excavation, followed in 2007 by a second phase of geological mapping and diamond core drilling on a 400 × 400 m grid (124 drill holes with an average depth of 26.6 m). The concession was acquired by PT Ceria Nugraha Indotama (PT CNI) in 2011, which has subsequently conducted HQ-size triple-tube diamond core drilling at various grid spacings from 200 m to 50 m. By the end of 2020, 1625 holes had been drilled for a total depth of 52,163 m.

7.2 Geology and Physical Setting

7.2.1 Bedrock Geology

The island of Sulawesi is a complex melange of litho-tectonic units bounded by large-scale tectonic dislocations and thrust faults. From west to east, these are (i) the west Sulawesi volcano-plutonic arc, (ii) the central Sulawesi metamorphic belt, (iii) the east Sulawesi ophiolite belt, and (iv) the continental fragments of Bangaii-Sula, Tukang Besi and Buton. Bedrock at Lapaopao (Fig. 9) is a fragment of the east Sulawesi ophiolite belt, one of the three largest ophiolites in the world, but separated from the main body to the north by intercalated metamorphic rocks. Ultramafic bedrock occupies about 90% of the concession, with 30% covered by alluvial, swamp and coastal sediments.

The ultramafic unit is in thrust contact in the NE with the Mekongga Complex, a Mesozoic metamorphic belt comprising muscovite schist, quartzite, meta-limestone and gneiss. The ultramafic rocks are mainly harzburgite with lesser dunite, hornblendite, lherzolite and pyroxenite. The level of serpentinization, estimated from loss on ignition (LOI), is low, with a median of about 4% LOI. Higher levels of serpentinization are present, particularly along fault zones such as the thrust contact with the metamorphic complex. Mafic rocks (diorite, diabase and gabbro) occur in places. The ultramafic complex is believed to be Cretaceous in age but was not emplaced until the Middle Miocene (Kadarusman 2019).

7.2.2 Geomorphology and Regolith Geology

The relief is strongly influenced by the underlying bedrock. The metamorphic complex to the east has the highest peaks in the region (up to 400 m above sea level) and an incised topography with steep slopes. The eastern NW-trending part of the ultramafic unit has an undulating topography, with elevations of mostly 150–200 m. The western part of the ultramafic comprises a narrow ridge trending southwest, bounded by steep slopes extending to the coast.

Nickel-rich lateritic regolith is formed to some degree over virtually the whole of ultramafic rock in the Lapaopao concession (Fig. 9), with bedrock exposed only in small areas on steep slopes and valley floors. The stratification in the lateritic profile is complex (Fig. 10), resulting from variable groundwater flow-throughs influenced by local factors such as geomorphology (slope), bedrock faults and shears, and the depth and fluctuation of the water table.

7.2.3 Climate

The region has a humid tropical rainforest climate. Although there are distinct dry and wet seasons, there is significant rainfall throughout the year (Table 2). The wet season (December—May) is influenced by continental Asian and Pacific Ocean air masses picking up moisture from warm oceans. The dry season (May to October) is influenced by drier Australian continental air masses.

7.3 Deposit Characteristics

7.3.1 Regolith Expression

The uppermost zone of the lateritic regolith represents the almost complete residual enrichment of various Fe oxides. The original rock fabric physically collapsed toward the top, where the zone is darker in colour and contains hematite. Ferruginous duricrust is commonly exposed at the surface. With increasing depth, primary rock fabrics are evident, and the colour is yellow-brown. Veins and films of black Mn oxides are present within the lower part of this layer. Although generally referred to as a single limonite zone, it is technically lateritic residuum grading downwards to ferruginous saprolite.

The hydrous silicate saprolite underlies a thin transition zone that separates it from the more

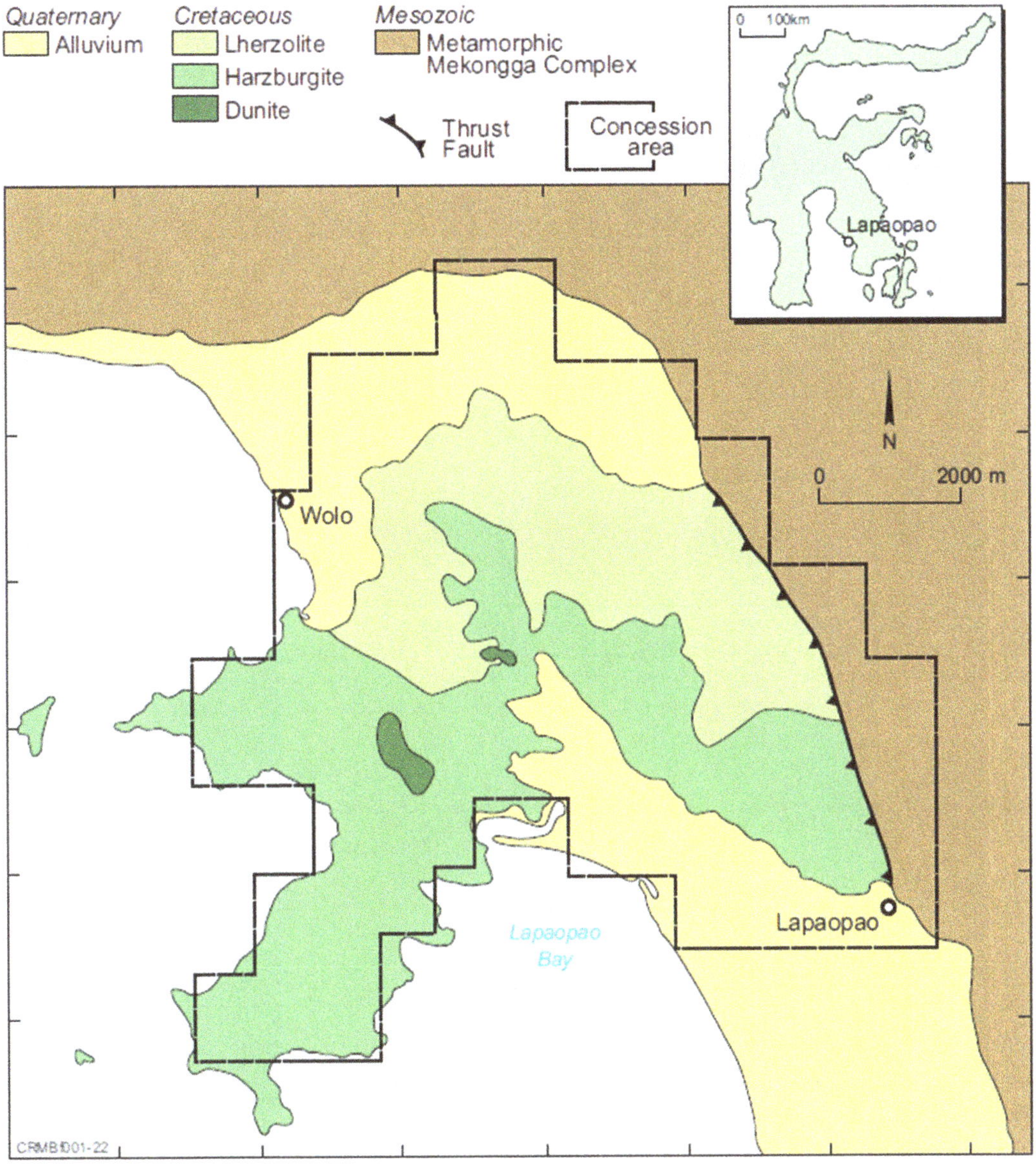

Fig. 9 Geological map of Lapaopao concession, Sulawesi. (Rusmana et al. 1993)

completely weathered ferruginous horizons above. The saprolite comprises various incomplete stages of formation and is composed of altered bedrock minerals combined with newly-formed hydrous silicate minerals (e.g., garnierite, smectite) and Fe oxides (e.g., goethite). The saprolite is not collapsed, and primary textures may be visible. Uneven leaching and weathering usually result in the saprolite containing boulders and fragments of fresh rock, surrounded by a matrix of highly weathered material. Garnierite veins are found near the base of the saprolite, commonly associated with silica (Fig. 4b). The underlying fresh bedrock can show traces of incipient weathering and can include thin veins of supergene mineralization.

The thicknesses of the main components of the lateritic profile were derived from a Ground

Fig. 10 Lateritic profile exposed in cutting, Lapaopao. The "limonite" zone, dominated by Fe oxy-hydroxide minerals, overlies saprolite containing Ni-rich hydrous silicate minerals and ultramafic bedrock. The zones are defined by distinctive chemical and mineralogical properties and are separated by highly irregular contacts

Table 2 Historical weather records, Lapaopao area. (From Kadarusman 2019)

	Jan	Feb	Mar	Apr	May	Jun	Jul	Aug	Sep	Oct	Nov	Dec
Mean temperature (°C)	28	28	28	28	27	26	26	26	27	28	28	28
Minimum temperature (°C)	23	23	23	23	23	23	22	22	22	23	23	24
Maximum temperature (°C)	33	33	32	32	31	29	31	31	32	33	34	29
Precipitation/rainfall (mm)	137	171	230	241	386	236	271	168	148	113	272	96

Penetrating Radar (GPR) survey (Fig. 11). The limonite varies from 0 to 27 m thick, and the saprolite from 0 to 36 m thick, with the top of bedrock detected to depths of more than 38 m. The profile is conspicuously thicker on the eastern arm of the ultramafic complex.

The geochemical variation downhole in the Lapaopao profile for major and some minor elements (Fig. 12) is typical of a complete Ni laterite profile. Although element distributions in individual drill holes may show significant deviations from those shown here, there are a number of features characteristic of lateritic regoliths. One such feature is the step-change variation in the abundance of many elements at the limonite-saprolite boundary, reflecting the sharp change in mineralogy across the transition zone—the Mg-discontinuity. Most notably, these are upward decreases in MgO and SiO_2 (reflecting the breakdown of silicate minerals and leaching of their components) and corresponding increases in Fe, Al_2O_3, and Cr_2O_3 contents (reflecting increased content of residual Fe oxide minerals). An Fe content of 30% was used as a geochemical indicator for distinguishing the ferruginous limonite zone from the siliceous hydrous silicate saprolite.

7.3.2 Mineralization

Nickel and, to a lesser extent, Co (the "ore" elements) are present in both the Fe-rich oxide (limonite) zone, and the Mg-rich hydrous silicate

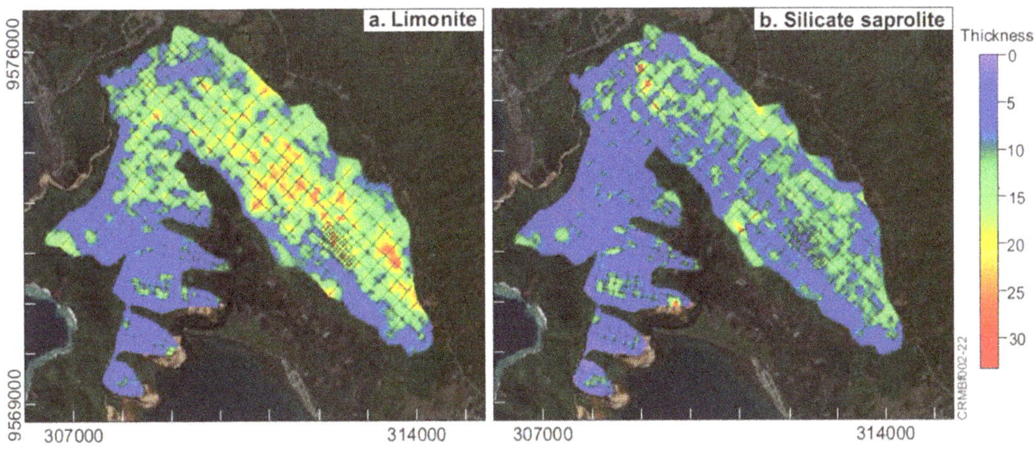

Fig. 11 The thickness of regolith zones is determined by ground-penetrating radar. (**a**) Limonite, surface to top of silicate saprolite. (**b**) Silicate saprolite, base of limonite to saprock. Coordinates and thickness scale in metres

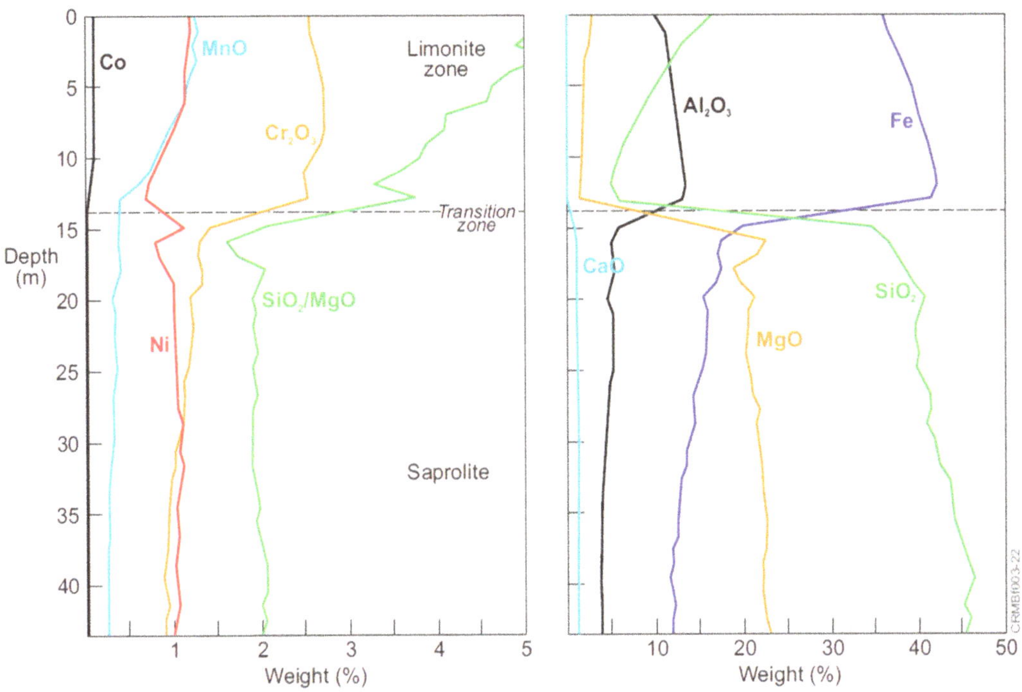

Fig. 12 Schematic geochemical profile of Ni laterite deposits at Lapaopao

saprolite, but their content is defined by the abundance of the host minerals and their capacity for accommodating the ore elements. The compositions of selected elements and oxides are shown in Table 3.

In the limonite zone, Ni occurs mainly in goethite (FeO(OH)) by substitution for Fe and/or by adsorption. Goethite is the final product of the breakdown and dissolution of olivine and pyroxene, in which all soluble components are removed

Table 3 Summarized chemical compositions of the regolith zones, Lapaopao

	Ni%	Co%	Fe%	SiO_2%	MgO%	Al_2O_3%
Limonite (oxide) zone						
No.	18,175	18,152	18,175	18,175	18,175	18,175
Mean	1.00	0.108	39.67	9.62	2.03	12.19
Std Dev	0.32	0.066	6.50	12.08	2.50	4.05
Q3 (75 pctile)	1.20	0.132	43.30	11.88	2.12	15.09
Median	0.98	0.100	41.00	5.23	1.51	13.11
Q1 (25 pctile)	0.77	0.070	37.62	2.63	1.18	10.12
Saprolite						
No.	28,997	28,994	28,994	28,996	28,997	28,980
Mean	1.05	0.028	13.51	43.36	21.76	4.35
Std Dev	0.55	0.044	7.60	11.44	11.74	2.78
Q3 (75 pctile)	1.41	0.036	17.77	45.81	32.07	5.79
Median	1.02	0.020	11.05	43.12	21.31	3.61
Q1 (25 pctile)	0.62	0.011	7.52	39.21	12.16	2.39

by groundwater, leaving only Fe. Early-formed, poorly crystalline Fe oxy-hydroxides have a high capacity to accommodate Ni: experimental studies have shown that goethite may contain up to 5.4 mole % Ni (Gerth 1990), but the relative proportions of Ni incorporated into the lattice, adsorbed onto the surface or present within its structure are still a matter of debate. Cobalt enrichment closely follows Mn content, reflecting the propensity for Mn oxides to scavenge Co from solution.

In the saprolite, Ni is hosted in partly weathered primary silicates by substitution for Mg, but also in newly formed hydrous silicate minerals such as serpentine, chlorite, sepiolite and garnierite. The high Ni content of some of the newly formed minerals is probably due to enrichment by Ni leached from higher in the profile, percolating downward in groundwater.

PT CNI has reported a total mineral resource estimate of 108 million dry metric tonnes at 1.38% Ni and 0.09% Co, based on cut-off grades of 1.10% Ni for the limonite zone and 1.30% Ni for the saprolite (Table 4).

7.4 Deposit Formation and Evolution

The formation of the Lapaopao deposit is a consequence of the combination of factors that influence the characteristics of Ni laterite deposits globally. The bedrock geology of the main harzburgite with lesser ophiolitic ultramafic rocks is fairly uniform throughout the property. The harzburgite is weakly serpentinized, as indicated by LOI data. Ultramafic rocks with low levels of serpentinization tend to form saprolite with a greater proportion of rocky material in the form of "boulders," and this is the case with Lapaopao. This attribute is well illustrated at Soroako, situated SE of Lapaopao within the same ultramafic belt, where highly serpentinized rocks are juxtaposed by faulting against poorly serpentinized rocks (Golightly 1981). The saprolite developed on the poorly serpentinized ultramafic rock has abundant rocky material, whereas it is absent in the serpentinized counterpart.

Tectonic activity during weathering can play a role in the development of Ni laterite. In certain cases, periods of tectonic uplift can reactivate previously mature profiles by lowering the water table and initiating renewed weathering. Sulawesi, in common with the surrounding oceanic-continental melange, has been subjected to massive tectonic activity since the late Palaeozoic. However, there is little evidence from the nature or complexity of the laterite profile at Lapaopao to indicate that there has been significant tectonism during its development since exposure in the late Miocene.

The difference in thickness of the limonite zone and saprolite between the western and eastern arms of the Lapaopao ultramafic (Fig. 11), although possibly due to tectonism, is more likely

Table 4 Lapaopao mineral resource estimate and production schedule

Mineral resource estimate (as of April 2021)

Resource category	Material type	Cut-off grade Ni %	Quantity million (DMT)	Ni %	Co %	Fe %	SiO_2 %	MgO %	Al_2O_3%
Indicated + measured	Limonite	1.1	37.30	1.26	0.12	38.96	10.82	2.33	11.58
	Saprolite	1.3	34.56	1.53	0.04	17.00	39.78	17.08	5.22
Subtotal indicated mineral resources			71.86	1.39	0.08	28.40	24.75	9.42	8.52
Inferred	Limonite	1.1	22.31	1.27	0.13	39.11	12.02	2.13	10.86
	Saprolite	1.3	14.29	1.54	0.04	15.89	40.89	18.90	4.43
Subtotal inferred mineral resources			36.61	1.37	0.09	30.04	23.30	8.68	8.35
Totals	Limonite	1.1	59.61	1.26	0.13	39.02	11.27	2.26	11.31
	Saprolite	1.3	48.85	1.53	0.04	16.68	40.11	17.61	4.99
Total mineral resources			108.47	1.38	0.09	28.96	24.26	9.17	8.46

DMT: Dry metric tonnes

Production schedule

	Mt/yr	Ni	Co	Fe	MgO	SiO_2	Fe/Ni	SiO_2/MgO
Limonite (HPAL)	3.6	1.26	0.12	39.29	2.13	9.97	31.06	–
Saprolite (RKEF)	3.75	1.59	0.04	17.45	16.31	39.13	–	2.40

the result of erosion. The thickness of a lateritic profile (or any component of it) results, initially, from the balance between the relative rates of downward weathering (the advance of the weathering front) and the removal of the top of the profile by erosion. The lesser thickness of limonite in the western arm compared to the eastern arm, relative to the approximate equivalent thickness of saprolite in both arms, appears to indicate that a significant part of the limonite zone has been removed. The more rugged topography of the western arm, with narrow ridges and steep slopes, is consistent with this interpretation.

7.5 Implications for Exploration, Evaluation, and Exploitation at Lapaopao

The most important consideration confronting further exploration and development of Lapaopao is to separate the two ore types that have strongly contrasting chemical and mineralogical compositions. Each type, dominated by either Fe oxyhydroxides or hydrous Mg silicates, has a specific method for extraction of the ore metals, each having different capital and operating economies and attendant risks. An important additional consideration at Lapaopao is that each type of mineralization is represented in sufficient proportion to be an individual development proposition. Many similar Ni laterite deposits show preferential development of one type over the other, with the result that only one is potentially economic for exploitation. PT CNI has carried out feasibility studies for both mineralization types. For the hydrous Mg silicate unit, rotary kiln electric furnace (RKEF) smelting was studied and found to be viable under a mine planning and smelter feed schedule. High-pressure acid leaching (HPAL) was determined to be the optimum processing method for the Fe oxyhydroxide limonite unit, which allows high recoveries of Ni and Co. The production schedules and chemical characteristics of each ore type are shown in Table 4.

PT CNI plans initially to commence a phased construction of up to four smelter lines to produce annually 55,600 t of Ni in ferronickel (FeNi) from 3.75 Mt. of hydrous Mg silicate ore grading $\geq$1.5% Ni (PT CNI 2021). Construction of an HPAL plant is scheduled to commence soon after, with an annual production capacity of 40,000 t Ni and 4100 t Co in a mixed hydroxide precipitate product from the treatment of 3.6 Mt. of ore.

Mining near the transition zone between the Fe- and Mg-rich ores, where there is an abrupt change in composition, presents the danger of creating a mixture that is not ideal for either treatment process. Because the contact between the two ore types can be quite irregular and visual control is not reliable, some form of careful grade control is necessary.

The mineral concession covers almost all of the ultramafic unit, so there is little opportunity to define additional resources. However, structural elements such as joints and shears can channel weathering to produce deeper ore development, including high-grade supergene minerals such as garnierite, possibly missed in areas of widely spaced vertical drilling. Closer drilling can potentially reveal such structures.

8 Sunrise Nickel-Cobalt-Scandium-Platinum Deposit, Fifield, NSW, Australia

8.1 Location, Discovery, and Development

The Sunrise Ni-Co-Sc-Pt project (previously known as Fifield and Syerston) is located ~440 km WNW of Sydney and ~ 5 km NW of Fifield, NSW, Australia (32°46′S, 147°26′E). Noble Resources NL identified a significant lateritic Ni-Co-Pt deposit developed over the Tout Complex in 1988. Subsequent drilling and evaluation by Clean Teq Holdings Ltd. (now Sunrise Energy Metals) has resulted in the definition of the Sunrise Ni-Co-Sc-Pt deposit, which remains undeveloped. The published mineral resource in 2016 is shown in Table 5.

8.2 Physical Setting

Sunrise is situated in gently undulating terrain at altitudes of 270 m to 320 m, bisected by a broad alluvial channel in the NE. The region is predominantly used for agriculture with scattered remnant native trees. The current climate is warm, temperate and subject to drought; mean annual daily temperatures range between 10.2 °C and 24.5 °C, with a mean annual rainfall of 458 mm (Condobolin Agricultural Research Station). Groundwater flow is insignificant, and very little perched water; the water table is below the drilling at >50 m depth. Shortwave infrared (SWIR) analysis of the drill core has shown increased levels of unbound water within the goethite-rich plasmic zone (GZ), associated with kaolinite and goethite.

8.3 Bedrock Geology

Regional sedimentary basement rocks of the Early to Middle Ordovician Girilambone Group have been intruded by Alaskan-type, mafic to ultramafic complexes that extend from Condobolin to S of Bourke. The Tout Intrusive Complex at Fifield, the protolith to the Sunrise deposit, has a serpentinized dunite core surrounded by pyroxenite and gabbro (Fig. 13). The complex has been dated at 439.6 Ma ±8.5 Ma, i.e., early Silurian (Webb 2014). The serpentinized dunite intrusive has mean contents of 30.0 wt% MgO and 0.15 wt% Al_2O_3. XRD analysis of 24 samples from 113 m to 136 m in SRC1257 shows no primary forsterite, with the mineralogy dominated by serpentine (mean 82%), brucite (mean 8 wt%) and spinels (8.5 wt %). The main serpentine minerals are lizardite (mean 52 wt%) and chysotile (mean 30 wt%) in a ratio of ~1.8:1. Where identified, the spinels are magnetite and chromite with similar abundances (means 3.2 and 3.6 wt% respectively). Trace amounts of carbonate (calcite, dolomite, magnesite), rutile, diopside, enstatite, quartz and hematite are also present. The surrounding pyroxenites include hornblende- and olivine-rich units.

8.4 Regolith Geology

The regolith has preferentially developed over the serpentinized dunite core, extending ~4 km E-W and ~ 2 km N-S to depths of 35–40 m thinning to 0–5 m over surrounding pyroxenites so that the base of the regolith over Sunrise has a basin-like

Table 5 Mineral resource estimate, Sunrise deposit (CleanTeQ 2016)

Cut-off NiEq	Resource category	Tonnage (Mt)	Ni%	Co%	Pt ppm	Sc ppm[a]	Fe%	Al%	Ca%	Mg%	Mn%
0.6	Measured	52	0.73	0.11	0.20	51	35	2.5	0.38	1.3	0.80
0.6	Indicated	49	0.58	0.10	0.21	56	35	2.3	0.40	1.2	0.70
0.6	Meas+Ind	101	0.65	0.10	0.21	54	35	2.4	0.39	1.2	0.75
0.6	Inferred	8	0.54	0.10	0.16	78	35	2.8	0.38	1.4	0.78

NB: Arithmetic discrepancies due to rounding. NiEq = nickel equivalent (Ni% + (Co% × 2.95)); Mt. = million tonnes
[a]Sc grades estimated by OreWin Pty Ltd., excluding resource overlying pyroxenites

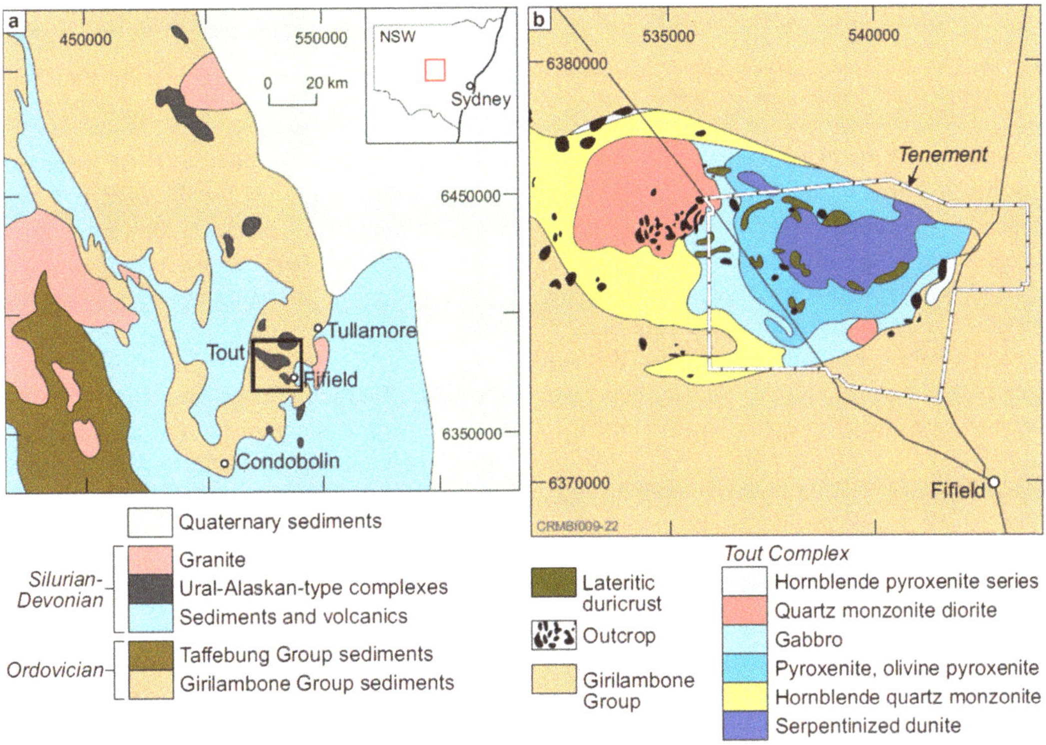

Fig. 13 Geology of the Sunrise deposit, NSW Australia. (**a**) Regional geology of the Tout Complex and other Uralian-Alaskan-type complexes of the Fifield Belt within the Lachlan fold belt. (After Keays et al. 2021). (**b**) Intrusive Complexes in the Project Area. (After Teluk 2001)

form (Fig. 14). High grades of Ni, Co and Pd occur over the dunite, whereas Sc is preferentially concentrated over the pyroxenites. The residual regolith has been subdivided into zones that reflect their geometallurgical (chemical and mineralogical) characteristics. Their relationship to conventional terminology is indicated in the following descriptions.

Alluvium (AZ) channel features that cross the deposit, which are clay (kaolinite) rich zones with very high Al_2O_3 and low Ni and Co grades.

Overburden (OVB) commonly characterized by clay, gravel and ironstone. The deep red maghemite hematite gravels are barren or contain very low grades of ore elements.

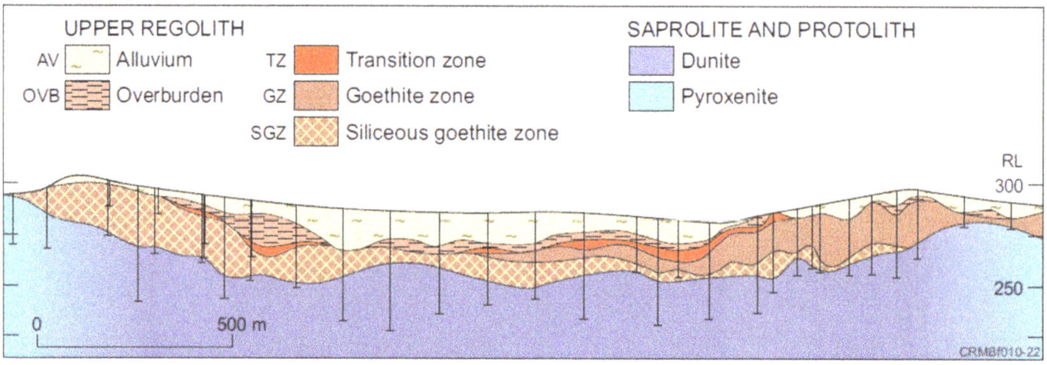

Fig. 14 Schematic section through Sunrise showing the thick regolith developed over the weathered dunite, shallowing over the pyroxenite. (After CleanTeQ 2016)

Transitional Zone (TZ) a mixed zone between overburden and residual regolith within the pedolith. It is strongly goethite-rich, with minor to major kaolinite and gibbsite and minor hematite and silica. Pisoliths are common in the upper parts of the zone, with a downward transition to orange-brown, goethite-rich clays. The TZ is variable in thickness (up to 15 m), with the upper part barren or with very low grades of ore elements.

Goethite Zone (GZ) a relatively uniform, very fine-grained, orange-brown, goethite-rich layer and a key economic zone. It typically contains more than 40% Fe and low abundances of Si (5–10%). The GZ contains black Co-rich Mn oxide wad and minor gibbsite, hematite, chromite, smectite/nontronite, kaolinite and magnetite. The zone is up to about 20 m thick, with a gradational upper boundary and a more abrupt and irregular lower boundary. This zone includes the plasmic zone within the pedolith above the pedoplasmation front and the collapsed ferruginous saprolite zone developed over weathered dunite.

Silicified Goethite Zone (SGZ) this zone is distinguished from the overlying GZ by an abrupt increase in silica, mainly as secondary quartz, minor smectite/nontronite, and a trace of Mn oxide wad, hematite, magnetite and chromite. Quartz is present as laminated veins or sub-horizontal bands and as irregular, coarse mesh boxworks where silica has precipitated on joints and fractures. Silica can be removed through beneficiation so that the SGZ is a key economic zone. This zone comprises ferruginous saprolite and lies between the pedoplasmation front and the Mg disconformity over weathered dunite.

Saprolite (SAP) this zone lies below the Mg disconformity and represents the saprolite and saprock of weathered ultramafic rock. Relict igneous fabrics are retained; some patches of relatively fresh bedrock are also observed. It is usually pale brown to grey and consists mainly of serpentine, magnesite, dolomite, smectite/nontronite, minor goethite and silica, with trace chromite and magnetite. Veins of magnesite and dolomite are very common. The saprolite rarely contains significant grades of Ni or Co.

Mineral distribution in the regolith at Sunrise is summarized in Fig. 15, based on drill core observations, chemical analysis and SWIR spectral analysis, providing an understanding of the weathering of ultramafic rocks (SRK Consulting 2018, and unpublished reports). SWIR analysis has demonstrated that kaolinite and gibbsite are important phases in the GZ zones overlying saprolite, becoming dominant in the TZ and above. As the dunite protolith contains negligible Al, their presence indicates they have been derived laterally and introduced into apparently residual zones by illuviation.

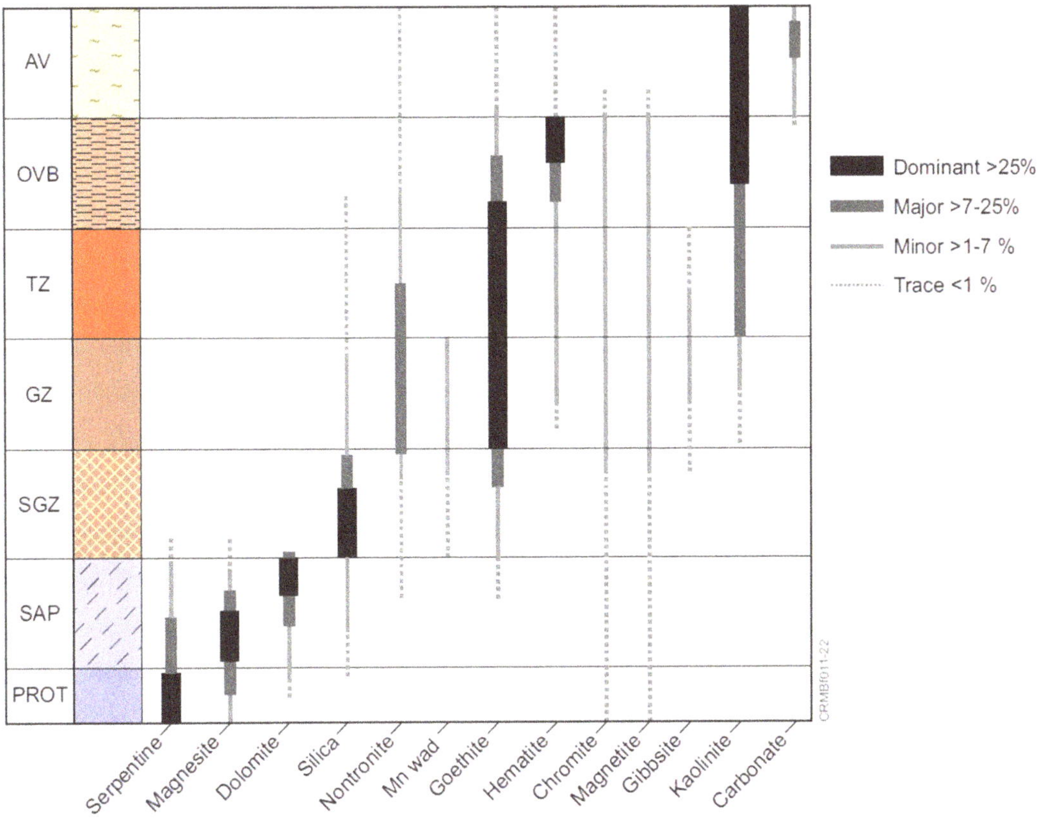

Fig. 15 Summarized distributions and relative abundances of the principal minerals in the Sunrise regolith profile based on existing mineral studies, drill core observations and SWIR analysis

8.5 Regolith Mineralization

Nickel Nickel concentration in the Sunrise protolith is ~0.25%, predominantly hosted by serpentine. Hydrolysis of serpentine releases Ni, along with Mg, Si and Fe. Nickel shows a gradual increase in concentration through the SAP and SGZ, largely due to relative enrichment, as indicated by the stable Mn:Fe ratio (Fig. 16). In the upper SGZ, Ni concentration increases through absolute enrichment associated with both Fe and Mn oxides. This enrichment continues through into the overlying GZ zone yet peaks at the SGZ-GZ boundary (partly associated with Mn oxides). Decreasing Fe and Mn abundances within the GZ are associated with a decrease in concentrations of Ni, indicating that both Mn and Fe oxides are hosts to Ni.

Cobalt The Co distribution in the regolith is controlled by its strong association with Mn oxides, which can be tracked from the saprolite through to the overlying Mn oxide-rich goethitic zones. The Mn-Co correlation in the SAP zone is $r^2 = 0.99$, with a Mn: Co ratio of 9:1 and a mean concentration of 190 ppm Co. Within the SGZ, the Co concentration increases (mean 610 ppm), with a Mn-Co correlation of $r^2 = 0.85$ (Fig. 17a). The Mn: Co ratios lie between 9:1 and 3:1, with the decreasing ratio indicating the addition of Co relative to Mn. Cobalt is most strongly enriched in the GZ zone (mean 1710 ppm Co) and maintains an overall Mn: Co correlation of $r^2 = 0.80$, with the Mn: Co ratio typically between 9:1 to 3:1 (mean 6:1). There is also a Mn:Co ratio trend at approximately 15:1 which requires further investigation with respect

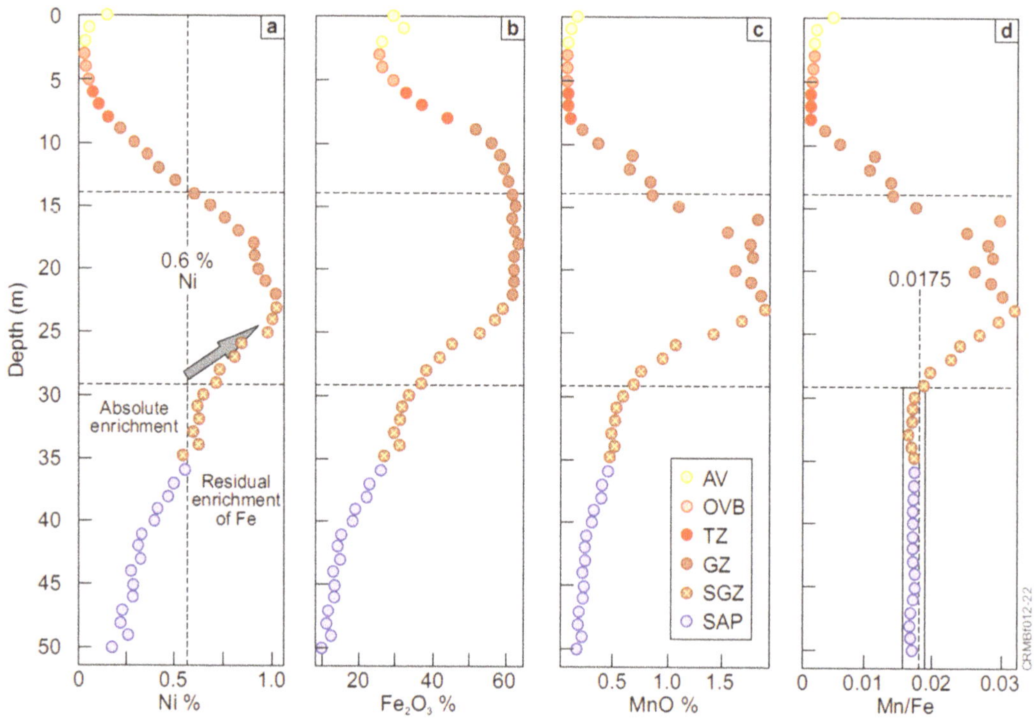

Fig. 16 Nickel distribution through the Sunrise profile and its association with Mn and Fe (average data from drill holes)

to its distribution, regolith location and control (Fig. 17b).

Platinum Although low grade (mean 0.21 ppm Pt), an extensive blanket of Pt enrichment is developed within regolith at Sunrise, particularly in the goethite zone and with localized steep dipping primary bedrock source, as indicated by the Phoenix Zone (Sunrise Energy Metals 2021).

Scandium The lateritic regolith over the Tout Complex has unusually high Sc abundances, preferentially developed over weathered pyroxenites (e.g., Chassé et al. 2017). The pyroxenite protolith contains 70–80 ppm Sc, whereas the dunite contains <10 ppm. The elevated Sc concentrations (>180 ppm) and the bulk of the Sc mineralization are present in the OVB, TZ and GZ, persisting downwards to the SGZ and SAP, generally at lower grades. The Sc mineral resource estimate at Sunrise, based on a 300 ppm Sc cut-off, is 45.7 Mt. at 421 ppm for 29,409 tonnes of Sc_2O_3 with 99.5% of the resource outside the dunite complex (SRK Consulting 2018, Table 14.23). The residual regolith hosts 53% of the resource, with the remainder in the overlying overburden and alluvium.

8.6 Development of the Regolith Profile

The weathering of serpentine has decreased its abundance upwards through the SAP, releasing metal ions and silicic acid. The silicic acid has precipitated as secondary silica (now quartz) along fractures and joints; it has also partly pseudo-morphed ferruginous serpentine, preserving the mesh texture. The removal of Mg in solution has resulted in residual accumulation of Si and Fe. The hydrolysis of serpentine within the SAP also forms secondary Mg silicates, which have lower Mg: Si ratios than serpentine. These secondary Mg silicates occupy micro-fractures

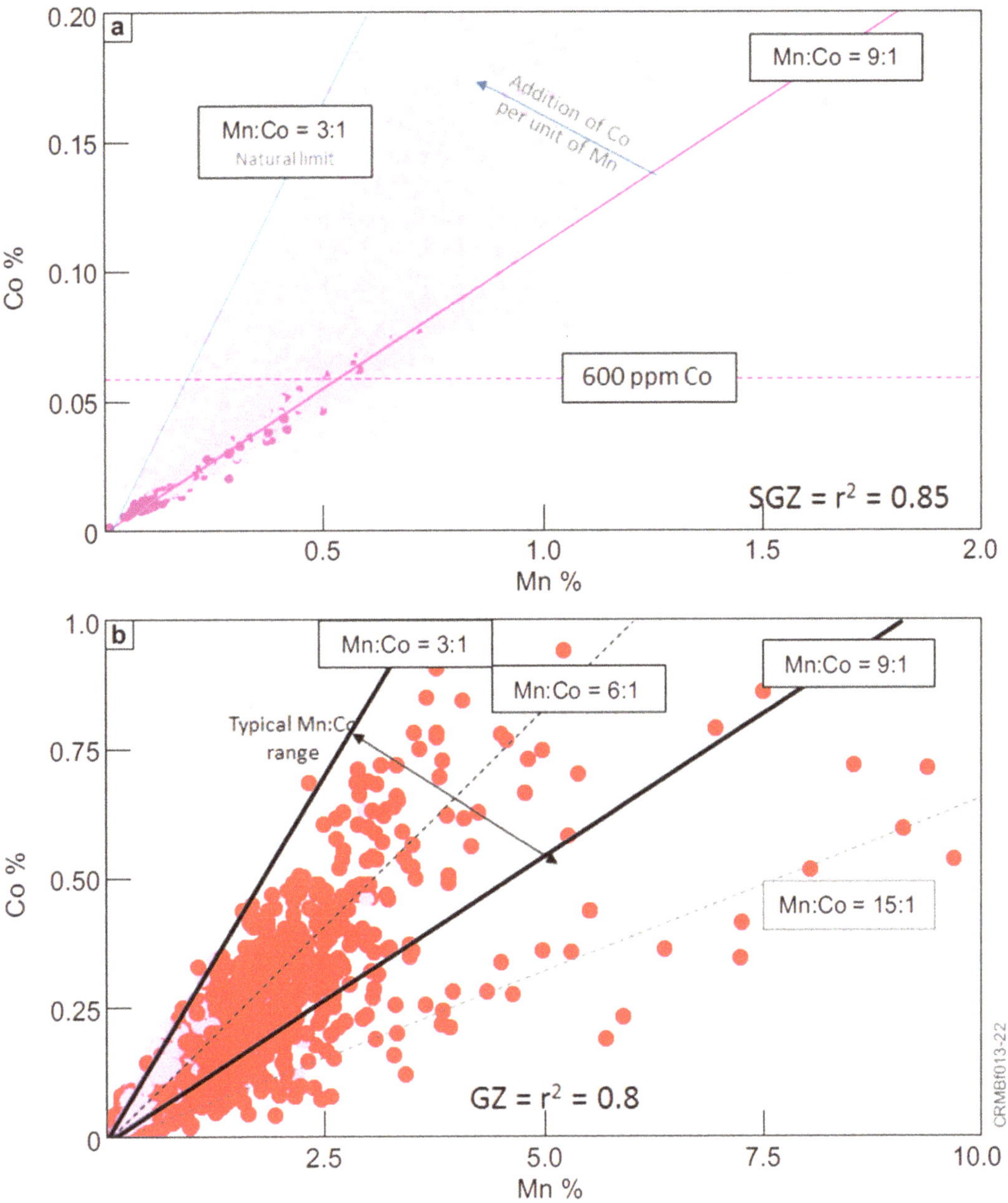

Fig. 17 Manganese vs cobalt scatterplots (**a**) showing the association within the SAP & SGZ, (**b**) showing the association within the SAP. SGZ and GZ zones

and cavities and pseudomorph serpentine grains, precipitating from components of primary minerals. They have a wide compositional variation due to increasing Fe and Ni substitution in the octahedral sites and a decreasing Mg: Si ratio through the saprolite.

The loss of Mg from the profile is typically sharp (the Mg- discontinuity), marking the top of the SAP and the base of the SGZ. The Mg-discontinuity can be identified by a MgO: Al_2O_3 ratio of 5, separating SAP (>5) from SGZ (<5), as shown in Fig. 18. The use of Al_2O_3 eliminates the need to account for the key mineral phases of serpentine and carbonates, as none contain Al_2O_3.

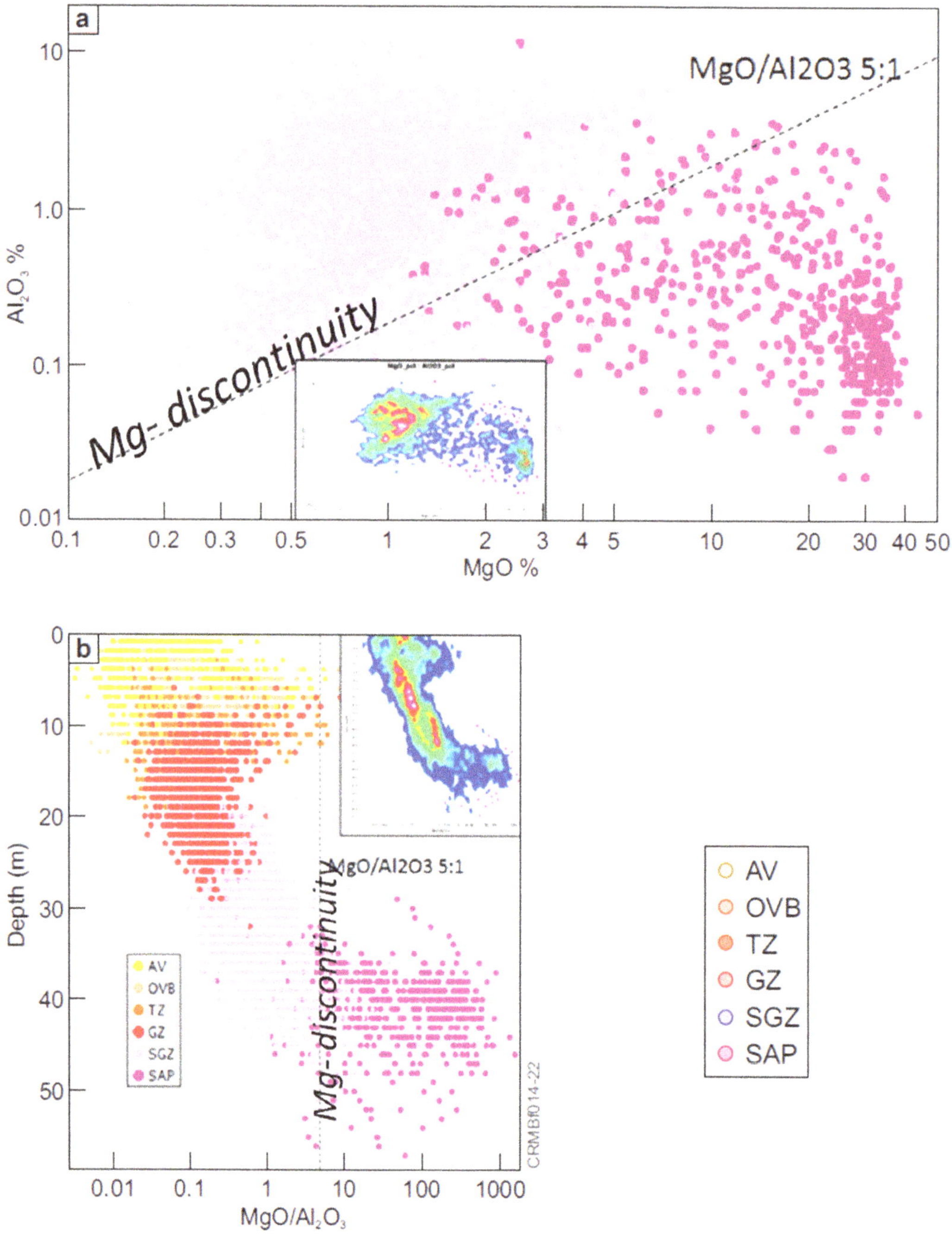

Fig. 18 MgO vs. Al2O3 (**a**) MgO/Al2O3 vs. depth (**b**) showing the separation of SAP from SGZ at the Mg discontinuity using a MgO/Al2O3 ratio of 5:1

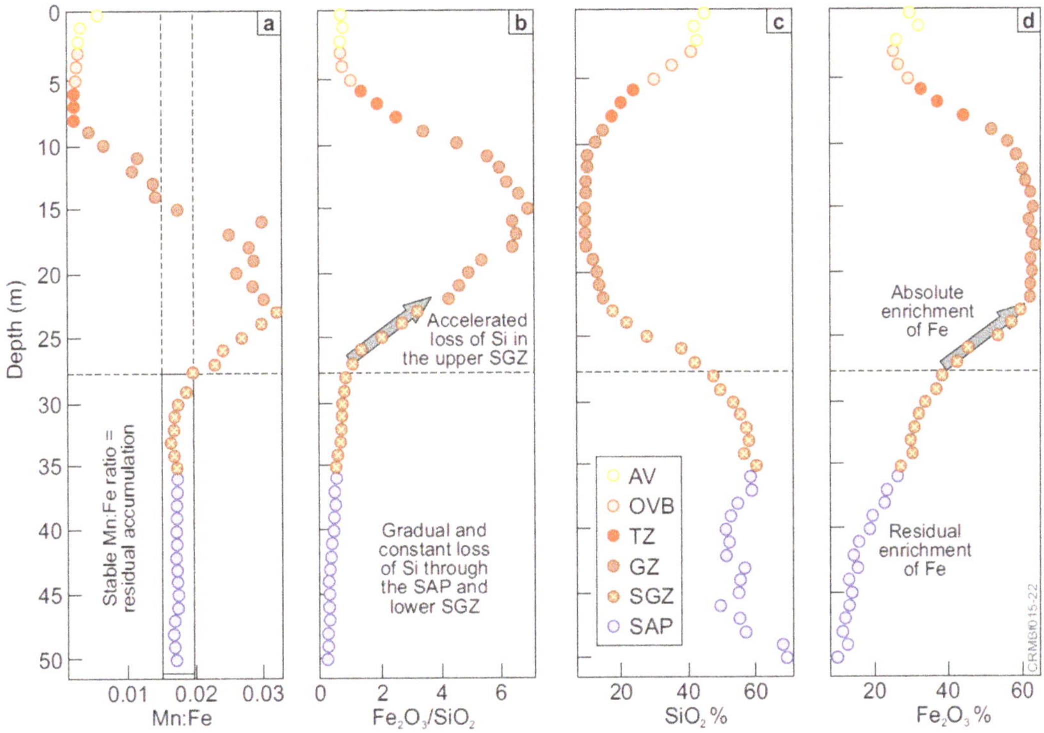

Fig. 19 Plots showing the residual and absolute accumulation of Fe in the Sunrise profile

Secondary Fe oxides initially precipitated along micro-fractures, in faults, and in zones within the SAP as a skeletal framework around the primary mesh texture. Their presence represents local transformation controlled by the hydrolysis of silicates and a Fe redox front within the regolith. Iron oxidation is the major control on the Eh and pH of weathering ultramafic rocks.

Within the SAP and lower SGZ, Fe has accumulated residually, as shown by the Mn: Fe ratio (Fig. 19), whereas silica (as silicic acid) has been removed, as shown by the Fe_2O_3:SiO_2 ratio. Removal of silicic acid is greater within the upper SGZ, extending into the GZ zone, resulting in the residual enrichment of Fe by the collapse of the lower GZ zone due to the absence of a supporting framework of silicified serpentine saprolite and silica fracture fillings. Iron concentration decreases in the upper GZ through the TZ. Goethite is the main Fe-oxide in SAP through to the TZ (Fig. 20), whereas in the overburden and alluvial zones, the dominant Fe oxide is hematite formed by dehydration of goethite.

8.7 Kaolinite Formation in the Regolith

The Al_2O_3 vs SiO_2 trend associated with GZ in Fig. 21 supports the presence of kaolinite ($Al_2Si_2O_5(OH)_4$) and gibbsite ($Al(OH)_3$). Separation of the SGZ and GZ zones at a SiO_2:Al_2O3 ratio of 3:1 demonstrates that kaolinite (SiO_2: Al_2O_3 ~ 1:1) is the dominant silicate mineral in the GZ zone and that there is negligible smectite (montmorillonite; SiO_2:Al_2O_3 ~ 3:1) or chlorite (SiO_2:Al_2O_3 ~ 2:1).

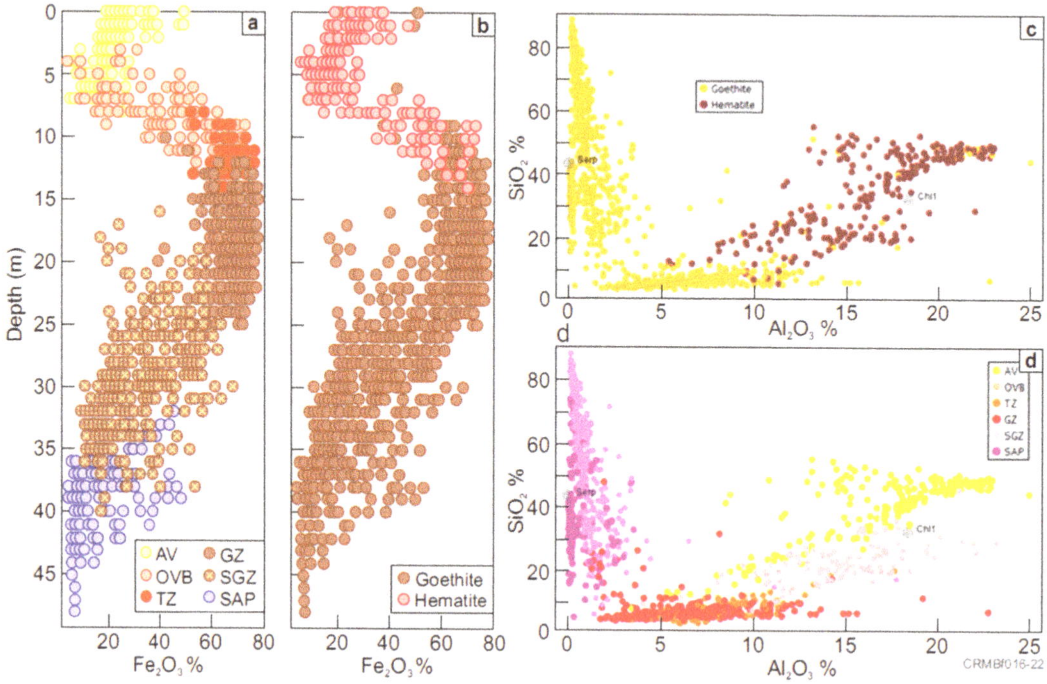

Fig. 20 Distributions of Fe_2O_3, Al_2O_3 & SiO_2 through the profile. (**a**) Fe_2O_3 distribution classified by geometallurgical zones. (**b**) Fe_2O_3 phases determined by SWIR. (**c**) Scatter plot of Al_2O_3 & SiO_2 classified by geometallurgical zones. (**d**) Fe oxide phases determined by SWIR

This conclusion is confirmed by SWIR analysis of samples from the GZ zones (Fig. 22), which shows that both poorly- and well-crystalline kaolinite are dominant, together with gibbsite. Neither kaolinite nor gibbsite are weathering products of ultramafic rocks; if alumina is released from the weathering of low-MgO ultramafic rocks, smectite preferentially forms relative to kaolinite.

The negligible Al contents of the dunitic protolith at Sunrise imply that these unusually high abundances of kaolinite and gibbsite in the GZ and overlying residual horizons of the regolith are due to the addition of alumina. The probable immediate sources are the overlying alluvium and overburden, originally derived from adjacent pyroxenite, gabbro and sedimentary bedrock. Within the profile, gibbsite dominantly lies below (deeper) kaolinite (Fig. 10), possibly indicating mobilization of Al in the profile. Gibbsite is considered a precursor to the formation of kaolinite, which can form in association with silicic acid, as shown in Eq. 1.

$$\underset{\text{silicic acid}}{2H_4SiO_4} + \underset{\text{gibbsite}}{2Al(OH)_3} \leftrightarrow \underset{\text{kaolinite}}{Al_2Si_2O_5(OH)_4} + \underset{\text{water}}{5H_2O} \qquad (1)$$

9 Conclusions

Nickel-cobalt laterites are supergene deposits formed exclusively on olivine-rich ultramafic protores. There are three mineralogically distinct deposit types, namely oxide, hydrous silicate and clay silicate, each requiring different processing. They are the product of the interaction of a variety of geological and environmental factors over long periods of geological time, during which the rocks have been exposed to weathering in seasonally humid tropical climates to form a deep,

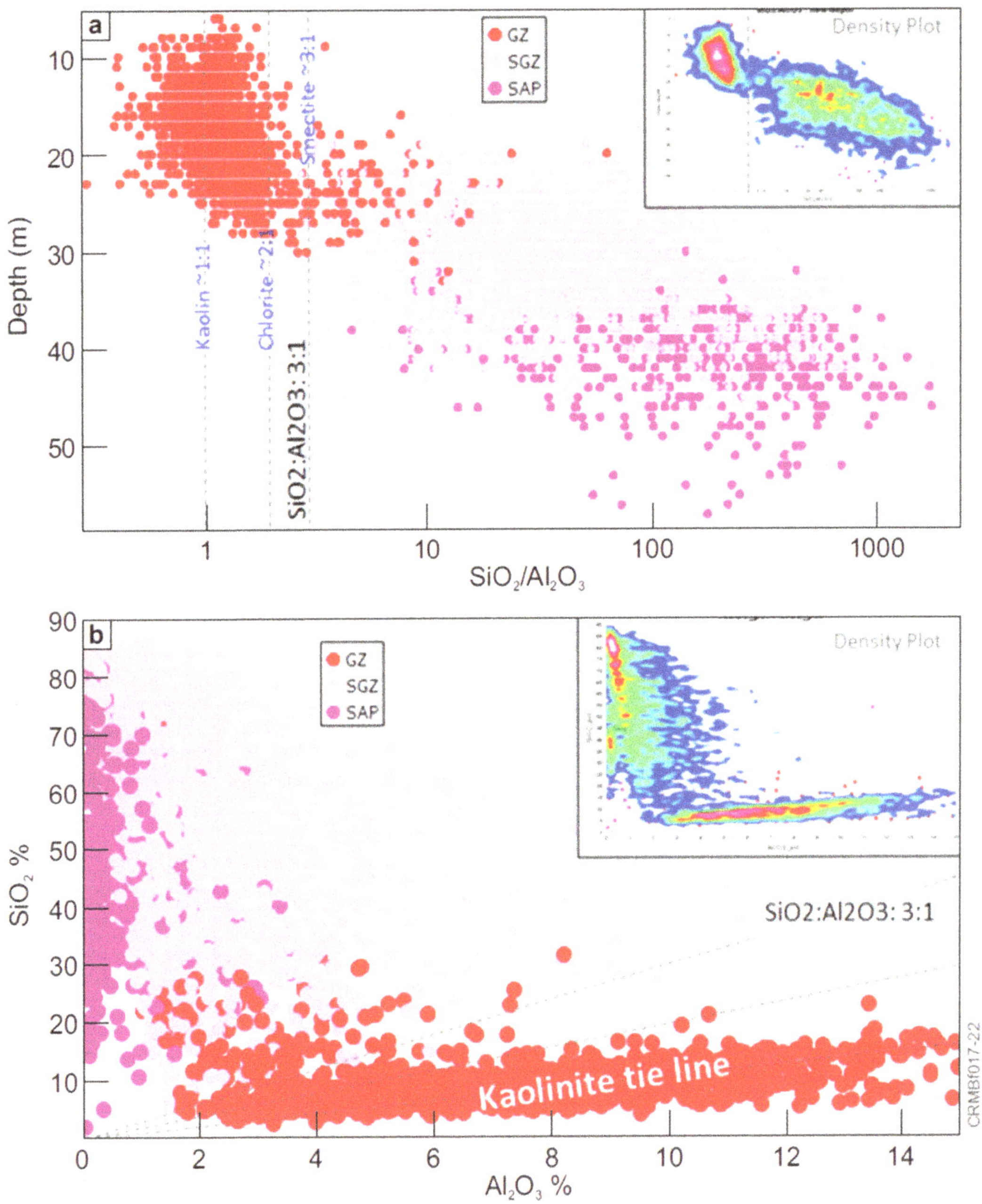

Fig. 21 SiO_2 and Al_2O_3 depth plots. (**a**) SiO_2:Al_2O_3 vs. depth showing the separation of GZ from SGZ and the stoichiometric ratios of key mineral phases, (**b**) Al_2O_3 vs. SiO_2 showing a separation between GZ and SGZ at an Al_2O_3:SiO_2 ratio of 3:1 reflecting a distinct change in mineralogy

intensely weathered regolith. Oxide deposits form in free-draining environments, with Ni (and Co) residually concentrated in goethite following leaching of Mg and Si. Hydrous silicate deposits form directly below the oxide horizon in slowly emergent terrains due to leaching of Ni from

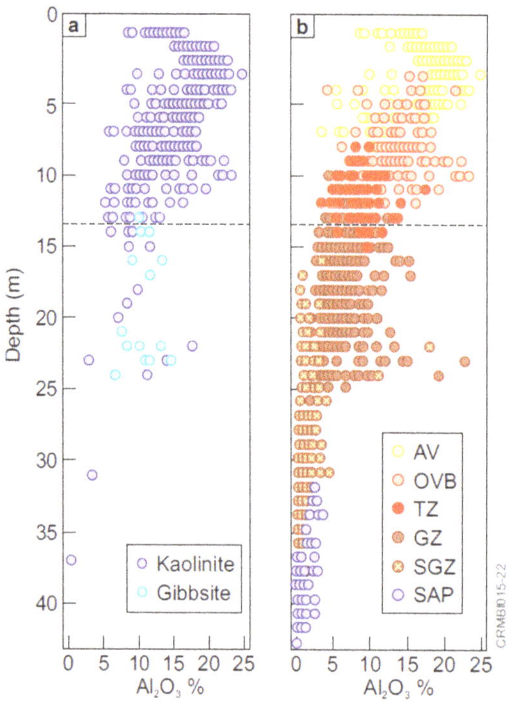

Fig. 22 Distribution of kaolinite and gibbsite detected by SWIR through the profile in relation to (**a**) depth, (**b**) geometallurgical zones

goethite and concentrating in saprolite as secondary hydrous silicates or by substitution for Mg in serpentine. Clay silicates are formed beneath the oxide horizon from peridotitic (Al-bearing) protores in poorly drained environments in which Mg and Si are partially retained during weathering. Thereafter, deposits in most regions have been subjected to tectonic and/or climatic changes. Where these caused little erosion, led to burial, or, in more arid climates, reduced the rate of weathering, modification has only been minor. Where there has been a significant uplift, losses by erosion increased, but under humid climates, this is offset, economically at least, by continued weathering and enrichment of Ni in hydrous Mg silicates forming in the surviving regolith. This sequential development is evident in the structure of the regolith. Each horizon in a lateritic regolith is considered to have developed from a precursor similar to the horizon that it overlies. However, ferruginous saprolite overlying hydrous silicate deposits appear to have no textural evidence of relic features such as veins and fissure fillings, typical of such ore, implying that the hydrous silicate mineralization is a late-stage modification of the profile.

All lateritic Ni deposits have an oxide ("limonite") resource, underlain by either a hydrous silicate or clay silicate resource, although in some, the oxide component may be partially or wholly eroded. In most mines, one ore type dominates, and because of the differences and the complexity of processing, only one is developed, and the other is undisturbed, stockpiled or used as a backfill. Lapaopao is an exception, with both oxide and hydrous silicate ores to be processed. Separating them will be crucial. Greater efficiencies in ore sorting and beneficiation may allow similar dual deposits elsewhere to be fully exploited.

Acknowledgments PT Ceria Nugraha Indotama is gratefully acknowledged for providing access to data from Lapaopao and permission to publish the case history. Sunrise Energy Metals (formerly CleanTeq Holdings Ltd) similarly provided access to data and permission to publish the Sunrise case history, which is acknowledged with thanks. All figures were drafted by Angelo Vartesi. Peter Gleeson is thanked for his comments on the manuscript.

References

Aiglsperger T, Proenza JA, Lewis JF, Labrador M, Svojtka M, Rojas-Purón A, Longo F, Ďurišová J (2016) Critical metals (REE, Sc, PGE) in Ni laterites from Cuba and the Dominican Republic. Ore Geol Rev 73:127–147. https://doi.org/10.1016/j.oregeorev.2015.10.010

Anand RR, Paine M (2002) Regolith geology of the Yilgarn Craton. Aust J Earth Sci 49:3–162

Berger VI, Singer DA, Bliss JD, Moring BC (2011) Ni-Co laterite deposits of the world—database and grade and tonnage models, Open file report 2011–1058. US Geological Survey, Reston, 26 pp. http://pubs.usgs.gov/of/2011/1058/ and http://mrdata.usgs.gov/laterite

Brand NW, Butt CRM, Hellsten KJ (1996) Structural and lithological controls on the formation of the Cawse nickel laterite deposits, Western Australia—implications for supergene ore formation and exploration in deeply weathered terranes. Australasian Institute of Mining and Metallurgy Publication series 6/96, Melbourne, pp 185–190

Brand NW, Butt CRM, Elias M (1998) Nickel laterites: classification and features. AGSO J Aust Geol Geophys 17:81–88

Butt CRM, Cluzel D (2013) Nickel laterite ore deposits—weathered serpentinites. Elements 9:123–128. https://doi.org/10.2113/gselements.9.2.123

Butt CRM, Williams PA, Gray DJ, Robertson IDM, Schorin KH, Churchward HM, McAndrew J, Barnes SJ, Tenhaeff MFJ (1992) Geochemical exploration for platinum group elements in lateritic terrain. AMIRA project 252. CSIRO Australia, Division of Exploration Geoscience report no. 332R. 3 vols, 413 pp. Re-issued 2001 as CRC LEME open file report 85. http://crcleme.org.au/Pubs/OFRSindex.html

Cathelineau M, Benoît Quesnel B, Gautier P, Boulvais P, Couteau C, Drouillet M (2016) Nickel dispersion and enrichment at the bottom of the regolith: formation of pimelite target-like ores in rock block joints (Koniambo Ni deposit, New Caledonia). Mineral Deposita 51:271–282. https://doi.org/10.1007/s00126-015-0607-y

Cathelineau M, Myagkiy A, Benoît Quesnel B, Boiron M, Gautier P, Boulvais P, Ulrich M, Truche L, Golfier F, Drouillet M (2017) Multistage crack seal vein and hydrothermal Ni enrichment in serpentinized ultramafic rocks (Koniambo massif, New Caledonia). Mineral Deposita 52:945–960. https://doi.org/10.1007/s00126-016-0695-3

Chassé M, Griffin WL, O'Reilly SY, Callas G (2017) Scandium speciation in a world-class lateritic deposit. Geochem Perspect Lett 3:105–114. https://doi.org/10.7185/geochemlet.1711

Chassé M, Griffin WL, O'Reilly SY, Callas G (2019) Australian laterites reveal mechanisms governing scandium dynamics in the critical zone. Geochim Cosmochim Acta 260:292–310. https://doi.org/10.1016/j.gca.2019.06.036

Chevillotte V, Chardon D, Beauvais A, Maurizot P, Colin F (2006) Long-term tropical morphogenesis of New Caledonia (Southwest Pacific): importance of positive epeirogeny and climate change. Geomorphology 81: 331–375

CleanTeQ (2016) Syerston nickel/cobalt mineral resource estimate. Updated Resource Estimate demonstrates potential for Syerston to become a leading global supplier of cathode raw materials to the lithium-ion battery industry. ASX announcement 22 Aug 2016

Cluzel D, Vigier B (2008) Syntectonic mobility of supergene nickel ores of New Caledonia (Southwest Pacific). Evidence from faulted regolith and garnierite veins. Resour Geol 58:161–170

Colin F, Nahon D, Trescases JJ, Melfi AJ (1990) Lateritic Weathering of pyroxenites at Niquelandia, Goias, Brazil. The supergene behavior of nickel. Econ Geol 85:1010–1023

Domènech C, Galí S, Villanova-de-Benavent C, Soler JM, Proenza JA (2017) Reactive transport model of the formation of oxide-type Ni-laterite profiles (Punta Gorda, Moa Bay, Cuba). Mineral Deposita 52:993–1010. https://doi.org/10.1007/s00126-017-0713-0

Dublet G, Juillot F, Morin G, Fritsch E, Fandeur D, Brown GE Jr (2015) Goethite aging explains Ni depletion in upper units of ultramafic lateritic ores from New Caledonia. Geochim Cosmochim Acta 16:1–15

Elias M (2002) Nickel laterite deposits—a geological overview, resources and exploitation. In: Cooke D, Pontgratz J (eds) Giant ore deposits: characteristics, genesis and exploration. Special publication 4. Centre for Ore Deposit Research, University of Tasmania, Hobart, pp 205–220

Freyssinet P, Farah AS (2000) Geochemical mass balance and weathering rates of ultramafic schists in Amazonia. Chem Geol 170:133–151

Freyssinet P, Butt CRM, Morris RC, Piantone P (2005) Ore-forming processes related to lateritic weathering. In: Hedenquist JW, Thomson JFH, Goldfarb RJ, Richards JP (eds) Economic geology, 100th anniversary volume. Economic Geology Publishing Company, New Haven, pp 681–722

Fritsch E, Juillot F, Dublet G, Fonteneau L, Fandeur D, Martin E, Caner L, Auzende A, Grauby O, Beaufort D (2016) An alternative model for the formation of hydrous Mg/Ni layer silicates ('deweylite'/'garnierite') in faulted peridotites of New Caledonia: I. Texture and mineralogy of a paragenetic succession of silicate infillings. Eur J Mineral 28:295–311. https://doi.org/10.1127/ejm/2015/0027-2503

Galí S, Soler JM, Proenza JA, Lewis JF, Cama J, Tauler E (2012) Ni-enrichment and stability of Al-free garnierite solid-solutions: a thermodynamic approach. Clays Clay Miner 60:121–135

Gantzmann MC, Schraut K, Vogel C, Gabler H-E, Huthwelker T, Adam C (2021) Investigation of scandium in bauxite residues of different origin. Appl Geochem 126:104898. https://doi.org/10.1016/j.apgeochem.2021.104898

Gaudin A, Decarreau A, Noack Y, Grauby O (2005) Clay mineralogy of the nickel laterite ore developed from serpentinised peridotites at Murrin Murrin, Western Australia. Aust J Earth Sci 62:231–241

Gerth J (1990) Unit-cell dimensions of pure and trace metal-associated goethite. Geochim Cosmochim Acta 54:363–371

Glazkovsky AA, Gorbunov GI, Sysoev FA (1977) Deposits of nickel. In: Smirnov VI (ed) Ore deposits of the USSR, vol II. Pitman Publishing, London, pp 3–79

Gleeson SA, Butt CRM, Elias M (2003) Nickel laterites: a review. SEG Newsl 54(1):10–16

Gleeson SA, Herrington RJ, Durango J, Velásquez CA, Koll G (2004) The mineralogy and geochemistry of the Cerro Matoso S.A. Ni laterite deposit, Montelibano, Colombia. Econ Geol 99:1197–1213

Golightly JP (1979) Nickeliferous laterites: a general description. In: Evans DJI, Shoemaker RS, Veltman H (eds) International laterite symposium. Society of Mining Engineers, New York, pp 3–23

Golightly JP (1981) Nickeliferous laterite deposits. Econ Geol 75th anniversary volume:710–735

Golightly JP (2010) Progress in understanding the evolution of nickel laterites. In: Goldfarb RJ, Marsh EE, Monecke T (eds) The challenge of finding new mineral resources: global metallogeny, innovative exploration, and new discoveries, Volume II: Zinc-Lead, Nickel-Copper-PGE, and Uranium Special Publication 15. Society of Economic Geologists, Inc., Littleton, pp 451–475

Ito A, Otake T, Maulana A, Sanematsu K, Sufriadin, Sato T (2021) Geochemical constraints on the mobilization of Ni and critical metals in laterite deposits, Sulawesi, Indonesia: a mass-balance approach. Resour Geol 71: 255–282. https://doi.org/10.1111/rge.12266

Kadarusman A (2019) Exploration, mineral resource update and ore reserves of Lapaopao nickel laterite project technical report for PT Ceria Nugraha Indotama (unpubl)

Kaiser Research online (2022) KRO Scandium Resource Center. https://secure.kaiserresearch.com/s3/Education.asp?ReportID=649798

Keays RR, Holwell DA, Prichard HM (2021) Platinum mineralisation in the Owendale Uralian-Alaskan-type complex, New South Wales, Australia: the effects of serpentinization on Cu-PGE-Ni sulfide. Ore Geol Rev 130:103928. https://doi.org/10.1016/j.oregeorev.2020.103928

Lewis JF, Draper G, Proenza JA, Espaillat J, Jiménez J (2006) Ophiolite-related ultramafic rocks (serpentinites) in the Caribbean region: a review of their occurrence, composition, origin, emplacement and Ni-laterite soils formation. Geol Acta 4:237–263

Llorca SM (1993) Metallogeny of supergene cobalt mineralization. New Caledonia. Aust J Earth Sci 40:377–385

Manceau A, Callas G (1985) Heterogeneous distribution of nickel in hydrous silicates from New Caledonia ore deposits. Am Mineral 70:549–548

Mao X, Retallack G (2019) Late Miocene drying of central Australia. Palaeogeogr Palaeoclimatol Palaeoecol 514: 292–304. https://doi.org/10.1016/j.palaeo.2018.10.008

McGowran B, Hill RS (2015) Cenozoic climatic shifts in southern Australia. Trans Roy Soc S Aust 139:19–37. https://doi.org/10.1080/03721426.2015.1035215

Mudd GM, Jowitt SM (2022) The new century for nickel resources, reserves, and mining: reassessing the sustainability of the devil's metal. Econ Geol 117: 1961–1983. https://doi.org/10.5382/econgeo.4950

Myagkiy A, Truche L, Cathelineau M, Golfier F (2017) Revealing the conditions of Ni mineralization in the laterite profiles of New Caledonia: insights from reactive geochemical transport modelling. Chem Geol 466: 274–284. https://doi.org/10.1016/j.chemgeo.2017.06.018

Myagkiy A, Golfier F, Truche L, Cathelineau M (2019) Reactive transport modeling applied to Ni laterite ore deposits in New Caledonia: role of hydrodynamic factors and geological structures in Ni mineralization. Geochem Geophys Geosyst 20:1425–1440. https://doi.org/10.1029/2018GC007606

Nahon DB, Tardy Y (1992) The ferruginous laterites. In: Butt CRM, Zeegers H (eds) Regolith exploration geochemistry in tropical and subtropical terrains, Handbook of exploration geochemistry, vol 4. Elsevier, Amsterdam, pp 41–55

Pelletier BG (1996) Serpentines in nickel silicate ore from New Caledonia. In: Grimsey EJ, Neuss I (eds) Nickel '96, Publication series 6/96. Australasian Institute of Mining and Metallurgy, Melbourne, pp 197–205

Pillans B (2005) Geochronology of the Australian regolith. In: Anand RR, de Broekert P, P. (eds) Regolith landscape evolution across Australia. CRC LEME, Perth, pp 41–61. http://crcleme.org.au/RegLandEvol/Geochron_of_%20Aust_Regolith.pdf

PT Ceria Nugraha Indotama (PT CNI) (2021). www.cerindocorp.com/. Accessed September 2021

Putzolu F, Balassone G, Boni M, Maczurad M, Mondillo N, Najorka J, Pirajno F (2018) Mineralogical association and Ni-Co deportment in the Wingellina oxide-type laterite deposit (Western Australia). Ore Geol Rev 97:21–34. https://doi.org/10.1016/j.oregeorev.2018.05.005

Putzolu F, Abad I, Balassone G, Boni M, Mondillo N (2020a) Ni-bearing smectites in the Wingellina laterite deposit (Western Australia) at nanoscale: TEM-HRTEM evidences of the formation mechanisms. Appl Clay Sci 196:105753. https://doi.org/10.1016/j.clay.2020.105753

Putzolu F, Abad I, Balassone G, Boni M, Cappelletti P, Graziano SF, Maczurad M, Mondillo N, Najorka J, Santoro L (2020b) Parent rock and climatic evolution control on the genesis of Ni-bearing clays in Ni-Co laterites: new inferences from the Wingellina deposit (Western Australia). Ore Geol Rev 120:103431. https://doi.org/10.1016/j.oregeorev.2020.103431

Qin HB, Yang ST, Tanaka M, Sanematsu K, Arcilla C, Takahashi Y (2020) Chemical speciation of scandium and yttrium in laterites: New insights into the control of their partitioning behaviors. Chem Geol 552:119771. https://doi.org/10.1016/j.chemgeo.2020.119771

Rusmana E, Sukido, Haryono R, Simanjuntak T (1993) Geological map of the Lasusua-Kendari Quadrangle, Sulawesi. Geology Agency, Bandung

Scotese CR (2000) Paleomap project. http://www.scotese.com/scotesepubs.htm. Accessed December 2020

Sevin B, Ricordel-Prognon C, Quesnel F, Cluzel D, Maurizot P (2012) First palaeomagnetic dating of ferricrete in New Caledonia: new insight on the morphogenesis and palaeoweathering of 'Grande Terre'. Terra Nova 24:77–85

Slack JF, Kimball BE, Shedd KB (2017) Cobalt. In: Schulz KJ, DeYoung JH, Seal RR, Bradley DC (eds) Critical mineral resources of the United States—economic and environmental geology and prospects for future supply: U.S. Geological Survey professional paper 1802, pp F1–F40. https://doi.org/10.3133/pp1802F

SRK Consulting (2018) Sunrise nickel cobalt project, New South Wales, Australia NI 43-101 technical report.

NI43-101 report prepared for CleanTeQ Holdings Limited by SRK Consulting (Australia) Pty Ltd. https://www.miningnewsfeed.com/reports/Sunrise_Technical_Report_2018.pdf
Sumitomo Metal Mining Co., Ltd, 2018. Establishment of scandium recovery business. https://www.smm.co.jp/en/news/release/uploaded_files/20160428en.pdf
Sunrise Energy Metals (2021) Bonanza grade platinum intersections at Phoenix Platinum Zone. ASX announcement 3 May 2021. https://wcsecure.weblink.com.au/pdf/SRL/02370372.pdf
Tauler E, Lewis JF, Villanova-de-Benavent C, Aiglsperger T, Proenza JA, Domènech C, Gallardo T, Longo F, Galí S (2017) Discovery of Ni-smectite rich saprolite at Loma Ortega, Falcondo mining district (Dominican Republic): geochemistry and mineralogy of an unusual case of "hybrid hydrous Mg silicate—clay silicate" type Ni-laterite. Mineral Deposita 52:1011–1030. https://doi.org/10.1007/s00126-017-0750-8
Tauler E, Galí S, Villanova-de-Benavent C, Chang-Rodríguez A, Núñez-Cambra K, Khazaradze G, Proenza JA (2023) Geochemistry and mineralogy of the clay-type Ni-laterite deposit of San Felipe (Camagüey, Cuba). Fortschr Mineral 13:1281. https://doi.org/10.3390/min13101281
Tavlan M, Thorne R, Herrington RJ (2011) Uplift and lateritization history of the Çaldağ ophiolites in the context of Neo-Tethyan ophiolites obduction and uplift; implications for the Cenozoic weathering history of Western Anatolia. J Geol Soc 168:927–940. https://doi.org/10.1144/0016-76492010-086
Teitler Y, Cathelineau M, Ulrich M, Ambrosi JP, Munoz M, Sevin B (2019) Petrology and geochemistry of scandium in New Caledonian Ni-Co laterites. J Geochem Explor, Elsevier 196:131–155. https://doi.org/10.1016/j.gexplo.2018.10.009
Teluk AJ (2001) Fifield platinum project, NSW, Australia. Platinum metallogenesis, Ural-Alaskan type complexes—technical report prepared for Rimfire Pacific Mining NL. Geodyne Pty Ltd., Sydney, November, 71 pp
Thorne RL, Roberts S, Herrington R (2012) Climate change and the formation of nickel laterite deposits. Geology 40:331–334
U.S. Geological Survey (2022) Mineral commodity summaries. https://doi.org/10.3133/mcs2023
Valeton I, Biermann M, Reche R, Rosenberg F (1987) Genesis of nickel laterites and bauxites in Greece during the Jurassic and Cretaceous, and their relation to ultrabasic parent rocks. Ore Geol Rev 2:359–404
Villanova-de-Benavent C (2015) Compositional and structural characterization of Ni-phyllosilicates in hydrous silicate type Ni-laterite deposits. PhD thesis. Faculty of Geology, University of Barcelona, 156 pp (unpublished)
Villanova-de-Benavent C, Proenza JA, Galí S, García-Casco A, Tauler E, Lewis JF, Longo F (2014) Garnierites and garnierites: textures, mineralogy and geochemistry of garnierites in the Falcondo Ni-laterite deposit, Dominican Republic. Ore Geol Rev 58:91–109. https://doi.org/10.1016/j.oregeorev.2013.10.008
Webb K (2014) The age and tectonometallogenic setting of the Fifield "Alaskan-type" intrusive suite, Lachlan Orogen. BSc (Hons) thesis, School of Earth & Environmental Sciences, University of Wollongong (unpublished)
Wells MA, Butt CRM (2000) Geology, geochemistry, mineralogy of the Murrin Murrin nickel laterite deposit. CRC LEME open file report 207, 2006, 193 pp. http://crcleme.org.au/Pubs/OFRSindex.html
Wells MA, Butt CRM (2017) Murrin Murrin and Bulong smectite-hosted Ni laterite deposits. In: Phillips GN (ed) Australian ore deposits. Monograph 32. Australasian Institute of Mining and Metallurgy, Melbourne, pp 149–152
Williams-Jones AE, Vasyukova VO (2018) The economic geology of scandium, the runt of the rare earth element litter. Econ Geol 113:973–988

Iron Ore

Erick R. Ramanaidou and Agnieszka Piechocka

Abstract

Supergene iron ores represent a large proportion of iron ores in the world and are particularly abundant in Australia, the largest iron ore producer and exporter, India and Brazil. The supergene iron ore includes the banded iron-formations (BIF) and the derived martite-goethite ores, the detrital iron deposits (DID) and the channel iron deposits (CID). The martite goethite iron ores are subdivided into supergene lateritic and mimetic types. The supergene lateritic ore results from in situ weathering of BIF under wet and warm tropical climates. The supergene mimetic enrichment of BIF shows a pseudomorphing of the gangue minerals by goethite. The upper part of the supergene mimetic iron ore is, in turn, subject to weathering, generating a hydrated zone. The DID represent a small percentage of supergene ores, and they are the result of the physical weathering and accumulation from adjacent pre-enriched in situ iron ores. The CID are giant Fe-rich deposits of fluviatile and lacustrine origin occurring in the Hamersley Province, the Yilgarn Craton and the Ashburton Basin of Western Australia, and in Kazakhstan.

E. R. Ramanaidou (✉) · A. Piechocka
CSIRO Mineral Resources, ARRC Kensington, Kensington, WA, Australia
e-mail: erick.ramanaidou@csiro.au

1 Introduction

We are made from stardust! The iron in the haemoglobin molecules in the blood in your right hand came from a star that blew up 8 billion years ago...The iron in your left hand came from another star (Jill Tarter).

Iron ore and its ultimate product, steel, is the most historically and economically significant commodity. In the last three millennia, iron ore has been mined from magmatic, skarn, hydrothermal or igneous magnetite-rich deposits, banded iron formations (BIF) and supergene deposits.

In the twentieth century, the largest iron ores were derived from the high-temperature enrichment of BIF, the so-called martite microplaty hematite iron ores with Fe content of more than 60 wt. %, occurring in Brazil, South Africa, India and Australia. Depletion of these direct shipping ores (DSO) led to the expansion of the BIF-hosted magnetite ores, requiring ore processing by removing the gangue minerals to produce a magnetite concentrate. The current high demand for iron ore by China has seen a similar impact in Australia and Brazil with the expansion of BIF-hosted magnetite deposits.

Finally, the currently mined supergene iron ore deposits represent the bulk of DSO in Australia and Kazakhstan, and to some extent in India and Brazil, consist of (1) the BIF-derived in situ martite-goethite ores including the lateritic and

R. J. Bowell, C. R. M. Butt (eds.), *Geology, Geochemistry and Formation of Supergene Mineral Deposits in Deeply Weathered Terrain*, Mineral Resource Reviews, https://doi.org/10.1007/978-3-031-75733-4_12

Table 1 Iron ore: world production by country

	Iron Ore Production (Mt)						Crude ore (CO)	Iron content (IC)	IC/CO
Country/Year	2015	2016	2017	2018	2019	2020	2020		
Australia	810	858	885	908	919	900	50,000	24,000	0.48
Brazil	431	421	454	460	405	400	34,000	15,000	0.44
Canada	46	47	50	52	58	57	6000	2300	0.38
China	46	366	345	335	351	340	20,000	6900	0.35
India	46	185	202	204	238	230	5500	3400	0.62
Iran	46	46	34	36	33	32	2700	1500	0.56
Kazakhstan	46	36	39	42	22	21	2500	900	0.36
Peru	46	11	13	14	15	15	N/A	1500	
Russia	46	101	95	96	98	95	25,000	14,000	0.56
South Africa	46	66	75	74	72	71	1000.00	640	0.64
Sweden	46	27	32	36	36	35	1300	600	0.46
Turkey	46	7	10	10	16	16	N/A	N/A	
Ukraine	46	63	61	61	63	62	6500	2300	0.35
United States of America	46	42	48	50	47	37	3000.00	1000	0.33
Other countries	46	99	98	88	81	75	18,000	9500	0.53
World Total	46	2374	2440	2465	2455	2386	180,000	84,000	0.47

mimetic iron ores; (2) the BIF-derived ex situ detrital iron deposits (DID) and (3) the terrestrial Phanerozoic ooidal ironstones or channel iron deposits (CID). Although both the DID and CID are of detrital origin, because of their size, importance and petrological differences, the DID sensu stricto and the Phanerozoic ooidal ironstones CID will be described separately.

In the last two decades, iron ore production and, in particular, supergene iron ore mining, has increased considerably, fuelled by very high demand from China's steel mills. In 2020, China used 61% of the world's total iron ore exports, sending worldwide iron ore production and price to unmatched levels. The major iron ore producers (Australia, Brazil, China, India, Russia, South Africa and Ukraine) have a variable proportion of iron ore content (IC) to crude ore (CO) as a result of ore quality (Table 1).

2 BIF-Derived Supergene Iron Ores

2.1 Banded Iron Formations

To fully understand the BIF-derived supergene iron ore deposits, covering both the in situ martite-goethite ores or the bedded iron deposits (BID) and the ex situ detrital iron deposits or DID, it is critical to understand the fresh BIF themselves as they are so closely intertwined.

Publications on Precambrian banded iron-formations are numerous and describe BIF from Australia (Morris 1983; Trendall 1980, 1983; Trendall and Blockley 1970; Rasmussen et al. 2012, 2015); Brazil (Dorr 1969, 1973; Rosiere et al. 2008; Spier et al. 2003); Mongolia (Ilyin 2009); North America (Bekker et al. 2010; Cloud 1973, 1983; Simonson 1987, 2003); South Africa (Beukes 1973, 1983; Beukes and Gutzmer 2008) and Ukraine (Belevtsev and Siroshtan 1983).

Banded iron formations are chemical sediments comprising a multiscale layered rock in which quartz (or its metamorphic equivalent) alternates with layers that are composed mainly of iron minerals such as magnetite or hematite (Fig. 1). Although earliest classifications of Precambrian banded iron-formations were defined for North America, the Australian classification of Trendall (1965) and Trendall and Blockley (1970) is now widely accepted; it is based on the scale of the banding (1) macrobanding (metre scale), (2) mesobanding (millimetre to centimetre), (3) microbanding (sub- to 5 mm) applied specifically to a varve-related type, and

Fig. 1 (**a**) BIF from the Western Australian Dales Gorge Member; (**b**) highly metamorphosed BIF of the Iron Quadrangle in Minas Gerais Brazil

(4) microlaminae (sub-mm) (Ewers and Morris 1981; Morris 1993).

The minerals occurring in BIF include, by order of importance, quartz, magnetite, hematite and stilpnomelane, carbonates, such as siderite and dolomite-ankerite series (Fig. 2) with additional minnesotaite, greenalite, pyrite and rare chamosite, ripidolite and ferri-annite. Fibrous riebeckite (crocidolite or blue asbestos) is found in the BIF of Hamersley and Transvaal (Beukes and Klein 1990) Provinces. In the BIF from Australia, Brazil, South Africa and North America, silica ranges from 30% and 52%, Fe varies between 20% and 40%, mostly as magnetite (see values of Fe^{2+} and Fe^{3+}) or hematite and alumina is less than 5% (Table 2). Calcium and Mg can reach 10% in carbonates, silicates and/or phosphates, whereas Na and K occur in silicates. Phosphorus, a deleterious element during the steel-making process, generally occurs as apatite. In Western Australia, a detailed compositional variation of BIF chemical composition from the South Flank deposit (Knight et al. 2018) shows that the Mount Newman Members N2 and N3 contain the most Fe% (32–37%) whereas Al, loss on ignition (LOI) and P are the highest in the West Angelas Member (Fig. 3).

Where metamorphic grade increases, the particle size of quartz and magnetite also increases (Dorr 1969). An increase in grain size and mineralogical changes were also observed by Pires (1995) from the Western greenschist facies to the Eastern amphibolitic facies in the Quadrilàtero Ferrìfero (Iron Quadrangle) in Brazil. The amount of hematite can also increase with metamorphism, for instance, in the fresh BIF of the Capanema mine, the amount of magnetite is not dominant, and most of the primary iron oxides are hematite (Ramanaidou et al. 1996; Ramanaidou 2009).

2.2 Martite-Goethite Iron Ores

2.2.1 Distribution

Supergene processes have affected BIF to considerable depths (Morris 1980, 1983, 1985) and more than 200 m (Ramanaidou et al. 1996; Ramanaidou 2009). Two types of supergene enrichment affect BIF to form Martite-Goethite (M-G) ores: (1) lateritic enrichment of BIF and (2) mimetic enrichment of BIF, which is, in turn, subject to lateritic weathering. By contrast, the supergene lateritic ores are residual ores resulting from the relative accumulation of Fe as other primary minerals are leached.

M-G supergene iron ores have been described from many countries; they are particularly common in Western and South Australia and also occur in India, South Africa and Brazil (Beukes

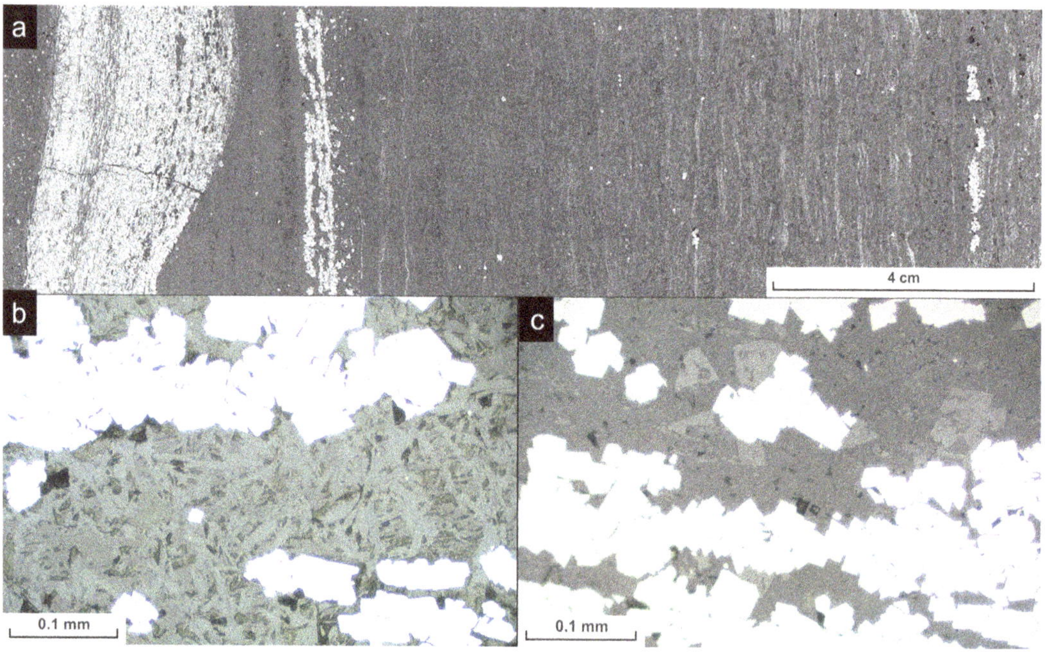

Fig. 2 (**a**) Polished sections of a Western Australia Dales Gorge BIF with alternating layers of iron oxides (white) and gangue minerals (grey) such as quartz, silicates and carbonates. (**b**) Magnetite rafts and silicates; (**c**) Magnetite rafts and carbonates and chert

et al. 2003; Chicarino Varajão et al. 1997; Dorr 1973; Freyssinet et al. 2005; Gagen et al. 2019; Melfi et al. 1988; Ramanaidou 1989, 2009; Ramanaidou et al. 1996; Rosiere et al. 2008; Roy and Venkatesh 2009; Spier et al. 2003; Taylor et al. 2001). The huge resource of M-G ores, estimated at 50 Gt, occurring in the Pilbara (Knight et al. 2018) is the dominant ore exported from Australia in 2021 (Fig. 4). Stratigraphically and geographically, the M-G ores (Fig. 5) occur in the Marra Mamba iron-formation (Marandoo, Mining Area C, West Angelas, Hope Downs, OB29, Christmas Creek) and in the Dales Gorge and Joffre members (Paraburdoo Eastern Range, Tom Price Section 6 and 7, Opthalmia Range and Rhodes Ridge) (Fig. 5). Morris and Kneeshaw (2011) described the Brockman iron-formation-hosted iron ores as generally richer in P than the Marra Mamba iron ores. M-G ores also occur in the Northern Pilbara (Abydos, Wodgina and Pardoo) and in the Yilgarn Craton (Koolyanobbing, Mt Jackson).

2.2.2 Supergene Lateritic Enrichment of BIF

About two-thirds of the Australian continent is covered by a lateritic regolith up to 200 m deep. The regolith consists of saprolite, preserving the primary fabric (texture) of the bedrock, overlain by a pedolith in which the bedrock fabrics have been destroyed, with the uppermost horizon commonly consisting of Fe-rich lateritic residuum. In many regions, the residual regolith is buried beneath continental sediments. The lateritic regolith is generally considered to have formed under seasonally humid tropical climates, although the present climate of much of the continent is semi-arid to arid. In the Brazilian Capanema mine, Ramanaidou et al. (1996) and Ramanaidou (2009) demonstrated that in situ weathering of metamorphosed BIF (itabirite) under heavy rainfall conditions caused the leaching of the BIF gangue and the oxidation of magnetite to martite with the development of thick horizons as a results of weathering fronts progressively

Table 2 Chemistry of Precambrian iron-formations

	1	2	3	4	5	6	7	8	9	10	11	12
SiO_2	40.85	42	43.51	44.34	49.5	40.85	42.5	45.9	51.1	38.8	42.84	45.78
Al_2O_3	0.41	0.3	0.36	0.89	2.4	0.41	0.4	2.23	1.2	2.5	0.2	2.04
Fe_2O_3	30.52	22		29.23	19.9	30.52	22.5	3.4	5.2	26.8	31.95	27.77
FeO		15.3	43.83	13.42	14.4		16.9	27.2	23.4		15.38	
MgO	4.63	2.88	3.03	2.3	2.9	4.63	3.78	4.95	4.25	3.8	1.89	5.21
CaO	10.26	6.7	1.81	1.78	3.5	10.26	5.81	1.46	1.86	8.2	1.7	2.85
Na_2O	0.196	0.13	0.03	0.53	0.22	0.196	0.43	0.26	0.19	0.09	0.42	0.02
K_2O	0.179	0.14	0.06	1.26	0.61	0.179	0.15	0.54	0.41	1.9	0.25	0.28
H_2O^+		1.1		0.98	2		1.3	3.58	2.86	2.7	0.92	
H_2O^-		0.12		0.17	0.6		0.17	1.26	0.21		4.12	
CO_2	12.09	8.37	5.81	4.62	3.1	12.09	5.25	7.73	6.47	16.5		14.3
TiO_2	0.016	0.06	0.03	0.05	0.14	0.016	0.2	0.12	0.06	0.1	0.02	0.09
MnO	0.269	0.14	0.07	0.17	0.07	0.269	0.12	0.12	0.07	0.3	0.06	0.24
P_2O_5	0.06	0.06	0.2	0.18	0.18	0.06	0.095	0.049	0.11	0.14	0.19	0.16
C		0.03		0.01	0.09		0.11	0.67	2.1	0.1		
S	0.102	0.14	0.07			0.102	0.09	0.47	0.7	0.45	0.03	0.32
FeS_2				0.12	0.15							
SO_2												
LOI						11.84*						14.34
Total	99.582	99.47	98.81	100.05	99.76	87.492	99.805	99.939	100.19	102.38	99.97	99.1
Fe as Fe_2O_3	30.52	38.7	43.83	43.99	35.7	30.5	41.3	33.6	31.2	26.9	49	27.77
Fe^{3+}	21.36	15.40	n.a	20.46	13.93	21.36	15.75	2.38	3.64	18.76	22.37	19.44
Fe^{2+}	n.a	11.93	34.19	10.47	11.23	n.a	13.18	21.22	18.25	n.a	12.00	n.a
Total Fe	21.36	27.33	34.19	30.93	25.16	21.36	28.93	23.60	21.89	18.76	34.36	19.44
Fe^{3+}/Fe^{2+}	n.a	1.29	n.a	1.95	1.24	n.a	1.19	0.11	0.20	n.a	1.86	n.a

	13	14	15	16	17	18	19	20	21	22	23
SiO_2	43.15	46.86	29.6	51.63	53.63	51.68	34.3	45.76	40.5	41.5	47.3
Al_2O_3	0.09	0.43	0.13	1.14	4.74	0.11	0.87	0.09	0.1	0.09	1.07
Fe_2O_3	46.37	24.68	2.71	16.87	29.44	39.79	56.6	49.17	58.7	57.7	43.1
FeO		17.19	36.58	17.68	3.24						
MgO	2.69	2.58	3.51	3.47	2.15	5.14	1.1	2.85	0.26	0.03	5.68
CaO	1.65	1.49	0.89	0.05	<0.01	5.65	2.63	1.75	0.05	0.02	3.03
Na_2O	0.04	0.16	0.08	0.16	0.08	0.03	0.32	0.04	0.01	0.01	0.33
K_2O	0.02	0.10	0.23	2.67	1.77	0.03	0.02	0.02	0.02	0.005	0.27
H_2O^+		0.57	1.02	1.4	3.42						
H_2O^-		0.08	0.05	0.1	0.95						
CO_2	4.47	5.81	24.75	4.83	0.03	0.07	n.m	n.m	0.07	0.58	n.m
TiO_2	0.01	0.05	0.01	0.07	0.24	0.05	0.09	0.01	0.02	0.017	0.4
MnO	0.05	n.dtm.	0.27	0.04	0.09	0.3	0.42	0.05	0.02	0.05	0.44
P_2O_5	0.21	0.25	0.36	0.02	0.11	0.09	0.29	0.22	0.04	0.02	0.1
C											
S	0.03										
FeS_2											
SO_2						0.05	n.a	0.03	n.a	n.a	n.a
LOI	4.06"					n.a	n.a		0.04"	0.04*	n.a
Total	94.31	n.a	100.19	100.13	99.89	102.92	96.64	99.99	99.82	99.442	101.72
Fe as Fe_2O_3	46.37										
Fe^{3+}	32.46	17.25	1.89	11.79	20.58	27.85	39.62	34.42	41.09	40.39	30.17
Fe^{2+}	n.a	13.36	28.43	13.74	2.52	n.a	n.a	n.a	n.a	n.a	n.a
Total Fe	32.46	30.61	30.32	25.53	23.1	27.85	39.62	34.42	41.09	40.39	30.17
Fe^{3+}/Fe^{2+}	n.a	1.29	0.07	0.86	8.17	n.a	n.a	n.a	n.a	n.a	n.a

1 Marra Mamba iron-formation, Paraburdoo (Ewers and Morris, 1981)

2 Iron-formation at base of Wittenoom Dolomite (Davy, 1975)

3 Brockman iron-formation, Dales Gorge Member (Ewers and Morris, 1981)

4 Brockman iron-formation, Joffre Member (Trendall and Pepper, 1977)

5 Weeli Wolli iron-formation

6 Marra Mamba iron-formation (Ewers and Morris, 1981)

7 Lower part of Marra Mamba iron-formation (Klein and Gole, 1981)

8 Lower part of Marra Mamba iron-formation (Klein and Gole, 1981)

9 Lower part of Marra Manba iron-formation (Davy, 1975)

10 Average for S bands within the Dales Gorge Member (Trendall and Pepper, 1977)

11 Average for 13 macrobands (Trendall and Pepper, 1977)

12 Average S macroband composition, Dales Gorge Member (Ewers and Morris, 1981)

13 Average BIF macroband compositions, Dales Gorge Member (Ewers and Morris, 1981)

14 Brockman iron-formation, Dales Gorge Member, BIF2 macroband

15 Brockman iron-formation, Dales Gorge Member, S13 macroband

16 Brockman iron-formation, Dales Joffre Member

17 Boolgeeda iron-Formation

18 Kuruman-Penge iron-formation (Klein and Beukes, 1992)

19 Urucum iron-formation (Klein, 2004)

20 Brockman iron-formation (Klein and Beukes, 1992)

21 Sandur iron-formation (Gutzmer et al., 2008)

22 Carajas iron-formation (Figueiredo E Silva et al., 2008; Klein and Ladeira, 2002)

23 Superior type -BIF (Mcclung, 2006)

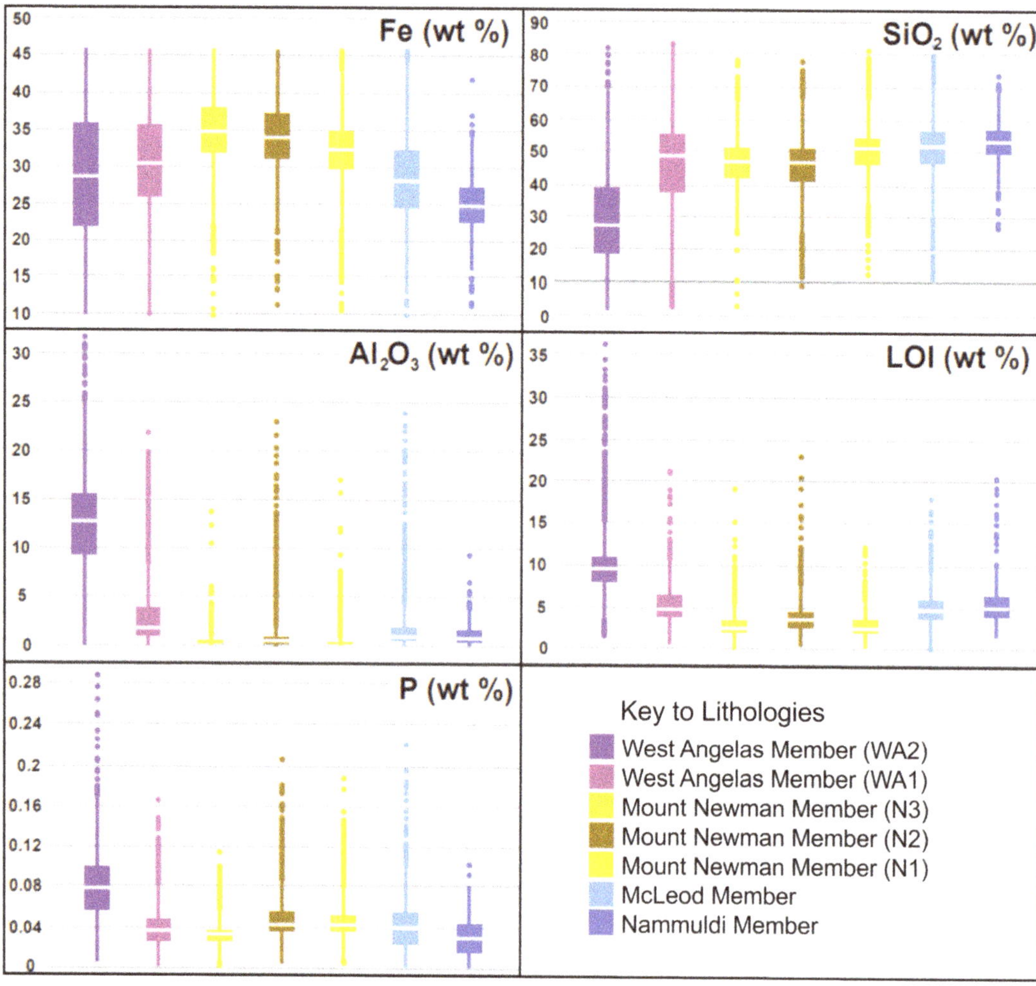

Fig. 3 Detailed chemical compositional variation of BIF from the Marra Mamba iron-formation and the West Angelas Member at South Flank in the Hamersley Province, Western Australia (Knight et al. 2018)

Fig. 4 Total high-grade iron ore resources (2016) in the Hamersley province categorised by major ore type

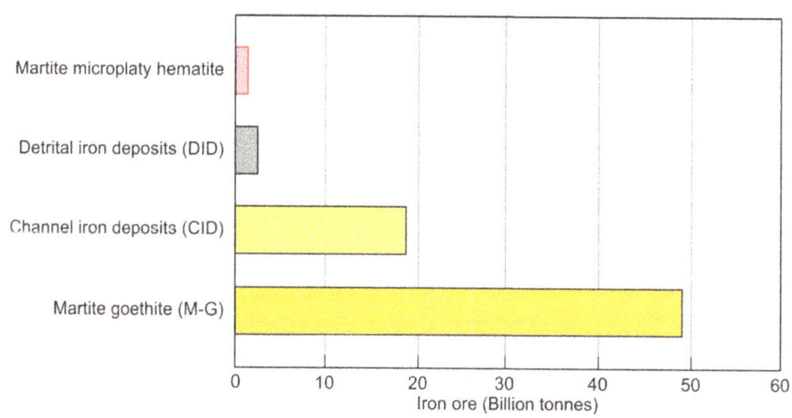

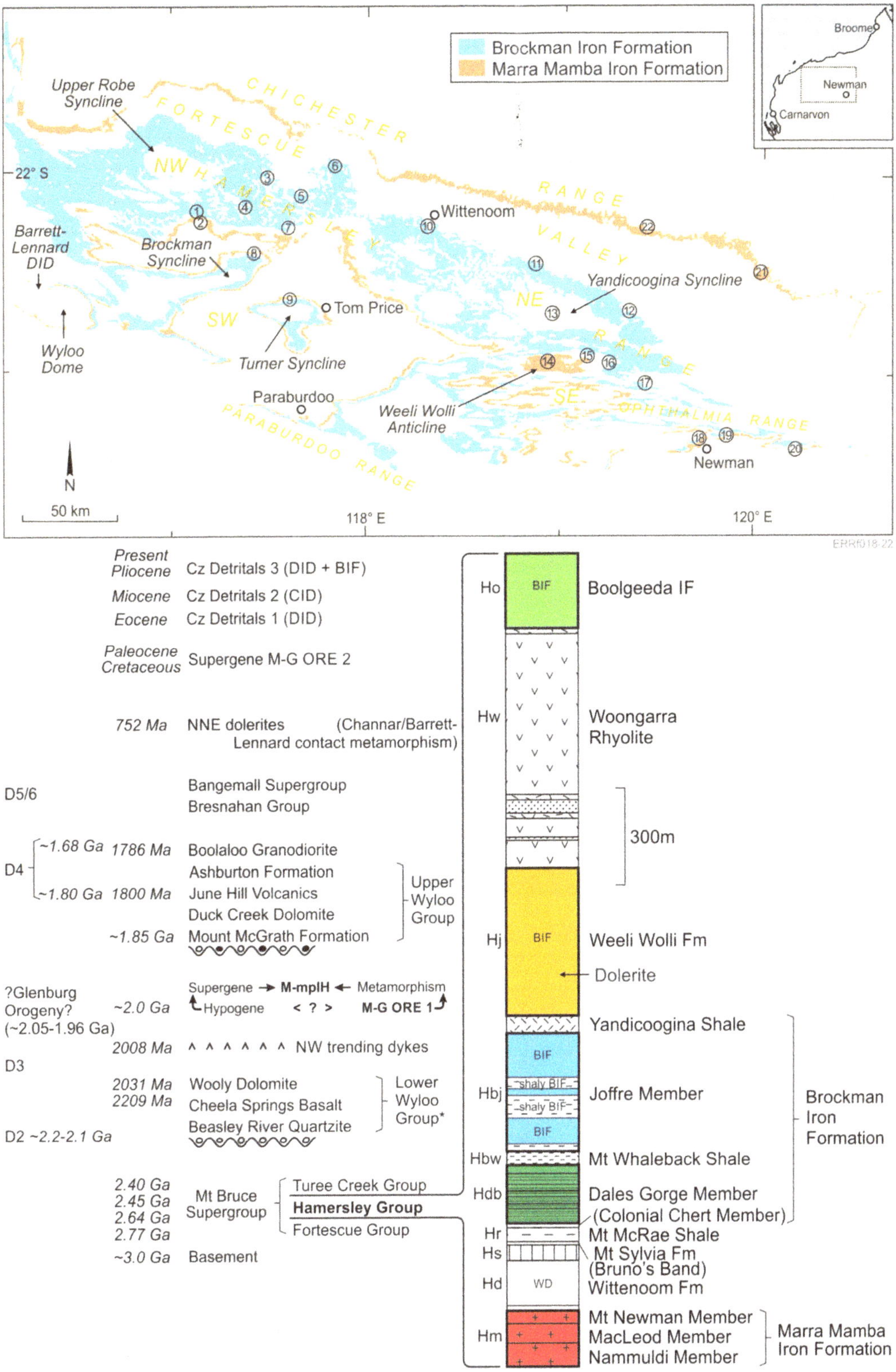

Fig. 5 Martite-Goethite Iron Ore in Western Australia: geographical location and stratigraphy

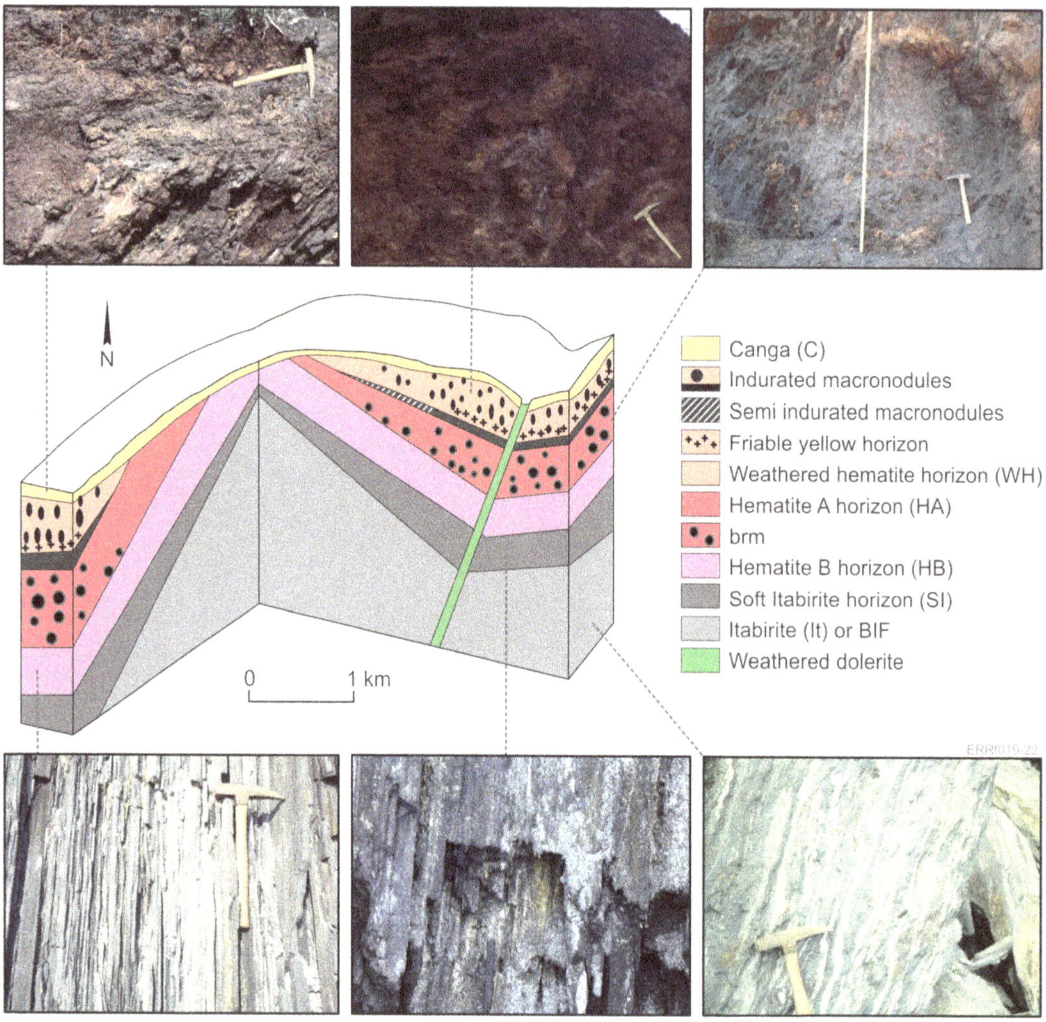

Fig. 6 Weathering fronts evolution in Capanema (Brazil)

descending (Fig. 6) and include from bottom to top (1) leaching of quartz and modest goethitisation in the lower horizons (Soft Itabirite and Hematite B) where weathering is subtractive and iron enrichment is relative, (2) pseudomorphic replacement of the primary iron oxides and oxy-hydroxides in the Hematite A horizon with an absolute accumulation of iron; (3) intense weathering in the aluminous rich Weathered Hematite horizon where three facies are characterised, (a) facies with preserved structure, (b) facies with macronodules and (3) facies where the structure has been totally destroyed and (4) a cellular Canga capping the weathering profile with hematitic nodules in a goethitic matrix. Through these various supergene processes, it was demonstrated that Al has a significant role and acts as an accelerant in a Fe-dominated setting. The common occurrence of alternating secondary Al-goethites and Al-hematites occurring in the Capanema Weathered Hematite horizon (Ramanaidou 1989, 2009) is identical to alternating goethite and hematite found in the Hamersley (Clout 2006) and in the Yilgarn (Anand and Paine 2002). In Carajas (Brazil), the prolonged weathering of BIF also resulted in a well-developed lateritic profile covered by a duricrust or canga (Gagen et al. 2019).

BIF-hosted Indian supergene iron ores also show very well-developed weathering profiles, and Mukhopadhyay et al. (2008) observed that all Indian iron deposits were subjected to chemical weathering that resulted in an enrichment of BIF and the production of blue dust ores and an intense goethitisation toward the surface. Similarly, Morris and Trendall (1988) showed that, at the Bailadilla mines in Chhatisgarh, the heavily leached hematite or 'blue dust' was overlain by a thick goethitic horizon capped by the highly weathered carapace.

Overall, supergene lateritic enrichment affecting the BIF is mostly subtractive, and the Fe enrichment is relative because the gangue material, principally quartz, is dissolved, and the primary iron oxides remain. The primary texture is significantly destroyed as weathering intensifies with the end-product a surficial ferruginous duricrust or canga with textural characteristics ranging from massive, vermiform, botryoidal and nodular to ooidal.

2.2.3 Supergene Mimetic Enrichment of BIF

In the M-G ores of Australia, India and South America, the supergene mimetic enrichment of BIF shows a replacement of the gangue minerals such as quartz, silicate and carbonate by goethite. The resulting texture shows an excellent preservation of all the precursor BIF petrological characteristics (Morris 1980; Ramanaidou and Morris 2010; Perring 2021). This goethite enrichment is an absolute accrual of iron and an almost elimination of the gangue chemical elements without the loss of the original BIF mineral textures. However, petrological studies have shown that the chemical composition of the goethite pseudomorphs after quartz, silicates or carbonates still retain small percentages of alumina, phosphorus and silica contents (Ramanaidou et al. 2008). Goethite pseudomorphed after carbonate contains the largest quantity of Al, P and Si; goethite pseudomorphed after quartz contains the smallest amount of Al, P and Si, whereas the goethite pseudomorphed after silicate contains intermediate amounts of Al, P and Si (Ramanaidou et al. 2008).

The age of the martite-goethite mimetic mineralisation was estimated to be Tertiary by MacLeod (1966) because of its association with the mature Hamersley surface. This was also postulated by Taylor et al. (2001) as Tertiary weathering on the ranges of the Brockman Iron Formation also formed typical martite-goethite mineralisation. Morris and Kneeshaw (2011) described a major second phase of M-G ores forming following general exposure by erosion in the Cretaceous–Paleocene. The first direct dating of the M-G ores using U-Th/(He) dating (Ramanaidou et al. 2019) showed that the Fe oxide's ages range from 45 Ma (Mid-Eocene) to 28 Ma (Early Oligocene).

A range of genetic models is currently hypothesised to describe the supergene mimetic M-G ores: Morris (1980, 1985) and Ramanaidou and Morris (2010) explained the M-G mimetic replacement genesis using an electrochemical cell model. They postulated that the deep-seated BIF-hosted ores extend to depths where oxygenated water is inaccessible. The mimetic goethite replacement of the BIF gangue minerals includes iron transported in the soluble Fe^{2+} form, followed by oxidation to Fe^{3+} in the anode area with precipitation as goethite with the cathode, the magnetite-rich BIF. This then freed the replaced gangue elements (e.g., Si, Al, Mg) to the water table and into the surface drainage. The oxidation process does not require free O_2 and is by electron exchange only. Consequently, the orebody formation ascends as erosion gets rid of the friable silica residue, forming a topographic low that, in turn, increases the groundwater flow through the system (Fig. 7). Lascelles (2002, 2007) offered a different model where, during compaction and diagenesis, portions of the BIF become magnetite-only BIF stratigraphy by removal of silica and subsequent supergene enrichment results in martite-goethite ore.

In 2018, Perring et al. proposed a genetic model of M-G mineralisation that includes (1) constant Cenozoic uplift of the BIF-rich Hamersley Province providing continuous BIF to the vadose zone; (2) acid and reduced fluids resulting from the interaction of meteoric water and organic matter during mid-Eocene temperate

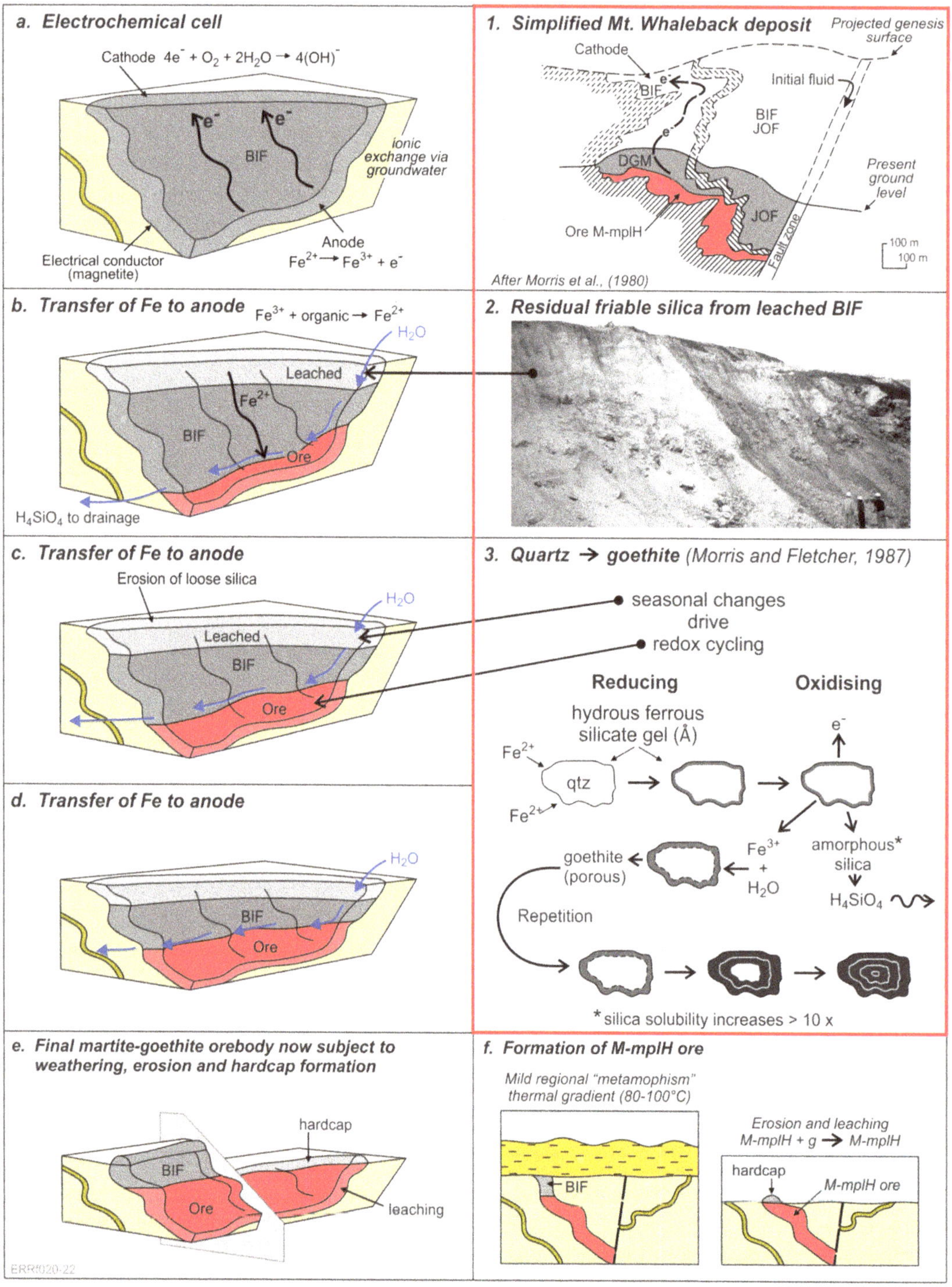

Fig. 7 Electrochemical Model for MG Genesis (Morris 1985)

climate; (3) leaching of the magnetite the ferrous ion into the vadose zone producing a supergene ore-fluid; (4) ore-fluid descending along the bedding planes producing the martitisation process and the mimetic replacement of the gangue minerals by goethite coupled; (5) alteration of the

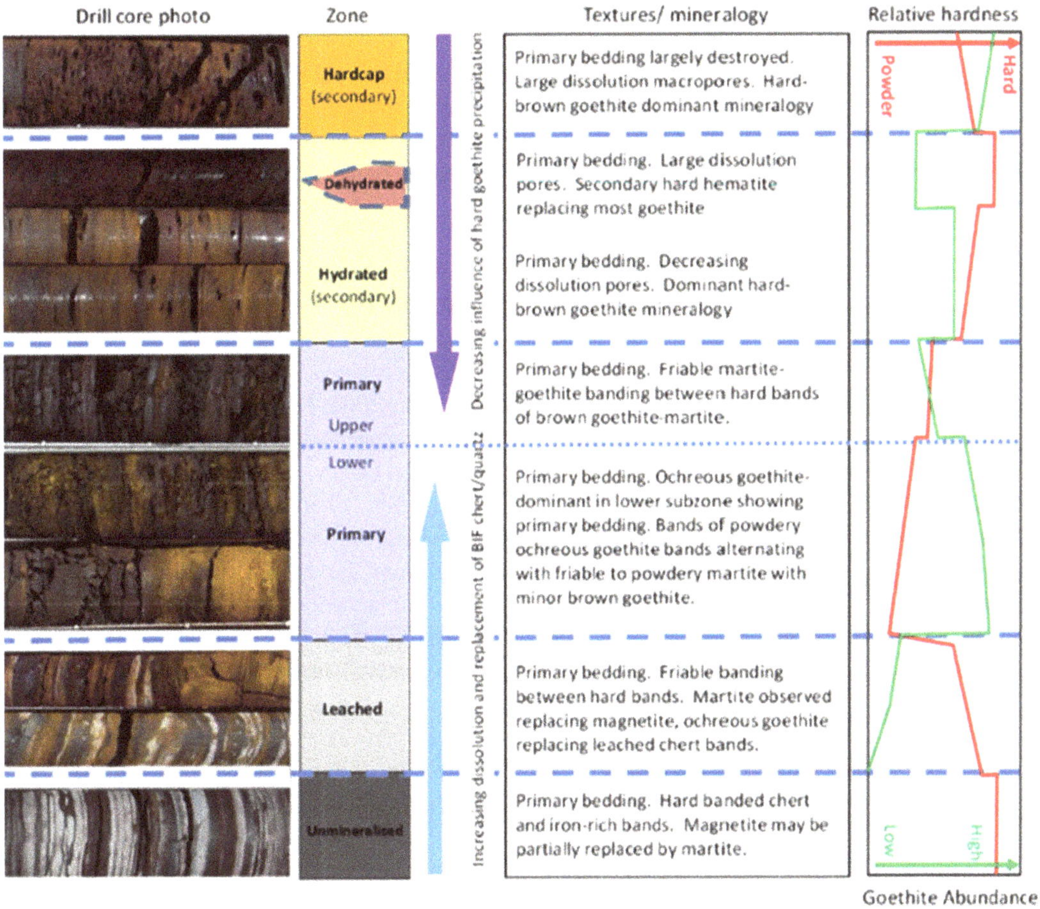

Fig. 8 Simplified vertical mineral/alteration zonation of a martite goethite deposit (Clout et al. 2017)

ore-fluid composition so silica deposition occurs until dilution.

Poulet et al. (2022) proposed a genetic model that showed that the paleo-stratigraphic permeability anisotropy (larger along strata than across) controls the supergene mimetic enrichment and distribution of the mineralisation. This non-continuous mineralisation is constructed from a simple 2D water flow model by analysing only one aspect of an otherwise very complex physical process.

2.2.4 Weathering of Mimetic Supergene Ore

The upper part of the supergene Fe iron ore is, in turn, subject to lateritic weathering. This is well documented in the Hamersley Province of Western Australia by Morris (1985), Harmsworth et al. (1990), Clout et al. (2017) and Perring (2021) but also in the Pilbara craton (Teitler et al. 2017).

In the Hamersley Province, Morris (1985) and Harmsworth et al. (1990) described weathering horizons including from surface to depth, the 'hardcap' consisting of (1) a metre-thick horizon of hematite-rich ore called the 'carapace', with preserved BIF texture resulting in part from the surface dehydration of the goethite from the ore matrix; (2) a 10–70 m thick 'hydrated zone' dominated by secondary aluminous vitreous and ochreous goethites resulting from the intense weathering of the parent BIF/Shale-band and (3) the ore proper which retains the original BIF texture and at times its primary ore hardness but is typically affected by groundwater leaching and secondary goethite reprecipitation.

Clout et al. (2017) proposed a more detailed classification for the Hamersley Province divided into six horizons that uses mineralogy, hardness and texture (Fig. 8) vertical zonation from the BIF to the surface, including (1) unmineralised BIF, (2) leached BIF, (3) lower primary ore, (4) upper primary ore, (5) secondary hydrated zone (including a dehydrated subzone) and secondary hardcap. Clout et al. (2017) also note that not all horizons will exist in the M-G deposits. Perring (2021) described a four-stage evolution in the M-G South Flank (Hamersley Province, Western Australia) deposit, including (1) martitisation and pseudomorph replacement of the carbonate–silicate–quartz by ochreous goethite; (2) micro-porosity development as a result of the dissolution of remaining quartz; (3) infilling of the micro-porosity by brown goethite and (4) intense leaching generating macro-porosity rimmed by a colloform intergrowth of vitreous goethite and (hydro)hematite.

In the Wodgina mine of the Pilbara craton, Teitler et al. (2017) described a supergene alteration sequence including (1) oxidation and martitisation; (2) silcretisation at 40 m below the modern surface, (3) goethite pseudomorph after quartz and (4) ferricrete with ochreous goethite rich in P, Zn, As, Ni and Co derived from local shales.

2.3 Detrital Iron Deposits

Detrital iron deposits or DID result from the physical breakdown and accumulation of alluvial or colluvial coarse gravels, with minor pisoids, derived from and mostly retaining the original textures of BIF-enriched iron ores such as Martite-Goethite or Martite-Microplaty types hardcap, or ferruginised surface BIF (Freyssinet et al. 2005; MacLeod 1966; Mohapatra et al. 2008; Morris 1994). The coarse gravels are cemented by a matrix of aluminous goethite and hematite. West Australian DID are generally lumpy, hematite-rich and low in P (Morris 1994).

DID resources in the Hamersley Province total 2 Bt (Fig. 4), with Fe ranging from 40 to 60 wt% but with negligible production (Morris and Ramanaidou 2007). Five Mt (57–62 wt% Fe) have been produced annually by Rio Tinto Iron Ore from the Brockman Syncline 2 Detritals, and smaller tonnages by BHP Iron Ore from Newman's Orebodies 25/23 and the Jimblebar W4 mine, both for blending with other lower lump iron ores into a saleable product (Butt et al. 2001). The lump product was extracted from the Brockman Syncline No. 2DID. The Barrett-Lennard (Wyloo dome in Western Australia) is a DID typically consisting of martite-microplaty hematite fragments (Dalstra and Guedes 2004; Macleod 1966; Morris 1985).

DID also occurs in the Rio Pardo area in Minas Gerais (Brazil) as well as in the Orissa state (India), where Chamakpur and Inganharan DID were mined (Mohapatra et al. 2008). A DID derived from the 1.7 Gt South African Sishen deposit 'laminated ore' accounts for 17% of the total reserve (Beukes et al. 2003).

The comprehensive study of DID in Western Australia by Kneeshaw and Morris (2014) described two types of detrital ores from the Eocene Cenozoic detrital deposits 1 (CzD1) and the Pliocene Cenozoic detrital deposits 3 (CzD3). Their chronological and geological classification of the DID recognised into two dominant geomorphological environments: (1) the southern Marra Mamba to Brockman Iron Formation strike valleys or MBSV, containing all DID sequences, and (2) the more recent northern Brockman IF plateau valleys or BPV only comprising the Miocene and Pliocene sequences. The different DID types result from the stripping of surface exposures due to major climatic changes. The CzD1 deposition was followed by an extended dry period that resulted in an extensively identifiable major dehydrated paleosurface. The CzD3 DID are the results of a revitalisation of exposure and erosion during the Pliocene—Quaternary period. The detrital Cenozoic stratigraphy in the Pilbara shows the climatic conditions required for the formation of DID (Table 3).

The genesis of DID (Kneeshaw and Morris 2014) has been exemplified in Area C, southern Marra Mamba to Brockman Iron Formation strike Valley and includes five phases (Fig. 9). They are (1) Eocene erosion at the contact between the

Table 3 Comparison of Cz Detrital stratigraphy of the Pilbara over time

CENOZOIC	Morris (1994)	Killick et al. (1996)	Kneeshaw et al., (2002)		Morris and Ramanaidou (2007)	Kneeshaw and Morris (2014)
	Tertiary Ore Detritals **TOD1, 2, 3**	**Detritals**	**Tertiary detritals** **TD1, 2, 3**		**CzD1, 2, 3**	**Cenozoic detritals** **CzD1, 2, 3**
			Yandi	**East Pilbara**	**Pilbara**	
	TOD3		**TD3**		**CzD3**	**CzD3**
Quaternary	Increasing BIF detritals	BIF detritus as crust stripped		Various BIF/ore detritals - Surface	Increasing BIF detritals	Increasing BIF detritals
Pliocene	Colluvial / alluvial ore detritals	Distal CID ↑ Proximal DID ↑ Stripping of ferruginous crusts	*(Not detailed)*	Canga Low lump Loose Compact Fine Aluminous *Lower five varieties grade distally into silts / clays*	Coarse hematitic colluvial/alluvial fragmentals ranging to pelletoidal	Coarse hematitic colluvial/alluvial fragmentals ranging to pelletoidal. *BIF-derived textures are typical*
	TOD2	Arid conditions	**TD2**		**CzD2**	**CzD2**
			Calcrete		Calcrete / silcrete	**Oakover** (Millstream) **Fm** Limestone (calcrete-silcrete)
Miocene	**CID** (pedogenic alluvials)	Ferruginous weathering crusts ↑	**Iowa Eastern & Barimunya Members CID**	Hardcap - earthy to vitreous goethite. **Clays dominant** with calcrete in upper zone.	**CID** or goethitic equivalents	**Barimunya & Robe Fm's** Billions of tonnes of pedogenic pelletoids and peloids => CID *Only rare BIF-related textures* Clay in many strike valleys where clay > pelletoid input Marillana (& Robe) Fmn
Oligocene		extreme stability of landscape	**Munjina Member** Clays / conglomerates		Mainly clays with minor organics **upper Munjina Member**	**Munjina Member** Basal clay minor CID + minor organics (+/- pyrite) Basal conglomerates
	TOD1		**TD1**		**CzD1**	**CzD1**
Eocene **Paleocene**	**Hardcap** on TOD1 Red ochre detritals, largely alluvial	↑	Not present ?	Red Ochre Detritals (ROD)	Clays/conglomerates ≡ **basal Munjina Member** Hematitic **hardcap** on mixed coarse fragmentals to silty red ochre detritals	Hematitic **hardcap** mixed coarse proximal fragmentals to distal silty red ochre detritals (ROD) BIF-derived textures typical
MESOZOIC **Cretaceous**		?				**MzD1** **Yaraloola Conglomerate** Rare M-G pebbles
PROTEROZOIC						**PzD2** **Eel Creek Formation** Late Proterozoic NE Pilbara Basal BIF marine conglomerate locally contains hematite pebbles (Yarrie Y10 deposit)
ERRf008-21						**PzD1** **Mt McGrath Formation** Upper Wyloo Group S Hamersleys Fluvial conglomerates frequently with M-(mplH) ore fragments (Barret-Leonard deposit)

Modified after Kneeshaw and Morris (2014)

Marra Mamba iron-formation and the friable West Angelas Member 1 of the Wittenoom Formation; (2) incision of a deep valley at the contact; (3) infilling of the valley by detrital clasts from neighbouring Marra Mamba M-G ore and induration of the clasts by dehydration of the goethitic matrix to form a carapace under dry conditions; (4) development of a Miocene deep weathering followed by erosion and deposition of ooids and pisoids in the meandering channels, formation of the goethitic matrix cementing the ooids and pisoids in the channels during the mid to late Miocene giving rise to the CzD2 detrital or CID; (5) Pliocene erosional phase forming the

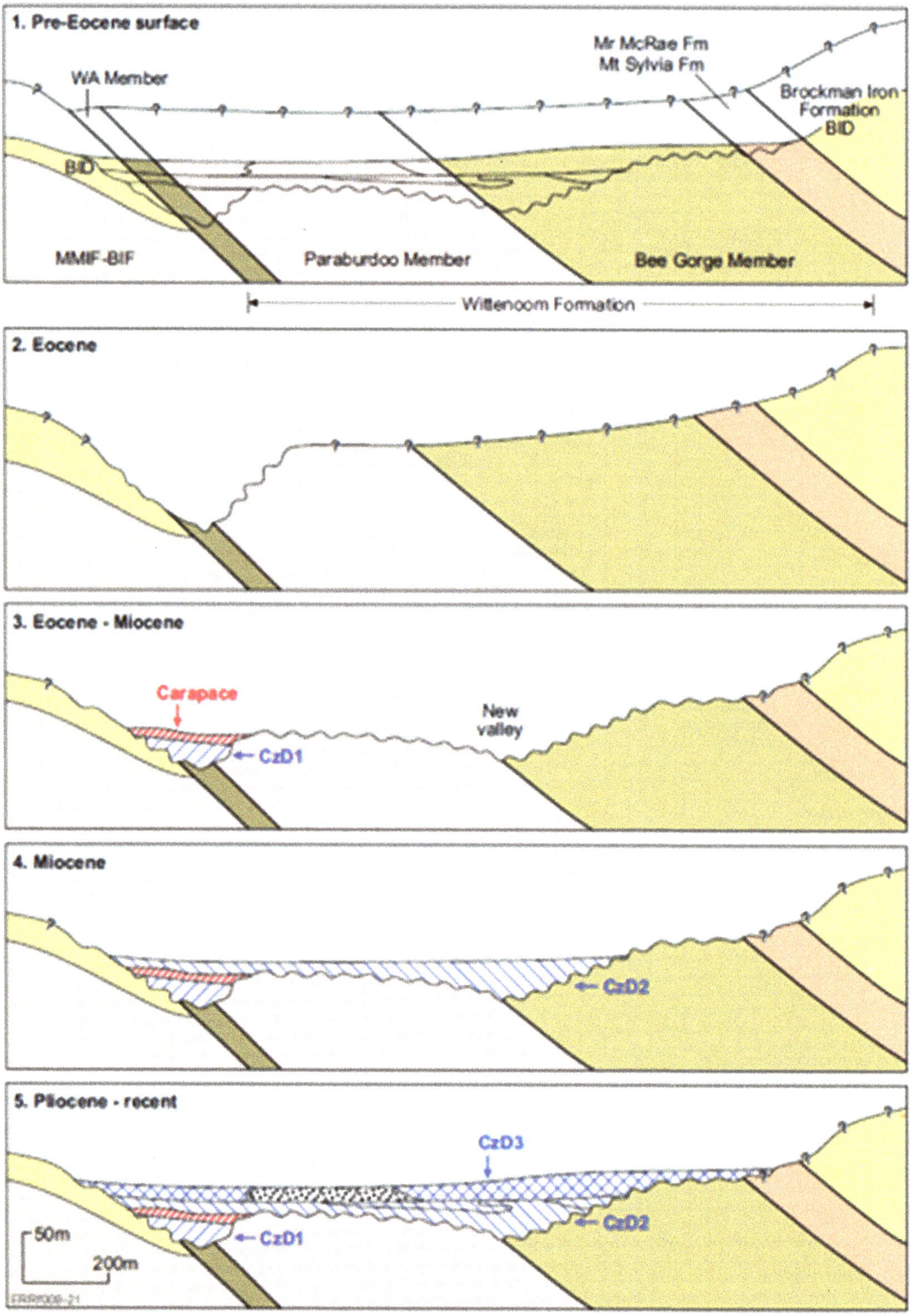

Fig. 9 Simplified interpretation of the growth history from pre-Eocene with Mesozoic–Paleocene Martite Goethite bedded iron deposits (M-G BID) Marra Mamba iron formation (MMIF) and banded iron formation (BIF) from Eocene to the current surface of the CzD1, 2, 3. (From Kneeshaw and Morris 2014)

CzD3 with clasts sourced partly from the Marra Mamba iron-formation but mainly from the taller Brockman iron-formation ridge on the other side of the valley (Fig. 9).

Alternative interpretations from Simpson et al. (2011) and Dalstra (2011) are based on local environments and settings. Simpson et al. (2011) proposed a genetic model for the Cenozoic DID of the Solomon area, and it suggested a model with three cycles of deposition with dominant interruptions and erosional periods. Dalstra (2011) also presented a genetic model for the DID that places the CzD3 type deposition in a period ranging from 45 to 5 Ma. However, neither CzD1 nor CzD2 occur in the sequence.

3 Channel Iron Deposits of the Pilbara Region

3.1 Occurrence, Distribution and Age

The channel iron deposits (CID) were first described (Woodward 1891) as '*hills capped with ferruginous sandstone*' in the lower Robe River and '*another line of flat-topped hills, but lower . . . capped by ferruginous sandstone containing large quantities of fossil woo*d' in the Nullagine area (Fig. 11). Harms and Morgan (1964) and Butler (1976) offer a summary of the discovery and early exploitation of the CID of the Robe River area. The CID are of terrestrial origin and differ from the marine Phanerozoic ooidal ironstones (POI) that have a worldwide distribution (McGregor et al. 2010) and have occurred since the late Cambrian (Fig. 10), such as the so-called 'Minette' ooidal iron deposits. Even though terrestrial POI from northwestern Nigeria (Kogbe 1978), Wealden sediments in southeast England (Taylor 1992), Kazakhstan (Yanitzkii 1960) and from the Hamersleys have been mined since the early 1970s (Ramanaidou et al. 2003), they received little attention until the early 1990s (Ramanaidou et al. 1991). Ramanaidou et al. (2017) provided a summary of the CID of the Pilbara Region of Western Australia. The term 'CID' was introduced by Ramanaidou et al. (1991) for deposits previously called 'pisolite ores' (MacLeod 1966), 'Robe Pisolite' and 'Goethitic Channel Ores'. The term is now generally recognised by both the mining and research communities to include all terrestrial, Fe-rich ooidal and pisoidal ironstone deposits of fluviatile, deltaic-lacustrine and lacustrine origin (Robe, Marillana and Poondano formations, the small deposits of the Yilgarn Craton and the Ashburton Basin (Butt 1979; von Perger 1991) as well as the fluvial deposit of Kazakhstan (Ramanaidou et al. 2003). Mining of CID commenced in the 1990s, and there is an estimated nearly 20 billion tonnes of CID in the Hamersley province (Knight et al. 2018). Although not as significant as the bedded resource, it presents as an additional ore type to contribute to the global market.

The Miocene Western Australian CID are Cenozoic detrital iron deposits but display significant dissimilarities with the minor Eocene and Pliocene detrital iron deposits or DID (Kneeshaw and Morris 2014 and Sect. 2.3 above). In the last 20 years, the volume of publications has been boosted by the significant increase in CID mining in Western Australia (Ramanaidou et al. 1991, 2003, 2017; MacPhail and Stone 2004; Heim et al. 2006; Morris and Ramanaidou 2007; Dalstra et al. 2010; Kepert et al. 2010; Ramanaidou and Morris 2010; Storkey et al. 2010; Danišík et al. 2011; Haest et al. 2011, 2012). CID mining commenced in 1972 in the Robe River—Bungaroo Province (Fig. 11), the longest valley sequence in the area, with CID partially preserved over 150 km from the Silvergrass Syncline to the Carnarvon Basin edge. The western CID of the Robe River (and Bungaroo), the Eastern Yandi CID (Fig. 11), the Central Hamersley CID Province of the Solomon East CID (Kepert et al. 2010), the Caliwingina deposit (Dalstra et al. 2010) and the Nullagine Iron Ore Project are part of the 18 Bt of the total Western Australian iron ore resource in 2020.

The CID occur mostly in meandering paleochannels that carve through Precambrian rocks (Ramanaidou et al. 2003), including granitoids, volcanics, metasediments and BIF (Figs. 12 and 13) and, in places, Paleogene detritals (Kneeshaw and Morris 2014). The

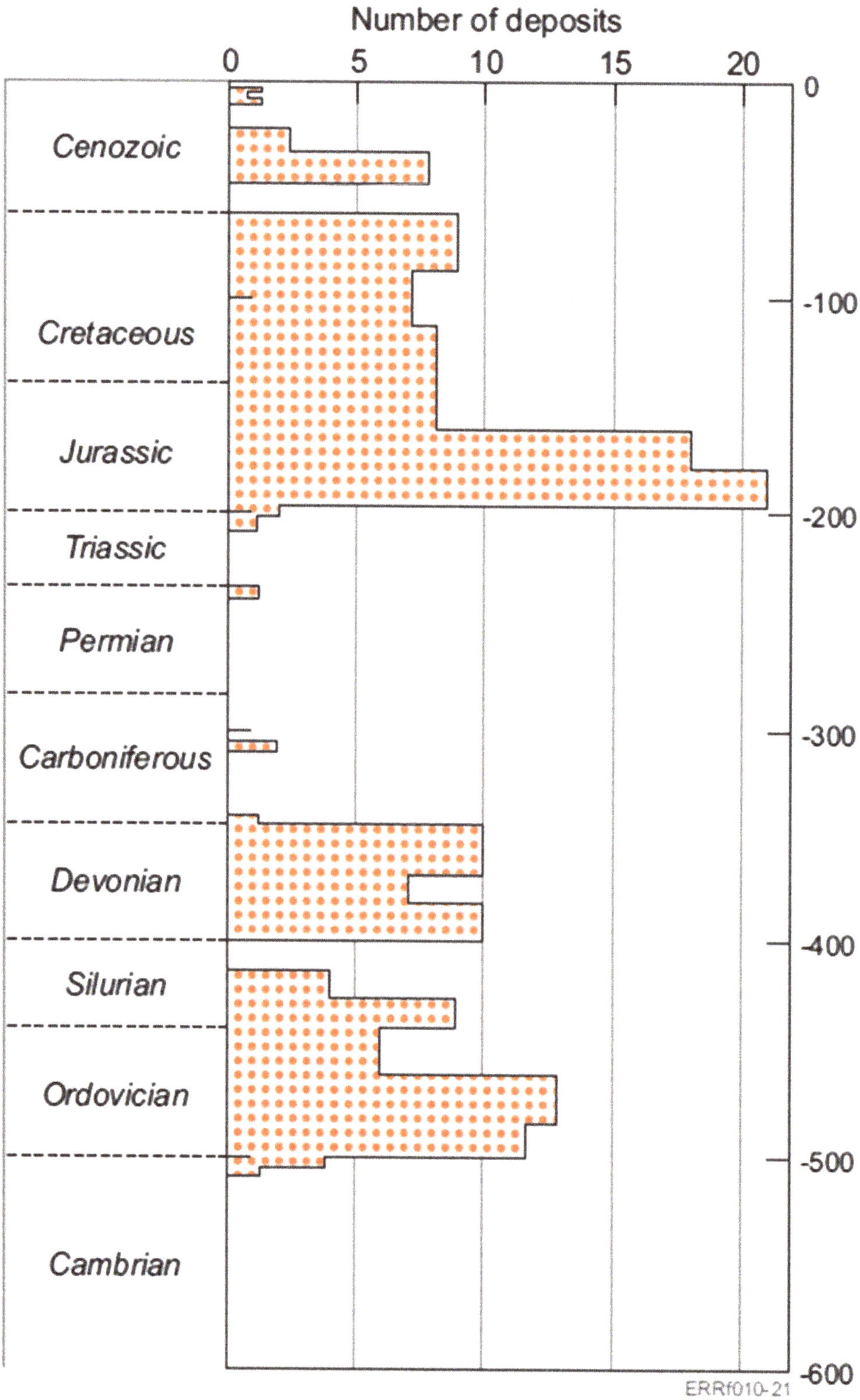

Fig. 10 Distribution of Phanerozoic ooidal ironstones (number of deposits) during the Phanerozoic

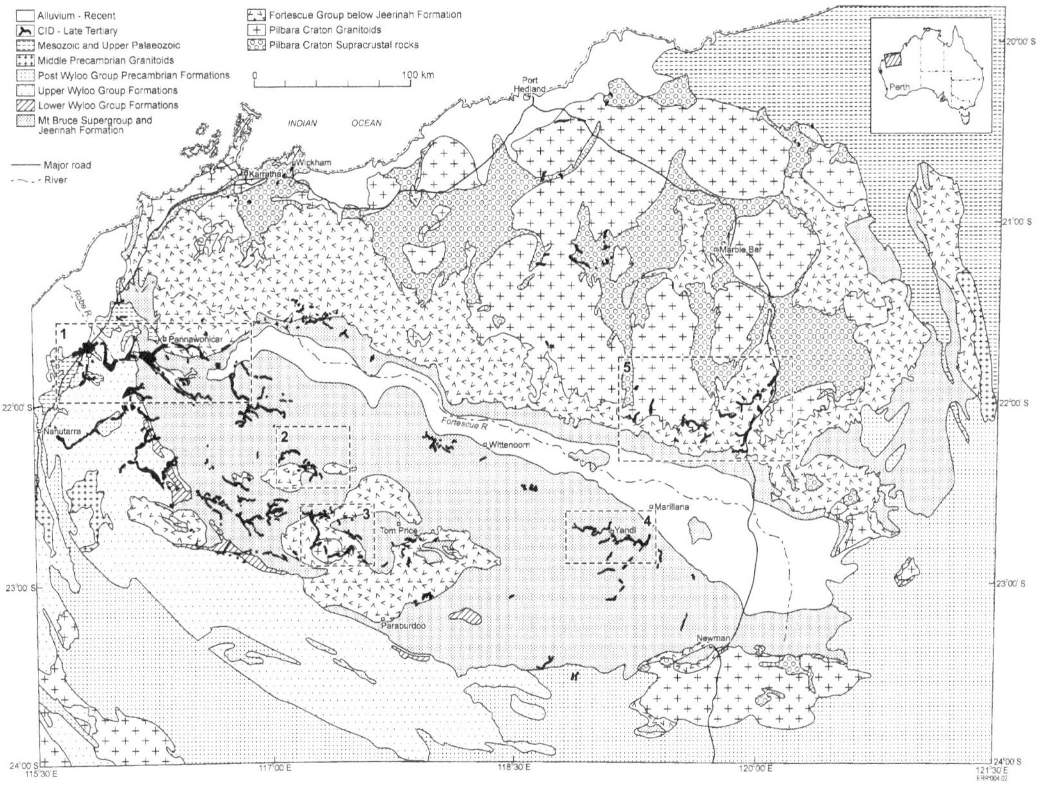

Fig. 11 Channel iron deposits (CID) distribution in the Pilbara region of Western Australia. Boxed outlines define the main CID provinces in the Hamersley Province: (1) Robe River and Bungaroo CID, (2) Central Hamersley (Caliwingina, Serenity—Cabbage Gum Bore, Solomon East), (3) Rocklea Dome, Beasley River CID, (4) Yandi CID and (5) Nullagine, Bonnie Creek

thickness of CID ranges between <1 and 100 m but can reach up to 200 m in the Bungaroo Valley. The CID have a width generally below 1 km but can be as wide as 5 km in Mesa A and Waramboo (Fig. 12). Recent incision has led to topographic inversion occurring as remnant mesas, in particular in the Western Robe River drainage system (Fig. 12), whereas the Eastern Marillana Formation (Yandi CID Province) occur in synclinoria on low-angle slopes or lean against the margins of the paleochannels (Fig. 13).

CID either rest on a lateritic surface or, uncommonly, overlie fresh bedrock. The CID can be classified into three facies (Ramanaidou et al. 2003) and include (1) the granular facies, the main source of CID ore, with an intraformational CID conglomerate subtype commonly found in the Robe and Yandi deposits; (2) the bedded facies; and (3) the weathered facies, including the leached, the silicified 'conchoidal' and the surface dehydration subtypes. All CID facies have variable Fe grades. The bedded, granular and intraformational conglomerate facies seem interbedded. The weathered CID facies are the result of weathering of the granular CID, and both the silicified 'conchoidal' and the surface dehydration subtypes are related to silcretisation and dehydration.

3.2 Petrology and Mineralogy

The CID texture is typically porous, consisting of Fe-rich granules, including peloids, pelletoids and

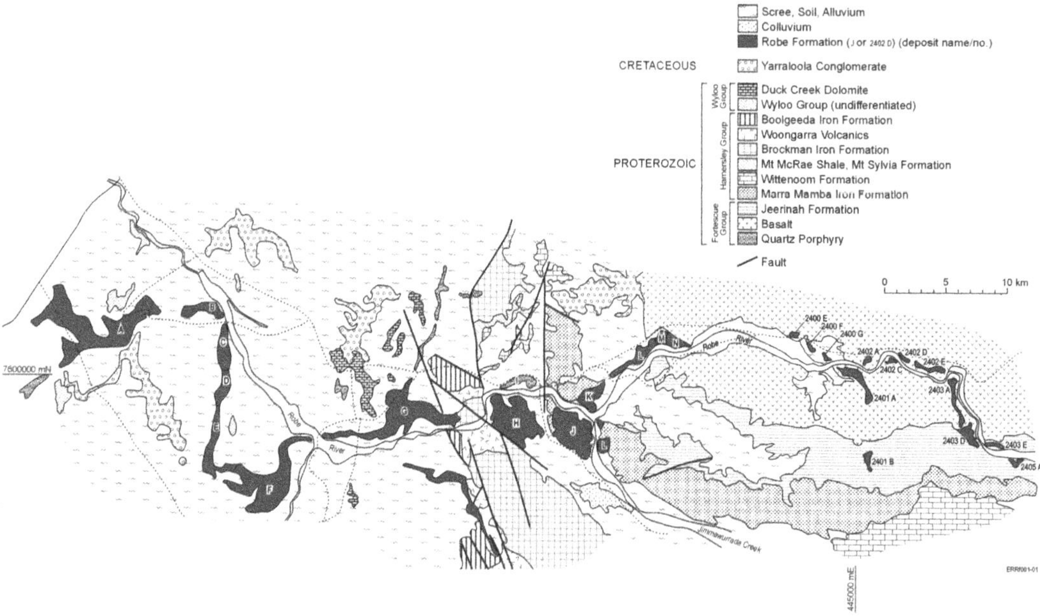

Fig. 12 Geological map of the Robe River area showing the distribution of channel iron deposits. (See Fig. 11 for location)

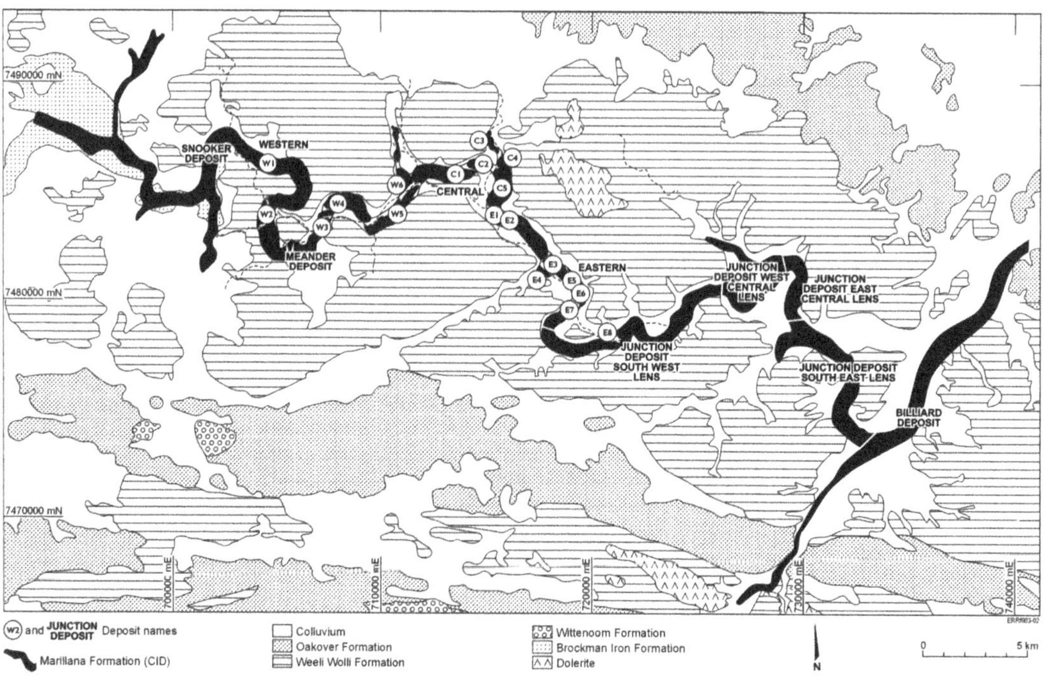

Fig. 13 Geological map of the Yandi area showing the distribution of channel iron deposits. (See Fig. 11 for location)

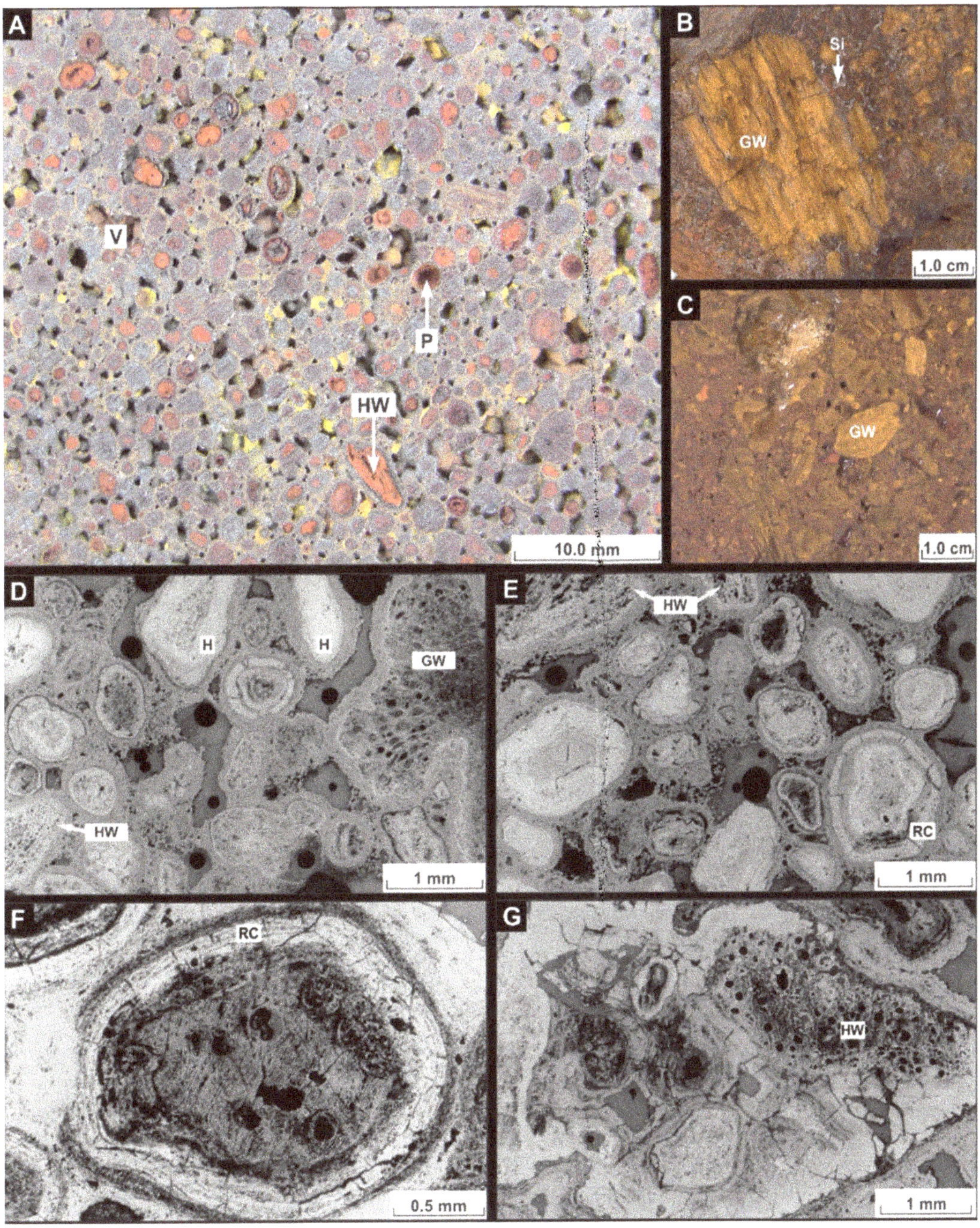

Fig. 14 Granular channel iron deposits with (**a**) Pelletoids with leached peloids (P), hematitised wood (HW) and minor matrix and voids (V). (**b**) Large goethitised wood (GW) and silica (Si). (**c**) Large goethitised wood (GW); (**d**) a range of granules including goethitic (GW) and hematitic (HW) wood fragments; (**e**) a range of simple nuclei; note radial cracking (RC) in the outer part of the hematitic nucleus; (**f**) hematite fossil wood nucleus with multiple hematitic layers showing RC; (**g**) large irregular granule, dominated by a complex nucleus including a HW fragment

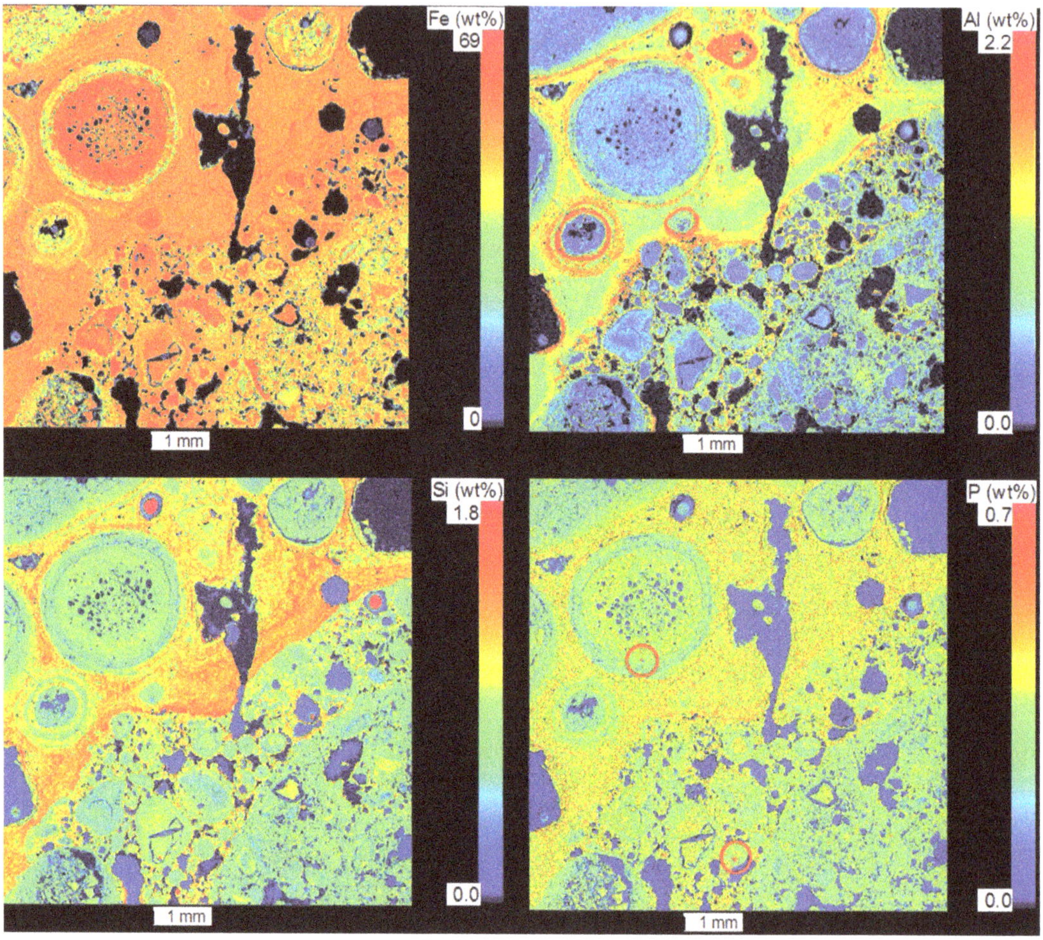

Fig. 15 Wavelength dispersive spectroscopic (WDS) X-ray mapping of the Fe, Al, Si and P distribution in a high-P (0.118 wt%) CID. Coincidental elevated concentrations of Al and Si infer the presence of kaolinite within the ooid cortex and matrix. 'Hotspots' of elevated P abundance (red circles) correspond to isolated occurrences of apatite

ferruginised wood cemented in a goethitic matrix (Figs. 14 and 15).[1] The mm-size peloids consist of astructural goethite fragments. The pelletoids, consisting of a nucleus surrounded by a cortex, include both ooids (diameter 0.25–2 mm) and pisoids (diameter 2–10 mm), with the ooids accounting for more than 80% of the total (Figs. 14 and 15). The nuclei are rounded or unevenly shaped, simple or complex and commonly contain a hematitic wood fragment (Ramanaidou et al. 2003); nucleus and pelletoid shapes are correlated (Ramanaidou et al. 2003). Fractures commonly occur in the hematitic nuclei and are developed radially. The Robe CID tend to have more spherical particles and more regularly developed layering with smaller cores than samples from the Yandi deposits. Ferruginised wood, with a maximum size of 50 mm, is a specific attribute of the CID (Ramanaidou et al. 2003; Morris and Ramanaidou 2007) and occurs

[1] Granules: include peloids, pelletoids and ferruginised wood. Peloids: fragments of silty goethite. Pelletoids include ooids (0.25–2 mm) and pisoids (2–10 mm) that are rounded bodies consisting of a nucleus surrounded by a cortex. The nucleus may be simple or complex.

as small hematitic nuclei of pelletoids and as discrete cm-sized goethitic fragments in the goethitic matrix. Stone et al. (2002) described 15% goethitised wood in the upper main ore zone of the Yandi CID. Goethite abundance varies from 75% (matrix-supported) to as little as 2% (grain-supported).

In the Robe CID, the alumina content of pelletoid nuclei ranges from 0.5% to 2.8%, whereas at Yandi, it is 0.2–0.7%. The goethitic cortex of a pelletoid is generally richer in alumina than its nucleus, up to 3% Al_2O_3 in the Robe CID, and at Yandi, up to 1.5% Al_2O_3. The goethitic matrix is lower in alumina (average 1.2% at Yandi and 1.85% at Robe) than the neighbouring cortices but higher in alumina than in the corresponding nuclei (Fig. 14). Phosphorus occurs in apatite grains or in association with goethite (Fig. 14).

In the CID, the two main ore minerals are Al-goethite and Al-hematite, whereas maghemite and/or kenomagnetite, two critical magnetic components, only account for a small portion of the Fe grade. The remaining minerals include kaolinite, Ca- and Fe-montmorillonite, gibbsite, chalcedonic silica, and calcite. Phosphorus is mainly associated with goethite (Wells and Ramanaidou 2011; Ramanaidou and Wells 2014) and apatite.

'Onion-skin' and more rounded pelletoids are richer in alumina than the simple and more irregularly shaped nuclei, mostly from the Yandi area, showing that aluminous country rocks have influenced their formation. Also, CID developed from the Jeerinah Formation and Marra Mamba Iron Formation tend to show higher amounts of goethite and an increase in titania. Trace elements in CID reflect the influence of regional rock sources. Zinc, for instance, is always higher in Robe ore than in Yandi ore, and in the Rocklea and Milli Milli Domes, the CID shows anomalous traces of Ni, Cr, Ti and V (Ramanaidou et al. 2003, 2017).

3.3 CID of Kazakhstan

The Kazakhstan CID has been described mostly in Russian studies (e.g., Zitzmann 1977) and Ramanaidou et al. 2003. Fossil wood fragments were reported in the Kerch Tobacco ores (Tsipurskii and Golubovskaya 1989) and Lisakovsk CID (Golubovskaya 2003). The Middle Oligocene Kazakhstan CID are located in river valleys with the largest deposit at Lisakovsk (total reserves of 2800 Mt) with a width of 2–8 km, a length of 100 km and ore horizons averaging 8 m in thickness. For the Kerch and Lisakovsk deposits, Fe ranges from 20 to 52 wt% (Zitzmann 1977), whereas for the limonitic Lisakovsk ooidal ore, Fe contents vary from 30.4 to 52.5 wt% (e.g., Golubovskaya 2003). Kerch brown ores comprise smectite, 'hydrogoethite', Mn-hydroxides, siderite, and Fe-Ca-phosphates, with a matrix comprising mainly 'hydrogoethite' (Golubovskaya 2001). The high P contents of the Lisakovsk CID are linked to the presence of discrete P-bearing phases such as chlorapatite, and Fe/Mn and Fe-Ca phosphates such as vivianite and kerchenite, mitridatite, bosphorite, and switzerite.

3.4 Genesis

MacLeod (1966) proposed that the CID were formed as bog iron-ore, whereas Schwann (2009) hypothesised that they form through the replacement of a greensand protore derived from volcanic ash. Lascelles (2007) proposed a model including sheetwash accretion of pisoids generated by the action of raindrop impacts on dry soils. However, physical modelling of raindrop impacts has proved to be inconclusive. The currently preferred model (Morris and Ramanaidou 2007) includes ferruginous ooids/pisoids derived from a lateritic profile associated with 'mud-bank' peloids as channel fill, followed by intraformational cementation by the goethitic matrix. The occurrence of magnetic components, such as kenomagnetite and maghemite, within the nuclei and cortices of ooids/pisoids is seen as the result of sporadic conditions caused by the reduction of goethite by organic matter during wildfires (Morris and Ramanaidou 2007; Anand and Gilkes 1987).

The timing of the formation of CID, originally based on fossil plant and pollen assemblages in the basal organic-rich material below the Marillana CID, indicated the Early Oligocene age (MacPhail and Stone 2004). (U-Th)/He dates for the goethitic matrix in the Yandi CID show ages ranging from 18 to 5 Ma (Heim et al. 2006), whereas Danišík et al. (2011) found ages between 18 and 7 Ma in the Robe River Mesa J CID. The CID ore formation process includes (1) hematitic cores in the ooids formed in the Early to Middle Miocene (18–14 Ma); (2) goethitic cortices formed in the late Middle to early Late Miocene (12 Ma); (3) ferruginisation of the wood fragments in the matrix are Late Miocene (9–7 Ma). Using (U-Th)/He dating and a three-component mixing model with one goethite and two hematites, Yapp and Shuster (2017) proposed that the age of crystallisation of the goethite in the Robe River Mesa J was 7 Ma. Al-adjusted δD values of this goethite, consistent with the formation of the goethitic matrix of the CID, were derived from a summer-season humid tropical climate that could have promoted multiple cycles of microbially mediated dissolution and recrystallisation of Fe (III) oxides (Yapp and Shuster 2017). Fritz and Yapp (2018) show that Miocene weathering environments that produced the Fe (III) oxides are present in the mole fraction (*X*) and δ13C of the Fe (CO_3) OH component in solid solution in the goethitic matrix and this could point to a formation in phreatic or near-phreatic conditions rather than vadose, suggesting a change in local climate condition of burial at 7 Ma (Fritz and Yapp 2018). This corresponds to the beginning of aridification in the Hamersley (Duclaux et al. 2013) and is consistent with the model of Morris and Ramanaidou (2007), which suggests that the goethitic matrix, the last event for CID genesis, is derived from modified chemically precipitated Fe hydroxyoxides sourced from leaching of Fe-rich soils in an organic environment. The Miocene period appears to be the peak period for CID formation at Robe River and Yandicoogina (Heim et al. 2006; Danišík et al. 2011).

4 Implications for Exploration, Evaluation and Exploitation

As discussed in the previous paragraphs, three supergene iron ore types have been recognised in the in situ martite-goethite ores, the ex situ detrital iron deposits and channel iron deposits. Despite the economic significance of the supergene M-G ores of the Hamersley Basin, their genesis remain equivocal (Morris 1980; Perring et al. 2018). Most outcropping M-G deposits have been discovered, meaning exploration efforts are now focused on undercover targets. Although no single supergene model exists for these ores, it is generally unequivocal that faults play a fundamental role in both styles (e.g., Morris 1985; Powell et al. 1999; Clout and Simonson 2005; Dalstra 2006, 2014; Bodycoat 2010; Perring et al. 2018; Piechocka et al. 2021). Harmsworth et al. (1990) stated an inconclusive genetic link between faults and martite-goethite deposits, and, despite their contrasting conclusions, structures continue to play an important role in targeting Fe ore and are used as critical exploration criteria for regional scale target generation. Therefore, although new challenges exist in exploration, the structure remains a favourable targeting criterion which (if identified) can be used to delineate potential targets undercover.

Similarly, most outcropping CID deposits have been discovered, so exploration will need to target buried CID deposits. In the last 50 years, the discovery of exposed CID in the Pilbara region depended on extensive mapping in combination with geophysics, drilling, and sedimentological and geochemical analyses. In 2013, Duclaux et al. (2013) used a forward numerical modelling method based on a hybrid hydrodynamic and sediment transport procedure to predict the location of undercover CID. The surface process modelling method was trialled at the regional scale to assess the dispersion of the Fe-rich weathered mantle over the Hamersley surface following the mid-Miocene aridification. The model successfully reproduced the deposition pattern of the exposed CID as well as predicted the occurrence of buried deposits (Fig. 16), particularly in the area near Bungaroo. Such modelling provides evidence that terrains with mafic and ultramafic lithologies, when subjected to intense deep chemical weathering followed by mechanical weathering, could be highly prospective for CID exploration, particularly for buried deposits. Furthermore, CID deposits are commonly overlain by the Oakover

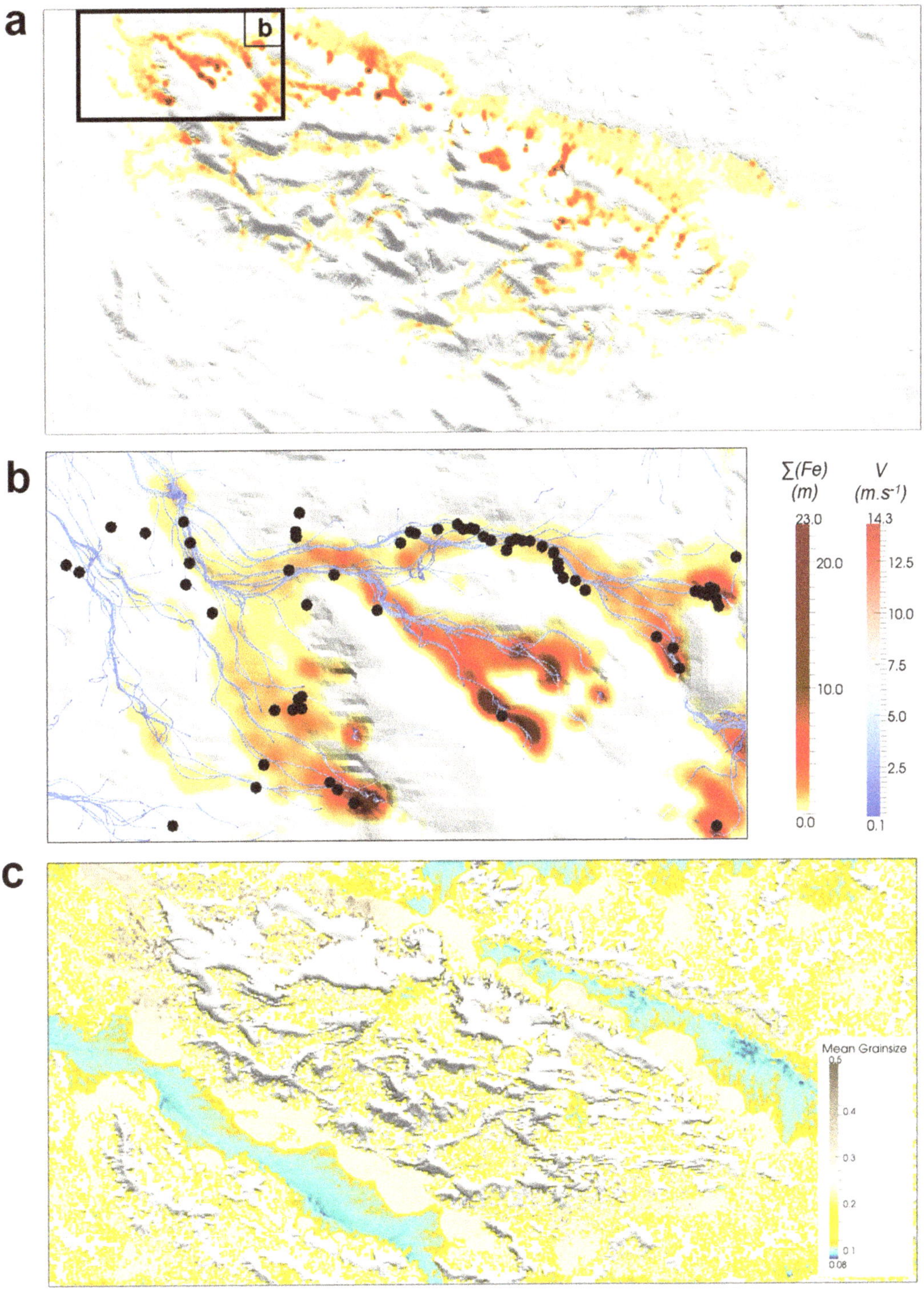

Fig. 16 Mid-Miocene surface dispersion modelling results: (**a**) predicted accumulation of ooids (sourced from a 10 m thick laterite layer overlying the Iron Formations. (**b**) Close-up view of the inset in (**a**) showing the known detrital iron ore (CID) occurrences in the area. The predicted pattern from the model is consistent with the known resource distribution. (**c**) Clastic sediments mean grain size distribution at the surface of the model after the Miocene climate aridification

Formation, a Tertiary silcrete/calcrete unit (Kepert et al. 2010; Ramanaidou and Morris 2010; Storkey et al. 2010) that preserves the CID. Identification of this formation could be an exploration guide for CID deposits. These alluvial fans concentrate sediments from the Hamersley Range watersheds and mark regions where buried CID can be expected.

The depletion of the premium M-G ores presents new challenges for the Pilbara iron ore producers as they are becoming lower in Fe grade and lower in lump fines ratio. Elevated P, Si and Al contents, as well as friable ochreous goethite, pose problems in the processing stages of the ore. The supergene Fe ores of the Hamersley Basin are informally divided into three types: (1) high P Brockman Iron Formation ore; (2) low P competent Marra Mamba Iron Formation ore; and (3) a low P friable Marra Mamba Iron Formation ore (Brent 2021). Type 2 also has a high Fe content (58–63%) with a reasonable lump yield and a favourable low P content (0.05–0.07%), whereas type 3 has the lowest Fe grade (56–62%) and lowest lump yield but with, a comparable low P content (0.05–0.07%) to type 2 (Brent 2021). The Marra Mamba ores typically have a higher proportion of goethite to martite as well as an increase in deleterious elements (such as Al_2O_3, SiO_2 and P). These differences in the chemical composition of Fe ores mean that new considerations must be made regarding how the ore will behave during processing and what the composition of the final product will be. This has led to research on characterising P and other impurities in goethite-rich iron ores (e.g., Pownceby et al. 2019), with the aim of increasing the understanding of these goethite-dominant ore bodies. The variable nature of the bedded iron ore deposits in the Pilbara presents an ongoing challenge for the industry to deliver a product that balances the lump yield, P content and Fe grade to produce a suitable blend for the customer. To overcome this challenge, the lower grade low P ores are blended with the higher Fe and higher P ores (Brent 2021).

In addition to M-G deposits, a second source of Fe ore comes from the CID. Two types of CID deposits are mined in Hamersley Basin and include a low and a high Al/Si CID (Brent 2021), with both types showing comparable Fe and P contents and elevated LOI. The low Al/Si type contains 1–2% Al_2O_3 and 4–7% SiO_2 and is a fines-only product, whereas the high Al/Si type contains 2–5% Al_2O_3 and 5–8% SiO_2, with a 20–35% lump yield.

Extraction of supergene iron ores via surface mining methods is relatively simple compared to more complex underground mineralised systems. However, there are several implications inherent to the M-G and CID deposits. Since most outcropping deposits have been discovered, future mining will be focused on buried deposits, which will result in a higher waste/ore strip ratio. The current strip ratios reported for Pilbara bedded Fe ore mines range from 0.75 to 4.1 (Aquila 2020; Roy Hill 2009) and 1:1 for CID (Inverarity et al. 2012); however, this ratio will increase for deposits with a deeper overburden. Although a low strip ratio means cheaper mining costs, there is insufficient material to back-fill the pits above the pre-mining water table once mining has been completed. A numerical and analytical modelling study by Inverarity et al. (2012) looked at the impact of mining on local groundwater and surface groundwater and concluded that strategic backfilling would have an acceptable impact on groundwater. Furthermore, with any exploration and mining activities, thought needs to be given to environmental implications. Iron ore producers must comply with environmental regulations and have waste management policies. For example, Green (2009) documents the waste management policy at Rio Tinto with thorough planning before exploration and mining to minimise the environmental impact as much as possible.

5 Summary and Conclusions

Iron ore and its ultimate product, steel, are arguably the most historically social and economically significant commodities of the last thirty centuries. Because of the increased demand for higher grade, production moved from early, low-grade bog ore and marine Phanerozoic ooidal ironstones to the giant BIF-hosted Martite-Goethite deposits, the large terrestrial Phanerozoic ooidal ironstones or CID, and the minor DID.

Although representing 60% of the Fe ore production in Australia, the geochemical composition and genesis of supergene M-G ores remain poorly understood. The low tonnage DID (3 Bt) are still a good source of lump in Australia and represents 17% of the Sishen iron ore deposit in South Africa.

CID are iron-rich detritals comprising coarse sand to fine gravel-size granules that include pelletoids (ooids and pisoids), peloids, ferruginised wood (goethitic in the matrix and hematitic in the nucleus), cemented by a porous matrix, and occur as paleochannels typically 1–5 km wide and 1–100 m thick. All Australian CID production is sourced from the Pilbara region and accounted for 30% of Australia's total Fe ore production in 2020, representing one of the two major mineable Fe ore types of the region. These almost uniquely Australian, Miocene-age ore deposits are hosted within the Robe Formation, the Marillana Formation and the Poondano Formation, along with minor deposits in the Yilgarn. Despite the considerable known CID ore reserves (18 Bt), future CID exploration and discovery will need to rely on increasingly more sophisticated exploration techniques, particularly for the discovery of buried deposits, as these become an increasingly important ore resource to the Australian mining industry.

Acknowledgements The authors thank Howard Baker, Rob Bowell and Charles Butt for their constructive reviews. We also express our gratitude to Angelo Vartesi for his help with the drafting.

References

Anand RR, Gilkes RJ (1987) The association of maghemite and corundum in darling range laterites, Western Australia. Aust J Soil Res 35:303–311

Anand R, Paine M (2002) Regolith geology of the Yilgarn Craton, Western Australia; implications for exploration. Aust J Earth Sci 49:3–162

Aquila (2020) Mineral resources statement. Available via the Aqulia website. https://www.aquilaresources.com.au/projects/west-pilbara-iron-ore/tables. Accessed 01 Sept 2021

Bekker A, Slack JF, Planavsky N et al (2010) Iron formation: the sedimentary product of a complex interplay among mantle, tectonic, oceanic, and biospheric processes. Econ Geol 105:467–508

Belevtsev YN, Siroshtan RI (1983) The Krivoy Rog basin. Elsevier, Amsterdam

Beukes NJ (1973) Precambrian iron–formations of Southern Africa. Econ Geol 68:960–1004

Beukes NJ (1983) Palaeoenvironmental setting of iron–formations in the depositional basin of the Transvaal Supergroup, South Africa. In: Trendall AF, Morris RC (eds) Iron–formation: facts and problems. 6. Elsevier, Amsterdam, pp 131–209

Beukes NJ, Klein C (1990) Geochemistry and sedimentology of a facies transition, from microbanded to granular iron-formation, in the early proterozoic Transvaal supergroup, South Africa. Precambrian Res 47:99–139

Beukes NJ, Gutzmer J (2008) Origin and paleoenvironmental significance of major iron formations at the Archean–Paleoproterozoic boundary. In: Hagemann S, Rosiere C, Gutzmer J, Beukes NJ (eds) Banded iron formation–related high–grade iron ore, vol 15. Society of Economic Geologists, Littleton, pp 5–47

Beukes NJ, Gutzmer J, Mukhopadhyay J (2003) The geology and genesis of high–grade hematite iron ore deposits. Transactions of the Institution of Mining and Metallurgy. Sect B Appl Earth Sci 11:18–25

Bodycoat FM (2010) Stratigraphic and structural setting of iron mineralisation at E Deposit (East), Area C, Hamersley Province, Western Australia. Appl Earth Sci 119(1):49–55

Brent, A (2021) Best in Class: Australia's Bulk Commodity Giants. Australian Iron Ore: when Quality meets opportunity. Available via Minerals Council of Australia. https://www.minerals.org.au/news/australian-iron-ore-where-quality-meets-opportunity. Accessed 01 Sept 2021

Butler RJT (1976) Geology of the Tertiary ironstones in the middle and upper Robe River area, Pilbara Region, Western Australia. MSc thesis, University of Western Australia, Perth (unpubl.)

Butt CRM (1979) Geochemistry of a pseudo-gossan developed on a black shale dolerite contact, Killara, Western Australia. J Geochim Explor 11:131–156

Butt A, Hawke P, Flis M (2001) Of detritals, derivatives and determination ? An example of detrital iron discovery. Explor Geophys 32:185–188

Chicarino Varajão CA, Ramanaidou ER, Colin F, Nahon D (1997) Gênese dos corpos de hematita compacta ("Hard hematite ores"): Metasomatismo, sedimentação ou alteração supergenica? REM: Revista Escola Minas, Ouro Preto 50:40–43

Cloud P (1973) Paleoecological significance of the banded iron–formation. Econ Geol 68:1135–1143

Cloud P (1983) Banded iron-formation—a Gradualist's dilemma. In: Trendall AF, Morris RC (eds) Iron-formation: facts and problems, vol 6. Elsevier, Amsterdam, pp 401–416
Clout J (2006) Iron formation-hosted iron ores in the Hamersley Province of Western Australia. Appl Earth Sci 115:115–125
Clout JMF, Simonson BM (2005). Precambrian iron formations and iron formation-hosted iron ore deposits. Econ Geol 100th Anniversary Volume, 643–679
Clout JMF, Counsell C, Simpson C (2017) Chichester and Solomon bedded and channel iron deposits. In: Phillips GN (ed) Australian ore deposits, AusIMM monograph 32, 6th edn. The Australasian Institute of Mining and Metallurgy, Carlton, pp 351–358
Dalstra HJ (2006) Structural controls of bedded iron ore in the Hamersley Province, Western Australia–an example from the Paraburdoo Ranges. Appl Earth Sci 115(4):139–145
Dalstra HJ (2011) From banded iron formation to iron ore—Genetic models and their application in iron ore exploration in the Hamersley Province, western Australia. In: Iron ore 2011. The Australasian Institute of Mining and Metallurgy, Melbourne, pp 73–82
Dalstra HJ (2014) Structural evolution of the Mount Wall region in the Hamersley province, Western Australia and its control on hydrothermal alteration and formation of high-grade iron deposits. J Struct Geol 67:268–292
Dalstra HJ, Guedes SC (2004) Giant hydrothermal deposits with Mg–Fe metasomatism: a comparaison of the Carajas, Hamersley, and other ores. Econ Geol 99:1793–1800
Dalstra H, Gill T, Faragher A, Scott B, Kakebeeke V (2010) Channel iron deposits, a major new district around the Caliwingina Creek, central Hamersley Ranges, Western Australia. Appl Earth Sci 119:12–20
Danišík M, Ramanaidou ER, Evans NJ et al (2011) (U–Th)/He geochronology of channel iron deposits, Robe River, Hamersley Province, Australia—implications for ore genesis. In: Iron ore 2011. The Australasian Institute of Mining and Metallurgy, Melbourne, pp 83–86
Davy R (1975) A geochemical study of a dolomite-BIF transition in the lower part of the Hamersley Group. West Aust Geol Surv Annu Rep 1974:88–100
Dorr JVN (1969) Physiographic, stratigraphic and structural development of the Quadrilâtero Ferrîfero, Minas Gerais, Brazil, U. S. Geological Survey professional paper, 641–A. U.S. G.P.O., Washington, DC
Dorr JVN (1973) Iron formations of South America. Econ Geol 68:1005–1022
Duclaux G, Salles-Taing T, Ramanaidou E (2013) Alluvial iron deposits exploration using surface processes modelling: a case study in the Hamersley Province (Western Australia). In: Iron ore 2013, Perth, Western Australia, 12–14 August 2013, pp 75–80
Ewers WE, Morris RC (1981) Studies of the Dales Gorge Member of the Brockman iron formation, Western Australia. Econ Geol 76:1929–1953
Figueiredo e Silva RC, Lobato LM, Rosiere CA et al (2008) A hydrothermal origin for the jaspilite-hosted, giant Serra Norte iron ore deposits in the Carajas Mineral Province, Para state, Brazil. In: Hagemann S, Rosiere C, Gutzmer J, Beukes NJ (eds) Banded iron formation-related high-grade iron ore, vol. 15. Society of Economic Geologists, Littleton, pp 255–290
Freyssinet P, Butt CRM, Morris RC, Piantone P (2005) Ore-forming processes related to lateritic weathering. Econ Geol 100:681–722
Fritz TO, Yapp CJ (2018) Miocene weathering environments in Western Australia—Inferences from the abundance and 13C/12C of Fe(CO3)OH in CID goethite. Geochim Cosmochim Acta 227: 186–210
Gagen EJ, Levett A, Paz A, Gastauer M, Caldeira CF, Valadares RBDS, Bitencourt JAP, Alves R, Oliveira G, Siqueira JO, Vasconcelos PM, Southam G (2019) Biogeochemical processes in canga ecosystems: armoring of iron ore against erosion and importance in iron duricrust restoration in Brazil. Ore Geol Rev 107:573–586
Golubovskaya EV (2001) Facies and geochemical features of the iron ore complex of the Kerch Peninsula. Lithol Miner Resour 36:224–235
Golubovskaya EV (2003) Some geochemical features of iron ores from the Lisakov deposit. Lithol Miner Resour 38:275–281
Green R (2009) Holistic management of sulfides at Rio Tinto Iron Ore's Pilbara mine sites. Min Technol 118(3–4):232–237. https://doi.org/10.1179/174328610X12682160006682
Haest M, Cudahy T, Laukamp C et al (2011) Characterisation of bedded and channel iron ore deposit using CSIRO's HyloggingTM Systems. In: Iron Ore 2011. The Australasian Institute of Mining and Metallurgy, Melbourne, pp 249–255
Haest M, Cudahy T, Laukamp C, Gregory S (2012) Quantitative mineralogy from infrared spectroscopic data; II, three-dimensional mineralogical characterization of the Rocklea Channel iron deposit, Western Australia. Econ Geol 107:209–228
Harms JE, Morgan BD (1964) Pisolitic limonite deposits in Northwest Australia. Proc Australas Inst Min Metall 212:91–124
Harmsworth RA, Kneeshaw M, Morris RC, Robinson CJ, Shrivastava PK (1990) BIF-derived iron ores of the Hamersley Province. In: Hughes FE (ed) Geology of the mineral deposits of Australia and Papua New Guinea: monograph 14. The Australasian Institute of Mining and Metallurgy, Parkville, pp 617–642
Heim JA, Vasconcelos PM, Shuster DL, Farley KA, Broadbent GC (2006) Dating paleochannel iron ore by (U–Th)/He analysis of supergene goethite, Hamersley Province, Australia. Geology 34:173–176
Ilyin AV (2009) Neoproterozoic banded iron formations. Lithol Miner Resour 44:78–86
Inverarity N, Dogramaci S, Rea I (2012) Combined numerical and analytical modelling as a means to assess closure options for open-pit mine, Pilbara

region. In: Proceedings of the International Mine Water Association (IMWA) Congress, Bunbury, Australia, pp 206A–206G

Kepert D, Clarke N, Simpson C, Edwards D (2010) Discovery of the Solomon iron deposits, Hamersley Province, Western Australia. Appl Earth Sci 119:21–27. https://doi.org/10.1179/037174510X12853354810345

Klein C (2004) Geochemistry and mineralogy of Neoproterozoic banded iron formation and their associated Fe and Mn deposit, with reference to the Griqualand West deposit, South Africa. J Afr Earth Sci 30:1233–1244

Klein C, Gole MJ (1981) Mineralogy and petrology of parts of the Marra Mamba iron formation, Hamersley Basin, Western Australia. Am Min 66:507–525

Klein C, Beukes NJ (1992) Proterozoic iron-formations. In: Condie KC (ed) Proterozoic crustal evolution, vol. 10. Elsevier, Amsterdam, pp 383–418

Klein C, Ladeira EA (2002) Petrography and geochemistry of the least altered banded iron-formation of the Archean Carajás Formation, northern Brazil. Econ Geol 97:643–651

Kneeshaw M, Morris RC (2014) The Cenozoic detrital iron deposits of the Hamersley Province, Western Australia. Aust J Earth Sci Int Geosci 61:513–586

Knight J, Perring C, Stephens D, Crowe M (2018) Discovery, geologic setting, and controls on iron mineralization, South Flank, Western Australia, Chapter 14. In: Metals, minerals, and society, SEG special publications, no 21. Society of Economic Geologists, Inc., Littleton, pp 321–346

Kogbe CA (1978) Origin and composition of the ferruginous oolites and laterites of North-Western Nigeria. Geol Rundsch 67:662–674

Lascelles D (2002) A new look at old rocks—and alternative model for the origin of in situ iron ore deposits derived from banded iron–formation. In: Iron ore 2002. Australasian Institute of Mining and Metallurgy, Melbourne, pp 107–126

Lascelles D (2007) A possible origin of pisolitic iron ore deposits from raindrops and sheetwash in a semi–arid environment. In: Iron ore 2007. The Australasian Institute of Mining and Metallurgy, Melbourne, pp 113–121

MacLeod WN (1966) The geology and iron deposits of the Hamersley Range Area, Western Australia, Geological Survey of Western Australia Bulletin, 117. B. Davies, Govt. Printer, Perth

MacPhail MK, Stone MS (2004) Age and palaeoenvironmental constraints on the genesis of the Yandi channel iron deposits, Marillana Formation, Pilbara, northwestern Australia. Aust J Earth Sci 51: 497–520

McClung CR (2006) Basin analysis of the mesoproterozoic bushmanland group of the Namaqua Metamorphic Province, South Africa. PhD thesis. University of Johannesburg, South Africa

McGregor F, Ramanaidou ER, Wells MA (2010) Phanerozoic ooidal ironstone deposits—generation of potential exploration targets. Appl Earth Sci 119:60–64

Melfi AJ, Trecases JJ, Carvalho A et al (1988) The lateritic ore deposits of Brazil. Sciences Geologiques, Bulletin 41:5–36

Mohapatra KB, Jena S, Mahanta K, Mishra P (2008) Goethite morphology and composition in banded iron formation, Orissa, India. Resour Geol 58:325–332

Morris RC (1980) A textural and mineralogical study of the relationship of iron ore to banded iron–formation in the Hamersley iron province of Western Australia. Econ Geol 75:184–209

Morris RC (1983) Supergene alteration of banded iron–formation. In: Trendall AF, Morris RC (eds) Iron–formation: facts and problems. 6. Elsevier, Amsterdam, pp 513–534

Morris RC (1985) Genesis of iron ore in banded iron–formation by supergene and supergene–metamorphic processes; a conceptual model. In: Wolf KH (ed) Handbook of strata–bound and stratiform ore deposits: Part II. Regional studies and specific deposits, vol 13. Elsevier, Amsterdam, pp 73–235

Morris RC (1993) Genetic modelling for banded iron–formation of the Hamersley Group, Pilbara Craton, Western Australia. Precambrian Res 60:243–286

Morris RC (1994) Detrital iron deposits of the Hamersley Province. CSIRO exploration geoscience restricted report, 76R

Morris RC, Kneeshaw M (2011) Genesis modelling for the Hamersley BIF–hosted iron ores of Western Australia—a critical review. Aust J Earth Sci 58:417–451

Morris RC, Ramanaidou ER (2007) Genesis of the channel iron deposits (CID) of the Pilbara region, Western Australia. Aust J Earth Sci 54:733–756

Morris RC, Trendall AF (1988) A model for the deposition of the microbanded Precambrian iron formations—discussion. Am J Sci 288:664–669

Mukhopadhyay J, Gutzmer J, Beukes J, Bhattacharya HN (2008) Geology and genesis of the major banded iron formation—Hosted high–grade iron ore deposits of India. In: Hagemann S, Rosiere C, Gutzmer J, Beukes NJ (eds) Banded iron formation–related high–grade iron ore. 15. Society of Economic Geologists, Littleton, pp 291–316

Perring CS (2021) Petrography of martite–goethite ore and implications for ore genesis, South Flank, Hamersley Province, Western Australia. Aust J Earth Sci 68:782–798

Perring CS, Crowe M, Hronsky J (2018) A new fluid–flow model for the genesis of banded iron formation–hosted martite–goethite mineralization, with special reference to the North and South Flank deposits of the Hamersley Province, Western Australia. Econ Geol 115:627–659 https://doi.org/10.5382/econgeo.4734

Piechocka A, Jessell M, Martin DMcB, Lindsay M, Ramanaidou E (2021). Preliminary 3D geological model and structural analysis of the Neoarchean to

Paleoproterozoic Hamersley Basin, Pilbara, Western Australia. In: Iron ore conference 2021

Pires FRM (1995) Textural and mineralogical variations during metamorphism of the Proterozoic Itabira iron formation in the Quadrilatero Ferrifero, Minas Gerais, Brazil. Anais da Academia Brasileira de Ciencias 67: 77–105

Poulet T, Giraldo JF, Ramanaidou E, Piechocka A, Calo VM (2022) Paleo stratigraphic permeability anisotropy controls supergene mimetic martite goethite deposits. Basin Res 35(2):572–591. https://doi.org/10.1111/bre.12723

Powell CM, Oliver NH, Li ZX, Martin DM, Ronaszeki J (1999) Synorogenic hydrothermal origin for giant Hamersley iron oxide ore bodies. Geology 27(2): 175–178

Pownceby MI, Hapugoda S, Manuel J, Webster NA, MacRae CM (2019) Characterisation of phosphorus and other impurities in goethite-rich iron ores–Possible P incorporation mechanisms. Miner Eng 143:106022

Ramanaidou ER (1989) Genese d'un gisement lateritique. Evolution supergene des itabirites Proterozoiques de la mine de fer de Capanema. PhD, Poitiers, France

Ramanaidou ER (2009) Genesis of lateritic iron ore from banded iron-formation in the Capanema mine (Minas Gerais, Brazil). Aust J Earth Sci 56:605–620

Ramanaidou ER, Morris RC (2010) Comparison of supergene mimetic and supergene lateritic iron ore deposits. Appl Earth Sci 119:56–59

Ramanaidou ER, Wells MA (2014) 13.13—Sedimentary hosted iron ores. In: Holland HD, Turekian KK (eds) Treatise on geochemistry, 2nd edn. Elsevier, Oxford, pp 313–355

Ramanaidou ER, Horwitz RC, Morris RC (1991) Channel iron deposits. CSIRO exploration geoscience restricted report, 62R

Ramanaidou E, Nahon D, Decarreau A, Melfi AJ (1996) Hematite and goethite from duricrusts developed by lateritic chemical weathering of precambrian banded iron formations, Minas Gerais, Brazil. Clay Clay Miner 44:22–31

Ramanaidou ER, Morris RC, Horwitz RC (2003) Channel iron deposits of the Hamersley Province, Western Australia. Aust J Earth Sci 50:669–690

Ramanaidou ER, Wells M, Belton D, Verral M, Ryan C (2008) Mineralogical and microchemical methods for the characterization of high-grade BIF derived iron ore. Rev Econ Geol 15:129–156

Ramanaidou ER, Wells MA, Morris R, Duclaux G, Evans N, Danišík M (2017) Channel iron deposits of the Pilbara. In: Phillips GN (ed) Australian ore deposits, 864 p. The Australasian Institute of Mining and Metallurgy: Melbourne, pp 375–380

Ramanaidou ER, Danišik M, Wells MA (2019) Preliminary dating of martite goethite iron ore in the Hamersley Province (Western Australia), ASEG Extended Abstracts, 2019:1, 1–2

Rasmussen B, Fletcher I, Bekker A, Muhling J, Gregory CJ, Thorne A (2012) Deposition of 1.88-billion-year-old iron formations as a consequence of rapid crustal growth. Nature 484:498–501. https://doi.org/10.1038/nature11021

Rasmussen B, Krapež B, Muhling JR, Suvorova A (2015) Precipitation of iron silicate nanoparticles in early Precambrian oceans marks Earth's first iron age. Geology 43(4):303–306. https://doi.org/10.1130/G36309.1

Rosiere CA, Spier CA, Rios FJ, Suckau EV (2008) The itabirites of the Quadrilatero Ferrifero and related high-grade iron ore deposits: an overview. Econ Geol 15:223–254

Roy Hill (2009) Roy Hill 1 iron ore mining project stage 2. Available via the Environmnetal Protection Authority. https://www.epa.wa.gov.au/sites/default/files/EPA_Report/3084_Rep1345RoyHillStage2ARIReportFINAL.pdf. Accessed 01 Sept 2021

Roy S, Venkatesh AS (2009) Banded iron formation to blue dust: mineralogical and geochemical constraints from the Precambrian Jilling–Langalata Deposits, Eastern Indian Craton. Appl Earth Sci (Transactions of the Institution of Mining and Metallurgy, Section B) 118:178–188

Schwann P (2009) An alternative genesis for pisolites and channel iron deposits in Western Australia. In: Iron ore 2009. The Australasian Institute of Mining and Metallurgy, Melbourne, pp 171–185

Simonson BM (1987) Early silica cementation and subsequent diagenesis in arenites from four early Proterozoic iron formations of North America. J Sediment Petrol 57:494–511

Simonson BM (2003) Origin and evolution of large Precambrian iron formations. Geol Soc Am Spec Pap 370: 231–244

Simpson C, Storey D, Kepert D, Boyd R, Nitschke N (2011) Solomon iron deposits—continued growth and development. In: Iron ore 2011. The Australasian Institute of Mining and Metallurgy, Melbourne, pp 649–660

Spier CA, Oliveira SMB, Rosiere CA (2003) Geology and geochemistry of the Aguas Claras and Pico iron mines, Quadrilâtero Ferrîfero, Minas Gerais, Brazil. Mineral Deposita 38:751–774

Stone MS, George AD, Kneeshaw M, Barley ME (2002) Stratigraphy and sedimentology features of the Tertiary Yandi channel iron deposits, Hamersley Province, Western Australia. In: Proceedings iron ore 2002. Australasian Institute of Mining and Metallurgy Publication Series 7, Melbourne, pp 137–144

Storkey A, Doecke A, Whaanga A (2010) Stratigraphy of the western channel iron deposits of the Marillana Creek operations, Western Australia. Appl Earth Sci 119:2–11

Taylor KG (1992) Non-marine oolitic ironstones in the lower cretaceous wealdon sediments of southeast England. Geol Mag 129:349–358

Taylor D, Dalstra HJ, Harding AE, Broadbent GC, Barley ME (2001) Genesis of high-grade hematite orebodies of the Hamersley Province, Western Australia. Econ Geol 96:837–873

Teitler Y, Duuring P, Hagemann SG (2017) Genesis history of iron ore from Mesoarchean BIF at the Wodgina mine, Western Australia. Aust J Earth Sci 64:41–62

Trendall AF (1965) Origin of Precambrian iron formations. Econ Geol 60:1065–1070

Trendall AF (1980) A progress review of the Hamersley basin of Western Australia. Geol Surv Finland Bull 307:113–131

Trendall AF (1983) The Hamersley basin. In: Trendall AF, Morris RC (eds) Iron–formation: facts and problems, vol 6. Elsevier, Amsterdam, pp 69–129

Trendall AF, Blockley JG (1970) The iron formations of the Precambrian Hamersley group, Western Australia; with special reference to the crocidolite. Bull Geol Sur West Aust 119:366

Trendall AF, Pepper RS (1977) Chemical composition of the brockman iron formation. Geol Sur West Aust 25

Tsipurskii SI, Golubovskaya EV (1989) Smectites of tobacco ores in Kerch deposits. Lithology and Minerals Resources 24:149–161

von Perger AD (1991) Exploration for concealed gold deposits, Yilgarn Block, Western Australia. CSIRO Division of Exploration Geoscience Restricted Report 238R

Wells MA, Ramanaidou ER (2011) Occurrence and mineralogical association of phosphorus in Australian bedded iron ore deposits. In: Iron ore 2011. The Australasian Institute of Mining and Metallurgy, Melbourne, pp 331–336

Woodward HP (1891) Annual general report of the government geologist for 1890 (Government Printer: Perth)

Yanitzkii HL (1960) Oligocene oolitic iron ores of northern Turgai and their genesis, vol 37. Academy of Sciences, Moscow

Yapp CJ, Shuster DL (2017) D/H of late Miocene meteoric waters in Western Australia: Paleoenvironmental conditions inferred from the δD of (U–Th)/He–dated CID goethite. Geochim Cosmochim Acta 213:110–136

Zitzmann A (1977) The iron ore deposits of the western U.S.S.R. In: Walther HW, Zitzmann A (eds) The iron ore deposits of Europe and adjacent areas, vol. 1. Federal Institute for Geosciences and Natural Resources, Hannover, pp 325–391

Supergene Origin of Kaolin Deposits

Ömer Işık Ece and Hatice Ünal Ercan

Abstract

The kaolin group minerals kaolinite, halloysite, dickite and nacrite have the same chemical composition, $Si_2Al_2O_5(OH)_4$, but different crystal structures. Supergene alteration is responsible for the formation of many economically operating kaolin deposits worldwide. Most supergene kaolin is formed by the incongruent dissolution of alkali feldspar in the presence of slightly acidic solutions; a combination of geochemical (chemical leaching) and biochemical (microbial nitrification and acidification) processes work together in almost all situations. The most important deposits have formed by complex multi-stage processes including (i) in situ weathering of granitic rocks under temperate to tropical climate conditions that produce deep soil profiles and lateritic regoliths containing residual kaolin, (ii) transport and deposition of sedimentary rocks, with synsedimentary and early- to late-diagenetic alteration of kaolin, and (iii) mixed-type kaolinization, including redeposition of kaolin and/or geologically recent weathering under temperate to tropical climatic conditions. Type (iii) deposits are the richest in kaolin group minerals. The largest and highest quality kaolin deposits known in the world have formed by supergene alteration of sedimentary rocks deposited on granite.

Dedicated to my spouse Kamuran Ece.

Ö. I. Ece (✉)
Department of Geological Engineering, Istanbul Technical University, Maslak, Istanbul, Turkey

Department of Geology, University of Georgia, Athens, GA, USA
e-mail: ece@itu.edu.tr

H. Ü. Ercan
Department of Chemical and Chemical Processing Technologies, Konya Technical University, Konya, Turkey
e-mail: hunalercan@ktun.edu.tr

1 Introduction

The term kaolin originated from Kauling (or Gaoling), a clay mine in the Jiangxi province of China (Schroeder and Erickson 2014). This mine was first developed in the eleventh century and remained operational until 1964 before its total depletion (Keller et al. 1980). Kaolin is a sub-group of the kaolinite-serpentine group of phyllosilicates and is characterized by one tetrahedral and one octahedral (1:1) sheet, a dioctahedral structure, and chemical composition of $Si_2Al_2O_5(OH)_4$. Due to the abundance of these elements, kaolin minerals are commonly found in soils, especially those in humid tropical regions. The influence of the continuous removal of K^+, Na^+, Ca^{2+}, Mg^{2+}, and, in part, SiO_2 by leaching in the presence of organic acids enhances the

R. J. Bowell, C. R. M. Butt (eds.), *Geology, Geochemistry and Formation of Supergene Mineral Deposits in Deeply Weathered Terrain*, Mineral Resource Reviews, https://doi.org/10.1007/978-3-031-75733-4_13

formation of kaolin group minerals in these climate belts. Kaolin minerals are commonly soft and earthy, usually white to cream, but their colour may change depending on the content of Fe or other impurities (Hinckley 1965). Kaolins can have mineralogical, particle size and morphological differences depending on their formation environment and the parent rocks from which they are derived. These properties control their industrial uses, such as ceramics, porcelain, paper coating, medicine, nanocomposites and plastics.

Kaolin minerals are formed by pH-controlled reactions under organic or inorganic acidic conditions by (1) weathering, as in situ or residual products, mostly of felsic (i.e., granitic, siliceous volcanic and pegmatitic) rocks, and (2) hydrothermal alteration of volcanic rocks and tuffs. *Supergene* kaolin deposits are those formed by weathering and include many sedimentary deposits derived by further alteration of kaolinized materials eroded from weathered source areas and deposited in continental or coastal environments. *Hypogene* kaolin deposits are formed by hydrothermal alteration. Supergene kaolin mineralization is commonly homogenous and characterized by oxidized species, whereas hypogene kaolin typically has more varied mineralogy, including reducing species (pyrite, chalcopyrite). However, it is not possible to distinguish all kaolin bodies as either supergene or hypogene as many have a mixed origin due to the supergene alteration (weathering) of hydrothermal deposits (Murray 1988; Murray and Keller 1993; Dill 2016).

Kaolin varieties include soft and hard types and flint clay, distinguished by Al_2O_3 and Fe_2O_3 concentrations, crystallinity (Hinckley 1965), and particle size and morphology. *Soft kaolin* has a higher Al_2O_3 content and crystallinity index, book-shaped to vermiform morphology, and larger particle sizes, whereas *hard kaolin* has a higher Fe_2O_3 content, a lower crystallinity index, parallel-aligned flakes and a much finer particle size, and hence are less permeable. *Flint clay* is a hard sedimentary kaolinite, commonly interbedded with coal seams; it has a very fine particle size, with high purity and high Al_2O_3 content, and is highly refractory and resists slaking in water (Jackson 1997).

2 Mineralogy, Chemistry and Thermodynamics of Kaolinite

Kaolin group minerals include; kaolinite, polytypes dickite and nacrite, and a polymorph halloysite. Kaolinite ($Al_2Si_2O_5(OH)_4$) has a triclinic crystal structure and consists of a tetrahedral (T–SiO_4) sheet and an octahedral (O–$AlO_3 + 3OH$) sheet. All kaolin group minerals consist of single-layer structures, but X-ray diffraction studies (Table 1) show that the stacking sequence of dickite and nacrite layers are different (Dixon 1989; Moore and Reynolds 1997). In the octahedral sheet, each Al ion is surrounded by six oxygen or hydroxyl (inner hydroxyl) ions arranged at the corners of an octahedron. The two sheets in each layer are strongly bonded via shared oxygen ions, whereas layers have hydrogen bonding between oxygen on the outer face of the *T* sheet of one layer and hydroxyl on the outer face of the *O* sheet of the adjacent layer (Nesse 2000).

Supergene kaolinite is a product of the weathering of heterogeneously distributed feldspars (e.g., plagioclase and orthoclase) in igneous and feldspathic sedimentary rocks (Ekosse 2000). Kaolinization of more mafic rocks mainly occurs in humid environments such as rainforests due to the abundance of organic acids. Kaolinite crystallization occurs as unstable feldspars dissolve and pore fluids become enriched in Al^{3+}-complexes and $HSiO_4^-$ after consuming H^+ (Eq. 1), as follows:

$$\begin{aligned}&2(\mathrm{K,Na})\mathrm{AlSi_3O_8} + 9\mathrm{H_2O} + 2\mathrm{H^+}\\&\longrightarrow \mathrm{Al_2Si_2O_5(OH)_4} + 4\mathrm{H_4SiO_4}\\&+2(\mathrm{K,Na})^+ \text{ Feldspar Kaolinite}\end{aligned} \quad (1)$$

The removal of K^+, Na^+, and dissolved silica from feldspars continues until the fluids become supersaturated in kaolinite. The formation of vermiform books of pseudohexagonal kaolinite results either from in situ formations of kaolinite or its alteration from the remaining feldspar. The morphology of kaolinite crystals and the fabric of kaolinitic clays have a genetic link to the environment in which they form by mechanisms controlled by the parent rocks, pH and ion

Table 1 Crystallographic data for kaolin minerals

	System	Space group	Cell parameters		References
Kaolinite	Triclinic	C1	$a = 5.154$ Å	$\alpha = 91.69°$	Neder et al. (1999)
			$b = 8.942$ Å	$\beta = 104.61°$	
			$c = 7.401$ Å	$\gamma = 89.82°$	
Dickite	Monoclinic	Cc	$a = 5.138$ Å	$\beta = 96.74°$	Joswig and Drits (1986)
			$b = 8.918$ Å		
			$c = 14.389$ Å		
Nacrite	Monoclinic	Cc	$a = 8.906$ Å	$\beta = 113.58°$	Blount et al. (1969)
			$b = 5.146$ Å		
			$c = 15.664$ Å		
Halloysite	Monoclinic	Cc	$a = 5.14$ Å	$\beta = 99.7°$	Kohyama et al. (1978)
Hydrated $d_{(001)} = 10$ Å			$b = 8.90$ Å		
			$c = 20.70$ Å		
Dehydrated $d_{(001)} = 7$ Å			$a = 5.14$ Å	$\beta = 101.9°$	
			$b = 8.90$ Å		
			$c = 14.90$ Å		

concentrations in the circulating fluids (Keller 1981).

Dickite and nacrite are rarely encountered in supergene kaolin deposits. Dickite, as a high-temperature mineral, is less common than kaolinite. It occurs mainly in pyroclastic units associated with pyrophyllite. Nacrite crystallizes as euhedral, pseudohexagonal or elongated crystals, and is generally found filling very thin, irregular veins associated with quartz and dolomite (Ruiz Cruz 1996). Nacrite mostly forms under high-stress conditions along the fault and shear zones at >50 °C. Unlike these two kaolin polymorphs, halloysite occurs as hydrated and dehydrated types in saprolite as a weathering product in temperate and tropical environments, and as a hydrothermal mineral where drainage systems discharge through underlying carbonate rocks (Murray 1988; Ece et al. 2008). Hydrated halloysite typically has a tubular structure and accommodates interlayer water (Joussein et al. 2005; Ece and Schroeder 2007).

Chemical data for kaolinite group minerals from various countries and deposit types are given in Tables 2 and 3. All the listed samples have lower molar SiO_2/Al_2O_3 ratios than the ideal ratio. There are almost no chemical differences between soft and hard kaolins of supergene origin, with only the Fe content slightly higher in hard kaolins. The particle size distribution shows that the <2 μm size fractions are about 65% or less in soft kaolin deposited in freshwater environments and about 90% in hard kaolin deposited in semi-marine environments (Hinckley 1965).

Thermodynamic modeling of halloysite, gibbsite, kaolinite and boehmite in the Al_2O_3–SiO_2–H_2O system can explain the simultaneous genesis and coexistence of both hydrated and non-hydrated mineral species as functions of the activities of water and dissolved silica (Trolard et al. 1990) (Fig. 1). When the activity of water decreases, the activity of dissolved silica at equilibrium also decreases and kaolinite becomes more stable than gibbsite; when the activity of water increases, it favours the stability of boehmite. Thus, when water activity is >0.724 or lower than <0.758, kaolinite equilibrates with gibbsite, but when water activity is <0.724, kaolinite equilibrates with boehmite (Trolard et al. 1990). When the silica activity increases, kaolinite is more stable than boehmite and gibbsite. The accepted view that halloysite gives way to kaolinite during long-term weathering is supported thermodynamically as halloysite is more soluble than kaolinite and has a less negative Gibbs free energy (Kittrick 1969). Halloysite is stable

Table 2 Chemical compositions of selected major kaolinite and halloysite ore deposits. (Ece et al. 2008; Pruett 2016; Hernández et al. 2019)

Country	New Zealand	China	Turkey	Turkey	Venezuela	Venezuela	Brazil	Brazil
Quarry name	Matauri Bay	Longyan	Kizildam	Kizildam	Los Pijiguaos	Los Pijiguaos	Capim River soft kaolin	Capim River hard kaolin
Mineral	Halloysite	Halloysite	Halloysite	Halloysite	Kaolinite	Kaolinite	Kaolinite	Kaolinite
SiO_2	50.40	48.00	39.26	42.43	42.99	50.90	44.6	45.0
Al_2O_3	35.50	38.00	34.41	34.32	39.81	37.78	38.7	38.2
Fe_2O_3	0.25	0.29	0.09	0.14	0.70	0.37	0.78	2.67
MgO	tr	0.30	0.06	0.03	0.01	0.09	0.04	0.03
Na_2O	tr	0.10	0.05	0.04	0.02	0.02	0.06	0.06
K_2O	tr	1.71	0.02	0.02	0.04	1.06	0.03	0.02
CaO	tr	0.16	0.06	0.03	0.03	0.01	0.02	0.02
TiO_2	0.05	0.02	0.02	0.03	1.63	0.76	0.57	1.95
MnO	0.01	tr	0.02	0.01	0.00	0.01	–	–
P_2O_5	0.06	–	0.52	0.12	0.02	0.02	0.09	0.21
SO_3	0.06	–	–	–	–	–	0.03	0.05
LOI	13.80	12.40	25.5	22.60			14.1	14.1
SiO_2/Al_2O_3	2.41	1.71	1.94	2.10	1.84	2.29	1.96	1.999

Table 3 Chemical compositions of dickite, nacrite, and kaolinite. (Nagasawa 1978; Wilson 2004)

Country	Japan	Japan	Japan	Japan	China	China	USA	USA
Quarry name	Hokugo-mura	Skokozan	Yonago	Kanagato	Datong-Shanxi	Yangquan-Shanxi	Middle Georgia	East Georgia
Mineral	Dickite	Dickite	Nacrite	Nacrite	Kaolinite	Kaolinite	Kaolinite	Kaolinite
SiO_2	45.62	45.99	41.52	45.17	42.28	44.78	45.30	44.00
Al_2O_3	38.92	39.34	37.51	40.24	39.37	39.05	38.38	39.50
Fe_2O_3	1.03	0.01	2.51	0.02	0.33	0.45	0.30	1.13
MgO	–	–	0.92	0.30	0.15	0.44	0.25	0.03
Na_2O	–	–	–	0.06	0.36	0.10	0.27	0.08
K_2O	0.08	–	–	0.12	0.94	0.15	0.04	0.06
CaO	0.12	0.06	0.64	0.40	0.58	0.66	0.05	0.03
TiO_2	–	0.21	tr	0.04	0.09	0.05	1.44	2.43
LOI	13.98	14.45	15.47	14.23	15.22	15.31	13.97	13.90
SiO_2/Al_2O_3	1.99	1.98	1.88	1.90	1.82	1.94	2.0	1.89

where water activity is >0.758, whereas kaolinite is stable when water activity is <0.758. The more hydrated minerals, halloysite and gibbsite, occur preferentially around large pores, where water activity is high, whereas kaolinite and boehmite form in microporous environments where water activity is low (Trolard et al. 1990).

3 Deposit Characteristics of Supergene Kaolinite

Supergene kaolin deposits are formed when parent rocks are exposed at or near the Earth's surface (average T_{atm} ~ 15 °C; P_{atm} ~ 1 bar; pCO_2 ~ 0.047–0.06 bar and pO_2 ~ 0.2 bar) and

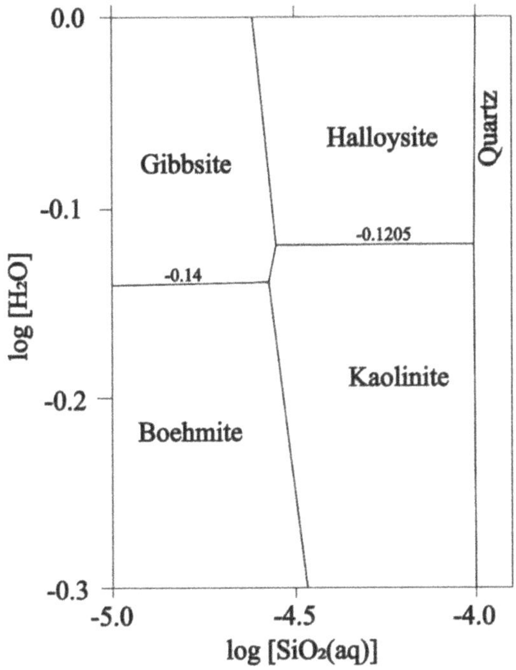

Fig. 1 Stability field distribution of halloysite, kaolinite, boehmite, and gibbsite as a function of water and dissolved silica activities. (Trolard et al. 1990)

eventually reach equilibrium under these new conditions. The partial pressure of CO_2 ($pCO_2 = 10^{-3.5} - 10^{-4.5}$) in soils is 10–100 times higher than in the atmosphere due to decomposition of organic matter and root respiration. The weathering processes that give rise to this secondary enrichment include oxidation, dissolution, chemical leaching, and erosion under the influence of surface and groundwaters. The enrichment of clays in the regolith is due to solid-state mass loss and differential mobility of elements relative to assumed conservative species such as Ti and Zr. The regolith is the permeable surface layer down to bedrock influenced by groundwater and has been described as a breathing, constantly evolving boundary layer where rock, soil, water, air, and living organisms interact with each other (Schroeder 2018).

3.1 Residual Kaolin Deposits, Products of In Situ Weathering of Crystalline Rocks

3.1.1 Characteristics and Environments of Formation

Deep soil profiles and lateritic regoliths, including lateritic bauxites, are the products of in situ rock weathering under humid tropical climates. The extreme rainfall and temperature conditions in tropical climatic belts affect the rate of decomposition and the development of soil-forming processes. A typical deep weathering profile is characterized by soils overlying discontinuous ferruginous units, mottled clays, saprolite and bedrock. This zonation displays the significance of shallow groundwater infiltrations that have had a major control on the weathering of the parent rocks and the formation of the regolith. Humid tropical climates and temperatures also enhance the growth of living organisms that contribute organic acids to the soil profile during weathering and through the addition of humus. Therefore, different climatic conditions result in different weathering processes that also vary depending on the bedrock type. Based on experimental studies, Al silicate minerals, mostly micas and feldspars, incongruently dissolve in solution to form clay minerals by the presence of humic and other organic acids (Huang and Keller 1970). In situ weathering of igneous rocks is more intense under tropical and sub-tropical rain forests due to high silica solubility caused by the combined chelating and acidic properties of organic acids. In organic systems, Si is more soluble than Al, whereas Al is more soluble than Si in inorganic systems at pH < 4.0.

Residual weathering of rock-forming minerals can be classified into two groups (Wilson and Jones 1983). (1) *Geochemical weathering* occurs under the influence of surface waters with atmospheric carbon dioxide (H_2CO_3) and only minor amounts of organic matter. It is characterized by the rapid crystallization of clay minerals, such as kaolinite and montmorillonite, and other accessory minerals, and is best developed in the soils and weathered rocks of humid tropical regions. (2) *Biogeochemical weathering* occurs under conditions strongly influenced by the presence of organic matter and living organisms. During biogeochemical weathering, crystallization is extremely slow, and the characteristic products are non-crystalline or poorly ordered materials and organo-mineral complexes. Biochemical weathering predominates in the soils of cool temperature climates. There are no calculated stability diagrams because mineral-water reactions are controlled by combinations of various organic acids.

The topography of the earth's surface influences soil profile development. Relatively flat land of low relief enhances the development of deep soils and regoliths. In addition, the retention of seasonal rainwater in forested watersheds carrying atmospheric CO_2 as H_2CO_3 results in topsoil development and the alteration of granitic rocks. Numerous residual medium-size kaolin deposits in Western Australia (e.g., Greenbushes, Gabbin, Jubuk, Ockley-Wickepin, Mount Gibson, Kerrigan); South Australia (e.g., Poochera, Calca, Mount Hope, Birdwood); Victoria (Hallum, Bulla and Lal Lal) and some of the Brazilian kaolins in the Amazon Region are the good examples of this type of weathering alteration.

3.1.2 Western Australia

The kaolin deposits in Western Australia are largely limited to the southwest terranes of the Archaean Yilgarn Craton and the Proterozoic Albany-Fraser Orogen. The largest recorded production of kaolin in this area is from the Greenbushes deposit, 250 km south of Perth, developed over a pegmatite dyke (Abeysinghe and Fetherston 1999). Additionally, the Wickepin kaolin deposit, about 220 km southeast of Perth, contains more than 100 million tons of extremely high-grade, easy-access kaolin. This represents one of the largest known remaining premium primary resources of kaolin in the world. The Wickepin kaolin deposit was developed on the weathered granite-gneiss basement of the Yilgarn terrane, locally intruded by younger (~ 2.8 Ga) dioritic and dolerite dykes. The dykes and granitic basement are overlain by a regolith, including transitional and saprolitic horizons, a mottled clay zone, and a colluvial lateritic horizon. The depth of weathering does not exceed 50 m, with the depth of the top of kaolin mineralization generally less than 12 m from the natural surface. The mineralization is a weathering product and is confined to the saprolite below the mottled zone. The thickness of the kaolin units varies from 3 to 16 m, whereas the average overburden varies from 4 to 10 m. The general assumption is that the formation of these kaolin deposits is associated with past deep weathering under humid tropical climates (e.g., see Ollier 1988; chapter "The Role of Regolith Formation and Evolution in the Development of Supergene Mineral Deposits"), The bleached (or "pallid") saprolite and plasmic horizons of these weathering profiles and nearby paleochannels generally contain thick sequences of variable grade clay. Kaolin has formed with weathering extending to a depth of 35–50 m below the surface (Abeysinghe and Fetherston 1999). Significant kaolin deposits of both residual and sedimentary origin are found elsewhere in this region. Sedimentary kaolin deposits occur in lacustrine palaeodrainage environments, such as the Mount Kokeby and Goomalling deposits.

3.1.3 New Zealand

Hydrated halloysites with book-like morphology occur in the Tauranga area, eastern North Island, New Zealand, in highly porous, poorly permeable, silt-dominated, and Si-rich pumiceous rhyolitic tephra deposits (Cunningham et al. 2016). The suggested alteration stages are solubilized volcanic glass–plagioclase → halloysite spheroids → halloysite tubes → halloysite plates → halloysite books. Halloysite is metastable because of the lower activation energy of nucleation than that required for kaolinite crystallization (Steefel and Van Cappellen 1990). Consequently, halloysite is expected to be the first mineral to precipitate from solution (Keeling 2015) if the rocks are constantly wet due to their low permeability in a humid climate. Such a situation arises where silicates interact with warm and/or corrosive (acidic) fluids or where less stable silicates, such as Ca-plagioclase, are exposed to weathering.

The Maungaparerua halloysite deposit in the North Island of New Zealand is about 210 km N of Auckland (Murray et al. 1977). The Early Pliocene (3.7 ± 0.04 Ma) rhyolite dome complex at Maungaparerua is bounded by a sequence of hydrothermally-altered basalt flows intercalated with non-marine siltstones and rhyolitic tuff units (Brathwaite et al. 2014). Several small quarries have been worked for halloysite-rich clay production in the rhyolite dome complex, which is overlain by younger, unaltered basalt flows. The rhyolitic rocks were initially hydrothermally altered and then subjected to intense surficial weathering under very humid climatic conditions to form a deposit of white halloysite. The deposit contains about 50% quartz and fine cristobalite and 50% a mixture of halloysite, kaolinite and allophane, with small amounts of feldspar in the coarse fractions. The upper part of the deposit contains relatively soft clay, 3 to 30 m thick, underlain by a hard-dense siliceous clay, up to 50 m thick, formed by surface weathering. This halloysite has high brightness due to low Fe and Ti contents and is a globally important resource. It has found a market niche as a component in quality porcelain manufacture with high whiteness and translucency. The isotopic $\delta^{18}O$ (21.1 to 22.4‰) and δD (−57.1 to −66.7‰) compositions of the halloysite indicate that the deposit is supergene rather than hydrothermal in origin. Brathwaite et al. (2014) concluded that the dominant process in the formation of halloysite was the deep weathering of sanidine rhyolite under water-saturated subtropical climatic conditions.

3.1.4 Brazil

Lateritic formations cover ~80% of the Amazon basin and have been developed under tropical humid climates over both the Precambrian basement and overlying sedimentary rocks. The lateritic cover was formed by intensive weathering that probably started in the early Mesozoic. They have a relatively monotonous and simple mineralogical composition but a complex structural organization within different pedological systems. The Amazon rain forests have a significant role in the development of abundant organic acids that intensify the lateritic weathering close to the ground surface (Murray 1988) and result in the development of kaolin deposits about 8 m thick. Kaolin deposits and extensive bauxite deposits coexist in Amazonia. However, the large and regional deposits of kaolin in the Amazon region were not formed by lateritization alone but are the result of a combination of supergene processes, importantly including alteration under subsequent reduced lacustrine and/or swamp environment (Fig. 2) (Costa and Moraes 1998).

Much of Brazil is underlain by a Precambrian basement of granites and migmatites associated with quartzites, gneisses, and calcareous rocks (Wilson et al. 2006). Brazilian kaolin deposits are classified as sedimentary with in situ weathering overprints. Source materials include granitic and pegmatitic rocks, anorthosites and volcanic rocks. Generally, kaolins derived from pegmatites are a mixture of kaolinite and halloysite with very low Fe and Ti contents; those from NE Brazil consist of very euhedral kaolinite. Kaolin from granites generally has a higher Fe content and invariably comprises mixtures of kaolinite and halloysite. Volcanic rocks produce fine, siliceous, generally halloysitic kaolin, whereas anorthosites commonly produce a mixture of kaolinite and halloysite (Wilson et al. 2006).

The three main kaolin deposits in the Paragominas—Capim kaolin bauxite district in Amazonia in N–NE Brazil have reserves of at least 1.0 billion tonnes of high-quality kaolin suitable for the paper industry. They are all Late Cretaceous–Early Paleogene sedimentary deposits, recrystallized by supergene processes, and most are similar to Georgia kaolins. The Cadam mining district contains hard kaolin of fine particle size in the Jari River area, south of the port of Belém on the Atlantic Coast. The Ipixuna (Imerys Rio Capim Caulim, IRRC) and Paragominas (Para Pigmentos SA, PPSA) Mines are N and S of the Rio Capim mining districts,

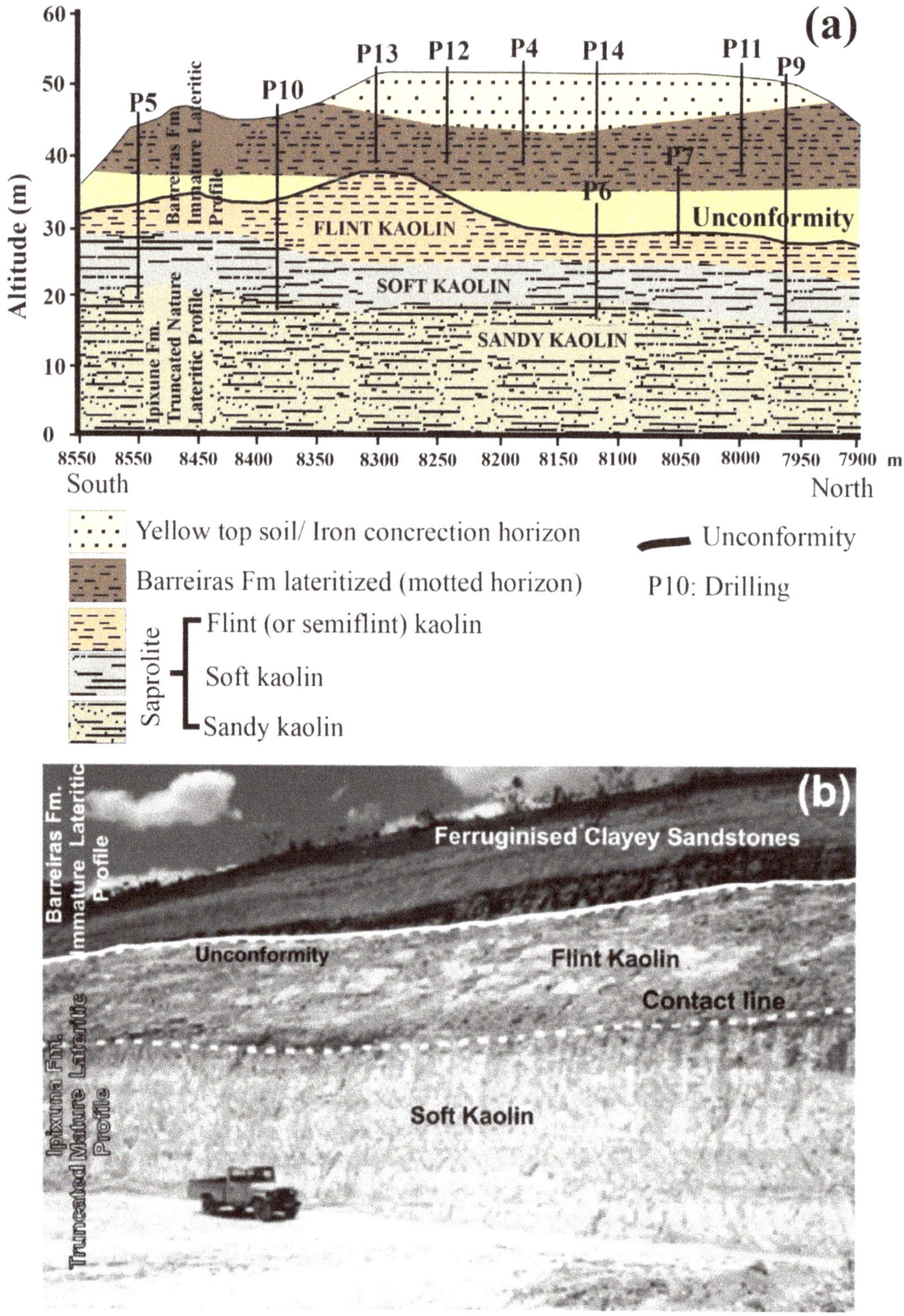

Fig. 2 (**a**) A geological cross-section in the area of Imerys-Rio Capim Caulim (IRCC) mine, Brazil, and (**b**) portion of the open pit IRCC kaolin mine showing the soft kaolin (ore), overlain by the flint kaolin and covered by the immature lateritized sediments of the Barreiras Formation. (After Costa et al. 2009)

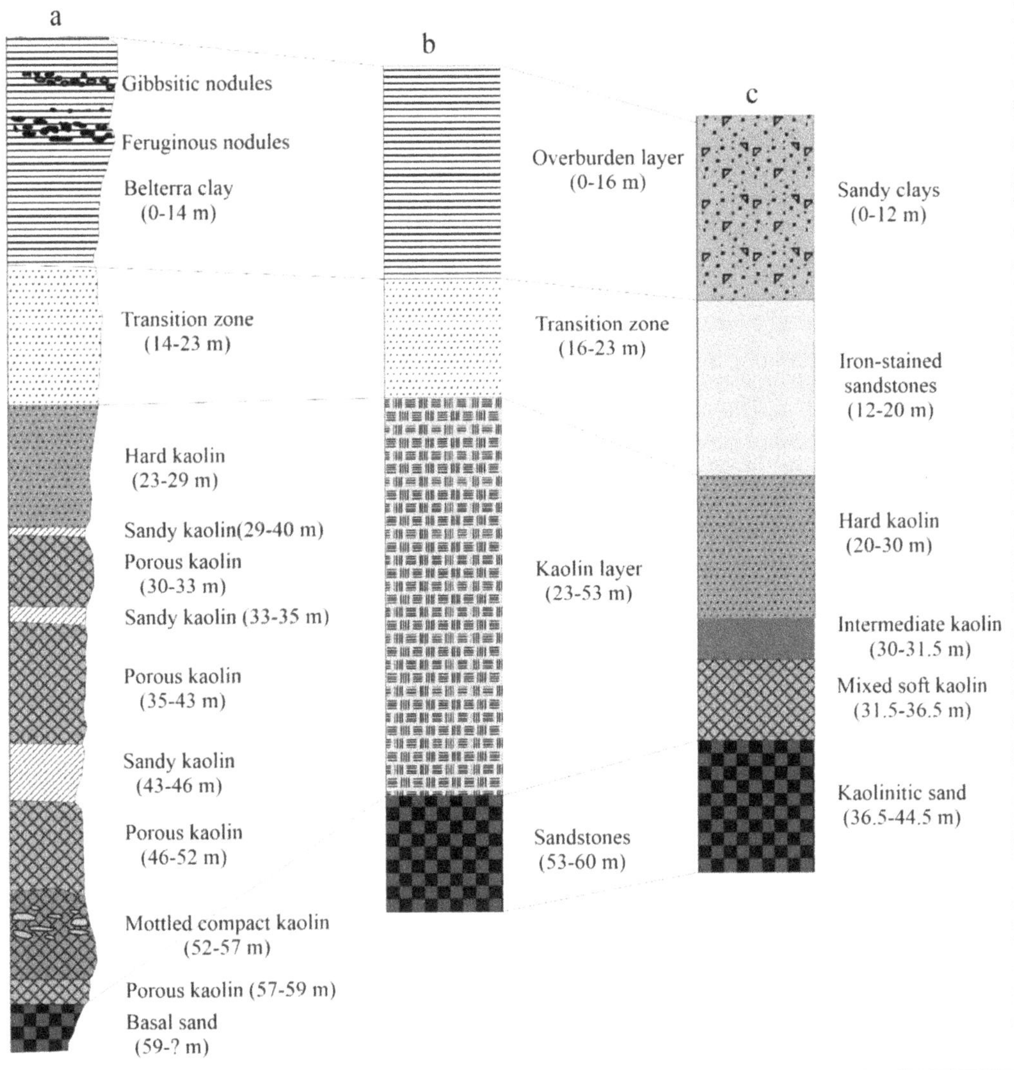

Fig. 3 (**a**) Typical cross-section of Morro do Felipe mine, Jari River, Brazil. (After Montes et al. 2002). (**b**) Profile from the Felipe II CADAM deposit, Jari River, and (**c**) geological cross-section of the Ipixuna kaolin deposit of PPSA, Capim River, Brazil. (After Wilson et al. 2006)

respectively (Murray et al. 2007). The generalized stratigraphic sections of kaolin deposits from the Jari River and the Capim River regions are shown in Fig. 3. The surficial geology of the region consists of a kaolin-bearing lower unit, which is a truncated mature lateritic succession derived from the Ipixuna/Itapecuru Formation and an upper unit, comprising the immature lateritized Barreiras Formation. The lower kaolin unit (Fig. 3) is characterized by sandy facies at the bottom and soft kaolin with flint facies at the top (Costa et al. 2009).

Kaolin mineralization in the Morro do Felipe deposit, located on the bank of the Jari River in the Cadam district (Fig. 3a), is about 60 m thick (Montes et al. 2002). Sedimentary features are present throughout the whole altered body, which indicates that the kaolin deposits were developed from kaolinitic clay sediments of the Cretaceous Alter do Chao formation (protore). These were originally derived from the erosion and transport of weathered granitic basement rocks and deposited during seasonal flooding of the meandering Amazon River. These sediments were then subjected to intense lateritic weathering in an area of relatively low topographic relief, resulting in enrichment of the kaolin, mainly through mineral dissolution, particularly of quartz.

The Ipixuna Formation (late Cretaceous—early Tertiary) in the Rio Capim area of the Cameta Sub-basin has been subdivided into three stratigraphic units: lower soft kaolin, intermediate kaolin and upper, indurated, semi-flint kaolin (Santos and Rossetti 2008). These units were affected by strong supergene alteration after deposition. Soft kaolin consists of well-structured, sub-angular to sub-rounded and, locally, angular kaolinized sandstones. The semi-flint kaolin was formed in sediments of variable origin but also includes a significant component of soft kaolin due to later pedogenic alteration by descending solutions rich in organic acid and bacterial activity. The soft kaolins are the principal ore and grade to a hard (flint) kaolin. Their high whiteness, fine particle size and good rheological properties make these a premium product extensively used in the paper industry. Six different facies of the kaolin deposits (sand kaolin, soft kaolin, lower transition facies, ferruginous crust, upper transition facies, flint kaolin) were studied to document the geochemical evolution of the deposits during lateritization and later weathering episodes (Sousa et al. 2006). The thick kaolin layers developed under the influence of major climatic changes from semi-arid to humid during the Cretaceous to Oligocene. Intense lateritic weathering produced the "soft" kaolin, which has the most favorable technical properties (Sousa et al. 2007). Ferruginization at the top of the soft kaolin led to the formation of a ferruginous plasma and disordered kaolinite with structural Fe in the kaolinite crystal lattice. Subsequent deferruginization under acidic reducing conditions led to the formation of the flint "hard" kaolin facies; silica solubility increases in acidic, organic-rich groundwaters (pH <4), resulting in almost total dissolution of quartz, degradation of the duricrust, dissolution of crandallite series ($CaAl_3(PO_4)_2(OH)_5(H_2O)$) minerals and the formation of poorly-ordered kaolinite (Sousa et al. 2006).

Kaolin deposits N of Manaus, located about 1000 km W of the Rio Capim, are developed in a ~ 20 m thick lateritic regolith formed in sedimentary sequences of arkoses, kaolinitic mudstones and kaolinitic sands of Cretaceous to early Tertiary age in the N-central Amazon Basin. Kaolinites have progressively evolved under humid tropical climatic conditions as weathering progresses downward in response to seasonally changing microenvironments (Giral et al. 1998).

3.1.5 Argentina

The Patagonian region of Argentina, in southern South America, has a Gondwana paleosurface of the Late Mesozoic age developed on Jurassic rhyolite volcanic units. The Patagonian residual kaolin deposits and epithermal precious metal deposits occur over a paleosurface of an extended Jurassic rhyolite province (Dominguez et al. 2016). During this time, the climate was wet and warm, allowing the development of extensive chemical weathering. Numerous deposits of kaolinite, halloysite and quartz up to 20 m thick occur along the Chubut River in the Chubut and Santa Cruz provinces in southern Patagonia, formed by in situ alteration of the Jurassic volcaniclastic rocks and by erosion of residual clays and their transport and deposition in small basins (Murray 1988; Dominguez et al. 2008).

Detailed micromorphological analysis of clay and non-clay minerals in these deposits indicates that they originated from the weathering of volcanic ash (Raigemborn et al. 2014) and that the climatic history provided favourable geochemical environments for the development of residual kaolin deposits. Cravero et al. (2001) reported that most of the residual kaolin deposits of Patagonia are related to a paleo-weathering surface developed over the rhyolitic rocks during the Upper Jurassic—Lower Cretaceous. However, the temporal trend of the clay mineral assemblages in the stratigraphic succession, coupled with evidence from sedimentological analysis, indicates a warm and seasonal climate for the basal part of the unit, a warm and humid climate with a relatively more perennial rainfall regime in the middle part of the unit, and a warm and less humid, probably sub-humid, climate in the uppermost section. Dominguez et al. (2008) divided the kaolins into three genetic groups. (1) Kaolins derived from weathering of pyroclastic sequences that are coarse-grained, composed of kaolinite + quartz ± halloysite, and have a very low specific surface area. (2) Kaolins formed by weathering of mostly crystalline pyroclastic rocks that yielded ordered kaolinite and illite (+halloysite) with a fine particle size distribution and intermediate specific surface area. (3) Kaolins formed by weathering of mainly vitreous pyroclastic rocks to form halloysite (+ kaolinite), with a fine-grained texture and moderately high specific surface area. All Patagonian kaolin deposits studied have isotopic values that plot in the supergene kaolinite field near the kaolinite line (Dominguez et al. 2016). For example, the $\delta^{18}O$ (19.3 to 19.1‰) and δD (−83.0 to −82.0‰) isotopic composition of kaolinites from La Esperanza and Cerro Rubio confirm that the main Patagonian kaolin mineralization was formed by weathering, with some localized sedimentary dispersion and redeposition.

3.1.6 Ethiopia, Africa

The Main Ethiopian Rift, which is the northernmost part of the East Africa Rift System, has been of interest to many geoscientists from different academic disciplines because it connects the Afar Depression to the north and the Kenya Rift to the south. The Belessa kaolin deposit is located on the western margin of the Central Main Ethiopian Rift. The district is underlain by Miocene to Quaternary rhyolite and pyroclastic tuff; the rhyolite is partially to completely kaolinized (Bedase et al. 2019). The deposit is elliptical in shape, about 3.5 km long, and has a maximum thickness of approximately 70 m. It is buried beneath 5 to 10 m. of silt- to sand-sized, reddish brown, eluvial sediments. The fresh rhyolite is dark and light grey to reddish brown and changes to yellowish to white where kaolinized. The host rock is composed of quartz and alkali feldspar phenocrysts set in a groundmass of very fine-grained quartz, alkali feldspar and plagioclase. Geological, mineralogical and geochemical studies indicate that in situ weathering played a major role in the formation of the Belessa kaolin deposit. The chemical alteration indices (CIA and CIW) for major elements on (A-CN-K and A-CNK-F) ternary diagrams and the high Ce + Y + La (>100 ppm) contents of all Belessa kaolin samples suggest a high degree of weathering and residual alteration (Dill et al. 1997).

3.1.7 Japan

The well-crystallized, pure and highly refractory Iwaizumi kaolin deposit of late Cretaceous and Oligocene age is located in NE Honshu, Japan. Kaolinitic clays occur in three units: from oldest to youngest, the Yokomichi, Komatsu and Nameiri formations (Fig. 4). The shale and flint clays have similar textures, with the uppermost shale grading into the lowermost clays. The fire-clay deposits of the Nameiri Formation, located higher in the section, are associated with coal seams. The fire clays of seams 1, 2, and 34 are of sedimentary origin, whereas the fire clay from seam 35 (Fig. 4) formed directly through in situ weathering of dacitic or rhyolitic tuff. The highly permeable character of these parent materials at Iwaizumi undoubtedly promoted the effective leaching of silica and other cations. The δD and $\delta^{18}O$ values of kaolinite from late Cretaceous and Oligocene deposits at Iwaizumi suggest that these clays formed during weathering rather than hydrothermal alteration. The Iwaizumi kaolinites are also depleted in D and ^{18}O relative to kaolinite formed during modern, tropical weathering, suggesting that the kaolinite developed under cool or cool-temperate conditions (Mizota and Longstaffe 1996). The oxygen-isotope compositions of the kaolinite increase slightly upward through the deposits, perhaps implying a modest increase in temperature from late Cretaceous to Oligocene time.

3.1.8 Ukraine

The Ukrainian Shield is a classic area for kaolin deposits, with a giant resource of about 1.2 billion tons. Forty-three deposits and 110 occurrences

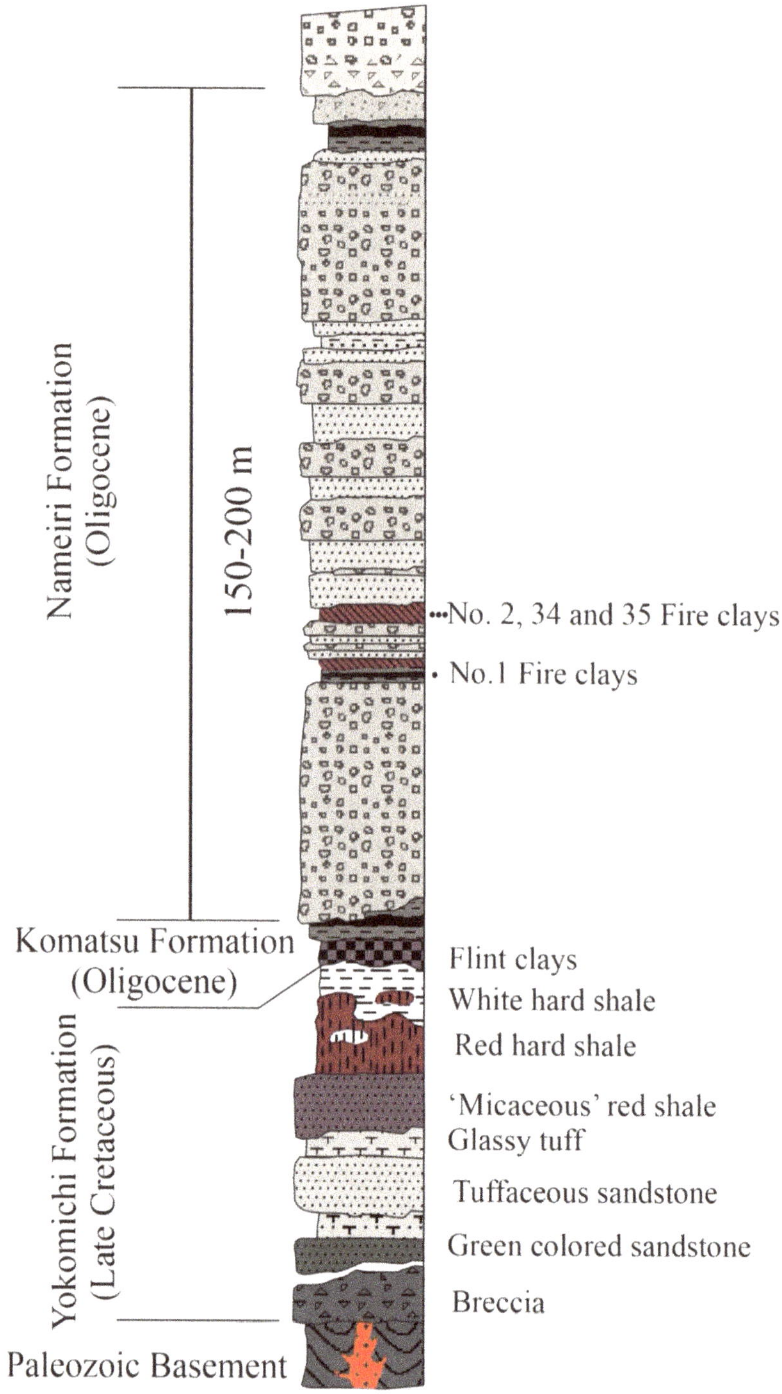

Fig. 4 Schematic section through the Iwaizumi clay deposits, Japan. (After Mizota and Longstaffe 1996)

are situated within a 950 km long and 350 km wide zone, extending from the Azov Sea to the Polish border (Sabelfeld and Greiling 2012). The kaolin deposits are classified into three groups: primary (in situ), secondary (sedimentary) and alkali. Primary deposits were formed by weathering under warm and humid climates during the Palaeozoic. Secondary kaolin is mostly redeposited from the primary occurrences. Primary deposits range from 1.4 to 120 m thick, with an average of 10–40 m. Secondary deposits are rare and 0.2–20 m thick, with an average of 5–8 m.

Primary and alkali kaolin (possibly a variant of primary kaolin) are the most economically significant deposits (Magidovich and Sivokin 1968) and generally occur as profiles with three subzones (Fig. 5):

1. Variegated kaolinitic clays unsuitable for industrial use due to high Fe_2O_3 and CaO contents.
2. Alkali kaolin ($K_2O > 2.5$–3%). The high K_2O content is due to the presence of fine-grained microcline and orthoclase feldspars and significant K mica.
3. Normal (basic) kaolin. The normal and alkali kaolins are characterized by high whiteness and low Fe_2O_3 and TiO_2 contents.

The largest single Ukrainian deposit is Veliko-Gadomezkoe, with reserves of 100 Mt. The Gluhivzi mining and processing plant, situated at the Gluhivzi Fault Zone, became a major producer with an annual capacity of 300,000 t of raw kaolin and 100,000 t of processed kaolin. The second largest is the alkali kaolin in the Dibrova Zapadnaya region, 8 km SW of the Prosyanovsk kaolin district (Magidovich and Sivokin 1968; Sonkin et al. 1983). The primary kaolin is formed by intense weathering of homogeneous leucocratic granites of the Dneprovsk magmatic complex, which are comprised of rose-colored microcline-rich granites and migmatites. There are no xenoliths of other rocks so the alkali kaolin unit is continuous. The total thickness of the overburden is about 25 m, with the kaolin beds between 14 and 25 m thick, including 7–12 m of alkali kaolin, a mixture of kaolinite and illite, commercially named "plastic clay" (Fig. 6).

3.1.9 Germany

The strong kaolinization of the New Aplite in the Kreuzberg Quartz Pegmatite of the Late Paleozoic Hagendorf-Pleystein Pegmatite Province, SE Germany, is supergene (Dill et al. 2015). The high heavy mineral and rare metal contents of the pegmatite complex are a guide to interpreting the origin of the kaolinite. The Nb-Ta-Ti heavy minerals in the overburden above the pegmatites are the same as those in the primary bedrock mineralization used to interpret the genesis of the felsic intrusive rocks. Supergene kaolinization continued from the Miocene through the Pliocene, and the stratigraphic succession can be correlated using mineral assemblages in the Tirchenreuth residual kaolin deposit. There were four stages of gradual alkalization of the meteoric pore fluids throughout the Neogene and the Quaternary, with the youngest at Tirschenreuth and the oldest, regolith, stage at Pleystein. Alteration of the Kreuzberg Pegmatite was hypogene in the west, but the kaolinization of the New Aplite in the east was supergene. The New Aplite is a relatively homogeneous unit within the Kreuzberg Pegmatite, with the basal massive aplite overlain by kaolinized aplite consisting of kaolinite with minor muscovite and feldspar; the kaolin body is about 300 × 40 m, orientated NW–SE.

The mineralogical and geochemical characteristics of the Tirschenreuth kaolin deposit indicate its granitic origin. Kitagawa and Koster (1991) reported: "The Tirschenreuth deposit is considered to have formed by *in situ* weathering because (i) the kaolinite has formed by the alteration of plagioclase and biotite, whereas microcline and muscovite are unaltered; (ii) the kaolin body has a very homogeneous composition and alteration zones cannot be recognized; (iii) there has been no addition of trace elements from sources other than plagioclase and biotite in the parent rock; (iv) particle-size distribution shows that the kaolinite crystals grew under uniform physical conditions; (v) quartz grains have smooth surfaces."

3.2 Kaolin Deposits Formed by Supergene Alteration of Sediments and Sedimentary Rocks

3.2.1 Characteristics and Environments of Formation

Sedimentary kaolin deposits are generally formed in shallow lacustrine, deltaic and coastal

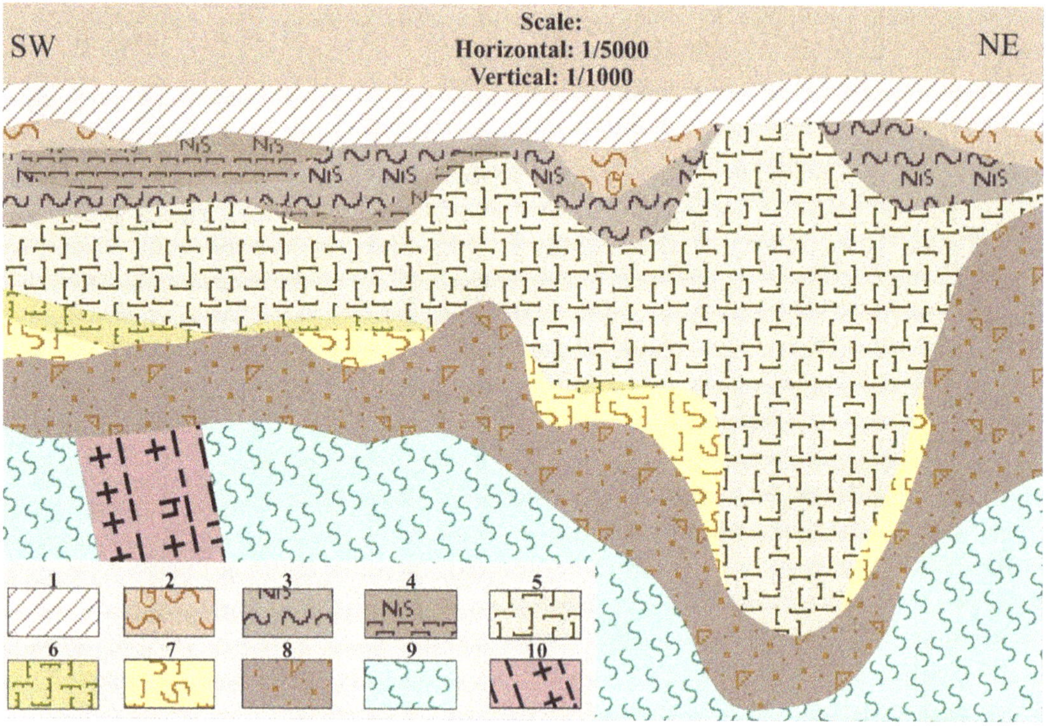

Fig. 5 Geological cross-section of Velykogadomenetsky primary kaolin deposit, Ukraine. 1- Quaternary loam; 2- Quaternary carbonate clays; 3- Miocene, Sarmatian clay; 4- Miocene, Sarmatian secondary kaolin. 5–8- Mesozoic-Cenozoic: 5- primary kaolin; 6- primary alkaline kaolin; 7- kaolinite-hydromica rock; 8- transition zone; 9–10- Palaeoproterozoic: 9- Berdychiv migmatite complex of 2 billion years; 10- gneisses and aplite-pegmatoid granites. No linear scale is given. (After Shcherbak and Goshovsky 2006)

environments. The quality, mineralogy, crystallinity, chemical composition and physical properties of each kaolin deposit depend on the provenance of the sediments and the physicochemical conditions of the freshwater movements in the depositional basin.

The kaolin deposits occur in sediments of terrigenous and pyroclastic origin that contain shales, sandstones, tuffs, lapilli tuffs, volcanic bombs and breccias. These are the products of erosion and transport of older rocks that have been deposited in shallow basins. Most sedimentary kaolin deposits are associated with coal, lignite and shales rich in organic matter that provide a combination of various organic acids (e.g., humic and fulvic acids) necessary for in situ chemical leaching of buried silicate minerals and amorphous materials as aqueous organic complexes. Sands and/or unconsolidated sandstone units within the stratigraphic succession provide suitably permeable beds or lenses for lateral and downward leaching and the formation of kaolin. Sedimentary kaolin deposits were formed mostly during the Cretaceous, Paleogene and Neogene Periods. The decomposition of organic matter in forests or semi-tropical swamp environments is the source of soluble organic acids. These have key impacts on soil profile development and kaolinization by the downward percolation of surface waters and leaching of basic cations, such as Ca, Mg, Fe, Na and K, from weathered silicate minerals. This process creates kaolinitic soils that tend to be slightly (pH 6.5) to very (pH 3.8) acidic at the surface and contributes to the development of distinct profile features associated with some forests (Morris 2004). The solubility of silicate minerals in organic and inorganic systems are very different; Si is more soluble than Al in the organic systems;

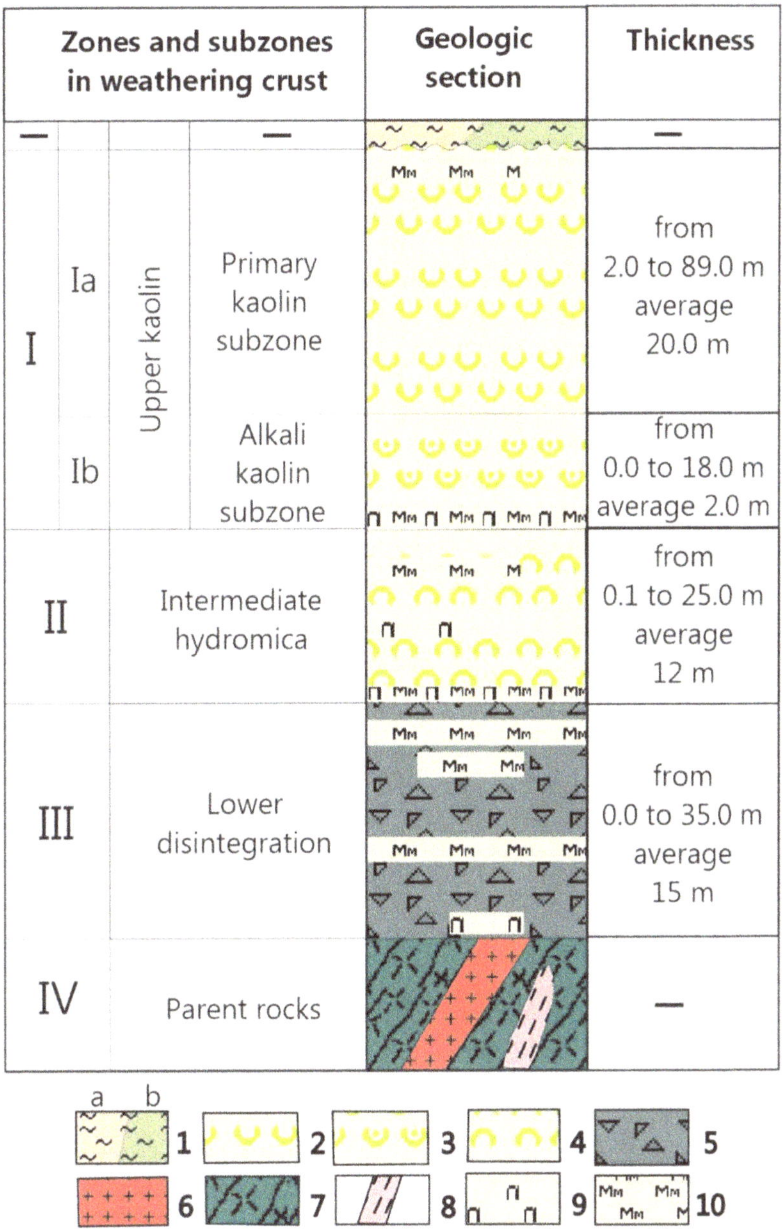

Fig. 6 Weathering profile of kaolin deposits in the southeastern Ukrainian Shield. 1a- variegated montmorillonite clays, 1b- sands and silts; 2- primary kaolin; 3- alkaline kaolin; 4- kaolinite-hydromica (illite); 5- claystone; 6- aplite-pegmatite-granite; 7- charnockites, granites and garnet-biotite migmatites; 8- biotite and garnet-biotite gneiss; 9- brecciated structures; 10- epigenetic montmorillonite. (After Sonkin et al. 1983)

hence, the presence of organic acid complexes is vitally important for the development of sedimentary kaolin deposits in ancient swamp environments.

Tonstein, a German term commonly used for some sedimentary kaolins, means "clay rock," a hard, compact rock composed of thin clay layers, mainly kaolinite, and a few other clay minerals interbedded with coal seams. Tonsteins are derived from the diagenetic alteration of volcanic ash deposited in acidic freshwater swamps or shallow lacustrine basins rich in organic matter. These anaerobic environments are partly responsible for the alteration of mineral phases in volcanic ash and kaolinite by dissolution and precipitation reactions. "*Tonstein*" is also used to refer to specific kaolinite-rich partings in coal in many other locations. For example, in the Permian and Jurassic of the Sydney Basin, Australia, coal seams within thick sequences of refractory clay contain claystone partings with characteristics similar to the Westphalian kaolinite tonsteins (Loughnan 1978). In the Paleocene Denver Formation, Colorado, USA, kaolinite-rich clay rocks are found both as thin, tonstein-like partings within a lignite seam and as thick irregular masses above it. The low-Eh, low-pH, and high concentrations of dissolved organic matter in pore fluids in the coal-forming swamp environments contribute to strong leaching of cations; geochemical and mineralogical data indicate that tonstein in Wyoming, USA, is derived from rhyolitic rocks (Zielinski 1985). Detailed studies of tonsteins as altered volcanic ash based on mineralogy, texture, radiometric age, and field relations were published by Bohor and Triplehorn (1993).

3.2.2 Şile, Turkey

The Şile Region, in NW Istanbul, Turkey, has thin, discontinuous, cyclic coal beds and kaolinitic clay deposits that were accumulated in lacustrine basins; the basins also received extensive volcanoclastic sediments from the erosion of highly weathered calc-alkaline volcanic rocks (Ece et al. 2003). This classic example of kaolinitic clay deposits with interbedded coal seams formed in Oligocene-Miocene swamp and shallow lacustrine environments under varying paleoclimatic conditions is illustrated in Fig. 7. The formation of the kaolin deposits is associated with cyclic post-depositional alteration and occurred in two stages: *firstly,* in situ, weathering of the Cretaceous andesitic tuffs, ash and agglomerates at the base of the lacustrine basin; *secondly*, surface weathering, with the enhanced transport of altered Paleozoic shale-sandstone-greywackes and Cretaceous tuffs, ash and other volcanic material from the surrounding hills into the lacustrine basin. Volcano-sedimentary-hosted clay formations in the north represent the freshwater deltaic plain, and clay formations hosted by limestones and Paleozoic rock represent sediments that have more fluvial input in the same sedimentary system (Demir 2021). Kaolinization developed in cyclic swamp environments, as indicated by the presence of thin- to thick-bedded coals; these provided the organic acids necessary for the post-depositional alteration of the redeposited shale-greywackes and volcanic rocks to kaolin. Andesitic volcanic rocks in the surrounding area were altered to smectite and, in turn, transported to the deltaic plain-swamp environments of the lacustrine basin, where they were leached to form kaolinite and amorphous silica under dysaerobic and low-pH conditions by organic acids generated by the degradation of thick vegetation. The kaolin deposits consist of disordered kaolinite, illite, smectite, gibbsite, quartz, pyrite, anatase, K-feldspar and goethite. The Şile region has relatively low annual temperature and high rainfall, which have induced intense weathering and regional groundwater movements in the shallow lacustrine basin. Depth-related changes in the distribution of clay minerals associated with coal beds are indicative of organic acid-mineral reactions (Fig. 7). Gibbsite has formed in the lower horizon of the stratigraphic section by dissolution of kaolinite during late-stage weathering. This transformation is associated with groundwater movement and increased weathering in permeable zones, as discussed for the Seto porcelain clay deposits in Japan (Takagi et al. 2021) (see Sect. 3.2.8).

3.2.3 Atlantic Coastal Plain, Georgia, USA

The sedimentary deposits in Georgia form the largest kaolin resource in the USA. They occur in a 300 km NE-SW trending belt on the Atlantic Coastal Plain, between Augusta and Macon, about 100 km SE of Atlanta. The kaolins and feldspathic sands were probably derived from the deeply weathered metamorphic rocks, granites and gneisses of the Piedmont Plateau of the southern Appalachians, and deposited near the Cretaceous—Early Tertiary shoreline in fluvial, deltaic or brackish to marine environments (Murray 1976). The Cretaceous kaolins generally form lenticular- to saucer-shaped bodies in relatively unconsolidated, variably kaolinitic, micaceous, and cross-bedded sands. The Tertiary kaolin occurrences are similar but are thicker and more persistent (Murray 1988).

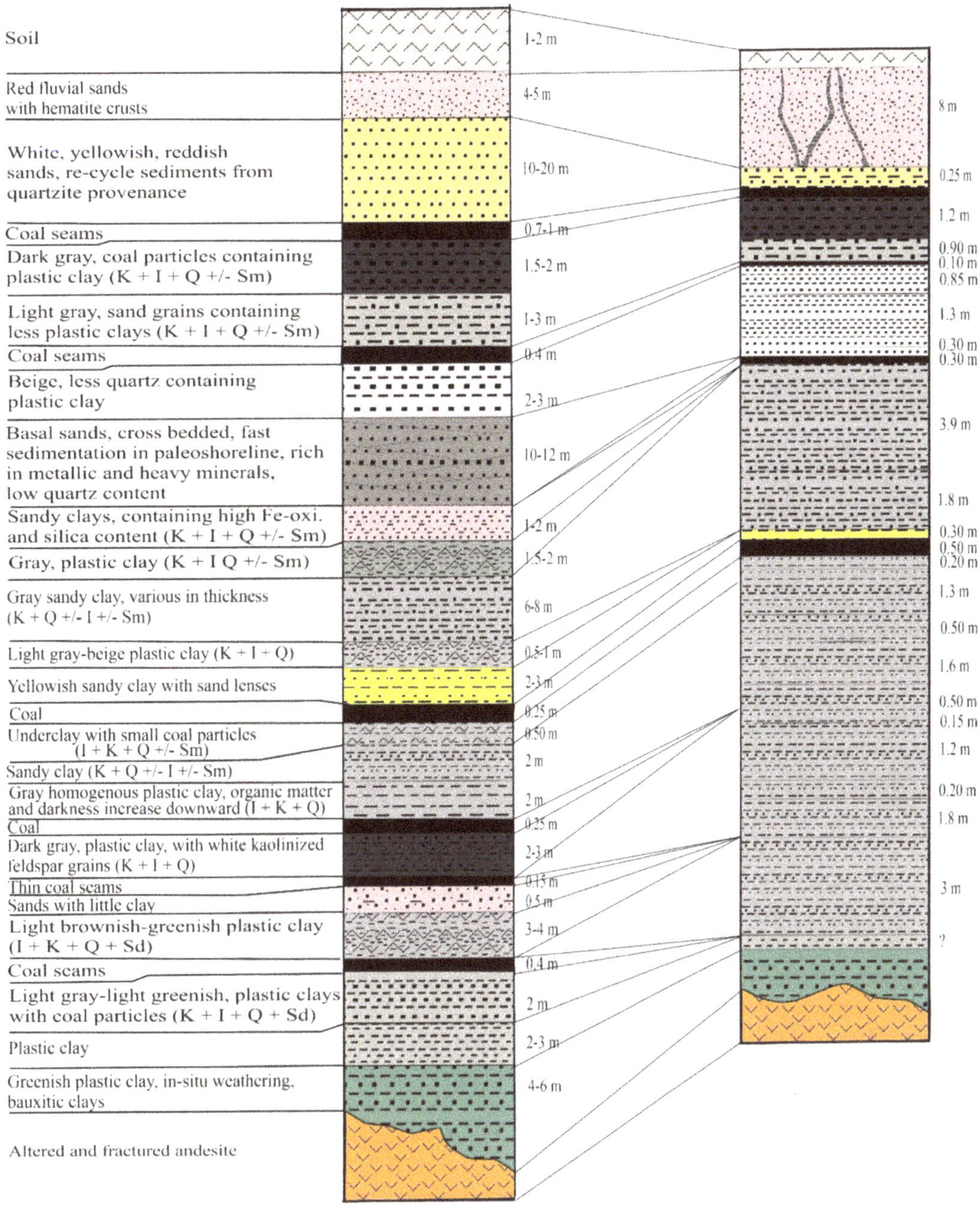

Fig. 7 *Left*, a stratigraphic section of Suluklu kaolinitic clay quarry, Istanbul, Turkey. K = Kaolinite, I = Illite, Q = Quartz, Sm = Smectite, and Sd = Siderite (Ece et al. 2003). *Right*, stratigraphic section of the same kaolin quarry after 20 years of mining activities. (After Demir 2021)

Two different types of kaolin deposits have been recognized: soft and hard. Mineralogy, crystallinity, Fe content, $\delta^{18}O$ and δD isotopic compositions distinguish Late Cretaceous soft kaolin from Early Tertiary hard kaolin (Hassanipak and Eslinger 1985). These differences probably reflect contrasts in the environments of deposition, particularly the salinity, pH, surface and shallow groundwater movements, and the effects of post-depositional overprinting. Essentially, the main difference between soft and hard kaolin deposits is related to the deposition in freshwater versus saline environments, respectively. In addition, the soft kaolins have a higher degree of structural order, slightly higher Al_2O_3 content, less exchangeable Ca^{2+}, more exchangeable K^{+}, and lower Fe_2O_3 contents than the hard kaolin deposits (Hinckley 1965). Soft kaolins were deposited in a freshwater environment with pH conditions suitable for kaolinite platelets stacking into booklets. They are also the result of extensive recrystallization of kaolin and other minerals deposited with an open face-edge orientation permitting higher permeability to infiltration of groundwaters. In comparison, the hard kaolins were deposited in seawater, with the saline conditions resulting in face-to-face flocculation and attendant higher compaction, lower permeability and, because of the higher pH, higher Fe_2O_3 contents than in freshwater environments. The $\delta^{18}O$ values and crystal sizes indicate that the Tertiary kaolins are not simply derived from the Cretaceous kaolins by winnowing but require post-sedimentation recrystallization of one or both. Hurst and Pickering (1997) concluded that sedimentary kaolins in Georgia not only had a sedimentary origin but, more particularly, have undergone intense post-depositional weathering in the recharge area of the regional groundwater system. This later weathering resulted in gradual differential leaching and attendant nucleation and growth of minerals compatible with the compositional changes.

3.2.4 Andoom-Weipa, Australia

The Weipa-Andoom kaolin deposit is located on the W coast of Cape York Peninsula, NE Queensland. The kaolin deposit overlies Cretaceous marine sediments of the Rolling Downs Formation (Andoom) and Cretaceous-Tertiary fluviatile sediments of the Bulimba Formation (Weipa), below pisolitic lateritic bauxite (Murray 1988; Taylor et al. 2008). These kaolins are formed by the alteration of sediments, as in Georgia and Capim. Weipa was a source of coating-grade kaolin, producing around 100,000 tpa of beneficiated kaolinite for export. This is a good example of fine-grained kaolinite with crystal shape and rheological properties suitable for paper coating, although the poor ordering and structural Fe content adversely affected its brightness. The profiles have the classic features of lateritic deep weathering, with gibbsite, boehmite, kaolinite, quartz, hematite and/or goethite, and ~ 2% anatase in the bauxite. The underlying kaolin resource, comprising the mottled zone and saprolite, consists mainly of kaolinite (up to 80 wt. %), quartz, with minor or trace amounts of muscovite, illite, smectite, halloysite, Fe oxides (hematite, goethite), Ti oxides (anatase, brookite) and poorly-diffracting phases (Ma 1996).

3.2.5 Bavaria, Germany

The Bavarian sedimentary kaolin deposits are situated near the western border of the Bohemian Massif close to Hirschau and Schnaittenbach. They are essentially an Early Triassic arkosic sandy kaolin that contains about 80% quartz and 10% kaolinite, with the remainder being partially kaolinized feldspar and a minor amount of smectite and illite (Murray 1988). These deposits are mined for glass sand, feldspar and kaolin products for use in the paper and high-quality ceramics industries. Kaolinization is considered to have occurred during the late Early Cretaceous whereas, in comparison, kaolinization of the residual Carboniferous granite-hosted deposits in the Tirschenreuth (Sect. 3.1.8) was Late Oligocene to Mid-Miocene (Gilg 2000). The sediments hosting the kaolins at Hirschau-Schnaittenbach are coarse to fine-grained arkoses that were laid down in upwards-fining cyclothems of different scales (Dill et al. 2016), with an original clay mineralogy similar to the limno-fluviatile sediments at the north of the Bohemian Massif. During the Early Cretaceous, the Late Jurassic carbonate platform was subjected to erosion and karstification, with a pre-Late Cretaceous relief of approximately 200 m (Gilg 2000).

3.2.6 Guadalajara, Spain

Kaolinitic sands cover a large area of NE and E-central Spain, but their kaolinite content is mostly relatively low. However, they host an important kaolin resource in an NW-SE belt extending from Guadalajara, E of Madrid, towards Valencia on the Mediterranean Coast. The kaolinitic sands were derived from deeply weathered granites and were deposited during the Early Cretaceous (Murray 1988). The sand layers are 10–30 m thick with a kaolinite content of 5–20%. The feldspar content decreases downwards, indicating increasing alteration of feldspars by percolating groundwater. Detailed studies of kaolin deposits in the Oliete basin have revealed the presence of both detrital and diagenetic kaolinites. Detrital kaolinites are anhedrals with low crystallinity and have formed by intense weathering in the source area during the warm climates of the Early Cretaceous (Bauluz et al. 2008). Such detrital kaolinites constitute the matrix of the claystone and siltstone. Three types of diagenetic kaolinites with high crystallinity and a high degree of order have also been recognized: kaolinite growing between "expanded" mica flakes, vermiform kaolinite, and euhedral kaolinite forming the matrix. These diagenetic kaolinites formed in situ in response to the incipient dissolution of K-feldspar and/or from organic acid-rich fluids derived from the maturation of organic matter in shales (Bauluz et al. 2008).

3.2.7 Suriname, South America

Four types of kaolin-containing bauxite deposits occur in northern Suriname, classified by their geomorphological characteristics. (Montagne 1964). From N to S:

(a) Young Coastal Plain: Holocene, flat, swamp-clay facies at about 1 m above sea level, between the Suriname River and the sea.
(b) Old Coastal Plain: Pleistocene to Late Tertiary, dry islands in swamps, marine and brackish sandy silty clays.
(c) Cover-Landscape: Tertiary to Mesozoic, the upper part White Sand Formation, unconsolidated sands resting on sandy clays, clays and lignites, and the lower part Berbice Formation, alternating sand and clay groups, with sandstones and conglomerates at the base.
(d) Hilly Interior: intensely weathered granites, schists and pegmatites of the Precambrian basement.

Cover landscape deposits are exemplified by the bauxite and kaolinite deposits of the Onverdacht region of northern Suriname (Fig. 8), where bauxite layers are interbedded with kaolinitic clay zones (Aleva 1965). Drilling data show a distinct layer of kaolinitic swamp clay in the uppermost part of the Onverdacht series. Paleocene and Lower Eocene arkoses and silty sedimentary beds, consisting of alternating layers of kaolinitic sand and swamp sediments, have been altered to a continuous bed of bauxite in the upper horizon. This presumably occurred during a period of humid tropical climate and good drainage. The boundary between bauxite and the underlying kaolinitic clay is mostly sharp but very irregular. The main bauxite unit encloses a series of kaolin lenses or beds and overlies several tens of metres of sedimentary kaolinitic clay.

3.2.8 Japan

The Seto porcelain clay formation in the Seto-Tono district, central Japan, is hosted by an intensely kaolinized, late Miocene, lacustrine arkose sediment with subordinate fluvial incursions (Takagi et al. 2021). These inland lakes were originally ephemeral bogs, with the sediments deposited during alternating episodes of stagnant water and completely dry conditions in a tropical to subtropical climate. At the Motoyama deposits, the formation has five sedimentary units: Kibushi clay, Gaerome clay, kaolin clay (white earth), sandy clay and silica sand (Fig. 9). The Kibushi clay is a dark brown clayey sediment, 1–6 m thick, composed of fine-grained kaolin and quartz, and very thin and irregular coal seams. The Gaerome clay, 5–15 m thick, is a compact sediment composed of fine-grained kaolin with a few organic fragments derived from the arkose sandstone and constitutes the main part of

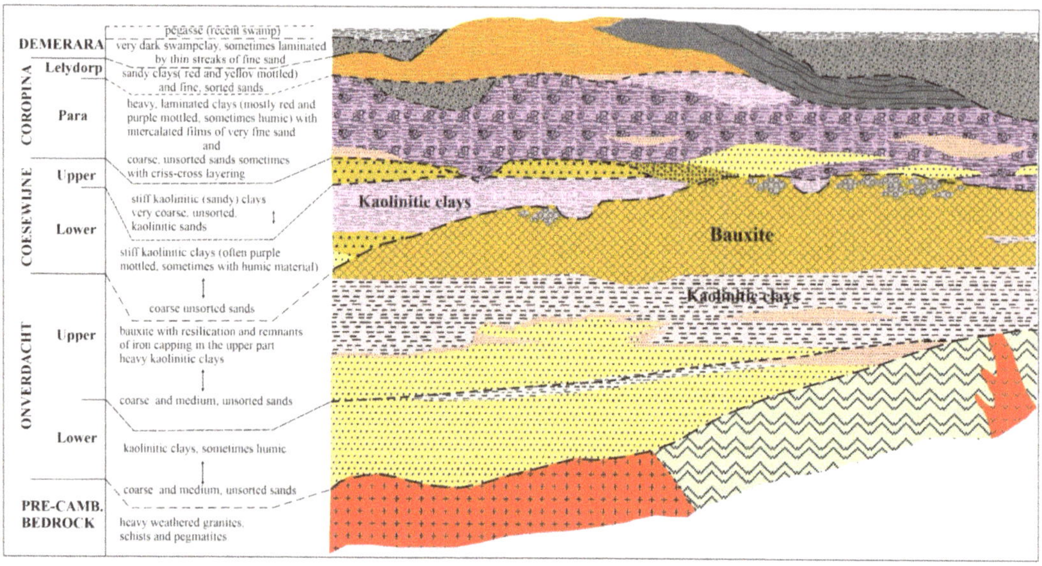

Fig. 8 Generalized stratigraphic section of the Onverdacht bauxite and kaolin district, Suriname. No scale is given. (After Montagne 1964)

the porcelain clay formation. The Gaerome clay contains plant fossils, indicating a calm and shallow lacustrine environment. The kaolin-rich white earth occurs as lenticular layers, irregularly distributed at a scale of several meters.

The deposits were probably formed by intense chemical leaching and kaolinization caused by microbial nitrification and acidification of lacustrine sediments beneath shallow inland lakes and highly vegetated swamps. In such environments, anaerobic nitrification of ammonia produced by bacterial decomposition of organic matter produces highly acidic conditions (pH < 4.0), resulting in kaolinization (Tagagi et al. 2021). Some NO_3^-, as dilute nitric acid, causes oxidation of Fe (denitrification). The acidity also leaches Fe that has previously migrated from the sediments to form limonite crusts. These reactions, summarized in Fig. 10, produce acid anions such as H_2CO_3, HCO_3^-, $SO4^{2-}$, NO^{2-}, and NO_3^- and result in kaolin formation. In addition, atmospheric CO_2 and rainwater provide H_2CO_3 in the soil, which dissociates with H^+ to HCO_3^- at pH <5.7.

3.3 Mixed Hydrothermal/Supergene Kaolin Deposits

3.3.1 Characteristics and Environments of Formation

This group of kaolin deposits was formed by multi-stage processes, with the initial hydrothermal mineralization overprinted by later weathering. It is likely that low-temperature hydrothermal fluids initiated the kaolinization of near-surface rocks close to quartz veins. Such partial alteration may increase the permeability of these rocks, opening them to surface weathering and the development of supergene kaolinization. Such overprinting has been proposed for kaolins in SW England (Sheppard 1977).

3.3.2 Cornwall, UK

Some of the largest and highest quality primary kaolin deposits in the world are located in Cornwall and Devon in SW England (Murray 1988), where kaolinite is derived from the late-stage magmatic or hydrothermal decomposition

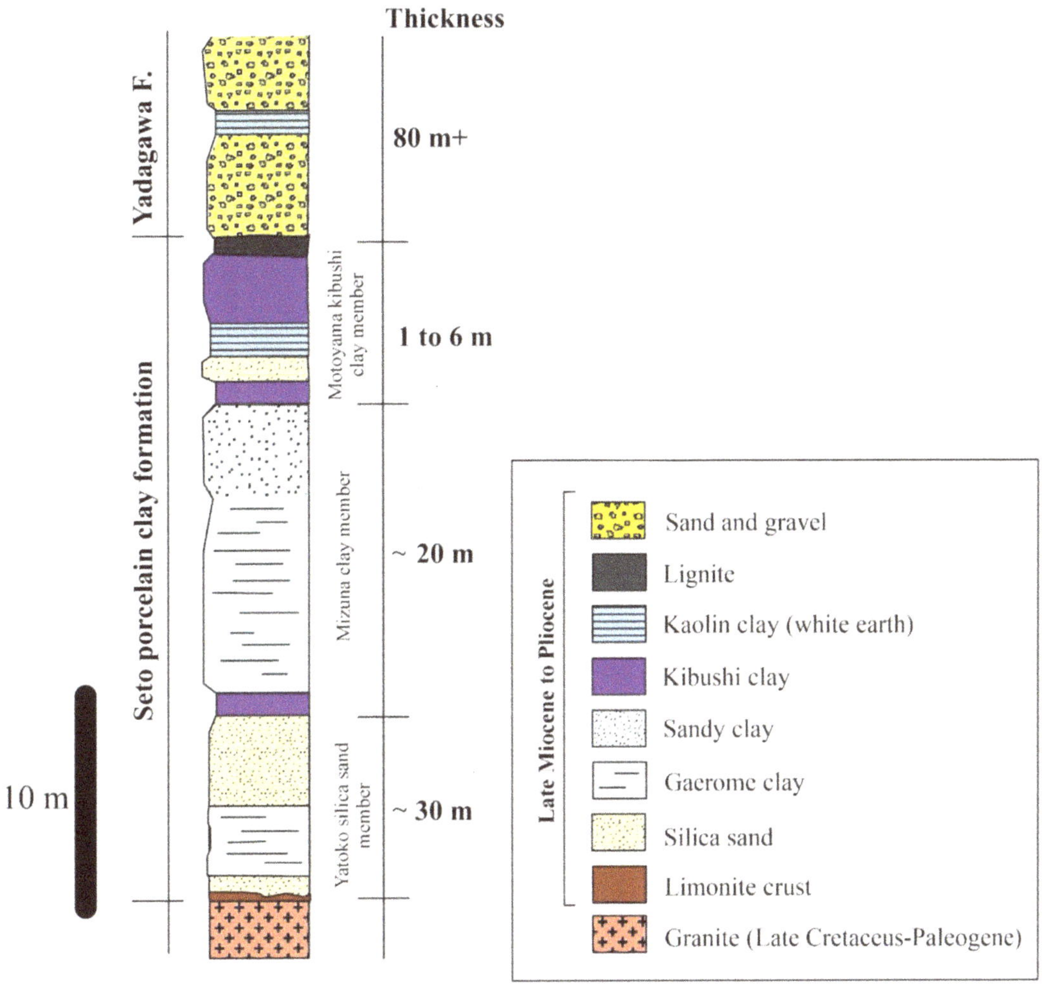

Fig. 9 Generalized stratigraphic section of the Motoyama sedimentary kaolin deposit, Japan. (Takagi et al. 2021)

of feldspar in granite (Bristow 1977). The Early Permian St. Austell granite in Cornwall has been subjected to two stages of magmatic activity: *firstly,* by the intrusion of several phases of biotite granite and, *secondly*, by the later intrusion of a non-megacryst lithium-mica granite that crystallized from a volatile-rich residual differentiate of the biotite granite magma (Manning and Exley 1984). Both intrusions were affected by at least two stages of hydrothermal alteration by a single aqueous fluid phase. Extensive Sn, Cu and W mineralization in the granite indicates that there was an intense multi-stage hydrothermal alteration that prepared the rocks for later supergene kaolin mineralization; the hydrothermal alteration transformed K feldspar to micaceous minerals ± illite, and Na plagioclase to mica, montmorillonite and kaolinite (Exley 1976).

The kaolinite content in the altered granite ranges from 10 to 20%, with a typical kaolin body exhibiting a funnel- or trough-like shape, narrowing downward to as deep as 230 m. Two distinct types of kaolinite occur in the granites: a fine platy type associated with feldspar and a vermiform type derived from muscovite. Because of the low Fe content of the granite, the resultant kaolinite is very white. Bristow (1977) and Murray (1988) summarized conditions for the evolution of the kaolin deposits in Devon and Cornwall as follows:

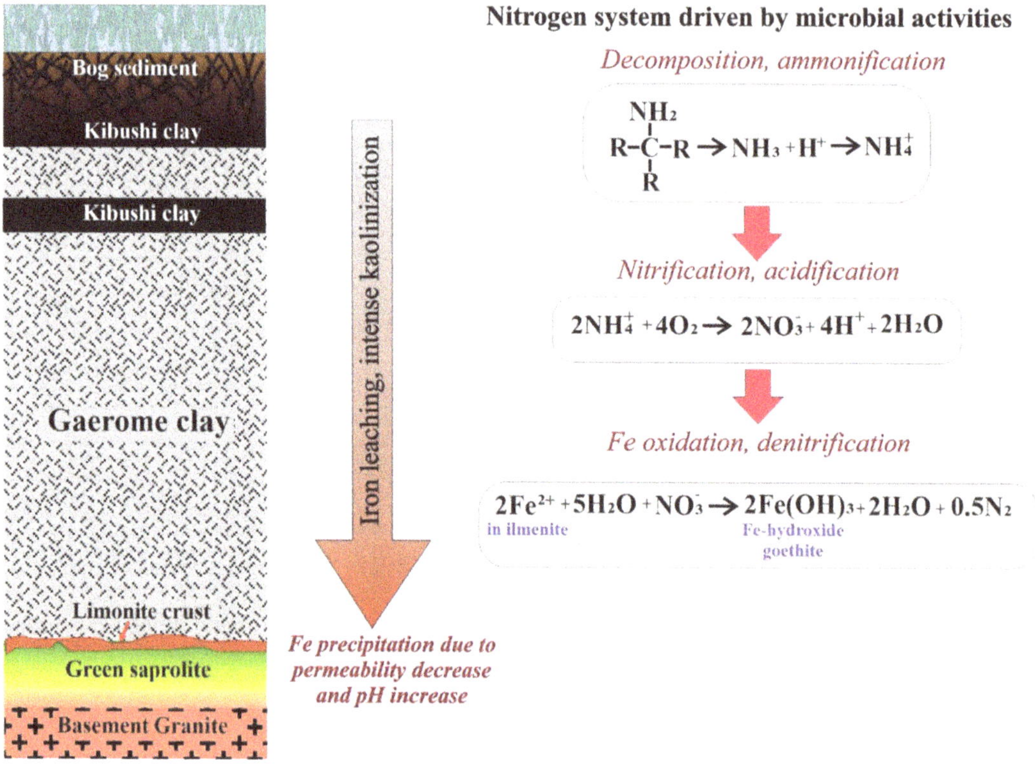

Fig. 10 Summarized cross-section of the Motoyama sedimentary kaolin deposits, Japan, showing proposed fluid flow and acidification and nitrification reactions. (Takagi et al. 2021)

(a) A parent granite with a low Fe content and virtually no biotite.
(b) An initial stage of hydrothermal alteration that served as a "preparation of the surface rocks," yielding a metastable matrix for later kaolinization,
(c) Supergene alteration sufficiently intense to transform the metastable matrix to kaolinite.
(d) A fairly gentle topographic relief with minor post-mineralization erosion, preserving the newly-formed kaolin body.

The earliest hydrothermal fluids (350 ± 100 °C) released from the crystallizing granite led to hydraulic fracturing and progressive feldspar alteration of the biotite granite host rock (Dominy and Camm 1998). This initial hydrothermal stage was characterized by quartz-tourmaline-cassiterite-bearing veins and pervasively greisenized wall rocks. In the late stage, low-temperature (100 ± 50 °C) and less saline geothermal solutions associated with active faulting produced the kaolinitic alteration assemblage and characteristic epithermal quartz veining. The alteration assemblage contains large (>10 μm) booklets of low disorder kaolinite set in a finer (<2 μm) kaolinite matrix. Alderton and Rankin (1983) proposed that low-temperature and low-salinity fluids were the most abundant in extensively kaolinized granite. These represent the latest phase of fluid activities and were probably responsible for the formation of the extensive kaolin deposits in the region at temperatures ranging from about 150 °C to below 70 °C. However, based on the isotopic values for $\delta^{18}O$ (17.9/ 20.4‰) and δD (−60/ −66‰) of the kaolinites and the general relationship of the deposits to present or recent erosion surfaces, Sheppard (1977) concluded that the major china clay deposits are of weathering origin, formed by the action of meteoric water during tropical weathering during the early Tertiary.

3.3.3 Kalabsha, Egypt

Three different kaolin occurrences (nodular, pisolitic and plastic) occur in Carboniferous and Cretaceous sedimentary rocks in the Kalabsha area, SW of the Aswan High Dam in southern Egypt. These rocks were predominantly derived from granitic plutons, with the remainder from metamorphic terranes. Textural, mineralogical and geochemical studies (Baioumy and Gilg 2011) suggest that the pisolitic kaolins are lateritic in origin. The lithostratigraphic succession shows nodular and pisolitic kaolins at the base, changing upwards to non-pisolitic plastic kaolins—the inverse of a typical laterite profile. It remains uncertain whether they were transported into the sedimentary basin as previously weathered clays, lateritized in situ after deposition, or formed by a combination of sedimentary and in situ weathering processes. Based on textural and geochemical variations, the overlying plastic clays, however, were not formed by reworking the pisolitic kaolins but were derived from saprolites underlying the lateritic crusts in the source region. A schematic diagram of possible sources, depositional mechanisms and early diagenetic changes of the pisolitic and non-pisolitic kaolins is shown in Fig. 11. The pisolitic kaolin is considered flint kaolin, based on its physical properties such as hardness, massiveness, conchoidal fracture, high density, very fine-grained nature, resistance to slaking in water, high crystallinity and homogeneous compositions (Baioumy and Gilg 2011).

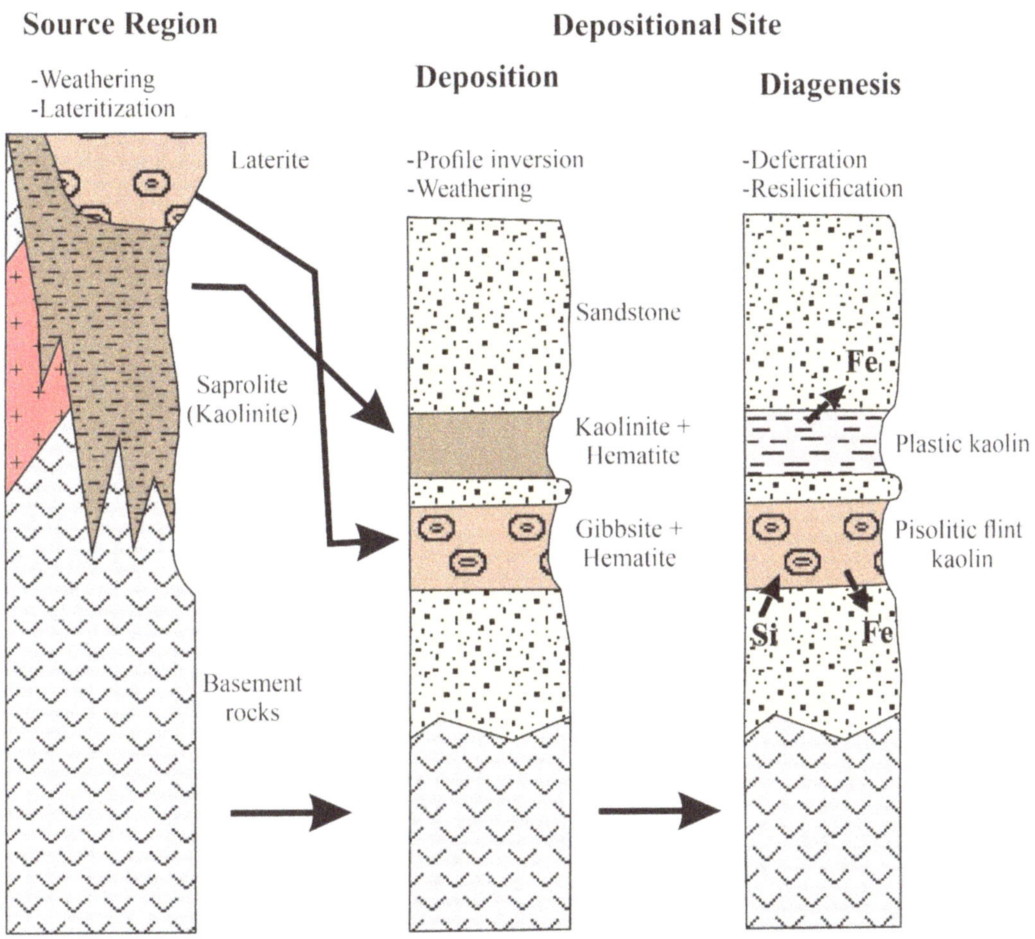

Fig. 11 Schematic diagram showing the possible sources, sedimentation, and in situ weathering of the pisolitic and plastic kaolins at Kalabsha area, Egypt. (After Baioumy and Gilg 2011)

An understanding of the processes of post-depositional alteration with the mobilization of Fe and Si would provide more information on the role of supergene activity on kaolin quality.

The δD and $\delta^{18}O$ values of the Cretaceous pisolitic flint kaolin deposit from Aswan (means 65‰ and 20.3‰, respectively) and plastic kaolin from the same area (means 66‰ and 19.5‰) are almost identical (Baioumy 2013). These studies suggest that the kaolinite of these Carboniferous and Cretaceous deposits formed under warm-temperate to tropical conditions. The O isotopic composition of Cenomanian rudist fossils (El-Shazly et al. 2011) similarly suggests the temperature ranging between 31° and 39 °C, i.e., that warm conditions prevailed during this period.

4 Summary and Conclusions

Kaolin group minerals are alteration products and are widely observed all over the world. The most abundant kaolins have formed by weathering of felsic (i.e., granitic, siliceous volcanic and pegmatitic) rocks. The deposits are characterized by their homogeneity and the presence of other oxidized minerals. The formation of supergene kaolin takes place under low pH conditions in the presence of organic and/or inorganic acids.

There are three broad groups of supergene kaolin deposits, although there are great differences within each group. (1) The most common are formed by in situ weathering of felsic rocks. (2) Sedimentary kaolins are the product of the erosion, transport and accumulation of previously kaolinized rocks, including felsic intrusive rocks, hydrothermally-altered volcanic rock tuffs and sedimentary rocks. (3) Post-depositional deposits formed by kaolinization in shallow continental or coastal basins, promoted by acidic anions generated by anaerobic decomposition of abundant organic matter. In addition, there are also mixed hydrothermal (hypogene) and supergene deposits.

The quality and abundance of the final products are dependent on the physico-chemical conditions of alteration. Elucidation of the genesis of the deposits and the possible relationship with adjacent host rocks is very important for determining the processes involved in their formation since these determine the particle sizes, crystallinity, chemical composition, mineral assemblages, alteration processes and intensities that control industrial applications. The many important deposits throughout the world were formed by different processes. A study of their origins provides important information about their genesis and assists in the exploration of new deposits.

Acknowledgments The authors would like to thank Charles R.M. Butt and Robert J. Bowell for inviting us to contribute to this book and for their very productive editorial help. Thanks also extend to Paul Schroeder of UGA and an anonymous reviewer who made significant suggestions during the preparation of the paper.

References

Abeysinghe PB, Fetherston JM (1999) Kaolin in Western Australia. Western Australia Geological Survey. Mineral resources bulletin, 19. Department of Minerals and Energy, Perth, 142 pp

Alderton DHM, Rankin AH (1983) The character and evolution of hydrothermal fluids associated with the kaolinized St. Austell granite, southwest England. J Geol Soc Lond 140:297–309. https://doi.org/10.1144/gsjgs.140.2.0297

Aleva GJJ (1965) The buried bauxite deposit of Onverdacht Suriname, South America. Geol Mijnb 44(2):45–58

Baioumy H (2013) Hydrogen and oxygen isotopic compositions of sedimentary kaolin deposits, Egypt: paleoclimatic implications. Appl Geochem 29:182–188. https://doi.org/10.1016/j.apgeochem.2012.11.013

Baioumy HM, Gilg HA (2011) Pisolitic flint kaolin from Kalabsha, Egypt: a laterite-derived facies. Sediment Geol 236:141–152. https://doi.org/10.1016/j.sedgeo.2011.01.002

Bauluz B, Mayayo MJ, Yuste A, Gonzalez Lopez JM (2008) Genesis of kaolinite from Albian sedimentary deposits of the Iberian Range (NE Spain): analysis by XRD, SEM and TEM. Clay Miner 43:459–476. https://doi.org/10.1180/claymin.2008.043.3.10

Bedase G, Getaneh W, Hailu B (2019) Geochemical and mineralogical evidence for the supergene origin of kaolin deposits—Central Main Ethiopian Rift. J Afr Earth Sci 149:143–153. https://doi.org/10.1016/j.jafrearsci.2018.08.005

Blount AM, Threadgold IM, Bailey SW (1969) Refinement of the crystal structure of nacrite. Clay Clay Miner 17:185–194

Bohor BF, Triplehorn DM (1993) Tonsteins: altered volcanic-ash layers in coal-bearing sequences, GSA special paper 285. Geological Society of America, Boulder, 44 pp

Brathwaite RL, Christie AB, Faure K, Townsend MG, Terlesk S (2014) Geology, mineralogy and geochemistry of the rhyolite-hosted Maungaparerua clay deposit, Northland. N Z J Geol Geophys 57:357–368. https://doi.org/10.1080/00288306.2014.920889

Bristow CM (1977) A review of the evidence for the origin of the kaolin deposits of southwest England. In: Proceedings of the eighth international kaolin symposium, Madrid-Rome, pp 1–19

Costa ML, Moraes EL (1998) Mineralogy geochemistry and genesis of kaolins from the Amazon region. Mineral Deposita 33:283–297. https://doi.org/10.1007/s001260050147

Costa ML, Sousa DJL, Angelica RS (2009) The contribution of lateritization processes to the formation of the kaolin deposits from eastern Amazon. J S Am Earth Sci 27: 219–234. https://doi.org/10.1016/j.jsames.2008.11.008

Cravero F, Dominguez E, Iglesias C (2001) Genesis and applications of the Cerro Rubio kaolin deposit, Patagonia (Argentina). Appl Clay Sci 18:157–172. https://doi.org/10.1016/S0169-1317(00)00021-1

Cunningham MJ, Lowe DJ, Wyatt JB, Moon VG, Churchman GJ (2016) Discovery of halloysite books in altered silicic Quaternary tephras, northern New Zealand. Clay Miner 51:351–372. https://doi.org/10.1180/claymin.2016.051.3.16

Demir H (2021) A new look at Şile clay deposits, eastern Istanbul: detail characterization and provenance. MS thesis, University of Georgia, Athens, 125 pp. ISBN: 9798516091599

Dill HD (2016) Kaolin: soil, rock and ore from the mineral to the magmatic, sedimentary and metamorphic environments. Earth-Sci Rev 161:16–129. https://doi.org/10.1016/j.earscirev.2016.07.003

Dill HG, Bosse HR, Henning KH, Fricke A, Ahrend H (1997) Mineralogical and chemical variations in hypogene and supergene kaolin deposits in a mobile fold belt–the Central Andes of northwestern Peru. Mineral Deposita 32:149–163. https://doi.org/10.1007/s001260050081

Dill HG, Dohmann R, Kaufhold S, Balaban SI (2015) Kaolinization—a tool unravel the formation and unroofing of the Pleystein pegmatite-aplite system (SE Germany). Ore Geol Rev 69:33–56. https://doi.org/10.1016/j.oregeorev.2015.01.016

Dill HG, Kaufhold S, Ehling A, Bowitz J (2016) Oxidized and reduced kaolin fan deposits: their sedimentological-mineralogical facies and physical-chemical regime (North-Bavarian Kaolin Mining District, Germany). Ore Geol Rev 72:459–484. https://doi.org/10.1016/j.oregeorev.2015.08.003

Dixon JB (1989) Kaolin and serpentine group minerals. In: Dixon JB, Weed SB (eds) Mineral in soil environments, vol 1. Soil Science Society of America Inc. (SSSA), Madison, pp 467–525. https://doi.org/10.2136/sssabookser1.2ed.c10

Dominguez E, Iglesias C, Dondi M (2008) The geology and mineralogy of a range of kaolins from the Santa Cruz and Chubut Provinces, Patagonia (Argentina). Appl Clay Sci 40:124–142. https://doi.org/10.1016/j.clay.2007.07.009

Dominguez E, Dondi M, Etcheverry R, Recio C, Iglesias C (2016) Genesis and mining potential of kaolin deposits in Patagonia (Argentina). Appl Clay Sci 131:44–47. https://doi.org/10.1016/j.clay.2015.12.031

Dominy S, Camm GS (1998) Geology and hydrothermal development of Bostraze-Balleswidden kaolin deposit, Cornwall, United Kingdom. Appl Earth Sci (Trans Inst Min Metall) Sect B 107:148–157

Ece OI, Schroeder PA (2007) Clay mineralogy and chemistry of halloysite and alunite deposits in the Turplu Area, Balikesir, Turkey. Clay Clay Miner 55:18–35. https://doi.org/10.1346/CCMN.2007.0550102

Ece OI, Nakagawa ZE, Schroeder PA (2003) Alteration of volcanic rocks and genesis of kaolin deposits in the Şile Region, Northern Istanbul, Turkey. I: clay mineralogy. Clays Clay Miner 51:675–688. https://doi.org/10.1180/0009855033840113

Ece OI, Schroeder PA, Smilley MJ, Wampler JM (2008) Acid-sulfate hydrothermal alteration of andesitic tuffs and genesis of halloysite and alunite deposits in the Biga Peninsula, Turkey. Clay Miner 43:281–315. https://doi.org/10.1180/claymin.2008.043.2.10

Ekosse G (2000) The Makoro kaolin deposit, southeastern Botswana: its genesis and possible industrial applications. Appl Clay Sci 16:301–320. https://doi.org/10.1016/S0169-1317(99)00059-9

El-Shazly S, Kostak M, Abdel-Gawad G, Klouckova B, SaberR SG, Salama FY, Mazuch M, Zak K (2011) Carbon and oxygen stable isotopes of selected Cenomanian and Turonian rudists from Egypt and Czech Republic, and a note on changes in rudist diversity. Bull Geosci 86:209–226

Exley CS (1976) Observations on the formation of kaolinite in the St. Austell granite, Cornwall. Clay Miner 11: 197–230. https://doi.org/10.1180/claymin.1976.011.1.06

Gilg HA (2000) D-H evidence for the timing of kaolinization in Northeast Bavaria, Germany. Chem Geol 170:5–18. https://doi.org/10.1016/S0009-2541(99)00239-9

Giral KS, Savin SM, Nahon DN, Girard JP, Lucas Y, Abel LJ (1998) Oxygen isotope geochemistry of kaolinite in laterite-forming processes, Manaus, Amazonas, Brazil. Geochim Cosmochim Acta 62:1865–1879. https://doi.org/10.1016/S0016-7037(98)00103-3

Hassanipak AA, Eslinger E (1985) Mineralogy, crystallinity and $^{18}O/^{16}O$, and D/H of Georgia kaolin deposits. Clay Clay Miner 33:99–106. https://doi.org/10.1346/CCMN.1985.0330203

Hernández AC, Sánchez-Espejo R, Meléndez W, González G, López-Galindo A, Viseras C (2019) Characterization of Venezuelan kaolins as health care

ingredients. Appl Clay Sci 175:30–39. https://doi.org/10.1016/j.clay.2019.01.003
Hinckley DN (1965) Mineralogical and chemical variations in the kaolin deposits of the coastal plain of Georgia and South Carolina. Am Mineral 50:1865–1883
Huang WH, Keller WD (1970) Dissolution of rock-forming silicate minerals in organic acids: simulated first-state weathering of fresh mineral surfaces. Am Mineral 55:2076–2094
Hurst VJ, Pickering SM Jr (1997) Origin and classification of Coastal Plain kaolins, southeastern USA, and the role of groundwater and microbial action. Clay Clay Miner 45:274–285. https://doi.org/10.1346/CCMN.1997.0450215
Jackson JA (1997) Glossary of geology, 4th edn. American Geological Institute, Alexandria., 769 pp. https://doi.org/10.4236/oje.2014.411060
Joswig W, Drits VA (1986) The orientation of the hydroxyl groups in dickite by X-ray diffraction. Neues Jahrb Mineral Monatshefte 1:19–22
Joussein E, Petit S, Churchman J, Theng B, Righi D, Delvaux B (2005) Halloysite clay minerals—a review. Clay Miner 40:383–426. https://doi.org/10.1180/0009855054040180
Keeling JL (2015) The mineralogy, geology and occurrences of halloysite. In: Pasbakhsh P, Churchman GJ (eds) Natural mineral nanotubes: properties and applications. Apple Academic Press, Oakville, pp 95–115. https://doi.org/10.1201/b18107-9
Keller WD (1981) Morphology of kaolinite weathered from a non-feldspathic mica-phyllite. Clay Miner 16: 289–295. https://doi.org/10.1180/claymin.1981.016.3.06
Keller WD, Cheng H, Johans WD, Meng CS (1980) Kaolin from the original Kaoling (Gaoling) mine locality, Kiangsi Province, China. Clay Clay Miner 28:97–104. https://doi.org/10.1346/CCMN.1980.0280204
Kitagawa R, Koster R (1991) Genesis of the Tirschenreuth kaolin deposit in Germany compared with the deposit in Japan. Clay Miner 26:61–79. https://doi.org/10.1180/claymin.1991.026.1.07
Kittrick JA (1969) Soil minerals in the Al_2O_3-SiO_2-H_2O system and a theory of their formation. Clay Clay Miner 17:157–167. https://doi.org/10.1346/CCMN.1969.0170304
Kohyama N, Fukushima K, Fukami A (1978) Observation of the hydrated form of tubular halloysite by an electron microscope equipped with an environmental cell. Clay Clay Miner 26:25–40. https://doi.org/10.1346/CCMN.1978.0260103
Loughnan FC (1978) Flint clays. Tonsteins and the kaolinite clayrock facies. Clay Miner 13:387–400. https://doi.org/10.1180/claymin.1978.013.4.04
Ma C (1996) The ultra-structure of kaolin. PhD dissertation. The Australian National University, 357 pp (unpublished)
Magidovich VI, Sivokin VI (1968) New deposit of feldspar—kaolin ore in the Ukraine. https://doi.org/10.1007/BF00680643
Manning DAC, Exley CS (1984) The origins of late-stage rocks in the St. Austell granite—a re-interpretation. J Geol Soc Lond 141:581–591. https://doi.org/10.1144/gsjgs.141.3.058
Mizota C, Longstaffe F (1996) Origin of Cretaceous and Oligocene kaolinites from the Iwaizumi clay deposit, Iwate, northwestern Japan. Clay Clay Miner 44:408–416. https://doi.org/10.1346/CCMN.1996.0440310
Montagne DG (1964) New facts on the geology of the "young" unconsolidated sediments of Northern Surinam. Geol Mijnb 43(12):499–515
Montes CR, Melfi AJ, Carvalho A, Vieira-Coelho AC, Formoso MLL (2002) Genesis, mineralogy and geochemistry of kaolin deposits of the Jari River, Amapa State, Brazil. Clays Clay Miner 50:494–503. https://doi.org/10.1346/000986002320514217
Moore DM, Reynolds RC (1997) X-ray diffraction and the identification and analysis of clay minerals. Oxford University Press, Oxford, 378 pp
Morris LA (2004) Soil development and properties. Nutrient cycling. Encyclopedia of forest sciences. Elsevier Academic Press, Oxford, pp 1227–1235
Murray HH (1976) The Georgia sedimentary kaolins. In: Proceedings of the 7th symposium, Congress of Kaolin International Geological Correlation Program, Committee on Correlation of Age and Genesis of Kaolin, Tokyo. University of Tokyo (pub), pp 114–125
Murray HH (1988) Kaolin minerals: their genesis and occurrences. In: Bailey SW (ed) Hydrous phyllosilicates (exclusive of micas), Reviews in mineralogy, vol 19. Mineralogical Society of America, Washington, D.C., pp 67–89
Murray HH, Keller WD (1993) Kaolins, kaolins and kaolins. In: Murray HH, Bundy W, Harvey C (eds) Kaolin genesis and utilization. The Clay Mineral Society, Boulder, pp 1–24
Murray HH, Harve C, Smith JM (1977) Mineralogy and geology of the Maungaparerua halloysite deposit in New Zealand. Clays Clay Miner 25:1–5. https://doi.org/10.1346/CCMN.1977.0250101
Murray HH, Alves CA, Bastos CH (2007) Mining, processing and applications of the Capim Basin kaolin, Brazil. Clay Min 42:145–151. https://doi.org/10.1180/claymin.2007.042.2.01
Nagasawa K (1978) Kaolin minerals. In: Sudo T, Shimoda S (eds) Clays and clay minerals in Japan. Development in sedimentology 26. Elsevier, Amsterdam, pp 189–219
Neder RB, Burghammer M, Grasl TH, Schulz H, Bram A, Fiedler S (1999) Refinement of the kaolinite structure from single-crystal synchrotron data. Clay Clay Miner 47:487–494. https://doi.org/10.1346/CCMN.1999.0470411

Nesse WD (2000) Introduction to mineralogy. Oxford University Press, Oxford, 356 pp
Ollier C (1988) Deep weathering, groundwater and climate. Geogr Ann Ser A Phys Geogr 70(4):285–290. https://doi.org/10.1080/04353676.1988.11880258
Pruett RJ (2016) Kaolin deposits and their uses: Northern Brazil and Georgia, USA. Appl Clay Sci 131:3–13. https://doi.org/10.1016/j.clay.2016.01.048
Raigemborn MS, Gomez-Peral LE, Krause JM, Matheos SM (2014) Controls on clay minerals assemblages in an early paleogene nonmarine succession: implications for the volcanic and paleoclimatic record of extra-andean patagonia, Argentina. J South Am Earth Sci 52:1–23. https://doi.org/10.1016/j.jsames.2014.02.001
Ruiz Cruz MD (1996) Dickite, nacrite and possible dickite/nacrite mixed-layers from the Betic Cordilleras (Spain). Clays Clay Miner 44:357–369. https://doi.org/10.1346/CCMN.1996.0440305
Sabelfeld A, Greiling RO (2012) Kaolin deposits in the Ukrainian Shield and their geological context. Z Dtsch Ges Geowiss 163(2):127–135. https://doi.org/10.1127/1860-1804/2012/0163-0127
Santos AEA, Rossetti DF (2008) Origin of the Rio Capim Kaolin based on optical (petrographic and SEM) data. J S Am Earth Sci 26:329–341. https://doi.org/10.1016/j.jsames.2008.08.004
Schroeder PA (2018) Clays in the critical zone. Cambridge University Press, Cambridge, 246 pp
Schroeder PA, Erickson G (2014) Kaolin: from ancient porcelains to nanocomposites. Elements 10:177–182. https://doi.org/10.2113/gselements.10.3.177
Shcherbak SV, Goshovsky SV (2006) Mineral deposit of Ukraine, volume II: industrial mineral deposits. Centre of Europe Publishing House, State Geological Survey of Ukraine, National Academy of Science of Ukraine, Kyiv-Lviv
Sheppard SMF (1977) The Cornubian batholith, SW England: D/H and $^{18}O/^{16}O$ studies of kaolinite and other alteration minerals. J Geol Soc 133:573–591. https://doi.org/10.1144/gsjgs.133.6.057
Sonkin LS, Vozianova TG, Fokanova NM (1983) Epigenetic montmorillonite in kaolin deposits in the southwestern Ukrainian Shield. Int Geol Rev 25(7): 797–802. https://doi.org/10.1080/00206818309466768
Sousa DJL, Varajão AFDC, Yvon J (2006) Geochemical evolution of Capim river kaolin, Northern Brazil. J Geochem Explor 88:329–331
Sousa DJL, Varajão AFDC, Yvon J, Da Costa GM (2007) Mineralogical, micromorphological and geochemical evolution of the kaolin facies deposit from the Capim region (Northern Brazil). Clay Miner 42:69–87. https://doi.org/10.1180/claymin.2007.042.1.06
Steefel CI, Van Cappellen P (1990) A new kinetic approach to modeling water-rock interaction: the role of nucleation, precursors, and Ostwald ripening. Geochim Cosmochim Acta 54:2657–2677. https://doi.org/10.1016/0016-7037(90)90003-4
Takagi T, Shin KC, Jige M, Hoshino M, Tsukimura K (2021) Microbial nitrification and acidification of lacustrine sediments deduced from the nature of a sedimentary kaolin deposit in central Japan. Nat Sci Rep 11: 3471. https://doi.org/10.1038/s41598-021-81627-4
Taylor G, Eggleton RA, Foster LD, Tilley DB, Le Gleuher M, Morgan CM (2008) Nature of the Weipa bauxite, northern Australia. Aust J Earth Sci 55:45–70. https://doi.org/10.1080/08120090802438241
Trolard F, Bilong P, Guillet B, Herbillon AJ (1990) Halloysite—kaolinite—gibbsite—boehmite: a thermodynamical modelisation of equilibria as function of water and dissolved silica activities. In: Geochemistry of Earth's surface and of mineral formation. 2nd international symposium, 2–9 July 1990, Aix en Provence, pp 294–297. https://doi.org/10.1016/0009-2541(90)90243-Z
Wilson LR (2004) Kaolin and halloysite deposits of China. Clay Miner 39:1–15. https://doi.org/10.1180/0009855043910116
Wilson MJ, Jones D (1983) Lichen weathering of minerals: implications for pedogenesis. In: Wilson RCL (ed) Residual deposits: surface related weathering processes and materials. Geological Society of London, Blackwell Science Publication, Boston, pp 5–12. https://doi.org/10.1144/GSL.SP.1983.011.01.01
Wilson IR, de Souza Santos H, de Souza Santos P (2006) Kaolin and halloysite deposits of Brazil. Clay Miner 41: 697–716. https://doi.org/10.1180/0009855064130213
Zielinski RA (1985) Element mobility during alteration of silicic ash to kaolinite—a study of tonstein. Sedimentology 32:567–579. https://doi.org/10.1111/j.1365-3091.1985.tb00471.x

Ion Adsorption Mineralisation in Regolith-Hosted REE Deposits

Charles D. Beard, Eva Marquis, Martin P. Smith, Martin Yan Hei Li, Guillaume Estrade, and Kathryn M. Goodenough

Abstract

Supergene deposits are an attractive source of rare earth elements (REE) because of the potential for reduced energy impact of mining relative to hard-rock deposits, for grade enrichment during the breakdown of rock-forming minerals, and for the potential separation of REE from deleterious radionuclides in the weathering environment. Ion adsorption deposits are a subclass of regolith-hosted REE deposits with the lowest cut-off grade of any currently producing REE deposit type (ca. 0.1 wt. % total rare earth oxides). They are formed via the breakdown of REE-bearing igneous, metamorphic and, less commonly, sedimentary rocks during supergene weathering, which releases the REE from rock-forming and accessory mineral phases. REEs are then carried within soil water, with the depth of maximum REE deposition commonly being localised by a sharp vertical gradient in soil water pH or redox, for example, where acidic surface-derived water interacts with higher pH groundwater. Here, REEs are adsorbed onto the surface of clay minerals, most commonly kaolinite and halloysite. A major advantage of ion adsorption deposits is the ease of REE extraction, with all currently active operations making use of relatively simple ammonium sulfate leach solutions. In situ, and heap or tank leaching approaches have been used. Today, extraction from ion adsorption deposits has a similar energy use and carbon emission potential per mass REE to operations at the hard-rock carbonatite REE deposits Mountain Pass and Bayan Obo. However, the 2D nature of IADs and low cut-off grade can result in greater risks of negative environmental impacts, such as slope instability, vegetation loss and groundwater contamination. Extraction from higher-grade IADs

C. D. Beard (✉)
School of Applied Sciences, University of Brighton, Brighton, UK

Department of Earth Sciences, University of Cambridge, Cambridge, UK

Department of Earth Sciences, Utrecht University, Utrecht, Netherlands
e-mail: c.d.beard@uu.nl

E. Marquis
Camborne School of Mines, University of Exeter, Cornwall, UK

M. P. Smith
School of Applied Sciences, University of Brighton, Brighton, UK

M. Y. H. Li
Department of Earth Science and Engineering, Imperial College London, London, UK

G. Estrade
Laboratoire Géosciences Environnement Toulouse (GET), Université Toulouse 3 Paul Sabatier, Toulouse, France

K. M. Goodenough
British Geological Survey, The Lyell Centre, Edinburgh, UK

R. J. Bowell, C. R. M. Butt (eds.), *Geology, Geochemistry and Formation of Supergene Mineral Deposits in Deeply Weathered Terrain*, Mineral Resource Reviews, https://doi.org/10.1007/978-3-031-75733-4_14

formed on REE-rich protoliths and the use of novel bioleaching cultures have the potential to increase recovery and significantly reduce environmental risk. Several prospects outside of China have now progressed to a detailed characterisation stage. It is, therefore, anticipated that ion adsorption deposits will continue to make important contributions to global REE production over the coming decade.

1 Introduction

The global transition to renewable energy infrastructure is driving rapid growth in the demand for 'critical raw materials' (Hodgkinson and Smith 2018), including rare earth elements (REE), which are required to manufacture permanent magnets for the generators of wind turbines and the motors of electric vehicles (Ballinger et al. 2020). In the past decade, supply networks for REE have undergone expansion and diversification (Fernandez 2017; Kynický et al. 2021). Production is presently dominated by mines on fresh or weathered carbonatites (Anenburg et al. 2021; Zhukova et al. 2021). However, extraction of REE from carbonatite deposits is complicated by a common association of the REE with radionuclides with associated financial, logistical and social acceptance pressures (Chakhmouradian and Wall 2012; Ali 2014; Jamaludin and Lahiri-Dutt 2017; Beland and Williams-Jones 2021).

Ion adsorption REE deposits (IADs), despite their relatively small tonnage and low grade, account for about 16.4% of 2023 global REE production (57,310 tonnes rare-earth element oxides [REO], Chinese production quota, Brazil, Malaysia, Myanmar (Burma)); U.S. Geological Survey 2024). Of the higher-value heavy REE (HREE; Gd to Lu + Y), these regolith-hosted IAD deposits in South China supplied the majority of global production (García et al. 2017). Ion adsorption deposits are an easily leachable type of REE mineralisation in which the REE occur in ionic form adsorbed onto mineral surfaces. They have previously been referred to as 'easily worked deposits of REE' (Neary and Highley 1984), 'weathered crust rare earth ore' (Chi and Tian 2007), 'weathered-crust elution-deposited rare earth' (Chi and Tian 2008), and 'exchangeable Rare Earth + Yttrium (MEX-REY)' (Bao and Zhao 2008). Recently, there has been a proposal by Li et al. (2017) to move away from the nomenclature that infers an ore-forming process, such as 'ion adsorption deposit' or 'weathered elution-deposited', to using 'regolith-hosted'. In some types of regolith-hosted mineralisation, the REE are hosted in forms where their extraction requires the chemical breakdown of ore minerals, for example monazite-(Ce) and xenotime-(Y) placer deposits (high radioactivity) and lateritically-weathered carbonatites (chapter "Carbonatite Hosted Residual REE Deposits"), the latter of which may have an adsorbed REE component. In addition to adsorbed REE, regolith deposits may also contain protolith REE minerals that survived weathering and secondary REE minerals that formed within the supergene environment, both of which can also represent resources of the REE but which may not be easily leachable (Li et al. 2017). 'Ion adsorption deposit' (IAD) is used herein to refer specifically to regolith-hosted REE mineralisation that is leachable by dilute electrolyte solutions. In this definition, the mineralisation within IADs does not necessarily need to be at economic grade (i.e., ore).

The relatively loosely-bound REE in ion adsorption deposits are amenable to extraction from the regolith by in situ, heap or tank leaching techniques using dilute electrolyte solutions, such as ammonium sulfate or magnesium chloride (Chi et al. 2005; Su 2009; Tian et al. 2010; Moldoveanu and Papangelakis 2012). The low economic cost of these leaching operations, compared with hard-rock mining, results in IADs having the lowest cut-off grades of all known REE deposit types (Kanazawa and Kamitani 2006; Simandl 2014; Su 2009). Presently active operations have total rare-earth oxide (TREO) grades of 0.08–0.15 wt.% for light REE (LREE)-rich deposits and 0.06–0.10 wt.% for heavy REE (HREE)-rich deposits (Ding and Deng 2013; Li et al. 2017). Whilst most technical reports state grade as rare-earth oxides (REO), a few report REE concentrations without oxygen, which in the case of neodymium is a 14.3% lower value than Nd_2O_3 for the equivalent amount of REE metal. Regolith profiles hosting ion adsorption ores are typically 6 to 30 m thick, with most less than 10 m thick, although some exceptional profiles can attain local thicknesses of 60 m (Hoshino et al. 2016). Ion adsorption deposits can avoid the radioactivity problems inherent to some other types of REE deposits (Wall et al. 2017; Beard et al. 2023) due to the separation of REE from U and Th during weathering. In the Ambohimirahavy deposit, Madagascar, leachable U and Th concentrations are several orders of magnitude lower than the REE and show little to no correlation with leachable REE concentration, indicating different host phases and behaviour during weathering (Estrade et al. 2019). The breakdown of rock-forming minerals additionally results in lower energy requirements for mining and often results in the enrichment of the REE in the regolith relative to its unweathered protolith by a factor of 2–5 times (Bao and Zhao 2008; Yang and Xiao 2011).

Several comprehensive reviews of ion adsorption deposits have been published in the past two decades, with significant contributions by Chi and Tian (2008), Bao and Zhao (2008), Hoshino et al. (2016), Sanematsu and Watanabe (2016), Li et al. (2017) and Zhou et al. (2020). Recent advances in the understanding of the characteristics and genesis of these deposit types, the importance of these deposits to global HREE+Y supply and outstanding questions around REE transport mechanisms, comparisons between deposit types (including HREE versus LREE enriched deposits), and global resource potential outside of China make this review timely. This article synthesises observations from ion adsorption deposits and presents them in a mineral systems framework.

2 Common Fundamentals of Ion Adsorption Deposits

Ion adsorption REE mineralisation occurs in regolith profiles formed via supergene weathering of granites and a variety of other rock types. The regolith profiles are generally subdivided into topsoil, pedolith, saprolite and saprock facies (Fig. 1, Eggleton 2001; Taylor and Eggleton 2001). However, some literature on ion adsorption deposits instead subdivides regolith profiles based on the extent of weathering and distribution of ion-adsorbed REE (Wu et al. 1990; Zhang 1990; Bao and Zhao 2008; Sanematsu and Watanabe 2016; Hoshino et al. 2016).

Ion adsorption REE deposits (IADs) form in temperate tropical climates (Sanematsu and Watanabe 2016), where the major and sometimes minor mineralogical components of rocks are chemically broken down in the near-surface environment. Released REEs are transported in soil waters, then loosely trapped via adsorption onto the surface of minerals, especially clays, oxides and oxyhydroxides, within the soil profile. Ion adsorption deposits commonly occur beneath mature pedolith profiles, from which REE have been leached by acidic soil waters during the breakdown of REE-rich protoliths (Fig. 1). Ion adsorption mineralisation is commonly localised by steep gradients in pH coincident with the pedolith–saprolith interface (Fig. 1).

Following the mineral systems framework (Knox-Robinson and Wyborn 1997; McCuaig and Hronsky 2017), the critical components required for the formation and preservation of IADs are as follows:

1. ENERGY: A humid temperate to tropical climate or palaeoclimate that facilitates weathering. The extent of weathering must release sufficient REE from major rock-forming minerals and minor REE-rich accessory phases.
2. LIGANDS: Ligand concentrations are typically low during weathering, but transport of the REE may be facilitated by the formation of humic acids in the soil zone (Bai et al. 1989; Bao and Zhao 2008; Chi and Tian 2008; Sanematsu and Watanabe 2016; Li et al. 2017).
3. SOURCE: Rocks that contain appreciable concentrations of REE, including granites, syenites and, potentially, carbonatites. The extent of protolith breakdown and the amount

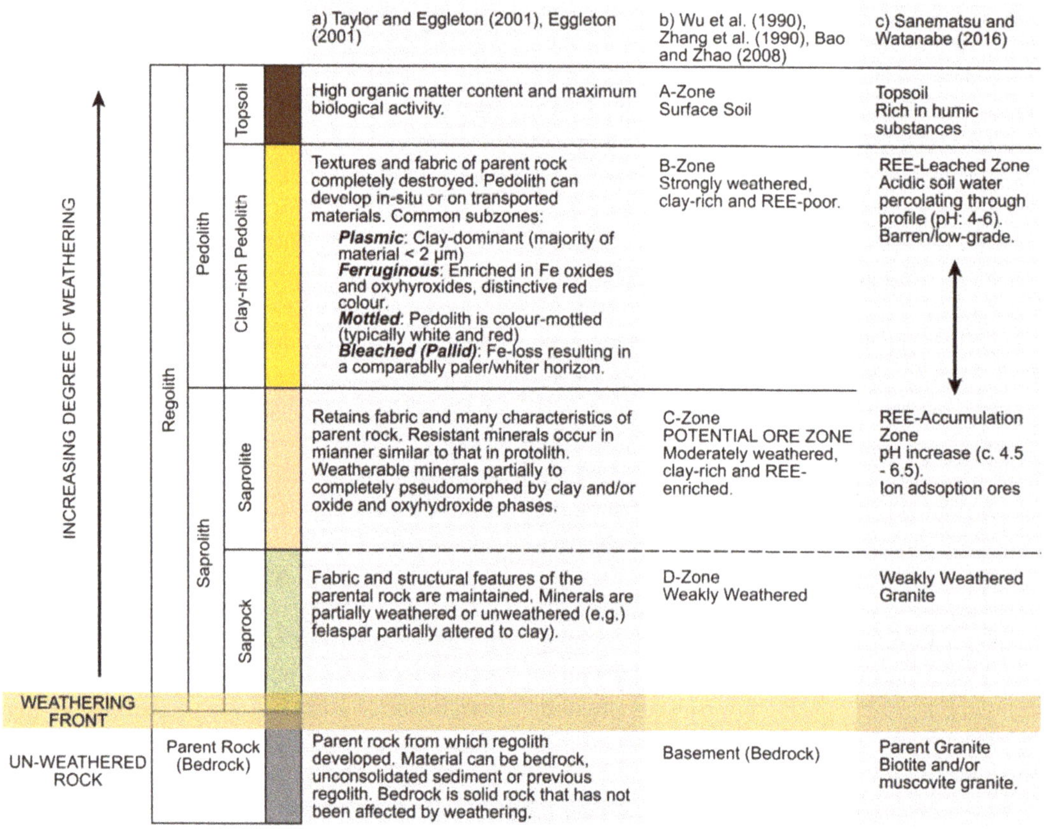

Fig. 1 Subdivisions of regolith profiles showing: (a) standard subdivision of regolith from Eggleton (2001) and Taylor and Eggleton (2001); (b) subdivision of Chinese regolith profiles containing ion adsorption ores as reported by Wu et al. (1990), Zhang (1990) and Bao and Zhao (2008); (c) subdivision of ion adsorption deposit regolith into REE leached and accumulation zones, which are controlled by changes in pH within the saprolite and pedolith proposed by Sanematsu and Watanabe (2016)

of REE released during weathering will be influenced by rock texture as well as the mineralogy and REE content of both rock-forming and accessory phases (Li et al. 2022). Hydrothermal alteration, particularly in the case of granites, can form REE minerals that are susceptible to weathering (Bao and Zhao 2008; Li et al. 2019).

4. TRANSPORT: Soil waters containing dissolved REE must flow from the 'source' to the 'trap'. Most ion adsorption mineralisation occurs relatively local to the source rocks. Vertical transport of REE on the order of 10s–100s metres has been reported within regolith profiles (see Sect. 6 below). Critically, the fluid containing REE sourced from weathering can be dispersed into the environment (no deposit) or may saturate secondary supergene mineral phases that sequester the REE and are not easily processable via leaching (potentially non-economic deposit, Li et al. 2017). Hydrology is controlled by the geomorphology of the landscape, which is, in turn, controlled by the geology. Large granitic batholiths tend to form shallow-angled slopes and plateaux, which are conducive to low throughput of groundwater (see also Sect. 7 below).
5. TRAP: The regolith must contain minerals with sorption capacity for REE. In the majority of case studies, the 'trap' minerals are halloysite or kaolinite clays (Wu et al. 1990; Kanazawa and Kamitani 2006; Bao and Zhao 2008; Sanematsu and Watanabe 2016; Li and

Zhou 2020), which are weathering products of feldspars, feldspathoids and micas in the same protolith that supplies the REE. REE sourced via the weathering of carbonatite may theoretically adsorb onto clay produced through the weathering of adjacent silicate rocks. Hydrogeology may also influence trapping resulting from vertical changes in permeability within laterite profiles (Bonsor et al. 2014).

6. OUTFLOW: The upper part of the regolith profile must be located above the water table, at least seasonally, to allow for the throughflow of acidic rainwater that is equilibrated with atmospheric CO_2 (Domènech et al. 2017). This is important for the breakdown of rock-forming and REE-rich accessory minerals of the protolith.
7. PRESERVATION: Relatively flat topography is required to hamper the effect of erosion. Most of the South China deposits are located on hills with slopes of less than 30 degrees (to a maximum gradient of 35 degrees, Li et al. 2017). Under certain conditions, granite batholiths may weather landscapes with km-scale plateaux and relatively shallow slope angles. These relatively large, structurally-simple geological features result in higher preservation potential for IADs than strongly heterogeneous geology, for example, ring dykes at the Tantalus Project, Madagascar (Estrade et al. 2019). However, weathering of granites to produce steep-sided pinnacles (inselbergs) will severely reduce preservation potential (Ruxton and Berry 1957). Ion adsorption deposits may also be degraded by overly extensive weathering. Under very humid tropical conditions, the formation of bauxite and the associated breakdown of clay minerals (Soler and Lasaga 1996) can reduce the sorption capacity of the soil profile and lead to the dispersion of REE into the environment.

3 Drivers of Weathering

Warm, humid conditions are a key driver for the development of thick regolith profiles that host ion-exchangeable REE (Bao and Zhao 2008; Sanematsu and Watanabe 2016). A tropical to temperate climate, or palaeoclimate, provides the necessary source of energy required to drive chemical weathering processes that break down rocks and minerals. The vast majority of actively mined IADs are restricted to South China and along the China-Myanmar border (Li et al. 2017; Sanematsu et al. 2016). Similar systems and potential IAD prospects have been recognised in the Americas, Africa and across SE Asia (see Fig. 2 and Sect. 6 below). Each of these occurrences is located in temperate to tropical climatic zones of the Köppen-Geiger classification (Beck et al. 2018; Fig. 2).

The influence of the timescale and intensity of weathering on IAD formation and seasonality are not yet well quantified. Field observations of basalt weathering rates on the Hawaiian Islands (Nelson et al. 2020), a reactive transport model for the formation of Ni-laterite from serpentinised peridotite in Cuba (Domènech et al. 2017) and a short-lived U-Th isotope disequilibria study on weathering of an alkaline granite in Amazonia, Brazil (Mathieu et al. 1995) all indicate laterite formation timescales of 100s of Ka to Ma for regolith with thickness greater than 10 m. This corresponds to a propagation rate of the weathering front into the protolith of about 5–10 cm/Ka. A sensitivity analysis of the reactive transport model (Domènech et al. 2017) highlighted the importance of porewater equilibration with atmospheric CO_2, the dissolution of which imparts acidity that facilitates the breakdown of primary minerals in the protolith. Where the soil profile is water saturated, there is limited input of dissolved atmospheric gas to the weathering system (Butt and Cluzel 2013), resulting in lower rates of chemical weathering. This highlights the importance of drainage and groundwater throughput in the formation of IADs.

Overly extensive weathering and the formation of bauxite regolith can degrade IAD mineralisation through the breakdown of clay minerals, such as kaolinite, to gibbsite [$Al(OH)_3$]. Reactive transport modelling by Soler and Lasaga (1996) highlights that acidic soil water (< pH 3.5) stabilises kaolinite at the expense

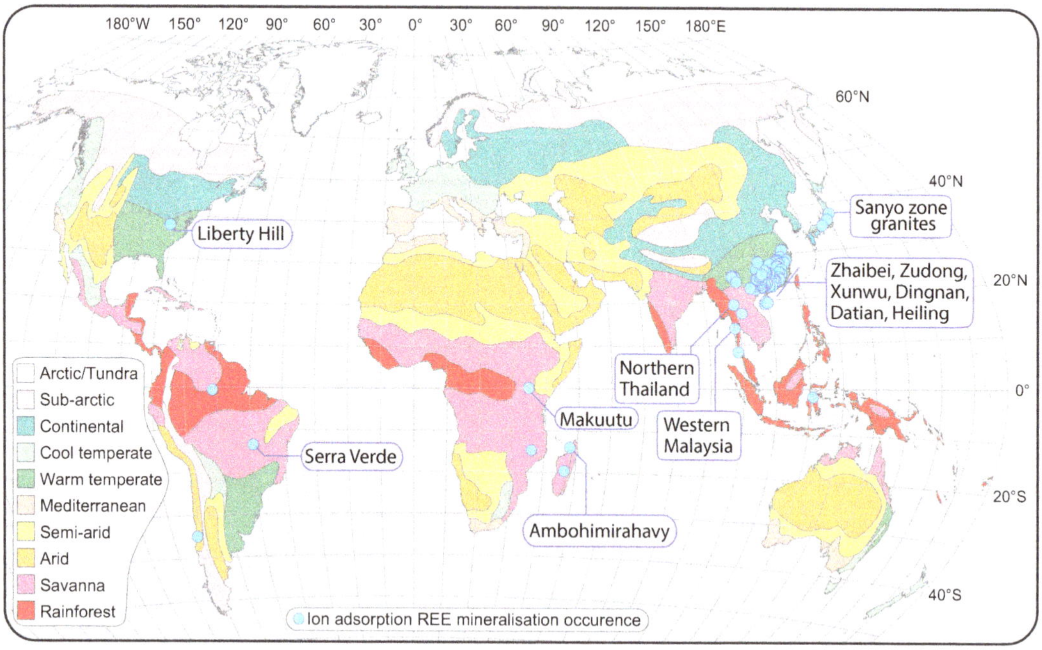

Fig. 2 A world map showing the global distribution of known regolith-hosted REE deposits with localities named in the text. The Köppen-Geiger climate zones are from Beck et al. (2018). (Winkel-Tripel projection)

of gibbsite, suggesting that bauxites cannot form via weathering by waters equilibrated with a CO_2-rich atmosphere or in regions with high soil humic acid contents as a result of enhanced plant activity. Rather, they reflect the weathering effect of weakly acidic groundwater in the presence of limited dissolved SiO_2 (<1–2 g/L; Bogatyrev et al. 2009).

4 Source: Protolith Composition, Texture and Structure

4.1 Protolith

The majority of economic IADs occur in regolith derived from granitic intrusions (e.g., Bai et al. 1989; Huang et al. 1989; Wu et al. 1990; Bao and Zhao 2008; Murakami and Ishihara 2008; Wang et al. 2015; Xu et al. 2017; Li et al. 2019, 2020), and felsic volcanic (cf. Li et al. 2017). These are mostly muscovite-, biotite- and muscovite-biotite-granite/granitoid intrusions of calc-alkaline to alkaline composition (Hoshino et al. 2016; Li et al. 2017, Table 1). Heavy REE-rich ion adsorption ores in southern China are situated on highly differentiated muscovite-biotite and muscovite granites with high SiO_2 (c. >75%), low P_2O_5 (c. <0.02%) and abundant alkali feldspar (Bao and Zhao 2008; Sanematsu and Watanabe 2016). A few examples are related to syenite intrusions (Le Couteur 2011), basalts (Zhang et al. 2016; Zhou et al. 2013), sedimentary successions (Ionic Rare Earths 2022) and metamorphic rocks (Huo 1992; Huang et al. 2021b).

There have been several investigations into how the mineralogy and texture of the source rock control the REE content of IAD deposits (e.g., Xu et al. 2017; Sanematsu et al. 2013, 2016; Li et al. 2019; Li and Zhou 2024; Estrade et al. 2019). The REE signature of IADs appears to be primarily controlled by that of the protolith, i.e., IADs enriched in HREE are developed upon granites that also exhibit HREE-enrichment (e.g., Zudong, S. China; Bao and Zhao 2008, Cheng et al. 2017; Li et al. 2019, Table 1).

Reviews by Bao and Zhao (2008) and Sanematsu and Watanabe (2016) both emphasise the importance of easily weatherable REE-bearing accessory minerals in the formation

Table 1 Summary of major (medium, large, or super-large) Chinese IAD variants. (Collated from Sanematsu and Watanabe (2016) and Li et al. (2017))

Type	Count	TREO grade [wt. %]	Parent rock type(s) [count]	Examples
Highly HREE-enriched	4	0.074–0.20	Muscovite granite [1] Muscovite-bearing granite [2] Two mica alkali granite [1]	Zudong, Datian, Xiawentian, Zhaibeiding
HREE-enriched	2	0.11–0.121	Two mica alkali granite [1] Biotite granite [1]	Xiazhuang, Guposhan
Highly LREE-enriched	25	0.06–0.38	Amphibole biotite granite [1] Biotite granite [10] Biotite monzo-granite [2] Biotite monzonite [5] Granite porphyry and rhyolitic lava [2] Granite porphyry and rhyolitic tuff [1] Granodiorite [1] Quartz syenite [1] Two mica granite [1] Two-mica monzo-granite [1]	Dingnan, Heling Bachi, Wuliting Anxi, Hanfang, Yangbu (Aozixia), Nanqiao, Xiache, Nantang, Hechang, Dabu, Danqian, Yangcun Nanshan, Gangxia Renju, Wujingfu Xiachuang, Laishi Shatian, Gonghe Datang, Jinjiling, Huashan
LREE-enriched	15	0.05–0.365	Alkaline syenite [1] Biotite granite [6] Biotite granite and fine-grained alkali granite [1] Biotite syeno-granite [1] Granite [2] Hornblende- and biotite-bearing monzonite [1] Monzo-granite [3]	Guanxi, Fukeng, Pitou, Dafuzu, Yijiang, Xishan Guposhan, Dazhou, Ancun Long'an, Longbazheng, Qinghu, Majiang, Xiaotung

of IADs. In general, the granitic protoliths of the deposits have undergone alteration, either due to deuteric fluids (Sanematsu and Watanabe 2016) or the infiltration of REE-enriched carbonate fluids (Xu et al. 2017). Also, multiple alteration would be necessary for progressive REE enrichment and replacement of refractory minerals by easily weathereable REE minerals in the granitic protoliths (Li and Zhou 2024). This alteration can replace common weathering-resistant primary magmatic REE-bearing minerals, such as allanite-(Ce), xenotime-(Y) and zircon, with REE-bearing phases that are relatively amenable to dissolution during weathering (Li et al. 2019; Sanematsu et al. 2013, 2016; Bern et al. 2017). High-temperature alteration appears to be an important precursor to the formation of many IADs. The thermodynamic stability of some REE minerals has been constrained experimentally, allowing for the prediction of optimal conditions for their dissolution during weathering (Li et al. 2022). In general, the less resistant a mineral is to dissolution, the more easily REE can be released into solution and, thus, the greater the potential for these REE to be fixed in an ion-exchangeable form.

4.2 Rock-Forming Minerals

During intense chemical weathering, rock-forming minerals decompose to a variety of weathering products (Fig. 3). While the REE content of most rock-forming minerals is low, their weathering can have a key influence on the formation of ion adsorption REE mineralisation.

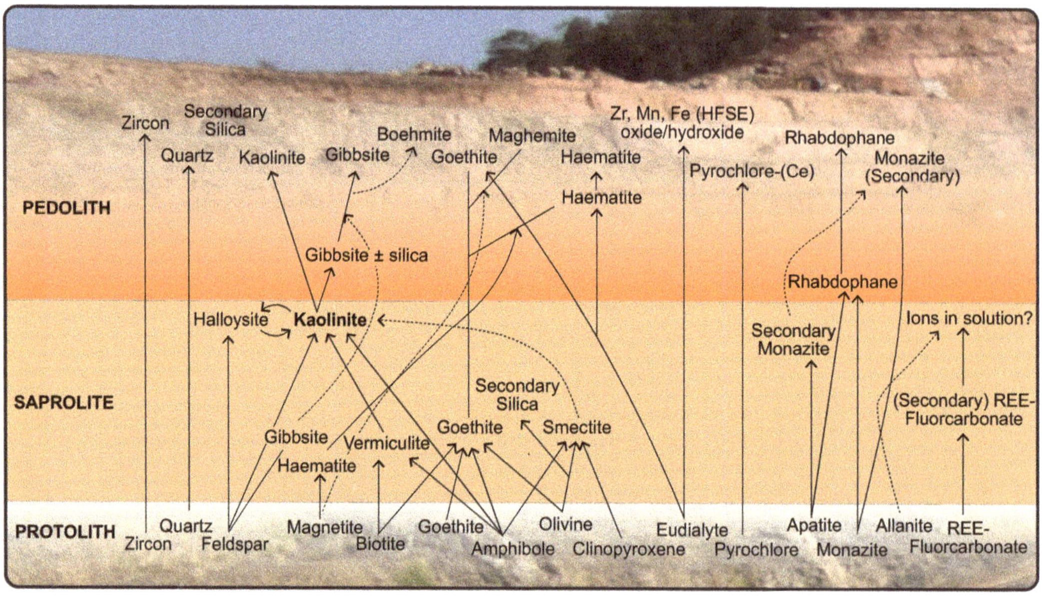

Fig. 3 Pathways for formation of secondary phases during chemical weathering. (After Anand (2005) and supplemented with data from Betekhtin (1956), Zakharov et al. (2011), Valeton et al. (1997), Lottermoser and England (1988), Lottermoser (1990) and Price et al. (2005))

Alumina-bearing phases such as feldspar, amphibole, clinopyroxene and biotite weather to form clay minerals, including kaolinite, halloysite and smectite (Wilson 2004; Anand 2005; chapter "The Geochemistry of Supergene Processes"; and references therein) that may subsequently adsorb REE. The modal proportion of rock-forming minerals is orders of magnitude higher than accessory minerals therefore, while their REE content may be low, their breakdown may still be a significant contributor to the resulting grade of easily leachable REE (e.g., garnet, epidote, clinopyroxene, amphibole all contain significant REE; Hermann 2002; Gregory et al. 2012; Beard et al. 2019; Vasyukova and Williams-Jones 2019).

4.3 Mineralogy of Accessory Phases

Most of the REE released during weathering is from the breakdown of accessory phases in the source rock (Sanematsu and Watanabe 2016; Li et al. 2017, Table 2). Weathering susceptibility varies with mineralogy; therefore, the mineralogy of the protolith, especially of its accessory phases, exerts a strong control on the mobility and fractionation of REE in regolith (e.g., Nesbitt 1979; Braun et al. 1990, 1993; Aubert et al. 2001; Laveuf and Cornu 2009; Stille et al. 2009). The kinetics of mineral dissolution can also be accelerated by radiation damage to the crystal lattice (Ewing and Haaker 1980; Ewing 1994; Price et al. 2005).

REE-fluorcarbonate minerals (the bastnäsite group) are particularly susceptible to dissolution in acidic soil water (Huang et al. 1989; Sanematsu et al. 2013; Sanematsu and Kon 2013; Sanematsu and Watanabe 2016). Easily-degradable REE-fluorcarbonates are typically precipitated along grain boundaries during hydrothermal alteration or, more rarely, during the last stages of magmatic differentiation in CO_2-bearing magmatic systems (e.g., Huang et al. 1989; Berger et al. 2008; Ishihara et al. 2008; Sanematsu et al. 2013).

Allanite-(Ce) and titanite are important sources of LREE and HREE, respectively, for

Table 2 Primary and secondary REE-rich mineral phases associated with the formation of ion adsorption-type deposits. The Levinson modifier [e.g., –(Y)] denotes the dominant element(s)

Phase	Formula	Environment
Aeschynite-(Y)	$REETiNbO_6$	Granite, syenite
Allanite-(Ce)	$(Ca,Ce)(Al_2Fe^{2+})(Si_2O_7)(SiO_4)O(OH)$	Granite
Britholite-	$(REE,Ca)_5[(Si,P)O_4]_3(F,OH,Cl)$	Nepheline syenite
Epidote	$Ca_2\{Al_2Fe^{3+}\}(Si_2O_7)(SiO_4)O(OH)$	Metamorphic rocks
Eudialyte	$Na_{15}Ca_6Fe_3Zr_3Si(Si_{25}O_{73})(O,OH,H_2O)_3(Cl,OH)_2$	Nepheline syenite
Euxenite-(Y)	$REENb_2O_6$	Granite, syenite
Fergusonite-(Y)	$REENbO_4$	Granite, syenite
Gadolinite-(Y)	$Y_2Fe^{2+}Be_2Si_2O_{10}$	Granite, pegmatite
Hingganite-(Y)	$(Y,REE,Ca)_2(\square,Fe^{2+})Be_2[SiO_4]_2(OH)_2$	Granite, pegmatite
Monazite-(Ce)	$Ce(PO_4)$	Various, incl granite
Parisite-(Ce)	$CaCe_2(CO_3)_3F_2$	Granite, syenite, carbonatite
Pyrochlore-(Ce)	$(Na,Ca,REE,Pb,U)_2Nb_2O_6(F,O,OH)$	Syenite, carbonatite
Synchysite-(Y)	$CaY(CO_3)_2F$	Granite, pegmatite
Titanite	$CaTi(SiO_4)O$	Granite, syenite
Waimirite-(Y)	YF_3	Hydrothermal
Xenotime-(Y)	$Y(PO_4)$	Various, incl granite
Yttrialite-(Y)	$(Y,Th)_2Si_2O_7$	Granite, pegmatite
Zircon	$ZrSiO_4$	Various, incl granite
Chernovite-(Y)	$Y(AsO_4)$	Secondary
Cerianite	CeO_2	Secondary
Rhabdophane-(Ce)	$Ce(PO_4).H_2O$	Secondary

ion adsorption ores (Sanematsu et al. 2013, 2016; Sanematsu and Watanabe 2016). The dissolution of allanite-(Ce) and titanite during acidic weathering is supported by the scarcity, or quantitatively reduced abundance, of these minerals in weathered profiles and clastic sediments derived from allanite-(Ce) and titanite-bearing protoliths (Deer et al. 1982; Banfield and Eggleton 1989; Braun and Pagel 1994; Condie et al. 1995; Harlavan and Erel 2002; Gieré and Sorensen 2004; Price et al. 2005; Sanematsu et al. 2013, 2016; Berger et al. 2014, Kalintsev et al. 2021).

Apatite group minerals are common REE-bearing phases, and their rate of dissolution has been shown experimentally to increase with decreasing pH (Valsami-Jones et al. 1998; Guidry and MacKenzie 2003; Chaïrat et al. 2007; Harouiya et al. 2007). The apatite dissolution rate is also promoted by increased CO_2 concentration due to the exchange of Ca between apatite and calcite (Middleton et al. 2013) and may be accelerated by the combined action of trees and fungi (Smits et al. 2012).

Eudialyte-group minerals (EGMs) are an important REE host in alkaline-silicate-associated REE deposits (Kalashnikov et al. 2016; Möller and Williams-Jones 2016; Borst et al. 2018; Beard et al. 2023). During the cooling of alkaline-silicate intrusive systems, the eudialyte-group minerals are susceptible to metasomatic and hydrothermal alteration and are commonly replaced by a variety of secondary phases, including zircon, which may lock up the REE (e.g., Estrade et al. 2014, 2015, 2018; van de Ven et al. 2019). Studies of supergene alteration of eudialyte-bearing phonolite (Lage's area, Santa Catarina, Brazil; Formoso et al. 1989) and nepheline syenite (Betekhtin 1961; Valeton et al. 1997; Zakharov et al. 2011) showed that eudialyte is not retained in regolith profiles. Betekhtin (1956) reported the accumulation of reddish black porous Zr and Fe hydroxides in the place of eudialyte, and Valeton et al. (1997) proposed that HFSE derived from the decomposition of EGMs are incorporated in oxide, hydroxide and amorphous phases of similar composition.

Ram et al. (2019) demonstrated using m-XANES that Zr can occur as an adsorbed ionic species on clays and Mn-oxyhydroxides alongside secondary oxyhydroxides.

Other uncommon accessory minerals, including gadolinite, hingganite and yttrialite, have been identified as important hosts of REE, especially the HREE, in granite and granitic pegmatite. For example, in the Zudong deposit in South China, these minerals are believed to host a significant amount of HREE in the parental muscovite granite (Li et al. 2019). Thermodynamic modelling suggests that these minerals are susceptible to weathering and would release the REE they host to the soil water, probably in incipient weathering (Li et al. 2022).

Accessory minerals resistant to chemical weathering include zircon, monazite-(Ce), xenotime-(Y), and niobate-group minerals. This resistance to dissolution is highlighted by the occurrence of these minerals in placer deposits (e.g., xenotime and monazite sands) and clastic sediments (Orris and Grauch 2002). Even in low pH aqueous solutions, the solubility of monazite-(Ce) and xenotime-(Y) is relatively low, although metamictisation may accelerate their dissolution (Petit et al. 1985; Cetiner et al. 2005). This resistance to weathering is evident at the Zudong intrusion, where Huang et al. (1989) showed that there was no notable reduction in the abundance of monazite-(Ce) and xenotime-(Y) during the weathering of the muscovite granite. Xenotime-(Y), fergusonite-(Y), and euxenite-(Y) (see Table 2) largely reside as relict minerals in the regolith (Li et al. 2019). Residual zircon crystals are also common in ion adsorption ores (Huang et al. 1989; Sanematsu and Kon 2013; Estrade et al. 2019).

5 Transport: Soil Water Chemistry and Bio-Geochemical Processes

In acidic solutions at ambient conditions, REEs are released into the weathering profile during the dissolution of REE-bearing minerals and are transported in solution downwards through the depth of the weathered profile (Sanematsu and Watanabe 2016), as well as downslope (Li et al. 2020, Liu et al. 2022). The REEs are believed to be transported primarily as complexes with humic, carbonate and bicarbonate ligands or as free ions in solution (Wood 1990; Sanematsu and Watanabe 2016; Li et al. 2022). Experimental studies indicate that carbonate, bicarbonate and humic ligands bind more strongly with the HREE than the LREE (Cantrell and Byrne 1987; Lee and Byrne 1993; Takahashi et al. 1997; Sonke and Salters 2006; Pourret et al. 2007). Bio-geochemical processes also influence REE and Y distributions in the regolith, including via solid-solution partitioning, oxidation-reduction reactions, bio-uptake and plant-influenced hydrological flow patterns (*c.f.*, Vázquez-Ortega et al. 2016). Although LREE/HREE fractionation is possible via variations in complexation behaviour in solution, the dominant control on the REE pattern of IADs appears to be the REE pattern of the source material and the susceptibilities of its REE-minerals to breakdown during weathering (see above; Sanematsu and Watanabe 2016; Estrade et al. 2019).

Studies of deposits in the Jiangxi area, China, have indicated that local groundwater is neutral to slightly alkaline and carbonate-rich, with HREE mainly transported as anionic dicarbonate complexes [$REE(CO_3)_2^-$] and LREE mainly transported as cationic carbonate complexes (Li et al. 2022). The stability of HREE-dicarbonate complexes in solution can potentially impede the adsorption of HREE onto kaolinite and halloysite. Therefore, in these deposits, the HREE preferentially stayed in the soil waters and were transported in groundwater towards the base of the slope, whereas the LREE were adsorbed on clay minerals in the regolith on the slope top. The regolith at the slope toe shows slight HREE enrichment, especially in the saprock, suggesting that the transported HREE have been fixed here. Because of their distinct aqueous solution chemistry, catchment-wide REE fractionation patterns are interpreted to result from a difference in aqueous complexation behaviour between the LREE and HREE (Li et al. 2020).

Fractionation of Ce relative to the other REE is common in weathered profiles, with positive Ce anomalies forming high in the regolith profile via oxidation of $Ce^{3+}$to Ce^{4+} (Fig. 4), which is responsible for the precipitation of cerianite [(Ce, Th)O_2], a phase that is relatively insoluble at surface conditions (Braun et al. 1990). The formation of cerianite can effectively separate Ce from the other REE and results in negative Ce anomalies lower in the regolith profile in which IADs may occur (Fig. 4).

The actinides (U + Th) may also be separated from the REE during weathering. Cations of Th^{4+} released during the breakdown of REE-Th bearing minerals are typically sequestered high in the weathering profile through precipitation of thorianite [ThO_2]. Uranium is typically oxidised during weathering from U^{4+} to U^{6+} and does not appear to accumulate significantly in the regolith (e.g., Estrade et al. 2019). Uranium is most probably diluted and dispersed into the wider environment.

6 Trap–Adsorption Mechanisms, Mineral Hosts and Preservation

The adsorption and retention of REE ions on mineral surfaces within regolith is controlled by a balance between their chemical activity on surfaces and their activity in solution. The former is controlled by the nature and surface area of the adsorption material, and the latter by the pH and ionic strength of the soil waters. Kaolinite and halloysite are the most common clays that adsorb REE in the IADs in southern China (Wu et al. 1990; Bao and Zhao 2008; Murakami and Ishihara 2008; Chi and Tian 2008; Li and Zhou 2020, 2023). Cation exchange of REE is also possible on other clay minerals [e.g., illite, particularly for syenitic protoliths (e.g., Wang et al. 2024)] and amorphous materials, which have permanent charge and/or pH dependent surface charge (variable charge; Schoonheydt and Johnston 2013). Alshameri et al. (2019) determined that the order of adsorption efficiency for the REE cations La^{3+} and Yb^{3+} from greatest to least was montmorillonite > muscovite > illite > kaolinite, but that kaolinite had the highest extraction efficiency; i.e., released the greatest proportion of its adsorbed REE when leached. Adsorption of REE in IADs is dominated by kaolinite and halloysite because these are the dominant clay minerals in the deposits. Thus, the boundary between the REE accumulation zone and the REE-leached zone is defined by a change in pH (Fig. 4). It has been theorised that this boundary shifts seasonally with changes in groundwater level (Sanematsu and Watanabe 2016).

As noted above, the saturation of cerianite [(Ce, Th)O_2] effectively removes Ce from the soil water high in the regolith profile. Fractionation among the remaining REE during adsorption processes is difficult to constrain through observations of natural systems. Sorption experiments on smectite and kaolinite indicate that the HREE are selectively adsorbed at high ionic strengths, but at the ionic strengths in soil and groundwater, there is minimal fractionation of the REE (Coppin et al. 2002). In the presence of humic acid, the adsorption of REE onto kaolinite appears to decrease with increasing pH. Furthermore, in the presence of significant humic acid, the LREE have a greater affinity for adsorption onto kaolinite than the HREE (Takahashi et al. 1997; Sonke and Salters 2006; Wan and Liu 2006).

The mechanism of adsorption of the REE onto clay mineral surfaces has been studied experimentally for many years, mainly focused on the behaviour of the lanthanides as proxies for actinides. Speciation studies of REE in aqueous fluids at ambient temperatures show that smaller heavy REE^{3+} cations (Gd–Lu) favour an eight-fold hydration sphere, whereas the larger light REE^{3+} cations (La–Eu) favour nine-fold coordination (Tanaka et al. 2008; Finck et al. 2017). On the basis of experimental sorption and desorption studies at low pH (< 6) and ambient temperature, the REE are proposed to adsorb to clays via electrostatic bonding as hydrated outer-sphere 8- or 9- co-ordinated cations (Coppin et al. 2002). At higher pH and at higher ionic strength than typical for soil pore waters during weathering, desolvation of hydrated REE ions may lead to

Fig. 4 A schematic model for the genesis of ion adsorption type REE mineralisation within weathered profiles developed on granite bedrock. (Modified after Sanematsu and Watanabe 2016). There is no distinction between pedolith and saprolite; instead, the key divisions are REE-leached and REE-accumulation zones. Chondrite-normalised plots of typical REE patterns for the REE-leached zone, REE-accumulation-zone and bedrock show changes in Ce-anomaly from positive to negative to none, indicating changes in cerium oxidation state. The background photo of the Catalão II mine in Brazil shows lateral variations in the thickness of regolith units. (Photograph: Charles Beard, May 2016)

the formation of permanently bound REE^{3+} complexes with bridging oxygens on the clay mineral surface (Takahashi et al. 1998, 2000, 2004; Coppin et al. 2003; Huittinen et al. 2010; Moldoveanu and Papangelakis 2012, 2013). The nature of adsorbed REE has been determined directly by extended X-ray adsorption fine structure (EXAFS) spectra of Y, and in some cases, Nd from sites in China, Japan and Madagascar (Yamaguchi et al. 2018; Ram et al. 2019; Borst et al. 2020). Each of these studies concluded that the REE^{3+} ions were dominantly held within IAD ore as easily leachable 8-to 9-coordinated outer-sphere hydrated complexes, dominantly onto kaolinite, with a small fraction of the REE potentially being adsorbed more strongly by the formation of inner-sphere complexes. Mukai et al. (2020) used nano-SIMS to demonstrate that La and Y were more concentrated on illite compared to kaolinite in a single sample from Jiangxi province, China and concluded that the REE were dominantly held as inner-sphere complexes that suppressed extraction by ammonium sulfate leaching. Feng et al. (2021) concluded on the basis of experimental studies and X-ray photoelectron spectroscopy that easily-leachable REE could also be adsorbed onto kaolinite at high pH (10) where the dominant form was as electrostatically bound hydroxides ($REE(OH)^{2+}$). Other phases that may hold adsorbed REE include Fe and Mn oxyhydroxides, although leaching studies suggest that this form does not predominate and is not recovered using standard lixiviant solutions (e.g., Sanematsu and Kon 2013; Li et al. 2019). In addition to mineral and solution chemistry, the texture of clay minerals within the regolith represents a strong control on their metal sorption capacity. Variable grain size, and hence surface area, of natural clays have a significant influence on their adsorption capacity (e.g., Churchman et al. 2006; Ugwu and Igbokwe 2019).

Taken together, these studies indicate that in the majority of IADs, the easily leachable fraction of REE^{3+} ions is held on the surfaces of kaolinite and halloysite clays as 8- or 9-coordinated outer-sphere complexes. The relatively weak electrostatic bonds holding the REE in place account for the ease of leaching. The remainder of the REE tenor of these deposits is in relatively resistant primary minerals that survived the weathering of the protolith and in secondary phases that formed within the weathering environment. Immobilisation of REE in secondary phases can hinder the development of ion adsorption ores. For example, secondary phosphate formation (e.g., crandallite-group minerals) can significantly hinder the mobility of REE within weathered profiles (Ishihara et al. 2008; Berger et al. 2014; Sanematsu et al. 2015; Sanematsu and Watanabe 2016).

7 The Preservation Potential of Ion Adsorption Deposits

Most laterites are geologically ephemeral features, being easily eroded and dispersed into fluvial systems (Ruxton and Berry 1957) or buried beneath younger sediments (e.g., Valeton et al. 1987), with rarer older laterites being preserved at surface only on stable cratons with relatively flat topography (e.g., 5.2–5.8 Ma; Wells et al. 2019, Cretaceous; Valeton et al. 1997; chapter "The Role of Regolith Formation and Evolution in the Development of Supergene Mineral Deposits"). The preservation potential of laterite-hosted ion adsorption deposits is, therefore, controlled largely by geomorphology, climate and tectonic processes. The dominant topographic features of the major ion adsorption deposit provinces of Jiangxi, Guangdong and Guangxi, China, are hills with low-angled slopes of <35° (Li et al. 2017; Liu et al. 2022). Gently sloping topography, coupled with favourable weathering conditions, results in low rates of denudation necessary for the development and preservation of regolith profiles that can host ion adsorption mineralisation (Bao and Zhao 2008; Li et al. 2017). Flat plateau landscapes are common on cratonic granite & granite-gneiss terranes in S. America, Australia and W. Africa, though as yet, no significant IAD mineralisation has been discovered in these locations.

Hydrological and hydrodynamic processes operating in ion adsorption mineral systems require further investigation. A recent study by Huang et al. (2021a) highlighted the influence that groundwater and the groundwater table have on the development of mineralisation at the Renju REE Deposit (Guangdong Province, China). Here, REE has been leached from the permeable vadose zone and transported downward through the regolith. The underlying water-saturated zone has relatively stable, high pH conditions that facilitated the accumulation of REE via adsorption onto kaolinite and halloysite. Macro- to micro-scale features in the protolith structure may also influence the hydrological system. Microcracks and intergranular permeability can be the main pathways for hydraulic conductivity in granites (Mosquera et al. 2000; Fu et al. 2019a). The presence of micro-cracks, in addition to the coarse grain size and greater plagioclase and biotite content, is suggested to be responsible for the development of thicker regolith profiles associated with the coarser-grained granites in the Darongshan batholiths compared to their finer-grained equivalents (Fu et al. 2019a).

8 Case Studies of Ion Adsorption Deposits

8.1 Distribution

Most well-characterised ion adsorption deposits are located in southern China (Fig. 2), with an estimated total defined tenor of 1.3 Mt. REO (Xie et al. 2016; Hoshino et al. 2016). A large fraction of production has shifted West to Myanmar (Burma) within the last five years (USGS 2024). The total reported tenor of Chinese IADs is comparable to that of a single medium-sized carbonatite or alkaline-silicate associated deposit, such as Mountain Pass, USA or Strange Lake, Canada (Beard et al. 2023). Ion adsorption mineralisation was first identified in the late 1960s in the south Jiangxi Province of China (Li et al. 2019). Further, IAD discoveries were made soon after across the Guangxi, Guangdong and Hunan provinces (Xie et al. 2016; Hoshino et al. 2016), with others located on the China-Myanmar border (Fig. 2; Li et al. 2017; Sanematsu et al. 2016). Only a few IADs are actively mined outside China. Exploration has identified occurrences and economic deposits in Brazil (Rocha et al. 2013; Santana et al. 2015), the USA (Foley and Ayuso 2015; Bern et al. 2017), Thailand (Sanematsu et al. 2013, 2015), Vietnam (Mentani et al. 2010), Laos (Sanematsu et al. 2009), Philippines (Padrones et al. 2017), Indonesia (Maulana et al. 2016), Malawi (Le Couteur 2011), Madagascar (Gilbertson 2013; Berger et al. 2014; Janots et al. 2015; Estrade et al. 2019; Ram et al. 2019; Borst et al. 2020), Uganda (Ionic Rare Earths 2022) and Chile (Bustos et al. 2022; Fig. 2a). Several deposits and occurrences are described below to highlight key processes described in the preceding sections.

8.2 South China

Until recently, regolith-hosted REE deposits in South China supplied the majority of global HREE production (García et al. 2017; USGS 2024). Ion adsorption mineralisation in China typically occurs south of 28°N in warm temperate climates with annual precipitation of over 1500 mm (Bao and Zhao 2008; Sanematsu and Watanabe 2016; Beck et al. 2018). The mineralisation is mainly hosted in the regolith of thickness varying from a few metres to up to 60 m that cover low convex-concave hills with slopes of <35°. The thickness of mineralised zones within the regolith varies from a few metres to up to a few tens of metres and becomes gradually thinner towards the slope toe in some deposits (e.g., Li et al. 2020). In general, the denudation rate is low and is in equilibrium with regional-scale exhumation, allowing for the preservation of the regolith profiles (Bao and Zhao 2008; Li et al. 2017). More than 170 regolith-hosted REE deposits have been discovered in Southeast China across the Jiangxi, Guangdong, Guangxi, and Fujian provinces, and recent discoveries of large-scale deposits have been reported in the Yunnan and Guizhou provinces in Southwest China (Table 1). This ion adsorption mineralisation is principally derived by weathering of muscovite-, biotite- and muscovite-biotite-granites of calc-alkaline to alkaline composition (Hoshino et al. 2016; Li et al. 2017). These granites are typically metaluminous to weakly peraluminous, with medium to high $Na_2O + K_2O$ (ca. 8–10 wt.%) and low P_2O_5 contents (<0.08 wt.%; Sanematsu and Watanabe 2016), but with variable REE concentrations from less than 100 to ~1000 ppm (Li et al. 2017). Large deposits have formed on rhyolite, felsic tuff and granite porphyry bedrock, for example, the Heling deposit in Jiangxi province (Bao and Zhao 2008; Hoshino et al. 2016). Notably, ion adsorption mineralisation was discovered recently in regolith derived from weathering of phyllite and schist (Huang et al. 2021b) and syenite (Wang et al. 2020). For South China IADs formed on igneous protoliths, the majority were emplaced in the Nanling Range during the Mesozoic Yanshanian magmatic event (Hu et al. 2017). Numerous granite IAD protoliths also formed during the Palaeozoic Caledonian magmatic event, especially in the Fujian province (Li et al. 2017).

Generally, the style of IAD in regolith shows the inheritance of REE patterns from the protolith, with HREE-enriched mineralisation developing over HREE-rich protoliths and LREE-enriched mineralisation over LREE-rich protoliths. The protoliths for all HREE-rich IADs described so far are highly differentiated granites with high SiO_2 (c. >75%), low P_2O_5 (ca. <0.02%) and contain abundant alkali feldspar (Bao and Zhao 2008; Sanematsu and

Watanabe 2016; Li et al. 2019). Principal examples of HREE-rich ion adsorption ores include the Zudong deposit (also called the Longnan deposit) of Jiangxi Province and the Zhaibeiding deposit of Guangdong Province. Examples of LREE-rich deposits include the Heling deposit and Zhaibei deposit of Jiangxi Province and the Renju deposit of Guangdong Province. In the following text, representative deposits from Southeast China are illustrated to demonstrate common features of ion adsorption mineralisation of different styles and from different protoliths.

The Zudong Deposit, China: A Representative HREE Deposit The Zudong deposit, also known as the '701 mine', was discovered in 1969 and was the first IAD identified globally. This deposit has a current resource of ~18,000 t REO at 0.1 wt.% REO (Li et al. 2019). The Zudong deposit developed via weathering of a two-feldspar, muscovite-bearing alkaline granite that was emplaced in the late Jurassic. The majority of the REE in the granite protolith are hosted in accessory REE minerals, including synchysite-(Y), gadolinite-(Y), hingganite-(Y) and yttrialite-(Y), and subordinately in xenotime-(Y), zircon and REE niobates (Bao and Zhao 2008; Li and Zhou 2024; Li et al. 2019). Within the regolith formed on this granite, the REE mineralisation mainly occurs in the lower pedolith and upper saprolite. The concentration of REE gradually decreases from a maximum of ~2000 ppm at the pedolith-saprolite interface towards both shallower and greater depths, except Ce, which has accumulated in the upper pedolith. The REE pattern of regolith generally resembles that of the protolith, except for Ce (Fig. 5b). In the regolith, about 60–70% of the REE are ion exchangeable and probably adsorbed on clay minerals, dominantly kaolinite and halloysite. Comparatively, a larger proportion of the LREE exists in ion exchangeable form. The rest of the REE are hosted in residual REE minerals, including xenotime-(Y), zircon and REE niobates, and in supergene minerals, mainly as chernovite-(Y). Proportions of REE hosted in secondary Fe-Mn oxyhydroxides and organic matter are negligible (Li et al. 2019). The major accessory REE minerals in the protolith, notably synchysite-(Y), gadolinite-(Y) and hingganite-(Y) are absent in the regolith and have probably been completely decomposed to release the REE they host. Their weathering susceptibility is consistent with thermodynamic calculations (Li et al. 2022): With progressive weathering, feldspars and muscovite decomposed to form various clay minerals, mainly kaolinite and halloysite, with a minor amount of smectite, illite and vermiculite (Li and Zhou 2020). Comprehensive characterisation of the clay minerals indicates that the adsorption capacity of the clay assemblage decreased significantly when poorly crystallised halloysite and kaolinite in saprolite re-crystallised to euhedral, well crystalline kaolinite in pedolith under progressive weathering (Li and Zhou 2020). The re-crystallisation process, coupled with the slightly acidic environment, allowed for the desorption of the REE from clay minerals in the upper pedolith. Hydrochemical analysis shows that the HREE can be efficiently transported in the soil waters and groundwater as REE-carbonate complexes to greater depth in the weathering profile (Li et al. 2019, 2022), where abundant halloysite of high adsorption capacity can efficiently fix and accumulate the HREE through adsorption. In typical regolith profiles, both the REE concentration and $(La/Yb)_N$ ratio gradually increase from the topsoil downwards until a maximum at lower pedolith-upper saprolite contact and then decrease with depth (Li et al. 2019).

Bankeng, Zhaibei (Dingnan), and Darongshan Deposits, China: Representative LREE Deposits The Bankeng deposit in the Jiangxi Province is a regolith-hosted LREE prospect and has a grade of ~0.15 wt.% REO (tonnage confidential). The protolith for the Bankeng deposit is a Triassic two-feldspar, biotite-bearing granite. The protolith is LREE-enriched with most of the bulk REE content hosted in monazite-(Ce), apatite, bastnäsite-(Ce) and parisite-(Ce). The IAD REE mineralisation occurs in the lower pedolith and upper saprolite, with 60–80% of the REE in ion-exchangeable form (Li et al. 2020). Cerium behaves differently to the other REE and is

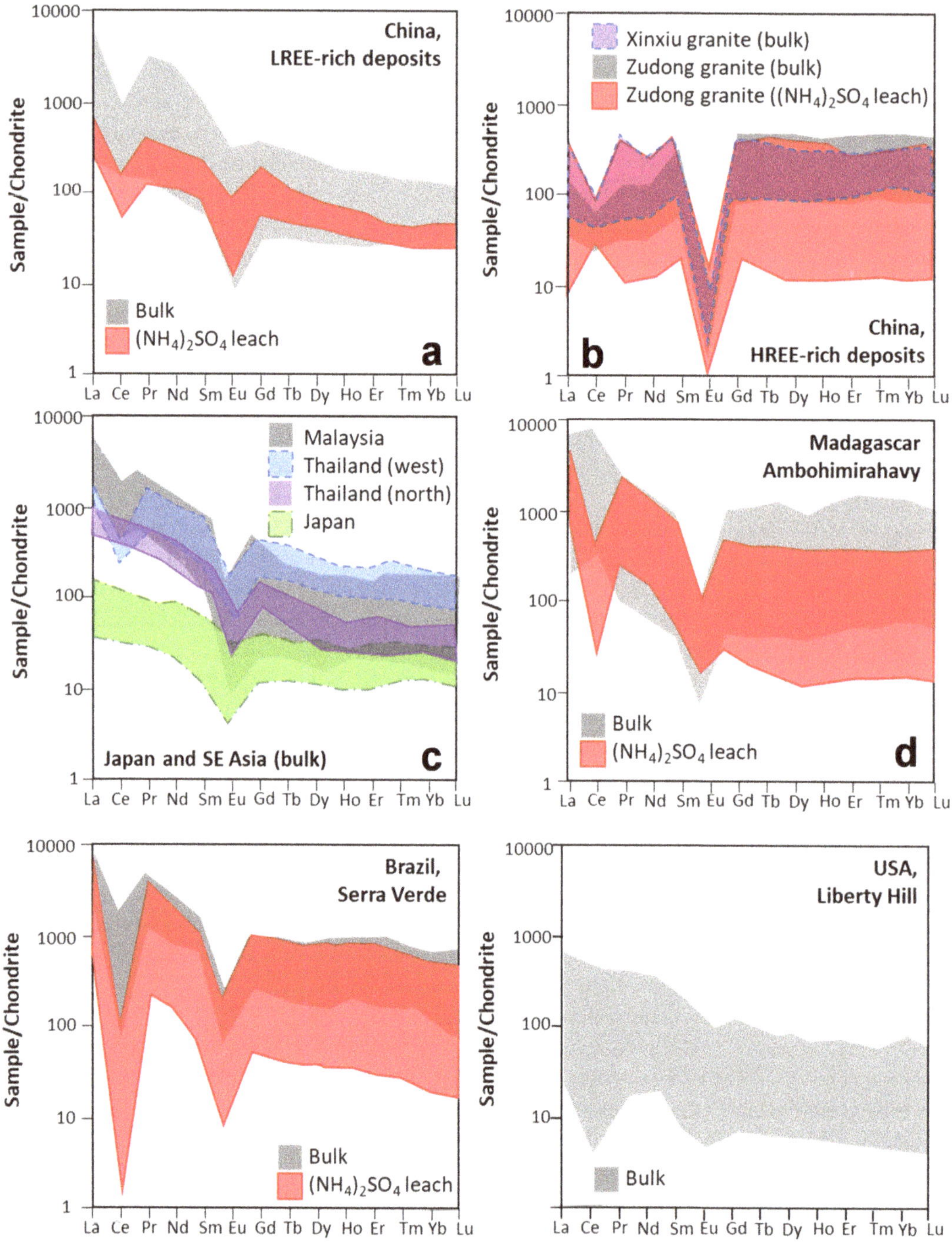

Fig. 5 Chondrite-normalised REE abundance diagrams for global regolith-hosted REE deposits. Summary data shown are bulk analyses of regolith and ammonium sulfate leachable REE fractions (where available). (Data are from Bao and Zhao (2008), Estrade et al. (2019), Murakami and Ishihara (2008), Sanematsu and Kon (2013), Zhou et al. (2013), Sanematsu et al. (2015), Bern et al. (2017), Pinto-Ward (2017), Fou et al. (2019), Li et al. (2019), Li and Zhou (2020), Yaraghi et al. (2020))

mainly bound to secondary Fe-Mn oxyhydroxides in the regolith, with peak concentrations at the pedolith-saprolite and saprolite-saprock interfaces. Kaolinite and halloysite are the main clay minerals and probably the main REE adsorbents in the regolith; however, illite constitutes up to 20% of the clay mineral assemblage in the lower saprolite, and vermiculite is up to 25% of the clay mineral assemblage throughout the regolith profile (Li et al. 2021). The relative amount of REE adsorbed on each of these clay minerals has not been determined, however it is likely that apart from kaolinite and halloysite, illite and vermiculite contribute to HREE adsorption in saprolite and pedolith, respectively (Li and Zhou 2023). Spatially across the ore-bearing catchment, the regolith at the top of slopes is relatively enriched in the LREE, whereas, at the base of hill slopes, the regolith tends to be enriched in the HREE. Carbonate complexation is shown to be the key factor for the preferential transport of the HREE downslope (Li et al. 2022), which additionally caused lateral fractionation of REE profiles along the strike of hillslopes. Vertical mobilisation has redistributed the REE from shallow pedolith to saprolite in the regolith profiles, and lateral groundwater transport may have further locally enriched the regolith at the base of hill slopes, especially in the lower saprolite and saprock units (Li et al. 2021).

Similar mineralogical and geochemical characteristics are also observed in other LREE deposits formed from the weathering of granite in the region (cf. Li et al. 2017). The Zhaibei (or Dingnan) deposit has 80,000 t REO at an average grade of ~0.09 wt.% REO. REE in the parent two-feldspar biotite granite is also mainly hosted in accessory minerals, including apatite, titanite, allanite-(Ce), synchysite-(Ce) and zircon (Ishihara et al. 2008; Xu et al. 2007). The protolith is LREE-enriched, and the REE pattern, except that of Ce, is inherited in the regolith, though the bulk concentration has been enriched two- to four-fold (Murakami and Ishihara 2008; Fig. 5a). The ion adsorption mineralisation mainly occurs in the saprolite and the ion exchangeable REE are 40–60% of the bulk content. Slight HREE enrichment in the regolith is due to the accumulation of residual zircon (Murakami and Ishihara 2008; Sanematsu and Kon 2013).

The Darongshan deposit in the Guangxi Province is another well-studied regolith-hosted LREE IAD (Fu et al. 2019a). The protoliths are granites of Hercynian (Devonian-Carboniferous) and Indosinian (Triassic) age. In this deposit, lithological variation between these granites controls the ion adsorption mineralisation. Regolith-hosted ion adsorption mineralisation is widely developed over the Hercynian granites, which have coarser mineral grains and more abundant apatite and monazite-(Ce). Weathering efficiently broke down the major minerals of these rocks to form kaolinite and illite, the major clay minerals in the regolith, whereas apatite released the REE. The highest grade is ~1500 ppm REE in the pedolith, with ~60% of the contained REE in ion-exchangeable form.

The Chongzuo Deposit, China: A Representative LREE Deposit from Weathering of Volcanic Rocks The Chongzuo deposit in Guangxi Province formed from weathering of Triassic, peraluminous, high-K dacite and rhyolite (Fu et al. 2019b). The REE concentration in these volcanic rocks is ~350 ppm, mainly hosted in accessory titanite, allanite-(Ce) and apatite. These accessory minerals were substantially decomposed during weathering, which released REE to regolith. Preserved regolith profiles have a maximum thickness (>10 m) on ridges. The regolith shows up to a five-fold enrichment of REE compared to the protolith, with the highest ion adsorption grades having developed in the pedolith. In general, the REE pattern of regolith follows that of the protolith, with the LREE being preferentially enriched. The highest REE grade is ~1700 ppm, and the ion exchangeable REE can reach up to 80% of the bulk REE content. Halloysite is the most important REE adsorbent, followed by kaolinite and illite. Up to ~10% and ~20% of the bulk REE content in regolith are hosted by Fe-Mn oxides and organic matter, respectively.

Regolith-Hosted REE Mineralisation from Weathering of Metamorphic Rocks, China Regolith-hosted ion adsorption REE mineralisation has been recently discovered in the Ningdu County of the Jiangxi Province (Huang et al. 2021b). The protolith is a metamorphosed sandstone with a REE concentration of ~200 ppm. Most of the REE in the sandstone are hosted by REE-epidote and rhabdophane-(Ce), and this accessory mineral assemblage is interpreted to result from post-metamorphic hydrothermal overprinting that has increased the propensity for the development of ion adsorption REE mineralisation. The regolith is usually less than 20 m thick, with up to ~1000 ppm REE, 70–80% of which is in ion exchangeable form. Clay minerals, including illite, kaolinite and hydrobiotite, are the main hosts of the ion-exchangeable REE. Compared to IADs formed from granite, the ore bodies in regolith derived from metamorphic rock at Ningdu are located at shallower depths in the middle or even upper pedolith. The shallow depth of the ion adsorption mineralisation probably results from the low permeability of the fine-grained metapelite-derived saprolite that hinders infiltration of the soil waters and transport of the REE to greater depth.

8.3 Southeast Asia and Northeast Asia

Ion adsorption REE mineralisation has been discovered in several countries across Southeast Asia, including Myanmar (Burma), Vietnam, Malaysia and Thailand (cf. Li et al. 2017). Most IADs in the region are formed on granites and have inherited the REE pattern of their protoliths (Fig. 5c). Occurrences in Myanmar are associated with two varieties of calc-alkaline granite: (1) hydrothermally-altered biotite-muscovite (HM) granites and (2) allanite-(Ce) and/or titanite-bearing (AT) granites (Sanematsu et al. 2016). Hydrothermal alteration of the HM granites resulted in the development of HREE-rich-mineral assemblages that include waimirite-(Y), synchysite-(Y) and xenotime-(Y), whereas the AT granites are relatively LREE-enriched, with allanite-(Ce) and titanite being the predominant hosts (Sanematsu et al. 2016). The mineralogical variation has resulted in differences in REE liberation between the HM and AT granites.

In Thailand, various granites across the country have been investigated for their potential for ion adsorption REE mineralisation (Sanematsu et al. 2013, 2015). In Phuket, ion adsorption mineralisation has been identified on the Kata Beach granite, which is a LREE-enriched transitional I- to S-type ilmenite-series biotite granite. The REE at Kata Beach are principally hosted in REE-fluorcarbonates that occur in cracks in feldspars, cavities and along grain boundaries of the parent granite (Sanematsu et al. 2013), indicative of a late- to post-magmatic origin. The breakdown of these REE-fluorcarbonates is the main source of REE to the regolith due to the ease of their dissolution in acidic soil water (Sanematsu et al. 2013). Regolith shows about two- to three-fold REE enrichment relative to the parent granite. The ion adsorption REE mineralisation mainly occurs in the lower pedolith, with 60–80% of the bulk REE content in ion exchangeable form. The ion exchangeable fraction is enriched in LREE (Fig. 5c), with HREE being sequestered in weathering-resistant minerals, including zircon and monazite-(Ce). Characterisation of Thai granites in the Western Province (WP) and northern Central Province (North Thai Granites) shows that a high abundance of monazite-(Ce) in the protolith and formation of secondary REE phosphates reduces the proportion of ion exchangeable REE in the regolith. In general, the WP granites have a relatively lower abundance of phosphates, and the regolith from the weathering of these granites shows a higher and wider range of bulk and ion-exchangeable REE concentration (Sanematsu et al. 2015).

In Malaysia, comprehensive investigations have been carried out on ion adsorption REE mineralisation developed over the Malaysian Western granite belt (Lumut and Telok Murok granites; Yaraghi et al. 2020). These Jurassic-Triassic aged S-type granites are peraluminous and contain ~400 ppm REE, with the majority hosted by accessory apatite, allanite-(Ce), monazite-(Ce), xenotime-(Y) and zircon. The REE are enriched in the regolith to a maximum concentration of ~2000 ppm in the saprolite, with preferential enrichment of LREE relative to the protoliths (Fig. 5c). The proportion of ion exchangeable REE ranges from 60–90%, except for Ce for which the proportion is extraordinarily low, probably due to retention in cerianite. REEs are mainly adsorbed on short-tube halloysite and smectite.

Weathered granitic rocks have been studied for their REE potential at a number of sites in Japan. Murakami and Ishihara (2008) showed that for granitic rocks of the Sanyo Belt, S.E. Japan, REE enrichment in regolith was relatively minor (Fig. 5c) and it was noted that enrichments could be greater in reworked, clay-rich Quaternary sediments capping the regolith profiles (Murakami and Ishihara 2008). However, a recent study indicates that up to 3 times enrichment has occurred in regolith and the REE speciation, mobilisation and adsorption have been greatly affected by soil pH, particularly when soil pH is >6 when inner-complexation of REE on kaolinite is significantly promoted (Nagasawa et al. 2024). A similar phenomenon is also observed in ion-adsorption mineralisation in South Korea (Jo et al. 2023).

8.4 The Tantalus Project, Madagascar

The Tantalus Rare Earth Ionic Clay Project (Tantalus Rare Earths AG Group) is located on the Ampasindava Peninsula of northwest Madagascar and was initially investigated due to the presence of REE-mineralized peralkaline granitic dykes, containing 0.1 to 4% TREO (Ganzeev and Grechishchev 2003; Desharnais et al. 2016; Marquis et al. 2023). In 2009, the presence of ion adsorption mineralisation was recognised in the regolith. By 2012, 4412 exploration pits (up to 10 m deep) and 359 boreholes (up to 130 m length) had been completed across the project area, which were used to build a 3D geological model of the regolith mineralisation. Reported measured plus indicated resources totalled 197.7 Mt. at 0.08% TREO and 0.01% Y_2O_3 (Desharnais et al. 2016). The Ampasindava Peninsula has a tropical monsoon climate, with annual mean temperatures ranging from 20 to 27 °C and annual rainfall of 1500 to over 2000 mm/annum. Lateritic weathering has developed regolith profiles ranging from 0 to over 40 m deep (average 13.5 m). Mineralogy consists mainly of kaolinite and halloysite clays, Fe and Al oxides and oxy-hydroxides, and quartz in typical lateritic profiles comprising bedrock, saprock, saprolite, mottled zone, ferruginous zone and soil (Estrade et al. 2019).

Protoliths of the Tantalus Project are plutonic to volcanic, silica-undersaturated to oversaturated, alkaline to peralkaline rocks of the Ambohimirahavavy Complex and its country rocks of Mesozoic mudstones, marls, limestones and sandstones (Estrade et al. 2014, 2019; Marquis et al. 2023). The Ambohimirahavavy Complex is part of the Cenozoic Alkaline Igneous Province of northern Madagascar, which comprises a series of intrusive to extrusive igneous complexes hosted within Mesozoic sedimentary successions and Precambrian basement. This highly heterogeneous sub-volcanic complex has experienced a diversity of pre- and post-emplacement magmatic and hydrothermal processes (Estrade et al. 2014, 2019). This has resulted in wide variations in bulk rock REE concentrations (100 to 15,000 ppm) and REE distribution in protoliths (Fig. 5d). The bedrock REE concentration is highest in the nepheline syenites and peralkaline granite dykes, which are the most evolved intrusive rocks of the complex. These evolved alkaline-silicate rocks contain a broad range of REE minerals, including eudialyte-group minerals, allanite-(Ce), REE-fluorocarbonates, fergusonite-(Y), monazite-(Ce), and pyrochlore group minerals (Estrade et al. 2019), which in the regolith have broken down partially during weathering. The proportion of easily recoverable REE within the regolith is 2 to 85% and varies according to (1) REE type, (2) depth in the weathering profile and (3) the nature of the bedrock protolith (Desharnais et al. 2016; Estrade et al. 2019). This is exemplified by, in some cases, relatively low REE content protoliths providing greater proportions of REE for adsorption, due to contained REE-host minerals being ameanable to decomposition (Estrade et al. 2019; Marquis et al. 2023).

The topography of the area is characterised by moderate to steep slopes and an annular ridge that surrounds a relatively flat central caldera. Elevations range from sea level to 713 m. Regolith profiles at topographic highs have thicker saprolite relative to pedolith compared to profiles on slopes and low flat-lying areas and constant positive Ce-anomalies (Estrade et al. 2019). This is indicative of the presence of thick unsaturated (vadose) zones with prevailing oxidising conditions at ridge/hill crests, indicating REE are leached and transported downslope.

8.5 Serra Verde, Brazil

The Serra Verde IAD entered production in 2024 and is hosted in the saprolite of a thick laterite profile situated on the Serra Dourada massif of the Goiás Tin Province, Brazil. Serra Dourada is the largest of several anorogenic-type granite bodies (~64 × 7.5 km) that make up the Palaeoproterozoic Rio Tocantins Sub-Province (Pimentel et al. 1999). It is dominantly a biotite granite, with minor amphibole granites at the borders and fine veins of porphyritic granite that cross-cut the main intrusive body (Bilal et al. 1997; Santana et al. 2015).

The biotite granite is rich in REE and HFSE and poor in phosphate (Pinto-Ward 2017; Fig. 5e). It consists of a Nb-REE rich portion to the NE, on which the Serra Verde deposit formed, and Nb-poor granite to the SW. Common REE-rich accessory minerals include zircon, bastnäsite-(Ce) (Nb-rich granite) and parisite-(Ce). Less common REE-phosphates include monazite-(Ce) and xenotime-(Y). Niobium is primarily hosted by pyrochlore in the Nb-rich granite and samarskite-(Y) in the Nb-poor phase. Additional accessory phases are fluorite, apatite, thorianite, ilmenite-rutile and titanite. The Nb-REE-rich phase of granite is characterised by enrichments in F and CO_2, suggesting a role for late-stage fluids in the generation of its elevated REE content (Pinto-Ward 2017).

The weathering profile that contains the Serra Verde deposit is preserved on plateaux within the Serra Dourada massif (Pinto-Ward 2017), the regolith and contained deposit having presumably been eroded and dispersed into the wider environment where the topography is steeper. Depth to bedrock is ~23–38 m, with an NI 43–101 compliant resource of 911 Mt. at 0.12% TREO (Serra Verde 2013/14, 2016/17, cited in Pinto-Ward 2017). Test leach data collected on a single profile show the recovery rate of the weakly-adsorbed REE ions (step-1) ranges from 16–31%, whereas REE associated with Fe-oxides (step-2), represent 44–57% of the bulk REE (Santana and Botelho 2022).

8.6 Liberty Hill, USA

The Carboniferous-age (305 Ma) Liberty Hill pluton comprises a biotite-amphibole granite with subordinate biotite granite and is situated in the highly-weathered and slowly-eroding South Carolina Piedmont (Bern et al. 2017). The granites contain a diverse assemblage of REE-bearing phases, including zircon, monazite-(Ce), allanite-(Ce), titanite, apatite and bastnäsite-(Ce). Hydrothermal alteration at the contact between these two granite facies resulted in locally high REE concentrations in the bedrock. The low phosphate content of the granites (< 0.11 wt.% P) inhibited the development of weathering-resistant REE-phosphate minerals during this hydrothermal event and, thus, the formation of secondary REE-phosphate phases (Bern et al. 2017). There has been variable development of ion exchangeable REE in the regolith, indicated by a broad range of leaching recovery rates from 3 to 77% REE + Y, with total regolith REE contents of 12 to 581 ppm (Fig. 5f). The highest ion adsorbed REE concentrations are observed on the hydrothermally-altered contact between the two granite facies, where smectite clays in the saprolite have developed through replacement of biotite (Bern et al. 2017). Lateral heterogeneity in bedrock mineralogy at Liberty Hill highlights the control of bedrock on IAD mineralisation potential, both from the point of view of bedrock REE content and mineralogy, as well as the adsorption capacity of weathering products formed within the regolith.

9 Extraction

Extraction of the REE from ion adsorption deposits is currently active only in China, Myanmar, Brazil, and Malaysia. All operations make use of ammonium sulfate ($(NH_4)_2SO_4$) leach solutions, which in experimental studies can achieve 60–90% REE extraction (Yang et al. 2013; Moldoveanu and Papangelakis 2013). Before 2008, extraction was dominated by heap or tank/pool leaching, and subsequently, in situ leaching has become more prevalent (Yang et al. 2013). Both technologies have a lower yield than laboratory leaching experiments, with high concentrations of REE remaining in tailings (Liang et al. 2014; Chao et al. 2016). In situ leaching operations use PVC-lined injection boreholes of 0.6–0.8 m diameter and between 2 and 6 m depth, typically with 5 m spacing. Leachate collection wells are drilled either at the base of slopes or in terraces. After a period of leachate injection, the system is typically flushed with fresh water to drive out any remaining leach solution (Xiao et al. 2003; Zhao et al. 2001; Yang et al. 2013; Vahidi et al. 2016). The REE are precipitated from the resulting solutions using either ammonium bicarbonate or oxalic acid and then calcined to produce rare earth oxides (Yu et al. 1990; Walters et al. 2011; Kynicky et al. 2012; Vahidi et al. 2016).

Recovery of REE can be improved by roasting the regolith prior to leaching (e.g., Moldoveanu and Papangelakis 2016), the use of novel lixiviants, including chelating ligands (e.g., Moldoveanu and Papangelakis 2021) or bioleaching with both fungi and bacterial cultures (Barnett et al. 2018). Relatively simple changes may have the potential to increase recovery and reduce environmental risk (see below). For example, a 1 molar ammonium chloride solution may be a more effective lixiviant than ammonium sulfate and results in lower concentrations of undesirable Al in the leach solutions (Stockdale and Banwart 2021).

10 Environmental Issues

The negative environmental impacts of REE extraction from IADs have received significant media attention but relatively limited scientific studies (Wall et al. 2017). To compare the impact of REE extraction from IADs via in situ leaching to mining of carbonatite-associated ores, Vahidi et al. (2016) carried out a life cycle assessment. The carbonatite systems were Bayan Obo, China and Mountain Pass, USA, with bastnäsite-(Ce) and monazite-(Ce) ores of 6–7.7 wt.% REO. Assessments of the ion adsorption mineralisation were made based on six sites in South China, containing 0.05–0.2 wt. % TREO. Despite the significantly lower grade of the IADs, the cumulative energy usage and carbon emissions were similar for both deposit types. The potential release of ammonium compounds to surface waters associated with in situ leaching of IADs carried a ~ 900% higher eutrophication risk relative to the mining of hard-rock occurrences. Vahidi et al. (2016) also noted that in situ leaching risked vegetation damage, landslide risk, and subsidence, all of which were difficult to quantify.

A detailed discussion of the potential impacts of surface mining and heap leaching of IADs was presented by Yang et al. (2013). They estimated that for the production of 1 tonne REE oxide, 300 m^2 of vegetation and topsoil were removed, 2000 tonnes of tailings were disposed of, and 1000 tonnes of wastewater with high concentrations of ammonium sulfate and heavy metals was produced. Changes to geomorphology by soil removal and waste disposal and to soil hydraulic properties by compaction and exposure of bedrock significantly increase the risk of storm flooding and land slippage. They concluded that uncontrolled mountaintop and hillslope mining of IADs had caused permanent loss of the ecosystem, severe soil erosion, air pollution, biodiversity loss and human health problems. In situ leaching has advantages over heap leaching in terms of reduced vegetation loss and soil removal. However, it still introduces significant nutrient contamination to the soil groundwater system, and changes in pore fluid pressures can result in landslides (Liu 2002; Yang et al. 2013). Reduced environmental impact from in situ leaching depends on the leachates used and the control of the groundwater hydraulic system.

Direct investigations of the redistribution of the REE around IAD sites are more limited. Liu et al. (2019a) detected elevated REE in tailing soils, stream waters and sediments, and farmland soils around mining sites. The REEs were distributed via transport in stream water as aqueous complexes of sulfate and carbonate and as colloidal minerals and organic matter (Liu et al. 2019b). However, there is currently little data on the ecotoxicology of the REE in the environment. Of more immediate concern for many REE deposits are concentrations of the actinides U and Th. These are typically low in lateritic deposits, and Th, in particular, is present as insoluble oxides that are not remobilised during leaching (Wall et al. 2017; Estrade et al. 2019).

In situ, leaching of IADs in South China has a similar energy usage and carbon emissions per mass of produced REE to mining operations on world-class carbonatite REE deposits. This is despite these IADs having an ore concentration two orders of magnitude lower. Responsible extraction from IADs should manage the significant risks of eutrophication and of landscape and ecosystem degradation, for example, via the use of heap or agitated leaching of ores and complete remediation once operations cease. If higher-grade IADs were identified, there is potential for significant energy efficiency gains relative to the mining of carbonatite REE deposits.

11 Summary and Conclusions

Ion adsorption REE deposits are a relatively understudied type of mineralisation that represents a small but important portion of defined global REE resources. They offer unique opportunities due to the potential for reduced energy impact of mining relative to hard-rock deposits and for the separation of REE from deleterious radionuclides in the weathering environment. The REE in IADs are sourced from the breakdown of rock-forming and accessory minerals in the bedrock. Consequently, the REE pattern of regolith is usually inherited directly from its source. Incomplete breakdown of the parent rock or the saturation of secondary REE-rich minerals in the weathering environment (especially phosphates) can sequester a portion of the REE in a non-leachable form, potentially biasing the REE pattern of the leachate relative to the bulk regolith. In producing IADs, the leach-recoverable REE fraction is commonly ~60–90% of the bulk. The easily leachable REEs are dominantly adsorbed as hydrated metal ions on the surface of kaolinite and halloysite clays, with smaller contributions from highly soluble secondary phases. The trapping of REE via adsorption is a balance between the chemical activity of the REE ions on mineral surfaces and their activity in solution. The depth of maximum REE deposition is commonly localised by sharp vertical gradients in soil water pH or redox, for example, through mixing of acidic surface-derived water with higher pH groundwater. Water saturation of the regolith can result in buffering of soil water pH to alkaline values. Under these conditions, the breakdown of primary minerals may be slowed because of the lack of acidic input as dissolved atmospheric CO_2 and kaolinite may be destabilised and break down to Al-oxyhydroxides (bauxite), releasing adsorbed REE into the environment. Seasonal variations in the depth of the water table may be important for IAD formation as they lead to alternation between water-saturated and unsaturated conditions and allow for periodic input of atmospherically equilibrated acidic water to the regolith. Lateritic soil profiles that host ion adsorption deposits are geologically relatively ephemeral features. Those discovered to date are best developed and preserved on gently sloping topography (<35°). Weathering of granitic terrain batholiths tends to produce landforms more conducive to laterite preservation relative to structurally heterogeneous geology of, for example, a faulted caldera system. Processing IAD ores via in situ leaching has advantages over heap leaching in terms of reduced vegetation loss and soil removal. However, in situ, leaching is associated with increased risks of eutrophication, slope instability and groundwater contamination.

A number of key questions remain regarding the genesis of ion adsorption deposits: which lithologies constitute optimal source material for IADs is still a matter for further study. Alkali granites, which commonly have high REE contents, do not necessarily produce high-grade IADs as much of their REE budget is retained during weathering by zircon, which can occur as either a primary phase or as an alteration product, for example, after eudialyte group minerals. Determining how much weathering is 'optimal' is of great importance, as the breakdown of kaolinite may lead to loss of adsorbed REE from the system. The hydrology of IAD systems, particularly the influence of seasonality on the spatial distribution of their REE tenor, remains largely unconstrained. An improved understanding of the development of porosity and permeability of regolith will be important to fully characterise the mechanisms by which the REE are trapped and redistributed in the regolith profiles, comparing both those having seasonally variable water tables with those that remain in steady-state. The influence of flora and fauna on the development of lateritic regoliths and their potential to host ion adsorption mineralisation will also require further study. Finally, most investigations so far have focussed on the characterisation of vertical profiles through regolith. The geomorphological factors that control lateral variations in regolith character and the preservation of thick laterite profiles have not yet been addressed in detail.

It is now clear that easily leachable, regolith-hosted REE deposits may develop globally on a variety of protoliths in sub-tropical and temperate soil systems. They may be preserved where geomorphology allows for the retention of thick, long-lived lateritic regoliths. Several prospects outside China have recently progressed to a detailed characterisation or active mining stage. Moreover, recent development on hyperspectral technique for regolith-hosted REE resources exploration would facilitate future resources development (e.g., Li et al. 2024; Tan et al. 2022). It is anticipated that regolith-hosted ion adsorption deposits will continue to make important contributions to global rare earth element production for some years.

Acknowledgements This work was supported by the WREED & SoS RARE projects [grant numbers NE/V008935/1, NE/M011429/1] funded through the UK Research and Innovation programme. We thank the WREED consortium members and attendees of the WREED project online workshops, which have guided this review. CB was further supported by a UKRI Future Leaders Fellowship [grant number MR/V02292X/1]. We thank Alfons Berger, an anonymous reviewer and editors Rob Bowell & Charles Butt for their constructive and helpful comments.

References

Ali SH (2014) Social and environmental impact of the rare earth industries. Resources 3(1):123–134. https://doi.org/10.3390/resources3010123

Alshameri A, He H, Xin C, Zhu J, Zinghu W, Zhu R, Wang H (2019) Understanding the role of natural clay minerals as effective adsorbents and alternative source of rare earth elements: adsorption operative parameters. Hydrometallurgy 185:149–161

Anand RR (2005) Weathering history, landscape evolution and implications for exploration. In: Butt CRM, Robertson IDM, Scott KM, Cornelius M (eds) Regolith landscape evolution across Australia: a compilation of regolith landscape case studies with regolith landscape evolution models. CRC LEME, Kensington, pp 15–45

Anenburg M, Broom-Fendley S, Chen W (2021) Formation of rare earth deposits in carbonatites. Elements 17(5):327–332

Aubert D, Stille P, Probst A (2001) REE fractionation during granite weathering and removal by waters and suspended loads: Sr and Nd isotopic evidence. Geochim Cosmochim Acta 65(3):387–406

Bai G, Wu C, Ding X, Yuan Z, Huang D, Wang P (1989) Genesis and spatial distribution of ion-adsorption type REE deposits in Nanling region. Institute of Deposit Geology, Ministry of Geology and Mineral Resources, Beijing, p 105, (in Chinese)

Ballinger B, Schmeda-Lopez D, Kefford B, Parkinson B, Stringer M, Greig C, Smart S (2020) The vulnerability of electric-vehicle and wind-turbine supply chains to the supply of rare-earth elements in a 2-degree scenario. Sustainable Prod Consumption 22:68–76. https://doi.org/10.1016/j.spc.2020.02.005

Banfield JF, Eggleton RA (1989) Apatite replacement and rare earth mobilization, fractionation, and fixation during weathering. Clay Clay Miner 37(2):113–127

Bao Z, Zhao Z (2008) Geochemistry of mineralization with exchangeable REY in the weathering crusts of granitic rocks in South China. Ore Geol Rev 33: 519–535

Barnett M, Palumbo-Roe B, Gregory SP (2018) Comparison of heterotrophic bioleaching and ammonium sulfate ion exchange leaching of rare earth elements from a Madagascan ion-adsorption clay. Fortschr Mineral 8: 236. https://doi.org/10.3390/min8060236

Beard CD, van Hinsberg VJ, Stix J, Wilke M (2019) Clinopyroxene/melt trace element partitioning in sodic alkaline magmas. J Petrol 60(9):1797–1823. https://doi.org/10.1093/petrology/egz052

Beard CD, Goodenough KM, Borst AM, Wall F, Siegfried PR, Deady EA, Pohl C, Hutchison W, Finch AA, Walter BF, Elliott HAL, Brauch K (2023) Alkaline-silicate REE-HFSE systems. Econ Geol 118(1): 177–208. https://doi.org/10.5382/econgeo.4956

Beck HE, Zimmermann NE, McVicar TR, Vergopolan N, Berg A, Wood EF (2018) Present and future Köppen-Geiger climate classification maps at 1-km resolution. Sci Data 5(1):1–12

Beland CMJ, Williams-Jones AE (2021) The genesis of the Ashram REE deposit, Quebec: insights from bulk-rock geochemistry, apatite-monazite-bastnäsite replacement reactions and mineral chemistry. Chem Geol 578:28. https://doi.org/10.1016/j.chemgeo.2021.120298

Berger A, Gnos E, Janots E, Fernandez A, Giese J (2008) Formation and composition of rhabdophane, bastnäsite and hydrated thorium minerals during alteration: implications for geochronology and low-temperature processes. Chem Geol 254(3–4):238–248

Berger A, Janots E, Gnos E, Frei R, Bernier F (2014) Rare earth element mineralogy and geochemistry in a laterite profile from Madagascar. Appl Geochem 41:218–228. https://doi.org/10.1016/j.apgeochem.2013.12.013

Bern CR, Yesavage Y, Foley NK (2017) Ion-adsorption REEs in regolith of the Liberty Hill pluton, South Carolina, USA: an effect of hydrothermal alteration. J Geochem Explor 172:29–40

Betekhtin AG (1956) Kurs mineralogii (Course of mineralogy). Moscow, Gosgeoltekhizdat Publ.

Betekhtin AG (1961) Kurs mineralogii (course of mineralogy). Gosgeoltekhizdat Publ, Moscow

Bilal E, Moutte J, Botelho NF, Marini OJ, Andrade GF (1997) Geochemistry of two Proterozoic A-type granites of Goiás state, Brazil: possible links with rapakivi series. An Acad Bras Cienc 69(3):349–365

Bogatyrev BA, Zhukov VV, Tsekhovsky YG (2009) Formation conditions and regularities of the distribution of large and superlarge bauxite deposits. Lithol Miner Resour 44(2):135–151

Bonsor HC, MacDonald AM, Davies J (2014) Evidence for extreme variations in the permeability of laterite from a detailed analysis of well behaviour in Nigeria. Hydrol Process 28(10):3563–3573. https://doi.org/10.1002/hyp.9871

Borst AM, Friis H, Nielsen TFD, Waight TE (2018) Bulk and mush melt evolution in agpaitic intrusions: insights from compositional zoning in eudialyte, Ilímaussaq complex, South Greenland. J Petrol 59(4):589–612. https://doi.org/10.1093/petrology/egy038

Borst A, Smith MP, Finch AA, Estrade G, Villanova-de-Benavent C, Nason P, Marquis E, Horsburgh NJ, Goodenough KM, Cheng X, Kynický J, Geraki K (2020) Adsorption of rare earth elements in regolith-hosted clay deposits. Nat Commun 11:4386. https://www.nature.com/articles/s41467-020-17801-5

Braun J-J, Pagel M (1994) Geochemical and mineralogical behavior of REE, Th and U in the Akongo lateritic profile (SW Cameroon). Catena 21(2–3):173–177

Braun J-J, Pagel M, Muller J-P, Bilong P, Michard A, Guillet B (1990) Cerium anomalies in lateritic profiles. Geochim Cosmochim Acta 54(3):781–795

Braun J-J, Pagel M, Herbilln A, Rosin C (1993) Mobilization and redistribution of REEs and thorium in a syenitic lateritic profile: a mass balance study. Geochim Cosmochim Acta 57(18):4419–4434

Bustos N, Marquardt C, Belmar Á, Cordeiro P (2022) Regolith-hosted rare earth exploration in the Chilean coastal range of the Central Andes. J Geochem Explor 234:106934. https://doi.org/10.1016/j.gexplo.2021.106934

Butt CRM, Cluzel D (2013) Nickel laterite ore deposits: weathered serpentinites. Elements 9(2):123–128

Cantrell KJ, Byrne RH (1987) Rare earth element complexation by carbonate and oxalate ions. Geochim Cosmochim Acta 51(3):597–605

Cetiner ZS, Wood SA, Gammons CH (2005) The aqueous geochemistry of the rare earth elements. Part XIV. The solubility of rare earth element phosphates from 23 to 150°C. Chemical Geology 217:147–169

Chaïrat C, Schott J, Oelkers EH, Lartigue J-E, Harouiya N (2007) Kinetics and mechanism of natural fluorapatite dissolution at 25 C and pH from 3 to 12. Geochim Cosmochim Acta 71(24):5901–5912

Chakhmouradian AR, Wall F (2012) Rare earth elements: minerals, mines, magnets (and more). Elements 8(5): 333–340. https://doi.org/10.2113/gselements.8.5.333

Chao YQ, Liu WS, Chen YM, Chen WH, Zhao LH, Ding QB, Wang SZ, Tang YT, Zhang T, Qiu RL (2016) Structure, variation, and co-occurrence of soil microbial communities in abandoned sites of a rare earth elements mine. Environ Sci Technol 50(21):11481–11490. https://doi.org/10.1021/acs.est.6b02284

Cheng X, Kynický J, Smith MP, Kopriva A, Brtnický M, Urubek T, Yang Y, Zhao Z, He C, Song W (2017) Origin of heavy rare earth mineralization in South China. Nat Commun 8:7. https://doi.org/10.1038/ncomms14598

Chi R, Tian J (2007) Review of weathered crust rare earth ores. J Chin Rare Earth Soc 25:641–650

Chi R, Tian T (2008) Weathered crust elution deposited rare earth ores. Nova Science Publishers, New York, 288 pp

Chi RA, Tian J, Li ZJ, Peng C, Wu YX, Li SR, Wang CW, Zhou ZA (2005) Existing state and partitioning of rare earth on weathered ores. J Rare Earths 23:756–759

Churchman, G.J., Gates, W.P., Theng, B.K.G., Yuan, G. (2006) Clays and clay minerals for pollution control. In Bergaya, F., Theng, B.K.G, Lagaly, G. (eds.) Handbook of clay science. Elsevier. Amsterdam pp. 625–675

Condie KC, Dengate J, Cullers RL (1995) Behavior of rare earth elements in a paleoweathering profile on granodiorite in the Front Range, Colorado, USA. Geochim Cosmochim Acta 59(2):279–294

Coppin F, Berger G, Bauer A, Castet S, Loubet M (2002) Sorption of lanthanides on smectite and kaolinite. Chem Geol 182:57–68

Coppin F, Castet S, Berger G, Loubet M (2003) Microscopic reversibility of Sm and Yb sorption onto smectite and kaolinite: experimental evidence. Geochim Cosmochim Acta 67:2515–2527

Deer WA, Howie RA, Zussman J (1982) Rock-forming minerals: orthosilicates, volume 1A. Geological Society of London, London

Desharnais G, Camus Y, Bisaillon C (2016) Resources for the Tantalus rare earth ionic clay project, Northern Madagascar, Updated NI 43–101 technical report. Montreal, SGS Canada Inc. 165 pp

Ding J, Deng G (2013) Main problems in the current ionic adsorption rare earth exploration specifications and their amendment proposals. Nonferrous Met Sci Eng 4:96–102. (in Chinese)

Domènech C, Galí S, Villanova-de-Benavent C, Soler JM, Proenza JA (2017) Reactive transport model of the formation of oxide-type Ni-laterite profiles (Punta Gorda, Moa Bay, Cuba). Mineral Deposita 52(7): 993–1010

Eggleton RA (ed) (2001) The regolith glossary. Surficial geology, soils and landscapes. CRC LEME, Perth, 144 pp. See also: https://publications.csiro.au/publications/#publication/PIlegacy:3643. http://crcleme.org.au/Pubs/Monographs/RegolithGlossary.pdf

Estrade G, Salvi S, Béziat D, Rakotovao S, Rakotondrazafy R (2014) REE and HFSE mineralization in peralkaline granites of the Ambohimirahavavy alkaline complex, Ampasindava peninsula, Madagascar. J Afr Earth Sci 94:141–155. https://doi.org/10.1016/j.jafrearsci.2013.06.008

Estrade G, Salvi S, Béziat D, Williams-Jones AE (2015) The origin of skarn-hosted rare-metal mineralization in the Ambohimirahavavy alkaline complex, Madagascar. Econ Geol 110(6):1485–1513. https://doi.org/10.2113/econgeo.110.6.1485

Estrade G, Salvi S, Béziat D (2018) Crystallization and destabilization of eudialyte-group minerals in peralkaline granite and pegmatite: a case study from the Ambohimirahavavy complex, Madagascar. Mineral Mag 82(2):375–399

Estrade G, Marquis E, Smith MP, Goodenough K, Nason P (2019) REE concentration processes in ion adsorption deposits: evidence from the Ambohimirahavavy alkaline complex in Madagascar. Ore Geol Rev 112:103027. https://doi.org/10.1016/j.oregeorev.2019.103027

Ewing RC (1994) The metamict state: 1993—the centennial. Nucl Instrum Methods Phys Res, Sect B 91(1–4): 22–29

Ewing RC, Haaker RF (1980) The metamict state: implications for radiation damage in crystalline waste forms. Nucl Chem Waste Manage 1(1):51–57

Feng X, Onel O, McAlister C-T, Noble A, Yoon R-H, Morris JR (2021) A study of rare earth ion-adsorption clays: the speciation of rare earth elements on kaolinite at basic pH. Appl Clay Sci 201:105920

Fernandez V (2017) Rare-earth elements market: a historical and financial perspective. Resour Policy 53:26–45. https://doi.org/10.1016/j.resourpol.2017.05.010

Finck N, Bouby M, Dardenne K, Yokosawa T (2017) Yttrium co-precipitation with smectite: a polarized XAS and AsFlFFF study. Appl Clay Sci 137:11–21

Foley N, Ayuso R (2015) REE enrichment in granite-derived regolith deposits of the southeastern United States: prospective source rocks and accumulation processes. Geological Survey Paper 3:131–138

Formoso MLL, Retzmann K, Valeton I (1989) Fractionation of rare earth elements in weathering profiles on phonolites in the area of Lages, Santa Caterina, Brazil. Geochim Bras 3(1):51–61

Fu W, Li X, Feng Y, Feng M, Peng Z, Yu H, Lin H (2019a) Chemical weathering of S-type granite and formation of rare earth element (REE)-rich regolith in South China: critical control of lithology. Chem Geol 520:33–51

Fu W, Luo P, Hu Z, Feng Y, Liu L, Yang J, Feng M, Yu H, Zhou Y (2019b) Enrichment of ion-exchangeable rare earth elements by felsic volcanic rock weathering in South China: genetic mechanism and formation preference. Ore Geol Rev 114:103120. https://doi.org/10.1016/j.oregeorev.2019.103120

Ganzeev AA, Grechishchev OK (2003) A new genetic type of rare-metal alkali granites of Madagascar. Russ Geol Geophys 44:539–553

García MVR, Krzemien A, del Campo MÁM, Álvarez MM, Gent MR (2017) Rare earth elements mining investment: it is not all about China. Resour Policy 53:66–76

Gieré R, Sorensen SS (2004) Allanite and other REE-rich epidote-group minerals. Rev Mineral Geochem 56(1): 431–493

Gilbertson J (2013) A competent persons report on the Tantalus project, Northern Madagascar, Prepared for Tantalus Rare Earths AG, Competent Persons report. SRK Exploration Services Ltd., Canada, p 146

Gregory CJ, Rubatto D, Hermann J, Berger A, Engi M (2012) Allanite behaviour during incipient melting in the southern Central Alps. Geochim Cosmochim Acta 84:433–458. https://doi.org/10.1016/j.gca.2012.01.020

Guidry MW, Mackenzie FT (2003) Experimental study of igneous and sedimentary apatite dissolution: control of pH, distance from equilibrium, and temperature on dissolution rates. Geochim Cosmochim Acta 67(16): 2949–2963

Harlavan Y, Erel Y (2002) The release of Pb and REE from granitoids by the dissolution of accessory phases. Geochim Cosmochim Acta 66(5):837–848

Harouiya N, Chaïrat C, Köhler SJ, Gout R, Oelkers EH (2007) The dissolution kinetics and apparent solubility of natural apatite in closed reactors at temperatures from 5 to 50 C and pH from 1 to 6. Chem Geol 244(3–4):554–568

Hermann J (2002) Allanite: thorium and light rare earth element carrier in subducted crust. Chem Geology 192:289–306. https://doi.org/10.1016/S0009-2541(02)00222-X

Hodgkinson JH, Smith MH (2018) Climate change and sustainability as drivers for the next mining and metals boom: the need for climate-smart mining and recycling. Resour Policy. https://doi.org/10.1016/j.resourpol.2018.05.016

Hoshino M, Sanematsu K, Watanabe Y (2016) REE mineralogy and resources. In: Handbook on the physics and chemistry of rare earths, Elsevier, vol 49, pp 129–291

Hu R-Z, Chen WT, Xu D-R, Zhou M-F (2017) Reviews and new metallogenic models of mineral deposits in South China: an introduction. J Asian Earth Sci 137:1–8

Huang D, Wu C, Han J (1989) REE geochemistry and mineralization characteristics of the Zudong and Guanxi granites, Jiangxi Province. Acta Geol Sin, English edition 2(2):139–157

Huang J, Tan W, Liang X, He H, Ma L, Bao Z, Zhu J (2021a) REE fractionation controlled by REE speciation during formation of the Renju regolith-hosted REE deposits in Guangdong Province, South China. Ore Geol Rev 134:104172

Huang Y, He H, Liang X, Bao Z, Tan W, Ma L, Zhu J, Huang J, Wang H (2021b) Characteristics and genesis of ion adsorption type REE deposits in the weathering crusts of metamorphic rocks in Ningdu, Ganzhou, China. Ore Geol Rev 135:104173

Huittinen N, Rabung T, Andrieux P, Lehto J, Geckeis H (2010) A comparative batch sorption and time-resolved laser fluorescence spectroscopy study on the sorption of Eu (III) and Cm (III) on synthetic and natural kaolinite. Radiochim Acta 98:613–620

Huo MY (1992) Distribution characteristics of the weathering crust type rare-earth resources in Nanling, China. J Nat Resour 7:64–70, [in Chinese with English abstract]

Ionic Rare Earths (2022) 100% success rate for phase 4 tranche 6 infill drill results, ASX release with JORC Code, 2012 edition report. https://wcsecure.weblink.com.au/pdf/IXR/02501473.pdf

Ishihara S, Hua R, Hoshino M, Murakami H (2008) REE abundance and REE minerals in granitic rocks in the Nanling range, Jiangxi Province, southern China, and generation of the REE-rich weathered crust deposits. Resour Geol 58(4):355–372

Jamaludin H, Lahiri-Dutt K (2017) Could Lynas make a difference in the global political economy of rare earth elements in future? Resour Policy 53 (September):267–273. https://doi.org/10.1016/j.resourpol.2017.07.005

Janots E, Bernier F, Brunet F, Muñoz M, Trcera N, Berger A, Lanson M (2015) Ce (III) and Ce (IV) (re) distribution and fractionation in a laterite profile from Madagascar: insights from in situ XANES spectroscopy at the Ce LIII-edge. Geochim Cosmochim Acta 153:134–148

Jo J, Yamanaka T, Shin D (2023) Nitrogen isotope geochemistry of ion adsorption-type REE mineralization: insights from the weathered granitoid rocks in the Sancheong district South Korea. Ore Geol Rev 157: 105429

Kalashnikov AO, Konopleva NG, Pakhomovsky YA, Ivanyuk GY (2016) Rare earth deposits of the Murmansk region, Russia—a review. Econ Geol 111(7):1529–1559. https://doi.org/10.2113/econgeo.111.7.1529

Kalintsev A, Brugger J, Etschmann B, Ram R (2021) An in situ, micro-scale investigation of inorganically and organically driven rare-earth remobilisation during weathering. Mineral Mag 85(1):105–116. https://doi.org/10.1180/mgm.2021.4

Kanazawa Y, Kamitani M (2006) Rare earth minerals and resources in the world. J Alloys Compd 408–412: 1339–1343

Knox-Robinson CM, Wyborn LAI (1997) Towards a holistic exploration strategy: using geographic information systems as a tool to enhance exploration. Aust J Earth Sci 44(4):453–463. https://doi.org/10.1080/08120099708728326

Kynicky J, Smith MP, Xu C (2012) Diversity of rare earth deposits: the key example of China. Elements 8(5):361–367. https://doi.org/10.2113/gselements.8.5.361

Kynický J, Smith M, Salvi S (2021) Editorial for special issue "Rare earth deposits and challenges of world REE demand for high-tech and green-tech at the beginning of the 3rd millennium". Minerals 11(4):378. Multidisciplinary Digital Publishing Institute

Laveuf C, Cornu S (2009) A review on the potentiality of rare earth elements to trace pedogenetic processes. Geoderma 154(1–2):1–12

Le Couteur P (2011) Geological report on the Chambe basin area of exclusive prospecting licence EPL 0325/11, Mulanje Massif, southern Malawi, East Africa. Vancouver, Micron Geological Ltd., p 101

Lee JH, Byrne RH (1993) Complexation of trivalent rare earth elements (Ce, Eu, Gd, Tb, Yb) by carbonate ions. Geochim Cosmochim Acta 57(2):295–302

Li MYH, Zhou M-F (2020) The role of clay minerals in forming the regolith-hosted heavy rare earth element deposits. Am Mineral 105:92–108

Li MYH, Zhou M-F (2023) Physicochemical variation of clay minerals and enrichment of rare earth elements in regolith-hosted deposits: exemplification from the Bankeng Deposit in South China. Clay Clay Miner 71:362–376. https://doi.org/10.1007/s42860-023-00250-8

Li MYH, Zhou M-F (2024) Hyper-enrichment of heavy rare earth elements in highly evolved granites through multiple hydrothermal mobilizations. Am Min 109: 1945–1959

Li YHM, Zhao WW, Zhou M-F (2017) Nature of parent rocks, mineralization styles and ore genesis of regolith-hosted REE deposits in South China: an integrated genetic model. J Asian Earth Sci 148:65–95

Li MYH, Zhou M-F, Williams-Jones AE (2019) The genesis of regolith-hosted heavy rare earth element deposits: insights from the world-class Zudong deposit in Jiangxi Province, South China. Econ Geol 114:541–568

Li MYH, Zhou M-F, Williams-Jones AE (2020) Controls on the dynamics of rare earth elements during sub-tropical hillslope processes and formation of regolith-hosted deposits. Econ Geol 115:1097–1118. https://doi.org/10.5382/econgeo.4727

Li MYH, Teng F-Z, Zhou M-F (2021) Phyllosilicate controls on magnesium isotopic fractionation during weathering of granites: implications for continental weathering and riverine system. Earth Planet Sci Lett 553:article 116613

Li MYH, Kwong HT, Williams-Jones AE, Zhou M-F (2022) The thermodynamics of rare earth element liberation, mobilization and supergene enrichment during groundwater-regolith interaction. Geochim Cosmochim Acta 330:258–277. https://doi.org/10.1016/j.gca.2021.05.002

Li MYH, Liu J, Tan W, Zhou J-X, Zhou M-F (2024) Characterization of regolith-hosted rare earth element deposits using reflectance spectroscopy: framework towards an efficient and reliable field exploration tool. J Geochem Explor 267:107578. https://doi.org/10.1016/j.gexplo.2024.107578

Liang T, Li KX, Wang LQ (2014) State of rare earth elements in different environmental components in mining areas of China. Environ Monit Assess. https://doi.org/10.1007/s10661-013-3469-8

Liu Y (2002) Soil erosion and conservation strategies of rare earth mining. Water Resources Development Research (in Chinese) 2(2):30–32

Liu W-S, Guo M-N, Liu C, Yuan M, Chen X-T, Huot H, Zhao C-M, Tang YT, Morel JL, Qiu R-L (2019a) Water, sediment and agricultural soil contamination from an ion adsorption rare earth mining area. Chemosphere 216:75–83

Liu W-S, Wu L-L, Zheng MY, Chao Y-Q, Zhao C-M, Zhong X, Ding K-B, Huot H, Zhang M-Y, Tang Y-T, Li C, Qiu R-L (2019b) Controls on rare-earth element transport in a river impacted by ion adsorption rare-earth mining. Sci Total Environ 660(2019):697–704

Liu W, Li Y, Wang X, Cui L, Zhao Z, Liu C, Xu Z (2022) Weathering stage and topographic control on rare earth element (REE) behavior: new constraints from a deeply weathered granite hill. Chem Geol 610:121066

Lottermoser BG (1990) Rare-earth element mineralisation within the Mt. Weld carbonatite laterite, Western Australia. Lithos 24:151–167. https://doi.org/10.1016/0024-4937(90)90022-S

Lottermoser BG, England BM (1988) Compositional variation in pyrochlores from the Mt Weld carbonatite laterite, Western Australia. Mineral Petrol 38(1):37–51

Marquis E, Goodenough KM, Smith MP, Estrade G, Rasolonirina F, Rakotoson RL (2023) Developing an integrated petrogenetic model for understanding REE deportment of the Ampasibitika intrusion and associated ion adsorption deposits. J Petrol 64(7). https://doi.org/10.1093/petrology/egad046

Mathieu D, Bernat M, Nahon D (1995) Short-lived U and Th isotope distribution in a tropical laterite derived from granite (Pitinga river basin, Amazonia, Brazil): application to assessment of weathering rate. Earth Planet Sci Lett 136(3–4):703–714

Maulana A, Sanematsu K, Sakakibara M (2016) An overview on the possibility of scandium and REE occurrence in Sulawesi, Indonesia Indonesian. J Geosci 3(2): 139–147

McCuaig TC, Hronsky J (2017) The mineral systems concept: the key to exploration targeting. Trans Inst Min Metall Sect B 2015:1. https://doi.org/10.1080/03717453.2017.1306274

Mentani T, Ohmura T, Watanabe Y, Urabe T (2010) So-called "ion-adsorption type" REE deposits found in weathered crust of ilmenite-series granite in Northern Vietnam [abs]. Geol Soc Am Abstr Programs 42: 336

Middleton AW, Förster H-J, Uysal IT, Golding SD, Rhede D (2013) Accessory phases from the Soultz monzogranite, Soultz-sous-Forêts, France: implications for titanite destabilisation and differential REE, Y and Th mobility in hydrothermal systems. Chem Geol 335:105–117. https://doi.org/10.1016/j.chemgeo.2012.10.047

Moldoveanu GA, Papangelakis VG (2012) Recovery of rare earth elements adsorbed on clay minerals: I. Desorption mechanism. Hydrometallurgy 117:71–78

Moldoveanu GA, Papangelakis VG (2013) Recovery of rare earth elements adsorbed on clay minerals: II. Leaching with ammonium sulfate. Hydrometallurgy 131:158–166

Moldoveanu GA, Papangelakis VG (2016) An overview of rare-earth recovery by ion-exchange leaching from ion-adsorption clays of various origins. Mineral Mag 80:63–76

Moldoveanu GA, Papangelakis VG (2021) Chelation-assisted ion-exchange leaching of rare earths from clay minerals. Metals 11:1265. https://doi.org/10.3390/met11081265

Möller V, Williams-Jones AE (2016) Petrogenesis of the Nechalacho Layered Suite, Canada: magmatic

evolution of a REE-Nb-rich nepheline syenite intrusion. J Petrol 57(2):229–276. https://doi.org/10.1093/petrology/egw003
Mosquera MJ, Rivas B, Prieto B, Silva B (2000) Microstructural changes in granitic rocks due to consolidation treatments: their effects on moisture transport. In: Proceedings of the 9th international congress on deterioration and conservation of stone. Elsevier, pp 601–607
Mukai H et al (2020) Microscopic analyses of weathered granite in ion-adsorption rare earth deposit of Jianxi Province, China. Sci Rep 10:20194. https://doi.org/10.1038/s41598-020-76981-8
Murakami H, Ishihara S (2008) REE mineralization of weathered crust and clay sediment on granitic rocks in the Sanyo Belt, SW Japan and the southern Jiangxi Province, China. Resour Geol 58:373–401
Nagasawa M, Shimizu Y, Yamaguchi A, Tokunaga K, Mukai H, Aoyagi N, Mei H, Takahashi Y (2024) Interpretation of vertical migration and enrichment processes of rare earth elements (REEs) in ion-adsorption-type mineralization in Japan based on REE speciation analyses. Chem Geol 670:122431. https://doi.org/10.1016/j.chemgeo.2024.122431
Neary CR, Highley DE (1984) Chapter 12—The economic importance of the rare earth elements. In: Henderson P (ed) Rare earth element geochemistry, vol 2. Elsevier, Amsterdam, pp 423–466. https://doi.org/10.1016/B978-0-444-42148-7.50017-4
Nelson ST, Barton B, Burnett MW, McBride JH, Brown L, Spring I (2020) The lateral and vertical growth of laterite weathering profiles, Hawaiian Islands, USA. Earth Surf Process Landf 45(12): 2940–2953
Nesbitt HW (1979) Mobility and fractionation of rare earth elements during weathering of a granodiorite. Nature 279(5710):206–210
Orris GJ, Grauch RI (2002) Rare earth element mines, deposits and occurrences, vol. 2, issue 189. US Department of the Interior, US Geological Survey
Padrones JT, Imai A, Takahashi R (2017) Geochemical behavior of rare earth elements in weathered granitic rocks in northern Palawan, Philippines. Resour Geol 67(3):231–253
Petit JC, Langevin Y, Dran JC (1985) Radiation-enhanced release of uranium from accessory minerals in crystalline rocks. Geochim Cosmochim Acta 49(3):871–876
Pimentel MM, Fuck RA, Botelho NF (1999) Granites and the geodynamic history of the Neoproterozoic Brasılia belt, Central Brazil: a review. Lithos 46(3): 463–483
Pinto-Ward C (2017) Controls on the enrichment of the Serra Verde rare earth deposit, Brazil. Unpublished PhD thesis, Imperial College London, 442 pp
Pourret O, Davranche M, Gruau G, Dia A (2007) Competition between humic acid and carbonates for rare earth elements complexation. J Colloid Interface Sci 305(1): 25–31
Price JR, Velbel MA, Patino LC (2005) Allanite and epidote weathering at the Coweeta Hydrologic Laboratory, western North Carolina, U.S.A. Am Mineral 90(1):101–114. https://doi.org/10.2138/am.2005.1444
Ram R, Becker M, Brugger J, Etschmann B, Burcher-Jones C, Howard D, Kooyman P, Petersen J (2019) Characterisation of a rare earth element-and zirconium-bearing ion-adsorption clay deposit in Madagascar. Chem Geol 522:93–107
Rocha A, Schissel D, Sprecher A, De Tarso P, Goode J (2013) Process development for the Serra Verde weathered crust elution-deposited rare earth deposit in Brazil. In: Proceedings of the 52rd conference of metallurgists (COM 2013), 27–31 Oct 2013, Montreal
Ruxton BP, Berry L (1957) Weathering of granite and associated erosional features in Hong Kong. Geol Soc Am Bull 68(10):1263–1292
Sanematsu K, Kon Y (2013) Geochemical characteristics determined by multiple extraction from ion-adsorption type REE ores in Dingnan County of Jiangxi Province, South China. Bull Geol Surv Jpn 64:313–330
Sanematsu K, Watanabe Y (2016) Characteristics and genesis of ion adsorption-type rare earth element deposits. Rev Econ Geol 18:55–79. https://pubs.geoscienceworld.org/books/book/1998/chapter/16271168/Characteristics-and-Genesis-of-Ion-Adsorption-Type
Sanematsu K, Murakami H, Watanabe Y, Duangsurigna S, Siphandone V (2009) Enrichment of rare earth elements (REE) in granitic rocks and their weathered crusts in central and southern Laos. Bull Geol Surv Jpn 60(11–12):527–558. https://doi.org/10.9795/bullgsj.60.527
Sanematsu K, Kon Y, Imai A, Watanabe K, Watanabe Y (2013) Geochemical and mineralogical characteristics of ion-adsorption type REE mineralization in Phuket, Thailand. Mineral Deposita 48(4):437–451. https://doi.org/10.1007/s00126-011-0380-5
Sanematsu K, Kon Y, Imai A (2015) Influence of phosphate on mobility and adsorption of REEs during weathering of granites in Thailand. J Asian Earth Sci 111:14–30
Sanematsu K, Ejima T, Kon Y, Manaka T, Zaw K, Morita S, Seo Y (2016) Fractionation of rare-earth elements during magmatic differentiation and weathering of calc-alkaline granites in southern Myanmar. Mineral Mag 80(1):77–102
Santana IV, Botelho NF (2022) REE residence behaviour and recovery from a weathering profile related to the Serra Dourada Granite Goiás/Tocantins States Brazil. Ore Geol Rev 143:104751. https://doi.org/10.1016/j.oregeorev.2022.104751
Santana IV, Wall F, Botelho NF (2015) Occurrence and behavior of monazite-(Ce) and xenotime-(Y) in detrital and saprolitic environments related to the Serra Dourada granite, Goiás/Tocantins State, Brazil: potential for REE deposits. J Geochem Explor 155:1–13

Schoonheydt RA, Johnston CT (2013) Chapter 5—Surface and interface chemistry of clay minerals. In: Bergaya F, Lagaly G (eds) Developments in clay science. Elsevier, Amsterdam
Simandl G (2014) Geology and market-dependent significance of rare earth element resources. Mineral Deposita 49(8):889–904
Smits MM, Bonneville S, Benning LG, Banwart SA, Leake JR (2012) Plant-driven weathering of apatite—the role of an ectomycorrhizal fungus. Geobiology 10(5):445–456. https://doi.org/10.1111/j.1472-4669.2012.00331.x
Soler JM, Lasaga AC (1996) A mass transfer model of bauxite formation. Geochim Cosmochim Acta 60(24): 4913–4931
Sonke JE, Salters VJM (2006) Lanthanide–humic substances complexation. I. Experimental evidence for a lanthanide contraction effect. Geochim Cosmochim Acta 70(6):1495–1506
Stille P, Pierret M-C, Steinmann M, Chabaux F, Boutin R, Aubert D, Pourcelot L, Morvan G (2009) Impact of atmospheric deposition, biogeochemical cycling and water–mineral interaction on REE fractionation in acidic surface soils and soil water (the Strengbach case). Chem Geol 264(1–4):173–186
Stockdale A, Banwart SA (2021) Recovery of technologically critical lanthanides from ion adsorption soils. Miner Eng 168:106921. https://doi.org/10.1016/j.mineng.2021.106921
Su W (2009) Economic and policy analysis of China's rare earth industry (in Chinese). China Financial and Economic Publishing House, Beijing
Takahashi Y, Minai Y, Ambe S, Makide Y, Ambe F, Tominaga T (1997) Simultaneous determination of stability constants of humate complexes with various metal ions using multitracer technique. Sci Total Environ 198(1):61–71
Takahashi Y, Kimura T, Kato Y, Minai Y, Tominaga T (1998) Characterisation of Eu(III) species sorbed on silica and montmorillonite by laser-induced fluorescence spectroscopy. Radiochim Acta 82:227–232
Takahashi Y, Tada A, Kimura T, Shimizu H (2000) Formation of outer- and inner-sphere complexes of lanthanide elements at montmorillonite-water Interface. Chem Lett 2000:700–701
Takahashi Y, Tada A, Shimizu H (2004) Distribution pattern of rare earth ions between water and montmorillonite and its relation to the sorbed species of the ions. Anal Sci 20:1301–1306
Tan W, Qin X, Liu J, Zhou M-F, He H, Wang CY, Huang J, Zhu J, Yao Y, Cudahy T (2022) Feasibility of visible short-wave infrared reflectance spectroscopy to characterize regolith-hosted rare earth element mineralization. Econ Geol 117:495–508
Tanaka K, Takahashi Y, Shimizu H (2008) Local structure of Y and Ho in calcite and its relevance to Y fractionation from Ho in partitioning between calcite and aqueous solution. Chem Geol 248:104–113
Taylor G, Eggleton RA (2001) Regolith geology and geomorphology. Wiley, Chichester, 375 p
Tian J, Yin J, Rao G, Jiang M, Chi R, Ouyang K (2010) Kinetics on leaching rare earths from the weathered crust elution-deposited rare earth ores with ammonium sulfate solution. Hydrometallurgy 101:166–170
U.S. Geological Survey (2024) Mineral commodity summaries 2024: U.S. Geological Survey, 212 p. https://doi.org/10.3133/mcs2024
Ugwu IM, Igbokwe OA (2019) Sorption of heavy metals on clay minerals and oxides: a review. In: Edibali S (ed) Advanced sorption process applications. IntechOpen book series. https://doi.org/10.5772/intechopen.80989
Vahidi E, Navarro J, Zhao F (2016) An initial life cycle assessment of rare earth oxides production from ion-adsorption clays. Resour Conserv Recycl 113: 1–11
Valeton I, Biermann M, Reche R, Rosenberg F (1987) Genesis of nickel laterites and bauxites in Greece during the Jurassic and Cretaceous, and their relation to ultrabasic parent rocks. Ore Geol Rev 2(4): 359–404. https://doi.org/10.1016/0169-1368(87)90011-4
Valeton I, Schumann A, Vinx R, Wieneke M (1997) Supergene alteration since the upper cretaceous on alkaline igneous and metasomatic rocks of the Poços de Caldas ring complex, Minas Gerais, Brazil. Appl Geochem 12(2):133–154
Valsami-Jones E, Ragnarsdottir KV, Putnis A, Bosbach D, Kemp AJ, Cressey G (1998) The dissolution of apatite in the presence of aqueous metal cations at pH 2–7. Chem Geol 151(1–4):215–233
van de Ven MAJ, Borst AM, Davies GR, Hunt EJ, Finch AA (2019) Hydrothermal alteration of eudialyte-hosted critical metal deposits: fluid source and implications for deposit grade. Fortschr Mineral 9(7): 422. https://doi.org/10.3390/min9070422
Vasyukova OV, Williams-Jones AE (2019) Tracing the evolution of a fertile REE granite by modelling amphibole-melt partitioning, the Strange Lake story. Chem Geol. https://doi.org/10.1016/j.chemgeo.2019.03.030
Vázquez-Ortega A, Huckle D, Perdrial J, Amistadi MK, Durcik M, Rasmussen C, McIntosh J, Chorover J (2016) Solid-phase redistribution of rare earth elements in hillslope pedons subjected to different hydrologic fluxes. Chem Geol 426:1–18
Wall F, Rollat A, Pell RS (2017) Responsible sourcing of critical metals. Elements 13(5):313–318. https://doi.org/10.2138/gselements.13.5.313
Walters A, Lusty P, Hill A (2011) Rare earth elements mineral profile series. British Geological Survey, Keyworth

Wan Y, Liu C (2006) The effect of humic acid on the adsorption of REEs on kaolin. Colloids Surf A Physicochem Eng Asp 290(1–3):112–117

Wang L, Xu C, Zhao Z, Song W, Kynicky J (2015) Petrological and geochemical characteristics of Zhaibei granites in Nanling region, Southeast China: implications for REE mineralization. Ore Geol Rev 64:569–582

Wang M, Zhou J, Liu F, Li Y, Li Y, Li M, Zhang Z, Wang C (2020) Mineralogy and elemental geochemistry of the Puxiong Nb-REE deposit in Jianshui City, Southeast Yunnan Province, SW China. Acta Geol Sin 94: 3746–3760, (in Chinese with English abstract)

Wang M, Li MYH, Zhou M-F, Zhou J-X, Sun G, Zhou Y, Li Y (2024) Enrichment of rare earth elements during the weathering of alkaline igneous systems: Insights from the Puxiong regolith-hosted rare earth element deposit SW China. Econ Geol 119:161–187

Wells MA, Danišík M, McInnes BIA, Morris PA (2019) (U-Th)/He-dating of ferruginous duricrust: Insight into laterite formation at Boddington, WA. Chem Geol 522:148–161

Wilson MJ (2004) Weathering of the primary rock-forming minerals: processes, products and rates. Clay Miner 39(3):233–266

Wood SA (1990) The aqueous geochemistry of the rare-earth elements and yttrium: 2. Theoretical predictions of speciation in hydrothermal solutions to 350 C at saturation water vapor pressure. Chem Geol 88(1–2): 99–125

Wu C, Huang D, Guo Z (1990) REE geochemistry in the weathered crust of granites, Longnan area, Jiangxi Province. Acta Geol Sin, English edition 3(2):193–209

Xiao ZZ, Tang XZ, Wang XM, Li JX (2003) Study on in-situ leaching stope of deeply soleplate ion-absorbed rare earth (first). Ind Miner Process 11:16–18, (in Chinese)

Xie Y, Hou Z, Goldfarb RJ, Guo X, Wang L (2016) Rare earth element deposits in China. In: Verplanck PL, Hitzman MW (eds) Rare earth and critical elements in ore deposits. Society of Economic Geologists, Littleton

Xu X, Lu W, He Z (2007) Age and generation of Fogang granite batholith and Wushi diorite-hornblende gabbro body. Sci China Ser D Earth Sci 50(2):209–220

Xu C, Kynický J, Smith MP, Kopriva A, Brtnický M, Urubek T, Yang Y, Zhao Z, He,C, Song W (2017) Origin of heavy rare earth mineralization in South China. Nature Communications 8:7. https://doi.org/10.1038/ncomms14598

Yamaguchi A, Honda T, Tanaka M, Tanaka K, Takahashi Y (2018) Discovery of ion-adsorption type deposits of rare earth elements (REE) in Southwest Japan with speciation of REE by extended X-ray absorption fine structure spectroscopy. Geochem J 52:415–425

Yang DH, Xiao GM (2011) Regional metallogenic regularities of the ion adsorption type of rare-earth deposits in Guangdong Province. Geol Resour 20: 462–468, (in Chinese with English abstract)

Yang XJ, Li AJ, Li XL, Wu YD, Zhou WB, Chen ZH (2013) China's ion adsorption rare earth resources, mining consequences and preservation. Environ Dev. https://doi.org/10.1016/j.envdev.2013.03.006

Yaraghi A, Ariffin K, Baharun N (2020) Comparison of characteristics and geochemical behaviors of REEs in two weathered granitic profiles generated from metamictized bedrocks in Western Peninsular Malaysia. J Asian Earth Sci 199:104385

Yu Q, Qiu D, Ma R, Zhou Z (1990) The techniques for extraction of rare earths from the ionic type rare earth ores and their possible improvements. Min Metall Eng 10(1):42–45, (in Chinese)

Zakharov VI, Maiorov DV, Alishkin AR, Matveev VA (2011) Causes of insufficient recovery of zirconium during acidic processing of Lovozero eudialyte concentrate. Russ J Non-Ferrous Met 52(5):423–428. https://doi.org/10.3103/S1067821211050129

Zhang Z (1990) A study on weathering crust ion adsorption type REE deposits, South China. Contrib Geol Miner Resour Res 5(1):57–71

Zhang Z, Zheng G, Takahashi Y, Wu C, Zheng C, Yao J, Xiao C (2016) Extreme enrichment of rare earth elements in hard clay rocks and its potential as a resource. Ore Geol Rev 72:191–212

Zhao J, Tang XZ, Wu C (2001) Status quo of mining and recovering technologies for ion-absorbed rare earth deposits in China. Yunnan Metall 30(1):11–14, (in Chinese)

Zhou L, Zhang Z, Li Y, You F, Wu C, Zheng C (2013) Geological and geochemical characteristics in the paleo-weathering crust sedimentary type REE deposits, western Guizhou, China. J Asian Earth Sci 73:184–198

Zhou M-F, Li MYH, Wang Z, Li X-C, Liu J (2020) The genesis of regolith-hosted rare earth element and scandium deposits: current understanding and outlook to future prospecting. Chin Sci Bull 65:3809–3824, (in Chinese with English abstract)

Zhukova IA, Stepanov AS, Jiang S-Y, Murphy D, Mavrogenes J, Allen C, Chen W, Bottrill R (2021) Complex REE systematics of carbonatites and weathering products from uniquely rich Mount Weld REE deposit, Western Australia. Ore Geol Rev 139: 104539. https://doi.org/10.1016/j.oregeorev.2021.104539

Channel and Playa Uranium Deposits

J. B. R. Drummond, A. R. Wilde, D. Layton-Matthews, N. P. James, and P. K. Pufahl

Abstract

Channel and playa uranium deposits are the most significant subclass of surficial uranium deposits globally. They form in arid/semi-arid regions within shallow (ca. 1–15 m depth) Cenozoic- to Quaternary-age paleochannel groundwater calcrete and associated smectitic fluvial/alluvial and playa-lacustrine sediments. The primary uranium mineral within these deposits is the potassium-uranyl-vanadate, carnotite ($K_2(UO_2)_2(VO_4)_2 \cdot 1–3H_2O$).

This chapter examines the global distribution and paragenesis of uranium mineralization in channel and playa uranium deposits, emphasizing the largest deposits in Western Australia and Namibia. Integrating geological, geomorphological, hydrogeological, and mineralogical observations from these deposits constrain the critical environmental and physicochemical factors essential to accumulating uranium into economically significant concentrations within a near-surface setting.

J. B. R. Drummond (✉)
IGO Ltd. South Shore Centre, South Perth, WA, Australia

Department of Geological Sciences and Geological Engineering, Queen's University, Kingston, ON, Canada

A. R. Wilde
Endell Ridge, Mount Claremont, WA, Australia

D. Layton-Matthews · N. P. James · P. K. Pufahl
Department of Geological Sciences and Geological Engineering, Queen's University, Kingston, ON, Canada

1 Introduction

The surficial uranium (U) deposit group defined by the International Atomic Energy Agency (IAEA) includes a diverse range of Tertiary to Recent U accumulations in near-surface sediments and soils (IAEA 2018). The largest deposits have been discovered in fluvial valley and lacustrine-playa environments and are thus the focus of this chapter. For simplicity, fluvial valley/channel and lacustrine-playa sub-types will be referred to herein as "surficial U deposits." Approximately 100 surficial U deposits and occurrences have been documented in 15 countries on five continents, with the majority in Western Australia (WA), Namibia, Jordan, Mauritania and Somalia (Fig. 1; Hall and Mihalasky 2017; IAEA 2018). Australia, Namibia, and Jordan collectively host 86% of surficial U resources. Surficial U resources are currently estimated to account for 4% of global conventional U resources classified as extractable for less than $130 USD/kg (Table 1; NEA/IAEA 2020). The Langer Heinrich deposit in Namibia is the only example to have been mined and accounted for ~3% and 47% of global and Namibian total annual U production between

R. J. Bowell, C. R. M. Butt (eds.), *Geology, Geochemistry and Formation of Supergene Mineral Deposits in Deeply Weathered Terrain*, Mineral Resource Reviews, https://doi.org/10.1007/978-3-031-75733-4_15

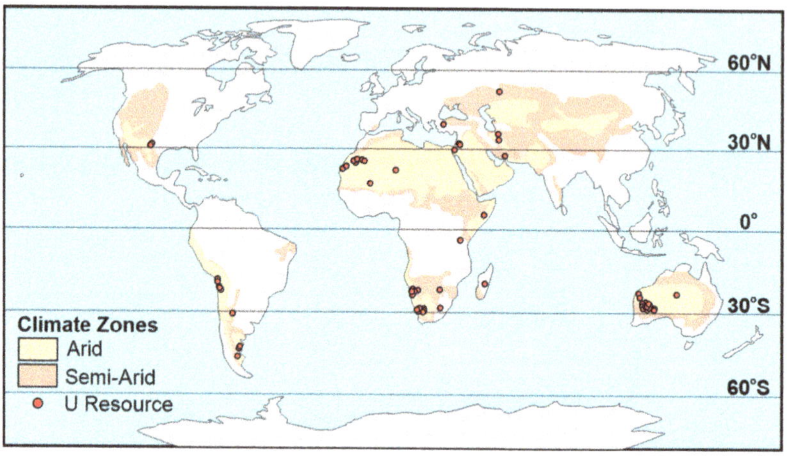

Fig. 1 Global distribution of channel and playa U deposits (NEA/IAEA 2020). Climatic zones modified from Beck et al. (2018)

Table 1 Summary of conventional channel and playa U resources and their proportional significance by country and world-wide

Country	C&P U Deposits[a]	Total Global C&P U Resource (tU)[a]	Total Country U Resource (tU)[a]	C&P U Resource proportion (%)
Namibia	9	89,693	448,300	20.0
Jordan	1	70,000	70,000	100.0
Australia	22	65,813	1,692,700	3.9
Mauritania	9	17,130	17,130	100.0
Tanzania	6	8917	58,200	15.3
Somalia	1	4309	7600	56.7
Argentina	5	3880	38,700	10.0
USA	2[b]	1027[b]	47,900	2.1
Botswana	1[b]	615[b]	87,200	0.7
Chile	1	51[c]	1400	3.6
India	1[b]	14[b]	195,900	0.0
Country Total	58	261,450	2,665,030	9.8
World U Resources (tU)	–	**261,450**	**614,7800**	**4.3**

[a]Data from NEA/IAEA (2020); [b]Data from Hall and Mihalasky (2017); C&P = channel and playa; Total Global C&P U Resource is comprised of assured and inferred conventional U resources (<USD 130/kgU); [c]Assured and Inferred conventional U resources <USD 260/kgU. C&P U Resource proportion = the relative proportion that C&P U deposit resources contribute to each countries total U resources

2011 and 2018, respectively (IAEA 2018; Paladin Energy Ltd. 2021).

2 Surficial U Deposit Characteristics

Surficial U deposits are also known as "calcrete-hosted U deposits" because of a close association between U and groundwater calcrete, defined as a non-pedogenic calcium and/or magnesium carbonate-rich rock formed through evaporative precipitation of groundwater at or near the water table: *calcrete* herein (Goudie 1972; Mann 1974; Mann and Horwitz 1979; Chen et al. 2002). Carnotite ($K_2(UO_2)_2(VO_4)_2 \cdot 1\text{–}3H_2O$; Anthony et al. 2005) is the predominant ore mineral, but various other uranyl minerals occur in minor amounts (See Table 2). The Daba-Siwaqa deposits of Central Jordan are a rare exception where the dominant U minerals are the Ca- and Na-uranyl vanadates, tyuyamunite and strelkinite (Table 2;

Table 2 Summary of uranium and vanadium minerals observed in channel and playa U deposits

Mineral	Formula	Observed locations	References
Uranyl-Vanadate phases			
Carnotite	$K_2(UO_2)_2(VO_4)_2 \cdot 1–3H_2O$	Widespread globally	**1, 2**
Tyuyamunite	$Ca(UO_2)_2(VO_4)_2 \cdot 5–8H_2O$	Trekkopje, Langer Heinrich, NA; Hasa-Qatrana Area, JO	**2, 3**
Metatyuyamunite	$Ca(UO_2)_2(VO_4)_2 \cdot 3H_2O$	Khan Az-Zabib, JO; Moab Bay, NA	**3, 4**
Strelkinite	$Na_2(UO_2)_2(VO_4)_2 \cdot 6H_2O$	Hasa-Qatrana Area, JO	**3**
Finchite	$Sr(UO_2)_2(VO_4)_2 \cdot 5H_2O$	Sulfur Springs Draw, SW, USA	**5**
Uranyl-Carbonate phases			
Rutherfordine	UO_2CO_3	Trekkopje, NA	**2**
Urancalcarite	$Ca(UO_2)_3(CO_3)(OH) \cdot 3H_2O$	Trekkopje, NA	**2**
Liebigite	$Ca_2UO_2(CO_3)_3 \cdot 10H_2O$	Trekkopje, NA	**2**
Swartzite	$CaMgUO_2(CO_3)_3 \cdot 12H_2O$	Trekkopje, NA	**2**
Andersonite	$Na_2K_3UO_3(CO_3)_3 \cdot 6H_2O$	Trekkopje, NA	**2**
Uranyl-Silicate phases			
Weeksite	$K_2(UO_2)_2Si_6O_{15} \cdot 4H_2O$	Silicified calcretes, Gascoyne Region, WA	**6**
Soddyite	$(UO_2)_2SiO_4 \cdot 2H_2O$	NE Goldfields, WA; Klein Trekkopje, NA	**7, 8**
Uraniferous Opal	Undefined	Gascoyne Region and NE Goldfields, WA	**7**
Uranyl-Phosphate phases			
Na-K Meta-Autunite	$Na_2(UO_2)_2(PO_4)_2 \cdot 10–12H_2O$	Boomerang Lake WA; Gascoyne Region, WA	**6, 9**
Vanadium-bearing phases			
Rossite	$Ca(VO_3)_2 \cdot 4H_2O$	Moab Bay, NA	**4**
Todorokite (Mn-Ox)	$(Na, Ca, K)_2(Mn^{4+}, Mn^{3+})_6O_{12}{\cdot}3–4.5H_2O$	Centipede-Millipede, Lake Way, Lake Maitland, WA	**10**

References: **1** – Mann (1974), **2** – Bowell et al. (2009), **3** – Khoury et al. (2014), **4** – Bowell and Davies (2017), **5** – Hall et al. (2019), **6** – Noble et al. (2011), **7** – Butt et al. (1977), **8** – Iilende (2012), **9** – Butt and Graham (1981), **10** – Drummond et al. (2021)

Khoury et al. 2014). Minor amounts of other vanadate minerals, such as rossite and V-rich Fe/Mn oxy-hydroxides, are common (Bowell and Davies 2017; Drummond et al. 2021). Thus, there is potential for these deposits to be significant sources of both vanadium (V) and uranium (Paladin Energy Ltd. 2020; Toro Energy Limited 2021; GSWA 2022).

More recent literature refers to these deposit types as channel and playa U deposits (Butt 1988; Noble et al. 2011; Hall and Mihalasky 2017; IAEA 2018). This preferred nomenclature attempts to encompass the broader environments in which these surficial U accumulations form rather than describing them by a particular lithological association. Channel U deposits occur in Tertiary to recent paleodrainages and are hosted in calcrete and in carbonate- and clay-rich clastic sediments (Mann 1974; Butt et al. 1977). The best-known and largest channel U deposits are Yeelirrie, WA, and Langer Heinrich, Namibia (49,284 tU and 73,306 tU, respectively). Playa U deposits occur in deltaic and transitional lacustrine-playa calcrete, carbonate- and clay-rich clastic sediments and, less commonly, in gypcrete (e.g., Lake Way, Centipede-Millipede, Lake Maitland, WA; Cape Cross, NA; WNA 2021). Shirtliff et al. (2017) proposed sub-dividing playa deposits into outflow (e.g., Centipede-Millipede) and playa (e.g., Lake Maitland); however, this classification has yet to be formally adopted by the IAEA.

2.1 Western Australian Surficial U Deposits

2.1.1 Introduction and Regional Distribution

All surficial U deposits in WA occur in Quaternary sediments within Tertiary-age paleochannels located in the Northern portion of the Yilgarn

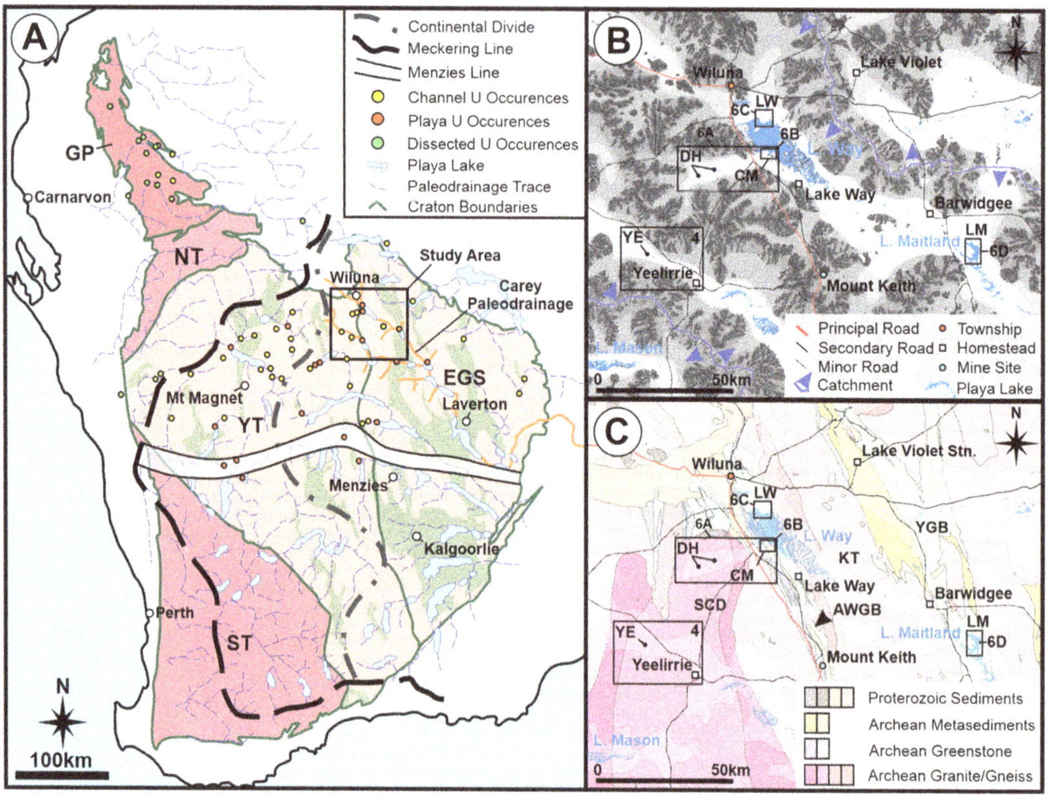

Fig. 2 (**a**) Locations for channel and playa U occurrences and deposits throughout the Northern Yilgarn and Gascoyne Province (GP) compiled and modified from Chudasama et al. (2018); Geographical information modified from Anand and Butt (2010). NT = Narryer Terrane; EGS = Eastern Goldfields Superterrane; YT = Youanmi Terrane; ST = Southwest Terrane. The black square indicates the study area. (**b**) 1 arc second Multi-resolution Valley Bottom Flatness (MrVBF; Gallant et al. 2012) imagery delineating a segment of the Carey paleodrainage examined in this study; Lake Way (LW), Centipede-Millipede (CM), Dawson-Hinkler (DH), Lake Maitland (LM), and Yeelirrie (YE). (**c**) 1:500 000 Precambrian geology underlying the study area. (Modified from GSWA 2018). SCD = Southern Cross Domain; AWGB = Agnew-Wiluna Greenstone Belt; KT = Kalgoorlie Terrane; YGB = Yandal Greenstone Belt. Black outline boxes in (**b**) and (**c**) correspond to deposit maps (Figs. 4 and 6)

Craton and the Gascoyne Province, between 23° 30' and 30°00' latitude (Fig. 2a; Mann 1974; Butt 1988; Hall et al. 2019). These deposits are geographically restricted by the onset of fluvial incision to the west (the Meckering Line; Mulcahy 1967) and to the south by a gradational shift in geochemical and hydrogeochemical characteristics of regolith, vegetation and groundwater (see Menzies Line; Fig. 2a; Butt et al. 1977). The Yeelirrie, Lake Way (including Lake Way and Centipede-Millipede) and Lake Maitland U deposits are the largest surficial U deposits found in WA (Fig. 2a–c; Butt et al. 1977; Hall and Mihalasky 2017; NEA/IAEA 2020). Uranium-series dating of carnotite within several WA deposits indicates mineralization ages are young, with most deposits exhibiting ages ca. <1 Ma (Dickson and Fisher 1980; Osmond et al. 1983; Briot and Fuchs 1984; Cuney and Kyser 2015).

2.1.2 Geological Setting

The Yeelirrie deposit is approximately 70 km southwest of Wiluna and 35 km west-southwest of the Lake Way deposits (Fig. 2b, c). The Yeelirrie deposit occurs along the central axis of a southeast flowing paleotributary that flows into Lake Miranda, eventually joining the main Carey paleodrainage at Lake Darlot. The Centipede-Millipede deposit is situated within another paleotributary of the Carey paleodrainage, known as the Lake Way Sub-Basin (LSWB), which flows into Lake Way from the west (Fig. 2b, c; Sanders 1972). The Lake Maitland deposit is located approximately 60 km southeast of Lake Way and resides within the Carey paleodrainage on the north-western margin of Lake Maitland (Fig. 2b, c).

Most Western Australian surficial U deposits are observed in drainages that are underlain by variable proportions of granite and greenstone of the Archean Yilgarn Craton (Condie 1981; Langford et al. 2000). Dissected calcrete U occurrences in the Gascoyne Province commonly overlie Proterozoic-age biotite gneiss (Fig. 2a; Butt et al. 1977). The Yeelirrie deposit occurs over Archean monzogranite and equivalent gneissic rocks of the Southern Cross Domain (SCD; Fig. 2c; Cassidy et al. 2006; Cameco 2015; Kneer 2015a). The Lake Way and Centipede-Millipede deposits occur over the Archean-age Agnew-Wiluna Greenstone Belt (AWGB; Fig. 2c). The Lake Maitland deposit overlies the Archean Yandal Greenstone Belt (YGB; Fig. 2c).

2.1.3 Regolith Development and Paleochannel Evolution

The Yilgarn Craton has been exposed to intense weathering for much of the Mesozoic to middle Cenozoic when the climate was hot and wet, forming a thick lateritic regolith over much of the landscape extending to ca. 100 m depth (Butt et al. 1984; Anand and Butt 2010). Fluvial processes during this time established numerous continent-scale drainage networks that remain preserved to this day (de Broekert and Sandiford 2005; Bell et al. 2012). Greenstone belts within the Yilgarn craton weathered to a greater extent than other rock types and controlled the location of sub-basins that currently manifest as playas. Subsidence of the interior of the Australian continent, through the early Eocene to late Miocene (due to the breakup of Gondwana), caused channel networks to fill with sediment, forming low-relief internal drainages (Magee 2009). The regional channel network was rejuvenated in the Miocene–Pliocene, incising fluvial channels into the regolith profile. Incision ceased with the onset of more arid conditions in the late Miocene associated with global cooling (see: Terminal Miocene Event; White 1994). Progressive aridification through the Miocene to the present resulted in the desiccation of rivers and lakes, forming extensive networks of playas and ephemeral drainages.

Tertiary paleochannel fill consists of immature and poorly-sorted fluvial and alluvial deposits derived from physical weathering of proximal exposed regolith and bedrock on catchment margins (Johnson et al. 1999; Brunner et al. 2011). Clastic material includes quartz, feldspar, kaolinite and illite, cemented by various authigenic phases, including carbonate minerals, smectite, haematite and goethite (Anand and Butt 2010). Late Quaternary deposits are predominantly aeolian-derived sandy silt and sheets of sand and gravel deposited in alluvial plains and outwash channels. These sediments have filled the older paleochannels and produced extensive low-relief alluvial sand plains. Present-day ephemeral streams and playas are situated along central axes and topographic depressions within paleodrainage networks. Playa sediments are composed of interbedded sand, silt, clay (kaolinite and detrital/authigenic smectite), and common gypsum and halite deposits.

2.1.4 Hydrology and Groundwater Evolution

Continued aridity throughout the late Quaternary to the present day has resulted in a stable, shallow and slow-moving groundwater regime along the central axes of paleochannels and in playas. Evaporation was (and is) particularly intense within playas and adjacent paleochannels, leading to the development of abundant calcrete and gypcrete deposits (Fig. 3a, b; Goudie 1972, 1973; Mann and Horwitz 1979). Such calcrete deposits have been repeatedly modified by phreatic and vadose zone meteoric diagenesis,

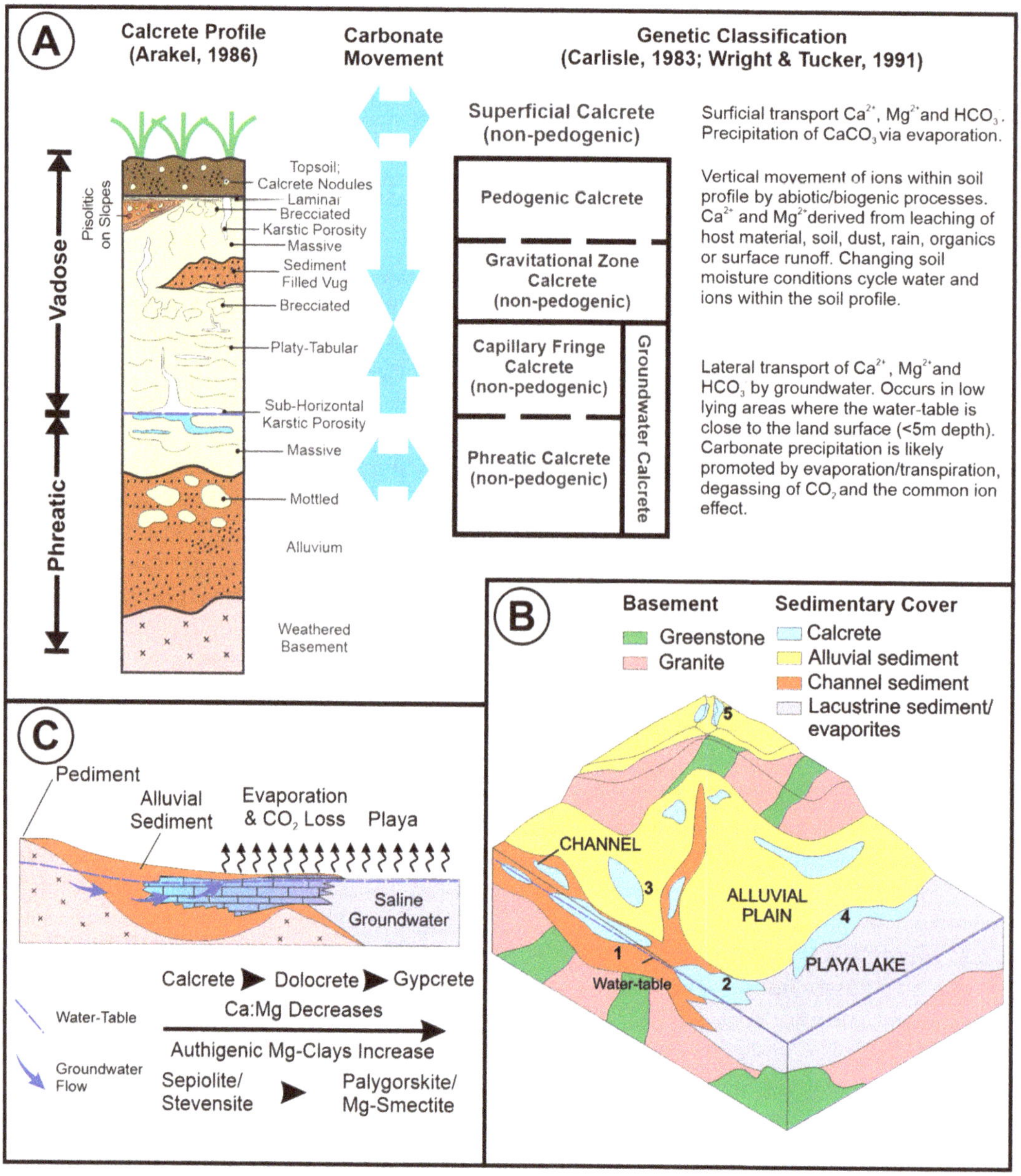

Fig. 3 (**a**) Generic groundwater calcrete profile with genetic classification and associated influence of diagenetic processes. (**b**) Geomorphological environments in which groundwater calcrete is observed. (1) Fluvial/paleochannel, (2) chemical delta, (3) alluvial plain/valley, (4) lake margin or playa, and (5) dissected/terrace. (Modified from Butt et al. 1977). (**c**) Schematic cross-section illustrating the evolution of groundwater and its authigenic precipitates across an arid alluvial basin. (Modified after Wright and Tucker 1991)

resulting in cyclic carbonate precipitation and dissolution, generating 5-to-10 m thick calcrete units with abundant karstic porosity (Fig. 3a).

Increasing aridity throughout the Quaternary has resulted in the chemical evolution of paleochannel groundwater from Ca-dominant to Mg-Ca-Si-dominant composition (Arakel et al. 1989; Drummond et al. 2021). This change has initiated extensive dolomitization and silicification of established calcrete bodies, as well as the precipitation of authigenic Mg-clays, opaline silica and sedimentary dolomite along with carnotite in fractures and pores of calcrete and associated alluvial sediments (Butt et al. 1977; Arakel 1986, 1988; Drummond et al. 2021). The degree of diagenetic alteration of calcrete bodies varies spatially and commonly increases toward playas, where dolomite, opaline silica and Mg-clays are more abundant (Fig. 3c). This late-stage authigenic mineral assemblage is observed in many surficial U deposits across WA, suggesting that their cogenesis is an essential control of the accumulation and distribution of U mineralization.

2.1.5 Channel and Playa Uranium Mineralization

The Yeelirrie orebody forms a 10 km long by 2 km wide ribbon-like zone of U mineralization in calcrete that occurs along the central axis of the paleochannel, where the valley floor has the lowest topographic gradient (Fig. 4; Western Mining Corp. 1975; Cameron 1984; Cameco 2015). Low-grade mineralization (100–1000 ppm U_3O_8) occurs between the surface and 10 m depth, with the locus of high-grade mineralization (>1000 ppm U_3O_8) at 4–7 m depth (Fig. 5a–c). Numerical groundwater flow and solute transport modeling conducted by Cameco (2015) confirms that the Yeelirrie deposit resides in a sub-basin situated over a pool of saline groundwater (10,000–30,000 mg/l TDS; Fig. 4). This sub-basin corresponds surficially where the valley bottom flattens at 500 m elevation (Fig. 4). Descriptions from Cameron (1984), combined with lithogeochemical data from Hudson (2010), indicate that carnotite is present in variably silicified calcrete/dolocrete and mottled, dolomitic/smectitic sediments that occur within and immediately beneath the main calcrete aquifer (Fig. 5a–c). Lithogeochemical interpretation also suggests that Mn oxides and celestine are variably coincident with U mineralization within the lowermost 1–2 m of the calcrete aquifer and in the underlying clay-rich sediments (Fig. 5a–c). The alluvial sediments within the paleochannel and beneath the ore body contain appreciable concentrations of V (100 to 300 ppm V; Fig. 5a–c) with no apparent correlation with U, suggesting that it is not present in carnotite, but most probably hosted by detrital phases and/or adsorbed onto clays and Fe/Mn-oxides. Gypsiferous sediment and gypcrete are also present in the top 1–2 m of the sediment profile and

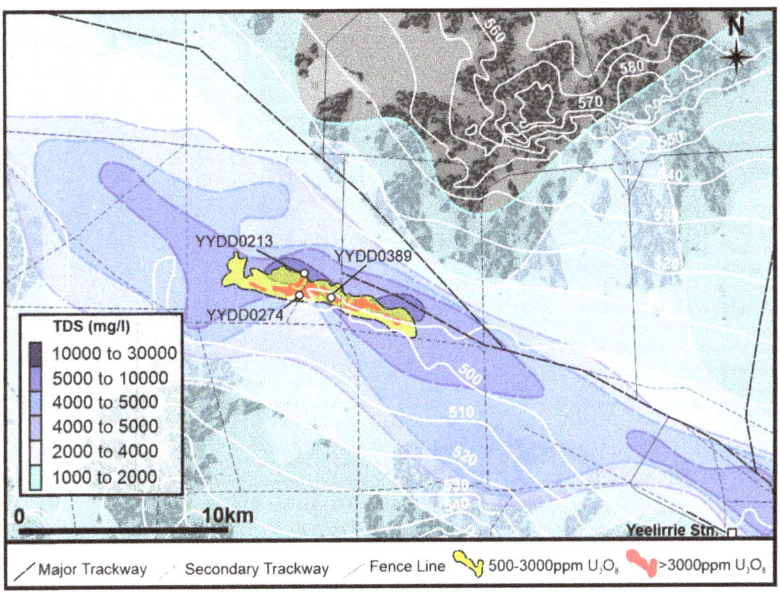

Fig. 4 Multi-resolution valley bottom flatness map overlain with the distribution of U mineralization at the Yeelirrie deposit. White dots indicate the location of representative air core drill holes summarized in Fig. 12. Groundwater salinity contours (Cameco 2015) reveal that the Yeelirrie orebody is situated over a saline pool of groundwater in a sub-basin within the primary catchment flowing southeast into Lake Miranda. Elevation contours derived from Shuttle Radar Topography Mission (SRTM) digital elevation product

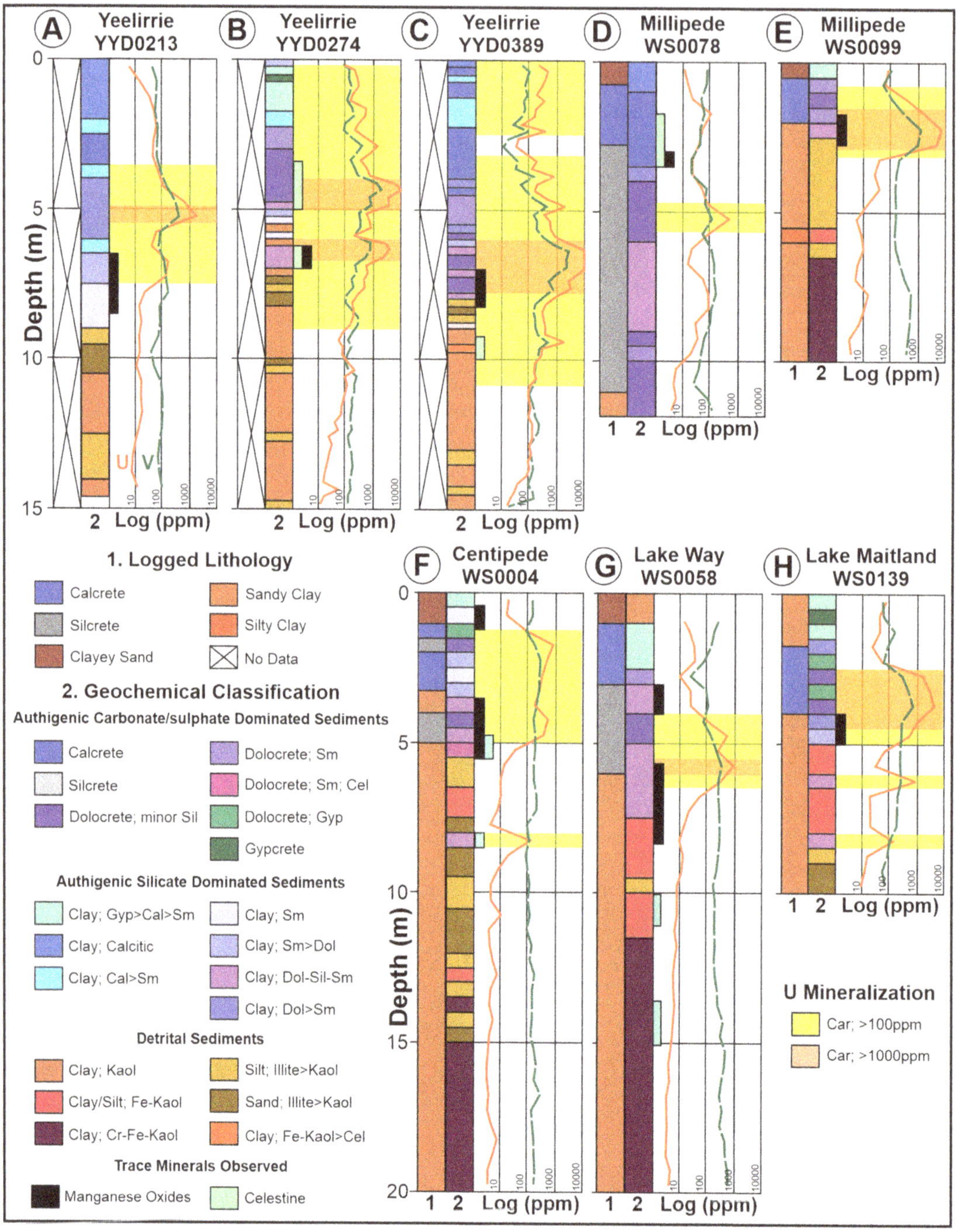

Fig. 5 (**a**–**c**) Representative air core drill holes from the Yeelirrie deposit. Geochemical classification interpreted with data from Hudson (2010). (**d**) – (**h**) representative sonic drill hole stratigraphy from Millipede (**d**, **e**), Centipede (**f**), Lake Way (**g**), and Lake Maitland (**h**). Lithological logging (1) compared against geochemical classification (2). The occurrence of celestine and Mn oxides was confirmed through a combination of logging descriptions and geochemical interpretation. Uranium and V were plotted on a log ppm scale to display zones of U mineralization. Mineral abbreviations: Sil = opaline silica; Cal = calcite; Sm = Mg-smectite; Dol = dolomite; Gyp = Gypsum; Cel = celestine; Kaol = Kaolinite; Fe = ferruginous (hematitic); Cr = elevated chromium

are uncommonly associated with lower-grade U mineralization (100 to 200 ppm U_3O_8; Fig. 5b).

The Centipede-Millipede U deposit occurs in chemical delta, clay pan and saltmarsh environments (Fig. 6a, b, d–f; Crabb et al. 1984). Carnotite occurs in the chemical delta at 4–5 m depth and is hosted in silicified dolocrete and silcrete with minor Mg-clays and amorphous Mg-silicate (AMS; Shirtliff et al. 2016b, 2017; Drummond et al. 2021). Silcrete replaces calcrete, whereas dolomite occurs as both a secondary replacive mineral after calcite in calcrete and as a primary precipitate in clay-rich sediments in the clay pan and salt marsh (Fig. 7a). Magnesium clays, AMS and opaline silica occur as late pore and fracture-lining cement and are commonly associated with sedimentary dolomite and carnotite (Fig. 7a–c). High-grade U mineralization at the Centipede-Millipede deposit occurs in saltmarsh and clay pan environments (Fig. 6b; Kneer 2015a) between 1–5 m depth and is associated with silicified and clay-rich dolocrete and dolomite- and smectite-rich sediments (Fig. 5d–f). Uranium mineralization is associated with minor Mn oxides, barite and celestine (Fig. 7a, b). The mineralized horizon commonly overlies highly oxidized, V-rich, ferruginous, illitic/kaolinitic sandy clay that contains elevated Cr and V, indicating a mafic provenance (200 to 1500 ppm Cr; Fig. 5e, f). Gypsum is commonly observed in the top meter of the sediment column and can be associated with minor carnotite mineralization. Downhole conductivity data collected by Toro Energy in 2015 across the Centipede-Millipede deposit identified the presence of a well-defined halocline that forms a hypopycnal groundwater estuary entering Lake Way (Humphreys et al. 2009; Shirtliff et al. 2016b; Drummond et al. 2021). Most U mineralization at Centipede-Millipede occurs where this fresh-brackish groundwater estuary is focused at the water table (1000–2500 mg/l TDS; 1–5 m depth). Significant U mineralization stops at the salt marsh-playa transition, which corresponds to an increase in groundwater salinity from saline in the estuary (ca. 30,000–50,000 mg/l TDS) to brine in Lake Way (>50,000 mg/l TDS). There is a conspicuous decrease in the abundance of carbonate minerals at this transition. Thermodynamic modeling of groundwater data suggests that calcite and dolomite become undersaturated at this interface (Mann and Deutscher 1978a, b; Drummond et al. 2021).

The Lake Way deposit occurs in saltmarsh at the terminus of the chemical delta and is coincident with the present-day water table between 4–10 m depth (Figs. 5g and 6a, c; French and Allen 1984; Shirtliff and Kneer 2014; Shirtliff et al. 2016a, b). The lithological and mineralogical associations are analogous to mineralization in the salt marsh at the Centipede-Millipede deposit, including being underlain by strongly oxidized V- and Cr-rich sandy alluvial clays (Fig. 5g).

Mineralization at Lake Maitland occurs over a very broad area, approximately 6 km long by 2 km wide, trending northeast to southwest, situated primarily in clay-rich, dolomitic saltmarsh and playa sediments with portions of the broad 100 ppm U_3O_8 ore grade shell occurring in chemical delta and lake margin calcrete (Fig. 6d; Cavaney 1984; Kneer 2015b; Shirtliff et al. 2016a, 2017). High-grade mineralization (>1000 ppm U_3O_8) forms a thin, arcuate shape with three branches that extend to the northwest (Kneer 2015b; Shirtliff et al. 2016b). The primary axis of high-grade mineralization is approximately positioned immediately on the up-flow side of the geomorphological expression of modern playa sediments (Czp; Fig. 6d). Carnotite is present as a thin, flat-lying horizon averaging 1.7 m in thickness, approximately 1–2 m below the surface (Fig. 5h). Mineral associations are akin to the Centipede-Millipede and Lake Way deposits in that carnotite is observed nucleating on Mg-clays (Mg-smectite, sepiolite, stevensite, palygorskite, and AMS), sedimentary dolomite and Mn oxides (Fig. 7d). Gypsum is common throughout the top 1–3 m of mineralized and barren playa sediments. In some cases, gypsum coincides with carnotite mineralization, but it is interpreted as genetically unrelated (Shirtliff et al. 2016a).

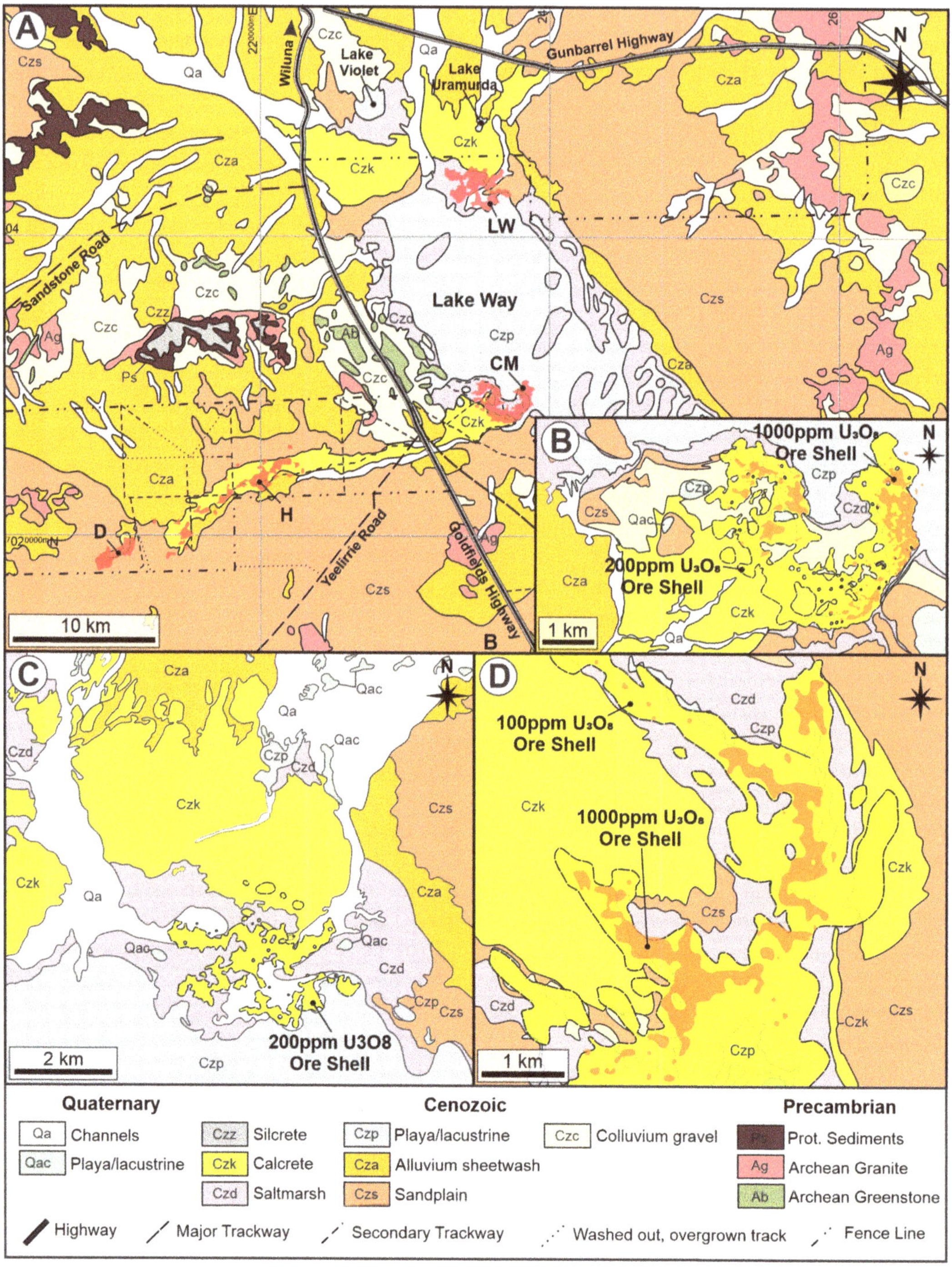

Fig. 6 (**a**) Location and distribution of channel and playa U deposits at Lake Way, WA. Uranium deposits are depicted as transparent red polygons. Deposit names: LW = Lake Way, CM = Centipede-Millipede, H = Hinkler, D = Dawson. (**b**) Centipede-Millipede deposit. (**c**) Lake Way deposit. (**d**) Lake Maitland deposit. Yellow transparent polygons in B and C depict the lower ore cut-off of 200 ppm U_3O_8 and 100 ppm U_3O_8 in D. Red transparent polygons in B and D delineate the high-grade ore zones of >1000 ppm U_3O_8 (Shirtliff and Kneer 2014; Shirtliff et al. 2016a, b)

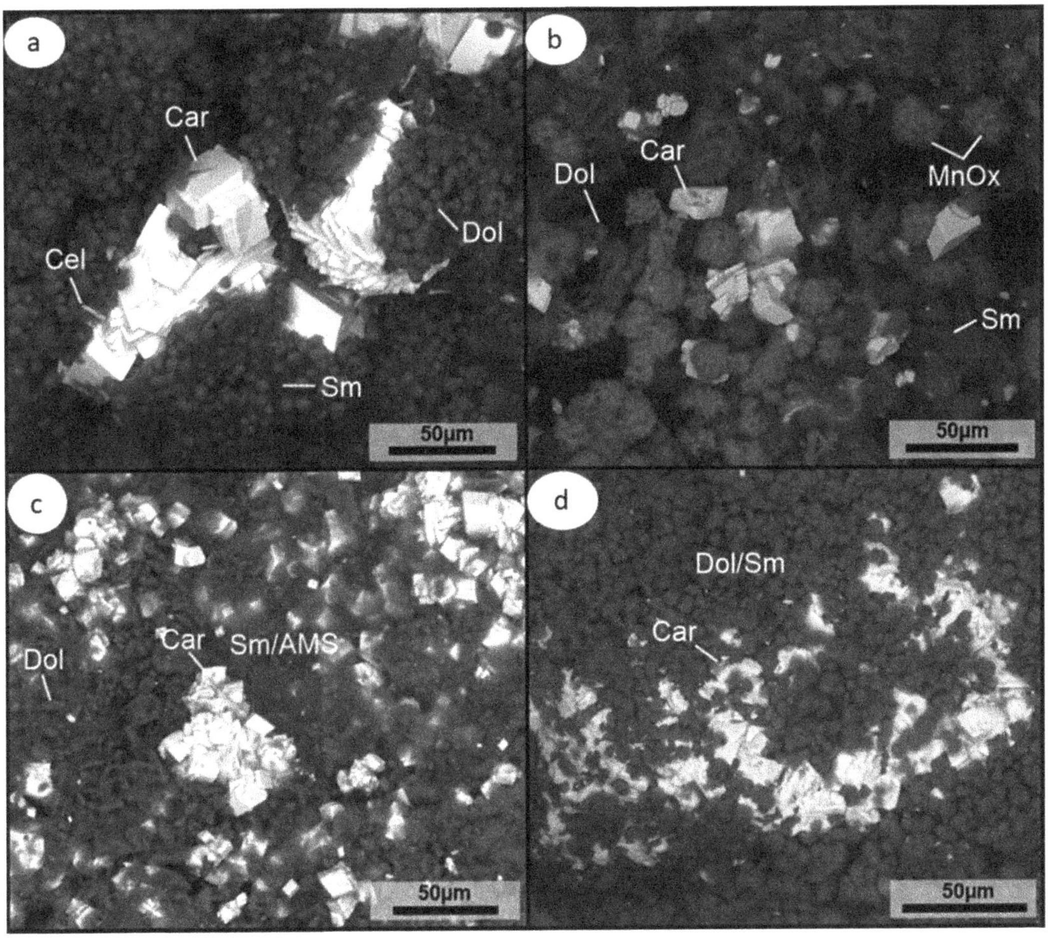

Fig. 7 (**a**) Carnotite (Car) and celestine (Cel) embayed in Mg-smectite (Sm) and spheroidal dolomicrospar (Dol) from a mineralized interval at the Centipede-Millipede deposit. (**b**) manganiferous (MnOx) and Mg-smectite-rich sediment with microcrystalline carnotite nucleating on both Mg-smectite and Mn oxide (todorokite) from the Centipede-Millipede deposit (Drummond et al. 2021). (**c**) Mg-smectite-rich sediment from the saltmarsh at the Centipede-Millipede deposit. Aggregates of carnotite are embayed in a mélange of Mg-smectite, AMS, opaline silica, and dolomicrospar. (**d**) Carnotite mineralized dolomite nodule from the Lake Maitland deposit. Dolomite nodules are composed of rounded dolomicrospar with a matrix of Mg-smectite

2.2 Namibian Surfical U Deposits

2.2.1 Introduction

The Erongo region in Namibia contains several significant paleochannel-hosted U deposits and minor occurrences hosted within carbonate-rich sediments that fill paleochannels developed on Proterozoic and Early Paleozoic (Damaran) metamorphic and igneous basement rocks. The contained U resource of these deposits amounts to nearly half a billion pounds of U_3O_8 (Table 3), making this region the most-endowed region in the world for this type of deposit.

2.2.2 Geological and Geomorphological Setting

The contemporary Kuiseb Basin contains four of the largest U deposits of the region, at Langer Heinrich, Tumas, Tubas and Aussinanis (Fig. 8). Much of this hyper-arid basin receives <100 mm

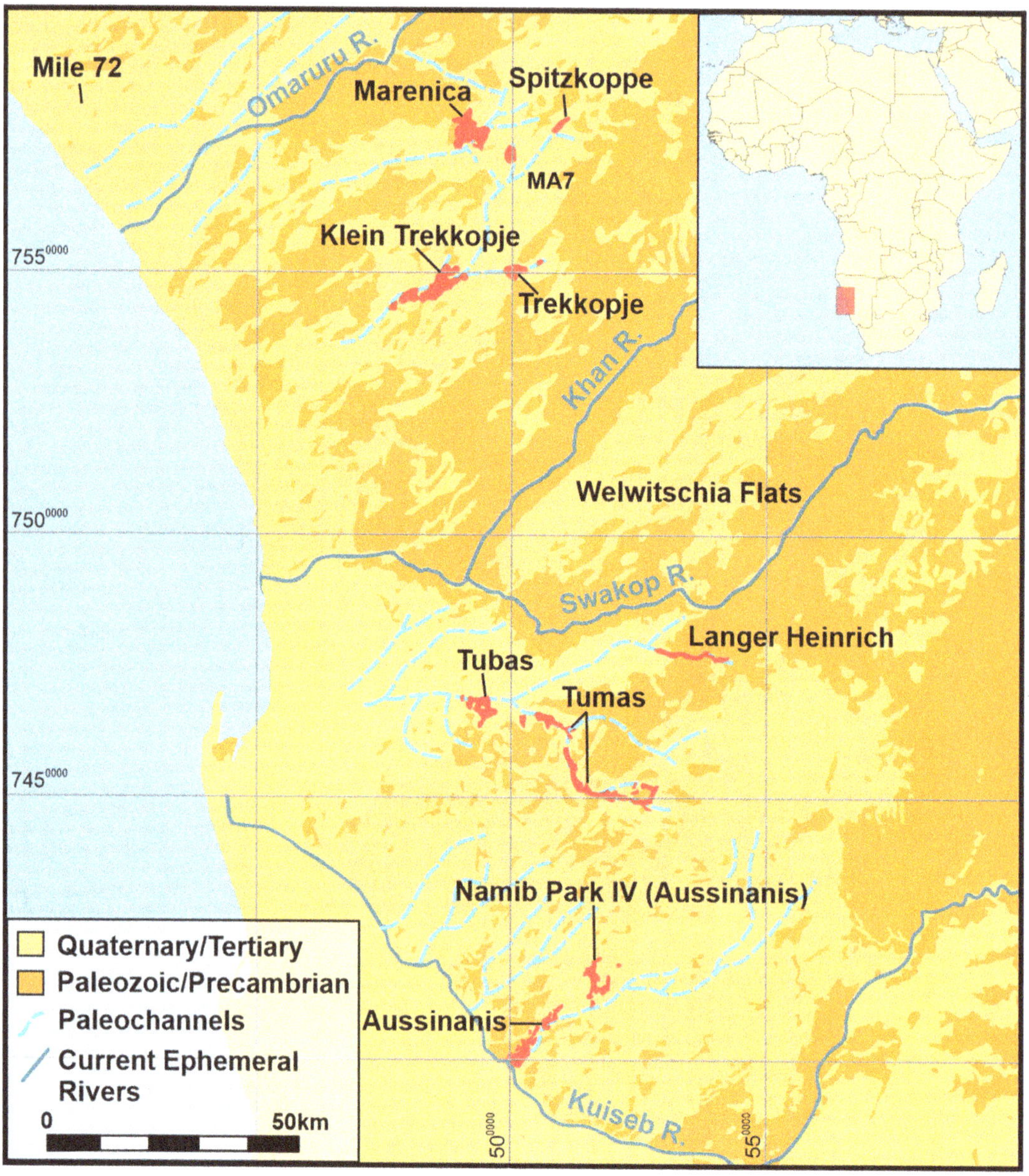

Fig. 8 Inset map – location of paleochannel-hosted U deposits of the Erongo region of Namibia (red rectangle). Main map – resent-day Kuiseb drainage basin, east of Walvis Bay. This basin contains four of the major U deposits of the region

of rainfall annually, and the Kuiseb River flows on average for only 16 days each year; water reaches the Atlantic Ocean only on rare occasions (Stone et al. 2010). Smaller active tributaries are generally spatially coincident with the paleochannels. Much of the rainfall in these tributaries probably evaporates before reaching the Kuiseb, and there are only a few small active playas, e.g., Tsondab Vlei and Zebra Pan (Eckardt et al. 2001). The paucity of active playas is a substantial difference between this region and Western Australia and is probably due to higher annual rainfall in WA (200–300 mm per annum).

Table 3 Uranium resources of palaeochannel type deposits of the Erongo region of Namibia

Deposit Name	Ore (Mt)	Grade (ppm U_3O_8)	Contained U_3O_8 (Mlb)
Langer Heinrich[a]	145.9	535	171.9
Tumas 1, 2 & 3[b]	200.0	258	114.1
Klein Trekkopje[c]	25.0	105	57.9
Marenica[c]	276.0	94	57.3
Aussinanis[c]	34.6	237	18.0
Koppies (Tumas)[d]	41.4	208	20.3
Trekkopje[c]	46.0	129	13.1
Tubas Red Sand[c]	34.0	170	12.7
Klein Spitzkoppe[e]	25.0	150	11.2
Welwitchia Flats[e]	35.0	120	9.2
Tubas Calcrete[c]	7.4	374	6.1
MA 7[c]	22.8	81	4.1
Hakskeen[e]	0.4	331	0.3
Mile 72[e]	0.4	230	0.2
TOTAL	**852.5**		**496.4**

[a]Paladin Energy Ltd. (2020), [b]Deep Yellow Ltd. (2023), [c]Hall and Mihalasky (2017), [d]Elevate Uranium (2024), [e]Roesener and Schreuder (1992)

The prominent paleosurface at the base of the paleochannels is referred to as the Namib Unconformity Surface and is believed to have formed during the Late Cretaceous (Ollier 1977). The morphology of this surface and the extent to which it has been modified (incised) during later events remains uncertain, although the paleochannels themselves have been reasonably well defined by a combination of airborne electromagnetic interpretation and drilling. There is little evidence of substantial weathering of the basement rocks below the Namib Unconformity Surface, probably a function of predominantly arid conditions since the Cretaceous (Ward 1987). Dating of secondary minerals at the Skorpion and Wolkenhauben base-metal deposits, however, indicates that some chemical weathering (and a relatively humid climate) occurred between 50–85 Ma (McInnes et al. 2006). A breccia of angular quartz vein fragments up to 3 m thick (Gomkaeb Conglomerate) is commonly seen at the basement contact and has been interpreted as a paleoregolith (Ward 1987).

Approximately 200 m of Early Tertiary to Miocene sediment occurs in the Kuiseb Valley (Fig. 8). This sedimentary sequence records fluctuating climatic conditions from the aeolian Tsondab Sandstone Formation and younger units (arid) to fluvial conglomerates of the Oswater and Karpfencliff Conglomerates (higher rainfall). Evaporites, including halite, sylvite and gypsum, occur at several stratigraphic levels, and some of the sediments were probably deposited in playas (e.g., Khommabes Carbonate Member, Tsondab Sandstone). Carbonate cementation is common throughout the sequence but is particularly intense in the 5 m thick Kamberg Calcrete that forms prominent scarps along the present-day Kuiseb valley. This unit is regarded as a stratigraphic marker, although intense carbonate cementation appears not to be confined to a particular stratigraphic interval (Ward 1987). Isolated outcrops of the tufa and travertine of the Hudoab tufa unit occur in the Kuiseb Valley. Tufa samples have been dated between 10.9–31 ka (Vogel 1989). Caves near Rossing Mountain contain "sinter" dated between 23–42 ka (Geyh and Heine 2014), indicating more humid conditions at this time.

2.2.3 Paleochannel Fill

Several major paleochannels in the Kuiseb Basin, east of Walvis Bay (Fig. 8), host significant U mineralization. These include Langer Heinrich and Tumas, which occupy 15 and 50 km of the respective paleochannels. The Hirabeb paleochannel hosts the Aussinanis deposit at its southern end, close to its confluence with the Kuiseb River. The depths of the paleochannels may vary markedly; at Tumas, the depth increases from a few metres in the east to over 100 m 50 km to the west.

The geology of the Langer Heinrich area is shown in Fig. 9. Unusually, the paleochannel sediments outcrop where they are not unconformably overlain by recent active sediments of the Gawib drainage (Trittschack 2008; Becker and Kärner 2014). Carbonate-cemented ("calcretized") sediments *above* the present level of the Gawib River and its associated deposits are evidence for substantial incision of the paleochannel before initiation of the Gawib River (Figs. 9 and 10). This incision and

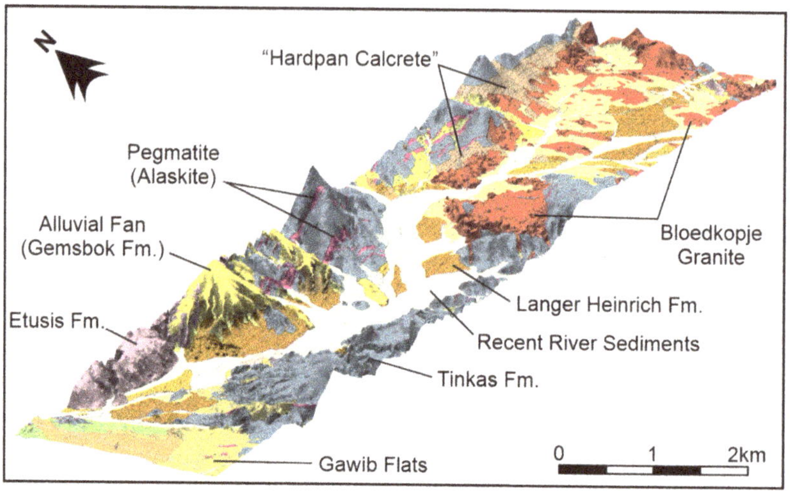

Fig. 9 3D geology of the Langer Heinrich deposit. Uranium mineralization extends from the Gawib Flats upstream to the Bloedkoppje Granite. Geology from (Trittschack 2008). Note the presence of "hardpan calcrete" above the present level of the Langer Heinrich Valley and recent sediments obscuring the older paleochannel sediments. Topographic relief is exaggerated 10:1

Fig. 10 Looking westwards down the Langer Heinrich Valley showing, relict calcrete deposits above the present level of the Gawib River and the uranium deposits

consequent exposure of deep paleochannel sediments facilitated the discovery of many surficial U deposits by airborne radiometric surveys (Becker and Kärner 2014).

The dominant paleochannel fill at the three deposits is poorly sorted, polymictic conglomerate (clasts generally up to 20 cm) and sand, cemented to varying degrees by carbonate minerals, chiefly calcite and minor clays. Minor silcrete has been identified at Trekoppje and Langer Heinrich. Fine-grained units consisting of muds and silts are occasionally present but generally constitute a negligible portion of the paleochannel fill, another significant difference to the Australian deposits. Clasts include monomineralic black quartz, which is probably derived from leucogranite pegmatites. Bedding can be correlated only for short distances, indicating transient high-energy flows, such as flash floods typical of desert climates (Hambleton-Jones et al. 1986; Hartleb 1988). The angular habit of detrital feldspar suggests deposition close to source.

At Tubas and in parts of the Hirabeb paleochannel, carbonate-cemented conglomerates are overlain by poorly consolidated coarse and rusty red sands of probable aeolian origin. At Tubas, this material hosts a significant U deposit, the Tubas Red Sand deposit. A veneer of sediment related to contemporary ephemeral drainage and sheet wash overlies the calcretized conglomerate and aeolian sand, typically reaching no more than a few metres thick. This material often obscures the courses of the paleochannels. The present-day land surface is generally defined by gypcrete, itself incised by recent drainage (Wilkinson 1990). Gypsum is only found within a few metres of the current land surface and contains carnotite in a few locations.

2.2.4 Uranium Mineralization

The U mineralization of the Kuiseb paleochannel, including Langer Heinrich, is poorly documented, despite intense exploration over the past five decades. Mineralization typically extends over many kilometres of paleochannel, occurring as irregular tabular bodies elongated

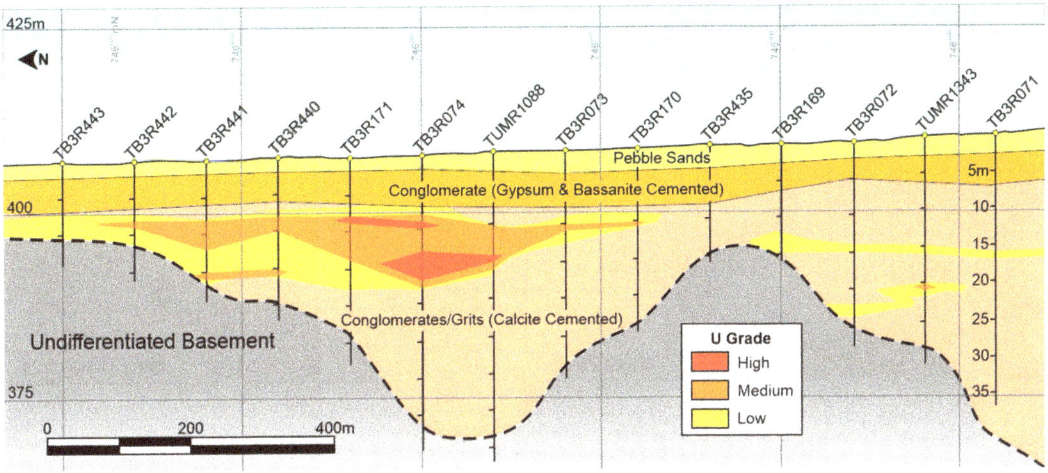

Fig. 11 Typical cross-section through the Tumas deposit

down drainage (Figs. 8 and 11), although there is some complexity due to the occurrence of some carnotite along fractures in the sediments and the underlying basement rocks. There appears to be no correlation between the intensity of U mineralization and the degree of calcrete deposition. Uranium minerals occur disseminated in the sediments and as coatings on clasts and fractures. The main ore mineral is usually referred to as carnotite, but tyuyamunite has been described at Langer Heinrich and appears to be at least as abundant as carnotite in the samples studied. Gangue minerals include quartz, feldspar and minor mica, magnetite, ilmenite, titanite, zoisite and apatite, plus post-depositional calcite and clays, with dolomite described at Langer Heinrich (Hartleb 1988; Trittschack 2008; Fleurances and Kinnaird 2011). Clay minerals include palygorskite, vermiculite, sepiolite, illite, smectite and kaolinite (Fig. 12a–d). Silcrete is also uncommon but is present in the upper part of the Trekoppje trial pit. Amorphous silica has been described at Langer Heinrich, but its abundance and distribution are poorly understood (Trittschack 2008). The paragenetic relationship between phases is poorly resolved. Recent studies suggest that U is associated with Mg-clays (e.g., palygorskite, smectite, amorphous Mg-silicates) and dolomite, and that these mineral phases post-date calcrete formation (Trittschack 2008; Iilende 2012; Drummond 2016). The absolute age(s) of U deposition remain uncertain, the only date being a U-Th age of 68 ± 2 ka (Fleurances and Kinnaird 2011). The source of U is usually regarded as granitic rocks. At Langer Heinrich, it is suggested that U was derived from the Bloedkoppie Granite, which outcrops immediately upstream of the deposits (Fig. 9). However, a correlation between U grade and the abundance of detrital black quartz points to a pegmatitic leucogranite ("alaskite") source. Indeed, pegmatites cutting the Bloedkoppie Granite contain significantly higher U content than the granite itself.

3 Environmental Factors Influencing Occurrence

Historical research and observations from the above case studies identify seven factors required to form economic surficial U deposits. These factors are: (1) geology; (2) geomorphology; (3) climate; (4) hydrology; (5) focus; (6) evaporation; and (7) time.

3.1 Bedrock Geology

The catchment geology of surficial U deposits is an essential parameter in providing paleochannel sediments and groundwater aquifers with the

Fig. 12 Back-scattered electron images of samples from Langer Heinrich. (**a**) – palygorskite (Trittschack 2008). (**b**) vermiculite (Fleurances and Kinnaird 2011). (**c**) and (**d**) association of carnotite with dolomite and Mg-silicates (AMS, Mg-Clay; Drummond 2016)

requisite ionic activities of U, V, and K to reach carnotite saturation (Mann 1974; Butt et al. 1977; Mann and Deutscher 1978a). Previous studies of Western Australian and Namibian deposits note minor abundances of carnotite are common in paleodrainage sediments and calcrete over granitoid rocks throughout semi-arid regions but with significant deposits spatially associated with a combination of U and V-rich bedrock lithologies (Butt et al. 1977; Toens and Hambleton-Jones 1980). In WA, the primary source rocks are the strongly weathered granites (source of U and K) and greenstones (source of V) of the Archean Yilgarn Craton. In Namibia, the primary sources of U are likely to be Neoproterozoic pegmatites and granites. These rocks generally have abundant V, suggesting that a separate source of V may not be required, although biotite schist has been proposed as an additional source of V (Toens and Hambleton-Jones 1980; Hartleb 1988; Youlton 2007). The paleochannel sediments may also provide a source of both U and V (Butt 1988; Anand and Butt 2010). In WA, U is hypothesized to be sourced from strongly weathered Precambrian granite, whereas Namibian deposits U is interpreted to be derived from the host paleochannel sediments (see below). The most cited sources of V are weathering of detrital magnetite, ilmenite and ferromagnesian minerals such as biotite and pyroxene, derived from granites and greenstones. In WA, mafic-derived sediments and reworked

V-rich lateritic ironstones deposited within paleochannels potentially serve as possible sources of V.

An important difference between the source geology in WA and Namibia is the degree of weathering. The Yilgarn Craton in WA has experienced multiple intense weathering and lateritization events since the Mesozoic and Cenozoic, resulting in the development of mature lateritic regolith profiles that affect not only the Precambrian basement but also the Tertiary paleochannel sediments that host surficial U deposits (Anand and Butt 2010). In Namibia, most paleochannels have formed over relatively fresh Precambrian and Paleozoic basement lithologies (Youlton 2007; Kinnaird and Nex 2016). The diversity of geological sources of U and V and variation in the degree of weathering between global occurrences demonstrates that several favorable geological source combinations can provide sufficient concentrations of U and V to form channel and playa U deposits and that heavily weathered source rocks are not always a prerequisite.

3.2 Climate

Most surficial U deposits worldwide are situated in semi-arid to arid climate zones, with the majority clustered between 20° and 30° latitude. Exceptions to this observation include equatorial deposits in Tanzania and Somalia and some smaller deposits 40° to 50° S in Argentina (Fig. 1). The distribution of surficial U deposits is interpreted to indicate that low annual precipitation (<300 mm per year) and high evaporation rates (>2500 mm per year) are an essential element in their genesis (Mann 1974; Butt et al. 1977; Carlisle 1980, 1983).

3.3 Geomorphology

Surficial U deposits are typically located along the central axis of low-relief paleodrainages and at the margins of playas, where the local water table is shallow, and groundwater flow is sluggish. It is interpreted that local basement constrictions in the paleodrainage, differences in sediment permeability, changes in hydraulic gradient, or contrasting salinity gradients focus U-bearing groundwater upward into the zone of evaporation (Mann 1974; Butt et al. 1977; Mann and Deutscher 1978a; Gaskin et al. 1981; Carlisle 1983; Drummond et al. 2021).

3.4 Hydrology

3.4.1 Aquifer Characteristics

Aquifer characteristics and groundwater composition are interrelated and together form the physical and chemical pathway which transports U-bearing fluids that form surficial U deposits. Previous workers have demonstrated through regional-scale hydrogeochemical surveys that groundwater composition is an essential factor controlling both U mobility and carnotite precipitation (Mann and Deutscher 1978b; Noble et al. 2011; Gray et al. 2016). For example, hydrogeochemical data from Gray (2001) and Gray et al. (2016) indicate that the distribution of surficial U deposits in WA is restricted to paleochannels in the Northern Yilgarn that possess oxidized groundwaters with circum-neutral pH. Thermodynamic modeling supports this interpretation, revealing that carnotite solubility is lowest between pH 6 and 8 (Mann 1974; Drummond et al. 2021). Previous research suggests that the development of paleochannel calcrete aquifers and their subsequent diagenesis is fundamental for establishing these favourable hydrological conditions.

3.4.2 Diagenesis of Groundwater Calcrete Aquifers

Diagenetic modification (dolomitization/silicification) of calcrete aquifers over time produces shallow, karstic aquifers that are highly permeable and open to the atmosphere. Equilibration of groundwater with calcite/dolomite in a calcrete aquifer is interpreted to result in groundwater with circum-neutral to slightly basic pH and high alkalinity, which maximize U mobility (Mann and Deutscher 1978a). Interaction of calcrete aquifer groundwater with circum-neutral saline groundwater within paleochannel sub-basins and groundwater estuaries entering playas results in the precipitation of abundant Mg-clays and opaline silica within calcrete and adjacent sediments. Capillary dilatancy associated with authigenic swelling clays (e.g., Mg-smectite) is interpreted to enhance the effect of evaporation within these shallow aquifers (Anand and Butt 2010; Bowell and Davies 2017; Drummond et al. 2021).

3.5 Focus

Although calcrete aquifers are common globally in arid/semi-arid climates, most are either unmineralized or contain only modest amounts of carnotite. It is apparent in many significant surficial U deposits that stable, persistent, and efficient hydrological focusing of groundwater into areas of high evapotranspiration over geologically relevant timescales (ca. >50 ka) is essential for the accumulation of large surficial U deposits (Batulin 1980; Gaskin et al. 1981; Cameco 2015; Chudasama et al. 2018).

Playa deposits such as Centipede-Millipede and Lake Maitland are interpreted to result primarily from vertical focusing of U-bearing groundwater through hypopycnal flow within a groundwater estuary and/or through changes in sediment permeability at the playa interface. It is suggested that the overall absence of lateral focusing mechanisms at the channel-playa transition results in expansive lobate zones of mineralization over a large spatial area with low to medium grades (i.e., 200–1000 ppm U_3O_8; Fig. 6b–d). High-grade, high-tonnage channel U deposits such as Yeelirrie typically exhibit a combination of a large catchment area and both vertical and lateral focusing in a relatively small saline sub-basin within a larger paleodrainage (Fig. 4). This three-dimensional focus of U-bearing groundwater establishes ideal conditions to maximize the influence of evaporation and, thus, the physicochemical processes governing carnotite precipitation (Fig. 4).

Less pronounced paleochannel sub-basins may be susceptible to breaching during wetter periods (i.e., climatic shifts associated with interglacial periods; e.g., Dawson-Hinkler deposit; Fig. 6a). The consequent rise in the water table is interpreted to disrupt the hydrology of the sub-basin, terminating the mineralization process. The numerous relatively small and low-grade nature of deposits in WA is interpreted to reflect ephemeral U accumulation events, punctuated with intervals of remobilization, as indicated by common U-series disequilibrium observed in most deposits (Dickson and Fisher 1980; Dickson 1984). In contrast, Namibian deposits such as Langer Heinrich and Tumas exhibit continuous ribbon-like zones of low-to-medium grade mineralization that are situated along paleochannel axes. It is hypothesized that the lack of discrete sub-basins within the Namibian paleochannel systems has resulted in spatially extensive, high tonnage deposits with relatively low to moderate grades.

3.6 Evaporation and Physicochemical Processes Influencing Carnotite Precipitation

Previous researchers hypothesized eight physicochemical processes that have the potential to facilitate carnotite precipitation. These processes include: (1) evaporation; (2) groundwater mixing; (3) pH change; (4) increase in K^+ activity; (5) dissociation of U and V complexes; (6) CO_2 degassing; (7) redox; and (8) sorption processes (Table 4). Research of numerous occurrences and significant deposits in WA and Namibia strongly suggests that appreciable carnotite mineralization cannot be attributed to a single physicochemical process; however, evaporation appears to be the fundamental parameter that is required to establish and maintain the physicochemical feedback loops that result in carnotite precipitation (Mann 1974; Mann and Deutscher 1978a; Bowell and Davies 2017; Chudasama et al. 2018; Drummond et al. 2021).

Table 4 Summary of potential U fixing mechanisms in channel and playa U deposits

Genetic model	Description
Groundwater mixing	U and V are transported by two separate aquifers, U as UO_2OH^+, V as $VO(OH)^+$ or $H_2VO_4^-$. When these fluids mix carnotite saturation is exceeded causing precipitation
pH change	Based on the solubility of carnotite as a function of pH, initial U and V bearing fluids are hypothesized to be at pH ~5.0, in equilibrium with atmospheric CO_2. U is carried as UO_2OH^+, V is carried as $H_2VO_4^-$. Species remain in solution below pH = 5.0. If pH rises to between 5.0 and 7.0 (ex: groundwater passes from clastic sediment into a calcrete aquifer) carnotite saturation is exceeded and precipitation ensues. If pH continues to rise above 7.0, carnotite may dissolve
Evaporation	Evaporation of aquifer water results in the increase in activity of K, U and V, achieving carnotite saturation. Evaporation can occur by surface evaporation from salt lakes and by sub-surface evaporation from the capillary zone above a shallow water table (<5 m). Groundwater calcretes are an ideal location for this process to occur because of their excellent porosity and permeability
Increase in K^+ activity	K activity is a function of the weathering rate of K-bearing minerals and by K^+ sorption onto negatively charged colloids (Fe-oxy(hydr)oxides, clays). K^+ activity may also increase sufficiently to induce carnotite precipitation where increased evaporation occurs in localized sumps and groundwater flow is sluggish
Dissociation of U & V complexes	Groundwater carries K^+, $H_2VO_4^-$ and U carbonate complexes. Change in pH, decrease in carbonate content or an increase in salinity may affect the stability of U and V complexes. Change in pH is unlikely to be significant; however, carbonate precipitation in a calcrete aquifer may induce dissociation of U complexes
Degassing	Rainwater or acidic groundwater percolate though sediment and interact with carbonate minerals. The greater confining pressure and pCO_2 of the deep aquifer allows the water to acquire significant concentrations of HCO_3^-. As the groundwater moves down flow it may rise closer to the surface, reducing the confining pressure causing degassing of CO_2 and the precipitation of carnotite
Redox	Vanadium is carried in the V^{4+} state ($VO(OH)^+$). In this situation, groundwater can contain high concentrations of U in V without carnotite precipitation. Groundwater enters a calcrete body and is exposed to an oxidizing environment which causes V^{4+} to oxidize to V^{5+} ($H_2VO_4^-$) allowing carnotite to precipitate. Thermodynamic modelling by Mann (1974) indicated that natural waters in WA calcrete environments have Eh able to accommodate V^{4+} and V^{5+} species. This model is the most cited of all. Revised thermodynamic modelling by Noble et al. (2011) suggests this model is less likely
Sorption processes	Calcretes commonly contain significant amounts of clay minerals and may be a controlling factor in the local activity of K^+. The cation exchange capacity of clays increases with increasing pH and salinity. In this setting, K^+ is preferentially adsorbed over Ca^{2+}, therefore the mineralogical composition of groundwater calcretes may influence carnotite solubility. Clays with adsorbed K^+ may also serve as carnotite nucleation sites

After Mann (1974), Mann and Deutscher (1978a, b), Boyle (1984), Bowell and Davies (2017)
Adopted from Drummond et al. (2021)

4 Genetic Model

4.1 Previous Genetic Models

The most widely cited genetic model of surficial U deposits is the redox-evaporation model proposed by Mann (1974) and Mann and Deutscher (1978a; Fig. 13; Table 4). This model postulates that U and V are sourced from deeply weathered source rocks (e.g., granite and greenstone; metasediment) and transported together through the paleochannel aquifer as uranyl hydroxide/carbonate (U^{6+}; UO_2OH^+ and $UO_2(CO_3)_2^{2-}$, respectively), and vanadyl hydroxide (V^{4+}; HVO_2^+) species. It was hypothesized that basement structures or permeability contrasts within

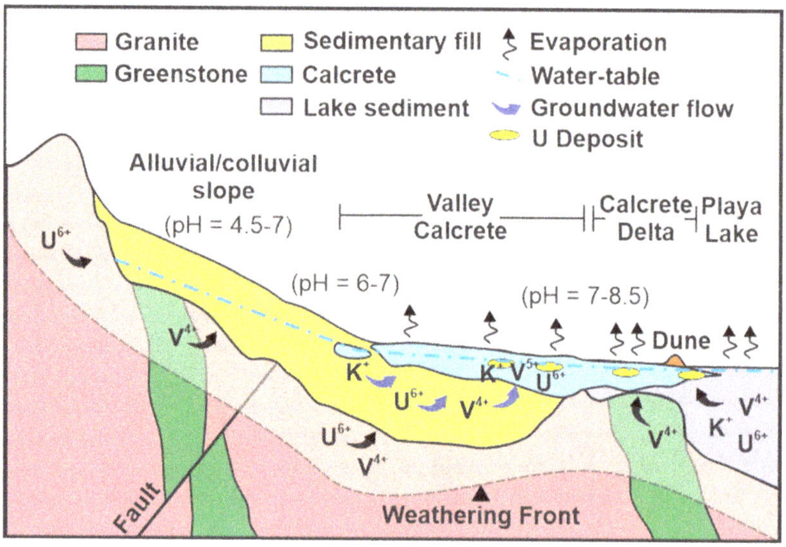

Fig. 13 Idealized model of calcrete-hosted U deposits after Mann and Deutscher (1978a). This model was previously established as the primary genetic model for channel and playa U deposits

paleochannel and playa sediments could direct groundwater flow upward and into the zone of oxidation and evaporation.

More recent thermodynamic studies of Northern Yilgarn groundwaters reveal the stability field of V^{4+} aqueous species is quite restricted, suggesting that this model is less plausible because V^{5+} ions ($H_2VO_4^-$ and HVO_4^{2-}; Fig. 14a; Noble et al. 2011; Drummond et al. 2021) are the most likely aqueous species. The presence of Mn oxides in mineralized calcrete and strongly oxidized hematitic alluvial-lacustrine sediments beneath these deposits is interpreted to indicate that V mobility is limited. The local activity of aqueous V in groundwater is potentially controlled by adsorption-desorption processes associated with pH fluctuations or Mn/Fe-redox cycling operating within the sediments and saprolite beneath calcrete aquifers (Fig. 14b; Jarvis et al. 1994; Drummond et al. 2021). Similar Mn/Fe oxides have not been recognized in Namibian deposits, and thus, the source and behaviour of V are still enigmatic. In WA and Namibia, carnotite is commonly associated with a gangue assemblage of authigenic opaline silica, Mg-silicates (amorphous Mg-silicate; palygorskite, sepiolite, saponite, montmorillonite, vermiculite and stevensite) along with calcite, dolomite, aragonite, celestine, barite.

4.2 Current Thinking

The most recent genetic model posits that carnotite precipitation is achieved in many surficial U deposits in WA through a series of physicochemical feedback loops initiated by hydrologic focusing and evaporation (Drummond et al. 2021; Fig. 15). In this model, alkaline, fresh/brackish groundwater (rich in U, K, Ca, Mg, Si and Sr) flows into a playa or sub-basin containing denser saline groundwater/brine (rich in Mg, Ca, K, V and SO_4) with near-neutral pH. The density contrast between the two groundwaters causes the lower-density, fresher water to be focused at the water table and into the influence of evaporation via hypopycnal flow (Fig. 15). The interaction of the two groundwaters with subtly contrasting pH initiates the precipitation of opaline silica, AMS and Mg-clays in the fractures and pores within the capillary fringe in calcrete and associated clastic sediments. The common presence of discrete intervals of celestine (+/- barite) with Mg-clay and dolomite in the calcrete and clay-rich sediments provides compelling evidence of limited groundwater mixing in the shallow aquifer setting (Hanor 2000). The accumulation of authigenic Mg-smectite in calcrete and underlying sediments is interpreted to increase the potential for "shrink-swell" capillary dilatancy that

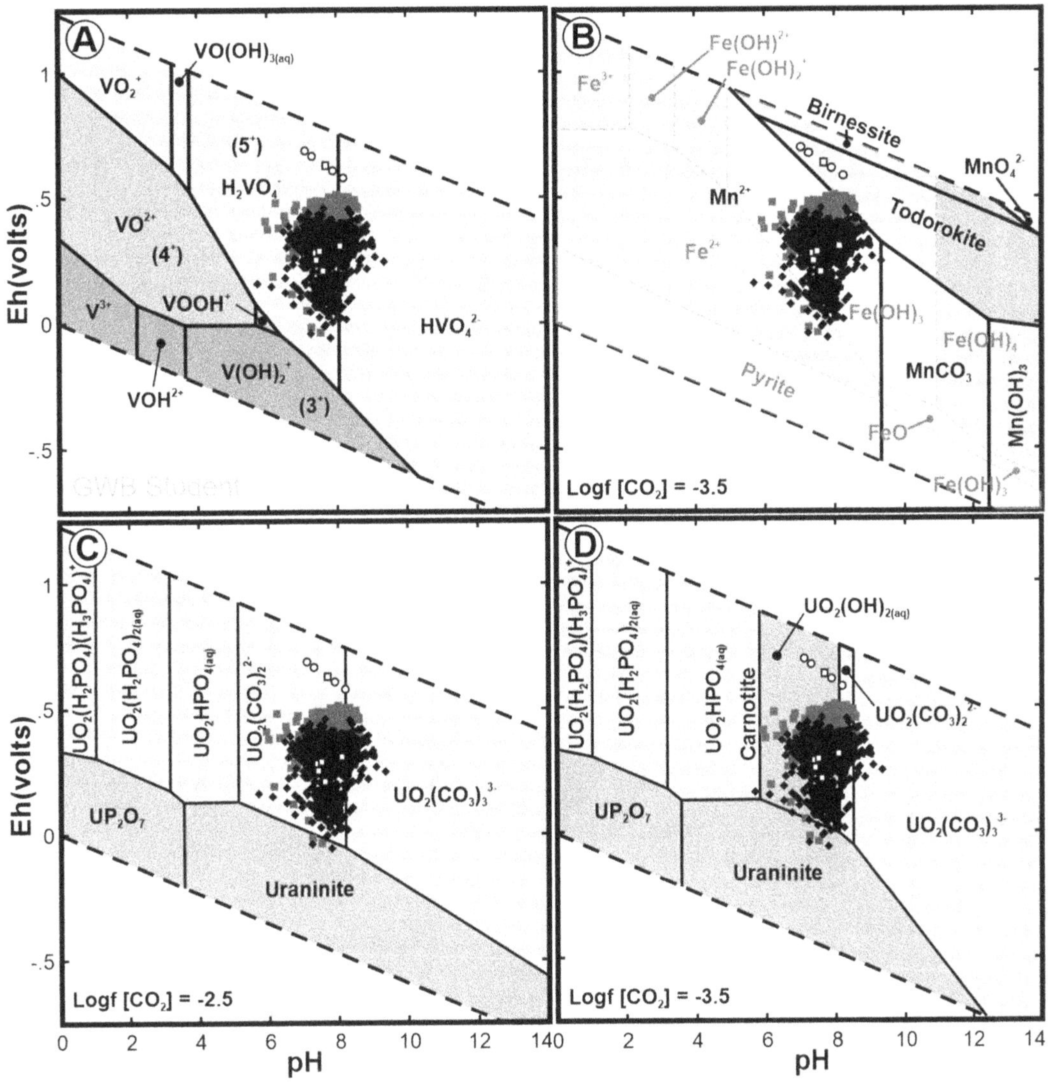

Fig. 14 (**a**) Modelled pH-Eh plots showing the composition of Northern Yilgarn groundwater modified from Noble et al. (2011). Dark grey squares are from stagnant wells; black diamonds are from flowing bores. Water samples from the LWSB are plotted for reference (white circles and squares; salt marsh and groundwater calcrete aquifers, respectively). (**a**) pH-Eh diagram illustrating V species with superimposed +5, +4, and +3 valence states indicated by light grey, medium grey, and dark grey fields, respectively. (**b**) pH-Eh diagram illustrating the stability of Mn and Fe mineral phases in a solution in equilibrium with the atmosphere. (**c**) pH-Eh diagram illustrating stable U species in a solution similar to water samples from the Northern Yilgarn. (**d**) pH-Eh diagram illustrating stable U minerals and aqueous U phases once a solution is equilibrated with the atmosphere

accentuates the effect of evaporation by facilitating the upward migration of groundwater into the capillary fringe. The Mg-rich and reactive substrate of authigenic AMS and Mg-smectites is interpreted to serve as a nucleation site for sedimentary dolomite; in addition, the negative surface charges of AMS, smectite and opaline silica are interpreted to serve as an adsorption sink for aqueous K^+. As discussed above, V is interpreted to have limited mobility in this setting, being adsorbed to Fe/Mn oxides and clays in the underlying sediments and the basal calcrete

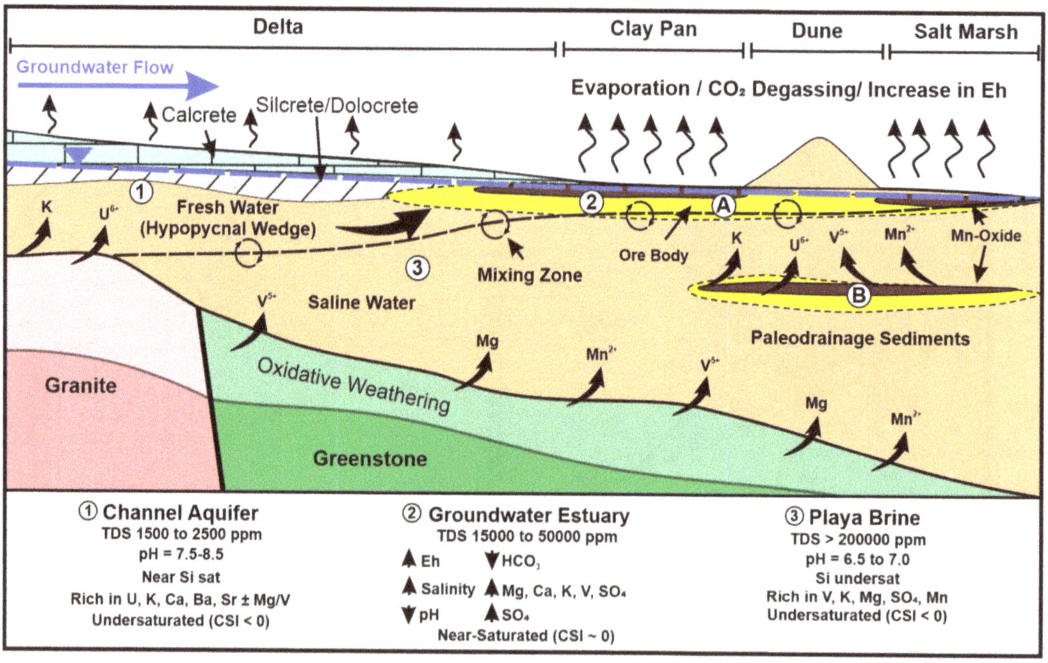

Fig. 15 Proposed genetic model for playa U deposits from Drummond et al. (2021). Locations 1–3 on the diagram correspond to the observed hydrogeochemical conditions listed. Location (**a**) – zone of present carnotite precipitation (yellow polygon). Dark brown polygons illustrate the approximate position of manganese oxides (Mn redox boundary). Location (**b**) – Hypothetical zone of relict U mineralization that is undergoing active dissolution. Similar hydrological conditions exist at Yeelirrie, suggesting this model is also applicable to channel U deposits

aquifer. The common presence of V-bearing Mn oxides with carnotite concentrated at or near the present-day water table in WA deposits suggests that Fe/Mn-redox cycling may also serve as an important mechanism for governing the local activity of V in these mineral systems. In this setting, U, V, and K are all concentrated at the water table and capillary fringe (Hambleton-Jones et al. 1986; Drummond et al. 2021). A combination of CO_2 degassing and evaporative precipitation of sedimentary dolomite at the water table is interpreted to decrease the local activity of HCO_3^- within porewaters, thereby inducing the dissociation of aqueous $UO_2(CO_3)_2^{2-}$ to $UO_2(OH)_{2(aq)}$, increasing carnotite saturation (Fig. 14c, d). Adsorbed K^+ on AMS, Mg-clays and opaline silica is interpreted to further increase the chances of carnotite precipitating by providing an abundance of reactive surface sites for it to nucleate.

5 Conclusions

With continued innovation in exploration, processing, and extraction techniques, surficial U deposits have the potential to become significant U and V resources in the future. To increase exploration success, an integrated mineral systems approach must be applied to refine exploration search space and direct the collection of exploration data. Investigations of surficial U deposits in WA and Namibia reveal seven mineral system parameters that must be present to form a significant deposit. These are:

1. *Geology* – Adequate sources of U and V. Uranium is highly mobile in paleochannel aquifers and may travel significant distances. Vanadium is interpreted to have limited mobility, restricted by adsorption processes in regolith and paleochannel sediments.

2. *Geomorphology* – Most non-dissected surficial U deposits occur on tectonically stable cratons containing vast paleochannel networks with shallow groundwater aquifers.
3. *Climate* – All surficial U deposits and occurrences are observed in arid to semi-arid climate zones with high precipitation deficits. This reflects the importance of evaporation in driving the physicochemical processes that facilitate carnotite saturation and precipitation.
4. *Hydrology* – Calcrete aquifer characteristics and groundwater composition are interrelated and form the physical and chemical pathways that transport U-bearing fluids. Diagenetic modification and karstification of calcrete produce shallow and highly permeable aquifers. Equilibration of groundwater in this setting with respect to calcite/dolomite results in circum-neutral pH and high alkalinity that maximize U mobility.
5. *Focus* – Sustained hydrologic focusing of U-bearing groundwater at the water table, along with both vertical and lateral focusing in a relatively small sub-basin within a larger paleodrainage, is essential to forming large deposits with economic grades.
6. *Evaporation* – With all the above factors present, evaporation drives the physicochemical processes that result in carnotite precipitation. The close association of carnotite with sedimentary dolomite and Mg-clays is convincing evidence that evaporative precipitation of carbonate minerals at the water table and capillary fringe facilitates the dissociation of uranyl carbonate complexes and increases the local saturation state of carnotite. When this process occurs in proximity to V-rich sediments, it is interpreted that the thermodynamic conditions for carnotite precipitation are achieved. Carnotite nucleation is interpreted to be facilitated by authigenic Mg-clays that act as adsorption sites for K^+.
7. *Time* – Stable hydrological focusing, stagnation, and evaporation must be present for significant periods of time (ca. >50 ka) to accumulate economic concentrations of U.

Acknowledgments JD acknowledges that the research into the Lake Way and Lake Maitland deposits discussed herein would not have been possible without the supervision and guidance of the late Dr. T. K. Kyser of Queen's University. Special thanks to Dr. Greg Shirtliff (Toro Energy Ltd.), who provided comprehensive data sets, geological samples, and valuable conceptual discussion. Thanks to Dr. Paul Morris (DMIRS) for arranging transportation and field equipment provisioning. Thanks to Marina Joury and Kiaran Smith (GSWA) for assisting with sample collection and field logistics. AW acknowledges discussions over the past 15 years with various geologists at Paladin and Deep Yellow, including but not limited to Jean-Christophe Corbin and Ed Becker and Prof. Judith Kinnaird of the University of the Witwatersrand for permitting access to unpublished research material.

References

Anand RR, Butt CRM (2010) A guide for mineral exploration through the regolith in the Yilgarn Craton, Western Australia. Austr J Earth Sci 57:1015–1114

Anthony JW, Bideaux RA, Bladh KW, Nichols MC (2005) Carnotite. In: Handbook of mineralogy, vol 4. Mineral Data Publishing, Tucson, p 680. http://www.handbookofmineralogy.org/pdfs/carnotite.pdf

Arakel AV (1986) Evolution of calcrete in palaeodrainages of the Lake Napperby area, Central Australia. Palaeogeogr Palaeoclimatol Palaeoecol 54: 283–303

Arakel AV (1988) Carnotite mineralization in inland drainage areas of Australia. Ore Geol Rev 3:289–311

Arakel AV, Jacobson G, Salehi M, Hill CM (1989) Silicification of calcrete in palaeodrainage basins of the Australian arid zone. J Geol Soc Austr 36:73–89

Batulin SG (1980) Regional factors of formation of uranium deposits in calcretes. Lithol Min Resour 15:437–442

Beck HE, Zimmermann NE, McVicar TR, Vergopolan N, Berg A, Wood EF (2018) Present and future Köppen-Geiger climate classification maps at 1-km resolution. Sci Data 5:180214

Becker E, Kärner K (2014) Geological setting of the Langer Heinrich uranium deposit, Namibia. In: Uranium Raw Material for the Nuclear Fuel Cycle: Exploration, Mining, Production, Supply and Demand, Economics and Environmental Issues (URAM-2009), IAEA TECDOC SERIES, IAEA-TECDOC-1739. International Atomic Energy Agency (IAEA), Vienna, pp 285–286

Bell JG, Kilgour PL, English PM, Woodgate MF, Lewis SJ, Wischusen JDH (2012) WASANT Palaeovalley Map – distribution of Palaeovalleys in Arid and Semi-arid WA-SA-NT. Geocat No. 73980, Thermatic Map, Geoscience Australia

Bowell RJ, Davies AA (2017) Assessment of supergene uranium-vanadium anomalies, Meob Bay deposit, Namibia. Geochem Explor Environ Anal 17:101–112

Bowell RJ, Barnes A, Grogan J, Dey M (2009) Geochemical controls on uranium precipitation in calcrete palaeochannel deposits of Namibia. In: The 24th International Applied Geochemistry Symposium, Fredericton Canada, 1st–4th June, 4 pp

Boyle DR (1984) The genesis of surficial uranium deposits. In: Surficial Uranium Deposits. IAEA-TECDOC-322, International Atomic Energy Agency (IAEA), Vienna, Austria pp 45–52

Briot P, Fuchs Y (1984) Processus de Minéralisation Uraniféres des Calcretes. In: Surficial uranium deposits, IAEA-TECDOC-322. International Atomic Energy Agency (IAEA), Vienna, pp 15–25

Brunner I, Wallis R, Elliott P, Thomson C (2011) Final report. Groundwater study proposed Yeelirrie development. Groundwater Study, URS Australia Pty. Ltd., Richmond, 116 pp

Butt CRM (1988) Major uranium provinces: Yilgarn Block and Gascoyne Province, Western Australia. In: Recognition of Uranium Provinces, Proceedings of a Technical Committee Meeting. IAEA-TC-450.5/7, International Atomic Energy Agency (IAEA), pp 273–304

Butt CRM, Graham J (1981) Sodian potassian hydroxonian meta-autunite: first natural occurrence of an intermediate member of a predicted solid solution series. Am Mineral 66:1068–1072

Butt CRM, Horwitz RC, Mann AW (1977) Uranium occurrences in calcrete and associated sediments in Western Australia. Technical Report, Commonwealth Scientific and Industrial Research Organization (CSIRO), Division of Mineralogy, Floreat. FP-16, 84 pp

Butt CRM, Mann AW, Horwitz RC (1984) Regional Setting, Distribution and Genesis of Surficial Uranium Deposits in Calcretes and Associated Sediments in Western Australia. In: Surficial uranium deposits, IAEA-TECDOC-322. International Atomic Energy Agency (IAEA), Vienna, pp 121–129

Cameco (2015) Numerical Groundwater Flow and Solute Transport Model of the Yeelirrie Uranium Deposit. Cameco, 73 pp. https://www.camecoaustralia.com/uploads/downloads/per_documents/Appendix_I1_Numerical_Groundwater_Flow_and_Solute_Transport_Model.pdf

Cameron E (1984) The Yeelirrie uranium deposit, Western Australia. In: Surficial uranium deposits, IAEA-TECDOC-322. International Atomic Energy Agency (IAEA), Vienna, pp 157–164

Carlisle D (1980) Possible variations on the calcrete-gypcrete uranium model. Technical Report, Bendix Field Engineering Corporation and University of California, Grand Junction, 76-022-E, 73 pp

Carlisle D (1983) Concentration of uranium and vanadium in calcretes and gypcretes. Geol Soc London Special Publ 11:185–195

Cassidy KF, Champion DC, Krapez B, Barley ME, Brown SJA, Blewett RS, Groenewald PB, Tyler IM (2006) A revised geological framework for the Yilgarn Craton, Western Australia. Technical Report, Geological Survey of Western Australia (GSWA), Perth, Record 2006/8, 8 pp

Cavaney RJ (1984) Lake Maitland uranium deposit. In: Surficial uranium deposits, IAEA-TECDOC-322. International Atomic Energy Agency (IAEA), Vienna, pp 137–140

Chen XY, Lintern MJ, Roach IC (2002) Calcrete: Characteristics, distribution and use in mineral exploration. Cooperative Research Centre for Landscape Environments and Mineral Exploration, Perth, p 160

Chudasama B, Porwal A, González-Álvarez I, Thakur S, Wilde A, Kreuzer OP (2018) Calcrete-hosted surficial uranium systems in Western Australia: Prospectivity modelling and quantitative estimates of resources. Part 1 – Origin of calcrete uranium deposits in surficial environments: a review. Ore Geol Rev 102:906–936

Condie KC (1981) Developments in Precambrian Geology 3 – Archean greenstone belts. Elsevier Scientific Publishing Company, Amsterdam, 425 pp

Crabb D, Dudley R, Mann AW (1984) Hinkler Well – centipede uranium deposits. In: Surficial uranium deposits, IAEA-TECDOC-322. International Atomic Energy Agency (IAEA), Vienna, pp 133–136

Cuney M, Kyser K (2015) Deposits Related to Low-Temperature Processes. In: Raeside R (ed) Geology and geochemistry of uranium and thorium deposits. Mineralogical Association of Canada, pp 251–261

de Broekert P, Sandiford M (2005) Buried inset-valleys in the Eastern Yilgarn Craton, Western Australia: geomorphology, age, and allogenic control. J Geol 113: 471–493

Deep Yellow Ltd. (2023) Mineral Resource and Ore Reserve. Company Website. https://deepyellow.com.au/projects/mineral-resource-and-ore-reserve/

Dickson BL (1984) Uranium series disequilibrium in the carnotite deposits of Western Australia. In: Surficial uranium deposits, IAEA-TECDOC-322. International Atomic Energy Agency (IAEA), Vienna, pp 165–170

Dickson BL, Fisher N (1980) Radiometric disequilibrium analysis of the carnotite uranium deposit at Yeelirrie, Western Australia. Aust Inst Min Metall 273:13–19

Drummond JBR (2016) SEM Mineralogical Studies on Langer Heinrich. Mineralogical Study, Queen's University Dept. of Geological Sciences and Geological Engineering, Kingston

Drummond JBR, Kyser TK, Bowell RR, James NP, Layton-Matthews D (2021) Diagenesis of paleodrainages in Lake Way and Lake Maitland, Western Australia, and the role of authigenic Mg-clays and dolomite in the genesis of channel and playa uranium deposits. Can Mineral 59:947–984

Eckardt FD, Drake N, Goudie AS, White K, Viles H (2001) The role of playas in pedogenic gypsum crust

formation in the Central Namib Desert: a theoretical model. Earth Surf Process Landforms 26:1177–1193
Elevate Uranium (2024) – U-pgradeTM. https://www.elevateuranium.com.au/u-pgrade/
Fleurances S, Kinnaird J (2011) Mineralogical, Geochemical and Isotopic Study of 6 samples from the Langer Heinrich Calcrete Deposit (Namibia). Mineralogical Study, Henry Poincare University, University of the Witwatersrand, Johannesburg
French RR, Allen JH (1984) Lake Way Uranium Deposit, Wiluna, Western Australia. In: Surficial Uranium Deposits, IAEA-TECDOC-322. International Atomic Energy Agency (IAEA), Vienna, pp 149–156
Gallant J, Dowling T, Austin J (2012) Multi-resolution Valley Bottom Flatness (MrVBF). Commonwealth Scientific and Industrial Research Organization (CSIRO), Version 3. https://doi.org/10.4225/08/5701C885AB4FE
Gaskin AJ, Butt CRM, Deutscher RL, Horwitz RC, Mann AW (1981) Hydrology of uranium deposits in calcretes of Western Australia. 177:311–316
Geyh MA, Heine K (2014) Several distinct wet periods since 420ka in the Namib Desert inferred from U-series dates of speleothems. Quater Res 81:381–391
Goudie A (1972) On the definition of calcrete deposits. Z Geomorphol 16:464–468
Goudie A (1973) Duricrusts in tropical and subtropical landscapes. Clarendon Press/University of Michigan, 174 pp
Gray DJ (2001) Hydrogeochemistry in the Yilgarn Craton. Geochem Explor Environ Anal 1:253–264
Gray DJ, Noble RRP, Reid N, Sutton GJ, Pirlo MC (2016) Regional scale hydrogeochemical mapping of the northern Yilgarn Craton, Western Australia: a new technology for exploration in arid Australia. Geochem Explor Environ Anal 16:100–115
GSWA (2018) 1:500 000 State interpreted bedrock geology. DMIRS-016, Dept. of Mines, Industry, Regulation and Safety (DMIRS), Perth. https://catalogue.data.wa.gov.au/dataset/1-500-000-state-interpreted-bedrock-geology-dmirs-016
GSWA (2022) Vanadium investment opportunities, Western Australia. Commodity Flyer, Department of Mines, Industry, Regulation and Safety (DMIRS), Perth
Hall SM, Mihalasky MJ (2017) Grade, tonnage, and location data for world calcrete-type surficial uranium deposits. U. S. Geological Survey (USGS) data release. https://catalog.data.gov/dataset/grade-tonnage-and-location-data-for-world-calcrete-type-surficial-uranium-deposits
Hall SM, Van Gosen BS, Paces JB, Zielinski RA, Breit GN (2019) Calcrete uranium deposits in the Southern High Plains, USA. Ore Geol Rev 109:50–78
Hambleton-Jones BB, Levin M, Wagener GF (1986) Uraniferous Surficial Deposits in Southern Africa, in Mineral Deposits of Southern Africa. In: Mineral deposits of Southern Africa. Geological Society of South Africa, Johannesburg, pp 2269–2287
Hanor JS (2000) Barite-Celestine geochemistry and environments of formation. Rev Mineral Geochem 40:193–275
Hartleb JWO (1988) The Langer Heinrich uranium deposit: Southwest Africa/Namibia. Ore Geol Rev 3: 277–287
Hudson S (2010) Mineral exploration report: Yeelirrie Project, TR53/6899H. Quarterly Mineral Exploration Report, BHP Billiton, Perth, 11 pp
Humphreys WF, Watts CHS, Cooper SJB, Leijs R (2009) Groundwater estuaries of salt lakes: buried pools of endemic biodiversity on the western plateau, Australia. Hydrobiologia 626:79–95
IAEA (2018) World distribution of uranium deposits (UDEPO), IAEA TECDOC Series. IAEA-TECDOC-1843, International Atomic Energy Agency, Vienna, 262 pp
Iilende, A. (2012) The source of uranium and vanadium at the Langer Heinrich and Klein Trekkopje uranium deposits-genesis and controlling factors for uranium mineralization, Western Namibia. Masters thesis, University of Namibia, Windhoek, 290 pp
Jarvis I, Burnett WC, Nathan Y, Almbaydin FSM, Attia AKM, Castro LN, Flicoteaux R, Hilmy ME, Husain V, Qutawnah AA (1994) Phosphorite geochemistry: state-of-the-art and environmental concerns. Eclogae Geologicae Helvetiae 87:643–700
Johnson SL, Commander DP, O'Boy CA (1999) Groundwater resources of the Northern Goldfields, Western Australia, Hydrogeological record series. HG 2, Resource Investigations Division, Water and Rivers Commission, East Perth, 57 pp
Khoury HN, Salameh EM, Clark ID (2014) Mineralogy and origin of surficial uranium deposits hosted in travertine and calcrete from central Jordan. Appl Geochem 43:49–65
Kinnaird JA, Nex PAM (2016) Uranium in Africa. Episodes 39:335–359
Kneer S (2015a) Combined Annual Mineral Exploration Report 170/2010 for the Period Ended 1 January to 31 December 2014: Wiluna Uranium Project. DMIRS Annual Report, Toro Energy Ltd. A105506, 76 pp
Kneer S (2015b) Combined Annual Report on Exploration for the Period Ended 31 December 2014: Lake Maitland Project East Murchison, WA. Annual Report, Toro Energy Ltd. A106289, 36 pp
Langford RL, Wyche S, Liu SF (2000) Geology of the Wiluna 1: 100 000 Sheet, 1:100 000 Geological Series Explanatory Notes. Geological Survey of Western Australia (GSWA), Perth, 26 pp
Magee JW (2009) Palaeovalley groundwater resources in arid and semi-arid Australia – a literature review. Literature Review, Geoscience Australia, Canberra, Australian Capital Territory (ACT), 224 pp
Mann AW (1974) Chemical ore genesis models for the precipitation of carnotite in calcrete. Technical Report, Commonwealth Scientific and Industrial Research Organisation (CSIRO), Minerals Research Laboratories – Division of Mineralogy, Floreat, Western Australia, FP-7, 20 pp
Mann AW, Deutscher RL (1978a) Genesis principles for the precipitation of carnotite in calcrete drainages in Western Australia. Econ Geol 73:1724–1737

Mann AW, Deutscher RL (1978b) Hydrogeochemistry of a calcrete-containing aquifer near Lake Way, Western Australia. J Hydrol 38:357–377

Mann AW, Horwitz RC (1979) Groundwater calcrete deposits in Australia some observations from Western Australia. J Geol Soc Aust 26:293–303

McInnes BIA, Evans N, Boni M, McDonald B (2006) (U Th)/He thermochronometry of supergene base metal ores and implications for Namibian Paleoclimate. Geochemica et Cosmochimica 70:A411

Mulcahy MJ (1967) Landscapes, laterites and soils in Southwestern Australia. In: Jennings JN, Mabbutt JA (eds) Landform Studies from Australia and New Guinea. ANU Press, Canberra, pp 211–230

NEA/IAEA (2020) Uranium 2020: resources, production and demand. Nuclear Energy Agency, Vienna. NEA No. 7551, 484 pp

Noble RRP, Gray DJ, Reid N (2011) Regional exploration for channel and playa uranium deposits in Western Australia using groundwater. Appl Geochem 26: 1956–1974

Ollier CD (1977) Outline geological and geomorphic history of the Central Namib Desert. Madoqua 10:207–212

Osmond JK, Cowart JB, Ivanovich M (1983) Uranium isotopic disequilibrium in ground water as an indicator of anomalies. Int J Appl Rad Isotopes 34:283–308

Paladin Energy Ltd. (2020) 2020 Annual Report – PDN. ASX Announcement, Australia Stock Exchange. Subiaco, 105 pp. https://www.paladinenergy.com.au/investors/financial-annual-half-yearly-reports/

Paladin Energy Ltd. (2021) 2021 Annual Report – PDN. ASX Announcement, Australia Stock Exchange, Subiaco, 100 pp. https://www.paladinenergy.com.au/investors/financial-annual-half-yearly-reports/

Roesener H, Schreuder CP (1992) Nuclear and fossil fuels – uranium. In: Mineral resources of Namibia. Republic of Namibia, Ministry of Mines and Energy, Geological Survey, Windhoek, pp 7.1-1–7.162

Sanders CC (1972) Hydrogeology of the Paroo Calcretes and Surrounding Areas, Wiluna District, Western Australia. Technical Report, Geological Survey of Western Australia (GSWA), Perth, Record No. 1972/7, 30 pp

Shirtliff G, Kneer S (2014) Combined Annual Mineral Exploration Report 170/2010 for the Period Ended 1 January to 31 December 2013: Wiluna Uranium Project. DMIRS Annual Report, Toro Energy Ltd, A101990, 34 pp

Shirtliff G, Kneer S, Cox R (2016a) Revisiting and renewing – a geological update on the shallow groundwater carbonate related uranium deposits of the Wiluna Project, NE Yilgarn, Western Australia, AusIMM Presentation, Adelaide

Shirtliff G, Kneer S, Cox R, Daniel G (2016b) Underestimation of uranium mineralisation by Gamma Probes at the Surficial Carbonate Associated Uranium Deposits of the Wiluna Project – Unrelated to Secular Disequilibrium. AusIMM Presentation, Adelaide. https://www.toroenergy.com.au/wp-content/uploads/2016/06/160608-AusIMM-Presentation-Adelaide_2.pdf

Shirtliff GJ, Kneer S, Cox RA (2017) Carbonate related uranium deposits of the north-east Yilgarn. In: Phillips GN (ed) Monograph 32 – Australian ore deposits. Australasian Institute of Mining and Metallurgy (AusIMM), Melbourne, pp 321–324

Stone A, Thomas DSG, Viles H (2010) Late Quaternary palaeohydrological changes in the northern Namib Sand Sea: New chronologies using OSL dating of interdigitated aeolian and water-lain interdune deposits

Toens PD, Hambleton-Jones BB (1980) Uraniferous surficial deposits. Atomic Energy Board, Pelindaba, Pretoria. PER-57, 18 pp

Toro Energy Limited (2021) Toro Initiates Lake Maitland Uranium and Vanadium Scoping Level Engineering Study. ASX Release, Australia Stock Exchange (ASX), 5 pp. https://cdn-api.markitdigital.com/apiman-gateway/ASX/asx-research/1.0/file/2924-02360448 6A1027303?access_token=83ff96335c2d45a094df02a206a39ff4

Trittschack R (2008) Geological identification and mineralogical characterisation of palaeosurfaces and channel fills at the Langer Heinrich uranium deposit, Namibia. PhD thesis, Martin-Luther-University Halle-Wittenberg, Germany: Martin-Luther-University, Halle-Wittenberg, 155 pp

Vogel JC (1989) Evidence of past climatic change in the Namib Desert. Palaeogeogr Palaeoclimatol Palaeoecol 70:355–366

Ward JD (1987) The Cenozoic Succession in the Kuiseb Valley, Central Namib Desert. Nambian Geological Survey, Department of Economic Affairs, Windhoek, 43 pp

Western Mining Corp (1975) A General Account of the Yeelirrie Uranium Deposits. Privately published and distributed to 25th Intern. Geol. Cong. tours, Sydney, p 14

White ME (1994) After the greening: the browning of Australia. Kangaroo Press, Kenhurst, 288 pp

Wilkinson MJ (1990) Paleoenvironments in the Namib Desert: the Lower Tumas Basin in the Late Cenozoic. University of Chicago, Committee on Geographical Studies, Chicago, 196 pp

WNA (2021) Uranium in Namibia. World Nuclear Association, 9 pp. https://world-nuclear.org/information-library/country-profiles/countries-g-n/namibia.aspx

Wright VP, Tucker ME (1991) Calcretes: an introduction. In: Calcretes (Eds Wright VP, Tucker ME) The International Association of Sedimentologists, Reading University, pp 1–22

Youlton B (2007) Controls on Uranium Mineralization at the Klein Trekkopje Prospect, Namibia. BSc Honours thesis, University of the Witwatersrand, Johannesburg, 63 pp

GPSR Compliance
The European Union's (EU) General Product Safety Regulation (GPSR) is a set of rules that requires consumer products to be safe and our obligations to ensure this.

If you have any concerns about our products, you can contact us on

ProductSafety@springernature.com

In case Publisher is established outside the EU, the EU authorized representative is:

Springer Nature Customer Service Center GmbH
Europaplatz 3
69115 Heidelberg, Germany

www.ingramcontent.com/pod-product-compliance
Lightning Source LLC
Chambersburg PA
CBHW081225130625
PP16633100001B/3
* 9 7 8 3 0 3 1 7 5 7 3 2 7 *